Rees's CLOCKS WATCHES AND CHRONOMETERS

Rees's Clocks Watches and Chronometers

(1819-20)

A selection from

The Cyclopaedia; or Universal Dictionary of Arts, Sciences and Literature

by

ABRAHAM REES

CHARLES E. TUTTLE COMPANY: PUBLISHERS
RUTLAND, VERMONT

Published by the Charles E. Tuttle Company, Inc.
of Rutland, Vermont and Tokyo, Japan
with editorial offices at
Suido 1-chome, 2–6 Bunkyo-ku, Tokyo, Japan

PUBLISHERS NOTE

This is a selection of articles and plates on the subject of horology, taken from Rees's *Cyclopaedia* – the first encyclopaedia to put primary emphasis on technological subjects.

The pages of the original were not numbered, and the pagination in this reprint, both of text and plate pages, has been given by the present publisher. References to the plates in the List of Illustrations on page vii, and in the Index specially provided for this edition, are to the new pagination; while in the original text the original plate numbers are referred to. For ease of cross reference the plates have been printed at the end of the appropriate sections of the text.

Rees's *Cyclopaedia* was first published 1819–20

First Tuttle edition published 1970

Library of Congress Catalog No. 75–116481
Standard Book No. 8048 0901–1

Printed in Great Britain

Contents

List of Illustrations

Chimes

CYCLOPÆDIA:

OR, A NEW

UNIVERSAL DICTIONARY

OF

ARTS and SCIENCES.

CHIMES, in *Horology*, is a ſpecies of muſic mechanically produced by the ſtrokes of hammers againſt a ſeries of bells, tuned agreeably to a given ſcale in muſic; the hammers are lifted by levers acted upon by metallic pins or wooden pegs ſtuck into a large barrel which is made to revolve by clock-work, and is ſo connected with the ſtriking part of the clock mechaniſm, that it is ſet in motion by it at certain intervals of time, moſt uſually either at every hour, or at every quarter of an hour. The muſic thus produced may conſiſt of a direct ſucceſſion of the notes conſtituting an octave frequently repeated, or otherwiſe may be a pſalm tune, or ſhort popular air in the key to which the bells are tuned. This ſpecies of mechanical muſic is by no means calculated to improve the taſte, and had its origin moſt probably, like clockwork itſelf, in ſome of the monaſtic inſtitutions of Germany, where, according to Dr. Burney, it prevails greatly, and where the ringing of changes on bells, as in England, is but little, if at all practiſed. The chime mechaniſm may be adapted either to act with the large bells of a church ſteeple, by means of wheel-work proportionably ſtrong to raiſe heavy hammers, or a ſet of bells of different diameters may be arranged concentrically, within one another

other on one common axis, ſufficiently ſmall to be introduced into the frame of a clock, or even of a watch, which we have ſeen performed by Margetts, who lately died in a ſtate of inſanity. The manner in which the hammers are moved by the pins of the chime-barrel muſt neceſſarily vary according to circumſtances, as the wires of different rooms are differently placed to produce the ſound of a houſe bell, but an ingenious workman will vary the length and ſhape of his levers of communication agreeably to the ſituation and diſtance of his bells from the clockwork, ſo that if we give a deſcription of the chime mechaniſm of a church clock, and alſo of a portable piece, he will not be at a loſs to know how to deviſe and adapt the neceſſary parts for any other conſtruction.—But before we proceed to our deſcriptions, it will be proper to give an account of the chime-barrel, and the method of inſerting the lifting pegs, or of what is called pricking it.

Chime-*Barrel* is that cylinder of braſs in ſmall clocks, or of wood in church-clocks, which gives motion to the hammers, that ſtrike on the bells, and produce a change or tune thereon, by means of pegs inſerted into certain points of its circumference at meaſured intervals, the pegs ſerving to lift the hammers in regulated ſucceſſion both of order and time.

In making a chime barrel for any given tune there are certain neceſſary conditions to be attended to: Firſt, the barrel muſt be well turned in a lathe upon its own arbor, ſo as to have the whole ſurface of its circumference concentrical, when revolving on its pivots. Secondly, the train of wheel-work, belonging to the chime-work, muſt make the barrel revolve in a ſpace of time exactly equal to what is required for playing the requiſite tune on any other inſtrument, which time may be exactly limited by proportioning the fly on the laſt wheel to the power of the weight or ſpring that urges the firſt wheel of the train. Thirdly, there muſt be as many bells as the compaſs of the propoſed tune contains muſical notes, and alſo as many rows of pegs inſerted into parallel circular lines on the circumference of the barrel as there are hammers to be lifted by them; and, laſtly, the whole circumference of the barrel muſt be divided into as many longitudinal parallelograms of equal breadth, as there are muſical bars in the propoſed tune, each of which muſt be again ſubdivided into as many parts as there are notes of the loweſt denomination, whether crotchets, quavers, or ſemi-quavers, in each bar; then the parallel dividing lines will correſpond to the bars, and the ſubdiviſions within will be the guides for placing the pegs in thoſe bars reſpectively: for inſtance, ſuppoſe that Pleyel's German hymn be required to be played by a chime clock of any deſcription, and that the points on the barrel where the pegs or pins are to be inſerted be required to be aſcertained?

In *fig.* 1. of *Plate* VII. *(Horology)*, it will be ſeen that this popular hymn is marked $\frac{2}{4}$, which implies, that the bar is meaſured by four, or crotchets, of which there are two in each bar; it is alſo obſervable that the quickeſt note in the tune is a ſemi-quaver, of which denomination eight conſtitute two crotchets, or one bar; conſequently the bar muſt be ſub-divided into eight. Now ſuppoſe the length of the barrel to be repreſented by the line, D D, or *d d*, becauſe the compaſs of the hymn is juſt one octave beginning and ending with D incluſively, the line, D D, or *d d*, muſt be divided into ſeven equal parts, which will require eight points in each to include them, viz. D, E, F, G, A, B, C, D, and *d*, *e*, *f*, *g*, *a*, *b*, *c*, *d*; then, becauſe the whole hymn contains 16 bars, the whole line, D *d*, which we aſſume as equal to the circumference of the barrel, muſt be divided into 16 equal parts, which we have made to fall oppoſite their reſpective bars, the better to elucidate our example; then each of theſe parts may eaſily be conceived to be ſub-divided into eight ſmaller diviſions, as from 1 to 2 in the lower line, D *d*. From the dividing points let ſtraight lines be drawn to complete the ſquare, or parallelogram as the caſe may be, and there will be a figure, D D *d d*, with 8 × 16 = 128 ſmall ſquares or parallelograms, which we will ſuppoſe to be on paper that will exactly cover the barrel when paſted round it, in order to convey a more diſtinct idea of the method of aſcertaining the true places of the pins on the barrel, which may be thus done:

The firſt note, B, is a crotchet, and therefore the pin, repreſented by a dot, that moves its hammer muſt be in the beginning of the line B *b*, and the next following pin muſt be at the ſpace of half a bar forwards to limit its continuance; accordingly the next note, D, is placed on the line, D *d*, at half a bar forwards; again this ſecond note, D, being alſo a crotchet, requires the next ſucceeding note of A to be removed another half bar, namely to the interſection of the lines, A *a*, with 2 2; but the third note, A, which we have juſt mentioned, is equal in length to three quavers, or ſix ſemi-quavers; the ſucceeding pin to repreſent B muſt conſequently be removed $\frac{6}{8}$ or $\frac{3}{4}$ of the bar, leaving $\frac{2}{8}$ or $\frac{1}{4}$ for the remaining quaver from B to C, the latter of which, being the firſt note of the third bar, falls on the interſection of the bar line, 3 3, with the hammer-tail line, C *c*: by the ſame rule the crotchet, C, once ſtruck continues half of a bar, and requires the following A to be half of a bar advanced, but A is here $\frac{3}{4}$ of a crotchet, and the following B only $\frac{1}{4}$ of the ſame or $\frac{1}{8}$ of the bar; the pin of B muſt therefore be at $\frac{7}{8}$ of the bar, and the ſucceeding one on the bar line to limit its duration to $\frac{1}{8}$. In the fourth bar there is a crotchet reſt, which is the reaſon why there is no dot or pin between its bar lines, as though the B of this bar had been a minim.—We might thus have analyſed the whole ſurface of the barrel, ſo far as relates to this hymn, but it is preſumed the preceding detail, clearly underſtood, will render the reſt perfectly intelligible without further explication.

When the hammers are very heavy, as in church clocks, thoſe bells which have quavers or ſemiquavers to be ſtruck immediately in ſucceſſion require to have each two hammers and each a pair of parallel circles pricked to perform within the limit of time; and when either the barrel is made adjuſtable, or the pins moveable, a number of tunes may be put on the ſame barrel, and where the bells are ſufficiently numerous, the tune may be played in two or even three parts.

This method of pricking a chime barrel for playing on bells differs from that of an organ-barrel in this reſpect, that in the former the length of the note is meaſured by the ſpace between the contiguous pins, whereas in the latter the limit of the note is produced by crank pieces, inſtead of pins, which pieces keep the pipes open, and therefore muſt cover the very ſpaces which lie between the pins of the other, the projecting parts of one barrel mutually correſponding to the vacant parts of the other.

Chimes *of a church clock. Plates* V. and VI. of *Horology* explain the chime mechaniſm in the clock-room of St. Margaret's church, Weſtminſter: A (*figs.* 1 and 3) is a barrel on which the rope B is wound; this rope, after paſſing over a fixed friction roller, has a heavy leaden weight faſtened to it, by the deſcent of which the mechaniſm is actuated: D is a large wheel at the end, and on the arbor of this barrel, which is worked by a pinion E, not ſeen, the arbor of which is ſquare at the projecting end for the key of the

the handle by which the weight is wound up; the barrel A is ſo contrived, by means of the click *b* and ratchet wheel *a*, (*fig.* 1) that, when the rope is in the act of being wound, the barrel ſlips round its ſtationary arbor, but when the weight is going down, it turns the large cylinder or chime barrel C attached to the arbor along with it; this chime barrel is made of wood, and has a number of pegs ſcrewed into it, which, as it turns round, at proper intervals take hold of the inner ends *d* of the horizontal and parallel levers F, raiſe them to a certain height, and then let them go ſuddenly; this motion at the ſame time depreſſes the outer ends *e* of the ſaid levers, and by means of the upright rods G, and other intermediate rods, raiſes the hammers which ſtrike upon the large bells in the belfry or ſtory above. On the end of the chime barrel next to barrel A, is a large wheel H (*figs.* 1 and 3) which takes into a ſmall pinion on the arbor of the wheel I; this wheel works a pinion on one end of the long arbor K, (*fig.* 1) on the other end of which arbor is a pair of fanners, or a fly L, the vanes of which catch the air, as they turn, and regulate the velocity of the machine. This fly is ſhewn ſeparate in *fig.* 2. A is a portion of the arbor, on which is looſely fitted an iron bar B B, having a vane D at each end to prevent the bar from ſlipping round the arbor in a retrograde direction; a ratchet wheel *b* is attached to the outer end of the arbor, and a click *c* is faſtened on the fly, which is kept to the teeth of the wheel by a ſpring, ſo that when the arbor turns the click ſtops againſt the teeth of the wheel, and turns the fly with it, but when the arbor is ſuddenly ſtopped (as hereafter deſcribed), the fly continues its forward motion for ſome time by means of its momentum, the click *c* in the mean time ſlipping round the ratchet *b*; by this contrivance all ſtrain upon the mechaniſm, by a ſudden check of the momentum of the fly, is avoided. On one of the diagonal or croſs bars of the wheel I (*figs.* 1 and 3), is a projecting piece of metal *f*, which piece, when the machine is to be ſtayed, is caught by a detent *g*, which detent may be moved towards the centre of the wheel (ſo as to clear the piece *f*) when the machine is to be put in motion; the upper end of the detent *g* is faſtened to an arbor M, ſo as to have a circular motion with it; to which arbor is alſo fixed another detent *h* (*fig.* 3) bent to avoid the bar of the frame. On the middle of this detent *h*, a piece of upright iron *i* is riveted, on the end of which a hammer N (*figs.* 1 and 3) ſtrikes; this hammer is raiſed at the proper hour by the church clock, and by its fall ſtrikes the piece *i*, (*fig.* 3) depreſſes the bent detent *h*, and conſequently moves the detent *g* from the ſtop of the piece *f*. O is a circular plate, having a notch in one part of its circumference; at the back of it are faſtened four arms *k*, *l*, *m*, (*fig.* 3) and another not ſeen. On the face of the wheel H there is a projecting tooth *n* (*fig.* 1) which takes hold of one of the arms on O, and at every turn of the wheel H, moves the wheel O round the ſpace of one quarter. Behind the arms *k*, *l*, *m*, are four knobs *p*, *q*, *r*, *s*, in a detached ſtate (ſhewn in *fig.* 4) againſt which a lever R (*figs.* 3 and 4) is preſſed, by a ſpring S; the uſe of this additional mechaniſm is to make the wheel O always deſcribe a complete quarter of a revolution whenever it moves at all, for ſuppoſe the arms *m*, *l*, *k*, (*fig.* 4) to have the poſition of the dotted lines *m'*, *l'*, *k'*, the pin *n* on the wheel H (*figs.* 1 and 3) takes the arm *m* and puſhes it downwards, during which courſe the knob *p* (*fig.* 4) raiſes the lever R into the poſition of the figure; the ſpring and lever then act upon the knob *p*, and quickly bring the arms to the dotted poſition, in which ſituation the letters of reference are advanced each to the next arm. This clock has moreover the ſtriking work, in which there is, as uſual, a wheel called the count wheel, which turns round once in 12 hours; in this wheel there are three pins, that, at the hours 4, 8, and 12, move the end of a lever that communicates, by means of an intermediate rod Q, (*figs.* 1 and 3) with the tail P of the hammer N, ſo as to elevate it; as the count wheel turns round, one of its pins lets go the lever, and the hammer N falls upon the detent *h*, (the end of which is ſuppoſed to be in the notch of the plate O) puſhes it down, and at the ſame time moves the detent *g* on the ſame arbor M inward; as ſoon as the hammer has made a ſtroke, it is partly lifted up again by the tail ſpring *t* (*fig.* 3) acting againſt the bar of the frame; but the end of the bent detent *h* is prevented from again entering the notch in the plate O, when the hammer riſes, by the lever R, which at the inſtant the lever *h* (*fig.* 4) is puſhed down, moves the wheel O a little round, by means of the ſpring S, into the poſition of the figure. When the hammer has thus ſtruck and removed the detent *g*, (*fig.* 3) the leaden weight pulls the barrel A, and ſets the machine in motion, during which motion the pegs in the chime barrel E ſtrike the lever, F, and move the bell hammers in the due ſucceſſion of time, the fly (*fig.* 2) in the mean time regulating the velocity of the barrel. At each revolution of the chime barrel E, the peg *n* in wheel H turns the wheel O round one quarter of a revolution, and by the time the barrel has turned four times, the notch in the plate O is brought again oppoſite to the end of the bent detent *h*, which therefore falls in by the weight of the detent *g*, the latter of which alſo falls at the ſame time, ſo as to catch the ſtop *f* which ſtays the machine. The ſaid notch in the plate O is ſo placed, that when the bent detent *h* is locked into it, the knobs *p*, *q*, *r*, *s*, on the back of this plate, are in the poſition ſhewn in *fig.* 4, ſo that this plate O moves, the inſtant it is permitted to do ſo, by the detent *h* being removed by means of the ſpring S, independently of the great wheel H. This machinery plays four different tunes, which are changed by turning the index *IV* (*fig.* 1) by the handle to the correſponding tune marked on the dial plate T; this handle has a pinion on its arbor, behind the dial plate, which works in a rack upon the crooked iron bar *v y*, ſo as to move it up or down when the handle is turned; the bar *v y* is made to move ſteadily, by having a crook *x* with parallel ſides at each end, ſliding againſt both ſides of correſponding ſteady pins ſcrewed into the frame; and the friction of a ſpring, 7, preſſing againſt it, prevents its being moved by accident: in the middle of the bar *v y* is a bend *y*, which acts like an inclined plane between the rollers 3, 4, fixed in a ſmall frame at the end of the bar W, ſo as to move that bar horizontally, while the crooked bar *v y* is moved vertically; to the horizontal bar W are fixed the centres of the keys, or ſhort levers, F, and a long iron plate, 5, with 16 notches to confine each lever to its own plane of action. The ſliding pins *g g*, moving in a cock with two perforations attached to the frame, confine the bar W to a horizontal motion. In ſetting out the pegs on the barrel, 64 parallel circles, four to each lever, are drawn round it, at equal diſtances; every four of theſe coincide ſucceſſively, by the rack-work adjuſtment, with one of the levers, ſo that the reſpective pegs upon the barrel in the firſt of each four circles may work its own lever, and play one tune; then by moving the levers, along with the bar W, the diſtance that two contiguous circles are apart, a ſecond ſet of pegs is preſented to the ſaid levers, which now play a ſecond tune; and in the ſame manner a third and fourth, ſucceſſively. The 16 rods G have each a ſcrew adjuſtment at their lower ends, and their upper ends are connected with revolving rods fixed to the ceiling, by which the motion is conveyed, under each bell, to the hammers, which are placed each in a line perpendicularly under the axis of its bell, fo as to ftrike near its lower extremity; (*Plate* vii. *fig.* 4); the weight of each hammer H is fupported by a fpring S, in fuch a manner that it rifes from the bell the inftant it has ftruck. There is moreover a long horizontal iron bar that goes acrofs the levers F (*Plate* V. *fig.* 1) not feen, the ends of which are connected with another lever alfo not feen, which, when it is pulled down, takes up all the hammers at once, fo as to clear the bells in the act of ringing. There are ten bells in this fteeple, fix of which have each two hammers, and the other four but one a-piece; the double hammers are ufed where the fame bell is required to be ftruck twice in fucceffion fo quickly, that the fame hammer could not be lifted up and be made to return twice in the requifite time.

In the mechanifm we have here defcribed, the chime barrel is fo conftructed as to be capable of playing only four tunes; but in fome fteeples which we have vifited, the barrels have parallel longitudinal apertures, which admit the lifting pegs to flide into any fituation, and which confequently render them difpofable for any tune within the compafs of the two extreme notes, the fliding pieces being fixed by thumb fcrews within the barrel.

CHIMES *of a common clock.* After the ample account which we have juft given of the chime mechanifm of a large church clock, we will fatisfy ourfelves with giving fo much only of the chime work of a common clock, as may fuffice to give the reader an idea how a communication is made from the barrel to the bells, which portion could not eafily be reprefented where the bells are at a diftance from the barrel, as is the cafe in a church fteeple. We beg leave, however, to refer our readers to our article CLOCK *with chimes*, for a more particular defcription of the mechanifm which connects the chime part with the going part of the clock, and which alternately locks and unlocks the train of wheel-work that regulates the chime-barrel, it being impoffible to defcribe in a complete manner the contrivance for making the chimes go, and repeat at every quarter of an hour, without defcribing at the fame time the ftriking mechanifm, which does not properly belong to this article. *Figures* 2 and 3 of *Plate* VII. (*Horology*) reprefent as much of the chime mechanifm of an ordinary clock as will ferve our prefent purpofe: A B C D conftitute a portion of the clock frame, of which one connecting pillar only is fhewn; E is the chime barrel of brafs, nearly equal in length to the pillar, and of a diameter fufficient to admit of diagonal rows of brafs or fteel pins to lift the hammer tails F, in the fucceffive order of a defcending octave, while the barrel is revolving, there being eight hammers G ufually ftriking fucceffively againft as many bells H, which are mounted, concentrically within one another, on a common axis held faft by the bearing piece of fteel, I, fcrewed to the face of one of the plates at its lower extremity. The mechanifm connected with the ftriking part of the clock is fo ingenioufly contrived and arranged, as may be feen under the article to which we have already referred, that at and after the firft quarter from any hour, it will repeat the octave once; at and after the fecond quarter, it will repeat the fame twice; at and after the third, three times; and at and after the hour itfelf, four times; alfo whenever the octave is repeated four times, the chime part unlocks the ftriking part, and allows the clock to ftrike the laft hour as foon as the chimes have ceafed, but not before. K is a brafs frame fcrewed at the lower end acrofs the fide of the fuperior part of the clock frame, which holds the works, and has a brafs plate N, of hammered metal attached to it by fcrews; this plate, N, is divided nearly to the top by eight parallel vertical flits, made by a faw, which form fo many fprings; the hammers all turn upon a common fixed arbor near P (*figs.* 2 and 3) by means of a perforation made in each folid part P above the tails, which folid part has a ftraight edge in each, againft the central point of which the refpective fpring bears a little above its inferior extremity (*fig.* 2); this mode of applying a fpring gives it the advantage of performing two offices, for it will not only keep the hammer tail in its proper fituation, to be taken by a pin of the barrel, but will give a fmartnefs to its blow, and then reftore it to its original pofition, by acting alternately, partly above and partly below the central hole; which kind of action will readily be underftood, by examining the opening joint of a pen-knife, and attending to the manner in which its fpring is applied. L is a cock fcrewed to the face of one of the plates of the frame, and holds the projecting pivot of the fly wheel, to which the fly M is attached with a fpring O preffing againft it, in fuch a manner as to hold it to the arbor of the wheel when the latter revolves, and yet to allow it to turn in a detached ftate, when the motion of the wheel is fuddenly arrefted at the conclufion of the chiming, which is a neceffary precaution for preventing the deftructive effect which might be produced by an inftantaneous check of the momentum of the fly, and anfwers the fame purpofe as the ratchet wheel and click, together with a fpring, on the large fly of the church chime work before defcribed.

The chime barrel which we have here defcribed, is pricked for the performance of a repetition of the eight notes of the octave by a defcending gradation, but it might as eafily have been made to chime the fame fcale by an afcending gradation, or to play any hymn or other tune within the compafs of the eight notes, or even with more notes, by the addition of a few more bells; and that without more than one hammer to each bell; for when the mechanifm is light, the hammers may be made to act with as much rapidity, as the hammers of a grand piano forte.

Chime Work in the Clock Room of S^t. Margaret's Church, Westminster.

Fig. 1.

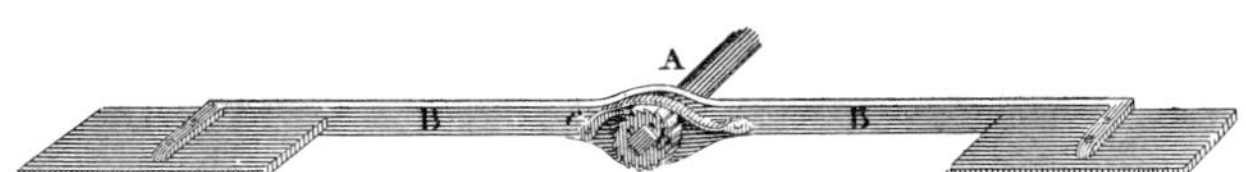

Fig. 2.

J. Farey Junr. delin.

Published as the Act directs, 1808, by Longman, Hurst, Rees & Orme, Paternoster Row.

Engraved by Wilson Lowry.

HOROLOGY.

Chime Work in the Clock Room

of S^t. Margaret's Church Westminster.

Fig. 3.

Fig. 4.

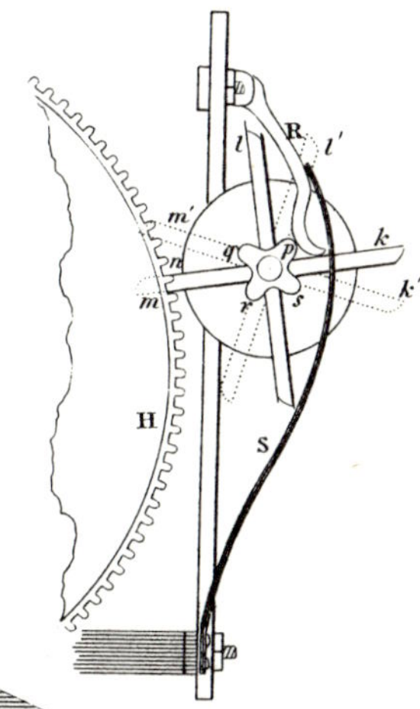

CHIMES.

Pleyel's German Hymn.

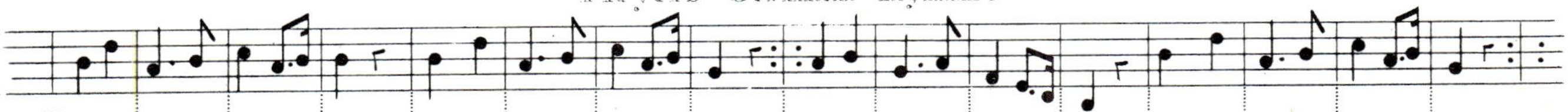

Fig. 1.

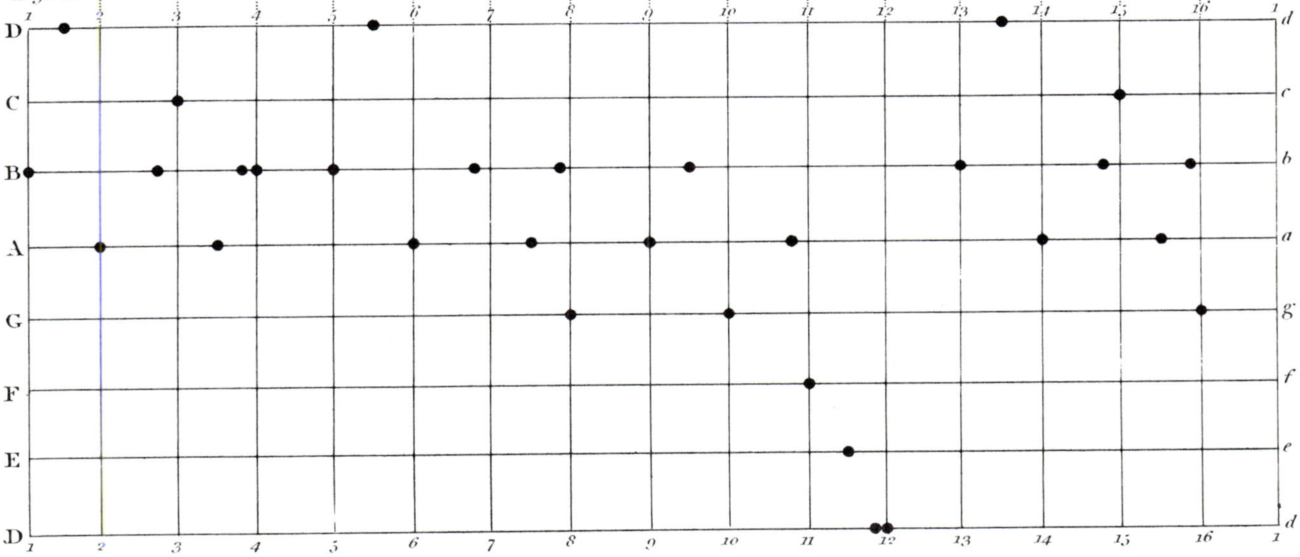

Fig. 4.

Fig. 3.

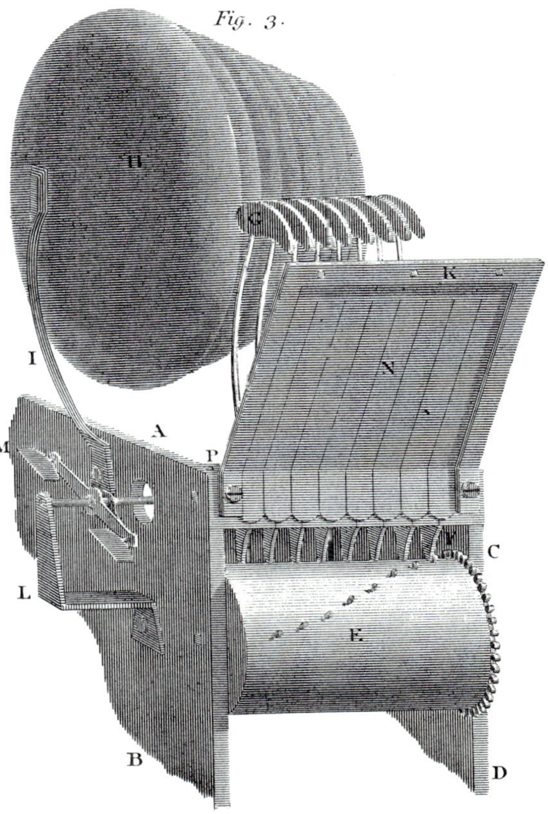

Fig. 2.

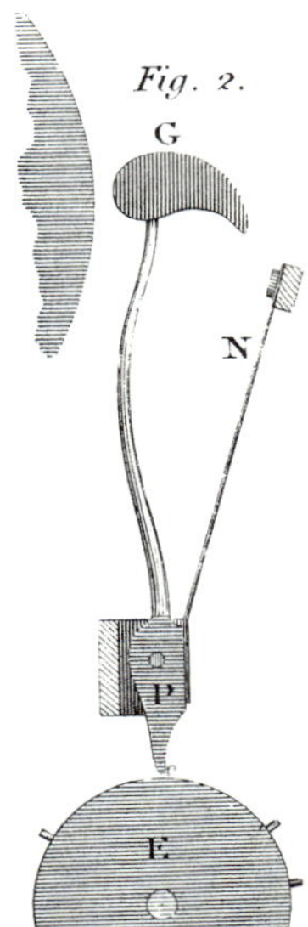

Published as the Act directs, April 1, 1807; by Longman, Hurst, Rees & Orme, Paternoster Row.

Engraved by Wilson Lowry.

Chronometer

CHRONOMETER, from *χρόνου*, *temporis*, and *μέτρον*, *mensura*, is a term in *Horology*, which, in its comprehensive signification, may mean any machine which *measures time*, of which there have been various kinds, such as *Clepsydræ*, *clocks*, *watches*, *regulators*, and *time-keepers*, or *time-pieces*, but the application of this term has been more particularly confined by mechanicians to two kinds of machines; first to such as measure very small portions of time only, by subdividing the second: and, secondly, to such as continue to measure long periods, with great accuracy under all the variations of temperature that arise out of the changes of season and climate.

The former of these two kinds of machines was originally constructed for philosophical purposes, such as measuring the time of the descent of a falling body through a given space, of the efflux of a fluid out of a given aperture under certain circumstances, of the passage of a heavenly body along the eye-piece of a telescope, of the velocity of sound compared with that of light, &c.

The description of a machine of this kind is given in Dr. Desagulier's Experimental Philosophy, and Dr. Hutton says he has seen one, that professes to measure so small a portion as the *fortieth* part of a second, but that it cannot be stopped with certainty within the tenth part of the proposed degree of accuracy. Mr. W. Nicholson, however, says, (vol. iii. p. 50, in a note of his Philosophical Journal, 4to series.) "that there have been instruments made to divide the second into a *hundred* parts; one of these, says he, made by Whitehurst, and regulated by a fly, repeatedly measured the time of the fall of a leaden bullet, in some experiments which I saw, with no greater variation than *one-hundredth* part of the second." The adoption of Mr. Atwood's machine has now superseded the use of such a nice measure of time in the above experiment, and for all the other purposes, the *beat* of a watch, counted by the ear while the eye observes the object of experiment, seems to be not only the most convenient chronometer of any, but is sufficiently accurate for any purpose when the beats are quick, and when their value is known. See our article Beats, in *Horology*.

The second kind of horological machines known by the appellation of chronometers, since Mr. Arnold the elder gave this name to his time-keepers, differs from an ordinary watch principally in the escapement and balance, and deserves our more particular notice, inasmuch as the act of navigating vessels over the extensive oceans of the universe, is greatly indebted to their accurate measurement of time, in all the variations of heat and cold, from the highest navigable latitudes to the equinoctial line. In steering a vessel over the trackless deep, the great desideratum is, to know at any given instant the relative longitude, or distance from the first meridian, and the latitude, or distance from the equinoctial line; a knowledge of these two guides will always suffice to direct on what point of the compass, where the variation is known, a vessel is to be steered, if no current interfere, in order to gain a given harbour. The latter of these two guides, viz. the present latitude, can always be obtained, independently of the ship's reckoning by the log line, by an observation of any of the heavenly bodies when at its greatest altitude, or even with sufficient accuracy by two successive altitudes taken at a distance from the meridian, provided the intermediate lapse of time be accurately noted. The other requisite, the present longitude, however, is not so readily obtained, the lunar method requiring tedious calculations not generally understood, and the occultations of the stars by the moon, and eclipses of the sun, moon, and Jupiter's satellites, not occurring with sufficient frequency to be of much benefit, even if they could be observed with accuracy and convenience on board a ſhip. The method by a good chronometer is, however, not only ſimple in its application, but at all times readily attainable, and therefore is coming into general uſe. We will premiſe a few obſervations reſpecting the utility of a chronometer at ſea, and the improvements it has ſucceſſively experienced, before we proceed to deſcribe the machine itſelf according to ſome of its more perfect conſtructions.

The earth, it is now generally believed, revolves on its axis, in every part of its annual orbit, in an uniform and equable manner, and, on this account, the period of its rotation has been fixed upon as the moſt proper ſtandard of our meaſure of time, and, indeed, is the only invariable ſtandard with which we are acquainted: this period, according to our mode of reckoning, is divided into 24 equal parts, as the rotation has a reference to the ſun; and each of thoſe 24 parts is called an hour, or a *ſolar* hour ſometimes, by way of diſtinction from the *ſidereal* hour, which is a 24th part of a rotation, as it regards a fixed ſtar; hence an hour, or ſolar hour, means one 24th part of the time elapſed ſince any ſpot on the globe paſſed the ſun on the meridian, or ſouth point, in regard to that ſpot: two hours mean twice that ſpace of time; and 24 hours the whole time of a ſolar rotation, which is longer than a ſidereal rotation by $3^{m}\ 55^{s}\ 54^{th}$ of ſolar time, by reaſon of the ſun having advanced apparently 59′ 8″ 10‴ in the ecliptic during a rotation; ſo that a ſolar, being only a relative rotation, is more than a ſidereal or abſolute rotation, by as much as, taken collectively, amounts to an entire rotation in each annual revolution of the earth, to which cauſe the apparent motion of the ſun in the ecliptic is owing: and, to this cauſe, is to be attributed the reaſon why there is a ſidereal more than a ſolar day in each year. But the period of a ſolar rotation of the earth, or any portion of it, may be, and frequently is, reckoned in other terms, implying ſpace paſſed through in a rotation, inſtead of the time occupied by the motion through that ſpace: mathematicians have long been in the habit of dividing a circle into 360 equal parts, one of which is called a degree, or 1°; and, as an equatorial ſection of the earth would be a circle, geographers and aſtronomers have ſuppoſed the equinoctial line divided into 360°, and each degree divided into 60′, (60 minutes) or geographical miles, which minutes are again ſubdivided into 60″ (60 ſeconds), as we divide an hour into 60^{m}, and each of thoſe again into 60^{s}: hence, as the whole 360° of the earth's circumference paſs the ſun's meridian ray in 24^{h}, we know that 15° muſt paſs the ſame in one hour, or 1° in four minutes of time, as alſo 1′ in 4^{s}; conſequently, when we know the time that has elapſed ſince any given ſpot on the globe has paſſed the meridian ſun, we know alſo, by allowing 1° to 4^{m} of time, how many degrees of the equinoctial have paſſed in the ſame time; hours, with their diviſions and ſubdiviſions, and degrees, with their diviſions and ſubdiviſions, being mutually convertible one into the other by direct proportion, or more readily by tables conſtructed on purpoſe. It is neceſſary, however, that we ſhould notice, that there is a difference between a *real* and an *apparent* rotation of the earth as it relates to the ſun, partly by reaſon of the earth's axis being inclined in an angle of nearly 23½° to its annual orbit, cauſing thereby a neceſſary reduction of apparent motion in the ecliptic, or earth's path, to real equable motion in the equator, and partly by the alternate acceleration and retardation of the earth's motion in her orbit at different times of the year, which irregularity requires a correction called the "Equation of the Center:" theſe two cauſes of apparent irregularity in the earth's rotations have their joint effects allowed for, by what is called "Equation of Time," uſually inſerted in a table with this title, and alſo placed in the column of "clock faſt," or "clock ſlow," in the almanacs; the quantity, therefore, correſponding to any given day in the year, in the equation table or almanac, muſt always be added to or ſubtracted from the time ſhown by an accurate chronometer, to make it agree with *apparent*, or what the French (and lately ſome of our Engliſh authors) improperly call *true* time; that time in our opinion being *true*, which is mean, and which correſponds to the real rotations of the earth; for theſe rotations, conſidered by themſelves, are equable, and not affected by the cauſes of thoſe two *apparent* irregularities we have noticed, as ariſing ſolely out of the relative poſitions and ſituations of the earth and ſun.

But we have not to regard only the period of the earth's ſolar rotation; its direction, alſo, muſt be conſidered, which is from that point of the horizon, which we call weſt, to that which is denominated eaſt; thereby cauſing the ſun, which is really a ſtationary body, or nearly ſo, to *appear* to move on the contrary from eaſt to weſt every ſolar day, and the ſtars likewiſe, in the ſame direction, once in every ſidereal day; but theſe latter bodies, being placed at an immenſe diſtance beyond the earth's orbit, have no apparent change of place, and, therefore, require no correction; conſequently, there is no difference at any part of the year between a *mean* or *true*, and an *apparent*, ſidereal day, which muſt have been the caſe, notwithſtanding the immenſe diſtance of the ſtars, if the rotations of the earth had not been uniformly equable. It is on account of this equality among the ſidereal days that aſtronomers have proportioned the pendulums of their regulators to vibrate ſidereal ſeconds, that the right aſcenſion of the heavenly bodies is given in ſidereal time, and that the late Margetts made his chronometers, with great ingenuity, to ſhow at the ſame time both ſolar and ſidereal time, and, conſequently, the ſun's mean right aſcenſion at any time, which is always equal to their difference: and we may add here, that the conſtant variation that is taking place between mean and apparent ſolar time, was the reaſon why a clockmaker in London, whoſe name is unknown, and after him on the continent, H. Sully, Alexander le Bon, Julien le Roy, Enderlin, l'Admiraud Paſſemant, Rivaz, Berthoud, and others, have made equation clocks on different conſtructions, to indicate both mean and equated, or apparent time.

From theſe introductory remarks on time, and its connection with, or rather dependence on, the earth's rotation on its axis, it is eaſy to conceive, that all places on the globe which paſs the ſun's meridian ray ſooneſt, count their 12 o'clock, or noon, earlier than thoſe which follow in ſucceſſion; but the eaſtern parts paſs firſt, and thence have their time more advanced, or earlier than the following more weſtern parts have; and the difference is, as we have ſaid, at the rate of 4^{m} for every degree of diſtance. This diſtance is called *longitude*, by reaſon of the equatorial diameter of the earth being *longer* than the polar diameter, in the direction of which latter, the breadth or *latitude* of the earth is counted both ways from the middle. The longitude may have its reckoning to commence at any aſſignable point on the globe, and all the other parts will be called eaſt or weſt of that point, which is called the firſt meridian, accordingly as they precede or follow it in each rotation of the earth, and the quantity will be either ſo many hours, minutes, and ſeconds of time, or ſo many degrees, minutes, and ſeconds of ſpace, as correſpond to that time. Now, it is very obvious, that, if we could at the ſame inſtant know the time accurately, as counted at each of two different places, ſituated reſpectively eaſt and weſt of one another, the difference

of

of those two times so indicated, would be their difference of longitude in time, which, converted into degrees, minutes, and seconds, would be their difference of longitude in this denomination, from which, in a known latitude, the actual distance of the two respective places may, by calculation, be ascertained. What, therefore, a chronometer has to do, is, *to tell at all times the hour, minute, and second, as counted at the first meridian*, whether London, Paris, or any other place, to the time of which it was accurately put previously to the commencement of a voyage; for, as the time at any island, or place of a ship, can be had by means of Hadley's quadrant, or sextant, or more accurately by means of Troughton's reflecting circle, from a celestial observation; the quantity that this time exceeds or falls short of the time indicated by the machine, as being the time at that moment at the first meridian, will be the island's or ship's comparative longitude in time, east, if the chronometer is behind, but west, if before the time by a celestial observation. In our English ships, the chronometer is a kind of travelling companion which tells, whenever consulted, what the exact time is at Greenwich; nor is it indispensably necessary that it should keep time exactly with the clock at Greenwich observatory, provided the daily gain or loss, called the *rate*, be ascertained and applied as a correction accordingly as it accumulates. It is, however, an indispensable requisite, that the daily gain or loss should not differ materially from itself at different periods, or under the changes of temperature experienced in different climates; and the fulfilment of this condition constitutes any portable horological machine, a marine *chronometer*, or time-keeper, whatever may be its construction or price. Any of the other methods of ascertaining the longitude may be occasionally put in practice with advantage, as a check upon the simple determination by the chronometer; for their operations will detect its daily errors, and ascertain nearly their amount at the time.

The first person who proposed to ascertain the relative longitude of any place or ship at sea, by means of an horological machine for indicating the time of the first meridian, was, as has been asserted, Gemma Frisius, about the year 1530; (vide "De Principiis Astronomiæ et Cosmographiæ.") This method was described and recommended in Carpenter's Geography so early as the year 1635; but the state in which horological machines was, at that time, prevented his accomplishing the design: the idea, however, once suggested, was valuable; and stimulated ingenious mechanists, in times succeeding, to attempt the accomplishment of an object of such national importance. The discovery of the isochronism of the pendulum turned the minds of ingenious men to the improvement of clocks; and we find that lord Kincardine tried a marine pendulum clock by Dr. Hooke in the year 1662; and that Christian Huygens, the celebrated Dutch mathematician and mechanician, contrived a time-keeper, actuated by a spring, and regulated by a pendulum, which was tried at sea by major Holmes in the year 1664, and spoken of by him in favourable terms. The escapement was of the crown-wheel kind, which, from its nature, is almost constantly under the influence of the maintaining power; but a small weight connected with the crown-wheel, was raised every half-second by the maintaining power, and gave an impulse to the pendulum, which, therefore, was not affected by the irregular transmission of the maintaining power through the train of wheel-work: this contrivance was ingenious, and obtained the name of *remontoir*. The pendulum, however, was not only unsteady in its action during the tossing of a ship, but was subject to a variation in its length by change of temperature, as well as to a change of weight depending on the parallel of latitude; the latter of which changes, indeed, was afterwards discovered. There was, moreover, a pair of cycloidal cheeks of brass so fixed, as that the thread of suspension, by being evolved from them alternately at each successive vibration, might make the bob of the pendulum describe the involute of a cycloid, which this author first proved, was itself a cycloid, possessed of the peculiar property of rendering the vibrations in long and short arcs of equal duration; this cycloidal doctrine was plausible in theory, but could not be reduced to practice, because it supposed; 1st, the pendulum invariable in length; 2dly, the collection of all the weight to be into one point; and 3dly, the absence of friction and other kinds of resistance, to which mechanism is subject. Huygens's contrivances, notwithstanding, together with the doctrines contained in his "Horologium Oscillatorium," may be considered as having laid the foundation of horological science.

The balance, which had preceded the pendulum, was again resorted to as a regulator of portable time-keepers; and though it was found incompetent to its office in a detached state, yet, by the aid of a slender spring to quicken and regulate its sluggish vibrations, it has ultimately turned out to be of eminent utility. It has been contested by Huygens and Dr. Hooke, which of these two skilful mechanicians first introduced the spring, called usually the pendulum spring, from the isochronal property which it possesses, like the pendulum, when of a proper strength, shape, and length; and F. Berthoud asserts, that though Hooke applied it first in a straight form, yet Huygens first adopted the spiral shape, as being more favourable to isochronism; while others are of opinion, with more probability, that Hooke actually applied it in a spiral form among the *twenty* several methods that he said, in his lectures at Gresham college, in the year 1664, might be used to answer the same purpose (vide "Lectiones Cutlerianæ," 1673.) Indeed the account of the spiral spring, adopted by Huygens, was not published in the Philosophical Transactions, until the year 1675, N° 112, whereas Hooke had discovered the isochronism of springs, and registered his discovery by an anagram composed of the Latin sentence "Ut tensio sic vis," in the year 1658.

But whoever was the inventor of the spiral form of the spring attached to the balance, and making with it a regulator for portable machines, they both still remained subject to alterations in their dimensions by the successive changes of heat and cold; for, by the former of these opposite temperatures, the spring became weaker, and the size and consequent momentum of the balance greater, in consequence of their enlargement, so as to produce a very sensible loss in the daily rate of going of a watch with such a regulator; also an acceleration beyond a mean rate was soon observed to be the consequence of increased cold, which, on the contrary, diminished the dimensions of the metallic parts, and thereby, at the same time gave additional strength to the spring, and likewise reduced the size and momentum of the balance. About the same period Leibnitz attempted to confine the vibrations of the balance with a spiral spring, to be of equal extent, by means of an additional spring to be applied to the balance wheel, and to be, like Huygens' remontoir, wound up by the maintaining power; but his endeavours, whatever ideas they may have suggested to Harrison, Mudge, or Haley, for their auxiliary springs, were not crowned with complete success. Hautefeuille also, in the year 1674, presented to the Academy of Sciences, at Paris, a balance with a straight spring, acting somehow instead of an escapement, but how far it resembled the spring detent of Arnold we know not, nor do we find that it was adopted in practice. This society, notwithstanding, thought the subject of such

 importance

importance, that, in the year 1720, they propoſed the following queſtion to be determined for a public reward: viz. "What is the moſt perfect method of preſerving on the ſea the equable motion of a pendulum, either by the conſtruction of the machine, or by the ſuſpenſion?" A memoir written by Maſſy, a Dutch clock-maker, obtained the prize, but he had not the ſatisfaction of ſeeing his plan executed. About a year afterwards (1724) Henry Sully, an Engliſh clock-maker, who had ſettled at Paris about eight years previouſly, preſented the ſame academy with a marine time-keeper, made in 1721, and publiſhed a deſcription of it in French, by the title of "Deſcription abrégée d'une horloge de nouvelle invention pour la juſtè meſure du temps en mer." Beſides the above, Sully made a ſecond marine time-keeper, which was tried at ſea in 1726, but the inventor died two years afterwards, a martyr to his horological ſtudies, before he had brought his machines to that ſtate of perfection which their object demanded. His pieces had vertical balances carrying cycloidal metallic pieces, round which a thread, or ſlender wire, was wound at the upper end, while the lower end was attached to a lever with an adjuſtable weight to effect the iſochroniſm of the balance, inſtead of a ſpiral ſpring: the horizontal pivots of the balance alſo moved on the angular point included between two large rollers, which method of leſſening friction, we believe, was the invention of this author. He alſo made a marine watch with a ſpiral pendulum ſpring, into which friction rollers, like Mudge's, were introduced, and had he lived longer, chronometry would no doubt have been greatly promoted by his labours.

It was about this period, that jewelling, another eſſential improvement in time-pieces, was introduced, according to Berthoud, by Mr. Fatio, a native of Geneva, who, not meeting with encouragement in France, came over into England, and brought his invention into notice.

The Academy of Sciences at Paris again propoſed a reward for the year 1747: the ſubject was "The beſt method of finding the hour at ſea, whether by day, by twilight, or at night, when the horizon cannot be diſtinguiſhed." The reward was obtained by Daniel Bernouilly's memoir, intitled, "Recherches mechaniques, et aſtronomiques," in which was diſplayed much ſcience, but the author's want of ſkill in mechanical operations prevented his labours being attended with complete ſucceſs.

In the mean time, the changes in the length of the pendulum began to be compenſated, firſt by means of quickſilver contained in a tubular rod, by Graham, and ſoon after by the oppoſite expanſions of different metals, by Harriſon, who, ſtimulated by the Britiſh parliamentary reward that had been previouſly offered to the public for marine time-keepers, applied the ſame principle to a watch to effect a ſelf-regulating kirb (or curb), for limiting the effective length of the ſpiral pendulum-ſpring to correſpond to the ſucceſſive changes of heat and cold, which changes were now known to alter the force of this ſpring, and the momentum of the balance. From this origin we may date the beginning of all the different kinds of compenſation-mechaniſm that have proved permanently uſeful in time-keeping; and if we add to Harriſon's invention of the metallic compenſation, his remontoir, and his addition of a ſecondary ſpring as an equivalent ſubſtitute for the maintaining power during the time of winding up, which is an eſſential requiſite in producing permanent motion, he may be fairly conſidered as the parent of modern chronometry.

The Britiſh parliament had, indeed, before the French academy, offered, ſo early as the year 1714, in the reign of queen Anne, a reward of 10,000l. for any method of determining the longitude within the accuracy of one degree of a great circle; of 15,000l. within the limit of 40 geographical miles, and of 20,000l. within the limit of 30 ſuch miles, or half of a degree, provided ſuch method ſhould extend more than 80 miles from the coaſt: and after this act, two others paſſed in the reign of George II. ſtat. 14 and 26, to promote the ſame purpoſe; but an act paſſed in the preſent reign, in the year 1774, repealing all the former ones, and offering ſeparate rewards to any perſon who ſhould invent a practical method of determining, within certain circumſcribed limits, the longitude of a ſhip at ſea: for a time-keeper, the reward held forth to the public is, 5000l. for determining the longitude to or within one degree; 7500l. for determining the ſame to 40 geographical miles, and 10,000l. for a determination at or within half of a degree. This act, notwithſtanding its abridged limits and diminiſhed rewards, has produced ſeveral candidates ſince Harriſon, who received the whole reward of the firſt act, for parliamentary remuneration, of whom Mudge, the two Arnolds, and Earnſhaw, have had their labours, as will be ſeen hereafter, crowned with partial ſucceſs. Beſides theſe, there have been various other chronometer-makers, whoſe pieces have performed with great accuracy, but whoſe names we omit to introduce here, leſt we ſhould ſeem partial to ſome at the expence of the reputation of others. Indeed, the art of conſtructing chronometers is lately become ſo general, that it is difficult to decide whoſe name ought to ſtand firſt on the liſt of excellent makers, and we hope that the ſpirit of competition for public fame will continue to entitle our Engliſh manufacturers to that preference among naval officers, which the excellence of their workmanſhip entitles them to expect. For, even in a commercial point of view, it was proved to the late Mr. Pitt, when he laid a tax on watches, by the committee of watch-makers convened in the pariſh of Clerkenwell, that a piece of the value of 500l. had been manufactured out of materials which did not, in their native original ſtate, coſt more than *ſixpence!* This repreſentation, we are credibly informed, induced the noble ſtateſman to abandon his plan of taxing an article, the value of which depended ſo much upon ingenuity and labour, and by the manufacture of which thouſands of ſubjects are entirely ſupported: nay, further, on learning that the French and Swiſs could afford to ſell three gold watches for the price of one Engliſh one, the ſame miniſter took off the duty of ſixteen ſhillings per oz. from watch-caſes of this metal, and ſubſtituted only one ſhilling, the price of the trial at Goldſmiths'-hall. We have juſt ſaid that Harriſon obtained the firſt and moſt ample reward for his inventions, but we are not to conclude from thence, that his pieces excelled all others; they were ſpecimens of great ingenuity and proofs of unwearied induſtry, which certainly were not overpaid; but the inventor himſelf was candid enough to confeſs that the balance, balance-ſpring, and compenſation-curb, were not contemporaneouſly affected by heat and cold, but that ſmall pieces of metal were ſooner affected than large ones, and alſo pieces in motion before pieces at reſt; whence he was led to conclude, that if the proviſion for heat and cold could properly be in the balance itſelf, as was the caſe with his gridiron pendulum in clocks, the time-piece might be made much more perfect.

Harriſon's ſuggeſtion of a compenſation-balance, in place of a compenſating curb for the balance-ſpring, found its way into France, and rouſed the attention of the watch-makers of that nation; and, to do them juſtice, we muſt allow that Peter, the eldeſt ſon of Julien le Roy, who was himſelf an eminent watch-maker, had the honour to be the firſt who accompliſhed the ſuggeſted deſideratum, by means of of two thermometers, one of mercury, and the other of alcohol, attached to and carried by the balance itself, which contrivance effected the compenfation, by bringing a portion of the mercury nearer to, or by removing it farther from the centre of the balance, according to the different ftates of the atmofphere. (See the Defcription under COMPENSATION-*balance*.) A chronometer on this conftruction was prefented by Peter le Roy to the king of France, on Aug. 5th, 1766, for which the prize of the Academy of Sciences was awarded him on the laft day of the fame month: he alfo publifhed an account to accompany the piece, entitled, "Memoire fur la meillure manière de mefurer le temps en mer, &c." in which memoir he afferts, a circumftance very worthy of notice, that he made another compenfation-balance entirely of pieces of different metals; viz. of brafs and fteel riveted together, like Harrifon's compenfation-curb, but bent into two feparate femi-circles in fuch a way, that each, carrying a metallic weight near its extremity, brought it alternately nearer to, or removed it farther from the centre of the balance thus formed, agreeably to the variations of the atmofpheric temperature; and though the inventor preferred at the time the thermometrical compenfation, yet the metallic one, fpoken of in the memoir in queftion, was, no doubt, the archetype of all our prefent compenfation-balances. It might now have been expected that a time-piece, with a movement aided by friction-rollers, or by jewels in the pivot-holes, and with a compenfation-balance regulated by a fpiral-fpring, would have performed alike under all circumftances, but ftill it was found that, however well the fufee was fhapen, and adjufted to the different intenfities of the main-fpring, yet fuch an unequal tranfmiffion of that power took place, even in the beft movements, in confequence of there being alternately favourable and unfavourable pofitions of the acting teeth of the wheels and pinions, and in confequence of the impediments to free motion occafioned by the variable denfity of the oil ufed, and by the acceffion of particles of extraneous matter, that the action of the pallet-wheel upon the pallets then in ufe was found fufficiently irregular to occafion an inequality in the impulfes given to the balance, and a confequent inequality in the magnitude of the arcs of vibration. This inequality in the magnitude of the arcs of vibration would not, indeed, have affected the rate of the going of the chronometer, if all lengths of the regulating-fpring had been found on trial to be equally ifochronal, but the fame Peter le Roy difcovered what Dr. Hooke knew long before, (vide his Poftfcript to "a Defcription of Heliofcopes, &c.") that there is a *certain length* in each good uniform fpring *which only* is ifochronal, or, in other words, which has the property of regulating the balance fo, that all arcs of vibration, long or fhort, fhall be performed in the fame time. This difcovery, or rather re-difcovery, of Peter le Roy, at a time when chronometry had made confiderable advances towards perfection, was calculated to do away the fanguine hopes that had been entertained of the good performance of time-pieces on Harrifon's conftruction, in which the effective length of the regulating-fpring was conftantly altering with the variations of temperature; and to this circumftance principally may be imputed, perhaps, the fmall number of time-pieces that were manufactured after Harrifon's model, notwithftanding the large premium which was awarded him. A remedy for the unequal tranfmiffion of the maintaining power had been, however, adopted by Harrifon, when he introduced the remontoir to produce equable action at the contrate-wheel of his pieces, a contrivance worthy of his genius, whether the idea was original, or borrowed by him from Huygens; but this was a remedy for only one of the two caufes of irregularity in the momentum of his balance; it might, and probably did, equalize the maintaining power nearly, but would not counteract the ifochronal defect produced in the regulating-fpring by the compenfation-curb, in all the various arcs of vibration which every piece is liable to experience in their different ftates of foulnefs.

The obftacles to equal tranfmiffion of force in a chronometer, led to the invention of various efcapements, both on the continent and in England; fome of which were intended to act ifochronally in concert with the regulating fpring, and others were fo conftructed as to give the impulfe almoft inftantaneoufly, and at the moft favourable inftant of the vibration of the balance, fo that the force derived from the maintaining power, to perpetuate the vibrations, might derange the natural ifochronal property of the balance, and its regulating fpring as little as poffible.

On confidering this fubject, it occurred to the moft fcientific artifts, that the regulating power of the balance and balance-fpring, which was found to be too much under the dominion of the maintaining power with the common efcapements, would be the leaft deranged if the impulfe derived from the maintaining power were momentary, particularly if it were applied at that point of the vibration where the momentum of the balance is a maximum; it alfo occurred, that the momentum of the balance itfelf ought to be as great as practicable, compared with the impulfe given to the pallet-wheel, and likewife that a momentum compofed more of velocity than of weight would be moft favourable for a balance with flender pivots. Thefe, and fimilar confiderations, the refult of much thought and reafoning, fuggefted a great variety of defigns for new efcapements, many of which have been brought into practice with an advantage correfponding to the importance of the object; the moft recent of which promifes to be of permanent utility in chronometry. Thefe efcapements have obtained the appellation of *free* or *detached*, from the circumftance of their being detached from the balance during the greateft part of its vibration. It has been matter of contention among horological writers, who was really the firft inventor of a detached efcapement, but it feems now to be pretty generally admitted, that Peter le Roy, whom we have mentioned as the inventor of the firft compenfation balance, was alfo the inventor of the firft detached efcapement. It would lead us far beyond the limits of our prefent article minutely to defcribe here all the variations in the fhape and mode of action of the different efcapements, by Julien and Peter le Roy, Berthoud, and others, that have been made on the continent, as well as of thofe made in Britain; on which account we fhall refume the fubject under the article ESCAPEMENT, and there give a detailed account of the fucceffive efcapements, as nearly in their order of time as can be afcertained. In the mean time, the reader will obtain, we prefume, a fufficient knowledge of the efcapements at prefent in ufe, from the defcriptions that are fubjoined to this article, of fome of the beft chronometers of modern makers, where we have given drawings and an account of an entire piece by Brockbank, and of fuch parts of the pieces of other modern makers as differ from it in conftruction. We wifh it, however, to be diftinctly underftood, that we difclaim all partiality to individuals, and give a drawing of Brockbank's chronometer in an entire ftate, for no other reafon but becaufe we are unwilling to diminifh the fale of the pamphlet containing the drawings and defcriptions of thofe by Arnold and Earnfhaw, lately publifhed by order of the commiffioners of longitude; and alfo becaufe Meffrs. J. Brockbank and Co. have been fo obliging as to allow us to take fuch original drawings as it is prefumed will fhew the relative fituations of the different parts in a favourable point of of view, for giving a clear idea of their relative offices and modes of action.

A method having been devifed of limiting the quantity of impulfe given to the balance, fo as to be juft fufficient to keep it vibrating when put in motion, but not fufficient to produce motion from a ftate of quiefcence, as will be feen hereafter, and a felf-compenfating mechanifm having been adapted to the balance itfelf, it may now be fairly inferred, that chronometers have arrived nearly at their *ne plus ultra* of perfection, and that they may fhortly be expected to find the permanent level of their price, uninfluenced by the recommendation of a fuppofed fuperiority arifing out of the *name* of the *maker*, or rather, as we might fay, of the *vender;* for it is a fact not to be controverted, that more chronometers have been fold by one individual than have been or could be made in the fame time under his own roof. Still however, the *adjuftment* of a chronometer for rate, temperature, beat, and pofition, is an object of the utmoft importance to the due performance of even the beft machine of this kind that ever was made; and if improvements are yet neceffary in this delicate but important branch of our manufactures, they are fuch as ought, and may reafonably be expected, to conduce to accuracy and expedition in completing thofe four kinds of operations; operations which have been found to be very troublefome in fome of the conftructions, inafmuch as one of the adjuftments, however delicately made, may not only be over or under-done, but may and frequently does derange another adjuftment previoufly made. Indeed, in the prefent advanced ftate of chronometry, one maker's excellence is often diftinguifhed from that of another by fome flight deviation from his contemporaries' method of fhaping certain parts, or of adjufting the balance according to fome fecret method peculiar to himfelf, which he does not choofe to difclofe, and for which, perhaps, he cannot give a good reafon; hence, an important trifle, no way dependent on or connected with fcientific principles, has been found, in certain inftances, to lay the foundation for pretenfions to exclufive merit.

We will conclude our narrative of the improvements in chronometry, by pointing out briefly fome of the diftinguifhing features of the different conftructions of our contemporaries; and leave to the public the exercife of their own choice in the felection of a particular maker. Without entering more minutely into a detail of the French chronometers in this place, it may be fufficient to ftate here generally, that their detached efcapements have detents, or pieces to fufpend the maintaining power for a certain time, moveable on an arbor with pivots, as will be feen under the article ESCAPEMENT; whereas the Englifh detents act by means of fprings without pivots, which confequently require no oil, and are alfo of a more fimple conftruction.

The late Mr. Arnold took out two patents for improvements in his chronometers, the one in April 1776, and the other in 1782. The former of thefe patents was for the invention and application of compenfation bars in the conftruction of his balance, together with the invention and application of what he calls the helical, but which is properly the cylindrical balance-fpring. The fecond patent was for three different ways of applying the compenfation-bars, for an improvement in the balance-fpring, particularly in the bending of the laft coil at the end of it, for his invention and application of the fpring-detent, and alfo for the cycloidal, or more properly epicycloidal fhape of the tooth of the balance-wheel. It has been already faid, that Peter le Roy was the firft who applied a metallic compenfation to the balance itfelf, but it does not neceffarily follow from thence, that Arnold did not alfo invent the one he adopted. We are perfuaded that a man of Mr. Arnold's known integrity and veracity would not make oath, in taking out a patent, of an invention which he did not at leaft confider as his own; and we fhall have occafion to fhew, under our article COMPENSATION-*balance*, a variety of different fhapes given by Arnold to his balances, and actually tried in practice *before* he adopted the one in prefent ufe; fome of which balances are yet in exiftence. At all events, Mr. Arnold muft be allowed the merit of having introduced the compenfation-balance into general ufe, when Peter le Roy preferred the thermometrical tube. The late Mr. Brockbank was the firft perfon who united the two metals by fufion, which le Roy had united by pins, that muft have interfered with the regularity of the flexure by different temperatures. Mr. Brockbank was alfo the firft who ufed the method of turning an expanfion-rim out of a folid compound plate, made by covering the fteel plate in a crucible of fufed brafs, and of cutting it into portions afterwards, thereby enfuring the uniformity both of figure and weight, which two properties are equally effential in any balance, as the name imports; fo that if it fhould be contended that the contruction of the compenfation-balance was not invented in this country, at leaft the practical application of the principle is our own; to which confideration we may add, that Arnold fenior invented and introduced the ufe of the fpring detent, which requires no oil. Mr. Arnold was likewife the firft watchmaker in England who laid much ftrefs not only on the fhape, but alfo on the particular length of the balance-fpring in practice, which Dr. Hooke and Peter le Roy had both fhewn was neceffary to be attended to in order to render the fpring ifochronal under all arcs of vibration, which is an effential object of adjuftment, and which no doubt influenced his choice of the fhape of the fpring; and we are informed that he was fo far fuccefsful in his attempt to afcertain the precife point for limiting the beft practical length of fome of his balance-fprings, that, after the example of Peter le Roy, to whofe contrivances, it muft be confeffed, he feems to have paid great deference, he fucceeded in making a movement go accurately without a fufee, by the mere regulation of an ifochronal fpring, which is an indubitable proof that the irregularities of the maintaining power can have but little influence on the rate of a chronometer with an ifochronal balance-fpring, efpecially when it has, moreover, a detached efcapement. Mr. Pennington affirms that there are many ifochronal points in every fpring, which difcovery accounts for the different lengths of the various balance-fprings that are made of the fame clue.

The late Mr. Mudge laboured to effect an equalization in the impulfe given to the balance by remontoirs of fpiral fprings, acting fo conftantly at each vibration, that the efcapement of his time-keepers could not properly be called a detached one; though the one introduced into her majefty's watch of his contrivance, and copied by the late Margetts, and by Emery, which gives an impulfe at every vibration, fufficiently great to produce motion from a ftate of reft, may be claffed among the detached kind.

Emery's balance had weights fliding on its croffes, and having their pofitions regulated under different degrees of heat by the variable flexure of compenfating bars compofed of two different metals, in the fhape of an S, as explained under our article COMPENSATION *balance*, where it will appear that this was one of the varieties invented by Arnold, whofe workman afterwards went to Emery. The Brockbanks have their chronometers diftinguifhed by the pofition of their locking and unlocking fprings, and alfo by a peculiar method of banking by means of the protrufion of the coils of the balance fpring, as will be hereafter explained.

Earnfhaw's

Earnfhaw's chronometers differ from Arnold's in the fhape and pofition of the detent and fprings, in the fhape of the balance wheel, and ftructure of the balance fpring and balance, all which will prefently be explained.

Recordon, fucceffor to *Emery*, at Charing Crofs, has a compenfation-balance perforated at the circumference with various tapped holes, into which the fcrews of adjuftment for temperature and pofition may be fucceffively removed, according to circumftances: this mode of adjuftment is practifed by Pennington, and was, we underftand, originally his contrivance.

Haley, a watch-maker at the corner of Wigmore-ftreet, Cavendifh-fquare, took out a patent on the 17th of Auguft, 1796, the particulars of which are given in the fixth volume of the "Repertory of Arts." The principle on which the patent was granted, confifted chiefly of a fecond cylindrical fpring and fpring-arbor with pallets, &c. interpofed between the efcapement-wheel and the balance to give an impulfe to the balance at each vibration, inftead of the impulfe ufually given by the force tranfmitted through the train. See ESCAPEMENTS *for Watches*.

Grimalde, in the Strand, who now makes a confiderable number of chronometers, fome of which, we learn, have the teftimony of naval officers in their favour, has informed us, that he places the cock fo conveniently, and adjufts for pofition fo readily, by a particular contrivance, which does not require the cock to be taken off, that it would, we think, contribute to the ftock of improvements already known, if he would make his method public.

Mr. *Hardy* has lately introduced a new mode of banking, by a lever attached to the exterior coil of an heliacal fpring, which is thrown out to catch a pin in the balance; and has propofed a new mode of making the fpring ifochronal, by making the ftud moveable on a fecond fpring; but experience muft prove their utility.

We might add a long lift of the other perfons who make chronometers, not in London only, but at Edinburgh and Liverpool, were we aware that there is any material difference in the conftruction of their mechanifm, or methods of adjuftment, from thofe, or fome of thofe, which we have noticed.

To aid the refearches of thofe readers who wifh to trace more minutely the rife and progrefs of the art of meafuring time by mechanical inventions, we fubjoin a lift of the principal authors who have, from time to time, written on this interefting fubject, viz. Hieronomi Cardani de Varietate rerum; 1557, fol. Bafileæ. Conrandi Dafypodii Defcriptio Horologii Aftronomici Argentinenfis; 1578, 4to. Argentorati. Guidonis Pancirolli Antiqua deperdit a & Nova reperta; 1607, 8vo. Ambergæ. L'ufage du Cadran ou de l'Horloge phyfique univerfel par Galilée; 1639, 8vo. Paris. Benedicti Haefteri Monafticæ Difquifitiones; 1644, fol. Antwerpiæ. Horloge magnetique, elliptique ou ovale nouveau, pour trouver les Heures du Jour & de la Nuit, par Pierre Georges; 1660, 8vo. Toul. P. Gafparis Schotti Soc. Jefu, Technica Curiofa, feu Mirabilia Artis; 1664; 4to. Herbipoli. Chriftiani Hugenii Zulichemii Horologium ofcillatorium; 1673. Parifiis. Lectiones Cutlerianæ. by R. Hooke; 1673. London. Gulielmi Oughtred Etonenfis Opufcula Mathematica hactenus inedita; 1677, 8vo. Oxonii. Matth. Campani de Alimentis Horologium, folo naturæ motu atque ingenio, dimetiens et numerans momenta temporis conftantiffime æqualia; 1677, 4to. Romæ. Pendule perpetuelle, par l'Abbe de Hautefeuille; 1678, 4to. J. J. Becheri Theoria et Experientia de nova Temporis dimentiendi ratione & Horologiorum Conftructione; 1680, 8vo. Londini. Gilberti Clark Oughtredus explicatus, de Conftructione Horologiorum; 1682, 8vo. Londini. Horological Difquifitions by Smith; 1698. London. Memoirs by De la Hire, Leibnitz, Saurin, Maffy, D. Bernoully, Romilly, J. and P. Le Roy, &c. &c., contained in the "Memoirs de l'Academie des Sciences à Paris," and in "Machines Approuves," in various volumes. Artificial Clock-maker, by Derham; 1714. London. Defcription d'une Montre de nouvelle Conftruction, par H. Sully; 1716. Regle artificielle du Temps, par H. Sully; 1717, (& 1737, a Paris, par Jul. Le Roy). Traité des Forces mouvantes, par Camus; 1722. Defcription abregée d'une Horloge d'une nouvelle Conftruction pour la jufte Mefure du Temps fur Mer, par H. Sully; 1726. à Bordeaux. Traité general des Horloges, par Alexandre; 1734. Traité d'Horlogerie pratique, par Thiout & Euderlin; 1741, 2 vols. 4to. à Paris. Traité des Echappemens, par Jodin; 1754, 12mo. à Paris. Traité d'Horlogerie, par Le Poute; 1755, 4to. à Paris. L'Art de conduire & de regler les Pendules & les Montres, &c. par Ferd. Berthoud, 1759, 12mo. à Paris. Etrennes Chronometriques, par Pierre Le Roy, 1759. Effai fur l'Horlogerie, par Ferd. Berthoud; 1763, 2 vols. 4to. & 2d ed. 1786, à Paris. Inftitutiones Horologicæ, in vol. ii. of Mathem. Inft. by Ben. Martin; 1764. The Elements of Clock and Watch-work adapted to Practice, by Alexander Cumming; 1766, 4to. London. The Principles of Mr. Harrifon's Time-keeper, with Plates of the fame; 1767, 4to. London. Dr. Mafkelyne's Obfervations and Calculations on ditto; 1766 and 1767, London. Journal du Voyage de M. le Marquis De Courtenvaux; 1768, 4to. a Paris. Voyage par M. Caffini; 1770, 4to. à Paris. Encyclopedie, Arts & Metiers. Voyage par M. d'Eveux-de-Fleurieu; 1773, 2 vol. 4to. à Paris. Traité des Horloges Marines, par Ferd. Berthoud; 1773, 4to. à Paris. Eclairciffemens fur l'Invention, la Theorie, &c. par Ferd. Berthoud; 1773, 4to. à Paris. An Introduction to the mechanical Part of Clock and Watch-work, by Thomas Hatton; 1773, 8vo. London. Precis des Recherches faites en France, &c. par M. Le Roy (fils de Julien); 1773, 4to. à Paris. Defcription concerning fuch Mechanifm as will afford a nice and true Menfuration of Time, by James Harrifon; 1775. Les Longitudes par la Mefure du Temps, par Ferd. Berthoud; 1775, 4to. à Paris. Voyage par Verdun, Borda, & Pingré; 1778, 2 vols. 4to. à Paris. De la Mefure du Temps, par Ferd. Berthoud; 1787, 4to. à Paris. Horlogerie Pratique, par M. Vigniaux; 1788, 8vo. a Paris. Traité des Montres & Longitudes, par Ferd. Berthoud; 1792, 4to. à Paris. Dr. Mafkelyne's Anfwer to Mudge's Narrative; 8vo. 1792. Atwood's Inveftigations for determining the Times of Vibration of Watch-balances, in the Philofophical Tranfactions of London; 1794. Mr. Mudge's Reply to Dr. Mafkelyne's Anfwer; 8vo. 1792. Suite du Traité des Montres a Longitudes, par Ferd. Berthoud; 1797, 4to. à Paris. Voyage de La Peroufe; 4 vols. 4to. à Paris, 1797. A Defcription, with Plates, of Mudge's Time-keeper; 1799, 4to. London. Ausfvhricke Gefchichte, &c. by John Henry Maurice Poppe; 1801, 8vo. Leipfic. The article WATCH-WORK in the 2d volume of the Supplement to the Encyclop. Britannica. Hiftoire de la Mefure du Temps, par les Horloges, par Fred. Berthoud; 1802, 2 vols. 4to. à Paris. A Journal of Natural Philofophy, Chemiftry, and the Arts, by W. Nicholfon; 4to. feries, and 8vo. feries; fee various volumes. Effai fur les Montres à Repetition, par François Crefpe; à Genéve, 1804, 8vo. p. 284. Explanations of Time-keepers conftructed by Mr. Thomas Earnfhaw, and Mr. John Arnold, publifhed by order of the Commiffioners

Commiſſioners of Longitude; 1806, 4to. with plates. London. Mr. Dalrymple's pamphlet; 1806. Appeal by Sir Joſeph Banks; 1806. Dr. Maſkelyne's Anſwer; 1806.

Chronometer, or *Time-keeper*, by Harriſon. It was our intention to have given a perſpective drawing of one of Mr. James Harriſon's chronometers, or time-keepers, with a correſponding deſcription, but on, an application by our draftſman to the aſtronomer royal to inſpect the interior parts of one of the machines, by the maker in queſtion, placed at the Obſervatory, he was informed that permiſſion to undo any of the covers or other parts of the mechaniſm could not be granted; we muſt therefore ſatisfy ourſelves with a verbal deſcription of the conſtruction, which, indeed, it is preſumed, will be deemed ſufficient, now that preferable conſtructions have been more recently adopted. We might, it is true, have copied the plans given in the ten plates of the pamphlet entitled the "Principles of Mr. Harriſon's Time-keeper," but ſome of them are ſo imperfectly explained, as to be unintelligible to any reader, except perhaps to ſuch as may have had occaſion to examine the original mechaniſm, and therefore would not be ſatisfactory to the public. The following particulars, relating to the dimenſions and other properties of the mechaniſm, apply more particularly to the *fourth* piece made by the inventor, and are extracted chiefly from the pamphlet juſt mentioned. The firſt obſervable diſtinction between Harriſon's train of wheel-work and that of an ordinary watch, is, that the numbers of his wheels and pinions are higher than had been uſual, as will appear from the ſubjoined arrangement, viz.

Firſt or great wheel 96
Center pinion acting
with it - - - . 21—120 ſecond wheel, which is concave
Its pinion 18—144 third wheel,
. Its pinion 16—120 contrate wheel,
Bal. wheel pinion 12—15 bal. wheel,
2 pallets

If theſe numbers be examined according to our mode of *notation* under the article Clock-*Making*, the value of the train will be $\frac{120 \times 144 \times 120 \times 15 \times 2}{18 \times 16 \times 12} = 1800$ vibrations in an hour, or juſt five vibrations in each ſecond. The firſt or great wheel will revolve in $\frac{96}{21}$ of an hour, or make one revolution, along with the fuſee, in $4\frac{4}{7}$ hours, ſo that $5\frac{1}{4}$ turns or ſpirals on the fuſee will maintain a motion of 24 hours, $6\frac{1}{4}$ will actuate the works $28\frac{4}{7}$, and $6\frac{9}{16}$ juſt 30 hours.

The balance-wheel does not differ eſſentially from that in a common watch, but peculiar care is neceſſary in ſhaping the pallets, as will be ſeen more particularly under the article Escapement.

The bad effect produced in ordinary watches by an irregular tranſmiſſion of the maintaining power through the train, is here guarded againſt by the introduction of a ſlender ſpring, or remontoir, which is wound up eight times every minute by the maintaining power, or main-ſpring, and which actuates the contrate and balance-wheels, and conſequently impels the pallets with a conſtant impulſe, independently of the maintaining power, the latter being employed for no other purpoſe but for winding up the former as before ſpecified. This ſlender ſpring, which is ten inches long, though it weighs only $3\frac{1}{2}$ grains, is coiled in a ſpring barrel, apparently concentric with the contrate-wheel, and has its outer end attached to a hook in the barrel, with its inner end attached to another hook on the contrate-wheel. In order to make this delicate ſpring anſwer its purpoſe perpetually, two wheels, and as many pinions, in addition to the foregoing train, called the fourth and fifth wheel, and fourth and fifth pinion, are introduced in connection with a fly; as is likewiſe a detent with five arms, reſembling a ſtar, turning on the pivots of a common arbor. The action of theſe parts is not eaſily underſtood, even from a reference to the original drawings, much leſs from a verbal deſcription; but the intelligent reader will form a general idea from being told, that there are eight pins placed equidiſtantly on the plane of the contrate-wheel in a ſmall circle round the arbor; that one of the five arms of the detent acts with theſe ſucceſſively as the wheel itſelf revolves; that a ſecond arm carries a ſmall roller acting againſt a piece of braſs on the fifth wheel; that a third is bent at the end ſo as to catch a pin in the rim of the fifth wheel; and that the fourth and fifth arms are mere counterpoiſes to the other three, to preſerve an equilibrium. The numbers of theſe wheels and pinions appear to be as follow: of the fourth wheel 112, which ſtands concentrically over the contrate-wheel and ſpring-barrel, of its pinion 14, of the fifth wheel 104, and of its pinion 12; and it is to the arbor of this laſt pinion, we preſume, that the fly is attached, though it does not appear evident either from the drawings or deſcription, at leaſt to our apprehenſion. The mode of applying the three effective arms of the detent bears ſome reſemblance to the action of the detents in the ſtriking part of a clock, in which the count-wheel is uſed inſtead of a ſnail, and where the locking and unlocking are alternately effected at meaſured intervals of time. The detached eſcapements, however, have now rendered this complex mechaniſm ſuperfluous, and, indeed, it is manifeſt that the propoſed object, of having a completely detached power to act alike at all times on the pallets, is not thus perfectly effected; for the locking and unlocking eight times in every minute, to be produced by the ſlender ſpring in queſtion, muſt be ſuppoſed to interfere in a certain degree with its regularity of action on the pallets; and if it ſhould be contended, that the eight deductions from the force of the ſpring are regular and periodical in every minute, and that therefore they produce an equable effect, yet the ſame argument may be uſed in favour of a well made train, in which the irregularities in the tranſmiſſion of force from the main-ſpring may likewiſe be periodical, the wheels and pinions not being compoſed of *prime* numbers.

But it was not enough for Harriſon, that nearly an equable force was applied to maintain the motion of the balance, while the balance itſelf and alſo the balance ſpring were ſubject to alterations in their dimenſions by changes of temperature; he well knew that the ſpiral ſpring uſed in watches, had more power when contracted, and leſs when elongated, than a mean power, and alſo that an enlarged balance has a greater momentum than a diminiſhed one, and *vice verſâ*; the mode alſo of effecting an adjuſtment to counteract the conſequent loſs or gain in the rate of going, was well known not only to him, but to all watch-makers, and mechanically applied with ſucceſs, as at the preſent day, by a moveable ſtud to limit the effective length of the ſpring; but this adjuſtment was not a *ſelf-acting* one, an index was required to be moved by a manual operation, which index was connected with the moveable ſtud, and pointed out the quantity of the adjuſtment; it remained for Harriſon to deviſe a mode of action, entirely dependent on the ſtate of the atmoſphere at any moment, which of itſelf, would produce the requiſite adjuſtment; this he did, as we have already ſaid, by riveting together a ſlip of braſs and a ſlip of ſteel, which two metals are of unequal expanſibilities, to the remote end of which compound bar he attached his clip to hold the exterior thread of his ſpring, after the

outer

outer end of it had paffed through and been pinned to a ftud of brafs attached to the upper plate of the frame; the confequence proved, what none but a real genius would have forefeen, that the brafs elongating and contracting alternately in oppofite temperatures more than the fteel, produced a curvature in the compound piece, the concave fide of which was always occupied by the metal leaft elongated, that is, by the fteel in hot, and by the brafs in cold weather; hence the compound piece, which carried the ftud backwards and forwards, was called a *kirb*, a Lincolnfhire word for *curb*, which contrivance is deftined to curb or command the effective length of the regulating fpring. The only material objection that experience has pointed out againft this felf-compenfating mechanifm, is, as we have before obferved, that fmall pieces of metal and pieces in motion do not alter their temperature at the fame time with large pieces and pieces in motion; nor yet fteel fo foon as brafs, even in fimilar circumftances; which confideration conftitutes an objection to the ufe of the thermometrical curb; an objection firft raifed by Harrifon himfelf, when it was greatly his intereft to have fuppreffed fuch a fuggeftion, as he was a candidate for the parliamentary reward. Vide COMPENSATION-*balance*.

Harrifon had befides remarked that in an ordinary watch the power which the main fpring has over the balance, through the medium of the train, compared with the power that the regulating fpring has over the fame, is as *one* to *three* generally; this power from the main fpring, he obferved, being fufficient to put the watch in motion from a ftate of quiefcence, muft be too imperious for the balance, fmall and light as it was, to control it; accordingly he propofed, reafoning thus *a priori*, to give an additional momentum to his balance, compared with his maintaining power: but momentum was to be attained in three different ways, namely, by additional weight given to a balance of the ufual diameter, by enlarging the diameter without increafing the weight, or laftly, by increafing both in a certain degree: the firft mode was objectionable on account of the friction likely to be produced on the balance-pivots by a heavy balance; the fecond was alfo objectionable on account of the refiftance of the air it was likely to experience; and therefore he fixed upon the laft mode of gaining momentum, by partly enlarging the diameter, and partly increafing the weight of the balance in ufe; by which means he conftructed a balance over which the force from the maintaining power of the remontoir has not more than *one-eightieth* part of the dominion that the balance-fpring has. Hence in the time-keeper there is not force enough in the maintaining power to excite motion from a quiefcent ftate, though there is power enough to overcome all the obftacles to continued motion, and to keep the piece going, when put in motion.

The balance of the time-keeper before us is defcribed to be of more than three times the weight of that of a large ordinary watch, and of three times its diameter; for, according to the notes taken by Dr. Mafkelyne, at the time of its examination, its diameter was $2\frac{1}{4}$ inches, and that of the plate $3\frac{8}{10}$; hence a point in its circumference will pafs through 24 inches, or about four times the fpace of an ordinary watch, in each fecond, as Harrifon calculates from the arc of its vibration; an advantage which has not been loft fight of by fome of the modern makers of chronometers; and it may be confidered as an axiom in chronometry, that the perfection of a balance, confidered fimply as a regulator, independently of its compenfation-mechanifm, confifts in its having the greateft poffible quantum of momentum with the leaft force from the train, and fmalleft quantity of friction and refiftance from the air

Notwithftanding, however, the above-mentioned contrivances in the conftruction of the balance, others were ftill wanting. Harrifon inferred that large arcs were defcribed in lefs time than fmall ones, from the circumftance of the piece going flower in a vertical pofition than in a horizontal one, where the vibrations were obferved to be vifibly longer, before any correction was applied; he therefore wanted, moreover, a compenfation for the errors of pofition. The time-piece, he obferved, did not go alike when in a vertical pofition, the hours III, VI, IX, and XIII, were fucceffively uppermoft, which defect he remedied by making the relative weights or dimenfions different, at different fides of the balance, thereby equalizing the arcs of vibration in each vertical pofition. Again, to render the time of an horizontal vibration equal to that of a vertical one, he introduced a contrivance which he calls a cycloid-pin, which, when in contact with the regulating fpring, quickens its vibrations; but as the longeft vibrations feemed to require fuch fecondary affiftance the leaft, in thofe the fpring left the pin for a longer time than in the fmaller arcs of vibration, and therefore they were lefs affected by it. This cycloid-pin is very imperfectly defcribed in the account of the plates, and, it appears, was added to the piece fent on trial to Jamaica, after its return; and it appears by no means certain, that any good end was obtained by its adoption. The idea of ufing adjuftments for pofition, however, has proved of permanent utility. It may be neceffary to add here, what will appear extraordinary, that there was no adjuftment for mean time in Harrifon's time-keeper, the compenfation curb having ufurped its place; but he profeffed to be able in general to afcertain its comparative rate of going by his regulator, or gridiron pendulum clock, near enough in three hours; and to apply the daily error in feconds to the time indicated, plus or minus, as the cafe might be; of courfe a longer trial was neceffary for a very nice rate, before a voyage was commenced. Still it was neceffary in a machine profeffing to meafure the conftant lapfe of time perfectly, that it fhould not ceafe to go while the main-fpring was in the act of being wound up; here was another field for our artifan's ingenuity; but with a great genius, as with a great warrior, or ftatefman, difficulties thrown in the way only tend to call forth the refources of an active and inventive mind; the auxiliary fpring was the refult of our inventor's fagacity, and the fufee itfelf, which before had been fhapen to equalize the varying force of the main fpring in its different ftates of tenfion, was now made to contain moreover within it a fecond fpring, fo bent in a contrary direction, that the force exerted by the firft or main-fpring wound it up to a refifting force exactly equal to the power neceffary to be exerted on the train, before this power began to be tranfmitted further along the train. The contrivance will not be eafily underftood by a verbal defcription, except by men converfant in watch-work; it is thus; a concave ratchet with 55 teeth, is fixed to the infide of the large end of the fufee, the concavity of which is fufficiently large in diameter to admit the fecondary fpring-barrel within it, into the cavity of the fufee; a fecond ratchet of 75 teeth, at the outer edge, inclined in an oppofite direction to thofe of the ratchet 55, and having the fecondary fpring-barrel attached to it, is placed contiguous to the great wheel of 96, plane to plane, and revolves on a tubular piece projecting from the plane of the great wheel; it does not appear how the click of the concave ratchet is placed to act with its teeth at the inner edge, but this click is no doubt placed on the plane of the large ratchet of 75, or on the barrel; the inner end of the fecondary fpring is hooked to the tubular piece on which the barrel and large ratchet attached to it revolve, and the outer end,

end, as uſual we preſume, to the ſide of the barrel within. There are two clicks to the large ratchet, one on each of its circumferences, which clicks are faſt to the plane of the frame-plate, together with their reſpective ſprings. The parts being thus connected, and the barrel inſerted into the cavity of the fuſee, the effect produced is this; ſuppoſe the key applied to the ſquare of the fuſee arbor to wind the piece up, the inclined teeth of the ratchet ſlide along the end of the click in this retrograde motion of the fuſee, during the act of winding, and no impediment occurs to free motion of the fuſee until the *garde* (or guard,) gives notice of the concluſion of winding; but let the key be withdrawn, and it will be obſerved, that the main-ſpring by its effort to relax, will urge the fuſee, in a direction contrary to the motion of winding, till ſome obſtacle oppoſe its motion; that obſtacle is the click of the concave ratchet, which if fixed in a ſtationary ſituation, would inſtantly arreſt the ſaid returning motion of the fuſee; but we have ſaid the click is fixed to the large ratchet or attached barrel, which contains, we will ſuppoſe, a relaxed ſpring; the returning motion of the fuſee, therefore, goes on, after the ratchet of 55 has caught its click, until the ſecondary relaxed ſpring in the little barrel is wound up as far as the ſtrength of the main-ſpring will wind it; the inner end of the ſecondary ſpring at that inſtant begins to act on the tubular projection of the great wheel, which may be called the barrel arbor, and urges it on as though the click had been fixed to the great wheel, and as though there had been no large ratchet, which Harriſon calls the perpetual ratchet, nor any ſecondary ſpring interpoſed between the firſt ratchet and the great wheel; and thus it is that the power is continually tranſmitted through the medium of an *intenſe* ſpring, after it has been firſt wound up from a ſtate of relaxation. Conceive again the key applied for the ſecond winding as before; the returning motion of the fuſee will now be but little, becauſe the ſecondary ſpring has been already wound up; but the effect of the contrivance here becomes evident; the ſecondary ſpring being previouſly wound up, the two clicks of the large ratchet of 75, to which the ſecondary ſpring-barrel is attached, keep it from relaxing, while the preſſure of the main ſpring is taken off by the act of winding; the force of this ſecondary ſpring, however, will exert itſelf ſomewhere to return to its natural ſtate, and becauſe the motion of the ſpring is ſtopped at the outer end, by the juſt mentioned clicks preventing the returning motion of the large or perpetual ratchet, the inner end of it will exert its whole force on the pin of the tubular part of the great wheel, conſidered as the barrel's arbor, and will thus impel the great wheel for a limited time with the ſame force with which it was acted upon itſelf, when wound up to the extreme, that is, with the whole force of the maintaining power to which it becomes a temporary equivalent. Thus we ſee four ſprings were uſed by Harriſon, which we are told were all made by Maberley, except the balance-ſpring that required rubbing away till it was found of a proper ſtrength to regulate for mean time nearly; the temper was given to this ſpring and the ſteel pinions by a melted mixture of one pewter and ſixteen lead; and to the balance ſpindle by a mixture of one pewter and twelve lead, the latter of which mixtures, in a ſtate of fuſion, the author ſays, is equal to 567° of Fahrenheit.

The effect of the thermometrical curb was increaſed by rubbing the ſides thinner, and decreaſed by thickening the edge with a burniſher.

The fuſee has 6¾ turns; the pivot-holes are all buſhed with rubies containing pieces of diamond at the bottom of each; and the pallets are of diamond. The fly at the fifth pinion is uſed to regulate the velocity with which the ſpring at the contrate-wheel is wound up every eighth part of a minute by the main-ſpring when unlocked. The dial-work is $\frac{112}{28} \times \frac{96}{32} = 12$, and the ſeconds are concentric by means of a wheel of 104 attached to the contrate-wheel driving another of the ſame number round an outer cannon in the centre.

The firſt time keeper which Harriſon made was in 1726, which Dr. Hutton aſſerts, did not err a ſecond in a month for ten years together; but the firſt time that one was publicly tried, was in a voyage to Liſbon, in the year 1736, which, being placed in a box, hung in gimbols, anſwered his expectation, and corrected the dead reckoning about a degree and a half; in conſequence of which, according to Dr. Mackay, the board of longitude granted him a gratuity, and deſired him to proſecute his labours.

In 1739, he finiſhed a ſecond piece more perfect than the firſt, and in 1749, according to Dr. Mackay, (but according to Dr. Hutton, in 1758,) a third, which was pronounced more ſimple in its conſtruction than either of the former; but his labours did not ſtop here; in 1761, his fourth piece, of which we have given a deſcription as nearly as we could without perſpective drawings, was produced for trial, and Mr. William Harriſon, the ſon, offered to take charge of it, in a voyage to and from Jamaica, which was accordingly performed in this and the following years. Mr. Robertſon, maſter of the Academy at Portſmouth, was fixed upon to take the rate of this piece, which he did, and reported that, on the 6th of November, 1761, at noon, it was 3^s ſlow after having loſt 24^s in nine days on mean ſolar time. The Deptford, in which ſhip the voyage was made, left Portſmouth on the 18th of the ſame month, and arrived at Madeira on the 9th of December following, when it was found that the reckoning was corrected by the time of the piece about a degree and a half. In the run from Madeira to Jamaica the reckoning was corrected 3°: and at the ſeveral iſlands, where the ſhip touched, the known longitudes agreed very nearly with thoſe given by the time-keeper. On Jan. 19th, 1762, the ſhip arrived at Jamaica; the time of mean noon was obſerved, by equal altitudes, at Port Royal, on the 26th of the ſame month, which, according to the piece, was $4^h\ 59^m\ 7^s.5$; but the original error on the 6th of November, 81 days, 5 hours before, was 3^s ſlow, this quantity, therefore, was to be applied as a correction, together with the accumulation of the daily error of $\frac{24}{9}$ in $81^d\ 5^h$, viz. $3^m\ 36^s.5$; this ſum of $3^m\ 39^s.3$ added to the time indicated, which, it has been ſaid, was $4^h\ 59^m\ 7^s.5$, make $5^h\ 2^m\ 47^s$ for the difference of longitude between Portſmouth and Port Royal; which determination was only 4^s of time leſs than the determination at Kingſton of the ſame, from the tranſit of Mercury over the ſun's diſc. This ſmall error in time correſponds to leſs than one nautical mile in the parallel of Jamaica.

The Merlin, on board of which the piece was now put, ſet ſail from Jamaica on the 28th of January, 1762, and experienced ſuch a violent ſtorm in the paſſage, as obliged young Harriſon to remove his piece into an expoſed ſituation; however, the ſhip arrived at Portſmouth on the 26th of March, and on the 2d of April the time of mean noon was found, from equal altitudes, to be $11^h\ 51^m\ 31^s.5$, to which, its former error of 3^s, together with the accumulation of the daily error, viz. $\frac{24^s \times 147^d}{9} = 6^m\ 32^s$ being added, make the time of mean noon by the time-keeper $11^h\ 58^m\ 6^s.5$. From

From this report it appears, that from Nov. 6th, 1761, to April 2d, 1762, though the piece had experienced many violent agitations at sea, and had been exposed to great changes of temperature, the whole error amounted to only $1^m\ 53^s.5$, or 28½′ of longitude on the equator, which quantity, (one second less than Dr. Hutton has stated), is not quite 18 nautical miles in the parallel of Portsmouth.

Though various objections were made to this trial, principally arising from the observations by which the longitude of Portsmouth and Jamaica had been ascertained, yet Harrison, we are informed, obtained a reward upon it from parliament of 5000l., and was ordered to make a second trial to Barbadoes. But previously to the second trial to the West Indies, the Board of Longitude, on the 17th of August, 1762, wished to place Harrison's piece in the hands of the astronomer royal, at that time Mr. Bliss, for trial at the Observatory, which wish was not complied with, by reason of some alteration to be made, probably by adding the cycloid-pin to aid the regulating spring; the same wish was repeated by the Board at their sitting on the 4th of August of the year 1763, which was again not complied with by the jun. Harrison, by reason of his not being yet sufficiently rewarded; however, on being desired to send the rate of going of his time-keeper, sealed up, to the secretary of the Admiralty, previously to his sailing, he consented to this request, and proposed to abide by the sealed rate on the trial to Barbadoes, which had been proposed. The annexed is a verbal copy of Mr. W. Harrison's declaration of the rate of going, or, more properly speaking, of the daily error of the time-keeper, to the Board of Longitude, dated Portsmouth, March 26th, 1764.

"My Lords and Gentlemen,

In obedience to your instructions, dated the 9th of Aug. 1763, I humbly certify that I do expect the rate of the going of the time-keeper will be as followeth; viz.

When the thermometer (Fahrenheit's, no doubt,) is at 42° it will gain 3 seconds in every 24 hours.

When the thermometer is at 52°, it will gain 2 seconds in every 24 hours.

When the thermometer is at 62°, it will gain 1 second in every 24 hours.

When the thermometer is at 72°, it will neither gain nor lose.

When the thermometer is at 82°, it will lose 1 second in every 24 hours.

Since my last voyage we have made some improvement in the time-keeper; in consequence of which, the provision to counterbalance the effects of heat and cold, has been made anew; and for the want of a little more time, we could not get it quite adjusted; for which reason the above allowances are necessary. This is its present state; and as the inequalities are so small, I will abide by the rate of its gaining, on a mean, one second a day for the voyage. I would not be understood, that it will always require so long time to bring those machines to perfection; for it is well known to be much harder to beat out a new road, than it is to follow that road when made. During the time of this experiment, the mean height of the thermometer shall be each day carefully noted down, and certified, which I will lay before the Board at my return. I am, &c.

WILLIAM HARRISON."

After having compared the time-piece with Mr. Short's regulator, in Surry-street, London, which had its error newly ascertained by an excellent transit instrument, Mr. Harrison junior went on board the Tartar on the 13th of February 1764, and proceeded to Portsmouth, where he again compared it with an astronomical clock in Mr. James Bradley's temporary observatory. It may not be foreign to our purpose to mention here, that the observatory just spoken of, was fitted up for the express purpose of observing the eclipses of Jupiter's satellites, as well as of keeping the clock in just time, in order that, to avoid future objections, the observations to be made by Messrs. Maskelyne and Green, at Barbadoes, on the same satellites, particularly the first, when compared with Bradley's, might ascertain the comparative longitudes of these two places, which was accordingly determined to be $3^h\ 54^m\ 20^s$.

Before leaving Portsmouth, which took place on the 28th of March 1764, Mr. Harrison took the rate of his time-keeper by equal altitudes, employed between the 29th of February and the 26th of March; and on April the 18th found, from comparing his observations of the sun at 4 P. M. at the ship, with the corrected time given by it, that the ship was at that instant only 43 miles eastward of Porto Santo, in consequence of which determination, sir John Lindsay, the master, steered accordingly, and saw the island in question before him at one o'clock the next morning, agreeably to expectation. On the 13th of May, the vessel arrived at Barbadoes, and on the four following days its error was ascertained by Messrs. Maskelyne and Green by equal altitudes of the sun, and also by a comparison with the astronomical clock at the observatory near Bridgetown; and it was found that the amount of the daily deviations from mean time was only 43ˢ in excess, or 10¾′ of a degree in longitude. Mr. Harrison set out on his return from Barbadoes in the New Elizabeth on the 4th of June; and, arriving at Surry stairs on the Thames, on July the 18th, found, from a comparison with Mr. Short's clock, the error of which had been ascertained on the very day, that the whole gain in the 156 days was only 54ˢ, allowing the sealed rate, of one second gain per day, as a correction: and it has been observed, that, if the allowances had been moreover made for the state of the thermometer, as stated in the declaration, the piece in that case would have been found to have been about 15ˢ only at variance with mean time, and this in the opposite extreme. Soon after this very satisfactory trial, a committee of seven scientific gentlemen and mechanicians were appointed by the Board of Longitune for examining the principles of Mr. Harrison's time-keeper, whose report was as follows: viz. "That Mr. Harrison has taken his time-keeper to pieces, in presence of us, and explained the principles and construction thereof, and every thing relative thereto, to our entire satisfaction; and that he also did, to our satisfaction, answer every question proposed by us, or any of us, relative thereto; and that we have compared the drawings of the same with the parts, and do find that they perfectly correspond." The committee were the Rev. N. Maskelyne, Rev. John Mitchell, Rev. William Ludlam, Mr. John Bird, Mr. Thomas Mudge, Mr. Larcum Kendal, and Mr. William Matthews. Mr. Harrison had then another 5000l. ordered him, with a promise that the residue of the whole parlimentary reward; which, by the act of queen Anne, was 20,000l., should be given him when a proper person could be found to execute his plan with equal success. Mr. Larcum Kendal, one of the committee already alluded to, undertook the task, and finished a time-piece on the same construction, or at least on the same principles, which was approved by Mr. Wales, in his voyage in company with captain Cook in the years 1772, 1773, &c. and which Dr. Hutton says performed even better than Harrison's, allowing for an acceleration in its rate. In consequence of this success, the parliament, to which an appeal was made, ordered the residue of the proposed reward

 to

to be paid; in addition to which, the gratuities of the Board of Longitude, of the Eaft India company, and of others, contributed to augment the whole fum to about 24,000l.

It fhould feem from Dr. Hutton's account, under the article *Longitude* of his Mathematical Dictionary, contrary to Dr. Mackay's authority in his book on the longitude, that the parliamentary rewards were made to the Harrifons by two equal payments of 10,000 l. each; to reconcile which accounts, we applied to Dr. Mafkelyne for authentic information, but, with his ufual referve, the Doctor declined giving us any information on the fubject. The difcrepancy in the dates of our two authorities, which agree in the whole amount, is not, however, of much importance. The compliment of the 20,000 l. was granted by parliament in the year 1774, and at the fame time the new act paffed for regulating the future rewards. For the credit of our ingenious countryman, we fhould have been happy to clofe our narrative of Mr. Harrifon's fourth time-piece here, but, as we profefs to give an impartial detail of all the facts that have come to our knowledge, relative to the different trials made of it, we are under the neceffity of adding a further notice, which is calculated to detract confiderably from its merit, that might feem to have been already indubitably eftablifhed. Though the piece had agreed with the longitudes of Portfmouth, Jamaica, and Barbadoes, to a great nicety, and had alfo accorded very well with the known longitudes of fome intermediate iflands, yet it was by no means certain that the refults would have been fo exact if taken on intermediate days; or, in other words, a coincidence of the rate with mean time at the end of the voyage, was no proof that there muft have been the fame coincidence in every part of it: to put the piece, therefore, to a more rigorous teft, the Board of Longitude, held April 26, 1766, came to a refolution to have it examined at the Royal Obfervatory from nearly the beginning of May of that year, to the end of February of the year following; accordingly, Dr. Mafkelyne received the piece from the hands of Philip Stephens, Efq. now Sir Philip Stephens, who was at that time fecretary to the Admiralty, on the 5th of May, and, in the prefence of Captain Thomas Baillie, of the Royal Hofpital, Greenwich; of Mr. John Ibbetfon, fecretary to the Board of Longitude; and of Mr. Larcum Kendal already mentioned, it was depofited in a deal box made purpofely to contain it, with a glafs cover made fecure with putty. The box had two locks of different wards, and two keys to each lock; alfo a pane of glafs in the fide fecured with putty; and whenever the piece was afterwards wound up, one of the keys was ufed by Dr. Mafkelyne, or one of his affiftants, Jofeph Dymond and William Bayly, and the other by Captain Baillie, or one of the officers at Greenwich Hofpital; and the conftant formality of a written teftimony was ufed, we are told, not only when the piece was wound up, but alfo when a comparifon of the rates of it and the obfervatory regulator was taken and regiftered. Dr. Mafkelyne afterwards publifhed the refults, under the title of "The original Obfervations of the going of the Watch from Day to Day," in a quarto pamphlet, from which we have extracted, or rather deduced, the fubjoined notices. The trial commenced on the 6th of May, 1766, and ended on the 1ft of March, 1767, including a fpace of 298 days, in which period the piece gained on mean folar time $1^h\ 10^m\ 27^s.5$; this accumulated error, divided by 298, gives $14^s.2$ very nearly, for the rate, or mean daily gain. On examining the different columns of the pamphlet, and the calculations grounded thereon, to afcertain the daily errors, both of the regulator and time-piece, as compared with the folar tranfits properly equated, we find a daily gain of the piece on mean time, on June 8th, amounting to $30^s.2$, or more than half a minute; which greateft deviation was when it was in a vertical pofition with XII. higheft, and the thermometer at 60°; the barometer being at the fame time at 29.9; but on fome other days the piece was, on the contrary, lofing, particularly in January 1767, when the thermometer was down at the freezing point; on one day in particular, the 4th, the daily lofs was as much as $6^s.5$, when the pofition was horizontal with the face upwards. Dr. Hutton, fpeaking of this trial, has faid in his Dictionary, that "the watch was now found to go fafter than during the voyage to and from Barbadoes, by 18 or 19 feconds in 24 hours;" this obfervation, however, is only accurate for the months of May, July, and about the end of October and beginning of November, and that when the pofition was horizontal and face upwards; but, even in this pofition, the temperature, in other months, produced confiderable irregularity in the daily rate. Whenever the pofition was vertical, or inclined 20° from a horizontal line, the rate depended greatly on the hour of the dial plate that was uppermoft, independently of temperature: from all which deductions, it is evident, that the time-piece wanted three adjuftments to be made at the time of its trial at the obfervatory; viz. firft an adjuftment for an error of $14^s.2$ in the mean rate, fuppofing the gains to have been equable at equal intervals of time; fecondly, an adjuftment for pofition; and, thirdly, an adjuftment for temperature; and, indeed, we find, according to Dr. Hutton, that Harrifon "had altered the rate of its going, by trying fome experiments, which he had not time to finifh before he was *ordered* to *deliver up* the watch *to the Board.*" Thus this trial, with all its precautions and formalities, was more calculated, perhaps, to prejudice the public opinion refpecting the future dependence to be placed in a time-keeper, than to appreciate the intrinfic value of a well regulated and well adjufted machine; and we cannot forbear adding, as our concluding remark, that Kendal's time-piece, tried and approved by Mr. Wales, was made after Harrifon's model during or after this trial; and that the inventor's fubfequent appeal to parliament was attended ultimately with the defired fuccefs.

CHRONOMETER, or *Time keeper*, by Mudge. *Plate* XIII. of *Horology*, contains fo much of the effential parts of Mr. Thomas Mudge's time-keeper, as will enable the reader to form a competent idea of its conftruction and manner of performance. In the "Defcription of the Time-keeper," publifhed in 1799, by Mr. Thomas Mudge, Jun. there are nine plates, eight of which are taken up in exhibiting the different plans and fections of the various parts feen in different points of view; they were drawn by Mr. Pennington, the original workman employed by the two Mudges, father and fon; but inftead of copying any of the plates, except our figures 6 and 7, which are from his Plate II., we have judged it to be more defirable to procure original drawings of a time-keeper, from one in the poffeffion of the faid Mr. Pennington, who now lives at the corner of Orchard Row, Camberwell, and who has obligingly given us his affiftance as well as permiffion to analyfe the piece. It is not neceffary to give a particular account of the cailiper of Mudge's time-keeper, as it does not differ in any material particular from that of the chronometers by other makers; nor is there any thing worthy of particular notice in the fufee, different from that of an ordinary watch, except that there is the auxiliary fpring making with it a *going fufee* of the ordinary conftruction. (See the defcription of Brockbank's *Chronometer.*) The barrel however, which contains the main fpring, is very fimilar to Harrifon's, and differs from ordinary barrels. *Fig.* 6, reprefents this barrel with the chain coiled round,

round, as it appears before it is wound up. A is the edge of the upper plate, and B of the pillar plate of the frame; the barrel is compofed of two diftinct parts, C and D, which together may be denominated an entire box, of which D is the body or box portion, and C the cover or lid; this lid C, which is attached to the arbor EF, is of itfelf, by Mr. Mudge, called the chain barrel, becaufe the chain is wound round it; and the box part, D, is alfo of itfelf called the fpring barrel, becaufe it contains within it the main fpring. A fection of this *fig.* 6. is feen parallel to it in *fig.* 7, where the fame letters imply the fame parts; from the latter of which it appears clearly that the portion C of the greater diameter may revolve with its arbor EF, feparately from the other portion D. G in both *figs.* 6 and 7, denotes a ratchet-wheel made in form of a rim or broad ring, with its inclined teeth on the outer or convex circumference, and is foldered, or fcrewed, to the lower end of the rim of the box D, in fuch a way, that a very fmall portion of the interior, or concave part, projects inwards into the box: this projecting part has a fhoulder turned away in the lathe from its upper plane, which makes the part within the box thinner than the part without, as may be feen in *fig.* 7; the bottom of the box *aa* is detached, and is of a diameter juft fufficient to allow it to enter the infide of the box; it has a correfponding fhoulder turned in the lathe at the circumference on its lower plane, which fhoulder refts on the above defcribed fhoulder in the concave projecting part of the ratchet; the ufe of this contrivance is this; the box, with its loofe bottom in it, is placed on the plane of the plate A, and the bottom *aa*, perforated in the centre for the revolving arbor, is fecured to it by two fcrews or more, entering from the plate, fo that though the box may be faid to be attached to the plate by its bottom, yet its rim is at liberty to turn along with the ratchet firmly faftened to it, while the bottom is at reft fcrewed to the plate, and the motion of the ratchet will be free or impeded in proportion to the quantity of preffure which the fixing fcrews occafion on the two fhoulders already defcribed. The parts being thus arranged, it is eafy to conceive that a click, fcrewed to the plate A at any convenient part of it, and taking into the teeth of the ratchet G, will hold the box, or fpring-barrel, from going in a retrograde direction, as well as if the box itfelf had been fcrewed to the plate A, but it will ftill be at liberty to revolve in the contrary direction on the application of any external force to it. Suppofe now the outer end of the fpring, hooked as ufual, to the interior fide of the barrel or box oppofite D, while the inner end is hooked on a pin in the arbor within the box alfo; the confequence will be, that the chain, which is coiled round its barrel or lid C, being wound away by the fufee, in the act of winding, will pull the arbor round, and its force will apply directly to the interior end of the main fpring firft, becaufe this is hooked to the arbor, and the coils of the fpring will follow one another fucceffively up to the centre, as the intenfity of the fpring is increafed by winding; whereas, in ordinary watches, where the fpring-barrel itfelf revolves without the arbor, the outer end of the fpring is firft acted upon in winding up, and the coils near the centre are laft affected by the force that winds. When the fpring requires to be fet up or down, it is done by turning the fpring-barrel and ratchet together with the contained fpring, while the click is raifed from the teeth of the ratchet, the arbor of the chain-barrel then being ftationary, or revolving no otherwife than according to the flow motion which the train regulates. The chain is wound round the fufee in a direction which actuates the wheels of the train in a backward manner, compared with the motion of an ordinary train, which direction of motion requires an additional wheel in the dial work; for a fmall wheel of 45 fixed to the arbor of the contrate wheel drives another of the fame number of teeth in a proper direction for indicating feconds, on a feparate graduated circle, at a diftance from the centre of the face; nor is the fecond, ufually called the centre wheel, placed in the centre of the hour and minute circles, as is common in watches; though oppofite the centre of the face; but its prolonged pivot bears a minute wheel of 52, driving a fimilar detached minute wheel, and alfo an attached pinion of 8 driving an hour wheel of 96, both which driven wheels have cannons to carry their refpective hands round two concentric circles, placed at the other fide of the centre of the dial. This method of placing the hands allows the contrate and centre wheels to have fmall pivots. The numbers of the movement publifhed by Mudge are as follow: viz. the great wheel has 100 teeth, the pinion acting with it 20, the centre wheel or fecond wheel 120, the next pinion 16, the third wheel 120, the pinion driven by it 15, the contrate wheel 120, the pinion driven by it 12, balance wheel 15, pallets 2; whence it feems that the fufee revolves in $\frac{100}{20}$ or 5 hours, and the contrate wheel in $\frac{120}{16} \times \frac{120}{15} = \frac{60}{1}$, or 60 times in the hour, and the value of the train is, according to our method of eftimating Harrifon's train,

$$\frac{120 \times 120 \times 120 \times 15 \times 2}{16 \times 15 \times 12} = 18000 \text{ vibrations in each}$$

hour, or five in each fecond. With this movement a fufee with $7\frac{1}{5}$ turns or fpirals in the groove, actuates the piece 36 hours; but there is a face laid down in Mudge's book for a continuance of eight days, the circles for the hours and minutes in which are at oppofite fides of the prolonged pivot of the centre wheel arbor, and the hour circle counts up to 24, which is certainly more convenient for aftronomical calculations, than one with only 12 hours. On fearching for an account of the train of this conftruction, we are only informed, by Mr. Mudge, jun. that fuch of the time-keepers as went eight days, were charged higher than thofe which continued at one winding only 36 hours; but the numbers of the movement are not given by him; on application to Mr. Pennington, however, we have been favoured with them, which are as follow, viz. great wheel 108, pinion on the centre wheel 12, centre wheel 128, pinion following 12, fecond wheel of the train, or third of the whole movement 120, pinion driven by it on the next arbor 12, contrate wheel 120, pinion following on the balance wheel arbor 12, crown wheel 15, pallets 2; fo that the value of the train is

$$\frac{128 \times 120 \times 120 \times 15 \times 2 = 55296000}{12 \times 12 \times 12 \ 1728} = 32000, \text{ on}$$

a fuppofition that the centre wheel revolves in an hour; but we find from the dial-work, which has what is called *borrowed minutes*, that this wheel revolves in $\frac{112}{63}$ of an hour or in 1.7 hours, therefore $\frac{63}{112}$ of 32000 = 18000 is the true train by the wheels and pinions before us; the hour hand revolves in $\frac{135}{10}$ of $1^h.7 = 24^h$; and the fufee revolves in $\frac{112}{63} \times \frac{108}{12}$ of an hour, which is = 16 hours; therefore a fufee with juft 12 turns will continue to go eight days, as was intended.

Fig. 1. of *Plate* XIII. is a perfpective view of the cock of Mudge's time-piece, fuppofed to be detached from the upper plate of the frame: the dotted fpiral round the centre is the balance regulating-fpring, to which there is a correfponding one underneath, called the compenfation-fpring.

Directly over these springs is placed a smaller concentric frame supported by three pillars, *a*, *a*, *a*, on the principal cock, and bearing four friction-rollers, of which the centres are marked by four points at the extremities of the dotted spiral; these points are here supposed to be demitted from the same plate to the large cock, as the former is taken off to show the spiral-spring; the pivot of the balance-arbor runs in the central point where the four rollers nearly meet, and touches the circumference of each, so as to produce a rotatory motion in each that lessens the friction at the pivot; *b b* is a metallic sliding-piece filed away on the side next the centre, to avoid the regulating-spring, and placed on the plane of the cock, to adjust one of the curb-pins for regulating the piece, which it carries underneath it; the cock has an oblong slit in it near the screw at *b*, for the pin to move backwards and forwards during the act of adjustment, and the slider itself has two oblong slits, one at the said screw, and the other at one of the little frame pillars, which slits not only admit of a longitudinal motion of the slider, but also keep it in its due line of position during the motion of adjustment: at *c* is an arbour, square at the end to admit a key, but round below, and has a fine screw at the remote end, which presses against a stud *d* in the cock, where is a third opening for the stud, thereby moving the slider in consequence of the part *e*, which is turned up at its end, being tapped to suit the screw of the arbor of adjustment: on the middle of the said arbor is a nut between *e* and *c*, divided into 30 on its edge, which is pointed to by a fixed line on a stationary piece adjoining, through which the arbor passes. There is, moreover, a scale, *f*, attached to the large cock, to which a line on the slider near *e* points as an index of adjustment, and moves one space for every entire turn of the screw, which is an addition of Pennington's to the original construction. *g* is an horse-shoe-spring placed on the cock to force back the slider, when the screw has a retrograde motion. These parts and their uses, it is presumed, will be easily apprehended from our drawing, without the variety of views which Mr. Mudge has given in his plates. The slider is held down to the plane of the cock by the screw near *b* at the curb-pin, and also by a collet going over the stud *d*, and pinned on. There is a second curb-pin fixed on a detent, or straight piece of steel, riveted to the under side of the piece *h*, and crossing the slider at right angles; this curb-pin is borne by the interior extremity of the detent, and the adjustment is made by the screw and studs *i* screwed into the cock, and fitted sideways into two notches in the piece *h*; the screw on *h*, which holds the whole down, passes through an oblong hole, covered by its head, to admit of the adjustment of distance. The mechanism of this second curb-pin, we fear, will not be very intelligible, as some of the parts are necessarily out of sight: *k* is the stud of the regulating spring, and *l* a piece made purposely to cover the second screw of the large cock, which screw, therefore, cannot be taken out for dismounting the cock until this cover is taken off, and the cover itself cannot be removed till the regulating spring-stud is previously taken off: this precaution was necessary to prevent the derangement or breaking of the balance regulating-spring, one of which would follow, if the cock were dismounted without the stud being first unscrewed; the nature of the precaution is this; the cover *l* moves on its centre, below *b*, as on a pivot, and the round part at the upper end covers one of the two screws of the cock, until the stud, embraced by the semi-circular fork of the other end, is removed, when the cover *l* then is at liberty: the other corresponding screw for fixing the cock is visible at the opposite projection above *c*. *Fig*. 2, represents the second balance-spring and mechanism of compensation for the effects of heat and cold, as they appear when the cock in *fig*. 1. is removed from the upper plate of the frame; *a a* is a piece of brass screwed to the plate by the three screws *d*, *d*, *d*; *e* and *f* are two similar compensation-bars, composed of brass and steel soldered together, with their positions reversed; that is, the piece *e* has the brass side next to *e*, and the piece *f* has the steel side next to *f*: these compound pieces, seen in a detached state in *fig*. 3, with their interior curved ends passing each other, are screwed, each by two screws, at their outer ends, to the similar pieces *g* and *g*, which are themselves screwed to the long piece *a a*, but in such a way that they are adjustable, when their fixing screws are not home; the adjustment is made by the two similar horizontal screws at *h* and *h*, which fit the studs that appear at *i* and *i* in the oblong apertures of the pieces *g* and *g* left for the studs; and by these, together with the narrow apertures, and two other smaller studs *k* and *k*, one at each interior end of the said pieces *g* and *g*, is preserved the parallelism of the motion of adjustment: *l l* is a lever or detent, carrying a curb, or piece with two pins at the upper end, a little distance from the stud *m* of the regulating spring, and having a little curve to avoid the balance-verge, together with two short levers or tails, near the rounded projection of the cocks *b* and *c*; this detent is fixed on an arbor, which is pivoted above into the cock *b*, and below into the foot *n* of the cock *c*, which is let down, out of sight, through an aperture in the plate; the lever of the detent, however, is prolonged back to a considerable distance behind the cock to the end *q* of the cross-piece of *a a*, on which is described a portion of a circle divided into 20 equal spaces, to indicate, as a thermometer, the situation of the curb at any particular time; which is another addition contrived by Pennington. Now the action of the compensation-bars in the time-keeper, denominated *blue*, is this; because brass is more elongated by heat, and also more contracted by cold than steel, the former will be longer in summer and shorter in winter than the latter; the consequence of which will be, (as we have already said, when speaking of Harrison's compensation mechanism, from which it should seem the principle of this is borrowed,) that the metal most elongated will become convex in hot, and concave in cold weather; for instance, the compound bar *e*, with the brass side next to *e*, will become concave in cold weather on the brass side, and, its remote end being fast by the screws to the fixed piece *g*, the interior end will move forward to form the curve and press below the remote tail-piece of the detent, which detent will therefore move and bring the curb towards *o*, the other compound bar *f* in the mean time becoming convex on the side *f*, will make way for the motion of the other tail-piece that points to *e*, by falling back at the same rate with which the other bar moves forward; but in hot weather the contrary will take place, i. e. the interior end of the piece *f* will come forwards and press on the other remote tail-piece, and the end of the piece *e* will give way for the backward motion of the tail-piece pointing to *f*, the curb in the mean time having its motion in a direction towards *p*: the former of these two contrary motions of the curb will shorten the spiral-spring, or, at least, the effective portion of it, as the curb limits its action, and the latter will lengthen the same: but it is well known, that a short spring has greater force than a long one, *cæteris paribus*; likewise, that the altered momentum of the balance retards the rate in hot weather, and *vice versâ*; hence, the increased force of the spring in cold weather thus becomes a compensation for the sluggish motion of the balance occasioned by the same natural cause, and the contrary in the opposite extreme of temperature. Mr. Mudge junior might seem to have been guilty of an error

error in his "Defcription, &c." page 169, of this part of his mechanifm, but he is there defcribing the time keeper *green;* the compenfation-bars in which were made to act with the tail-pieces of the detent, which point refpectively *towards* them inftead of thofe which point *from* them : the nature of the action of the tail-pieces in *blue,* is, on the contrary, as we have defcribed, and as is clearly feen by the view we have given of the pofition of thefe bars in *fig.* 3. There is an error in the account by Mudge of this part of his mechanifm, where he afferts, that "the natural effect of heat is to leffen the momentum of the balance: and of cold to increafe it." We beg leave to obferve on this, perhaps unguarded, affertion, that whatever caufe enlarges the dimenfions of the balance muft undoubtedly increafe its momentum; and that heat has this effect is evident from the curve formed by the fuperior elongation of brafs compared with the lefs expanfible metal, fteel. Hence, the fpiral is fhortened by the curb in fummer and lengthened in winter.

In *Fig.* 4 (*Plate* XIII.) *a* is a cock in which the upper pivot of the upper pallet runs; the cock *b* is that in which the lower pivot of the lower pallet runs; *c* is the cock, on the nofe of the potence, on which both thofe pivots run that are near the axis of the balance-wheel; *d* is the cock in which one of the balance-pivots runs; *e* the cock for the other pivot of the balance-wheel; and *f* the potence fcrewed to the under fide of the upper plate, and bearing the fmall cocks, nearly in the relative pofitions in which they ftand in the figure; it may be proper to add, that the femicircular vertical excavation in *e* is to admit the arbors of the pallets within it, and that the pivot holes on thefe cocks are all jewelled.

Fig. 5 (*Plate* XIII.) exhibits a perfpective view of the balance, on an enlarged fcale, which is in the original of half an inch diameter, together with the auxiliary fprings which are fubftituted for the maintaining power, the balance-wheel, and the pallets. A B is the balance, and E F its verge, fhaped like a crank, of which the pivots I and K run in two fyftems of rollers, that have each a little frame, one of which has already been defcribed under our defcription of *fig.* 1, to which the lower one, placed at K, and fupported by a pillar fixed to the upper plate, is fimilar: one of the croffes of the balance bears a piece, M, to counterpoife the crank of the verge, and to keep the balance in equilibrio. C is the ftud of the regulating fpring, placed, as we have already feen, on the cock of the balance; and D is the ftud of the compenfation-fpring, fcrewed to the upper plate of the frame; thefe two fprings are both attached by their interior ends to the upper part of the verge above the crank, and have thin helices wound in the fame direction. The reafon why the verge has the fhape of a crank, is, that the little cocks or bearing pieces, reprefented in *fig.* 4, may have room for being fixed in their refpective fituations to catch the pivots of the balance-wheel arbor, and of the two pallet arbors, *c d* and *g h*, which arbors are exhibited in *fig.* 5, without their cocks, in order that the acting parts of the efcapement may be prefented to view; *a* is a pin, projecting from the upper bend of the crank verge, which in its motion meets with another longer pin, *b*, made faft to the arbor of the upper pallet, which pallet is feen a little curved with a fmall bend or hook at the outer end, between the letters *c* and *d*: this pallet-arbor has the upper auxiliary fpring fixed to it by its interior end, the outer end being pinned to the ftud at H; G is the pallet-wheel with its arbor difcontinued near the place where the contrate-wheel pinion is fixed. Mr. Nicholfon has given a very good account of the action of Mudge's efcapement in the 2d volume of his Journal, 4to feries, page 57, but has reprefented the fhape of the tooth in this balance-wheel fimilar to that in an ordinary watch, which, indeed, Mr. Atwood had done before him in the Philofophical Tranfactions of 1794; thefe authors have alfo omitted to mention the fpring of compenfation, which we here notice, becaufe their readers would otherwife be at a lofs, from the view of the figure, to know how the compenfation for temperature is effected, or, indeed, whether there is any at all; their object being only the account of the fprings as they related to the balance and efcapement. Pennington, who made the drawing for Mr. Atwood's paper, tells us, however, that the fhape given to the teeth is not of much importance. There is another pin, *e*, at the bottom of the crank, fimilar to *a* above, and alfo a fecond pallet-arbor, *g h*, carrying another pin, *f*, and a fecond auxiliary fpring, the outer end of which is pinned into the ftud I: the helix of this fpring is wound in a direction contrary to that of the upper auxiliary fpring. We have endeavoured to give fuch a perfpective reprefentation of the mechanifm of efcapement, together with the balance and compenfation, in one figure, as might exhibit to view all the parts of action; and the effect produced when the time-keeper goes, may be thus defcribed.

There are fix fprings employed in the time-keeper: the main fpring, the fpring of the going-fufee, the regulating or pendulum-fpring pinned at C, the fpring of compenfation pinned at D, and the two auxiliary fprings on the two pallet-arbors; let us fuppofe all thefe fprings exactly in their natural ftate of quiefcence, and that the crank verge, and pins on the two bends of it, are fo adjufted as to reft quietly againft the pins *b* and *f* of the pallet-arbors: alfo let the two pallets be at liberty, one remaining in the fpace diametrically oppofite the acting tooth which refts againft the other, which will be the cafe as the number of teeth is an odd number (15); in this fituation all the arbors will remain at reft till fome force is impreffed on the balance-wheel, provided the piece be fuffered to lie on a table; now wind up the main-fpring, and the fpring of the going-fufee will be firft put into a ftate of intenfity, and then the power will be tranfmitted through the train, with fome occafional variations of intenfity, arifing from the unavoidable friction of the teeth of the train, up to the balance-wheel, which will be impelled forward, till, pufhing againft one of the pallets, fay the upper one, it meets with fome refiftance from the auxiliary fpring on the pallet-arbor; however, the power of the main-fpring is fufficient to wind this auxiliary fpring up till the tooth of the balance-wheel, fliding on the concave fide of the pallet, is ftopped by the little hook at its extremity, which hook of the pallet now acts as a detent, and arrefts the further motion of the pallet-wheel as urged by the main-fpring; the quantity that the pallet-arbor revolves during this fhort action of the pallet-wheel, we are told, is about 27°. Now all is at reft again, and would continue fo, if no external force were to put the balance in motion: let therefore an impulfe be given to the balance fufficient to make it move through a femi-vibration, which in one of the original machines was 133° with one of its balances, and 119 with another; during this femi-vibration, the pin *a* of the crank catches the pin *b* of the pallet-arbor at the 27th degree from the ftate of quiefcence, and confequently difengages the pallet, and goes on winding up the auxiliary fpring the remaining 106°, the balance-wheel in the mean time running on, on being difengaged from the detaining pallet, by the impulfe of the main-fpring, till it arrives at the hook of the lower pallet; but during this unreftrained motion, it has no connection whatever with the balance; on which account it is that the efcapement is called a detached one; detached, as it relates to the main-fpring, except for the moment of unlocking,

locking, which is not more than $\frac{1}{25}$th part of the time of a vibration, but not detached, as it relates to the auxiliary ſpring, which is uſed as a ſubſtitute for the maintaining power; for we have ſeen that the pin in the crank of the balance verge impels the pin in the pallet-arbor all the time that the auxiliary ſpring is wound through a ſpace of 106°. The balance wheel by the impulſe of the main-ſpring having now urged the lower pallet till it is detained by the bend in its extremity, the pin *f* in the lower pallet-arbor has alſo in its turn been moved 27° from the place of reſt, in which ſituation it is now ready to be caught by the pin *e* in the lower bend of the crank verge; at length the balance begins to return from the point at 133°, the extremity of its vibration, and in its turn is impelled by the pin *b*, acting againſt the pin *a* of the crank, during the whole ſemi-vibration of 133°, ſo that it is itſelf impelled 27° more than it impels the pin, and the difference of the continuation of the two alternate impulſes conſtitutes the quantum of maintaining impulſe, that keeps the piece in continual motion; which quantum, in fact, is a deduction from the retarding force in the latter ſemi-arc of vibration, rather than an addition to the impulſive force in the firſt ſemi-arc. The balance-crank in its return on paſſing the point of original quieſcence, has done with the ſpring of the upper pallet, which now remains in the poſition we at firſt found it; but when the crank has paſſed the point of original quieſcence the ſpace of 27°, its pin *e* meets with the pin *f* of the lower pallet-arbor, which is in its turn impelled by the ſaid pin *e*, the ſpace of 106°, exactly as the former one was; the former, or upper pallet, in the mean time being again carried forward 27° by the next following tooth of the wheel; at length this ſemi-vibration is complete, and the balance-crank returns, driven by the pin *f* through the whole 133° to the point of its ſpring's quieſcence as it was before by the pin *b*, and thus the proceſs is finiſhed through two ſucceſſive excurſions; which proceſs may be conceived to continue for months, or even years, without interruption. Some of the balance-wheels were made of tempered ſteel and ſome of braſs, and the acting portions of the pallets of flint, agate, ruby, or ſapphire. The two ſlender ſprings within the crank, which have been juſt deſcribed, are denominated *auxiliary* ſprings, becauſe they *aid* the two other ſprings in the reſpective returning parts of each vibration, and in this point of view may be conſidered alſo as regulating ſprings, but, being only $\frac{1}{20}$th of the ſtrength of each of the others, are too weak of themſelves to perform this office without the addition of the other more powerful ones, one of which has its effective length conſtantly, though very ſlowly, changing by the ſucceſſive changes of atmoſpheric temperature, which is the reaſon why we called it the compenſation-ſpring.

It is of the utmoſt importance that all the four ſprings of the balance ſhould be at their reſpective points of quieſcence when the balance is at reſt, otherwiſe the law of their forces, which is aſſumed to be directly as their tenſions, will not be the ſame in all, and ſome may be accelerating while others are retarding, whereas they ought all, like one ſpring, to accelerate and retard the balance conjointly in every part of the vibration, unleſs, indeed, the large ſprings are found not to be iſochronal, and the ſmall ones have their points of quieſcence ſo placed as to effect a compenſation, which Mr. Atwood, in his ingenious paper of Feb. 27, 1794, in the Philoſophical Tranſactions, has ſhewn to be a feaſible adjuſtment. For inſtance, if the point of quieſcence of each of the auxiliary ſprings, or even of one of them, be placed ſomewhere in the firſt ſemi-arc of vibration of the balance, the acceleration effected thereby will be leſs in this than the retardation will be in the following ſemi-arc; in conſequence of which the time-keeper will go ſlower; but if the point of quieſcence of the auxiliary ſpring be in the latter ſemi-arc of vibration, the contrary will be the caſe, that is, the time-keeper will accelerate its rate; nay, if the quantity of deviation from the exact point of quieſcence be only one degree of a circle, the ſame author has calculated that where the balance is of one inch radius, and of a correſponding given weight, the daily gain or loſs will be about $19\frac{1}{2}$ˢ, and the quantity increaſes with the diminution of the arc of vibration; ſo that if the arc of vibration were to become 60°, in one caſe the daily loſs would be 44ˢ.33, and in the other the daily gain would be 43ˢ.6; which conſideration ſhews the imperious neceſſity of having the points of quieſcence of all the ſprings exactly adjuſted; and alſo that the pins in the pallet-arbors be ſteadily fixed; for an alteration in their poſitions is in fact an alteration in the quieſcent points of the auxiliary ſprings.

If we were to reaſon from theory, we ſhould be diſpoſed, from the conſideration we have given this mechaniſm, to ſay, that the principle upon which it is conſtructed is excellent, inaſmuch as the impulſes given to the balance, and the oppoſitions made to it during its vibrations, are in proportion to the diſtance from the point of reſt at all times, agreeably to the laws of gravity in a pendulum; the diſcontinuance of the retarding force for a ſpace of 27° in each excurſion, deranges as little as may be the ſcale of varying forces by which the regulation is effected; whereas in the other detached eſcapements, where the impulſe from the train is momentary, ſuch impulſe, to be powerful enough to maintain the continual motion of the balance, may be ſuppoſed to partake of the nature of a *jerk*, and to derange the regularity of the ſcale of forces, according to which, the balance-ſpring ought naturally to act; however, in practice it has not been proved that this conſtruction exceeds ſome of the more ſimple ones; but we will not undertake to affirm, that the aſſertion is quite accurate which ſays, that the great number of pivots requiring oil, and the various ſprings that require adjuſtments for time, temperature, power, and poſition render it extremely difficult to obtain a permanent rate. It is ſaid by the younger Mr. Mudge, that his father conceived the idea of making a time-keeper, and had organized his notions reſpecting its conſtruction, ſo early as Auguſt 1771, before he knew the conſtruction of Harriſon's; which ſeems probable, from various obſervations made in his "Thoughts on the Means of improving Watches," publiſhed in 1763, and written ſome time before; but were we diſpoſed to judge alone from a compariſon of the principles of his and Harriſon's time-keepers, we ſhould at leaſt ſay, that there is a remarkable reſemblance, almoſt too great for accidental coincidence; the principle of the compenſation mechaniſm is preciſely the ſame, and the auxiliary ſprings, though greatly different in their mode of acting, the one being before and the other behind the balance-wheel, the one wound up eight times in a minute, and the other once at each vibration, yet the object is the ſame in both, namely, to give a regularly modified impulſe to the balance, which balance again has a great momentum in both conſtructions, in conſequence of its enlarged dimenſions. Mr. Mudge, in a letter dated Plymouth, Oct. 5, 1775, and addreſſed to his excellency count Bruhl, his friend, ſays, "I do not think it difficult to deduce from reaſons, *a priori*, that there is one diameter (of a balance), with a proportionable weight, by which a greater momentum will be procured than by any other; and that you will loſe momentum either by increaſing or diminiſhing that diameter;" he does not, however, give thoſe reaſons, but has recourſe to experiment. The original balance was ſuppoſed to be too heavy in proportion to its diameter, a larger but lighter one was

was therefore made and tried, but the alterations neceſſary to be made in the cock rendered the trial doubtful, though the opinion entertained was in favour of a diminiſhed momentum being occaſioned by the alteration; a third was then made, one-tenth of an inch bigger than the firſt, and nearly four-tenths leſs than the laſt, "and I find (ſays our author) upon comparing the diameters, vibrations, and weights of the two (laſt made) that the momentum of the leaſt is to the momentum of the biggeſt as 9 to $7\frac{2}{3}$ nearly." The ſubjoined are the data and calculations on which the above concluſion was founded; viz.

leaſt balance:

Inches
Diameter 2.15 × 266° = 571.9 = the velocity
Vibration 266°
Weight 56.5 grains
Then 571.9 × 571.9 = 327069.61 × 56gr.5 = 18479433 = momentum.

Inches
Diameter 2.47 × 238° = 587.86 = the velocity
Vibration 238°
Weight 45.5 grains
Then 587.86 × 517.86 = 345579.3796 × 45gr.5 = 15723862 = momentum.

Let it be remarked here, that Mr. Mudge has ſuppoſed all the weight at the circumference of the balance, and has aſcertained the velocity from the diameter, which we apprehend ſhould have been from the radius or diſtance from the centre of motion, which mode would not indeed affect the relative, but would double the real velocities; alſo the ſquare of the velocity is multiplied by the weight, contrary to the opinion of thoſe who aſſert that the ſimple velocity ſhould be uſed as a multiplicand for the weight uſed as a multiplier, in order to effect a product equal to the momentum. According to this latter rule, if we take the double velocity as before, the reſpective momenta will be 371.9 × 56.5 = 32312.35 and 587.86 × 45.5 = 26747.33, which reſults are very nearly as 6 to 5.

It does not appear, notwithſtanding the above calculation, what was the relative power derived from the auxiliary ſprings, conſidered as a maintaining power, compared with the regulating power, which we have ſeen was 1 : 80 in Harriſon's time-keeper; nor is it quite certain, though very probable from the account, that the balance, with the greateſt momentum, was uſed in the trials of the going of the firſt time-keeper, on which an application was made for the parliamentary remuneration. Pennington found, from ſome experiments on a time-keeper that ſtopped in a voyage, that the neceſſary force of a main-ſpring ſhould be to the force juſt ſufficient to keep the piece in motion at firſt, as $17\frac{1}{2}$ to $10\frac{1}{2}$: i. e. when $10\frac{1}{2}$ oz. will produce a continuance of motion at firſt, 7 oz. more muſt be added for foulneſs in wear. Theſe were the exact proportions in *green*, though in *blue* they were found ſomewhat different.

Though Mr. Mudge had made and approved his firſt time-keeper, previouſly to Harriſon's having obtained his laſt 10,000l., the complement of his rewards, which ſum was granted by parliament in 1774, yet he was unfortunate enough to omit making application for the trial of his time-keeper till the act had paſſed, which limited the whole reward to 10,000l., or one half of that propoſed by the act of queen Anne, and alſo rendered the limits of trial more circumſcribed; the latter of which circumſtances rendered the attainment of even a portion of the diminiſhed reward more difficult, than the attainment of the whole great reward which Harriſon was fortunate enough to obtain by his indefatigable perſeverance. The act of queen Anne allowed to Harriſon's trial a voyage of only ſix weeks, at the end of which, if his time-keeper was found to have kept time within four minutes of error, or one degree of longitude, he might claim a portion of the reward thereon; but the time ſpecified in the act of 1774 was ſix months, during which, the error was not to exceed four minutes, for the ſmalleſt portion of the reward. In the four trials made of the time-keepers of Mr. Mudge by Dr. Maſkelyne, from the years 1776 to 1790, the firſt of which trials was of N° 1, in 1776, 1777, and 1778, and the ſecond, third, and fourth of the pieces denominated *blue* and *green* in 1779 and 1780, again in 1783 and 1784, and again in 1789 and 1790, it appeared from the Doctor's reports to the Board of Longitude, that none of the time-keepers had kept time within the limits preſcribed by the act of Geo. III. On the 1ſt of March, 1777, however, the aſtronomer royal reported to the Board, that the (1ſt) watch made by Mr. Mudge had gained only 1m 19s in 109 days, in conſequence of which, it was reſolved, that a letter be written to the Navy Board to pay Mr. Mudge 500l. to enable him to finiſh two more watches on a ſimilar conſtruction, which was deemed preferable to any other that had been previouſly tried; but after this, the main-ſpring of this piece broke, and on being replaced, the piece was found, on a trial of 15 months, commencing on Nov. 11, 1776, to have gained daily 8s.6 at the end, more than at the beginning of this ſecond portion of the trial. With reſpect to the three trials of the time-keepers, *blue* and *green*, we ſhall have occaſion to ſpeak of them hereafter. Since the publication of the "Narrative" by Mr. Thomas Mudge junior, it is well known what the difficulties were that Mr. Mudge encountered from the oppoſition which the Board of Longitude raiſed againſt him in his appeal to the Houſe of Commons, after his memorial to the Board had been unſucceſsful on the 11th of June, 1791. To inquire into and aſcertain the comparative merits of Mr. Mudge's time-keeper, however, a committee was appointed by the houſe in conſequence of the appeal conſiſting of the following honourable members: viz. Mr. Pitt, Mr. Fox, Mr. Ryder, Mr. Bragge, Sir Gilbert Elliot, Mr. Gregor, Sir George Shuckburgh, and Mr. Windham, of whom Sir George Shuckburgh was conſidered as the friend of the Board of Longitude, and was therefore candidly propoſed by Mr. Windham to witneſs all the meaſures about to be taken. The committee very properly directed their attention, in the firſt inſtance, to two material objects; 1ſt, to appoint a ſub-committee of ſcientific gentlemen, and men of practical ſkill in mechanics, to examine and report their opinion of the principles upon which the conſtruction was founded; and ſecondly, to aſcertain by a compariſon with other chronometers the accuracy which Mr. Mudge's time-keeper had given proof of in its actual meaſurement of time. The ſub-committee nominated was compoſed of the following liſt: viz.

The Biſhop of St. David's, Mr. Atwood, Mr. De Luc.	Scientific Gentlemen.
Mr. Ramſden, Mr. Ed. Troughton,	Mathematical Inſtrument Maſters.
Mr. Holmes, Mr. Haley, Mr. Howells.	Watch Makers.

This ſub-committee after due examination made the ſubjoined report to the ſelect committee, viz.

"We whoſe names are underwritten, to whom it hath been referred, by a Select Committee of the Honourable Houſe of Commons, to receive confidentially from Mr. Mudge, a communication of the principles of his time-keeper, and to report thereon, being aſſembled at the

houſe

houſe of his Excellency Count Bruhl, in Dover Street, on Tueſday the 14th inſtant, did cauſe a time-keeper, which was produced to us as one of the two made by Mr. Mudge, and ſubmitted to trial at the Royal Obſervatory at Greenwich, for twelve months, from June 1789 to June 1790, to be taken to pieces in our preſence, by Mr. Matthew Dutton; and having carefully examined the parts of the ſame, we find a contrivance in it for deſtroying the inequalities of the maintaining power derived from the main ſpring, which as far as we know and are informed is altogether new; and having conſidered the ſame, and put many queſtions concerning it to Mr. Mudge, jun. and Mr. Matthew Dutton, we are of opinion, that the ſaid contrivance is well calculated for producing the deſired effect, and that a diſcloſure of it may conduce to conſiderable improvements in the art of making time-keepers. We moreover declare that we find great ſkill and ability diſplayed in every part of the workmanſhip, as well as much ingenuity in this particular invention. At the ſame time it is our opinion, and we think it our duty to declare it ſo to be, that no judgment can be formed of the exactneſs of any time-keeper by theoretical reaſoning upon the principles of its conſtruction, with ſuch certainty as with ſafety to be relied upon, except it be confirmed by experiments of the actual performance of the machine.

Samuel St. David's,
George Atwood,
Jesse Ramsden,
Edward Troughton,
John Holmes,
Charles Haley,
Dover Street, William Howells,
the 20th of May, 1793. J. A. de Luc."

Notwithſtanding this report and the teſtimonies produced by Mr. Mudge in favour of his three time-keepers, one made in 1774, and the other two in 1777, from regiſters kept by Doctor Maſkelyne, Doctor Hornſby, Count Bruhl, Doctor (now Baron) Zach, of Saxe Gotha, and Mr. Dutton, who had been partner with Mr. Mudge, yet the committee, at a loſs for an unerring rule by which to aſcertain at all times an exact rate to be depended upon in future trials, declared it as their opinion that ſome of Mr. Arnold's chronometers, particularly No. 36 and No. 68, "had gone with a degree of accuracy greater than could be ſhown on any correſponding trial of Mr. Mudge's;" but in another part of the report of the ſelect committee, which is too long to be copied at full length, it is ſaid, alluding to the foregoing report of the ſub-committee that "in virtue of this report, and of ſuch other evidence as the inquiry has furniſhed, your committee have no difficulty in declaring, that they conſider the improvement in queſtion ſufficiently aſcertained, and as likely to conduce to advantages ſufficiently important to attract the notice of Parliament;" then, after the attention of parliament had been directed to the circumſtances of a life, ſpent in hope of benefiting the public, more than of enriching the individual, the report concludes with theſe words; viz. "For theſe conſiderations, joined to thoſe above ſet forth, your committee think themſelves authorized to recommend the petitioner to the attention of the Houſe, conceiving that the circumſtances attending his caſe give him a ſtrong plea to favour; and that the invention of which he is the author, contains an important improvement in the art of conſtructing time-keepers, ſuch as the Houſe might well wiſh to ſecure to the public, as well as to reward the perſon by whom it was produced." Accordingly, in the year 1793, the Houſe of Commons, after the examination of various witneſſes, notwithſtanding the oppoſition of the Board of Longitude, granted to Mr. Mudge, in addition to the 500l. previouſly received by way of encouragement, the further ſum of 2500l. under the act of 1774.

We have ſaid that the ſelect committee were without a certain rule that would apply in all caſes to aſcertain the true rate of a time-keeper at each ſucceſſive period of trial, a ſtandard rate previouſly obtained being in point of accuracy no longer applicable than while the piece conforms to that rate in going. The rule called "Dr. Maſkelyne's method" is this; when the period of obſervation, or trial, continues many months, he takes a mean rate from the going in the *firſt month*, and applies it as the ſtandard to any ſix ſucceſſive months after, mediate or immediate, which method allows of ſix periods in a year of ſix months each, or twelve in a year and half; and when a mean rate taken from the firſt month was thus applied, whether it was additive or ſubtractive, the aggregate amount of the daily errors thus equated, taken on any day of a period, is called *the error* of that day, and the evidences brought againſt the going of Mudge's watches were the *greateſt error*, and alſo the *mean error*, on an average of the whole period. Mr. Mudge, on the contrary, wiſhed the mean rate to be taken on a period of at leaſt ſix months, as Harriſon's had been on ſix weeks, or whole time of trial, and produced as evidence in favour of the going of his watch, the rate taken from a mean of the daily errors during ſix months of the trial, and alſo the greateſt deviation from that rate on any two ſucceſſive or remote days, without regarding the aggregate of the daily errors, or what is called *the error* at any particular part of the trial, it being contended that *the error* would never be very great if the rate were properly taken. In conſequence of theſe different opinions in reſpect to the mode of judging of the performance of a time-keeper, ſome of the members put queſtions to the witneſſes, particularly to Dr. Maſkelyne, tending to aſcertain, whether or not a rate could be taken at ſea as well as by land? and, on being anſwered in the negative, in what time, on touching at any known point of land, the longitude of which is known, a rate could be obtained? to which the Doctor's reply was "a month:" and his reaſon for requiring ſo long a time to get a rate, was, that in a voyage only the rates on the firſt and laſt days can, he ſays, be aſcertained accurately, but by land the ſucceſſive daily errors can be compared together, ſo that if a ſudden change takes place in the rate, the date of that change can thus be aſcertained, which he contends is not the caſe on a trial at ſea. It would ſwell our preſent article too much were we to inſert here all the trials in favour of and againſt Mudge's time-keepers, and alſo of thoſe with which they were contraſted; ſuffice it therefore that we refer the reader, who wiſhes to know all the particulars, to the "Narrative" publiſhed by Mr. Mudge, to Dr. Maſkelyne's "Anſwer to the Narrative," and Mudge's "Reply to the Anſwer." It will ſatisfy the ordinary reader, we preſume, to know the annexed particulars reſpecting the moſt and leaſt favourable trials. In the firſt trial by Dr. Maſkelyne, made during 15 months; from April 20, 1779, to July 17, 1780, the errors obtained by Dr. Maſkelyne's method of the time-keepers denominated *green* and *blue*, eſtimated from eight periods of ſix months each, were theſe; viz.

	′	″
Of *green* the mean error of eight periods was	11	1
The greateſt - - - -	16	25
The leaſt - - - -	7	20
Of *blue*, the mean error of the ſame periods	22	30
The greateſt - -	36	26
The leaſt . - -	5	53

In

In *green*, the daily error, or rate, was altered at the end of the trial 11^s.3 which was in excefs.

In *blue* the fame was more in excefs, viz. 18^s.

This trial was the leaft favourable to both time-keepers.

The moft favourable trial of *Green* feems to have been from 21ft July 1783, to 12th Sept. 1784, by Dr. Mafkelyne, in which its mean error on feven periods was 3^m 8^s, the greateft 7^m 10^s, and the leaft 1^m 25^s, the rate at the end having become fafter by 3^s.

But the moft favourable trial of *Blue*, feems to have been its laft under Dr. Mafkelyne in the years 1789 and 1790, in which the leaft error, eftimated on fix periods of fix months each, was 3^m 5^s, and the greateft 6^m 0^s, the acceleration of the rate being at the end 2^s.5 per day. The proof in favour of Arnold's No. 36 was, that the greateft variation in its daily rate on a trial of 15 months by Dr. Mafkelyne was 7^s; that its greateft error on eleven periods of fix months each, was 2^m 31^s, and its mean error only 54^s. The proof of Mr. Arnold's No. 68 was from a trial of Mr. Everard, of Lynn, in Norfolk, from which it appeared, from 46 periods of fix months each, that the mean error was, according to Dr. Mafkelyne's method, only 2^m 33^s, in only eight of which periods the error exceeded 4^m, and that the greateft variation in its daily rate was 5^s taken on any two parts of its trial. To thefe and other comparative proofs againft Mr. Mudge's time-keepers, the inventor oppofed various trials, as we have before faid, under Doctor Hornfby, Count Bruhl, Admiral Campbell, and Baron (then Doctor) Zach, as well as fome under Mr. Dutton, but as the difference in the daily rates, and not the aggregate, or what is called *the error* is chiefly given as the refult of each trial, we cannot well make a comparifon of refults comprifed in terms of different denominations.

Mr. Mudge junior, previoufly to his father's death, eftablifhed a manufactory for time-keepers, and employed Meffrs. Howells, Pennington, Pendleton, and Colman, to make them for him; fome few of which performed in a way that merited the approbation of certain naval officers of great refpectability, particularly lord Keith Elphinftone, and lord Hugh Seymour, but the difficulty of making the adjuftments fo accurately as Mr. Mudge fen. had done, and the high price put upon them, about 150 guineas each, (which indeed was too fmall, as the younger Mudge was a great lofer by the manufactory) induced the Admiralty to decline giving any other than occafional orders for his majefty's navy; the chronometers of Arnold and Earnfhaw, which were deemed equally good by the Board of Longitude, being fold at an inferior price; though Kendal's price for making a time-keeper after Harrifon's model was 400l. We decline accompanying Mr. Mudge through his complaints againft his opponents, particularly againft Dr. Mafkelyne, whom he has accufed of being too partial to his own darling child, the *lunar method*, to do juftice to any method purely mechanical for anfwering the fame important purpofe; a ferious complaint this, which is corroborated by an affertion, that Meffrs. Harrifon and Arnold fenior, made fimilar complaints: but be this as it may, it is a fact that the younger Mudge had only fold *eleven* time-keepers of his father's conftruction at the time his book was publifhed in 1799, though he had others finifhed, or nearly fo, and that Barraud and Jamefon, as we are informed, made propofals to the Admiralty for finifhing Mudge's time-keepers at the reduced prices of ninety guineas each, though it is well known that they cannot be afforded for this fum. In juftification of the Aftronomer Royal, we will conclude our account of Mudge's time-keeper with the concluding paffages of his own "Anfwer" to the "Narrative;" "the ufefulnefs of the Board of Longitude is too well known to the public, and acknowledged by all but a difappointed artift, to require my pointing out inftances in which they have materially ferved the public and done honour to the nation. Doubtlefs they deferve commendation in another refpect for having been careful concerning the diftribution of the public money. They might indeed have been properly cenfured, if they had given it away to a perfon not legally entitled to it by the act of parliament, or by a partial preference of the lefs deferving perfon to the more deferving ones."

Chronometer by *Meffrs. John Brockbank and Co.* Among other chronometer makers of reputation we have already had occafion to mention the Brockbanks, of No. 6, Cowper's Court, Cornhill, London, fome of whofe chronometers have performed with a degree of accuracy equal to that of perhaps any other maker, though they never applied for any parliamentary remuneration; the original firm of the houfe was Meffrs. John and Miles Brockbank, the former of whom lately died, and the latter is juft gone out of the bufinefs, and is fucceeded by his nephews, Meffrs. John and William Brockbank, fo that the firm is now Meffrs. John Brockbank and Co. On our applying for permiffion to take one of their chronometers to pieces, in order to give a full defcription of all its parts, the requeft was politely and readily granted. *Plate* XV. contains perfpective views of the different portions of the Brockbanks' chronometer, with the exception of the cafe or box, dial-work, face and hands, guard, fpring-ratchet, and fuch other fubordinate parts as are common to an ordinary watch, which may be feen more particularly under the article Watch-work. *Fig.* 1, exhibits the upper plate of the frame, feen a little obliquely, together with the balance, regulating cylindrical fpring, and the three different cocks for holding refpectively the balance verge, upper pivot, the ftud for the outer end of the balance-fpring and the mechanifm for banking. A, B, C, and D, are four circular holes perforated through the plate, juft large enough to receive the ends of the four pillars, denoted by the fame letters in *fig.* 2; E is a hole fomewhat lefs, to receive the end of the fpring-barrel arbor, and is at the place where the ratchet and click are placed to regulate the main-fpring, and preferve its intenfity when regulated: *a* is the compenfation balance, at prefent feen in the form of an ellipfe, as placed on its verge; but in *fig.* 3 is feen in its true fhape of a circle, where the eye is fuppofed to be perpendicularly over it; *b* is the cylindrical balance fpring, with its coils equal in diameter to the radius of the balance, having ufually from four to nine folds, attached at its lower extremity to the balance verge, by means of a collet and pin, and at its upper extremity to a ftud at the extreme end of cock *d*, which is compofed of two parts, fcrewed together by two fcrews, the heads of which appear near *d*; this cock is then fcrewed to the plane of the upper plate by a fingle fcrew, which fcrew, by the aid of two little fteady pins, faft in the cock near the fcrew hole, one of which is reprefented by a diminutive circle, and inferted into fmall holes in the plate, holds the cock firm in its proper fituation; *c* is the principal cock, the end of which has a jewelled hole that receives the upper pivot of the balance, the correfponding pivot is hid, being fupported by the crank part of the potence, which is reprefented by D E F in *fig.* 6, and is fcrewed to the under fide of the upper plate by a fcrew at E, aided by three fteady pins; the pivot hole in this potence is alfo jewelled. The third cock, *e*, is feen on an enlarged fcale in *fig.* 4, where it will be defcribed by and by. The cock, *c*, alfo appears to be compofed of two pieces joined together by two fcrews above the bend, the heads of which appear near *e*, but it is in one piece, and these screws fix it more firmly to the plate. The fusee guard, as in common watches, is placed under this upper plate, and is therefore out of sight.

Take the upper plate from the frame now, by unscrewing the four screws that go into the ends of the pillars, a contrivance of Messrs. Brockbank's, and *fig.* 2 will be presented to view, in which are contained the main-spring, fusee, and chain, with the mechanism of the perpetual ratchet, seen separately in *figs*: 7 and 8, the movement of the piece, and the lower or pillar plate with the four pillars, one of which has indeed been purposely left out in the drawing, that it might not intercept the view of the movement. A, C, and D, are the three pillars alluded to in our description of *fig.* 1, and B the place in the lower plate, where the fourth pillar ought to be screwed, or riveted, which we have said is left out of the drawing; E is the spring barrel, containing a well tempered main-spring, stronger than the spring of an ordinary watch, because the escapement is a detached one, requiring the momentary impulse given to it to be pretty strong; there is nothing particular in the construction of this barrel, except its size, which is proportioned to the spring it contains; one of which, for a large box chronometer, has been found equal to support a weight of 11 lbs. or upwards; the chain is also of the usual construction, except that it is stronger than is necessary in a common watch; it is represented as partly wound on the fusee, and partly on the barrel, which is generally the case when a watch or chronometer is going, so that, in this situation, if some obstacle to motion were not presented to some one of the wheels of the train, before the balance is taken out, the piece would run rapidly down, and be in danger of breaking some of the pivots or other delicate parts, on which account a bristle is usually put through the crosses of the third or fourth wheel to prevent such accident, while the balance is taken out, a small obstacle being sufficient to arrest the motion when applied near the top of the train, where the maintaining power is diminished in proportion to the number of wheels and pinions in the train that it has passed through; the reason of which is more particularly explained under our article CLOCK MOVEMENT, where the observations we have made are equally applicable to the movement of a clock, chronometer, or ordinary watch. F is the fusee grooved in such a way, after being made of the shape of the frustum of a paraboloid, that the decrease of the acting radius is always inversely proportional to the intensity of the main-spring; by which admirable contrivance the effective power of this spring is at all times very nearly alike; the adjustment of the varying levers, or points of action, of the fusee, is made very conveniently by a long lever with a moveable weight, like a steel-yard, being inserted on the square of the fusee arbor made for the key, as will be explained more particularly in its proper place. (See FUSEE and CLOCK-TOOLS.) The number of turns of the fusee, it has been already said, depends on the number of hours it is intended to be actuated by the spring at one winding up, and this number again depends on the ratio between the great wheel and the centre pinion, as has been mentioned in our descriptions of Harrison's and Mudge's time-keepers. The angular point of the cap, at the smaller end of the fusee, near F, is to catch the shoulder of the guard, when the chain has filled all the turns of the spiral groove; otherwise the chain would wind back again a little way, and the power of the main-spring would become too great, or perhaps the spring might even break by being over-strained. Concentric with the fusee at the large end, contiguous to the pillar plate, are the great wheel, two ratchets, and a secondary spring to keep the chronometer going while it is wound up; this idea, and also the mechanism proper for effecting such purpose, originated, as we have seen, with James Harrison; but the present application of his principle is much more simple than his was, inasmuch as the spring is so constructed, as not to require a barrel or box to contain it. The particulars will be more minutely detailed when we come to *figs.* 7 and 8. In the pocket chronometer before us, the great wheel, which is moveable round the fusee arbor in one direction only, like that of a common watch, has 60 teeth: *a* is the centre wheel arbor, on which is a pinion of 12 leaves revolving in an hour; this arbor is stronger than any other arbor in the train above, not only because more power is impressed on its pinion than on any other pinion higher in the train, but also because the minute hand is borne by it, and also motion given to the hour hand from it; the centre wheel fixed on this arbor has 64 teeth, which impel the pinion of eight leaves on the second arbor *b*; the second wheel of 60 teeth is fast also to this arbor, and impels the third pinion of eight leaves fast on the third arbor, *c*, which is the arbor for the seconds hand, which hand moves in a circle of 60 out of the centre of the face, therefore cannot be mistaken for the minute hand, which is a matter of some importance in a chronometer: on this arbor of the seconds hand, a contrate wheel is usually placed in a common watch, for the sole purpose of altering the direction of motion, that the balance or crown wheel may have an horizontal arbor, but here the case is different, the balance wheel, or more properly speaking the escapement wheel, has a vertical arbor like the rest of the train; the third wheel therefore is here like the rest in position, and has 80 teeth driving its pinion of eight on the fourth arbor, *d*, which also carries the escapement wheel with 15 teeth. From this account of the movement, we now know that the fusee revolves in $\frac{60}{12}$ of an hour, or in five hours, and also that, as there are $6\frac{1}{4}$ turns in it, the whole period of continued motion, at one winding up, will be 32 hours; also by our former method of estimating the value of a train exemplified in Harrison's and Mudge's trains; we know that $\frac{64 \times 60 \times 80 \times 15 \times 2}{8 \times 8 \times 8} = \frac{921600}{512} = 18000$ are the number of vibrations in an hour, or five in a second; but as it will be seen presently that there are only half as many audible beats as vibrations in this detached escapement in a second, this circumstance would induce any person, not previously acquainted with it, to conclude that the train is a lower or slower one than is used in an ordinary watch, though the reverse, in point of fact, is the truth. In some of the best box chronometers, the pinions are 10, and the train = 14400, owing to the large size of the balance. At *e* is a slender spring screwed to the pillar plate, and pressing against the tail of a lever or detent, fixed to the arbor, *f*, and resting like a click in the inclined teeth of the large or perpetual ratchet, the edge of which ratchet wheel is just visible above the large wheel under the fusee, but is seen better in *fig.* 7; the little ratchet and its click are not seen in this 2d figure, but is seen also within the perpetual one in *fig.* 7.

Fig. 3 represents the balance detached from its verge and regulating-spring after being taken from its collet on the verge; there are three radial arms meeting and uniting at the centre, which carry each a third part of a circle; this circle is at first turned in a lathe out of a solid compound plate of steel and brass, a circular plate of steel of the required diameter being covered with brass by immersion in a crucible of this metal melted previously by heat, after which it is divided into three equal portions, which will, by these operations, be of similar dimensions and weight when nicely divided;

divided; this method of uniting the metals, and of forming the balance in a lathe, was the invention of Mr. J. Brockbank senior, though never before publicly known; at the outer end of each of the three radial bars, which are of steel, are three screws of adjustment for time, and also for position, at *a*, *a*, and *a*, where the three portions of the compound circle are supported by their respective radial bars, or sometimes a little at one side. When the chronometer gains considerably, each of these three screws are screwed back or outwards, to increase the momentum of the balance by enlarging its effective dimensions, but the quantity of adjustment must be precisely the same in each of the three screws, supposing the balance previously in equilibrio in all positions. On the contrary, when there is a considerable loss in the rate, the three screws must be alike screwed in a certain quantity, depending on the quantity of the daily error, which a little practice only will determine; but when one of the three screws is screwed in for position, each of the other two must be screwed half the same quantity out, and *vice versâ*, the taps being the same in all. The convex side of the rim of the balance is brass, and the concave side steel; the superior expansibility of the former, therefore, will occasion the remote ends of each of the three expansion pieces, forming the circle, to approach the centre of the balance in high temperatures, by reason of those ends being at liberty, and the opposite ends being fixed to the radial bars; but in cold weather the contrary will be the case; the brass, being more liable to contract with cold, as well as more liable to elongate with heat, than steel, exerts a natural effort, in low temperatures, to straighten the expansion bar, and therefore makes the remote end, in each of the three, recede from the centre of the balance; again, the same heat that makes the remote end of each expansion piece approach the centre of the balance, makes, at the same time, the radial levers elongate, and consequently removes the fixed end of the same expansion pieces from the centre a certain quantity, and *the compensation* consists in a due balance of the two opposite and contemporary motions of the opposite ends of each of the three expansion bars. It is found from experience that the momentum of the balance is best preserved in an equable state by loading each expansion piece with a metallic weight of similar dimensions, and so contrived that they may slide along the expansion pieces, by adjustment, till the exact point is found such in each, that the equipoise of the whole is not destroyed, and yet that the momentum will remain the same in the opposite extremes of temperature, as well as at a medium. The additional weights are denoted by the letters *b*, *b*, and *b*, which have a groove and a screw each, to fix them to the expansion rims. If the additional weights of compensation, *b*, *b*, *b*, were screwed close to the screws of adjustment for rate and position at *a*, *a*, and *a*, it is evident that, as the radial bars are there elongated by heat, these loads would thereby be carried outwards so as to increase the momentum of the balance sufficiently to make the chronometer retard, being an effect similar to that produced by turning the screws *a*, *a*, and *a*, back from the centre; also if the said weights were fixed at the remote ends of the expansion rims, which ends move inwards with the greatest velocity, while their fixed ends are moving but slowly outwards, it is equally evident that, if the weights are too large, which are thus made to approach the centre with the greatest possible velocity, the momentum of the balance will thereby be diminished too much, and the chronometer will have an accelerated rate; there is, consequently, a certain load, which, being fixed in a certain point between the opposite ends of each of the three expansion rims, will render the momentum of the balance least liable to vary in the different degrees of temperature; for if the weight is comparatively small, it must necessarily be fixed near the end at liberty, but if comparatively large, it must necessarily be fixed at some distance from the said end; but until it has been determined in practice, or by experiment, what diameter is best for a given weight of a compensation-balance, with a given escapement and maintaining power, it is not easy to assign the precise quantity of each weight, and the corresponding point in its expansion piece where it ought to be fixed, in order to produce the best practical effect, i. e. the greatest quantum of invariable momentum with the least possible friction and resistance from the air. Various experiments, no doubt, have been tried in the adjustments of the balance, to effect this purpose in the best manner, taking into consideration, moreover, the relative thickness of the brass and steel of the expansion rims; though scarcely any thing has yet been published on the subject; to aid the views of those who are thus laudably employed, we presume to suggest, that the *shapes* of the loading pieces, and of the adjustment screws at present in use, are calculated to meet with too much resistance from the air; and if clock-makers wish to avoid such resistance, by a proper shape given to their balls or bobs in a pendulum, where the momentum is very great, surely the same object is worthy of the notice of the makers of chronometers. Mr. Brockbank once used weights shaped like a double cone, but laid them aside again, supposing it to be a matter of indifference. How far our suggestion may prove useful may be proved experimentally by trying the variation of rate of any machine, small enough to admit of being put under the receiver of an air-pump, whilst it is kept in a complete vacuum, or nearly so; this experiment, however, should not be attempted where there is not a syphon or other gage, to indicate to the eye the state of exhaustion during the trial. A gentleman who tried this experiment, informs us, that the variation in the rate of one of Brockbank's chronometers, was only about one second per day minus.

In the balance before us there is a pin screwed into the remote end of each expansion-piece, exactly opposite to as many projecting pieces of metal borne by the radial arms at *c*, *c*, and *c*, which additions might appear to have something to do with the compensation, but they are only precautionary contrivances, called guard-pieces, to prevent the bending of the expansion-rims too much inward or outward by any jerk or other accidental cause. In the most modern of Messrs. Brockbanks' chronometers, the load of the expansion-pieces is shaped as at *d*, in *fig*. 3, where a third additional screw is put in the direction of a tangent, to adjust for small errors in temperature, where the screws only are moved without the loads, which otherwise might be moved too much, so as to effect more than the desired quantity. *Fig*. 4, is the cock of the banking, marked *e*, in *fig*. 1, but on an enlarged scale; it is screwed down to the plane of the upper plate by the screw A, and kept immoveable by the steady-pin, denoted by a diminutive circle near the screw: B is a small piece of metal borne by a light arm that has a joint, the centre pin of which is seen represented by a dot on the end of the cock near the screw C; this screw is so placed on a side-piece fastened to the cock, that the end of it bears against the side of the joint-piece, and acts as a rest to prevent the joint from opening further than the adjustment requires: D is a very slender straight piece of balance-spring, not blued, with one end in the piece B, and the other in the cock, so that if any slight force is applied on the side of piece B, this piece will have a motion on its centre towards D, but will return to its original position by the force of the flat spring D. E is an erect pin attached to the radial arm, marked *f* in the balance,

as

as exhibited in *fig.* 1, which pin has a triangular head that remains contiguous to the projecting pin in the piece B, and very nearly touches it in the present position in both figures: now the cylindrical spring of the balance may be so adjusted, that the pin E shall be a quarter of a circle, or any given number of degrees, from the piece B, when this spring is in its quiescent state. Suppose it to be a quarter of a circle; and suppose the balance put into a forward motion by any accidental external force; if this force should be great, the balance would revolve, perhaps, two whole revolutions in its circle without any obstacle to stop it, which would overset it, and endanger the teeth of the escapement wheel, which would hobble on irregularly; to avoid such effect, Mr. J. Brockbank sen., having observed that the coils of his spring enlarge their diameters, particularly near the upper end, when wound up by one entire revolution of the balance, ingeniously hit upon the idea of making this protrusion of the coils of his spring subservient to the purpose of limiting the extent of the vibration of his balance, or of constituting the contrivance called the *detached banking*. In an ordinary watch a pin is put into the rim of the balance, which projects sufficiently to prevent its passing two little studs, placed at equal distances from the point of quiescence of the said pin, which studs, therefore, limit the quantity of the whole arc of vibration: this quantity, consequently, when two studs are used, must necessarily be less than a circle, even if the common escapement would admit of such a long vibration; and if one stud only were placed just a semicircle each way from the pin's quiescent point, the limit would then be an exact circle, after deducting the thickness of the stud; but in a chronometer one entire vibration generally exceeds a circle, when the piece is clean; a circumstance which constitutes one of its excellencies, by augmenting its momentum; hence the contrivance wanted was to allow the balance to vibrate more than an entire circle, and then to stop it at a given part of the second revolution: to effect this purpose, the late Mr. J. Brockbank invented the mechanism already described in *fig.* 4, which acts thus; when the balance first begins to vibrate, the coils of the spring do not alter their shape, and the pin E of the balance passes the piece B attached to the cock untouched, the position of B not having been altered by any external force; but when the balance comes round a second time, the protrusion of the coils becomes great enough to strike against the piece B, placed near the spring; it now yields to the impulse of the protruding part of the spring, and moves towards D, taking its pin along with it; and in this new situation it is that the pin of B is presented to the head of the pin E, carried by the balance, and acts as a banking-stud to prevent the further vibration, or, perhaps, we should say rotation, of the balance; in the contrary motion of the balance it cannot be overturned, by reason of the locking-spring not being moved by the backward motion of the lifting pallet. The contrivance before us is certainly an ingenious one, and, we understand, answers its intended purpose very well. The weight of the pin E is, of course, counterpoised in the adjustment for position.

Figs. 5 and 6 represent the escapement-wheel, pallets, and springs used in the business of alternately locking and unlocking the teeth of the pallet-wheel, on an enlarged scale, and as they would be seen by an eye placed over the principal cock, if the upper plate were transparent; but they are hid by this plate in *fig.* 1; the wheel, however, may be seen on its arbor projected on the fusee in *fig.* 2, and the pallets may easily be conceived to be put under the upper plate, through an aperture in the plate, and to be attached to the lower part of the balance-verge above the lower pivot, which pivot, we have said, rests on the crank part of the potence D E F, beneath the said plate, which potence is also supposed transparent. The shorter, or inner slope of the tooth is directed, not to the centre of the wheel, but to a point in the radius, about $\frac{1}{4}$ from the centre, which mode of shaping is called under-cutting. The same letters of reference apply to both the *figs.* 5 and 6, in which A is the escapement-wheel of 15 teeth, (but of 13 in a box-chronometer,) and B the large pallet of polished steel, first made circular, but afterwards notched, to make way for the teeth in passing, and to receive a piece of finely polished diamond, *c*, at that part of the notch where the end of tooth 1 nearly touches it in *fig.* 5; the little dotted piece *d*, like a bird's head, placed on the balance-verge, concentric with the large pallet, but below it, so as to be covered by it, is the lifting or unlocking-pallet made of steel, set with sapphire: C is a long slender spring, screwed to the under side of the upper plate, carrying a protuberance, *a*, called the locking-pallet, and embracing at its loose end, by a semicircular bend, the verge of the balance, the bent part of which, being hid under the large pallet B, is represented by a curve line of double dots: D is another very slender spring fixed by a pin at *c* into a hole in the end of the bar *c d*, which is elastic, and screwed to the potence at *d*; the screw *c* is tapped into a stud in the potence, and bears against the bar *c d*, and adjusts the length of D, which reaches a little beyond the curved end of the locking-spring, called also sometimes the detent-spring; this slender spring D is denominated the unlocking-spring, because the lifting or unlocking pallet, hitting it on the end, drives it against the crooked end of the locking-spring, and thereby forcing the pallet *a* from a tooth of the wheel, unlocks the wheel and leaves it under the controul of the maintaining power; lastly, *b* is a screw, which, supported by a little tapped cock, constitutes a rest for the detent-spring, when it has returned, in consequence of its elasticity, from the situation it was driven to by the lifting pallet. The two slender springs C and D point in the same straight line through the centre of the balance-verge; and the action of the escapement mechanism is this: suppose the balance to be at rest in the first place, and the respective positions of the parts as in *fig.* 6, the circular side of the large pallet nearly in contact with tooth 1, and the tooth preceding it; the lifting pallet must be as in *fig.* 5, a little short of the end of the unlocking-spring D, and tooth 2 resting on the locking pallet *a*; while every thing remains in this state, the maintaining power is suspended by the tooth 2 being detained by the locking-pallet *a*, and the machine will not have power to put itself into motion; but let some external force be applied to put the balance in motion, by turning the chronometer suddenly round horizontally, or otherwise, and, supposing the motion impressed to be in a direction from B towards the lifting-spring D, the angular point of the lifting-pallet, moving along with the balance and large pallet B, strikes the extreme end of the locking-spring D, and drives it, and also the bent end of the locking-spring C, against which the other rests, far enough to disengage the locking-pallet *a* from tooth 2 of the escapement-wheel; which wheel, therefore, instantly runs on by the impulse through the train, in a direction towards *b*; but in the mean time the sapphire face of the pallet, moving with the balance, has got before tooth 1; this tooth, therefore, now strikes the said face of the large pallet and urges it on, we will suppose, till it arrives at the line joining the centre of the wheel and great pallet, as is represented in *fig.* 5; in this situation, it will be seen, the lifting-pallet has quitted the end of the lifting or unlocking-spring, and the

the locking-pallet *a* is about the mid-way between tooth 2 and tooth 3: during the continuance of this portion of the wheel's impulfe on the face of the large pallet, the extreme end of tooth 1 has flided forwards along the plane of the pallet's face, which we have faid is fapphire, and, therefore, produces but little friction; but now, as the tooth 1 advances in its revolution beyond the line joining the centres of the wheel and large pallet, it proportionably returns along the face of the fapphire, till, at length, it completely efcapes the pallet, in which fituation it would run on violently, but by this time the locking-fpring has returned to the end of fcrew *b*, and has prefented its locking-pallet to receive tooth 3, which has juft arrived at this pallet, when tooth 1 drops from the large pallet; the train, confequently, is again detained, and the balance proceeds in its vibration, together with the large and lifting pallets, in a ftate completely detached from every obftacle, except the balance-fpring, which fpring, by being wound up, oppofes its free motion, and, at length ftops it; the ftation, however, is but of fhort duration, as the balance-fpring conftantly exerts its power to bring back again the balance; it returns; the back or curved part of the lifting-pallet ftrikes the end of the lifting fpring D at the inner fide, which being very flender gives way, and the locking fpring, not being difturbed, remains in *ftatu quo*. The returning force of the flight fpring, D, is exhaufted at the inftant of its arrival at its point of former quiefcence, and the balance proceeds in its retrograde vibration, by the accelerated force it has acquired on its arrival at this quiefcent point, until the oppofition of its fpring renders it again ftationary for an imperceptible moment, after which the original procefs, which has been defcribed, is refumed, and an alternation of backward and forward vibrations is perpetuated by means of one little impulfe, given by each fucceffive tooth of the wheel, as long as the maintaining power continues in a ftate of fufficient intenfity. With a given maintaining power and a given efcapement-wheel, the momentary impulfe given to the balance, in any chronometer of this conftruction, has its intenfity meafured directly by the effective length of the large pallet, in any fituation of the acting tooth; whence the intenfity of the impulfe would be a minimum in the fituation exhibited in *fig.* 5, in a point lying in the line of the centres, if this fituation were not the moft favourable to the tranfmiffion of the impulfe; but as the wheel here impels the pallet in the direction of a tangent from its point of action, and as this tangent is alfo perpendicular to the face of the pallet, the effect on the balance is here the greateft, independently of the force from the accelerated velocity. The lifting-pallet makes its ftroke at that part of the balance's vibration, when it has come within 20° of its original point of quiefcence, and as the fcale of forces may be confidered as changing foon after this point from an increafing to a decreafing one, there will here be very little difturbance produced in that fcale; befides, the continuance of the faid ftroke will be fmall as well as its force great by being near the point where the velocity is a maximum.

Again, in any chronometer of this conftruction, the arc of continuance of the wheel's action on the great pallet is inverfely proportional to the diameter of it, compared with that of the efcapement-wheel; therefore the larger the pallet the fmaller is the arc of action, and *vice verfâ*; but then, we have faid, that the impulfe is proportionably larger or more intenfe; it is probable, therefore, that there is a medium between a large impulfe given in a fmall arc of action and a fmall impulfe continuing during a larger arc of action, which medium may produce an effect moft fteady and moft in concert with the varying fcale of momentum in the different parts of the total vibration. Mr. Earnfhaw has laid much ftrefs on the circumftance of his large pallet being of half the diameter of the efcapement-wheel; but the inference does not appear to have been the refult of any nice calculations or varied experiments of his; for before him his mafter, the fenior Mr. Brockbank, ufed a pallet in his pocket chronometer, exactly fimilar, though it was a trifle lefs in his box one: but Mr. Arnold's large pallet, in general, is much fmaller, though fome of his chronometers have had as large pallets as thofe of either of the other two. Mr. Miles Brockbank informs us, that his brother and he found from experience, that a fmall pallet does not produce fo large a vibration as a large one with the fame maintaining power.

Fig. 7 exhibits a view of the great wheel, and two ratchets with their clicks fitted to the large end of the fufee, and conftituting, with it, what is called the *going fufee*; the conftruction is more fimple than that of Harrifon's, which we endeavoured to defcribe without a drawing, and anfwers its purpofe equally well. The ratchet, *a*, or fmall ratchet, is fixed by two fcrews or pins to an excavation in the large end of the fufee, the central part being left perforated for the fufee-arbor to pafs through, and its plane lying in the plane of the end of the fufee; its click and fpring, *b*, are fcrewed to the plane of the large or perpetual ratchet, *c*, which has its teeth inclined in a direction contrary to thofe of the fmall ratchet, *a*, and are bedded in a groove turned in the end of the fufee, between the fmall ratchet and great wheel; we have fhewn but one click for the fmall ratchet, but generally there are two, one at each fide of it; the click-fpring of the perpetual ratchet is fcrewed to the upper plane of the pillar plate, as feen at *e* in *fig.* 2, where it will be feen that the click turns on an arbor, and acts as a detent; *e* (*fig.* 7.) is the great wheel, and *fig.* 8 is a horfe-fhoe fpring, bedded between the perpetual ratchet and the great wheel, a circular groove being turned in the plane of the large wheel, or it might be in that of the large ratchet, or partially in both, to form a bed for this fecondary fpring; the pin, *a*, at one end of the fpring, is inferted into a correfponding hole in the bed of the wheel, and the other pin, *b*, into a fimilar hole perforated through the perpetual ratchet at *f*; this fpring, thus connected with both the great wheel and perpetual ratchet, would produce no other effect than to attach them together, and make them like one wheel, if the horfe-fhoe piece were not elaftic, in which cafe the large ratchet would be fuperfluous, and the effect produced would be that of an ordinary fimple ratchet; but the piece in *fig.* 8 is of a fpring temper, and its elafticity fmall enough to be acted upon by the main-fpring, fo as to make the two pins, *a* and *b*, at the ends approach each other; and in this fituation it is that the fecondary fpring is faid to be wound up, and in which it continues whenever the chronometer is going. When the key is applied to the fufee-arbor to wind up the piece, the click, *b*, will flide over the ferrated teeth of the fmall ratchet, *a*, which revolves with the fufee, and the large ratchet, *c*, would alfo revolve a little way with it, to let down the intenfity of the fecondary-fpring, which now exerts its force to remove its ends apart from each other to their natural ftate as they are feen in *fig.* 8.; but the click or detent, held to its teeth by the fpring *d*, prevents this little motion from *b* towards *a*, which otherwife would have taken place from the pulling of the pin *b* inferted into its little hole; in confequence of this oppofition to the great ratchet's temporary motion by the action of its detent, the pin, *a*, at the other end of the fecondary-fpring pulls at its hole in the great wheel, and draws it towards *b*, or, in other words, draws the great wheel round in a contrary direction, and with a force equal for a time to that of the original maintaining power by which the two pins were made to approach each other. The reafon of pin *b*, in *fig.* 8, being made to project both ways acrofs the end of the fecondary

condary ſpring, is, that the remote end beyond *b* may move in a circular little aperture made through the plane of the great wheel behind *f* in *fig.* 7, which aperture allows the two ends of the ſpring to approach and recede ſteadily, and the length of the aperture is determined by the quantity that pin *b* is drawn by the main-ſpring towards pin *a* before there is an equipoiſe in their intenſities.

Some years ago Mr. Ed. Troughton contrived a jambol or gimbol for preſerving the horizontal poſition of a pocket chronometer at ſea, which is loaded with a weight, turning on the point of a pin, like the card of a compaſs, and continues to be made by Meſſrs. Brockbank and Co. with conſiderable advantage to the going of the chronometer. The order of the adjuſtments is this: firſt the adjuſtment for temperature is made in heat of from 90° to 120° of Fahrenheit, and alſo in as great a degree of cold as can be obtained; ſecondly, the chronometer is cleaned anew, and has freſh oil put on; thirdly, the adjuſtment for rate and poſitions is made; and, laſtly, the rate is taken. There are, beſides the three pallet-faces of ſapphire, eight jewelled holes in the beſt chronometers; viz. two at the balance verge pivots in the cock and potence, two at the pivots of the eſcapement-wheel, and two for each of the two next wheel pivots, called the fourth and third wheels of the movement, beginning with the great wheel as firſt, but the third and ſecond of the train, if we count from the centre or hour-wheel. The pivots are tapering, in the form of a cone, and bear on their ends in action, which ſhape gives ſtrength, without adding to the friction.

It cannot be expected that the beſt chronometers will ever be manufactured at ſo low a price as watches without the compenſation and detached eſcapement; but when, as many different hands are employed in making their ſeveral parts, as are employed in the making and finiſhing of an ordinary watch, we may expect that the price will be proportionably reduced; the loweſt price that we have been informed of, as the price of any good maker for a pocket chronometer, is forty guineas, but in general they cannot be afforded for ſo little when all the adjuſtments are well made, which take up much time as well as patience. At preſent the *movement*, that is, the frame containing the barrel, fuſee, wheels, and pinions, all but the eſcapement-wheel, is made, like the movement of a watch, by the different workmen employed for this purpoſe in Lancaſhire; the motion or dial-work is next added by a workman in London, who has the main-ſpring, chain, face, and hands, from the reſpective makers in town; then the eſcapement-maker and the jeweller are employed to finiſh their departments; and, laſtly, the *maker*, as he is called, finiſhes the adjuſtments, and puts the works into the box, or caſe, or both, as may be required. In the progreſs of theſe different ſtages, or even after the adjuſtments are begun, it is frequently neceſſary to alter, and many times to change certain parts, as the ſprings, compenſation-rims, adjuſtable weights, &c. which neceſſarily enhance the price the chronometer might otherwiſe be afforded for.

The teſtimonies, both public and private, of the accuracy of ſome of Meſſrs. Brockbanks' chronometers in meaſuring time, are ſufficiently nnmerous to eſtabliſh their credit. Several letters from naval officers and rates were delivered to the Board of Longitude during the time of Mr. Earnſhaw's application for remuneration, which cannot be reſtored without an order from the Board; otherwiſe it would have been in our power to have laid ſome of them before the public. Governor Hunter, in his publication, has given a teſtimony ſo extraordinary, that we cannot, in juſtice, with-hold it; he has aſſerted, that, from the time of his ſetting out from Port Jackſon in New Holland, on a twelve months voyage round a great part of the globe, in the courſe of which he remained ſome time at the Cape of Good Hope, to the time of his return to the ſame port, a time-keeper or chronometer by the Brockbanks was found to have gone ſo well, that the error, at the end of the voyage, did not exceed *one ſecond* of time. This, however, will be conſidered rather as an extraordinary coincidence of the ſtate of the watch at the beginning and end of the voyage, than as a proof that its accuracy was thus great at all the intermediate parts of the voyage. Mr. Gavin Lowe, of Iſlington, has a pocket chronometer, the rate of which was given to Sir Joseph Banks to be laid before the Board of Longitude, which, we have heard, exceeds, in accuracy, the rate of any other pocket chronometer that has been made, inaſmuch as that the rate in it is not ſenſibly affected by cleaning in the courſe of many years wear. But, as we have ſaid, it is not our intention to preſs upon the public attention the merits of any individual maker excluſively, we ſhall ſatisfy ourſelves with copying only one additional teſtimony in favour of our preſent maker, from the report of lord Hugh Seymour, who tried three of Mr. Mudge's time-keepers againſt one of Mr. Earnſhaw's and one of Meſſrs. Brockbanks' chronometers, in a cruiſe from the 18th of May, 1796, to the 19th of Auguſt following.—The report was this: viz.

"At noon, May 31ſt, the town of St. Mary, on the iſland of that name, bore N. 10° W. diſtance 30 miles. The longitude of the ſhip, at that time, taken from the requiſite tables and corrected by the above bearings, gave 25° 3′ 15″ W. Mr. Mudge's watch, called No. 4, gave 1′ 45″ W. Mr. Mudge's watch, called *green*, gave 1′ 45″ E. Mr. Mudge's watch, called *blue*, gave 9′ 15″ E. Theſe watches were taken from the academy at Portſmouth, May 10. A watch made by Mr. Earnſhaw gave 3′ 45″ W. and one made by Mr. Brockbank gave 15′ 45″ E. of the ſhip's place. The mean of all gave 4′ 18″ E. Theſe two laſt watches were too ſhort a time on board at Spithead to obtain their rate exactly, *but were given a new rate this day.*

"At noon, June 4th, the town of Delgada, on the iſland of St. Michael's, bore N. 5 miles. The ſhip in the longitude of 25° 42′ W. No. 4 at that time gave 11′ 15″ W.; *green*, 5′ W.; *blue*, 5′ E.; Earnſhaw, 2′ 15″ E.; and Brockbank, 1′ 45″ E. The mean of all 1′ 30″ W. of the ſhip's place.

"At noon, July 16th, Cape St. Vincent bore N. 81° E. 4 miles. The ſhip in the longitude of 9° 7′ W. No. 4 at that time gave 3′ 15″ E.; *green*, 10′ 30″ W.; *blue*, 13′ 30″ E.; Earnſhaw, 16′ 30″ W.; and Brockbank, 2′ E. The mean of all 1′ 39″ W. of the ſhip's place.

"At four in the afternoon, Auguſt 12th, the light-houſe of St. Agnes was ſeen bearing N. diſtance 19 miles. The longitude of the ſhip at that time was 6° 28′ W. and ſuppoſing Scilly to be in *that* longitude. No. 4 gave 27′ 30″ E.; *green*, 8′ 30″ W.; *blue*, 18′ 15″ E.; Earnſhaw, 56′ 30″ W.; and Brockbank, 6′ 15″ E. The mean of all 2′ 30″ W. of the ſhip's place.

"At noon, Auguſt 13th, the Start bearing N. 28° E. diſtance 19 miles: the longitude of the ſhip was 4° 5′ 15″ W. No. 4 at that time gave 28′ 15″ E.; *green* 9′ 15″ W.; *blue* 17′ 15″ E.; Earnſhaw 59′ 15″ W. and Brockbank 6′ 45″ E. The mean of all 3′ 9″ W. of the ſhip's place.

"At Spithead, Auguſt 18th, the ſhip in the longitude of 1° 7′ 20″ W. No. 4 gave 30′ 20″ E.; *green*, 12′ 10″ W.; *blue*, 15′ 50″ E.; Earnſhaw, 1° 11′ 10″ W.; and Brockbank, 7′ 20″ E. The mean of all, 5′ 55″ W. of the truth."

From this report it is evident, that, after a proper rate was aſſigned, Meſſrs. Brockbanks' chronometer performed with a degree of accuracy which far exceeded any one of the other four, indeed, we may ſay, which has ſeldom been equalled by any other chronometer.

After being apprifed of thefe and other teftimonies, we inquired why the Brockbanks never applied to parliament for a public trial of any of their chronometers, with a view of obtaining the premium under the act of Geo. III., and received for information, that, after having been refufed a private trial at the Royal Obfervatory, and on finding that much trouble was likely to attend the application to the Board and fubfequent public trials, the idea at one time entertained was given up; which circumftance is our reafon for having troubled the reader with lord Hugh Seymour's report at full length.

Chronometer by *Mr. Arnold.* After the minute defcription of all the parts of Meffrs. Brockbanks' chronometer, it would be fuperfluous to repeat here an account of fuch parts as are common in all the modern chronometers; we therefore propofe to omit the drawings of the movement and other portions of the mechanifm contained within the frame of both Mr. Arnold's and Mr. Earnfhaw's chronometers, and beg leave to refer thofe readers who wifh to fee all the individual portions of each of thefe two, to a pamphlet lately publifhed by the Hon. Commiffioners of Longitude, which is charged five fhillings, and in which are contained three plates of each author, together with the defcriptions of the plates, and the queftions put by the Board of Longitude relative to each conftruction, together with the anfwers. This pamphlet is entitled, "Explanations of Time-keepers conftructed by Mr. Thomas Earnfhaw, and the late Mr. John Arnold," Payne and Mackinlay, Strand, 1806. The movements are made with pinions of 8 or 10, according as they are intended for pocket or box-chronometers, to which the correfponding wheels for trains of 14,400 or 18,000, may be had by infpection in our tables under our article Clock-movement, calculated on purpofe for the workmen who are movement-makers, and who chiefly refide in Lancafhire. It may be neceffary to mention, that Mr. Arnold's box-chronometers have very ftrong main-fprings requiring a deeper barrel than is neceffary for the length of the frame-pillars, on which account there is a cap fixed on the plane of the upper plate, to receive the lower pivot of the barrel arbor, and to hold the click and click-fpring of the ftrong ratchet, as placed on the fquare of this projecting arbor; but there is no occafion for fuch addition in the pocket-chronometer. The efcapement-wheel A, fhown in *fig.* 1, of *Plate* XIV. of *Horology*, is Mr. Arnold's, on an enlarged fcale, and, like Meffrs. Brockbanks', is placed near the lower end of its arbor, within the frame, fo as to have only a fmall portion of it feen by an eye placed over the cock, when the piece has its natural pofition reverfed: this wheel is what is called a *funk* one; that is, it has its teeth, like thofe of a cylinder-efcapement-wheel in this refpect, projecting from the plane of the wheel, as feen in *fig.* 2, which reprefents a fide-view of the fame wheel. The fhape alfo of the teeth of the wheel before us differs from that of Meffrs. Brockbanks' and Mr. Earnfhaw's in another refpect: the triangular acting part of each tooth, which is raifed from the plane of the wheel, it bounded by two ftraight lines and a curve; the curved portion, which acts with the jewelled face of the large pallet B, and which Mr. Arnold jun. in his defcription calls a *cycloidal curve*, is defcribed as being generated by the revolution of a fmall circular piece of metal with a tracing-pin in its circumference, while it rolls on the circumference of a larger metallic circle, as a bafe, and is, therefore, properly fpeaking, *epicycloidal*; a cycloid being generated by a circle rolling on a ftraight line: the proportions of the generating circle, and its bafe are ftated to be as the diameter of the large pallet to that of the efcapement-wheel; but, by confulting what we have faid on the proper fhape of acting teeth in an impelling-wheel in our article Clock-Movement, the reader, we prefume, will agree with us, that, to have as little friction as poffible, the fmall circle with the tracing-point, called the generating circle, ought to be equal to the *radius*, not the *diameter*, of the large pallet; which pallet may properly be confidered as a pinion for the fhort time it is impelled by a tooth of the efcapement-wheel; in the fame place, above referred to, it will be feen, indeed, that the difference of the curves, generated by a tracing-piece equal to the diameter, and by a tracing-piece equal to the radius of the fame wheel or pinion, will fcarcely be fenfible till the two curves have been carried on farther than is neceffary for forming a fmall tooth; fo that the diftinction in theory makes no confiderable difference in practice, which we here mention, left an unfavourable opinion fhould be entertained of the teeth in queftion: but, what may feem to fome of our readers a curious circumftance, when the epicycloidal tooth in any wheel is formed by a generating circle of double the fize it ought to be from theory, the friction of the parts in wear will ultimately produce the curve that ought to have been originally formed by a generating circle of a due fize, provided the tooth of the wheel acting with it is of a proper fhape, and of a more durable metal; becaufe, when the teeth of any two wheels that act together are both formed trufy from proper generating circles rolling on proper bafes, thefe teeth will roll over one another without, or nearly without friction, and preferve their original figure unimpaired; but if one of the two fhall happen not to have the exact curve, the friction in the action will wear away the fuperfluity of fubftance beyond what ought to have been there to conftitute the true curve; and as a large generating circle gives a fuller tooth than a fmall one, there will neceffarily be that fuperfluity in its fize which we have fuppofed to exift, when the diameter is ufed for radius in the generating circle. Thus a flight deviation from the true epicycloidal fhape of the tooth of any wheel will be rectified by its action with a lefs deftructible body of an exact fhape for true action, and it will be feen, under the article to which we have already referred, that a *ftraight line*, the particular fhape of the jewelled face of the large pallet, is one of the varieties of an interior epicycloid generated by any generating circle revolving on the concave fide of a circle of *twice its diameter* ufed as a bafe of generation.

We made thefe obfervations under an impreffion that Mr. Arnold's tooth *rolls* over the furface of the face of the pallet during the time of its impulfion, in which cafe the friction and deftruction of the parts of contact would have been the *leaft poffible*; but on examining his drawing more minutely, and on adverting to the queftions put to Mr. Arnold, junior, by the Board of Longitude, we find that the tooth of the wheel always continues to act at the extreme point of the pallet's face, over which angular point, probably rounded a little, the curved part of the tooth *flides* inftead of rolling, by reafon of every acting part of the tooth coming in fucceffion to the fame point of the pallet. This mode of action does certainly require that the ratio between the generating circle and its bafe fhould be exactly as ftated by Mr. Arnold; it being that particular cafe where a wheel drives a lantern pinion with fmall fpindles, as may be feen by the reference we have more than once given to Clock-movement. However, we are ftill perfuaded, that to give an impulfe to the large pallet without friction, would be more defirable than the method before us, if it is equally practicable. An adoption of the mode we have propofed would, we think, on mature confideration, require either

either the wheel to have more teeth than has been usual, or the pallet to be larger, to which we see no objection, as the arc of action would in either case be diminished.

"The size of the pallet depends upon the number of teeth in the escapement wheel, says Mr. Arnold, in his Description of his father's escapement. The radius of the pallet should be equal to the distance between any two teeth of the wheel, and then their relative motions will be equal. If the wheel has twelve teeth, the radius of the pallet will be thirty degrees, measured on the circumference of the wheel, and its diameter sixty degrees (nearly), measured in the same manner, which will make it half the size of the wheel. If it has thirteen teeth, the pallet will, in diameter, measure fifty-five degrees and a half; if fourteen teeth, fifty-one degrees and a half; and if fifteen teeth, which is the number generally applied to pocket time-keepers, it will be forty-eight degrees. The marine (or box) time-keeper is made to beat half-seconds, the balance making 240 vibrations both ways in a minute; for if the balance-wheel has 15 teeth, the fourth wheel 80 teeth, and the balance pinion 10 teeth, there will be 120 beats, or half seconds, in one minute. It is also made with the escapement wheel of 12 teeth, the balance pinion having 7, and the fourth wheel 70 (counting from the great wheel); consequently there will be 120 beats or half seconds in one minute, as before. It has been already remarked that the pallet for 12 teeth must be half the diameter of the wheel, and for 15 teeth five-twelfths, or fifty degrees.

"The pocket time-keepers, that they may not be disturbed by motion, have what is called a quicker train, the seconds hand making 150 beats upon the dial, or 5 beats in two seconds. The escapement wheel has 15 teeth, the balance pinion 8 teeth, and the fourth wheel 80; consequently there will be 150 beats in one minute, the pallet being 50 degrees in diameter, measured upon the diameter [ought to be circumference] of the balance-wheel. No mention has been made of the numbers of the teeth in the other wheels and pinions, as they are of little or no importance, and may be varied considerably."

We beg leave to differ here from Mr. Arnold, being decidedly of opinion that pinions of 6, 7, and even 8, are by no means so well calculated to transmit the maintaining power equally, as pinions of 10 and upwards; indeed Camus has demonstrated in his chapter on the proper shape of teeth of "*Cours de Mathematique*," lately translated into English, that no strong pinion with a number of teeth under 10 will act with a wheel of ordinary size, entirely on one side of the line joining the centres of the acting wheel and pinion; therefore will not act without much friction; besides, the more numerous the teeth are in a given wheel, provided the strength of the tooth be sufficient for its purpose, the less the teeth take into those of the pinion, and consequently the less the friction, in this second point of view; if pinions of 12 or upwards were to be adopted in watch-movements and clock-movements, the advantage accruing from the adoption would be considerable, provided the weight of the wheels, towards the third and fourth wheels, were as little as their requisite strength will admit.

Mr. Arnold has given four positions of his escapement wheel in *Plate* III. of his description, delivered to the Board of Longitude; figure one shows the wheel locked, and the balance returning in a detached state from its second excursion; the second figure shows the situation of the three pallets at the instant of unlocking; the third, which is our *fig.* 1, of *Plate* XIV., shows the situation of the different parts when the impulse is half given, the jewelled face of the large pallet being in the line that joins the centres of the escapement wheel and pallet; and his figure four exhibits the parts at the moment when the impulse ceases to be given by the acting tooth of the wheel. After what we have said of the mode of acting in Messrs. Brockbanks' escapement, it will easily be apprehended what those relative positions are from a verbal description; particularly as we have put the same letters to the corresponding parts of both escapements, to assist the reader in comparing what we have before said with our present description. A is the escapement-wheel of Mr. Arnold's marine or box chronometer, made of brass, and having 12 teeth, with their triangular ends projecting upwards, or rather downwards, when the face is up, from the plane of the wheel; B is the large impelling pallet of steel, at first made circular, like Messrs. Brockbanks', but having its notch terminated by two straight lines pointing to its arbor, in one of which its jewel *c* is fixed; C is the locking spring, screwed at its remote end to the under surface of the upper plate, and playing in a notch, or straight groove made in the plane of the plate to receive it; its weakest part is about C, or between C and the fixing screw, about which point it may be said to turn as on a centre, but having no pivot, it requires no oil; about the middle of this spring, C, is a second weaker spring, D, attached to it, which, in Messrs. Brockbanks', is a detached spring lying at the opposite side of the great pallet, and pointing to its centre; the end of this slender spring, which is called the unlocking spring, comes nearer towards the centre of the large pallet than the same end of the spring C; the said two springs, thus attached together, have a great resemblance, as they are seen in the figure, to a metallic pen in a pocket case of instruments, when one of the nibs is longer than the other; nearly at one third of the spring C, from this interior end, and on the side next to the wheel at *a*, is the locking pallet, the acting portion of which is a jewel; this jewelled pallet rests against the heel of the tooth, or nearest angular point towards the centre of the wheel, and in the act of unlocking is driven *inwards*, to allow the projecting portion to move *behind* it, when the wheel is unlocked; but, lest the locking spring should yield to the pressure of the wheel when locked at any time, a screw *b*, tapped into a stud in the upper plate opposite the pallet, *a*, bears against the exterior side of the locking spring, to prevent its falling back beyond a certain limit. In the drawing, the pallet appears to be a continuation of the screw through the spring, on account of being placed over, or very nearly over it: the centre of motion of the unlocking spring, D, is near the screw we have just described, from which it is free, by being narrower than the locking spring, C, and consequently also weaker, when equally thin; the unlocking spring, D, therefore, is at liberty to move back towards the screw-head without affecting the position of the locking spring, C, but when, by an impulse received, it moves in a contrary direction, it must necessarily take the locking spring along with it, and consequently the pallet, *a*, attached to this locking spring also. The lifting, or unlocking pallet in Mr. Arnold's construction, is a straight piece of steel, *d*, carrying a jewel, pointing not in a direction nearly opposite to the face of the large pallet, as is the case in Messrs. Brockbanks', but in one vibration follows the face of the large pallet a very little, and in the other precedes it as much: if two lines were drawn along the acting faces of these two pallets, which are fixed by friction on the verge of the balance, the angle contained would be very small in the figures given by Mr. Arnold, in consequence of his locking on the second tooth, but we have not exact data whereby to calculate it, as will be seen more particularly in our account of Mr. Earnshaw's chronometer, which

which follows, because we are not informed by Mr. Arnold where the unlocking pallet refts when the regulating fpring of his balance remains quiefcent. Meffrs. Brockbanks' efcapement wheel, like Mr. Arnold's, locks, as we have feen, at the neareft tooth behind the face of the large pallet, whereas Mr. Earnfhaw locks at the third, counting the tooth of action one, as will be feen by and by; but, unlike all the others that we have feen, Mr. Arnold's locking fpring receives an impulfe *inwards*, to ftrike the locking pallet from the tooth of its wheel: we pretend not to affert, from theory, that this kind of action is either more or lefs favourable to the efcapement than when the locking pallet is driven *outwards;* in either cafe, if the face of the tooth is fuch, that the efcapement wheel has no recoil during the difengagement, the total refiftance will be nearly fimilar; for though that part of the tooth, which is neareft the arbor of the wheel, preffes on the pallet more than the extreme points would do of the *fame wheel*, under the fame circumftances, in confequence of which, the preffure is here greater near the termination of the impulfe given to the locking pallet than at its beginning, yet we do not conceive this to be any difadvantage, becaufe the locking pallet may thus be fuppofed to be unlocked gradually, rather than by a fudden jerk, which muft be partly the cafe, when the preffure of the wheel's tooth againft the pallet is a maximum at the commencement of the impulfe that detaches the locking pallet. When, however, a comparifon of this kind is made between two efcapement wheels, we ought to take into the account their relative diameters, the relative maintaining powers as exerted at thefe wheels, the relative ftrengths of the regulating fprings, as well as of the locking fprings, and alfo the relative points in the arcs of vibration where the unlocking pallets ftrike, compared with the relative points of quiefcence of the balance fprings; all of which are data, involving a complexity of calculations not entered into, we prefume, by the chronometer maker, when he feels difpofed to prefer one conftruction to another, on views more fuperficial.

We have already faid, that in our *fig.* 1., the wheel is at the middle point of its arc of action, confequently, the lifting pallet *d* has let go the contiguous end of the unlocking fpring D, and the pallet *a* has returned to the fcrew point, to be ready to receive the next following tooth 3.; the direction of motion being fuch, that the teeth, 1, 2, 3, &c. follow one another in fucceffion. In this fituation, it appears to us, that the heel of tooth 2, which is now moving, is too near the pallet *a* in the figure; for, when the other half of the impulfe has been given, and the tooth 1 has efcaped the pallet, the tooth 3, which is now nearly twothirds of a fpace from the pallet *a*, will have one-third drop before it arrives at it, which muft be prejudicial to the fteady motion of the balance. It is eafy to fee, that, during this vibration of the balance from D towards *c*, the unlocking pallet *d* muft have hit the projecting end of D before it paffed it, and alfo muft have carried it and the fpring C, together with the pallet *a*, towards the wheel, until the pallet *d* cleared the extreme end of D; alfo that the impulfe given in paffing muft have been of a continuance depending on the quantity that the end of the fpring D prefents of its length to the paffing pallet *d*; in the returning vibration, there will be the fame continuance of the impulfe given by the pallet *d* to the fame fpring D in a contrary direction, but then the flender fpring is the only one to be moved, and it yields to the flighteft impulfe, thereby occafioning no fenfible derangement in the fcale of forces by which the regulating fpring controuls the balance: hence one impulfe received from the maintaining power through the medium of the train, by the face of the great pallet during the angle of its action, or rather of the efcapement wheel's action on it, is fufficient to overcome all the friction and refiftance the balance meets with, from whatever fource, and to perpetuate its vibrations. From the prefent pofition of the refpective parts of action, it is alfo eafy to fee, that the impulfe given to the end of fpring D muft have taken place a very fhort time before the tooth 1 caught the face *c* of the large pallet, on which time depends the quantity of *drop* of tooth 1, before it comes into action; which drop ought to be very fmall, left an accelerated force fhould impel the large pallet with a jerk, and endanger fome of the finer pivots. Mr. Pennington tells us that the prefent pofition of pallet *d* is very nearly that in which the balance fpring ought to be quiefcent, in order that the chronometer may be well in beat, a circumftance not noticed in Mr. Arnold's account, but a very effential circumftance to be known, when we judge of the fitnefs of the efcapement for anfwering its purpofe for a long continuance; for, as the momentum of the balance is always a maximum when it paffes the quiefcent point of its regulating fpring, or, in other words, when the force of this fpring ceafes to be accelerated, and begins to be retarded, it is acknowledged, we believe, univerfally, that the nearer this point of greateft momentum the balance is, when the unlocking pallet makes its ftroke, the lefs is the derangement in the uniformity of the balance's motion, which its regulating fpring is intended to produce: but, it has been faid, that the prefent pofition is that of original quiefcence of the regulating fpring; and it will be obferved, that there is an arc of feveral degrees contained between the extreme ends of pallet *d*, and fpring D, if a circle were defcribed to touch them both; whence, it might be concluded, that the pallet *d* ought to touch the end of fpring D, when the regulating fpring is at its quiefcent point, for then the ftroke would be made exactly at the moment of its having the greateft momentum: fuch conclufion might be good in theory, and, indeed, this is the pofition which Mr. Earnfhaw has given in his drawings, when he defcribes his balance as being in a ftate of quiefcence; but we fhall referve what we have to fay further on this important point, till we come to fpeak more particularly of Mr. Earnfhaw's efcapement in our following fection. In the mean time, we will only generally obferve, that, to put a chronometer, with a detached efcapement, *into true beat*, the pallet *d* muft be in the middle of the arc of efcapement, which we fuppofe to be nearly in the pofition of our figure, when the balance-fpring is quiefcent. Any further notice here on the mode of acting in this efcapement, we think unneceffary. *Figs.* 3 and 4, exhibit Mr. Arnold's balance; the former fuppofes the eye placed over the centre, and the latter at one fide in a line paffing through its plane. The circular bar of metal *a*, *b*, carrying three weights of adjuftment for pofition, within the expanfion pieces *c*, *d*, in *fig.* 3, we underftand, were added to the original balance by Mr. Arnold junior; but the generality of his chronometers have not had fuch addition, being capable of the neceffary adjuftment without.

The expanfion rims, which are about the third portion of a circle each, were originally foldered together by an intermediate mixture, and bent into the requifite fhape by a pair of pliers fhaped on purpofe to give the defired curve regularly; and, in fome of the *beft* chronometers, Mr. Arnold informs us himfelf, that he ftill continues this practice, which is greatly reprobated by Mr. Earnfhaw; we will not undertake to decide the practical queftion, which, in theory, we fhould have lefs difficulty to decide; the regularity of weight and fhape enfured by turning in the lathe, which was at firft the fenior Brockbank's practice, and is fince that of

Mr.

Mr. Earnſhaw, who formerly worked under him, ſeems to promiſe fairly for anſwering the purpoſe beſt, particularly when two metals only are united by fuſion; but, it is contended by Mr. Arnold, that the true figure given in the lathe no longer remains when the ring is cut into portions, ſuch as halves or thirds; for he ſays, the ſeparate parts aſſume, by their elaſticity, an inſtantaneous alteration in their figure, generally becoming portions of a circle of ſmaller radius than that of the original ring, and the ſeparate portions do not always undergo a change exactly ſimilar, owing to circumſtances which cannot be eaſily detected. To ſatisfy our doubts reſpecting this objection to turning an expanſion ring in a lathe, we applied to Mr. Pennington, who is allowed to be inferior to none of his contemporaries in practical ſkill, in all the different conſtructions of modern date, and we find it to be his practice, like Mr. Arnold's, not only to bend his expanſion pieces with pliers, but alſo, to ſolder them previouſly, as he formerly did thoſe of Mr. Mudge's; and he is convinced that his balances, though perforated in many places to receive various ſcrews of adjuſtment, are as ſenſibly and regularly obedient to the changes of temperature as thoſe of any other maker: it may be proper to add here, that Mr. Pennington has an excellent regulator, with a compenſation pendulum of a peculiar conſtruction; and that he has lately fitted up a little room with an inſulated brick and ſtone pillar for his tranſit inſtrument; of which he is fully acquainted with the uſe and neceſſary adjuſtments.

The ſcrews *d*, *d*, in *figs.* 3 and 4, are for rate; the cylindrical pieces, *c* and *c*, tapped for the ſcrews at the ends of the expanſion pieces, and having little holes at their exterior ends for a fork ſcrew-driver, are for the adjuſtment for temperature; and the two additional ſcrews, *e* and *e*, are for the adjuſtments for poſition, particularly when the interior ring *a b* is not introduced: theſe laſt ſcrews, *e*, and *e*, when uſed for poſition, will alter the momentum of the balance, and conſequently the rate of going, if one of the two is not juſt as much ſcrewed in as the other is ſcrewed out, when the rate is previouſly adjuſted, unleſs, indeed, there is a difference in their relative dimenſions and weights, which, in this caſe, would interfere with the adjuſtment for temperature.

Mr. Arnold uſes a cylindrical ſpring with his balance, and is very particular in aſcertaining the exact effective length that ſhall produce the ſame rate, whether the arc of vibration be long or ſhort; the trial of this adjuſtment is made by uſing the main-ſpring greatly relaxed, or let down by its ratchet, and again when it is ſet high, or has its intenſity increaſed; which alteration is equivalent to an addition or ſubtraction of weight in the maintaining power of a clock in order to increaſe or diminiſh the arc of vibration in a pendulum. When the chronometer is new, or clean if uſed ſome time, the ſemi-arc of vibration varies according to circumſtances from 180° to 230°, making in the whole vibration from a circle to 460°; but, when the oil grows thick, or when dirt has obtained admiſſion into the upper part of the train, the arc will ſometimes be reduced to 240°; ſo that, if the long and ſhort arcs were not performed in the ſame time preciſely, an alteration would take place in the rate, which might be conſiderable enough to do away all dependence on an accurate meaſurement of time. The balance-ſpring is uſually made by Mr. Arnold of ſteel wire hardened and tempered, though he ſays that wire hard rolled, or wire made of gold with a mixture of from one-eighth to one quarter of copper, will do; but the two latter are leſs permanently elaſtic, and the hard rolled wire will ſometimes require to be tapered, at that end which is next the balance ſtud: the beſt length of a balance-ſpring, which is longer in box than in pocket chronometers, lies between 5 and 20 inches to become iſochronal; but Mr. Arnold does not ſeem to apprehend, that there are various intermediate lengths, as Mr. Pennington aſſerts, which are equally iſochronal. In the box chronometer, which has a heavier balance than the pocket one, and a ſlower train, the weight is taken from the ſupporting pivot of the verge, by an ingenious application of the cylindrical regulating ſpring thus; when the ſpring is wound round a cylinder, to be blued, the coils are put contiguous, which ſhape would be afterwards preſerved from the elaſtic temper then given to it, if no force were to ſeparate them; but when one end of the ſpring is attached to the verge collet, and the other to the ſtud, theſe oppoſite ends are forcibly removed from each other to ſuch a diſtance, that the coils are ſeparated from contact, and the effort, exerted to bring them again into contact by their elaſticity, lifts nearly the whole weight of the balance. Formerly, there was a contrivance for banking in Mr. Arnold's chronometers, but the banking pin or lever rubbing between two coils of the ſpring, was found to be injurious, and was therefore laid aſide; and we do not learn that any other banking has been ſubſtituted. Mr. Arnold lays great ſtreſs on his ſtud being placed ſo as to fix the end of his balance-ſpring at half the diſtance between the centre and circumference of the coils, ſo that the laſt coil, at each end, is made ſo much ſmaller than the reſt, as prevents any protruſion of the large coils, and preſerves the cylindrical ſhape apparently unaltered by the action.

We underſtand, that neither of the two Arnolds ever placed a chronometer for trial at the Royal Obſervatory, for the expreſs purpoſe of applying for parliamentary remuneration; but we have ſeen, under our account of Mr. Mudge's time-keeper, that Nos. 36 and 68 were tried againſt his, and pronounced to be ſuperior, both with reſpect to their accurate going, and alſo in regard to the ſimplicity and practicability of their conſtruction; ſo that 1322 l. were given to Mr. Arnold ſenior by the Board of Longitude at different times by way of encouragement, and 1678 l., its complement to 3000 l., were given in December 1805 to Mr. Arnold junior, being at the ſame time that 2500 l. were given to Mr. Earnſhaw, in addition to the 500 l. which he had previouſly received. When the queſtion was difficult to decide, whether Mr. Arnold's or Mr. Earnſhaw's chronometers were moſt worthy of public reward, the Board of Longitude very properly rewarded the labours of both theſe makers.

Mr. Arnold, like Meſſrs. Brockbanks, was deſired to lay before the Board of Longitude ſome of the rates of chronometers made by his father and himſelf, which have not been returned to him, but the public are in poſſeſſion of the certificates of ſeveral rates which Mr. Arnold publiſhed in the year 1791; which rates are copied into Mr. Dalrymple's publication, together with ſome additional teſtimonies, which are too long for us to introduce here, and which, therefore, we muſt requeſt thoſe readers to refer to, who wiſh to be informed of all the particulars ſtated in the certificates; it is ſufficient, for the purpoſes of the general reader, that we have given him the peculiarities of the conſtruction of the chronometer itſelf, together with an hiſtorical notice of Mr. Arnold's inventions, and an account of the mode of action of the eſcapement; the general merit of chronometers of this conſtruction no diſintereſted perſon, that we know of, has yet denied.

CHRONOMETER by *Mr. Earnſhaw.*—For the ſame reaſon that we have not given the movement and parts of the chronometer contained in the frame of Mr. Arnold, we think it

6 not

not neceffary particularly to defcribe all the parts of Mr. Earnfhaw's that are common to the other modern chronometers. The peculiarities of the conftruction are confined to the balance, the balance-fpring, the efcapement-wheel, and the acting parts of the efcapement. To thefe parts, therefore, we propofe to confine our account. Mr. Earnfhaw fays, that his train is the beft poffible, viz. 18oo vibrations of the balance in an hour, which, we have feen, has always been the train of the other chronometer-makers, particularly for the pocket ones; his box pieces have 13 teeth each in the efcapement-wheel, and his pocket ones 15, like thofe of Meffrs. Brockbanks'; we are not told what pinions are ufed in the movements; nor is it faid, in Mr. Earnfhaw's account delivered to the Board of Longitude, that this is a confideration of any moment. *Fig.* 5 of *Plate* XIV. is an exact copy of Mr. Earnfhaw's *fig.* 1 of his *Plate* III., which was taken from his model laid before the Board of Longitude on June 7th, 1804, at the fame time that Mr. Arnold's model was produced; the plate and cocks, being not neceffary for explaining the mode of action, are in our figure omitted. Mr. Earnfhaw has given fo minute an account of his efcapement accompanying the model, that we cannot defcribe it better than by copying his own words, which are nearly thefe: viz.

"The fmall wheel M S K is called the large pallet; it is a cylindrical piece of fteel, having a notch or piece cut out of it at *l h r*; againft the fide of this notch is a fquare flat piece of ruby, or any hard ftone, *h l*, ground and polifhed very fmooth, and fixed faft into the pallet. The cylinder is fo placed, with refpect to the balance-wheel, that it may not be more than juft clear of two adjoining teeth. E F is a long thin fpring, which is made faft at one end, by being pinned into a ftud, G, and made to bear gently againft the head of an adjufting fcrew, *m*: the other end is bent a little into the form of a hook; to this fpring there is fixed another very flender fpring at *y*, which projects to a fmall diftance beyond it. This fmall fpring lies on the fide of the thick fpring neareft to the balance-wheel. The adjufting fcrew, *m*, takes into a fmall brafs cock, at *a p*, which is fcrewed faft to the upper plate by a ftrong fcrew. Upon the fpring E F there is fixed a femi-cylindrical pin, which ftands up perpendicular upon it, and of a fufficient length to fall between the teeth of the balance-wheel A B C D. This pin is called the locking-pallet, and is placed on the oppofite fide of the fpring reprefented to view. Through the centre of the cylindrical pallet M S K, a ftrong fteel axis paffes, called the verge; the pallet is made faft to this axis, which alfo paffes through the centre of the balance, and is made faft to it; it has two fine pivots at its extremities, upon which it turns very freely, between two firm fupporting pieces of brafs, fcrewed firmly, and made as permanent as poffible, by fteady pins, to the principal plate. A little above the cylindrical pallet M S K is fixed a fmall cylindrical piece of fteel, *i n*, having a fmall part projecting out at *i*, through which the verge alfo paffes; this is called the lifting pallet, (and is from ⅓ to ½ the diameter of the large pallet); it fixes upon the verge like a collar, and is made faft by a twift, fo as to be fet in any pofition with refpect to the large pallet M S K. The end E G of the long fpring E F being made very flender, if a fmall force be applied at the point *o* to prefs that end out from the wheel A B C D, it eafily yields in that direction, turning, as it were, upon a centre at G; it is alfo made to flide in a groove made in this ftud, in fuch a manner that the end *o* may be placed at any required diftance from the centre of the verge. Having defcribed the feveral parts as they appear in the figure, we next come to their connection or fituation with refpect to each other. Let the long fpring E F be fuppofed to be fo placed, that the end of the flender fpring *y i* may project a little way over the point of the lifting-pallet *i n*, but not fo clofe but that the point of the pallet may pafs by the hooked end of the fpring E F without touching it; the head of the adjufting-fcrew *m* is alfo fuppofed to bear gently on the inner fide of the faid fpring E F, or that neareft to the wheel, and at the fame time the locking-pallet is fo placed, that one of the teeth, D, of the balance-wheel may juft take hold of it. This pallet is not vifible in its proper place in the figure, being covered from fight by the fcrew *m*, and part of the fpring E F; its pofition is therefore reprefented by the dot *k*, on the oppofite fide of the wheel, having the tooth A juft bearing up againft it. From the above defcription of the feveral parts of the efcapement, and their connection with each other, it will be eafy to fee the mode of its action, which is as follows:

"A force being fuppofed to be applied to the balance-wheel, fo as to caufe it to move round in the direction of the letters A B C D, one of the teeth, as D, will come againft the locking-pallet (as reprefented at A, and the locking-pallet by *k*). The wheel is then faid to be locked, being prevented from moving forward by this pin. Let the balance be now fuppofed to reft in its quiefcent pofition, and it will have the fituation reprefented in the figure; the lifting-point *i* of the pallet *i n* will be juft clear of the projecting end of the flender fpring, the face *h l* of the large pallet M S K will fall a little below the point of the tooth B, and the balance having its fpiral or helical (meaning cylindrical) fpring applied to it remains perfectly at reft in this pofition. Now, as the balance and the two pallets M S K and *i n* are fixed faft to the verge, it is plain they muft all move together; let, therefore, the balance be carried a little way round in the direction of the letters M S K; by this motion the end *i* of the lifting-pallet *i n* will be brought to prefs up againft the projecting end of the flender fpring, and as this fpring is fixed on the fide of the fpring E F, neareft to the balance-wheel, the point *i* will prefs the two fprings together out from the balance-wheel; then, as only the point of the tooth D (fee its pofition at *k*) touches the locking-pallet, when the fpring E F was at reft againft the head of the fcrew *m*, it will, by the fpring being preffed out from the tooth, have flipt off (for the locking-pallet which was before fuppofed at *k*, will now be at *a*, clear of the tooth A of the balance-wheel); the wheel being now at liberty will move round by the force fuppofed to be applied to it; but as the point *i* of the lifting-pallet moves on and preffes out the fpring, the point *l* of the large pallet approaches towards the point of the tooth B of the balance-wheel, fo that when the fpring E F is fufficiently pufhed out to unlock the wheel, the point *l* of the large pallet will be got to *d*, and in this pofition the point of the tooth B of the balance-wheel will fall upon it, at the fame time the point of the tooth D has juft dropped off from the locking-pallet *m*; the force of the wheel being by this means applied to the top of the pallet *h l*, gives an increafed momentum to the balance, and affifts it in its motion in the fame direction, and by the continued motion of the large pallet in the direction M S K; the point of the tooth B, which keeps preffing and urging it forward, moves up towards the bottom of the face of the pallet towards *h*, until the plain flat furfaces of the tooth and pallet come into contact, by this time the end, *o*, of the flender fpring has dropt off from the point, *i*, of the lifting-pallet, and the two fprings have returned again into their quiefcent pofition, the fpring, E F, gently bearing againft the head of the adjufting fcrew, *m*, and the locking-pallet in a pofition to receive the next tooth,

C, of the balance-wheel. When the two ſurfaces of the tooth and pallet are thus in contact, the greateſt force of the wheel is exerted upon the pallet, and of courſe upon the balance moving with it. The tooth ſtill preſſing againſt the face of the pallet, and the pallet moving in the direction MSK, it at laſt drops off, leaving the balance at perfect liberty to move on in the ſame direction in which it was going. Juſt as the point of the tooth B, which has been preſſing the large pallet round, is ready to leave it, the next tooth, C, of the wheel is almoſt in contact with the locking pallet *m*, ſo that the inſtant the tooth B drops off, the wheel is again locked, and the action of that tooth upon the balance is finiſhed. As the balance moves with the greateſt freedom upon its pivots, the force of the tooth has given it a conſiderable velocity, ſo that the balance ſtill keeps moving on in the ſame direction, after the preſſure of the tooth is removed by ſlipping off from the pallet, until the force of the pendulum-ſpring (which is not repreſented in the figure) being continually increaſed by being wound up, overcomes the momentum of the balance, which, for an inſtant of time is then ſtationary, but immediately returns by the action of the pendulum-ſpring, which exerts a conſiderable force upon it in unwinding itſelf. As the balance returns, the point *i* of the lifting pallet *i n* paſſes by the ends of two ſprings, EF, *y o*, and, in paſſing by, puſhes the projecting end, *o*, of the ſlender ſpring in towards the balance-wheel, until it has paſſed it; after this, the projecting end *o* again returns and applies itſelf cloſe to the hooked end of the ſpring EF, as before. The ſpring *y o* is made ſo ſlender, that it gives but little reſiſtance to the balance, during the time the point *i* of the lifting pallet is paſſing it, and of courſe cauſes but little (if any) decreaſe in its momentum. During the time the point *i* of the lifting pallet is paſſing in the ſmall ſpring *y o*, the long ſpring EF remains ſteadily bearing againſt the head of the adjuſting ſcrew *m*, as the hooked end at *o* juſt lets the end of the lifting pallet paſs by without touching of it. As the ſpring has now been continually acting upon the balance, from the extremity of its vibration in the direction MSK, it has given it the greateſt velocity, when the point *i* of the lifting pallet is paſſing the end *o* of the ſlender ſpring; for at this inſtant the ſpring which was wound up by the contrary direction of the balance, is now unwound again, or in the ſame ſtate as it was in its quieſcent poſition at firſt, and of courſe has no effect upon the balance at all in either direction; but the balance, having now all the velocity it would acquire from the unwinding of the ſpring, goes on in the direction SMK, until the force of this ſpring again ſtops it and brings it back again, moving in the ſame direction as at firſt, with a conſiderable velocity. By this return of the balance, the point *i* of the lifting pallet comes up again to the projecting end *o* of the ſlender ſpring, puſhes back the long ſpring EF, and unlocks the wheel; and another tooth falling upon the face of the pallet *h l* gives freſh energy to the balance; and thus the action is carried on as before."

In this quotation we have given the original letters of reference, but as the balance and cocks are left out in our *fig.* 5, the direction of the balance has been indicated by the letters referring to the large pallet, which, being placed on the ſame verge, has the ſame motion. We have added two dotted lines in the large pallet to ſhew that the direction of the lines, bounding the notch in the circumference, is towards a point in the radius of it, equally diſtant from its centre and circumference; alſo we have added two ſimilar dotted lines from two ſeparate teeth, to ſhew that the undercutting or ſloping of the interior ſide of the tooth is likewiſe directed to a point equally diſtant from the centre and circumference of the wheel, which point is at double the diſtance from the centre compared with the point guiding the undercutting ſlope in Meſſrs. Brockbanks' chronometers; for the interſecting dotted lines form tangents to the dotted circle deſcribed with half the radius of the wheel. The locking ſpring points directly to the balance arbor, but is laid in ſuch a way as not to be an exact tangent to a radial line drawn from the centre of the wheel to the locking pallet, the angle formed at the pallet being ſomewhat leſs than a right angle, in order, as Mr. Earnſhaw ſays, "that the wheel may have a tendency to draw the ſpring into it," for ſafe locking. The eſcapement wheel, balance, balance-ſpring, pallets and ſprings for locking and unlocking are at the outſide of the upper plate, and are preſented to view when the cock is taken off. We hardly underſtand Mr. Earnſhaw's reaſoning, when he ſays that his ſcape wheel "unlocks in a ſimilar circle which the wheel makes, which renders it a perfect dead ſcape," and that "Mr. Arnold's is locked on the other ſide of the wheel, and in the act of unlocking the ſpring moves in towards the centre of the wheel, which is a different direction to that which the wheel takes, and produces a recoil." In the firſt place we are at a loſs to conceive how the pallet on the detent-ſpring and the point of a tooth in the wheel can move in a *ſimilar circle*, to ſize, or direction of motion, unleſs the radii are alike, and the centres of motion coincident; and in the next place, we ſee no reaſon to conclude that a pallet carried *directly from* the centre of the wheel's motion ſhould be conſidered in the ſame direction as that of a tooth in the wheel, any more than a pallet carried *directly towards* the centre of the ſame, in the act of unlocking; in both caſes, if the detent ſpring is a tangent to the tooth of the wheel that holds the pallet, the motion *in* or *out* will be at right angles to the circumference of the wheel, and if the tooth is conſiderably undercut there muſt neceſſarily be a recoil in both caſes. The only correction for this is, to make the angle formed by the detent ſpring and radial line of the wheel at the reſting pallet, to be as much leſs than a right angle, as the angle of undercutting is; accordingly Mr. Earnſhaw ſays he has made this angle a little leſs than a right angle, but then he has undercut his tooth much more than the difference, and therefore, as appears to us, there muſt be recoil, or a little backward motion in the ſcape wheel, in the act of the detent pallet's eſcaping; whereas Mr. Arnold's locking pallet reſts againſt a ſtraight line directed towards the centre, which is alſo the direction of the pallet's motion in unlocking, conſequently there can be no ſenſible recoil in his eſcapement. In our deſcription of Mr. Arnold's eſcapement the reader may recollect that he was referred to this place for ſome further account of the proper *adjuſtment for beat* in a chronometer of the modern inſtruction, we here reſume the ſubject with an obſervation of Mr. Earnſhaw himſelf which we have already quoted; "let the balance, ſays he, be now ſuppoſed to reſt in its quieſcent poſition, and it will have the ſituation repreſented in the figure," alluding to the figure which we have copied: we preſume not, merely upon the ſtrength of our theoretical reaſoning, to aſſert, that the adjuſtment for beat is improper, if the balance ſpring is quieſcent when the face of the lifting pallet *i* is in contact, or nearly in contact with the end of the unlocking ſpring *i y*, but we are aſſured by workmen well qualified to judge, and who give ſcientific reaſons for the aſſertion, that the lifting pallet is not only about 24° on one ſide of the requiſite point of poſition, but is actually at the wrong ſide of the unlocking ſpring; the reaſon given us is this, whenever a chronometer is in *true beat*, the quieſcent point from which the excurſions of the balance commence

 is

is in the *mean point* of, the angle of efcapement; this is obvioufly the cafe in the anchor efcapement of a pendulum clock, and requires a neceffary adjuftment when the clock is at firft fixed up; and a little confideration will prove that an attention to this adjuftment, though not equally obvious, is equally defirable in a chronometer. What is called the angle of efcapement differs from what is called the angle of action, or angle of impulfe, in a chronometer; the former is included in the arc comprehended between the point where the unlocking fpring lets go or efcapes the face of the lifting pallet, and the point where the impelling tooth of the wheel drops off or efcapes from the face of the large pallet; but the fecond, or arc of impulfive action, is comprehended, between the point where the impelling tooth commences, and the point where it terminates its action; this fecond arc is fmaller than the former, and is always contained in it; the two have indeed one common termination, but each has a feparate commencement; for the wheel muft neceffarily be unlocked before its action on the large pallet can begin, and the unlocking takes place previoufly to the unlocking or lifting pallet's quitting the end of the unlocking-fpring: if we fuppofe that the flender fpring $i\,q$ bends back, the fpace of 4°, more or lefs, before the unlocking-pallet $n\,i$ quits it, after the wheel is unlocked, this quantity will conftitute the difference between the arc of efcapement and the arc of the wheel's action, provided there be no drop of the impelling-tooth on the heel of the large pallet after the wheel is unlocked, and before the action commences; but if we fuppofe the faid drop to be alfo 4°, then will the whole arc of efcapement exceed the whole arc of action on the pallet by 8°. Admit now that the efcapement wheel and large pallet have their diameters to each other as 2 : 1, which is the ratio Mr. Earnfhaw in general adopts, and that this wheel have 13 teeth, which is the number in his box-chronometer; then $\frac{360°}{13.} = 27°.7$ nearly, is the diftance between the points of two fucceffive teeth; admit alfo that there be a fecond drop of 2° from the tooth to be locked to the locking-pallet after the arc of action is finifhed; then the fum of the two drops, $4 + 2 = 6°$, that precede and follow the arc of action, being fubtracted from 27°.7, or whole diftance between the two neareft teeth, leaves 21°.7 meafured on the circumference of the wheel for the total arc of action on the pallet; but the wheel is double the diameter of the pallet, therefore, the fame quantity is $21°.7 \times 2 = 43°.4$ meafured on the circumference of the large pallet, and the arc of efcapement exceeds the arc of action, on our fuppofition, by 8°, confequently, $43°.4 + 8° = 51°.4$ will be the arc of efcapement: now it will be feen, on referring to the plate, that Mr. Earnfhaw has placed his lifting-pallet in the figure within lefs than 4° of one of the two extremities of the arc of efcapament, and on that fide of the unlocking-fpring $i\,\gamma$, towards which this fpring bends at the inftant of the lifting-pallet's final efcape; and yet he fays that this is its fituation when the balance-fpring is in a ftate of quiefcence; let us try what will be the confequence of fuch a pofition; the chronometer will meafure time very well, and the impulfe of the lifting-pallet, we allow, will be given at the moft favourable inftant, namely, at the inftant of the fpring's quiefcence, or point of the balance's greateft momentum, which, it is evident, was Mr. Earnfhaw's reafon for fixing his lifting-pallet fo; but then, which is our objection, as one excurfion of the balance from the point of reft muft neceffarily be 47°.7 (51°.4 — 4°) before it clears the arc of efcapement, and as the excurfions muft be fimilar to the right and left, when the balance vibrates freely, the whole arc of vibration can never be diminifhed to lefs than 95°.4, in this cafe, without the chronometer's ftopping, which, as the chronometer has not the power of commencing motion of itfelf, is a very ferious objection to the prefent pofition of the lifting-pallet; it may, indeed, be faid, in reply, that the chronometer muft be very dirty before its arc of total vibration becomes fo much diminifhed as to be little more than a quarter of a circle, but this is not the only predicament in which the ftoppage will take place; any fudden check, or quick horizontal motion given to the vibrating balance, that makes it return even once before it has exceeded this limit, will bring it to reft; that is, nearly the fame effect will be liable to be produced as if the arc of the efcapement had been almoft double the prefent quantity with a pallet of ½ the diameter of the wheel, which is Mr. Earnfhaw's cogent objection to Mr. Arnold's conftruction. In making this comparifon with a double arc of efcapement, we, of courfe, fuppofe the point of the lifting-pallet's quiefcence to be in the middle point of the arc of efcapement, as Mr. Arnold's appears to be in his figure, which we have copied, and which fituation we now proceed to fhow, is moft favourable to the continuance of the chronometer's going without interruption with any given pallet. We propofe to place the lifting-pallet's face exactly in the middle point of the arc of efcapement, which is not a theoretical propofition of ours, but we have examined chronometers adjufted fo for beat, and have found that, when the balance is drawn round a trifle beyond either extremity of the efcapement-angle, they have the power of commencing motion within a limit very little exceeding the efcapement-angle, which, in Mr. Earnfhaw's box-chronometer, we have fhown, may be about 51°.4. Hence the chance is almoft two to one in favour of that chronometer's avoiding ftoppage by dirt or accident, which has its quiefcent lifting-pallet in the middle of the arc of efcapement, compared with that which has the fame pallet at one end of the fame arc.

The only objection which carries the femblance of argument againft our mode of adjuftment for beat, an adjuftment we believe too generally overlooked, is, that the impulfe of the lifting-pallet is not at the moment of the balance's greateft momentum, but about 24° before the momentum is a maximum, allowing the pallet to be quiefcent at the middle of the arc of efcapement, which we contend for; our reply is, that the momentum is fo nearly a maximum at 24° from it that the difference produces no fenfible bad effect on the balance, compared with the probable effect from ftoppage when the adjuftment for beat is nearly as bad as poffible. Indeed, Mr. Atwood has calculated, and his calculations feem to have been verified by practice, that when, in the cafe of Mr. Mudge's time-keeper, the quiefcent point of his auxiliary fpring is at one fide of the quiefcent point of his ftronger fpring, the effect produced is a daily gain or daily lofs in the rate, accordingly as the diftance from coincidence of the refpective points of reft fell on the right or left hand fide of the quiefcent point of the ftrong regulating fpring, and the daily gain or lofs thus to be effected is ingenioufly propofed, in our author's excellent paper in the Philofophical Tranfactions of 1794, to be ufed as an adjuftment for rate, or even as a compenfation for want of ifochronifm in the balance-fpring. The deduction from the momentum of the balance, occafioned by the impulfe of the lifting-pallet, being fomewhat analogous to the want of adjuftment in the quiefcent point of Mr. Mudge's auxiliary fpring, might indeed produce a flight lofs in the daily rate, if the fcrews of adjuftment for rate did not compenfate fuch lofs; but when it is confidered that the impulfe from the wheel fucceeds the deduction from the momentum almoft inftantaneoufly, and that the deduction we we ſpeak of is preciſely the ſame in each vibration, the uniformity of the going of the chronometer will not be ſenſibly altered thereby.

Let us ſee, in the next place, what will be the arcs of action and of eſcapement in Mr. Earnſhaw's pocket chronometers, in which the eſcapement-wheel has 15 teeth; $\frac{360°}{15} = 24°$ is the diſtance between two contiguous teeth, from which ſubtract, on our former ſuppoſition, 6° as the ſum of the preceding and following drops, and the angle of action will be 18°, or 36° meaſured on the circumference of the large pallet, if it be half the ſize of the wheel; but the angle of eſcapement will be greater than the angle of action by 8°, therefore $36 + 8 = 44°$ is the whole arc of eſcapement, ſuppoſing, as before, the data for the drops to be accurate. This angle of eſcapement turns out to be leſs, on our ſuppoſition, than that with only 13 teeth in the eſcapement by 7°.4, and therefore will admit a lifting-pallet ſmaller than the box chronometer has, in order to have the ſame angle of eſcapement, or the ſame liability to ſtop by dirt or accident; it being ſelf-evident that the ſame ſubtending line, conſidered as a chord, will meaſure more degrees on the periphery of a ſmall circle than on the periphery of a larger. This comparifon of the two arcs of eſcapement ſhews, that the arc in queſtion depends not entirely on the relative diameters of the eſcapement and large pallet, as Mr. Earnſhaw ſuppoſed, when he objected to Mr. Arnold's proportions, but on their relative diameters conjointly with the number of teeth in the wheel, the latter of which has been overlooked by Mr. Earnſhaw, but particularly inſiſted on by Mr. Arnold. However, we agree perfectly with Mr. Earnſhaw, that in any wheel with a given number of teeth, a large pallet for receiving the impulſe of the wheel will require a ſmaller arc of action as well as a ſmaller arc of eſcapement, than a pallet that has a ſmaller diameter, under any given adjuſtment for beat, which is a conſideration worthy of general notice.

On counting the teeth in Mr. Earnſhaw's eſcapement-wheel, which we have copied, we were ſurpriſed to find only 12 teeth in it, particularly as we are informed, in his deſcription, that his numbers are 13 and 15 reſpectively; we can only account for this circumſtance by ſuppoſing that the draftſman miſtook the number; which we here notice, leſt the reader ſhould ſuppoſe the fault to lie in our figure excluſively.

Fig. 6 of *Plate* XIV. repreſents Mr. Earnſhaw's balance, which, like Mr. Arnold's, has only two compenſation pieces, and thoſe much ſhorter than his, being little more than a quadrant each: *a* and *a* are the ſcrews of adjuſtment for rate, ſcrewed in to make the chronometer go faſter, and out to make it go ſlower. The ſliding pieces *b* and *b* have each a circular groove, turned in a lathe or turning-frame, deep enough to form a bed for the expanſion bars, in order that the interior ſide ſcrews *c* and *c* may preſs againſt the edges of the expanſion pieces, and retain the ſliding pieces in any given ſituation. *Fig.* 7 is a lateral view of this balance and verge without the pivots, which Mr. E. ſays ſhould be conical except very near the ends, which ſhould be cylindrical; and ſhould run in a "jewel hole as ſhallow as poſſible, ſo as not to endanger cutting the pivot; and the part of action of the hole ſhould be made quite back, with only a very ſhallow chamfer behind to retain the oil." The ſliding pieces *b* and *b* are the weights of adjuſtment for temperature; in making theſe weights, a braſs ring is firſt made in the lathe, and is then cut through into fourteen equal parts in a clock-engine, by an operation ſimilar to that of cutting a wheel into fourteen teeth, ſo that each piece is the fourteenth part of a circle, after deducting the thickneſs of the cutter, which is not mentioned. One of the weights in a box chronometer, is about twenty grains generally. The expanſion rims are turned out into a ring from a circular compound plate, which is made after the method of Meſſrs. Brockbanks, by fuſing the braſs in a crucible, containing the ſteel circle, held in a horizontal poſition during the time of fuſion. The conſtruction is very ſimple, but, as it appears to us, if the adjuſtments for poſition were made by means of the ſcrews of rate, and weights for temperature only, one adjuſtment would derange another, and that for poſition would be very difficult to make nicely, as there is no tangent ſcrew, or other contrivance, to move the weights *b* and *b* by gradual ſlow motions. As though to get rid of this objection, which was too obvious to be overlooked, Mr. E. conceives that the cauſes of a difference in the rate in different poſitions, are large balance pivots in part, but principally the badneſs of the balance-ſpring; his words reſpecting the latter cauſe, are theſe; "much difficulty has fallen to the lot of watch-makers in the endeavour to make time-keepers go nearly the ſame in the different poſitions: I have had my ſhare of this, but it is now over; by far the greateſt part of this difficulty ariſes from the balance-ſpring not being properly made. But if the ſpring is made as I ſhall deſcribe hereafter, you have only to make the balance of equal weight, and it will go within a few ſeconds per day in all poſitions alike; and if it vibrates not more than one circle and a quarter, by applying a ſmall matter of weight to that part of the balance, which is downward, when in the poſition that it loſes moſt, (it) will correct it with great accuracy."

The Board of Longitude not knowing what this "ſmall matter of weight" implied, nor how it is to be attached to the balance, were induced to put the following queſtion, viz. "When the weight is wanted to adjuſt the watch in the poſitions of 3 and 9, by what means do you obtain that weight, in the manner your balance appears to be made; if you know any thing more that is material concerning the making time-keepers go nearly the ſame in different poſitions, communicate it."

Anſ.—"To adjuſt the watch in the poſitions of 3 and 9, (ſays Mr. E.) I fix on to one of the compenſation-weights that is downwards, when in that poſition that it loſes moſt, a ſmall piece of braſs, not larger in diameter than a common pin-head, and nearly as thin as foolſcap paper. I fix it on with a very ſmall particle of *bees wax*, not larger than the common dot of an *i*: that is, if the watch is gaining on mean time. But if the watch is loſing, I then take out the balance, and with a drill drill out a ſmall matter from that compenſation-weight, that is uppermoſt when in the poſition that the watch loſes moſt; this I have found to correct it without ſo many ſcrews and fans as I have ſeen in ſome time-keepers." After having read the above anſwer, we were almoſt induced to examine whether the balance itſelf might not alſo be fixed to the verge-collet by bees wax inſtead of ſcrews, but we recollected that in the drawing there is a little ſcrew-head at each ſide of the centre of the balance, which appear to be uſed for this purpoſe. Mr. E. we truſt, will pardon the obſervation.

The rule for adjuſting the balance for temperature is this; put the watch into about 85 or 90 degrees of heat by the common thermometer, mark down exactly how much it gains or loſes in 12 hours, then put it into as ſevere a cold as you can get, for 12 hours, and if it gains one minute more in 12 hours in cold than in heat, move the compenſation weights farther from the arm of the balance about $\frac{1}{8}$ of an inch; and if it gains one minute more in 12 hours in heat than

than in cold, move the weights ⅝ of an inch nearer to the arm of the balance, and fo on in the like proportion, trying it again and again, till you find the watch go the fame in whatever change of heat and cold you put it. Of courfe this adjuftment precedes the adjuftment for pofition, fo that moft probably the pofition, whatever it is, during thefe trials, is always the fame, a circumftance not fpecified. The total vibration at firft is confined to a circle and a quarter, which quantity, it is faid, will produce the moft fteady performance; but we are not informed what the banking is that limits the greateft vibration, nor indeed whether there is any banking at all, which we have been informed there is not in general.

Mr. Earnfhaw's greateft difficulty in the progrefs of his labours feems to have been to find out what he calls the *invifible properties* of the balance-fpring, and to render it not exactly ifochronal, but fo nearly fo, that its deviation from ifochronifm may compenfate for a *relaxation* in the fpring, real or imaginary, arifing from conftant ufe, which is *fuppofed* to affect the permanency of the rate. After complaining bitterly of much difappointment in his firft labours, our author fays, "I found, in the courfe of reafoning on bodies, that watch-fprings *relax* and *tire* like the human frame, when kept conftantly in motion; and this may be proved by the following experiment: let a watch, that has been going a few months, go down, let it be down for a week or two, or more, then fet it going, and if it be a good time-keeper, fo as not to be affected by the weather, it will go fome feconds per day fafter than it did when it was let down, but it will again lofe its quicknefs in a gradual manner, gaining lefs and lefs, till it comes to its former rate. Therefore, finding that ifochronal fprings would not do, and likewife having made fprings of fuch fhape as would render long and fhort vibrations equal in time, (which) conftantly lofe (loft) the longer the watch went; I then made them of fuch fhape as to gain in the fhort vibrations about five or fix feconds per day more than the long ones; this quantity could only be found by long experience; and the way I proved this, was to try the rate of the watch with the balance vibrating about ⅓ of a circle, then tried its rate when vibrating a circle and a quarter, and if the fhort vibrations go flower than the long ones, that watch will lofe on its rate, and if they are equal, it will likewife lofe, but that only from *relaxation;* and if it gains in the fhort vibrations more than five or fix feconds in 24 hours, it will in the long run gain on its rate, but if not more than that quantity, and if the time-keeper is perfect in heat and cold, and every other part, the above properties will render it deferving of the name of a perfect time-keeper: and this is a principal caufe of my time-keepers excelling all others; and this is the principal caufe of fome of my time-keepers going better than others; though by me, the fprings of them being made to accord more exactly to the above proportions; and this is the caufe which has enabled me to foretel what my time-keepers would do, which Dr. Mafkelyne, Mr. Crofley, and others can teftify. The above effect is produced as follows: I find the common *relaxation* of balance-fprings to be about five or fix feconds per day on their rates in the courfe of a year; therefore if the fhort vibrations are made, by the fhape of the fpring, to go about that quantity fafter than the long ones, and as the fpring relaxes in going by time, fo the watch accumulates in dirt (dirt accumulates in the watch more properly), and thickening of the oil (takes place), which fhortens the vibrations, the fhort ones then being quicker compenfated (compenfates) for the evil of relaxation of the balance-fpring." (See Mr Earnfhaw's "Explanation," pages 8 and 9.)

The Board of Longitude having put the following queftions on this fubject, "Explain what you mean by this? and how it is performed?" Mr. E. replied: "All watchmakers know how to draw and taper balance-fprings, though they did not know how much they were to be tapered to that certain degree which could only anfwer the purpofe of a complete time-keeper. I perform it in the following manner: take a length of balance-fpring-wire, fay about twelve inches for box time-keepers, and draw it between two fmooth potence files, beginning from the end about ⅙ of its length, make one draw, the next about ⅖, and fo on, advancing ⅙ every draw till you come to the top, preffing the files juft fo hard together as will make them bite or take hold of the fpring; do the fame with two oil-ftones, only give 12 ftrokes inftead of 6, which will take off all burrs which the file left on." ("Explanation," p. 29 and 30.) We are further told, that when the fpring is *tired* with long continuance of work, its vigour may be reftored by *reft*.

Thefe fprings, that have the invifible properties defcribed, "are made of foft fteel, rolled hard, and not hardened and tempered with heat and cold, that procefs not being at all neceffary." Their length varies in box chronometers from 12 to 20 inches, and in pocket ones from 5 to 7, like Mr. Arnold's, and the fhape is cylindrical, with the two extreme coils, each about half the diameter of the other coils, agreeably to the fpecification in one of Mr. Arnold's patents. The thicknefs of the fteel in the expanfion-pieces of the balance is about $\frac{8}{1000}$ of an inch, and that of the brafs twice, or nearly three times as thick; the diameter of the balance without the weights and fcrews $1\frac{1}{5}$, and with them $1\frac{2}{5}$ inch; the total weight of it with weights and fcrews 3 dwts. 10 gr. The mainfpring barrel is $1\frac{3}{16}$ inch, and the depth ½ inch, and the fpring in it has from 4½ to 5 turns. The weight of each of the two balance fcrews is from 5 to 6 grains; and the figure which the expanfion pieces affume when cut from the original compound ring is fuffered to remain unaltered, on a fuppofition that bending by any mechanical means will injure the regularity of their obedience to the changes of temperature; the locking and unlocking fprings are made as weak at the bending parts as are practicable to afford the neceffary refiftance, as a detent, to the efcapement-wheel; they are brought to fhape by filing, then fmoothed with a piece of fteel and oil-ftone powder, and laftly hardened and tempered. The pallets, which are of fteel, have each a jewel fet in them, and thofe on the verge of the balance are twifted round and fet to their fituations refpectively by the friction of their central holes. In his laft anfwer to the queftions of the Commiffioners of Longitude, when afked, "How fhould the lifting and large pallet be placed with refpect to each other?" Mr Earnfhaw's words were, "the fmall lifting-pallet muft be moved round to fuch pofition that when the wheel is unlocked the face of the large pallet fhould be juft within the compafs of the wheel-tooth which is to act on it;" but, what appears to us remarkable, no queftion was afked by the Board, nor any particular notice taken, that we can find, either by Mr. Arnold or Mr. Earnfhaw, refpecting the relative pofitions of the lifting-pallet and the quiefcent point of the regulating fpring, excepting in the inftance of Mr. Earnfhaw, that we have had occafion to notice; which notice, we truft, will benefit not only him, but moft of the other makers of chronometers, with the detached fpring-detent efcapement.

The mention of this efcapement brings to our recollection a patent which was faid to have been taken out for the invention of a detent-fpring in the name of Wright, a Quaker in the Poultry, and which was, till lately, confidered to be

Mr. Earnſhaw's patent, but which was never actually taken out, as we have been lately informed from good authority; what part of the ſpring conſtituted the pretended invention, and how it differed from Mr. Arnold's ſpecification, we are not informed; the only difference that we are aware of is, that one ſpring-detent unlocks inwards, and the other outwards; but Meſſrs. Brockbanks' have always unlocked outwards, though they have not the two ſprings inſerted into one another, as Meſſrs. Arnold and Earnſhaw have: neither do we ſee any good reaſon for preferring one of the three conſtructions to either of the other. They have all been found to anſwer the deſired purpoſe; and the three makers, who have ſeparately ſold each a thouſand chronometers or thereabouts, have turned out of their hands ſome machines that have performed but indifferently, as well as others, that have done them great credit, as well as the navy great ſervice.

"Two time-keepers, ſays Dr. Maſkelyne, in his preface to the "Explanations," publiſhed by order of the Commiſſioners of Longitude, conſtructed by Mr. Thomas Earnſhaw, were tried three ſeveral times at the Royal Obſervatory, by order of the Commiſſioners of Longitude, for a twelvemonth or more at a time between 1798 and 1802, as candidates for ſome of the great rewards held out by the act of parliament of the 14th of his preſent Majeſty; but were adjudged not to have gone within any of the limits preſcribed by the act, and, therefore, not thought proper to be ſent to ſea, to undergo the ſubſequent trial required by the act. However, as they appeared to have gone with ſufficient exactneſs, in the two laſt trials, to be of conſiderable uſe in navigation, the commiſſioners, on the 3d and 17th of March 1803, reſolved unanimouſly to grant to Mr. Earnſhaw the ſum of 2500 l. in addition to 500 l. which they had given him before." Sir Joſeph Banks baronet, however, entered *a proteſt* againſt this *unanimous* reſolution, to which Dr. Maſkelyne has replied in a private pamphlet; which, therefore, we cannot further notice; but Mr. Earnſhaw, elated, probably, with his ſucceſs, publiſhed an advertiſement in the public papers by no means calculated to conciliate his very reſpectable opponent; in reply to which, Mr. Dalrymple, a gentleman well known to the world as a geographer, has recently written a pamphlet for public circulation, which we here notice, as impartial by-ſtanders, that we may ſet the author right in ſome of his obſervations, which we conſider as the produce of his zeal for his friend's cauſe, rather than of his mature judgment. The pamphlet we allude to, is intituled "Longitude;" the author of which, ſpeaking of the effects of heat and cold on the balance-ſpring, and the mechanical means uſed by Harriſon and others, ſays thus: "Various modes of correction were practiſed; but, as *all* theſe were by *conſtraint*, checking the natural effects of heat and cold on the ſpring, they were found ineffectual. Mr. Arnold, imitating the ſimplicity of nature in her operations, ſo conſtructed *his balance*, that *heat* and *cold* ſhould *enlarge* or *diminiſh* it, in the *ſame degree* as they operate on the *ſpring* of a *watch*, ſo that the *effect* muſt be *uniform*; the balance, by becoming *larger* or *leſs*, exactly counteracting the effect of *heat* and *cold* on the ſpring." (page 13 and ſeq.). What Mr. Dalrymple's term *conſtraint* here means, we do not exactly conceive; our opinion of Mr. Harriſon's curb, or kirb, is, that its inward and outward motion alternately affected by changes of temperature, limited the effective portion of the ſpring to the *ſame dimenſions*, in point of length, under all circumſtances; which limit in the length of the ſpring *conſtrained* it, to be ſure, to become an uniform regulator, or very nearly ſo: but that the *enlargement* of the balance ſhould counteract the effects of *heat*, or the *leſſening* of it ſhould counteract the effects of *cold*, is to us a new doctrine, juſt the reverſe of what we have been in the habit of propagating; we beg leave to repeat, on this occaſion, that the expanſion pieces of the balance are intended to preſerve the effective diameter of it unchanged by heat or cold; the ſame heat which elongates the radial arms brings the expanſion weights nearer to the centre, and preſerves the momentum unaltered; which effect produced on the diameter of the balance is a *conſtraint*, as much as the alternate lengthening and ſhortening of the balance-ſpring is, by artificial means, ſimilar in principle, and alike in oppoſition to the direct effect of natural cauſes. Of the ſame nature is the following opinion, in p. 80: "Before I quit the ſubject, I muſt obſerve, that the confidential committee, to whom Mr. Mudge's time-piece was explained, repreſented the *eſpecial merit* of it to be in "a contrivance for deſtroying the *inequalities* of the *maintaining power* derived from the *main-ſpring*." Before that time, the late Mr. Arnold had *entirely* deſtroyed the *inequalities* proceeding from the *main-ſpring*; this was demonſtrated by an *experiment*, in preſence of ſeveral perſons, at Mr. Aubert's houſe in Auſtin Friars," &c. The experiment here alluded to, was that of the iſochronal length of the balance-ſpring, which made the long and ſhort arcs of vibration to be performed in the ſame times, or, in other words, which compenſated the great inequalities of the maintaining power where no fuſee was uſed; ſo that the contrivance did not "entirely deſtroy the inequalities" proceeding from the main-ſpring, but allowed them to be as great as poſſible, and then compenſated thoſe *increaſed inequalities* by the iſochroniſm of the regulating-ſpring, as we have before ſtated. It appears to us, that the author in queſtion, at the time he made thoſe ſeveral obſervations on chronometry, had not given himſelf time to form a diſtinct apprehenſion of the difference between cauſe and conſequence.

Nearly of a ſimilar kind does Mr. Earnſhaw's reaſoning appear on the tapering of the balance-ſpring, where he ſays, "if a man is to go four miles in the ſame time as he has gone one mile, he cannot do it with the ſame power; no, he muſt have impelling force to quicken his motion, or he will be four times as long in doing it. Therefore, inſtead of the ſpring being equal in all its parts, it muſt be made to increaſe in thickneſs to the outer end," &c. (Page 8, of the "Explanations.") Mr. Earnſhaw ſeems here to have forgotten, that in every uniform ſpring, of whatever ſhape, the law of its action, as aſcertained long ago by Dr. Hooke, is "ut tenſio ſic vis," *i. e.* the impelling force is directly as the tenſion, or diſtance from the point of reſt; and if he knows any thing of the law of accelerated forces, he will find, on conſideration, that there is no occaſion to apply two ſeparate cauſes to produce an exact effect, when one of them, ſingly, is competent to the purpoſe; if, indeed, more than a due effect is to be produced, then an auxiliary cauſe may be had recourſe to, but ſtill we muſt look to the primary cauſe, for regularity as well as for the continuance of the effect, particularly as, in the caſe before us, the primary cauſe is the natural one, and the auxiliary one only artificial. The ſpring of uniform thickneſs will produce a ſcale of accelerated velocities ſimilar to that produced by the uniform action of gravity alone. Another of Mr. Dalrymple's remarks is this; "The *ſpiral ſpring* was found to be another ſource of irregularity; as, by action, it changed its elaſticity, and its *power* was not the *ſame* in all parts of the *volute*;" (p. 14, "Longitude"): and he ſtrengthens the force of this obſervation by adding Mr. Arnold's well known reply to the committee of the Houſe of Commons, when they aſked him,

What

What objections are there to the common ſpiral-ſpring? which reply was, " that it is never a ſpiral but when it is at reſt." Our reply to theſe remarks is this, that no ſpring can act at all without motion, and, conſequently, without change of figure: nor would any ſpring anſwer the purpoſe of a regulating-ſpring, if its power or elaſtic force were the ſame at every degree of tenſion, as both Mr. Earnſhaw and Mr. Dalrymple ſeem to have perſuaded themſelves is uſually the caſe. The beauty of the law of a ſpring's tenſion is, that it does not depend on the ſhape, provided its dimenſions be uniform: indeed Mr. Earnſhaw himſelf has found out from practice, that " the only advantage attending the cylindrical ſhape is, that it is rather eaſier (more eaſily) made; a ſaving of about one hour of time." Mr. Earnſhaw has particularized ſix advantages which his eſcapement has over Mr. Arnold's, in p. 12 of his " Explanations;" but, as the reader will probably think we have detained him too long, we refer him to the original information, in reading which, he will now be able to form an opinion for himſelf as to the real exiſtence of thoſe advantages; indeed, when all reaſoning on the ſubject is at an end, that particular conſtruction of a chronometer will no doubt ultimately prevail in practice, which ſhall be found to perform the beſt, and at the ſmalleſt expence.

We might have added to our article an account of different chronometers by French makers, but, as a chronometer differs from an ordinary watch principally in the *compenſation-balance* and *eſcapement*, their properties and peculiarities of conſtruction deſcribed under theſe two heads will be amply ſufficient, after what we have here ſaid on the ſubject: and, with reſpect to our preſent article, which has grown upon our hands to a greater length than we intended it ſhould, the reader, however he may differ from us in ſome particulars, will at leaſt acknowledge, that we have, in general, taken the liberty of thinking for ourſelves, and have, moreover, freely expreſſed thoſe thoughts, uninfluenced by the authority of names, even of thoſe to whoſe aſſiſtance we are indebted; which we conceive to be the only way of doing juſtice to an intereſting ſubject, that has never before gained general circulation in the Engliſh language.

Mr. Mudge's. TIME KEEPER.

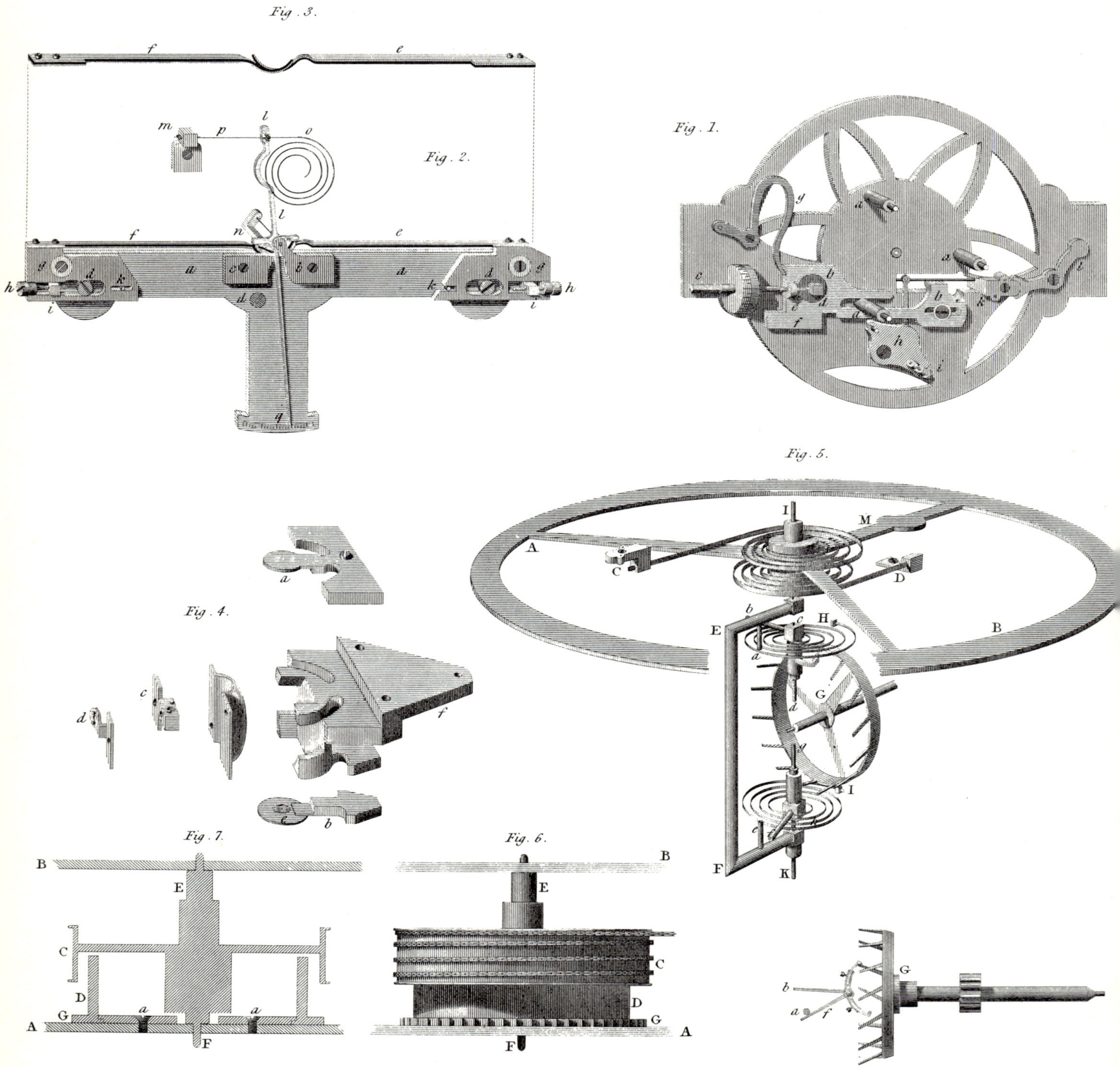

Farey, delin. *Published as the Act directs, 1819, by Longman, Hurst, Rees, Orme & Brown, Paternoster Row.* *Lowry, Sc.*

ARNOLD'S AND EARNSHAW'S CHRONOMETERS.

Fig. 1.

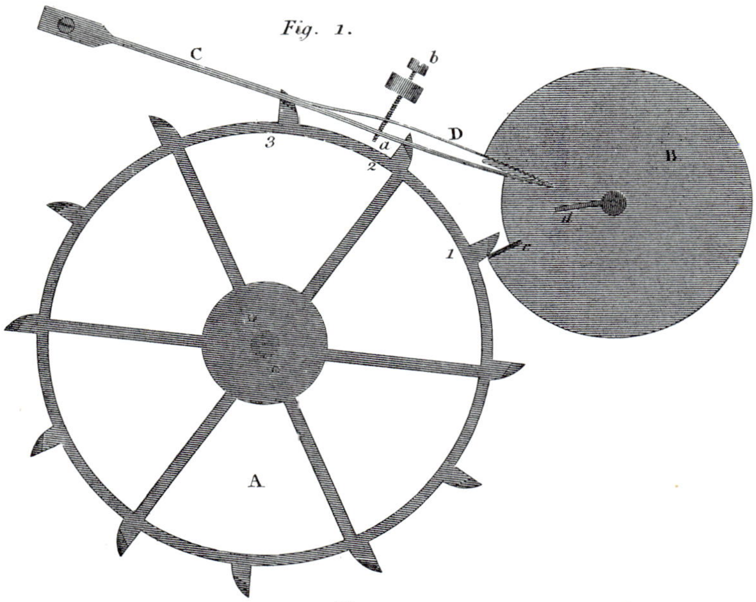

Fig. 2.

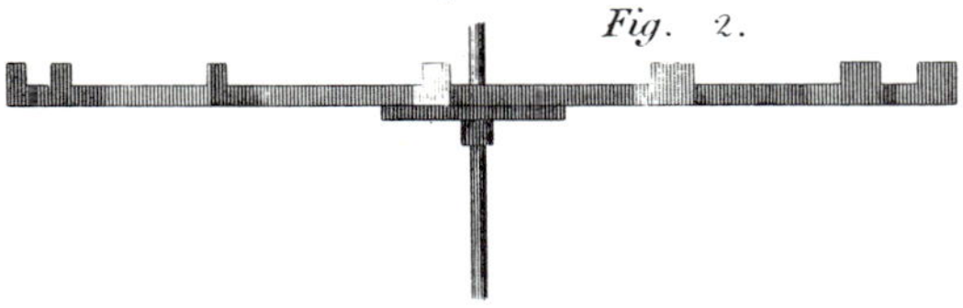

Fig. 3.

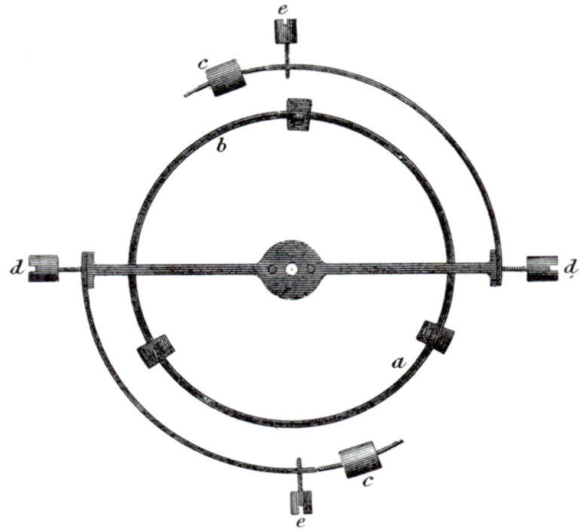

Fig. 4.

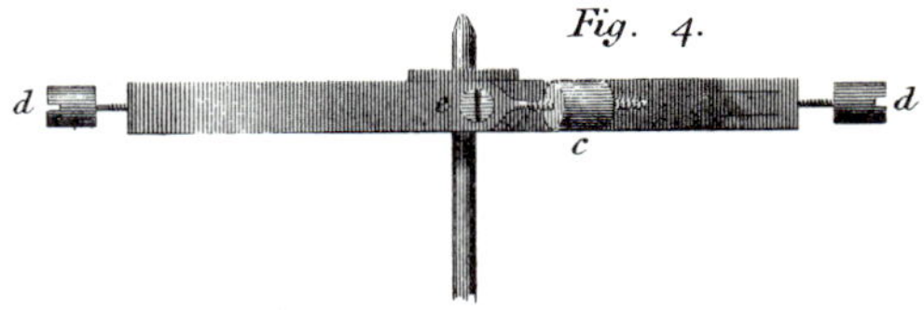

Fig. 5.

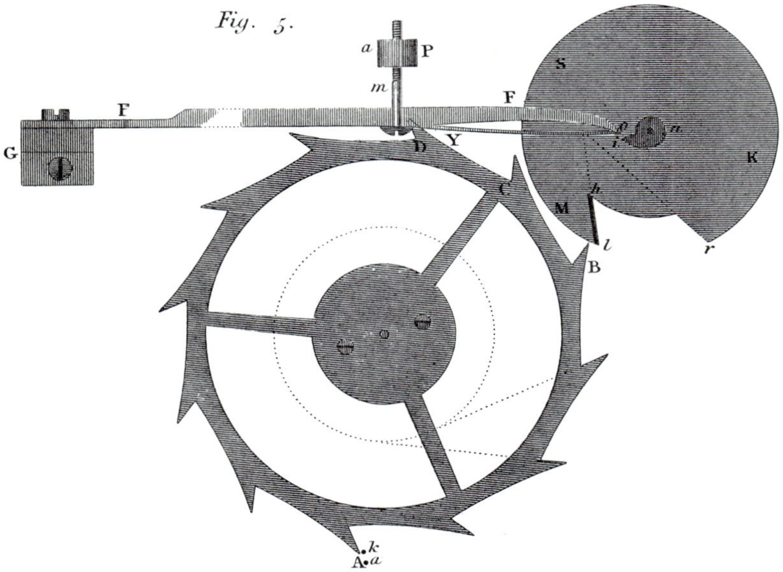

Fig. 6.

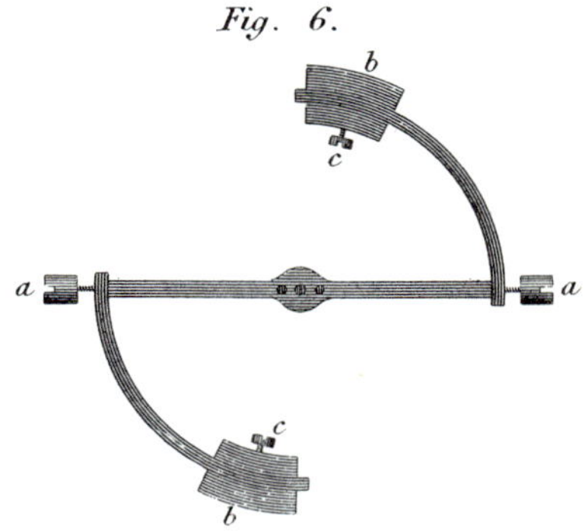

Fig. 7.

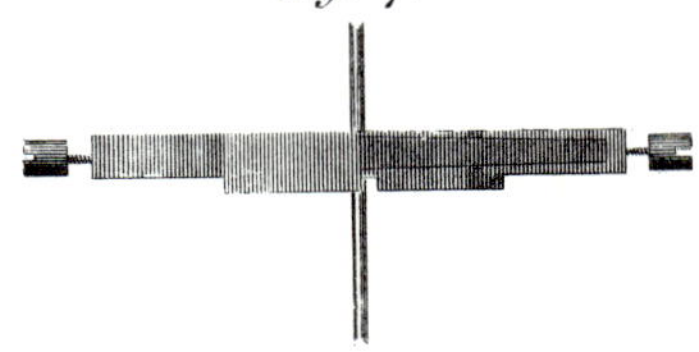

J. Farey Jun.r delin.

Published as the Act directs, Dec.r 1807, by Longman, Hurst, Rees & Orme, Paternoster Row.

Engraved by Wilson Lowry.

BROCKBANK'S CHRONOMETER.

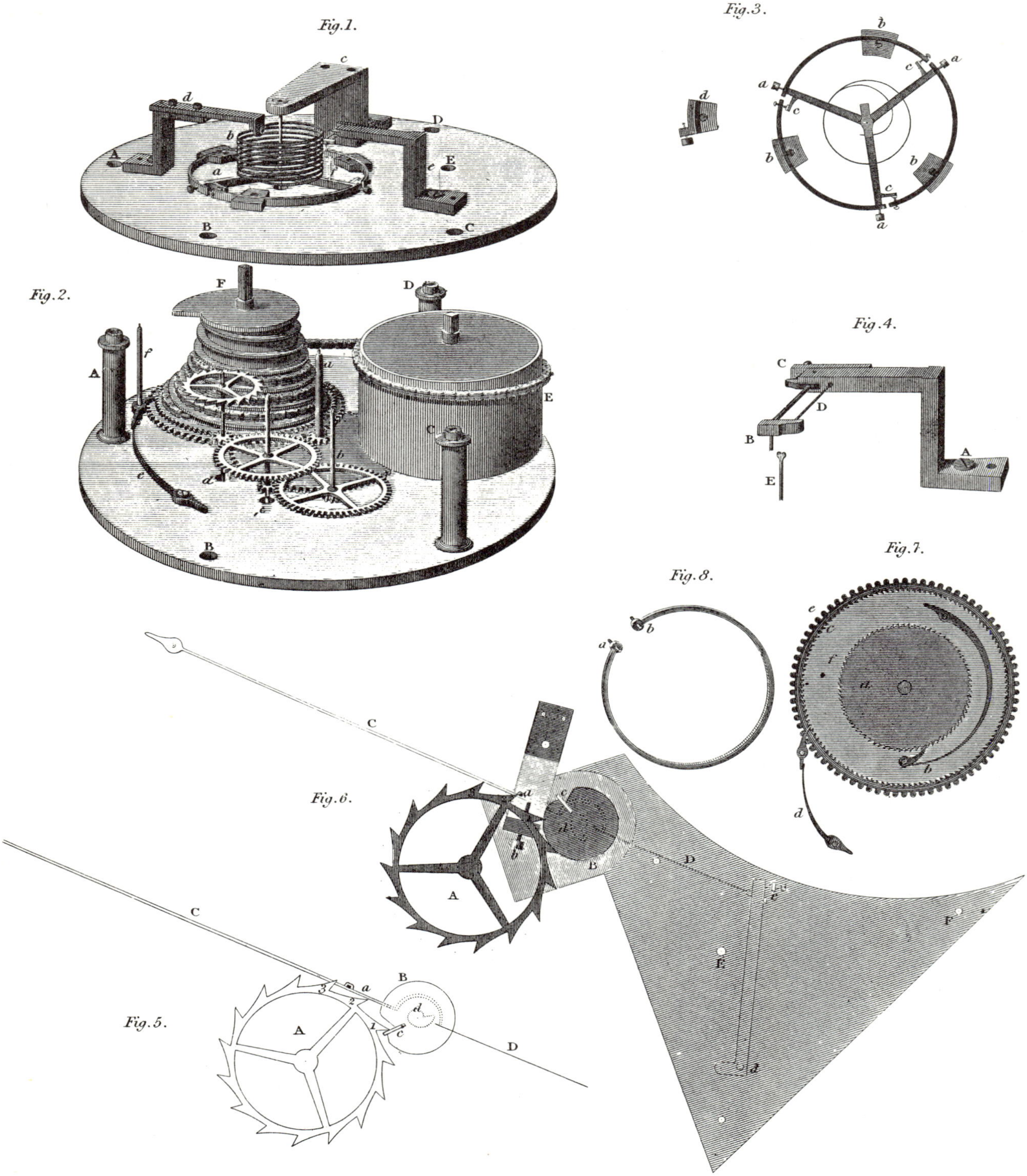

J. Farey Jun. del.
Published as the Act directs, 1817, by Longman, Hurst, Rees, Orme, & Brown, Paternoster Row, London.
Lowry sculp

Clepsydra

Clepsydra, so called from κλεπτω, *surripio*, and ὑδωρ, *aqua*, was an horological instrument of great antiquity, among the Egyptians and other eastern nations, probably before sun-dials were invented; though the name of the original inventor is not handed down to us; the construction has been varied in different ages and countries, according to the variation of the different modes of reckoning time, but one principle is the basis of all the forms it has undergone, namely, the constant dropping, or running of water through a small aperture, out of one vessel into another. At first the indication of time was effected by marks corresponding to either the diminution of the fluid in the containing vessel, during the time of emptying, or to the increase of the fluid in the receiving vessel during its time of filling; but it was soon found, that the escape of the water was much more rapid out of the containing vessel, when it was full, than when it was nearly empty, owing to the difference of pressures at different heights of the surface; this irregularity in the dropping, presented an obstacle which required much ingenuity to correct. In our account of the different constructions of clepsydræ, we will class them under the two heads of *ancient* and *modern*.

Ancient Clepsydræ.—According to M. Vitruvius Pollio, the first improver of the ancient clepsydra, or water-clock, was Ctesibius of Alexandria, the son of a barber, who, about 245 years before Christ, spent much time in devising mechanical contrivances for removing not only the obstacle in question, but also another equally formidable one, which arose from the daily inequality of the Egyptian hours. As one-twelfth part of the time elapsed from sun-rise to sun-setting on any day, was called an hour of that day; and as one-twelfth part of the time that passed from sun setting to sun-rise, was called an hour of the night; not only did the hours of day differ from the hours of night, but from one another, at all times, except at the vernal and autumnal equinoxes; hence it became necessary, either to make the water fall irregularly into a receiving vessel, with equidistant hour-marks, or to have varying hour marks for a regular efflux; the first of these methods (which probably preceded that of Ctesibius) was thus effected, *viz.* 1. A conical hollow vessel, A, was inverted, or placed like a funnel in a frame C C (*Plate* I. *fig.* 1. of *Horology*) there being a very small aperture at the apex of the cone, and another solid cone, B, every way similar as to dimensions, was plunged into the hollow one when filled with water to a greater or a smaller depth, accordingly as the efflux was wanted to be more or less rapid, and then adjusting marks, corresponding to every day and night in the year, were put on a long stem D, inserted into the broad end of the solid cone B, and kept in its position by the frame, as represented in the figure, to show how much the inner cone was to be depressed or elevated, to accelerate or retard the issue of the fluid for the corresponding time; H was the spout which supplied a constant influx of water, and I the waste pipe, connected with the top of the conical vessel, which carried off the superfluous water; hence the constant influx of water preserved an unvarying height of the surface from the aperture, which aperture was varied at pleasure, by the elevation or depression of the inner cone; it if now we ſuppoſe the ſubjacent veſſel to be a cube, cylinder, or any other regular figure, and equidiſtant hour-marks to be properly made on its ſide, the ſurface of the water, or an index borne by it on a piece of cork, would, as it roſe, indicate the hours correſponding to thoſe marks.

The imperfections of this clepſydra were theſe: 1. It required two daily manual adjuſtments, one in the morning, and the other in the evening; and, 2. It made no allowance for the variation of fluidity, in different degrees of temperature, which, it is aſſerted (but perhaps without proof), greatly influenced the iſochroniſm of the drops. As an improvement, or rather appendage, to this conſtruction of the clepſydra, a bar, E E, with rack-work at the upper end, as ſhewn by the dotted lines, was made to float on the ſurface of the lower veſſel by means of an affixed piece of cork, F, ſo that as the cork and its bar roſe in the veſſel, the teeth of the bar turned a ſmall wheel, G, fixed to the upper part of the frame by a cock, on the arbor of which wheel a hand was put, which revolved and indicated the hours on a fixed dial-plate. This addition, however, did not render the inſtrument a more accurate meaſure of time, but only indicated the hours, ſuch as they were, in an improved manner. It may be worthy of remark here, that water was at once the regulator and the maintaining power of the inſtrument before us; the interval between two ſucceſſive drops was to the clepſydra what one vibration of the pendulum is to a clock, or one oſcillation of the balance is to a watch; and the floating of the indented bar was in place of a weight or ſpring to move the wheel to which the hand was attached; conſequently it might be ſaid to be an horological machine of the ſimpleſt conſtruction poſſible. The adjuſtment of the two cones was regulated by the latitude of the place, owing to the manner in which the hours were divided; at Alexandria, for inſtance, the greateſt and leaſt velocity of the drops were required to be to each other as 70 to 50, the longeſt and ſhorteſt hours in that latitude being reſpectively $1^h\ 10^m$ and 50^m of equable time; and in higher latitudes the diſparity is ſtill greater.

2. The next attempt to improve the clepſydra was by conſtructing it ſo that its aperture was adjuſted, as the year advanced, by the putting of an index to the ſun's place in an ecliptic circle; which attempt, of courſe, rendered the inſtrument more complex. Perrault conceives the parts to have been thus adapted, according to the deſcription given of it by M. Vitruvius Pollio, in his book "De Architecturâ" (cap. ix. lib. ix.).

Fig. 2. of *Plate* I. repreſents an ancient clepſydra with an horary circle and a variable aperture: A is a reſervoir, to the top of which is attached a water-pipe, not ſeen in the drawing, to preſerve an equal preſſure by carrying off the ſuperfluous water; B is a pipe projecting from the reſervoir into the upper part of the drum, M N, on the front of which drum the ecliptic circle is marked; O D L is a ſmaller inner drum, which revolves on a tubed arbor, F, and which is repreſented as drawn out of its place; this ſmall drum has a thorough groove, *a b*, varying in breadth all round it, like a hoop tapering throughout from the broadeſt part both ways to its oppoſite point, and is of ſuch a diameter that the middle of the groove juſt reaches to, and coincides with, a perforation under the tube, B, at the upper part of the great drum, ſo that, as the little drum, which carries the diurnal index, L, and nocturnal index, O, oppoſite to the former, is turned round by hand, the variation in the breadth of the groove occaſions a correſponding variation in the velocity of the efflux of water, by making a larger or ſmaller aperture, accordingly as the ſun's place is more or leſs advanced in the ecliptic, the largeſt aperture being when the diurnal index is at the beginning of Capricorn; a little baſon or funnel attached to the upper part of the fixed tube or hollow arbor, F, (not viſible), receives the water in its fall within the drum, and tranſmits it through the ſaid tube by G into the receiving veſſel, H, in which is floated the piece of cork, I; this floating-piece is connected, by a chain, with the counterpoiſe, K, after it is folded round the arbor, P, which carries the hour-hand of the dial-plate; conſequently, as the water riſes in the veſſel, H, the piece, I, is raiſed, and its counterpoiſe, K, at the ſame time falling gives motion to the arbor and hour-hand, and the hours are longer or ſhorter according to the breadth of the groove which is at any time under the perforation of the tube, B, *i. e.* according to the place in the ecliptic to which the proper index is put.

This clepſydra, like the preceding one, compoſed of two cones, requires two manual adjuſtments, one in the morning and the other in the evening, and makes no allowance for the (ſuppoſed) variation of fluidity occaſioned by the different ſtates of the weather; and the variation in the breadth of the groove or ſlit, it is preſumed, was more plauſible in theory, than feaſible in practice; the contrivance, however, was ingenious, and beſpoke the inventor's acquaintance with aſtronomy.

3. The next improvement in the ancient clepſydra was probably that of Cteſibius, which was an automaton, or ſelf-adjuſting machine, and is repreſented by *fig.* 3, which, according to Perrault and Ferd. Berthoud, exhibits the interior conſtruction of this machine; A is the end of a tube over which an image ſtands, which is connected with a full reſervoir, and from the eyes of which, conſidered as invariable apertures, the water continually flows or drops in a regulated manner into it; this tube conveys the water from M towards B into the top of a long regular veſſel, B C D F, which it gradually fills, and raiſes the cork, D, with its attached light pillar, C D; on the top of this pillar is ſurmounted another image holding an index which points to the diviſions on the large column above. Now, when the water riſes in the veſſel that contains the cork, it alſo riſes in the ſmall tube, F B, which conſtitutes one leg of a ſyphon, F B E, that is connected with the bottom of the cubic veſſel; conſequently, when the index has mounted to the uppermoſt diviſion on the large column of hour lines, conſiſting of twice twelve, the water flows over the bent part, B, of the ſyphon, and immediately empties the veſſel into one of the ſix troughs or diviſions of the water-wheel, K, which is thus turned one-ſixth part of a revolution, during which time the image falls with its index to the bottom of the column, to be ready for the next day. This portion of the mechaniſm would have been ſufficient to conſtitute the machine, if the hours had been conſidered as of equal length throughout the year, but the Egyptian mode of dividing and reckoning time made it requiſite that the hour lines ſhould ſlope out of an horizontal direction on the ſurface of the column, ſo as to make variable ſpaces, and alſo that the column ſhould revolve once in a year, to preſent all the variations of ſpace to the index. This annual motion of the column is ſaid to be effected by wheel-work in the following manner:—on the arbor of the water-wheel, K, is fixed the pinion, N, of ſix leaves, which impels the contrate-wheel, I, of 60 teeth in $6 \times \frac{60}{6} = 60$ days, then on the perpendicular arbor of I is another pinion, H, of ten leaves, which drives the wheel, G, of 61 teeth round in $60 \times \frac{61}{10} = 366$ days, and along with it the horary column, into which its arbor is inſerted at L. On the bottom of the column is marked an ecliptic circle; and 12 perpendicular lines drawn lengthwiſe down the column divide it

it into the refpective figns, which are ferviceable for afcertaining the requifite flope of the hour lines in any month. The writer of this article, however, fufpects, that the above train of wheel-work is only what Perrault, the tranflator of Vitruvius, fuppofed to be that of Ctefibius; for, on referring to the original account of Vitruvius, the year in which the column revolved is ftated to be 365 days, a period which might be effected thus:

Let the water-wheel have only five compartments inftead of fix, and let an endlefs fcrew be cut on its arbor to impel a wheel of 73 teeth, with a perpendicular arbor, to be inferted into the column of hours, which will, by fuch a fimple conftruction, revolve in $5 \times \frac{73}{1} = 365$ days, agreeably to the original account.

The clepfydra, in one of its earlier forms, was ufed as an aftronomical inftrument, by the help of which the equator was divided into twelve equal parts, before the mathematical divifion of a circle was underftood; it was deemed of more value than a fun-dial, on account of its dividing the hours of the night as well as of the day. It was introduced into Greece by Plato, and into Rome by P. Cornelius Scipio Nafica, about 157 years before Chrift.

Pliny fays (lib. xxxvii.) that Pompey brought a valuable one among his fpoils from the Eaftern nations; and Cæfar is faid to have met with an inftrument of this kind in England, by the help of which he obferved that the fummer nights of this climate are fhorter than they are in Italy. The ufe which Pompey made of his inftrument was to limit the fpeeches of the Roman orators; which Cicero alludes to when he fays "latrare ad clepfydram."

4. Befides the ancient clepfydra, above defcribed, F. Berthoud mentions another (Hiftoire de la Mefure du Temps, tom. I. p. 20.), which was called the *anaphoric*, on the dial-plate of which were projected the circles of the fphere, including the parallels of the fun's altitude, with the femi-diurnal and femi-nocturnal arcs, to which an adjuftable bead, as the fun's reprefentative, pointed as an index to fhew the hours, parallels, &c. as the dial-plate revolved daily by means of wheel-work, which was impelled by water. It does not feem certain at what period this inftrument was invented and ufed; but Berthoud thinks that tables of the fun's motion muft have exifted previoufly to its invention, and alfo a knowledge of projections of the fphere on a plane furface, whence he fixes the date pofterior to the time of Hipparchus, who, according to Pliny, died about 125 years B.C. The name *anaphoric* is evidently derived from *anaphora*, which was the fecond houfe in the heavens, according to the doctrine of aftrology, which prevailed about the time here fpecified.

In Athenæus, lib. iv. p. 174, we have a hiftory and defcription of an ancient inftrument. He tells us that it was invented in the time of the fecond Ptolemy Euergetes, by Ctefibius, a native of Alexandria, and by profeffion a barber: or rather, that it was improved by him, for Plato furnifhed the firft idea of the hydraulic organ, by inventing a night-clock, which was a clepfydra, or water-clock, that played upon flutes the hours of the night at a time when they could not be feen on the index.

The anecdote in Athenæus concerning the mechanical amufements of the great ideal philofopher, is curious. What a condefcenfion in the *divine* Plato to ftoop to the invention of any thing ufeful! This mufical clock muft have been wholly played by mechanifm.

In defcribing it, Athenæus fays, it refembled in appearance a round altar; but was not to be ranked with ftringed but wind inftruments, compofed of pipes; the orifices of which being towards the water, when it was agitated, produced from the pipes, by its fall, a foft and pleafing found.

Modern Clepfydræ.—The modern method of dividing the natural day into 24 folar hours of equal length, has rendered the preceding conftructions of the clepfydra ufelefs for fome centuries back; and, notwithftanding the fcience of hydroftatics is much better underftood by the modern than it was by the ancient philofopher, fo that a fcale of altitudes correfponding to the variable velocities of the efflux of a fluid out of a given aperture can be afcertained by calculation for a containing veffel of any capacity or figure, yet, fince the happy inventions of the balance and pendulum, as regulators of watches and clocks, horological machines, actuated by the motion of water, have become fo rare, as to be confidered as objects only of curiofity.

Beckmann, in his "Hiftory of Inventions," vol. i. p. 136. attributes the contrivance and introduction of a water-clock to fome time between 1643 and 1663, and gives nearly the fame brief account of one as we meet with in "Bion, on Mathematical Inftruments," and alfo in "Ozanam's Recreations," edited by Dr. Hutton; the laft of which authors faid, in the year 1693, that the firft water-clock brought to Paris about that time was from Burgundy. He alfo fays, that father Timothy, a Barnabite, had given the machine all the excellence it was capable of, by conftructing it fo as to make it go a month at one winding up, and to exhibit not only the hours on a dial-plate, but alfo the fun's place, day of the month, and feftivals throughout the year.

How thefe and fimilar particulars might be indicated, will be eafily apprehended from the following defcription, which is agreeable to the accounts given of a water-clock of the 17th century by the authors already named.

1. In *fig.* 1, of *Plate* II. of *Horology*, A B C D is an oblong frame of wood, to the upper part of which two cords, A *a* and B *b*, are fixed at their fuperior extremities, and at their inferior, to the metallic arbor, *a b*, of the drum, E, which contains diftilled water; this water is confined in cells fo peculiarly conftructed, that they regulate the velocity with which the drum fhall defcend by the force of gravity from the top to the bottom of the frame, and the ends of the arbor indicate the hours marked on the vertical plane of the frame during the time of defcent. An obferver, who knows not the nature of the interior cells of the drum, is furprifed to fee that its weight does not make it run down rapidly, when mounted to the top of the frame by merely folding the ftrings round the arbor, there being apparently no mechanical impediment to the natural action of gravity. To explain how this phenomenon is produced, we muft refer to *fig.* 2, which is a fection of the drum at right angles to its arbor; this circular plane we will fuppofe to be fix inches, which is about the ufual fize, in diameter, and to reprefent the inner furface of either of the two ends of the drum, which may be made of any of the unoxidable metals; then, if we conceive feven metallic partitions, F *f*, G *g*, H *h*, I *i*, K *k*, L *l*, and M *m*, to be clofely foldered to both ends of the drum, in the floping direction indicated by the figure, where the black lines are equidiftant tangents to the fmall dotted circle of an inch and half diameter at the points *f*, *g*, *h*, &c.; it is evident, that any fmall quantity of water introduced into the drum would fall into two, or at moft three, of the lower compartments, and would remain there until fome external force fhould alter the pofition of the drum, fuppofing in this cafe the cords tied faft to the arbor; but we have faid that they are wound round the circumference of an arbor, that has a fenfible diameter, fuppofe one-eighth of an inch; therefore,

they are removed one-fixteenth of an inch, or upwards, if we take their thicknefs into the account, from the centre of the drum, which would alfo be its centre of gravity, if it were empty, on which account it would, in that cafe, revolve to the left, in the direction F G H downwards, from the cord being at the remote fide of the centre, as reprefented by N O; but conceive the water to be included now and then, it would be elevated to the right, till its weight became a counterpoife to the gravity of the heavier fide of ths drum, in which fituation all motion would ceafe, and the drum would remain, fufpended, indeed, by the cords, but in a ftate of equilibrio. Conceive again a fmall hole perforated in the partition preffed upon by the water near the circumference of the large circle, and alfo at the points F, G, H, I, K, L, M, and the confequence will be, that the water will firft force its way flowly through the perforation at K, from the more elevated to the lower compartment, which effect will diminifh its power as a counterpoife, and give fuch an advantage to the heavy fide, F G H, of the drum, confidered as empty, as will occafion a fmall degree of motion towards the left, and confequently carry the water once more towards the right; but now the water paffes through the perforation of the next partition alfo at I, and produces again the fame effect, as has been defcribed with refpect to K, and will continue to do fo, at the fucceffive perforations, till all the compartments have been filled and emptied by means of thefe perforations, in fucceffion, which kind of motion of the drum, contrary to that of the water, it is now not difficult to conceive will be pretty regular, if all the partitions are perforated exactly alike. The difference of the preffures of the water in cells, nearly full and nearly empty, will occafion fome little deviation from regularity; but thefe will be periodic, and muft be allowed for in the hour divifions, which ought to be made by a comparifon of the fpaces fallen through, with the time indicated by a clock or watch. About nine ounces of diftilled water will fuffice for a clepfydra of fix inches diameter, and two inches depth, and the velocity of the fall may be limited, either by varying the quantity of water, or by hanging a fmall metallic cup F, to receive weights, by a cord wound in a direction contrary to the cords of fufpenfion, to act as a counterpoife in aid of the water, if the fall be too rapid, or *vice verfâ*.

It is abfolutely neceffary that the arbor fhould fit the central fquare hole fo well as to prevent the efcape of water from the drum, otherwife the inftrument would continue to gain velocity, till at length it would no longer afford a true indication of time.

Sometimes a cord, *c d*, with a weight, P, is made to pafs round a pulley fixed to an arbor at the top of the frame, with a noofe paffing over the axis near *a*, as is feen in the fame figure, which arbor, projecting through a dial-plate or face, turns round and carries a hand to indicate the hours like an ordinary clock; when this conftruction is preferred, it is an indifpenfable requifite that the circumference of the pulley's groove be exactly of the fame dimenfions as the fall of the drum in 12 or 24 hours, accordingly as the dial is divided.

This clepfydra, it is faid, goes fafter in fummer than in winter, which is owing to the drum being relatively heavier in rarefied than in denfe air; we can hardly fuppofe that any alteration in the fluidity of the water, as formerly fuppofed, would make any difference. The minute hand and alfo the ftriking part of a common clock might eafily be fuperadded to this clepfydra.

2. Another form, and that a very fimple one, of the modern clepfydra has derived its origin from that law in hydroftatics by which the efflux of water out of an orifice is influenced under different preffures, or which is the fame thing, at different depths from the furface, the velocity being directly as the fquare root of the height of the furface from the aperture. If a glafs veffel, like that in *fig.* 3, therefore be taken, out of which all the water will flow in exactly 12 hours, from a fmall aperture in its lower extremity, the whole height muft be divided, or fuppofed to be divided, into the fquare of 12 or 144 equal parts, of which parts 11 × 11, or 121 meafured from the bottom, or 23 meafured from the top, will give the divifion for the hour 11, 10 × 10 or 100 from the bottom will give the line for 10, 81 for 9, 64 for 8, and fo on down to the bottom, as reprefented in the figure; which fcale is in the inverted proportion of that according to which heavy bodies fall in free fpace by the fole force of gravity.

If, inftead of the veffel itfelf being divided by hour-lines as above directed, the ftem of a floating piece like an hydrometer were to have a fimilar fcale kept in a perpendicular direction, by paffing through the central hole of a cap or cover of the veffel, the indication of time would be made on the ftem at the furface of the cap, which conftruction would admit of the veffel being of wood or metal.

3. But fuch a figure might be given to the containing veffel as would require the dividing marks to be equi-diftant, which Dr. Hutton, in his recent edition of "Ozanam's Recreations," has afferted to be a paraboloid, or veffel, formed by the circumvolution of a parabola of the fourth degree, the method of defcribing which, he has given thus:

Let A B S, *Plate* II. *fig.* 4, be a common parabola, the axis of which is P S, and the fummit S. Draw, in any manner, the line, R *v* T, parallel to that axis, and then draw any ordinate of the parabola A P, interfecting R T in R; make P Q a mean proportional be tween P R and P A, and let *p q* be a mean proportional alfo between *p r* and *p a;* and fo on. The curve paffing through all the points Q *q*, &c. will be the one required, which, being made the mould for a veffel to be caft by, will produce an inftrument, which, when perforated at the apex, will have the fingular property of equalizing the fcale, fo as to correfpond to equal times while the water is running out. Mr. Varignon has given a geometrical and general method of determining the fcale for a clepfydra, whatever may be the fhape and magnitude of the veffel. (See "Memoires de l'Academié Royale des Sciences," p. 78, 1699.)

4. Another method of making a water-clock with equi-diftant hour lines in any regular veffel, is effected more fimply than in the preceding one, by means of the fyphon fixed faft in the centre of a broad piece of cork, which is floated in any regular veffel, as the cylindrical one at *fig.* 5, for as the power of a fyphon to empty any veffel filled with water depends upon the difference of atmofpheric preffures at the furface of the water and at the orifice of the longer leg, it is clear that while the fhorter leg finks with the furface of the water in the veffel during its time of emptying, the relative preffures, depending on the diftance from the furface of the water to the orifice of the lower leg, will continue unaltered in any ftate of the atmofphere; hence equal portions of water will be difcharged in equal times; and a light cock cemented on the lower orifice would afford a means of adjufting its aperture to the fize of any veffel that may be fixed upon; or otherwife a fecond receiving veffel may be divided into equal fpaces for the hours, which would in this cafe be indicated by the furface of the rifing water.

Befides the preceding methods of meafuring time by means of water, there are others nearly fimilar, fuch as the double jet d'eau, which, like the fand-glafs that may be claffed with thefe, requires to be inverted as foon as empty, and it is eafy to conceive a variety of ways of applying any liquid to anſwer the purpoſe of meaſuring pretty nearly a given number of hours, but we do not learn that the moſt accurate of the clepſydra is comparable to an ordinary clock, though it has been aſſerted, that Amontons conſtructed one in ſo accurate a manner, that he hoped to find it uſeful in aſcertaining the longitude at ſea by means of its accuracy; we regret that it is not in our power at preſent to procure the pamphlet in which the account of it was publiſhed. "Remarques & Experiences Phyſiques ſur la Conſtruction d'une nouvelle Clepſydre," &c. Paris. Jombert. 1695.

5. We ſhall conclude our account of theſe horological inſtruments with detailing the conſtruction and action of a clepſydra, publiſhed in the 44th volume of the Philoſophical Tranſactions by the Hon. Mr. Charles Hamilton.

A B and C D are two ſimilar oblong veſſels attached to a frame of wood, which may eaſily be conceived to ſurround figure 6, which ſhews only the interior mechaniſm; *a b* and *c d* are two columns of wood ſo floating in water, that their counterpoiſes, F and G, juſt keep their ſuperior ends equal with the ſurface of the water by means of connecting chains paſſing over the pulley *f*, and another hid by the dial plate; the former of theſe pullies, *f*, has a click which puſhes the ratchet on the barrel, *i*, when the counterpoiſe, F, falls, but ſlips eaſily over the ſlopes of the teeth when the ſaid counterpoiſe riſes; the latter pulley has alſo a ſimilar click acting in like manner, with a ſecond ratchet at the oppoſite end of the barrel, *i*, which ratchet is alſo hid in the drawing, ſo that whichever of the two counterpoiſes ſhall at any time be falling, the barrel, *i*, will move forwards in the ſame direction; and carry the minute hand along with it on the dial-plate; the hour hand goes round by means of dial-work, as in an ordinary clock or watch, where a diminution of velocity is effected by two wheels and two pinions. The action is thus produced by means of five ſyphons and two balances.

The water enters with an unvaried influx, drawn from a reſervoir, by a ſyphon of ſmall bore, the longer leg of which is ſeen at J, into the middle of what may be called a horizontal trough, ſupported like a balance by a fulcrum at K, in ſuch a manner, that either end of the balance may be elevated accordingly as the long veſſels A B and C D require to be alternately filled; near the top of each of theſe veſſels is inſerted a long ſyphon or tantalus, *l* and *m*, the lower legs of which reach down to two ſmall cylindrical veſſels, *n* and *o*, which are poiſed by another balance at the fulcrum *p*; theſe cylindrical veſſels have, in like manner, each a ſmall ſyphon, *q* and *r*; laſtly, a ſilken thread tied to the upper end of the cylinder, *n*, is carried up round a ſmall pulley faſt to the frame at *s*, and is faſtened to the end of the trough under it, and a ſimilar thread is faſtened in like manner to the cylinder *o*, and end of the trough under the ſmall pulley *t*. Now it is eaſy to conceive, that when the veſſel, A B, is filled to nearly the head of the tantalus *l*, the bore of which is larger than of the feeding ſyphon J, the water will be diſcharged into the cylindrical vaſe *n*, which conſequently will preponderate, and by means of the ſilken chord elevate the end of the trough higher than the horizontal line, and make its oppoſite end under the ſmall pulley, *t*, to be depreſſed, which will therefore conduct the water into the other long veſſel C D; during this action the counterpoiſe, F, riſes, and its pulley, *f*, produces no effect on the ratchet by reaſon of the click, *h*, ſliding over the ſloping ſides of its teeth, but the counterpoiſe, G, falls, and the click of its pulley (not ſeen) puſhes the ſecond ratchet forwards in the direction of the figures of the face I. II. III. &c.

When C D is nearly full, the long ſyphon, *m*, begins to diſcharge its water; makes the cylindrical vaſe, *o*, preponderate, and again elevates by means of its ſilken ſtring the end of the trough under the ſmall pulley *t*, and depreſſes the oppoſite end to fill the veſſel, A B, again, during which time the click, *h*, of the pulley, *f*, acts with its ratchet; and thus the alternate increaſe and decreaſe of the water in the two veſſels are continued without interruption, ſo long as the feeding ſyphon continues to ſupply a ſufficient quantity of pure water. We think, however, that the mechaniſm is nearly as complex as that of a clock itſelf, and conſequently ſhould prefer a water-clock, ſuch as that made by Perrault in the year 1699, where a pendulum is uſed as the regulator, and water only as the firſt mover. For the account, ſee "Machines Approuvées," tome i. p. 39.

The ſame Perrault alſo made a water-clock with a balance and ſtriking part, an account of which is given in the volume of "Machines Approuvées," which we have juſt referred to; and in the ſeventh volume of the ſame work, is a deſcription of a regulator going by water, invented by Peronnier, and improved by Le Roy, the ſon, in 1746. (See page 335.)

Clepsydra is alſo uſed for an hour-glaſs of ſand.

Clepsydra is alſo applied to a chemical veſſel perforated in the ſame manner.

ANCIENT CLEPSYDRÆ.

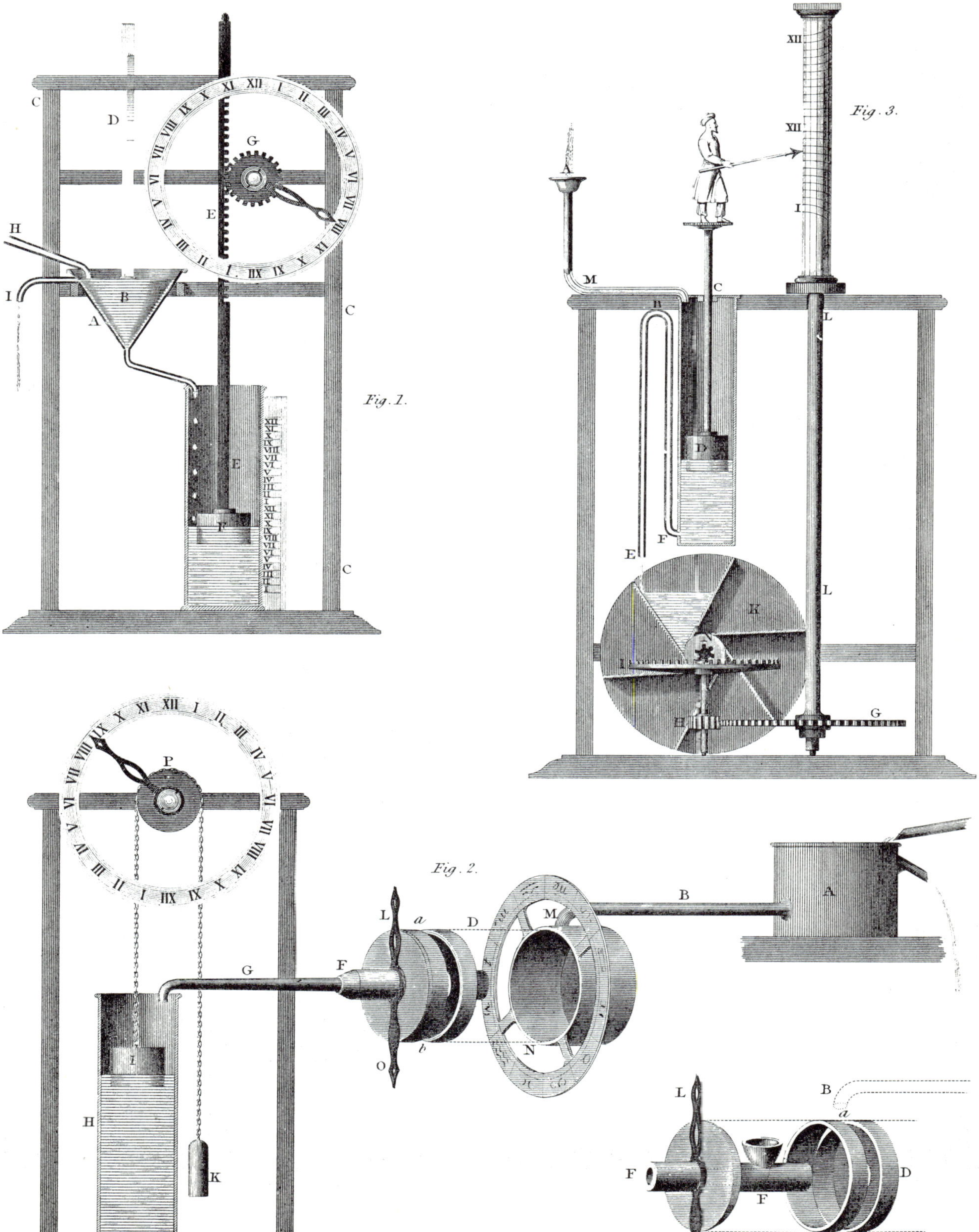

Farey, delin. Published as the Act directs, 1819, by Longman, Hurst, Rees, Orme & Brown, Paternoster Row. Lowry, Sc.

HOROLOGY.

MODERN CLEPSYDRÆ.

page 60

Clock

Clock, in *Horology*, is a machine which meaſures, ſubdivides, and indicates the ſucceſſive portions of time with a degree of accuracy that has deſervedly given it the preference over the clepſydra, and brought it into general uſe for civil, domeſtic, philoſophical, and aſtronomical purpoſes; it is conſtructed of various materials, and after different models, to accommodate the views of various individuals; its value, therefore, varies from three half-crowns to one hundred pounds and upwards; but though every houſeholder almoſt is now in poſſeſſion of a clock, few individuals, comparatively ſpeaking, know either the theory of its action, or the ſubſerviency of the ſeparate parts of its mechaniſm to produce their deſtined effect; and yet, no inſtrument has been more the object of ingenuity, or the ſubject of ſcience, for two or three ſucceſſive centuries, than the machine in queſtion, before it attained its preſent moſt improved conſtruction. Its name is derived either from the German *die gloke* or *die kloke*, a clock, or from the French *la cloche*, a bell, againſt which it uſually ſtrikes the hour indicated.

In tracing the records of antiquity for the origin of the firſt horological machine, that had a ſuſpended gravitating body as a maintaining power, and a regulator of ſome determined ſhape and dimenſions to check its velocity at ſmall equidiſtant intervals, different authors have fixed upon different ingenious men as its inventor, and have quoted with ſome confidence paſſages from the more ancient writers in confirmation of their reſpective opinions: hence Archimedes and Poſſidonius before the Chriſtian æ a, Boethius in the fifth century, or about the commencement of the ſixth, Pacificus about the middle of the ninth, Gerbert at the end of the tenth, Wallingford near the beginning of the fourteenth, and Dondi at the end of the fourteenth, have ſeverally been aſſerted to have been the firſt contrivers of a clock. The difficulty of aſcertaining the exact period when a regulated machine for meaſuring time without the aſſiſtance of water was firſt invented, ariſes from this conſideration, which has not been ſufficiently attended to; *viz.* the appellation *horologium* has by ancient writers been indiſcriminately applied to all inſtruments that had any thing to do with hours, whether regulated or not, ſuch as ſun-dials, clepſydræ, and inſtruments for merely repreſenting the motions of the heavenly bodies, like our orreries and planetaria; and, laſtly, clocks, or machines with a bell, to ſtrike the hours, placed in the ſteeple *(le clocher)*, of ſome abbey.

The ſphere of Archimedes, made two hundred years before Chriſt, as mentioned by Claudian, was evidently an inſtrument with a maintaining power, but without a regulator, and therefore would not meaſure time in any other way than as a planetarium, turned by a handle, meaſures, or rather exhibits, the reſpective velocities of the planetary bodies. The ſame may be ſaid of the ſphere of Poſſidonius eighty years before Chriſt, as mentioned by Cicero (De Natura Deorum), notwithſtanding the deference which is due to the opinion of Dr. Derham.

When Bernardus Saccus (Hiſt. Ticin. lib. vii. c. 17.) aſcribes the invention of clocks to Boethius in the year 510, he paſſes over that part of the quotation from Caſſiodorus, which ſays, "that the hours were determined *guttis aquarum*; by drops of water;" hence his horologium was evidently nothing more than a clepſydra. The authority upon which Pacificus, archdeacon of Verona, has been deemed the inventor of clocks in the year 850, is the ſubjoined portion of his epitaph:

"Horologium nocturnum nullus ante viderat,
En invenit argumentum et primus fundaverat;
Horologioque carmen ſpheræ cœli optimum
Plura alia graviaque prudens invenit."

Profeſſor Hamberger, in a paper read to the Society at Gottingen in 1758, (vide Beckmann, vol. i. page 419, & ſeq.) has given reaſons for believing this "Horologium nocturnum," as it is called in oppoſition to a ſun-dial, or horologium diurnum, to have been a clepſydra, notwithſtanding the "nullus ante viderat," for it was in the ninth century that Hildemar, ſpeaking of the Monks, in regard to their obſerving the hours, ſays, "he who wiſhes to do this properly muſt have "horologium aquæ," which is conſidered as a proof that clocks did not then exiſt. (Commentar. in Reg. S. Bened. cap. 8.) On the contrary, Bailly in his "Hiſtory of Modern Aſtronomy" (vol. i. p. 321.) aſſerts, that Pacificus was the inventor of a clock going by means of a ſuſpended weight, an eſcapement, and a balance; if this information be accurate, no doubt he was the firſt who is known to have made a clock; but F. Berthoud (Hiſtoire de la Meſure du Tems, tom. i. p. 49.) very properly obſerves, that the celebrated author has adduced no authority for the aſſertion he has uſed, and that he believes it to be inaccurate. With reſpect to Gerbert, who was made pope Silveſter II. in the year 999, Dithmar (Chron. lib. vi. p. 83. fol. 1580.) believes that his horologium was only a ſun-dial which he made at Magdeburg; his words are "Gerbertus, a finibus ſuis expulſus, Ottonen petiit imperatorem, et cum eo diu converſatus, in Magdeburg horologium fecit, illud recte conſtituens, conſideratâ per fiſtulam quadam ſtellâ nautarum duce." Here no wheels or weights are mentioned, but, on the contrary, his obſervation only of the pole ſtar, which aſſiſted him in fixing the horologium or dial; nay, in Gerbert's book "De Aſtrolabio" he explains the method of erecting dials for all latitudes, but ſays nothing of any other kind of horologia. To what has been above remarked, concerning the different ſignifications of the word horologium, may be added what is ſaid by the anonymous author of the life of William Abbot of Hirſhau, who lived in the eleventh century, *viz.* "naturale horologium ad exemplum cæleſtis hæmiſpherii excogitâſſe," which expreſſion alludes to ſome piece of aſtronomical mechaniſm at that time invented; and according to the account of Trithemius reſpecting the horologium made in 1232, and ſent by the ſultan of Egypt to the emperor Frederic II. (Chron. Hirſaug. ad. h. a.) as well as from Leland's deſcription of Richard Wallingford's horologium, called Albion (all by one) made in 1326, theſe muſt have been inſtruments claſſing rather with our orreries than our clocks; for the motions of all the heavenly bodies appear to have been conducted by the maintaining power, whatever it was, without any controlling or regulating mechaniſm. (Vide Tanneri Biblioth. Brit. Hibern. p. 629.) It may be further remarked here, that the words, "horologium

gium dirigere," "ordinare," temperare," &c., which denote the office of the facriftan, in the writings of the faid William Abbot of Hirfhau, evidently allude to the adjuftments of the clepfydra. Hence it may be fairly inferred, that no one of the preceding contrivers of horologia was alone the inventor of a regulated clock: the invention was, no doubt, of monaftic origin, or, at leaft, intended firft for monaftic purpofes, when the ftated periods of prayer required the attendance of the monks by night as well as by day. That the horologium of Dondi, however, was a clock can admit of but little doubt; it was conftructed at Padua about the end of the fourteenth century, by order of Hubert, prince of Carara; and the defcription given of it by Petrus Paulus Vergerius (in Vit. Princip. Carrar. ap. Murator. tom xvi. p. 171.) is this; "Horologium quo per diem et noctem quatuor et viginti horarum fpatia fponte fuâ defignarentur, in fummâ turri conftituendum curavit." This fpontaneous defignation of 24 hours of day and night, by an horologium, placed on the fummit of a turret or fteeple, correfponds exactly to our church-clock; but ftill it remains uncertain whether or not Dondi, who was afterwards called Horologius, was the original inventor. About the middle of the fourteenth century feems then to be the time that affords the firft certain evidence of the exiftence of a clock, or regulated horological machine, notwithftanding the frequent mention of horologia in preceding ages, as applied to other horological inftruments; and the following are fome of the earlieft authentic notices that hiftory has recorded on the prefent fubject.

1. In the "Chronica Mifcella Bononienfis" (in Muratori. tom. xviii. p. 444.) it is faid that the firft clock at Bologna was fixed up in the year 1356.

2. About the year 1364, Henry de Wyck or Henri de Vic, a German artift, placed a clock in the tower of Charles V.'s palace. (Vide Moreri. Diction. art. Horloge du Palais.)

3. In Rymer's Fædera is mentioned the protection of Edward III. to three Dutchmen, Orlogiers, who were invited from Delf into England in the year 1368, from which time we may probably date the introduction of clock-work into England.

4. Conradus Dafypodius has given an account of a clock erected at Strafburg about the year 1370.

5. According to Froiffard, Courtray had a clock about the fame period, which was carried away by the duke of Burgundy in the year 1382.

6. Lehmann fays that at Spire was a clock in the year 1395.

7. Nuremberg had a clock in 1462, Auxerre in 1483, and Venice in 1497; and it appears from a letter of Ambrofius Camaldulenfis (lib. xv. Epift. 4.) to Nicolaus of Florence that clocks began to be common in private familes on the continent, about the end of the fifteenth century. (Beckmann.) It is probable therefore, that clocks began to be general in England too about the fame period, for we find in Chaucer, who was born in 1328, and died, as is fuppofed, in 1400, the following diftich: *viz.*

"Full fickerer was his crowing in his loge,
As is a clock, or any Abbey orloge."

The honourable Daines Barrington (Archeologia, vol. v. p. 415.) is indeed difpofed to believe that clocks were in ufe at the beginning of the fourteenth century, and quotes the following paffage, as a proof, from the Italian poet Dante, who was born in 1265 and died in 1321.

"Indi come horologio che ne chiami,
Nel hora che la fpofa d'Idio furge
Amattinar lo fpofo, perche l'ami."

But we have feen that clepfydræ were called horologia, which had alfo fometimes ftriking mechanifm; therefore the paffage before us will apply with equal propriety to that inftrument. The fame author alfo fuppofes that the clockhoufe, near Weftminfter Hall, was furnifhed with a clock in the fixteenth year of Edward I., or in 1288 (vide Selden's Preface to Hengham), out of a fine impofed on Radulphus de Hengham, chief juftice of the king's bench, but there appears to be more of conjecture than proof in the detail. From the teftimonies which have been here adduced refpecting the origin of a clock, the conclufion to be drawn is, that this machine is neither of fo ancient a date as fome writers fuppofe, nor yet among thofe more recent inventions which are placed in the laft two centuries; and that the inventor is not certainly known. Fer. Berthoud, who has written more volumes on the fubject of clock-work than any other man, concludes his refearches with a belief, founded in ftrong probability, that a clock, fuch as that of Henry de Wick, is not the invention of any one man, but an affemblage of fucceffive inventions, each of which is worthy of a feparate contriver: for inftance—1. Wheel-work was known in the time of Archimedes; 2. the weight applied as a maintaining power had at firft a fly, moft likely fimilar to that of a kitchen jack; 3. the ratchet wheel and click, for winding up a heavy weight without detaching the teeth of the great wheel, was found the next indifpenfable contrivance; 4. the regulation of the fly, depending on the ftate of the air, was abandoned, and a balance fubftituted; 5. an efcapement confequently became neceffary, which, in conjunction with the balance, conftituted a more regular check upon the tendency which a falling weight has to accelerate its velocity, than a fly ufed as a regulator could of itfelf be; 6. the application of a dial-plate and hand, to indicate the hours, was the confequence of the regularity introduced into the going part of the mechanifm; and, laftly, the ftriking portion, to proclaim at a diftance, without the aid of a watching-man, the hour that was indicated, completed the lift of inventions. Such a fucceffion of ingenious contrivances, introduced by different men, to improve upon the firft rude inftrument, is perfectly analogous to the fucceffive improvements which the prefent clock has experienced, at different periods, fince Henry de Wick's clock was conftructed, which is the moft ancient one of which we have a particular defcription.

Defcription of the Ancient Clock made by Henry de Wick.

Going Part.—*Fig.* 1, of *Plate* VIII. of *Horology*, reprefents, in profile, the movement of this clock, and *fig.* 2, the front view of the fame inftrument: A is a weight, fufpended by a chord, which is folded round a metallic barrel B, the arbor, *a a*, of which has its ends or pivots, *b*, *b*, paffing through fmall circular holes in the plates C C and D D, which, in this ancient clock, were of iron, the former being bent at right angles at the ends, and fcrewed to the latter, inftead of having pillars; thefe plates, when thus attached, conftitute what is called the frame.

The action of gravity of the weight, A, has a natural tendency to fall towards the earth, and to carry along with it the whole mechanifm, but the frame being fixed firmly on a fupport is fuftained in its place; inftead, therefore, of all the mechanifm being pulled from its fituation, the barrel, B, is pulled round to allow the cord to efcape, and would move with an accelerated velocity, according to the laws of falling bodies, if it had not the wheel, F, with inclined teeth, beft feen in *fig.* 5, pinned to it, which is called the ratchet wheel, and which has its motion ftopped by the fmall pointed piece of metal, *c*, called a click, inferting its pointed end into a fpace between two teeth, and kept to its place by a flender

flender fpring, *d*. This fpring and alfo the click are fcrewed to the wheel G, as may be feen in *figs.* 1, 2, and 5, which is not faftened to the arbor of the barrel, confequently, when the weight pulls downwards, the ratchet wheel pufhes the click, and with it the wheel G, but when the weight is taken off, and the ratchet wheel turned backwards, the wheel, G, keeps its pofition, by reafon of the click fliding eafily along the floping fides of the ratchet wheel, which it is allowed to do by being made to move freely on its fcrew as a ftud ; this contrivance for moving the wheel, G, when the weight falls, and for leaving it at reft when the weight is raifed, is the mechanifm both for winding up the weight, which is called the maintaining power, and alfo for tranfmitting this power to the wheel, G, when wound up : but, as the fufpended weight of this palace clock, of Charles V. of France, was too heavy to be wound up by the ftrength of one man, the wheel Q, *figs.* 1 and 2, was fixed on the fame arbor with the barrel, and the pinion, *n*, was applied to it, to be turned by a handle or key taking hold of its fquare arbor at P, which additional wheel and pinion formed a kind of crane for elevating the weight, and the click held it in any ftate of elevation.

The wheel, G G, having the motion of the barrel communicated to it by means of the click, tranfmitted it to the pinion, *e*, *fig.* 1, and confequently to the wheel, H H, riveted on the fame arbor *f f*; this wheel, H H, again tranfmitted the motion received to a lantern pinion, *g*, and confequently to the wheel, I I, fixed on its arbor, which was called the crown wheel or efcapement wheel (roue de rencontre) : this laft wheel by giving the pallets or fhort levers, *h*, *i*, each a pufh alternately by two teeth, at oppofite fides of its circumference, and moving in oppofite directions, one forward and the other backward, gave a vibratory motion to the vertical arbor, K, moving on two pivots, *l*, *k*; and as the regulator or balance was fixed on this arbor, it was thus made to vibrate backwards and forwards at every pufh of the crown wheel upon the pallets, which alternate motions of the balance were called vibrations or ofcillations, and their frequency was regulated by the fmall weights, *m*, *m*, better feen in *fig.* 2, placed at their correfponding diftances on the arms, L, L, of the balance, by means of little equidiftant notches : if the vibrations were too quick the momentum of the balance was increafed by removing the fmall fufpended weights farther from the centre of motion, and *vice versâ*. As the balance was neceffarily heavy to counterpoife a large maintaining power, a flender cord, M, was fufpended from a fmall cock, T, placed on the large cock, S, that preferved the perpendicular direction of the arbor K, which cord bore the principal weight of the balance, and fuffered it to vibrate without the whole friction which its weight would have occafioned on the lower pivot.—The pallets were placed at about 90° from each other on the arbor or verge of the balance, fo that when one of them was parting with its tooth of the crown wheel, the other was in a fituation to receive the oppofite one immediately ; this action of the crown wheel upon the pallets, which efcaped the teeth alternately, was called, and is ftill denominated, the efcapement, or by abbreviation, 'fcapement of the clock ; it is by means of it that the large weight is prevented from falling with rapidity to the ground, by receiving a check at every impulfe of the wheel of efcapement on one or other of the two pallets ; and the interval between two fucceffive impulfes was regulated by the time of a vibration of the balance, which period we have feen was adjuftable. Thus the whole duration of a vibration was the meafure of time, and the wheels and pinions were employed, firft to tranfmit the maintaining power in order to overcome the obftacles to motion which the balance met with from friction and refiftance of the air ; and fecondly to number the vibrations and indicate them in a vifible form by a hand, O, *fig.* 2, on a dial plate, not given in the figures, as they amounted to hours. The former of thefe offices of the wheel-work we have already defcribed without attending to the numbers of the teeth into which the wheels and pinions were divided ; but to comprehend the nature of the latter office, this confideration muft be taken into the account. The efcapement wheel, I I, has one tooth completely efcaping the pallets at two vibrations of the balance, which we will fuppofe adjufted to be exactly equal to as many feconds of time ; then on this fuppofition, the efcapement wheel of 30 teeth will make one entire revolution in 60, or one minute, as will alfo its pinion, *g*, of 8, by reafon of its being faft on the fame arbor ; but as this pinion acts with the wheel, H H, of 60 teeth, only 8 of thofe teeth will pafs the pinion in every minute, therefore as often as 8 are contained in 60, fo many minutes will this wheel take to revolve in, which are $7\frac{1}{2}$; its pinion, *e*, of 8 alfo revolves in $7\frac{1}{2}$ minutes, on account of being on the fame arbor, which pinion acts with the great wheel G G, fo that eight of its teeth pafs in $7\frac{1}{2}$ minutes ; but its number is 64, or eight times eight, it therefore revolves in eight times $7\frac{1}{2}$ minutes, *i. e.* in one hour exactly : hence as often as the cord is wound round the barrel, fo many hours will fuch a clock continue to go after winding. Now as the wheel, G G, revolves once in every hour, it is evident that the pinion, *d*, on its arbor, which projects through the frame, will turn round alfo in an hour. Its number of teeth is 8 ; therefore $8 \times 12 = 96$ is the number of teeth of the dial wheel, N N, which goes round along with the hand, *o*, placed on its arbor, in 12 hours. Thus while the train, as it is called, of wheels and pinions is tranfmitting the power of the fufpended weight forwards to the balance, it is alfo counting back again the number of vibrations to conftitute an hour, and pointing out on the dial plate, by the addition of a hand, each fucceffive hour, as the vibrations accumulate. Should it be afked here, why a large weight is neceffary to actuate the wheel-work, the anfwer is, that the power of it is gradually diminifhed as it approaches the regulator according to that law of dynamics, by which the force is known to be equal, whether a large weight moves with a fmall velocity, or a fmall weight with a proportionate great velocity, if there were no friction in the train, one fixtieth part of the weight, hung on a fimilar barrel placed on the crown wheel, would have the fame effect on the balance, as the large weight itfelf in its prefent fituation on the hour arbor ; but then it would want winding up fixty times in the fame period in which it requires only once winding up with a large weight falling with diminifhed velocity ; fo that convenience was the object to be attained in fixing upon a large maintaining power.

Clock Part, or *Striking Part*.—The ftriking, or what is properly called the clock part of the antique piece of mechanifm, which we are defcribing, is, perhaps, the moft ancient of any which is tranfmitted to us, and which therefore deferves a particular detail, efpecially as the ordinary Dutch clocks retain very nearly the fame conftruction to this day. We cannot give a better idea of this portion of the inftrument, than by tranflating F. Berthoud's account of it, from his "Hiftoire de la Mefure du Temps," as we have done, with fome verbal deviations, in our defcription of the going part.

In *Plate* VIII. *figs.* 3 and 4 reprefent the wheel-work of the ftriking part, adopted in Henry de Wick's clock : A and B, *fig.* 3, are the principal plates, or rather bars of iron ; C and D, the connecting parts inftead of pillars ; the weight F, fufpended by the cord, that furrounds the cylinder G, is is the maintaining power; and this cylinder has a rachet wheel fixed to it, like that in *fig.* 5, in the going part, and connected with a click faſtened to the wheel, H, *figs.* 3 and 4. for the purpoſe of winding up the weight; for performing which, a handle is put on the ſquare arbor of the pinion *a*, which acts with the remontoir wheel, I, faſtened to the cylinder. Wheel H has eight pins projecting from its plane at *b*, *c*, &c. which pins lift ſucceſſively a hammer that ſtrikes againſt the bell, which parts are not repreſented in the plate, but their principle may be readily conceived from the deſcription, which we ſhall hereafter give, of the ſtriking part of a modern clock. Wheel H drives the pinion *d* fixed on the ſame arbor with the wheel K, which again drives the pinion *e*, on the projecting arbor of which is fixed the fly L, the office of which fly is to regulate the interval between each blow of the hammer, which it does by means of the reſiſtance that its revolving wings, or fanners, meet with from the air.

The manner in which the reſpective number of blows of the hammer for each hour is regulated, is an ingenious invention, and is thus effected; the arbor of the firſt wheel, H, is made to project through the frame, ſo as to take a pinion, *f*, of eight leaves, to drive the wheel M of 78 teeth, which number is equal to 1 + 2 + 3 &c. + 12, or whole number of ſtrokes in 12 hours. On the wheel M is fixed another wheel, N, *figs.* 3 and 4, called the *count* wheel, which has 12 notches on the edge, at unequal diſtances, *viz.* at $\frac{1}{78}$, $\frac{2}{78}$, $\frac{3}{78}$, &c. of the circumference, correſponding to the hours 1, 2, 3, &c. to regulate the number of ſtrokes at each hour. Q, *fig.* 4, is a detent fixed on the arbor R of *fig.* 3, with its claw reſting on the edge of the count-wheel N; to which arbor a lever, T, is alſo fixed, reaching to the pins of the 12 hour wheel N of *fig.* 2, and likewiſe a ſhorter lever, V, within the frame directly above a ſingle tooth, *o*, on the arbor of the fly L. Now as the 12 hour wheel revolves by the going part, one of its pins catching the end of the lever T at every hour depreſſes it, and at the ſame time raiſes the claw or catch of the detent from the notch of the count-wheel, and alſo the ſhort lever V from its tooth *o*, by reaſon of the detent, and two levers, being all faſt to the ſame common arbor R; the weight F in this ſituation, makes the wheels run on till another notch of the count-wheel comes to the claw of the detent Q, when it again falls by the gravity of the load P, placed alſo on the ſame common arbor R, and filling the notch ſtops the count-wheel; at the ſame time, the ſmall lever V falls in the way of the tooth *o* of the fly, and arreſts the motion of the wheels within the frame; ſoon after which the fly alſo comes to reſt, partly by the reſiſtance of the air, and partly by a ſpring preſſing on the end of its arbor, round which it can revolve in a detached ſtate, like the fly of the chime mechaniſm, which we have before deſcribed.—The number of ſtrokes which the hammer makes when raiſed by the pins *b*, *c*, &c. depends on the diſtance between the two notches of the count-wheel, on which the claw of the detent reſts at the time. The ſame proceſs is repeated every time that one of the pins of the 12 hour wheel, depreſſes the end of the lever T, and detaches thereby the claw of the detent and tooth *o* of the fly arbor, ſo as to permit the weight F to actuate the wheel-work for the limited time of ſtriking.

We are not informed what are the numbers of teeth in the wheel-work of the ſtriking part contained within the frame, nor is it of much importance, as this movement has no other uſe, than that of regulating the reſpective velocities of the fly and pin wheel H, that raiſes the hammer; but when the wheel M has 78 teeth, as we have ſtated, it is neceſſary that the pinion *f*, that drives it, ſhould have juſt as many leaves as there are pins on the wheel H; otherwiſe one tooth of the wheel M would not correſpond to one ſtroke of the hammer againſt the bell, which is a neceſſary condition; the pinion *f* therefore has eight leaves, correſponding to the eight lifting pins of the wheel H, and 78 ſtrokes are given in a progreſſion increaſing by unity, during each twelve hours, after equal hourly intervals of ſilence.

Hiſtory of the ſucceſſive Improvements in Clocks.

The preceding account of Henry de Wick's rude clock muſt have prepared the reader for a deſcription of the ſucceſſive alterations and improvements which ingenious mechanicians and artiſts have deviſed in the clock during the laſt two centuries; for it muſt have occurred to him in the peruſal, that large iron wheels continually expoſed to the oxiding influence of the atmoſpheric air, in which unequal and ill ſhapen teeth were cut with the inaccuracy of a manual operation, were by no means calculated to tranſmit the maintaining power with perfect regularity to the balance, ſuppoſing it to have been a good regulator; but when it is further conſidered, that the alternate direct puſhes of the balance-wheel againſt the pallets muſt have produced jerks, and deſtroyed or greatly diſturbed the regularity of this moſt eſſential part of the mechaniſm, great accuracy was not to be expected in the indication of time; ſo that, as we ſee, even minutes were deemed too ſmall portions of time to be indicated by ſuch a machine. We find, notwithſtanding, that ſo early as the year 1484, Walther made uſe of a balance-clock for heavenly obſervations, as did the landgrave of Heſſe after him; and ſuch ſeems to have been the utility of the clock, thus early, for aſtronomical purpoſes, that Gemma Friſius propoſed a portable one to be uſed at ſea for aſcertaining the longitude by ſo ſoon as the year 1530. About the year 1560 Tycho Brahé was in poſſeſſion of four clocks, which indicated hours, minutes, and ſeconds, the largeſt of which had only three wheels, one of which was three feet in diameter, and had 1200 teeth in it, a proof that clock-work was then in a very imperfect ſtate. Tycho however obſerved, that there was an irregularity in the going of his clocks, which depended upon the changes in the atmoſphere; but he does not appear to have known how ſuch effect was produced. In the year 1577 Moeſtlin had a clock ſo conſtructed as to make juſt 2528 beats in an hour, 146 of which were counted during the ſun's paſſage over a meridian, or azimuth line, and determined his diameter to be 34′ 13″, ſo that, the ſcience of aſtronomy began thus early to be promoted by the aſſiſtance of clock-work; and as clocks firſt promoted the ſtudy of aſtronomy, it will be ſeen by and bye that aſtronomy in its turn gave riſe to ſome of the moſt eſſential improvements in clock-work, and that, as the arts and ſciences were more and more cultivated, improvements in clock-work kept pace with them, and employed the talents of the moſt ingenious men of each ſucceeding age.

One of the firſt additions to the mechaniſm we have before deſcribed was what we call an alarum or larum, ſtill uſed in the Dutch wooden clocks, which contrivance took its origin from the circumſtance of prayers being uſed at ſtated periods in monaſteries by night as well as by day; the fervors of devotion were not found always unfettered by ſleep at the hour proclaimed by the bell, an invention conſequently became neceſſary to rouſe the ſleepy prieſt to his duty by a continued ringing in his ſtunned ears: for a deſcription of which mechaniſm we muſt refer to the article *Thirty-hours* CLOCK *with a Larum*.

It is not quite certain at what time exactly the bulky ſize of the ancient clock was reduced to a ſtate of portability, which muſt have conſtituted a real improvement, as the main ſpring muſt have been invented previouſly; the ſubſtitution of of which for a large heavy body, as a firſt mover, conſtituted a ſecond æra in horology, from which we may date the origin of the fuſee, or mechaniſm for equalizing the variable power of a coiled ſpring; and from which our modern watches and chronometers of great value derive both their form and principle.

F. Berthoud (Hiſtoire de la Meſure du Temps, i. 79.) ſuppoſes that a portable clock muſt have been invented ſome time before the year 1544, which was that in which the Corporation of Maſter Clock-makers at Paris had a ſtatute enacted in their favour by Francis I. to this purpoſe, *viz.* "No one, of whatever ſtation, if he be not admitted a maſter, ſhall make, or cauſe to be made, clocks, alarums, watches, large or ſmall, or any other machine for meaſuring time, within the ſaid town, city, and precinct of Paris, on pain of forfeiture of the ſaid works, and of arbitrary penalty, &c." This ſtatute, however, inſtead of proving the origin of portable clocks, only proves that in the year 1544 they had begun to be commonly made in France, and probably had been introduced from Germany, which country, no doubt, was the nurſe as well as mother of horology. Indeed we have lately been favoured with the ſight and examination of a portable clock, at preſent in the poſſeſſion of Mr. Peckitt of No. 50, Old Compton ſtreet, which, from an inſcription engraven in the Bohemian language, appears to be much older than the ſtatute above quoted: it was made in the year 1525 by Jacob Lech of Prague, and differs in its conſtruction from De Wick's clock, in that it has a ſpiral ſpring, with a fuſee of ſoft metal, and a ſcrew inſtead of notches at the ends of the double levers of the balance, with tapped weights of lead for the adjuſtment to time, with the addition of ſome wheel-work, to ſhew the motions of the ſun and moon, in an engraven ecliptic, and alſo a contrivance to ſtrike one at every hour. The wheels are of iron, and retain certain punched marks of diviſion which prove that they have been cut with a file by hand; and the levers being faſt to the arbor or verge of the pallets, will vibrate either in a horizontal or vertical poſition. A catgut was originally the band of the fuſee; but the introduction of a modern metallic chain unfortunately has deſtroyed nearly three out of the eight ſpiral threads at the ſmaller end, ſo that inſtead of going 48 hours, *viz.* ($8^{thr} \times 6^{h}$), the remaining five threads will allow the piece to go only 30 hours, or $5^{thr} \times 6^{h}$, which it now does, though irregularly, with vibrations of nearly one ſecond each by eſtimation. Beckmann, in his "Hiſtory of Inventions," mentions this clock as having been the property of Mr. Ferguſon, at whoſe ſale the preſent proprietor purchaſed it in the year 1777, and he intends, we underſtand, to bequeath it to the Britiſh Muſeum, where, no doubt, it will be conſidered as a great curioſity. On examination of the wheel-work, we found the moon's train to be $\frac{12}{6} \times \frac{82}{6} = \frac{984}{36}$ of a day $= 27^{d}\ 8^{h}$ for a periodic revolution, and the ſun's $\frac{984}{36} \times \frac{81}{6} = \frac{79704}{216} = 368$ days, the inaccuracies of which trains prove that planetary mechaniſm in clocks had at that period made no great progreſs towards perfection.

But the portable clock of which we have here given a ſhort notice, we have reaſon to believe, was not among the firſt that were made, for at firſt the ſpiral ſpring was folded in a box, the arbor of which had the great wheel on it, and the irregularity of its action was in ſome meaſure equalized by a ſecond ſpring, which, being diſpoſed in a certain curve, oppoſed the principal ſpring when wound up, and acted in the ſame direction with it when its intenſity began to remit; this piece of mechaniſm, which was a German invention, preceded the invention of the fuſee, and was called *ſtack-freed.* Berthoud has given a drawing and deſcription of a portable clock, probably by Gourdain, without a fuſee; and ſome of the modern French watch-makers have probably borrowed from it their idea of making a watch go well without a fuſee. For the ſhape and uſe of which equalizer of irregular power, ſee the article FUSEE.

When the ſize and weight of a clock were reduced, the quantity of friction was ſo diminiſhed, as to allow the thread of ſuſpenſion, on the verge of the balance, to be abandoned, which circumſtance gave riſe to a new poſition of the balance, the verge of which was now placed horizontally upon its pivots, which were a little flattened, ſo that the weight might be ſupported by its edge; and this contrivance was called the *knife edge* ſuſpenſion, a ſpecies which F. Berthoud has pronounced ſuperior to the ſuſpenſion with a ſlender ſpring, to which the Engliſh artiſts are much more partial, though it does not appear that their partiality is founded on any thing like impartial experiment.

Such was the ſtate of clock-work when Galileo, the celebrated philoſopher and mathematician to the duke of Florence, obſerved, that two lamps, or other heavy bodies, ſuſpended by ſtrings of the ſame length, made their vibrations in long or ſhort arcs, very nearly, if not exactly, in the ſame ſpace of time; this iſochronal property he publiſhed at Paris in a treatiſe, called "L'Uſage du Cadron ou de l'Horloge phyſique univerſelle," in the year 1639, which explained, in 15 chapters, its uſe in philoſophy, aſtronomy, muſic, phyſic, &c. and though he never applied the pendulum as a regulator to ſuperſede the balance in clocks, yet we may date from his diſcovery a third æra in clock-work, namely, the origin of the pendulum clock, which continues in uſe in our day; the idea of the iſochroniſm of a detached pendulum at leaſt was his, and the inveſtigation of the law, by which bodies fall in free ſpace, was applied by him to determine the lengths of two pendulums, that ſhall vibrate in times that are in a given ratio to each other.

It has been a ſubject of great contention, who had the honour to be the firſt artiſt, or mechanician, who applied a pendulum to clock-work; without pretending to decide the conteſt, which, at this diſtance of time, can be determined only by hiſtorical evidence, we will briefly lay before the public the pretenſions of different men to this invention, and leave the reader to judge, from ſuch facts as we have been able to collect on the ſubject. Bernard, one of the profeſſors of aſtronomy at Oxford in the laſt century, has aſſerted, that the Arabians, beſides having clepſydræ and ſun-dials, made uſe of pendulums in aſtronomy long before this period, as we know Riccioli, Tycho Brahé, Langrenus, Vendelin, Merſenne, Kircher, Hevelius, Mouton, and Galileo himſelf, did in a detached ſtate; but we do not find that any of them uſed it in conjunction with wheel-work. According to Becker (Bailly's Hiſt. of Modern Aſtron.), Juſte Birge, a native of Switzerland, in the year 1552, and who was Rothman's ſucceſſor at the obſervatory at Caſſel, from 1590 to 1597, was the firſt who applied a pendulum to a clock, which application, however, he never publiſhed to the world, that we can learn, and which, therefore, if the fact be true, never benefited the world.

According to profeſſor Venturi (Eſſai ſur les Ouvrages phyſico-mathematiques de Leonard de Vinci), Sanctorius applied a pendulum to clock-work ſome time before the year 1625, in which he publiſhed his "Commentarii in Avicennam," and deſcribed ſeveral inſtruments which he had explained to his auditors 13 years before, in his lectures read at Padua.

Vincentio Galilei, ſon of the famous Galileo, is alſo ſaid

(Exper. de l'Acad. del Cimento) to have made a pendulum clock, fuggefted by his father's difcovery, fo early as the year 1649, at Venice; but Chriftian Huygens. the juftly celebrated mathematician and philofopher of Zuylichem, contefted the honour of priority of the application with him, which conteft gave rife to that excellent treatife on clock-work, "De Horologio Ofcillatorio," which laid the foundation of moft of the fubfequent improvements in horometrical machines, and in which it appears indubitably, that he made, or directed the making of, a pendulum clock before the year 1658. From comparing what has been adduced by thefe two contending mechanicians, an impartial reafoner is led to conclude, that Vincentio Galilei may have applied, in a rough mechanical way, not made generally known, fuch a pendulum to clock-work as proved no mean fubftitute for the balance; but that Huygens applied it in a more fcientific and mafterly manner, and probably without any knowledge of what his opponent had previoufly done; hence the honour of the invention, as it has been called, has been generally attributed to Huygens.

But whilft we have ftudioufly avoided entering into a minute difcuffion of all the proofs that have been adduced in favour of each of thefe foreigners, to fubftantiate their refpective claims to originality of project, with refpect to the pendulum, we fhould act in oppofition to our feelings as Englifhmen, as well as to our profeffions of impartiality, if we did not here avail ourfelves of fome documents that have recently fallen into our hands, which bear the marks of authenticity, and from which it appears, that neither Vincentio Galilei, nor yet Huygens, was the firft who adapted a pendulum to a clock, but that an artift in London, named Richard Harris, invented and made a pendulum clock about eight years before either of them dated their claims; an engraven plate, bearing date "Great Ruffel Street, Dec. 21, 1798," is at this time hanging in the veftry room of St. Paul's church, Covent Garden, of which the annexed is a verbal copy, *viz.*

"The (new) turret clock, and bells of this church, were made A.D. 1797, by Thomas Grignon, of Great Ruffel Street, Covent Garden, the fon and fucceffor of Thomas Grignon, who, A.D. 1740, brought to perfection what the celebrated Tompion and Graham never effected, *viz.* the horizontal principle in watches, and the dead-beat in clocks, which dead-beat is a part of the mechanifm of the turret clock. Thomas Grignon, fenior, made the time-piece in the pediment at the eaft end of this parifh church, deftroyed by fire A.D. 1795. The clock fixed in the turret of the faid (late) church, was the firft long pendulum clock in Europe, invented and made by Richard Harris of London, A.D. 1641; although the honour of the invention was affumed by Vincenzio Galilei, A.D. 1649, and alfo by Huygens in 1657. This plate is here affixed by Thomas Grignon of this parifh, the fon of the above Thomas Grignon, as a true memorial of praife to thofe two fkilful mechanicians, his father and Richard Harris, who, to the honour of England, embodied their ideas in fubftantial forms that are moft ufeful to mankind."

In order that the reader may judge for himfelf, what degree of credit is to be attached to the preceding memorial, the writer of the prefent article pledges his honour to the public, that the following extract, which, on inquiry, is Mr. Grignon's authority for attributing the invention of the pendulum to Richard Harris, is copied verbatim from a manufcript marginal note, written in page 12 of an old book, at prefent in his poffeffion, called "Effayes of Natural Experiments made in the Academie del Cimento, under the protection of the moft ferene prince Leopol, of Tufcany, written in Italian by the fecretary of that academy, and tranflated by Richard Waller, F.R.S. London, 1684."

(Copy of the Extract.)

"The great clock belonging to Covent-Garden, has a long pendulum, and was made by Richard Harris of London, in the year 1641, which was eight years before Vincenzio Galilei put his father's obfervations into practice, as appears by the date 1649.

"The ingenious Mr. Huygens applied the pendulum to a clock in the year 1657, and attributed the invention to himfelf, which created a difpute between him and Vincent Galileo; this laft affirming that he had put it in practice in 1649; and the reafon of Richard Harris's not appearing, (which would have decided the controverfy,) in all probability was, that he being only a private workman was entirely unacquainted with any difpute which might happen between Vincent Galileo and Mr. Huygens; or he might be dead before the difpute arofe, it being fixteen years after he made the faid church clock."

The manufcript of which this is an exact copy bears every mark of the antiquity afcribed to it by the prefent Mr. Grignon, who is reputed to be a man of veracity, and who moft pofitively afferts it to be his father's hand-writing, as indeed it appears to be, when compared with fome other of his papers which we have feen. It may not be foreign to the fubject to add here, that the late Mr. Grignon, who died on April 4, 1784, aged 71, was a good mathematician as well as an excellent workman, and was patronized by Archibald, duke of Argyle. He was an intimate friend of James Fergufon and other fcientific men of his time, and one of the firft members of the Society of Arts in London, to which fociety he prefented a regulator in the year 1759, which is yet in one of the rooms at the Adelphi, and which has the improved dead-beat efcapement, and very high numbers in the wheel-work to avoid friction, which was another idea of his own. The peculiar properties of this clock, however, feem not to have been noticed by, or even perhaps known to any of the prefent members of that numerous fociety, fo as to become an object of particular attention. Mr. Grignon, notwithftanding, affures us, that it will keep the fame time whether its maintaining power be four or twelve pounds, which property he afcribes principally to his father's improvement of Graham's dead beat efcapement.

From this account of Harris's clock, we confefs ourfelves difpofed to contradict the generally received opinion, afferted by other authors, that Fromantil, the Dutchman, was the firft who made pendulum-clocks in England fo late as the year 1662; one of which was given by bifhop Ward to Grefham College, Oxford. Indeed Dr. Hooke, as great a genius as any we have mentioned, made a pendulum-clock, we find, in 1658, for Dr. Wilkins, afterwards bifhop of Chefter, which was prior to Fromantil's clocks, and fome authors have been difpofed to make him the inventor of the pendulum as a regulator.

The pendulum being once applied to clock-movements, was found to be a regulator fo much fuperior to the old balance, that Gemma Frifius's idea of making a marine or nautical time-piece was attempted to be realized by the ingenious Huygens, of which attempt we have already fpoken under the word CHRONOMETER, where we entered into a detail of the merits and conftruction of various pieces of mechanifm ufed for the determination of a fhip's longitude. It may not be improper, however, to fay further of Huygens' marine-clock, that its pendulum vibrated more flowly as it approached

proached the equator, proving, as Picard has fince afferted from experiments made in 1669, the fhape of the earth to be that of an oblate fpheroid. But Huygens, whofe induftry was equalled only by his ingenuity, foon difcovered that the ifochronal property afcribed to the pendulum by Galileo, was only true in circular arcs, when the length of the arc of vibration remained the fame; for that long circular arcs required fomewhat more time for a vibration than fhort ones, fo that variations in the maintaining power produced variations in the time indicated, by altering the length of the arcs of vibration. This difcovery prefented a ferious difficulty, and gave rife to one of the moft ingenious contrivances ever introduced in the mechanifm of a clock, though practical experience foon proved it to be of little or no ufe; we mean the contrivance of two cycloidal cheeks at the point of fufpenfion confidered as evolutes of a cycloid, round which the thread by which the pendulum was fufpended was bent, and occafioned, as we have before noticed under the word CHRONOMETER, the ball to move in the involute of the faid curve, which Huygens firft demonftrated to be itfelf a cycloid, poffeffing theoretically this peculiar property, amongft others, that a heavy body defcending along it from any given point therein, will defcend to the loweft part in the fame time that any other heavy body will fall from any other higher or lower point thereof; whence it was concluded, that a pendulum, vibrating in fuch a curve, would regulate a clock in the beft poffible manner with any maintaining power whatever. The idea was truly ingenious, and the mechanifm equally fo, as we have faid, in theory; nay, the pendulum really moved in a cycloid by this contrivance, but then the alternate pufhes of the balance-wheel againft the pallets were communicated by means of an intermediate lever, called the fork, (which was another invention of our author) to the pendulum, which additional force fo greatly difturbed the natural power of gravity, that the pendulum attached to clock-work was no longer poffeffed of the ifochronal property which it poffeffes when moving in a cycloid in a detached ftate, with all its weight collected in one point, as the theory fuppofes; befides, the moifture of the atmofphere affected the pendulum's ftring of fufpenfion, and what had not been obferved fufficiently at that time, the metallic rod itfelf elongated with heat, fo that the cycloidal cheeks were at length found to be of no benefit as an auxiliary to the natural pendulum, and have therefore been abandoned.

After the length of a fecond's pendulum was afcertained by Huygens to be 3 feet 8½ lines, old French meafure, and after he had propofed it as an univerfal ftandard of meafure, we find that aftronomers began to afcertain the right afcenfion of the ftars by it, and alfo the equation of time, that Hipparchus had fpoken of long before, which application made it neceffary that there fhould be fome contrivance for winding up the maintaining power of the clock during the time that it continued to go, otherwife the quantity of the earth's rotation, that might take place during the act of winding up, muft have been allowed for by conjecture; hence the endlefs chain with a detached ratchet, was invented for this purpofe by Huygens, which is ftill in ufe, we believe, in fome few aftronomical clocks. The fame author was probably the contriver of our prefent dial-work, for changing the hour into 60 minutes by the addition of another hand at the centre of the clock-face; and a circular pendulum was likewife invented and applied, as he fays, with fuccefs to fome of his clocks, the principle of which was, that as the ball of the pendulum revolved in an horizontal circle, when fufpended by a vertical axis, the centrifugal force and power of gravity fo counteracted each other, as to produce an equable motion, while the arbor revolved without an efcapement. We are not informed why this pendulum has not been imitated.

The next invention worthy of notice in clock-work, was the mechanifm of *repetition*, by means of which the clock is made to obey the pull of a perfon in bed, who, from any caufe, may wifh to know the hour which was laft ftruck; this ingenious contrivance, in a great meafure, has fuperfeded the ufe of the ftriking-work with the count-wheel, and was the invention of Barlow, a London clock-maker, about the latter end of the reign of Charles II. *viz.* in the year 1676.

The noife which this curious piece of mechanifm made in the world, fet feveral ingenious artifts to making repeating apparatus, particularly Quare, in London, and Julien Le Roy, Collier, Lançay, and Thiout, on the Continent (Machines approuvées, tomes v. and vi.). Nearly about the fame period, we find that the comparifon of the earth's rotation with the regular motion of a pendulum clock, produced a defire to indicate not only *mean* but *equated* time on the dial-plate, by various contrivances, which, by the bye, were by no means calculated to improve the regularity of the going part, but only added to the complexity of the machine, and introduced unneceffary and variable friction. We know not certainly who was the firft to execute the equation work, but, according to Sully, an Englifh clock-maker, who fettled at Paris (Regle artificielle du Temps, ed. 1717), the firft equation clock which is recorded, was fent from London to Charles II. king of Spain, before 1699. This contrivance, which was more curious than ufeful, caufed a great number of attempts to produce the fame effect; father Alexander, a Benedictine, prefented a project of this kind to the Academy of Sciences at Paris, in 1698: Le Bon, a Parifian clock maker, prefented another, which was much admired, to the fame fociety, in 1717; which was alfo done by Julien Le Roy, in the fame year, at Paris; and to thefe names we may add thofe of Thiout, a curate of St. Cyr, Duchefne, Kriegfeiffen, Enderlin, L'Admiraud, Paffemant, Rivaz, Berthoud, and others, who have beftowed much time on a fubject which requires only the infpection of an equation table, and which, therefore, our Englifh artifts now think an expenfive trifle, fcarcely worth their notice, any further than as a matter of curiofity.

The equation mechanifm naturally led to the practice of making clocks to go a month, three months, and even a year, with one winding up; likewife to the introduction of mechanifm to fhow the fun and moon's rifing, culminating, and fetting, and planetary motions in general. Indeed Oronce Finée, mathematician to the French kings, Francis I. and Henry II., had made a planetary clock fo early as 1553, concerning which, and the theory of the planetary motions, he publifhed a Treatife, at Paris, in 1557; and the automaton of Huygens, made in 1703, was properly a planetary clock, and one of a very ingenious conftruction.

We come next to an important improvement, which, like moft other real improvements in the art of meafuring time accurately, belongs to the Englifh, we mean the introduction of the anchor pallets, which, even Berthoud confeffes, was a contrivance of Clement, a London clock-maker, in the year 1680 (fee Smith's Horological Difquifitions, London, 1698); the advantage of this efcapement, with an horizontal arbor for the fwing-wheel, as it is called, over that of the crown-wheel efcapement, is, that it will admit the efcape to take place with a fmall angle of vibration, fo as to prevent the maintaining power from acting on the pallets a long time by a direct pufh; as was the cafe with the crown-wheel efcapement; befides, a fhort arc of vibration in a circle is ſo like a cycloid at the loweſt point, that a ſhort arc, with a heavy ball on the pendulum, ſoon began to be adopted generally, to the excluſion of the cycloidal cheeks, which, had they poſſeſſed their theoretic property in practice, would now have been of little value with a vibration in ſhort arcs, even if the ſuſpenſion thread had been ſtrong enough to bear a heavy weight, and at the ſame time ſufficiently flexible. This change in the clock eſcapement introduced the cuſtom of ſuſpending the pendulum from a cock, by means of a piece of watch ſpring, which was another important invention of Clement, according to Smith's authority. It may be proper to obſerve here, that the truly ingenious Dr. Hooke too claimed the reputation of being the inventor of the anchor eſcapement, and affirmed, ſays Sully, that he had ſhown to the Royal Society a pendulum with a ſimilar eſcapement, ſoon after the fire of London, in the year 1666, as we have ſeen, when we treated of *Chronometers*, that he alſo conteſted with Huygens and Hantefeuille the honour of having been the inventor of the balance-ſpring of watches.

But whether Clement or Hooke were the inventor of the anchor eſcapement, it was ſoon found that a clock of this conſtruction gained time conſiderably by an addition to the maintaining power, or, which may be conſidered as the ſame thing, by diminiſhing the weight of the pendulum ball, the cauſe of which is the recoil, or retrograde motion, occaſioned in the ſwing wheel, which, oppoſing the natural vibration, ſhortened the arc in the aſcent; the ſeconds pendulum, with this eſcapement, was called for the firſt time, the royal pendulum.

Such was the ſtate of clock mechaniſm, and ſuch were the alterations and improvements therein, at the concluſion of the ſeventeenth century, or beginning of the eighteenth, which period conſtitutes a fourth epoch in the hiſtory of clock-making. The expanſion of metals by heat had been known ever ſince the year 1648, but the ingenuity of art had not yet deviſed a remedy for the alternate elongation and contraction of the pendulum rod in ſummer and winter reſpectively, though the uſe of the clock in aſtronomy imperiouſly demanded ſome compenſation: this honour fell to the lot of our celebrated George Graham, who, in the year 1715, ſucceeded in his attempt to preſerve the diſtance from the point of ſuſpenſion to the centre of oſcillation of a pendulum unaltered, notwithſtanding it was expoſed to all the variations of temperature incident to our climate; the compenſation was produced, as will be more particularly deſcribed under the word Pendulum, by means of mercury encloſed in a cylindrical glaſs veſſel, guarded by a frame, and ſubſtituted for the ball of a metallic pendulum rod, ſo that while the rod of metal, conſtituting the verge of the pendulum, lengthened downwards, the column of mercury lengthened upwards, which was an original and truly ingenious idea. Pendulums of this conſtruction, when well adjuſted, have been found to meaſure time with a degree of accuracy far beyond what the former pendulum had any pretenſions to, for a continued length of time: the principal objection to its general adoption was its liability to break in carriage, or by other accident:—but Graham, feeling the force of this objection, took a wide view of the ſubject, and, looking round for other reſources, ſuggeſted the idea of uſing the oppoſite expanſions of different metals as a means of compenſation in a pendulum, which idea was immediately adopted by Harriſon, at that time an obſcure carpenter in Lincolnſhire, at a village called Barton; who, overcoming all the difficulties attending his retired ſituation, which is the property of a great genius, aſtoniſhed the world by the production of the grid-iron pendulum; a pendulum which is not liable to the objection of want of portability, and which has till lately been generally adopted in aſtronomical clocks of the beſt conſtruction. For its deſcription we muſt again refer to the article Pendulum. Still, however, a better eſcapement than the common anchor eſcapement was wanting; for it was not ſufficient to have obtained an invariable pendulum in point of effective length, while its natural iſochronal property, ariſing from gravity alone, continued to be greatly diſturbed; *viz.* while alternate additions and ſubtractions of force were derived, and that often unequally, through the medium of the train of wheel-work to the eſcapement, from the maintaining power firſt, and thence improperly modified to the pendulum itſelf. Accordingly, we find Graham, about this time, becoming the inventor of the dead-beat eſcapement, and alſo Harriſon of a ſilent eſcapement, which will go without oil; both of which are ſpecimens of great originality of contrivance, as will be ſeen when we come to deſcribe them in their proper place: the former of them has proved itſelf of ſuch great utility, that it has been adopted in the generallity of regulators or clocks for aſtronomical obſervations, in conjunction with Harriſon's grid-iron pendulum, moving with a heavy ball in a ſmall arc of vibration: of this kind is the clock at the Royal Obſervatory, at Greenwich, of the accuracy of which it has been affirmed, that it ſeldom gains or loſes, on an average, more than one ſecond of time in five days. To avoid the wearing out of the parts moſt in action, and the influence of friction, the beſt clocks of this conſtruction, like the one we have juſt mentioned, have pallet wheels of hardened and poliſhed ſteel with pallets of ruby or agate, which require little or no oil: this kind of ſubſtance, we believe, was firſt uſed in time-pieces by a Frenchman, De Baufre, about the year 1704. The pivot holes alſo are ſometimes buſhed with jewels, to avoid the production of verdigris, and the clamminneſs of thickening oil; but the conſequent additional expence ſeems to be hardly compenſated by ſuch a refinement as the laſt.

Harriſon was alſo, as we have ſeen before, the inventor of the auxiliary ſpring and additional ratchet on the barrel arbor, which is a much neater way of making the clock go during winding, than that of Huygens by means of an endleſs chain, or of the forcing elaſtic bolt, which has been made to ſlip into a tooth of one of the wheels, and to puſh it forwards during the act of winding and for ſome minutes after.

We might now ſuppoſe that the clock has arrived at its *ne plus ultra*, with reſpect to further improvements, but ſtill we find ſo many ſucceſſive alterations, if not improvements, in the eſcapement, and mode of compenſating the pendulum's expanſion, that to notice all the fanciful minutiæ that have been magnified into importance by different modern artiſts, would be to write a long volume on the ſubject. We will, however, out of the great variety of horological contrivances in this century, ſelect thoſe which have originality and utility to recommend them to public notice, and paſs in ſilence, for the preſent, over the inventions of chimes, organs, cuckows, planetary motions, atmoſpherical winding-up, mercurial mover, &c. &c. as objects of fancy more than of real uſe in clock-work. (For the two laſt ſee "Machines Approveés.") Among the contrivers of detached and other clock *eſcapements*, as will hereafter be ſeen under this word, we may place Grignon, Mudge, Cummins, Nicholſon, &c. in London; and on the Continent Julien le Roy, Peter le Roy, Sully, Du Tertre, De Bethune, Le Paute, Amant, Robin, Berthoud, &c all of whom have merited commendation for ingenious contrivances in their conſtruction of this moſt ſcientific part of the clock. The principal inventors of compenſation pendulums, (which will be deſcribed under the

the word PENDULUM,) after Graham and Harrison, have been, in France, Regnauld, Deparcieux, Julien le Roy, Cassini, and Berthoud; and in England, Ellicot, Cummins, Nicholson, and Troughton, the last of whom has very recently made such a disposition of the grid-iron pendulum, by means of concentric tubes of brass containing within them the rods of iron, as gives it the appearance of an ordinary simple pendulum rod; and his accurate mode of adjusting for temperature by a delicate spirit-level pyrometer, lately invented, bids fair for introducing this elegant pendulum into general use in clocks which profess to be accurate; indeed, we know, that many pendulums of this construction have already been made for observatories both public and private.

We might mention here various curious as well as useful engines, tools, pyrometrical, and other instruments, which owe their origin to the successive improvements in clock-work, but we think the subject of sufficient importance to demand a separate account of these. We must not, however, pass in silence over some other real improvements connected with the pendulum, one of which is, fixing the ball at the centre of oscillation, instead of supporting it by a tapped nut at the lower extremity, the latter of which methods is only useful when a simple substance, such as metal, wood, glass, &c. is used for a rod; for in these cases a species of compensation is effected by the upward expansion of the ball; in some of which cases a pin, passing through a point of the ball between the centre and lower extremity, will be still better. In those constructions where the ball itself is not easily adjustable, a second light ball is made to screw up and down a projecting piece of the rod below the heavy ball, to adjust for time, and sometimes a micrometer-screw is adapted for measuring the quantity of the adjustment; this is an idea, no doubt, borrowed from Huygens, whose clock had a second adjustable weight on the body of the rod to answer the same purpose, which circumstance we omitted to mention before. The improved suspensions, however, have rendered these secondary balls in many clocks superfluous, which requiring the clock to be stopped for their adjustment were inconvenient; in these suspensions the adjustment for rate is made by a micrometer-screw at the cock or top of the pendulum, even while going, to which mechanism also is added a lateral adjustment for putting the pendulum into beat, instead of bending the fork as is practised in ordinary clocks. Lastly, the modern practice of fixing the cock to the solid wall or other steady situation, and allowing the pendulum to find its own perpendicular line before it be fixed in that situation, is among the real improvements of this art, and cannot be too much recommended. But, notwithstanding all these, and doubtless other improvements in the art of clock-making, one desideratum yet remains to be discovered; *viz.* a simple substance that is *not expansible* by heat, for the rod of the pendulum; we think that pyrometrical experiments, on different substances of nature and combinations of art, have not yet been sufficiently extended, and we beg leave to suggest an opinion, for which we are indebted to Mr. Troughton, that tobacco-pipe-clay, or the composition of Wedgewood's thermometer, if properly baked, may prove on trial to afford a very simple rod, at least for a half-seconds pendulum; particularly if a metallic cap, cemented or otherwise fastened to its upper extremity, should be furnished with a knife-edge suspension, which the celebrated Berthoud, as we have said, affirms has less friction than the slip of watch spring; but if the latter be preferred, we presume the compensation for it, and also for the argillaceous rod, if it should require any, may be made by pinning the ball a little below the centre to the lower extremity of the rod, and then the adjustment for time may be as usual at the cock, with lateral screws to close the slit when the adjustment is complete. From the experiments which we have made with wooden pendulums, we find that they are more affected by moisture than by heat, and are therefore not to be depended on.

Turret Clock of the Royal Palace at Hampton Court.

After having given a history of the principal improvements made from time to time in clock-work, we propose to describe, individually, so many clocks of different constructions, in successive order, as will enable the reader to form a competent judgment of their properties and relative merits, which office we undertake with the more pleasure when we reflect that the English language is yet without a book descriptive of the various constructions of a machine, which, in one form or other, is now in the possession of almost every house keeper.

According to Dr. Derham, the oldest English clock extant is in a turret of the royal palace at Hampton, constructed in the year 1540, which time was in the reign of king Henry VIII. by a maker whose initials are N. O. When we consider that this clock contains mechanism for representing the motions of some of the heavenly bodies, and that the celebrated Copernicus was living at the time of its date, and had not published his book "On the Revolutions of the Celestial Orbs;" when we reflect also, that more than a century elapsed after this time before the pendulum was applied as the regulator of clocks, these considerations appear sufficiently interesting, to induce a minute examination of the wheel-work of this ancient clock, particularly of that part of it which constitutes its celestial mechanism.

Fig. 1, of *Plate* X. is the calliper of the wheel-work which produces the celestial motions, taken from Dr. Derham's "Artificial Clock-maker," published in 1714 (third edition), whose account we will first copy in his own words, and then make some observations on the value of the wheel-work. The Hampton Court clock, says the Doctor, "shews the time of the day, and the motion of the sun and moon, through all the degrees of the zodiac, together with the matters depending thereon, as the day of the month, the sun's and moon's places in the zodiac, moon's southing," &c.

"To shew how completely (for that age) the wheel-work is laid under the moving part of the dial-plate, I have given the calliper thereof, which represents the several wheels and pinions only which lie under the dial-plate, and drive the several motions in this manner. In the centre of all, both the dial-plate and its wheel-work are placed on a fixed arbor, which hath a pinion on the end of it, which drives both the solar and lunar motions, by means of a large wheel of 288 teeth turning round upon it once in 24 hours, which large wheel is drawn round by a pinion of 12, fixed on the arbor of the great wheel within the clock, which turneth round once in an hour. The wheel 288 thus turning round in 24 hours, carries about with it the wheel 37, and its pinion of 7 leaves, as also the other prickt (dotted) wheel, and its pinion, on the other side. The pinion 7, of the wheel 37, drives another wheel of 45 teeth, which carries round the moon's ring or circle. On the opposite side the aforesaid pinion 8 drives round the prickt wheel, whose pinion drives a wheel of 29 teeth, whose pinion, of 12 leaves, drives round the wheel 132 that carries the sun and the zodiacal matters."

"These were the numbers of the wheel-work (continues the Doctor) remaining in the year 1711; but the prickt wheel and pinion were taken out formerly by some ignorant workman that was not able otherwise to amend the clock; but

but were ſupplied, and the whole movement repaired lately, by that ſkilful artiſt, Mr. Lang. Bradley, in Fenchurch Street, London." (p. 121, 122.)

This deſcription gives a clear idea how the movements were actuated; but the numbers of the dotted wheel and pinion being unknown, leave the ſolar movement incomplete, thus $\frac{*}{8} \times \frac{29}{*} \times \frac{132}{12}$, ſo that the exact value of this original combination of wheels and pinions for the annual motion unfortunately cannot be with certainty known; though there is no difficulty in aſcertaining a wheel and pinion to be ſubſtituted, to complete the wheel-work for either a ſolar or a civil year.

The wheel-work for a lunation, however, is entire, namely, $\frac{37}{8} \times \frac{45}{7}$ or $\frac{1665}{56}$ of 24 hours, which will be found to be equal to $29^d\ 17^h\ 34^m\ 17^s.14$, which is too long a period by $4^h\ 50^m\ 14^s.31$. This exceſs, in the ſhort ſpace of one lunation, it will be remarked, is very conſiderable; and it will be inferred from hence, either that a ſynodical revolution of the moon was not at that time well aſcertained in England, or that the inventor of the aſtronomical movement was unable to calculate numbers more accurate to repreſent it by wheel-work. We are of opinion, notwithſtanding, that neither of theſe was the caſe, and our reaſon for ſuch an opinion is founded in theſe three conſiderations; firſt, great ingenuity is ſhewn in the diſpoſition of the wheel-work, ſo great indeed, that it will be ſeen hereafter, that Mr. Ferguſon has evidently copied it in the dial-work of his principal aſtronomical clock; ſecondly, the calculations have evidently been made for a lunar day, and not for a lunation, becauſe the moon's age is indicated by the difference between the ſolar and lunar apparent daily motions, that is, by the ſolar hand pointing to the moveable lunar dial-plate, and not by a lunar index pointing to a graduated fixed plate; and, thirdly, the length of the lunar day was not aſcertained with accuracy by any means ſo remotely as a lunation, on account of a deception that exiſted in the determination. Let us examine theſe conſiderations a little more cloſely; as the large wheel of 288, which revolves in the ſpace of 24 hours, carries both the ſolar and lunar movements along with it, except the pinion of 8, which remains always immoveable in the centre, it follows, that, becauſe this great wheel revolves from weſt to eaſt, the wheel of 37 borne by it, and connected with the fixed pinion, will revolve in the ſame direction the ſpace of 8 teeth out of 37 in every 24 hours; alſo the pinion of 7, fixed on the centre of the wheel 37, will, by the ſame connection, be carried $\frac{8}{37}$ of 7 teeth in the ſame time, but the ſecond wheel of 45 will revolve in a contrary direction, or from eaſt to weſt in each 24 hours, the ſpace $\frac{8}{37}$ of $\frac{7}{45}$ parts of a ſolar day, which, reduced into a ſimple fraction, is $\frac{56}{1665}$ of 24 hours, or 48.432432, &c. minutes, ſo that the falling back of the wheel of 45 in every revolution of the large wheel occaſions the moon's daily ſpace on the dial-plate to be equal to 48.432432 minutes on the contiguous ſolar face of 24 hours, and, if we divide the whole of 24 hours, or 1440 minutes, by this quantity, we ſhall have $\frac{1440}{48.432432} = 29^d\ 17^h\ 34^m\ 17^s.14$ for the lunation, as before.

In order now to ſee how large the day's ſpace for the moon's age ought to be on the moon's plate, we muſt divide 1440, the minutes in a day, by 29.53058, the days in a lunation, and the quotient 48.76301 will be the number of the minutes that the moon's plate ought to loſe, with reſpect to the ſolar index, in a natural day, which quantity differs from what the wheel-work effected only .33057, or very nearly ⅓d of a minute's motion in every 24 hours, which would be deemed very inconſiderable at a time when the clock itſelf muſt have been without ſuch a regulator as would make it go truly for any length of time; the inventor, therefore might not think it neceſſary to calculate a more accurate movement.

But there might be another cauſe of error in the calculation ariſing from the aſſumed data; it appears, as we have intimated, that for a conſiderable time after the period in queſtion, though the length of a lunation was known pretty accurately, yet the length of an exact lunar day was not properly apprehended; even Benjamin Martin among other aſtronomical writers (Inſtitution of Clock-work, vol. ii. p. 412.) who had lunar tables before him, has inconſiderately but erroneouſly aſſerted, that the moon paſſes the meridian on each day at a mean rate about $48\frac{7}{10}$ minutes later than on the preceding day; it is true, indeed, that the difference between the ſun's and moon's daily mean motions in the ecliptic is 12° 11′ 27″, which ſpace anſwers very nearly to $48\frac{3}{4}$ of time, but then, as we have already ſeen, this is the diſtance between theſe two heavenly bodies at the end of a ſolar day after new moon, and not at the inſtant when the moon tranſits the meridian the firſt time after, as has been falſely concluded; for during the time that the earth is revolving on its axis the ſpace of 12° 11′ 27″ or $48\frac{3}{4}$ minutes of time, the moon continues her advance from the ſun, and gets as much farther as correſponds to 1.643 minutes more; and laſtly another ſmall advance is made whilſt the earth is revolving this laſt ſpace; and ſo on till the moon is found on the meridian: the aggregate of the moon's daily advance from the ſun and of all the ſucceſſive proportional parts may be thus aſcertained at one operation; becauſe a ſynodical revolution of the moon is performed in 29.53058 days, and becauſe the moon paſſes the meridian leſs frequently by once than the ſun does in that time, let 1440 be divided by 28.53058, and the reſult will give the lunar day equal $50^m.472$ or 50 minutes and 28.32 ſeconds. Mr. Ferguſon, however, has diſcriminated between a lunar day, and the time anſwering to the moon's daily motion added to a ſolar day, one being a corrected and the other an incorrect period; in conſequence of which we find the annexed note in page 116 of his "Tables and Tracts," *viz.* "It is generally believed that the moon revolves from the meridian to the meridian again in 24 hours 48 minutes, but that is a miſtake: for if ſhe did, there would be 30 complete days from change to change."

We repeat, therefore, that a miſapprehenſion of this nature was very likely to be productive of an error in the calculation of a movement which was intended to repreſent the moon's meridian paſſages: and let it be recollected, that what has been here ſaid reſpecting the two lunar periods $48^m.76301$, and $50^m.472$, will be ſerviceable in aſcertaining the moon's age and the tides at any place; the former, being the moon's daily motion in ſolar time, applies to calculations of the moon's age; and the latter, being the exact length of a lunar day, applies to the aſcertaining of the tides, which diſtinction, we think, is not generally made.

After the writer of the preſent article had made the preceding calculations and obſervations on the original conſtruction of the aſtronomical clock at Hampton Court, he felt an inclination to inſpect the preſent clock, that he might know what the ſolar movement is, which Bradley is ſaid to have ſubſtituted, when a wheel and pinion of the original movement were loſt; accordingly, on the 8th of May 1802 he embraced an opportunity which occurred, of gaining permiſſion to aſcend to the lofty ſituation in which this clock is placed, which enterprize was attended with ſome perſonal danger; but it proved on a minute and careful

ful examination, that the whole of both the annual and lunar movements are different from the original ones recorded by Dr. Derham.

The lunar movement is $\frac{59}{10}+\frac{45}{9}=29\frac{1}{2}$ days and the annual one $\frac{73}{12}+\frac{42}{7}+\frac{150}{15}=365$ exactly. The preſent central pinion is a double one conſiſting of a 10 and a 12, fixed as the former one of 8 is deſcribed to have been, and pinned together; they are made of box, as are alſo the pinions of 7 and 9, to prevent their cankering, or oxiding, as it is now called, and both the trains are arranged as already deſcribed. The wheel of 42 is made of braſs; but the reſt, being very large, are made of iron. The great wheel of 288, which connects the clock-work with the aſtronomical movements, appears to be the only portion of the original work, both by its marks of antiquity and the number of its teeth, which are cut on the inner edge of its circular part; there are two croſs bars riveted to this indented rim to carry the heavenly movements, and as there is no counterpoiſe to theſe, it was ſuſpected at the time that their riſing and falling weight would alternately accelerate and retard the going of the clock, which is connected with it by means of an horizontal arbor about 3 yards long by eſtimation: accordingly, on inquiry, it turned out that the time of the day indicated is ſometimes 5 minutes or more too back, and again as much too forward on the ſame day, every day, which circumſtance had not before been accounted for. Indeed our author had made a memorandum, before he ſaw this clock, merely from conſidering its conſtruction in Dr. Derham's account, that moſt probably this would be the caſe, unleſs there ſhould be ſome counterpoiſe.

The inſcription "L. Bradly, 1711," was marked on the frame of the going part of the clock, which has evidently been new, either all of it at that time, or ſome part of it ſince, ſo that what the original regulator was, does not appear, nor are the initials "N. O." to be found at preſent. There are three barrels and weights, one for the going part which has a very heavy long pendulum; one for the ſtriking part; and one for ſtriking the quarters: the preſent eſcapement is a pair of pallets acting alternately into pins projecting from the plane of a wheel with an horizontal arbor, which kind is now pretty common in England, and, according to Berthoud, was invented by M. Amant, a clock-maker of Paris, late in the eighteenth century. (See ESCAPEMENT.)

In Groſe's Antiquities it is ſaid that this clock, conſidered as an aſtronomical clock, was invented by Tompion; but this account cannot be true, becauſe that famous artiſt lived in Dr. Derham's time, a century and half after the original conſtruction; he may in all probability have been employed in making ſome of the alterations either in the aſtronomical or going part, which circumſtance has given riſe to this account.

The hand and divided circles are in the following order on the face, *viz.*

1. or inmoſt ſmall circle—twice XII for ☾'s ſouthing:
2. Moon's age $29\frac{1}{2}$:
3. Ecliptic with ſigns and days of the month:
4. Sun and hour hand revolving in 24 hours:
5. 24 hours marked I. II. &c.

The moon's phaſe is ſhewn by a circular hole cut on the hour index covering more or leſs of a blackened plate placed under it on the lunar dial.

The going and ſtriking parts of the clock before us have nothing particular in them to require our further notice.

A Thirty-hours Clock, with a 'Larum and Count-wheel ſtriking Work.

An ordinary 30-hours houſehold clock has uſually a ſeconds pendulum, 39.14 inches long, in the latitude of London, as meaſured from the point of ſuſpenſion to the centre of oſcillation, according to a mean of five determinations, and is contained in a caſe of wood of correſponding length reſting on the floor of the room where it is placed; its going and ſtriking parts are actuated by two ſeparate ponderous bodies, the cords or chains of which go reſpectively round two diſtinct pullies, with pins inſerted into their grooves, to hold the cords or chains from ſlipping in a detached ſtate, without taking the pullies along with them; and in the act of winding up every morning, or every evening, as may be the cuſtom, each ponderous body, uſed for actuating the wheel work, is drawn up ſeparately and ſucceſſively; the ſtriking mechaniſm in moſt of the modern clocks is of the kind deſcribed hereafter and ſeen in *Plate* XII. known by the name of the rack and ſnail ſtriking part; but as the older clocks and ſome few of the modern ones are made with the count-wheel ſtriking mechaniſm, we propoſe to deſcribe, in the firſt place, a clock with ſtriking mechaniſm of this kind in its moſt improved ſtate, together with an alarum, or 'larum, for rouſing a perſon from ſleep at a given hour, and alſo a ſimple method of making the clock go while it is under winding. *Plate* IX. of *Horology* ſhows a perſpective view of a clock of this kind in *fig.* 1, where the covering parts are removed, ſo as to allow an exhibition of all the parts of the mechaniſm in their reſpective ſituations, which, we believe, has never been properly done before. The frame of this clock is nearly a cube made of ſix plates of braſs, with four pillars at the corners connecting the top and bottom plates, and the bell is mounted over the frame, as repreſented in the figure; there are beſides, within the frame, two pair of croſs bars of braſs, one pair holding the going part, and the other holding, ſeparately from the going part, the ſtriking part of the mechaniſm; all which are too plainly ſeen to require particular reference to the drawing. In a common 30-hours clock that indicates ſeconds, the centre wheel arbor uſually carries the minute hand, and revolves in an hour, and with pinions of 8 this wheel has 64 teeth, and the ſecond wheel 60, with a ſwing wheel of 30; but with pinions of 6 the centre wheel has 48 teeth, and the ſecond wheel only 45, with a ſimilar ſwing wheel for the common anchor pallets; but the clock before us does not ſhew ſeconds, and conſequently has higher numbers in its train, and a pendulum ſhorter than a ſeconds pendulum: the firſt wheel, *a*, on which the pulley for winding up the going part is fixed faſt without a ratchet in this conſtruction, for a reaſon to be explained by and bye, is aſſumed as revolving in two hours, in order that the fall of the ſuſpended weight may not be ſo great, as it would have been if it had revolved once in an hour; this firſt wheel of the train has 80 teeth, and drives the pinion of 8 on the arbor of the ſecond wheel, *b*, which has alſo 80 teeth, impelling a ſecond pinion of 8 on the arbor of the ſwing wheel, *c*, which has 45 teeth; the ſwing wheel therefore makes 100 revolutions $\left(\frac{80}{8}\times\frac{80}{8}\right)$ while the firſt wheel revolves once; *i. e.* it revolves once in 50 minutes, and therefore is not proper for carrying a ſeconds hand; but the 45 teeth of the ſwing wheel do not all completely eſcape the pallets of the anchor

anchor in less than 90 (45 × 2) vibrations of the pendulum; hence 50 × 90 = 4500 are the vibrations per hour, and as the lengths of all pendulums are inversely to each other as the squares of their vibrations per hour, the pendulum before us must necessarily be 20 inches long, for as the square of 4500 : the square of 3600 :: $39^{in}\cdot14$: $20^{in.}$, very nearly. The pallets are of the recoil kind (see ESCAPEMENT), and have an arbor of great length, with a crank in the middle of it nearly, to avoid the fly of the striking part which comes through an aperture in the top plate; the inner end of the pallets arbor has its pivot inserted into a piece of metal, *d*, on the top plate, and the outer end of the same has its pivot passing through a cock, *e*, on the same plate, which cock also bears the pendulum by a slip of watch main-spring; the pendulum receives its impulse from the crutch, not seen, attached to the protruding portion of the said pivot. What we have so far described constitutes what is called the going part, sometimes also called the watch part of the clock, from its watching the lapse of time. The first wheel, *a*, which revolves once in two hours, is not placed in the centre, but a little below, and has its strong arbor passing through the cross partition bar, *ff*, of the interior frame work, and receiving a wheel of 40 teeth, and also a pinion, *b*, of ten leaves, which are attached together and inserted on this arbor by friction; the wheel of 40, which, we have seen, revolves in two hours, drives a concealed pinion with a long tube, called the cannon pinion, the place and size of which are ascertained by dots at *k*; on the tube of this pinion, which has 20 teeth, and which therefore revolves in an hour, is placed the minute hand, *n*, the end of the tube being squared to admit the square aperture of the hand; the pinion of ten leaves, which also revolves in two hours, drives the wheel, *i*, of 60 teeth in twelve hours, the tube of which admits on its circular part the hour hand, *m*, which consequently revolves in 12 hours $\left(2 \times \frac{60}{10}\right)$; this work is denominated the dial-work, being that which regulates the relative velocities of the two hands as seen in *fig.* 3. The next portion that offers itself for description is the 'larum portion, which has an immediate connection with the dial-work, and has the time of its going off limited thereby. On the tube of the 12-hours wheel, *i*, is placed loose, or at least so tight only as friction will fix it, the small plate, *o*, pointed to by the tail of the hour hand, *m*, which small plate has 12 hours engraven on it, and a pin inserted into it behind, which comes in the way of the lever *p* every 12 hours; this pin is put into a certain situation with respect to the hour of 12, and the end of the lever, *p*, also, in order that the pin may catch the said lever at a certain hour placed under the tail of the hour hand at any time previously to the hour intended; the consequence is, that when the 12-hour wheel has revolved far enough to present the pin of the small 'larum dial borne by it to the end of the lever *p*, this lever is elevated a little, and as its arbor, *q*, has its pivots running in the interior frame work of the going part, a second lever, *r*, on the same common arbor, is also at the same time elevated from the pin, *t*, of an escapement crown wheel, *s*, better seen detached in *fig.* 2, at which instant the small weight of the 'larum pulls the pulley on the back of the escapement wheel, *s*, of the 'larum, and gives it a rotatory motion as long as the weight continues to fall. The escapement wheel here mentioned has coarse teeth of the serrated kind, which act with two pallets on the perpendicular arbor *v*, in the same way that the pallets of a common watch act, except that the latter are on a horizontal arbor, and have their frequency regulated by the vibrations of the balance; whereas here there is no regulator, but the pallets go and come alternately as fast as the impelling weight can force them to move; on the top of the perpendicular arbor of these pallets is fixed a hammer with two faces within the bell, represented by dots, which moving backward and forward from one interior side of the bell to the other, with the force communicated to the pallets by the pallet-wheel, make a reiterated noise, the intensity and continuance of which are sufficient to disturb the repose of a sound sleeper. When the weight has drawn up all the cord, it is only necessary to pull it up again, and the lever, *r*, acts as a detent with the pin, *t*, of the pallet-wheel, till the pin of the 'larum dial, set to any given hour, shall again detach it, when the same continued noise will be resumed.

The last portion of the clock is the striking portion, which also has a connection with the dial work; the wheel, *g*, which revolves in the space of two hours, has two pins at the distance of a semicircle from each other, behind the wheel as seen in the figure; one or the other of these two pins at the end of each hour seises the end 1 of a tail piece attached to the long horizontal arbor 2 3, which reaches the whole depth of the two internal frames; this long arbor has another bar 4, or detent, which reaches far enough to fall in the way of a pin in the wheel 7, or warning wheel, so as to arrest the motion of the striking movement when in its quiescent position. Parallel to the long arbor, and above it, is another but shorter arbor 5, turning by its pivots in the interior frame of the striking part; this short arbor 5 has a bent lever 6 by which it may be raised by the contact of the detent 4 of the long arbor 2 3, and also two catches or detents 7 and 8, all fixed at right angles to the axis of motion; the detent 7 falls into a notch made in a hoop placed fast to the wheel marked 13, thence called the hoop-wheel or detent-wheel sometimes, which revolves once at every blow of the hammer, and the second catch 8 falls successively into 12 notches cut at unequal distances on the edge of plate 11, called the locking plate, which is fixed fast to the wheel marked 12, called the count-wheel, because its teeth count the stroke-spaces between the notches of the locking-plate that are placed respectively at $\frac{1}{78}$, $\frac{2}{78}$, $\frac{3}{78}$, &c. of the circumference of the plate from each other, as explained when we described the ancient clock of Henry de Wick. The wheel 9, first actuated by the cord or chain passing round the third pulley fastened to it, has 12 pins for raising the tail-piece 10, of the hammer, the arbor, *l*, of which is seen in the figure; the spring of the hammer tail is concealed from view, but the shaft of the hammer is seen passing through an opening in the top plate at 16, and the head of the hammer is represented within the bell, where it strikes, by a dotted design above 16. The pin-wheel or striking-wheel 9 has 60 teeth, and drives a pinion of 10 leaves on the arbor of the hoop-wheel 13, behind the cross piece of the frame work; the hoop-wheel has 70 teeth driving a pinion of 7 leaves on the remote end of the arbor of the warning wheel 7, of 56 teeth; which wheel again impels the pinion 14, of 7 leaves, on the arbor of the fly, one half of which is seen through an opening in the top plate at 15. On that end of the arbor of the pin-wheel which passes through the back part of the frame-work is inserted a pinion of 12 leaves, called the pinion of report, driving the counting wheel 12 of 78 teeth. The action of the striking part is thus:—

One of the pins in the two-hour wheel, *g*, first lifts the tail 1, of the long lever 2 3, and with it the detent 4, the warning wheel 7 is not yet at liberty, but begins to revolve the instant that this detent has raised the curved arm 6 over it, which arm raises with it both the catches or detents 7 and 8, that leave the hoop-wheel 13, and also the count-wheel 12, under

12, under the command of the fuſpended weight; the motion of the wheel 7, however, does not proceed far till its detent 4 is raiſed into the way of its pin, and the motion of it is arreſted; the noiſe of this temporary motion of wheel 7 is called the warning, and the wheel itſelf the warning-wheel; preſently the pin of the 2-hours wheel, *g*, drops from the end of the tail 1 of the arbor 2 3, and this tail as well as detent 4 reſume their quieſcent poſition; during the temporary motion of the warning wheel 7, the hoop-wheel, and alſo the locking-plate attached to the count-wheel, had moved far enough to take the notches from the claws of their reſpective catches or detents 7 and 8, the moment, therefore, that the detent 4 takes its quieſcent poſition, the warning-wheel is again at liberty, as are alſo the hoop-wheel and count-wheel; the whole movement, conſequently, now proceeds, and the pin-wheel raiſes the hammer tail as often as the pins meet with it, until the detent 8 meets with a notch to receive it into the locking-plate, at which inſtant all motion is at an end, and the detent 7 of the hoop falls alſo into its notch, and holds the whole movement firmly in a quieſcent ſtate, till the ſecond pin of wheel *g* again detaches the detents, and renews the ſame proceſs; which happens at the concluſion of every hour.

From this account of the movement of the ſtriking part, and of the other auxiliary parts of this mechaniſm, it is eaſy to apprehend the reaſon of the numbers of teeth fixed upon in their different wheels and pinions; for firſt, becauſe there are 12 pins in the ſtriking or pin wheel, it is neceſſary that the pinion of report on the ſame protruding arbor ſhould have 12 leaves, in order that every tooth of the count-wheel, which counts a ſtroke of the hammer, ſhould have its motion correſponding with that of the pins reſpectively; but in twelve hours there are 78 ſtrokes, therefore 78 teeth in the count-wheel, one of which meaſures the firſt interval on the locking plate; two of which the ſecond; three, the third, and ſo on till the laſt ſpace between the notches is meaſured by twelve teeth of this wheel; again, as the pin-wheel has 60 teeth and twelve pins, each pin is removed from the next $\frac{60}{12} = 5$ teeth; if the hoop-wheel, or detent-wheel were neceſſarily obliged to have an exact revolution at every ſtroke of the hammer, the pinion on its arbor driven by the pin-wheel muſt neceſſarily have five leaves only; but when the teeth are not laid very deep into one another, the play will allow the hoop-wheel to have only one revolution in two ſtrokes, which is the caſe before us, where the pinion has ten leaves; the pin-wheel, however, might very well have had 96 teeth, and the pinion in queſtion 8 leaves, and then there would have been an entire revolution of the hoop-wheel at each ſtroke. We have ſeen that the count-wheel revolves once in 12 hours and that the pin-wheel revolves in $\frac{78}{12}$ of this time, or makes $6\frac{1}{2}$ turns; but if the number of pins had been 13, and the pinion of report alſo 13, the time of a revolution of theſe would have been $\frac{78}{13}$ of 12 hours, or one in two hours, which is the caſe with the great wheel of the going part, and the two movements would, in that caſe, have been more uniform with reſpect to the calculations of continuance. It is of but little importance what the numbers of teeth be in the two remaining pinions and warning wheel as they only regulate the velocity of the fly, provided the teeth are numerous enough to act without much friction.

As we mean not to introduce any other clock with the count-wheel mechaniſm after the preſent one, it may be proper to notice here, that when this kind of ſtriking work is uſed in an eight-days clock, or other clock of longer continuance than 30 hours, another wheel and pinion, called in ſuch caſe the great wheel and its pinion, muſt be placed on the end of the barrel, or fuſee, accordingly as a weight or ſpring is uſed for the maintaining power; and the number of teeth of this additional wheel will depend partly on the number of leaves of the pinion and partly on the number of coils of the cord or chain on the barrel, or fuſee, as the caſe may be; which never can be difficult to determine when it is known in what time the pin-wheel revolves, and what the continuance of the clock is required to be; for inſtance, if the pin-wheel revolves with 13 pins in two hours, $\frac{72}{12}$ will be the wheel and pinion to make the barrel or fuſee turn once in 12 hours, and 16 turns will be ſufficient for a continuance of eight days.

The manner in which the clock in queſtion may be placed to go well, is ſhown in *fig.* 3, where A is a bracket fixed to a ſolid wall, to hold the frame containing the works, which are here ſuppoſed to be without a caſe; the face or dial requires no explanation after what we have ſaid of the ſmall innermoſt circle or 'larum dial; nor have we any particular remarks to make on the pendulum, or its rod, in this clock, which profeſſes not to meaſure time alike under the variations of atmoſpheric temperature; but the manner in which one ſuſpended weight impels both the going and ſtriking movements, the former even while winding, is ingenious and deſerves attention. Huygens long ago propoſed and indeed actually uſed an endleſs cord with a detached ratchet, ſo applied that one half of the weight continued always to impel the going movement of his clock; and a detached ratchet gathered up the expended portion of the cord once in 24 hours. (See Clock-*work*) The preſent contrivance is evidently borrowed from his invention, but its application ſerves two diſtinct purpoſes, one half of the weight impels the going part, and the other half the ſtriking part of our preſent clock, in the manner following;—The pulley of the going part we have ſaid is fixed faſt to the arbor of the great wheel and always revolves along with it, but the pulley of the ſtriking part, like the pulley of the 'larum, has a rachet formed of a click taking hold of one of the croſſes of the pin-wheel. ſo that it will move back without this wheel, but not forwards; the cord, which has its ends nicely united, goes over both pullies, firſt over the ſtriking pulley, as Huygens's does over the detached ratchet, and down the loweſt plate of the clock frame; it then paſſes under the running pulley, *a*, which has the weight hung to it, and up again over the going pulley; thirdly, it comes down again through the ring of lead *b*, which is only a dead weight to ſtretch the cord over the pulley of the ſtriking part; and, fourthly, aſcends again over the ſtriking pulley, where the ends meet; the clock is wound up by pulling the cord, *c*, downwards, till the principal weight is raiſed to the bottom of the bracket, while one of the cords of the running pulley continues to act with half of the whole weight on the pulley of the going part the whole time; ſo that this power, when duly drawn up, is not only equivalent to both the movements, but is perpetual, as well as invariable in its intenſity.

Eight-days portable Clock with repeating Mechaniſm.

A portable eight-days clock differs from an ordinary 24-hours clock principally in five reſpects: in the firſt place, it is actuated by a ſpring; ſecondly, it has a ſhorter pendulum; thirdly, it has conſequently a higher train; fourthly, it requires a fuſee; and, laſtly, it has frequently a crown-wheel eſcapement; in ſhort, it may be conſidered as a watch on an enlarged

enlarged ſcale, except that it has uſually the ſtriking mechaniſm, which is introduced in the repeating watches only. In the clock, which we propoſe here to deſcribe, we have ſubſtituted the ſwing-wheel with an iſochronal eſcapement, for the crown-wheel eſcapement, which we are of opinion ought to be baniſhed from all clocks and watches entirely, as being too much the ſlave of the maintaining power. It will not be neceſſary to render our deſcription of the clock before us very long, as we mean to make it the ſubject of analyſis under our ſubſequent article CLOCK-*movement.*

Plate XI. of *Horology* preſents a perſpective view of a portable eight-days clock, deſigned and callipered agreeably to the numbers and directions given in the article juſt mentioned, which conſtruction we have preferred deſcribing here, in order that the reader may there ſee our reaſons for every part of the mechaniſm, if he wiſhes for ſuch explanation: let it however be underſtood, that we by no means hold out the preſent clock as a model for others to follow, but give it as one of the hundred different varieties or more that might be deviſed to anſwer the ſame purpoſe.

Plate XII. exhibits the dial-work and ſtriking part, including the repeating mechaniſm of the ſame clock, which we ſhall deſcribe in its turn.

Fig. 1, of *Plate* XI. expoſes to view the interior face of the back or pillar plate A A A A, and the wheel-work contained within the frame; B, B, B, B, B, are five pillars ſcrewed into the back plate, and tapped at their projecting ends to receive five fixing ſcrews, when the front plate is put in its place to complete the frame: there are two barrels with main-ſprings, C and D, of which C is for the going part, and is cloſed with its cap; but D, which is the barrel for the ſtriking part, is left open, to ſhow the coils of the main-ſpring contained in it. There are alſo two fuſees, E and F, attached to their reſpective great wheels; E the fuſee of the going part, and F that of the ſtriking part. G is the centre or hour wheel of 64 teeth placed partly behind the great wheel of 96, by which its pinion of 8 is actuated; hence the fuſee revolves in $\frac{96}{8}$ of an hour, or once in 12 hours, ſo that the 16 ſpiral grooves, filled by the gut, allow the continuance to be juſt eight days. The ſecond wheel H, of 60 teeth, has its pinion of eight impelled by the centre wheel, and in its turn impels the pinion of 8 on the arbor of the third wheel of the train, I, which is here alſo the eſcapement wheel; the revolutions therefore of the arbor of wheel I are $\frac{64}{8} \times \frac{60}{8} = 60$, while the hour-wheel revolves once; conſequently this arbor, which revolves in $\frac{1}{60}$ of an hour, is proper for carrying the ſeconds-hand round in a minute. The eſcapement-wheel has 60 teeth, and the pendulum vibrates twice in every ſecond: but one tooth does not eſcape the pallets of the anchor K, until two vibrations have been completed; conſequently 60 teeth eſcape, *i. e.* the eſcapement-wheel makes one revolution in 120 vibrations, or in the ſpace of one minute: hence the clock before us indicates half-ſeconds. The ſquare ends of the two fuſees are oppoſite two holes in the clock-face, at each ſide of the centre of the circles of indication, (which are too well known to need deſcription), but a little below it, and the ſame handle fits both ſquares: L is a jointed lever fixed to the interior ſide of the front plate called the guard-gut, or ſimply the guard, the uſe of which is to prevent the chain or gut from doing more than juſt fill the ſixteen grooves of the fuſee in winding; M is a ſpring alſo faſt to the interior ſide of the front plate, which preſſes the lever L towards the middle of the fuſee, and keeps it there till the chain or gut, meeting with it, drives it back again, in the act of winding, ſo far till the claw *o* is preſented to the beak or catch of the end piece N, which then ſtops the further motion of the fuſee and limits the quantum of chain to be wound up. Theſe pieces L and M conſtituting the guard, being attached to the front plate, are taken off with it, when the frame is diſmounted, but we have put them into their places in a detached ſtate in our figure, to ſhow more evidently the nature of their office. O is the arbor of the warning piece, which will be deſcribed in its place preſently. The wheels P, Q, and R, with their reſpective pinions, conſtitute the movement of the ſtriking part, and the fanner S regulates the velocity with which they move. The wheel P has eight pins which lift the croſs piece *t*, of the arbor T, eight times in each revolution of the wheel P; theſe elevations of the piece *t* occaſion ſo many portions of a revolution of the arbor T, which arbor has its pivot projecting to V, behind the frame, and carrying on its ſquared projection the hammer V, *fig.* 2; the hammer is conſequently raiſed every time a pin of wheel P moves the piece *t;* W is a long and ſtrong ſpring, called the hammer-tail-ſpring, attached to the back plate of the frame, and preſſing with its upper extremity under the croſs pin, paſſing through a hole in the arbor T, near the face of the back plate; ſo that when the hammer is raiſed at any time, the ſpring W urges it back again with a ſmart blow, and makes it ſtrike the bell behind the frame, which is concealed from the ſight in our figure; but leſt the blow ſhould be too ſtrong, a counter ſpring, U, is fixed to the contiguous pillar, which breaks the violence of the blow, and makes the hammer return ſmartly to its place when the blow is made; this ſpring, U, alſo ſerves as a guard, in caſe a ſtroke of the hammer ſhould be made when the bell is taken off at any time. The fuſee F is provided with a guard ſimilar to that of the fuſee E; and the vane of the fly or fanner S is kept to the arbor, on which it is placed, by the friction of a preſſing ſpring, ſo that it will go round either with or without the arbor, the latter of which is the caſe only after the ſtriking has ceaſed, till the momentum of the fly has been annihilated by the reſiſtance of the air.

Fig. 3, is a portion of the top of the exterior face of the back plate of the frame, conſtituting the ſuſpenſion of the pendulum: *a* is a bridge, or double cock, in which the projecting pivot of the pallet's arbor is ſupported; *b* is a ſmall cock over this, or may form a part of the ſame, the protruding part of which is ſlit with a ſaw, to allow the ſlip of watch main-ſpring, *f f*, to paſs; by which the pendulum rod is ſuſpended; and *c* is a thumb-ſcrew to cloſe the ſlit, and clamp the piece of ſpring when the true length of the pendulum is aſcertained; the upper end of the ſpring, *f f*, is borne by the bearing piece *d* above the frame, to which it is uſually made faſt by a pin paſſing through the piece *d*, and a hole in a piece of braſs pinned to the extreme end of the ſpring; the apparatus for raiſing and lowering the piece *d* is better ſeen in *Plate* XII. to which the reader is now deſired to turn for the remainder of the deſcription; only let him bear in mind that the effective length of the pendulum is meaſured from the inferior edge of cock *b*, in *Plate* XI. to the centre of oſcillation of the bob, and that the quantum of this meaſure is adjuſted by the elevation or depreſſion of the piece *d*, while the ſlit of the ſtationary cock *b* is not cloſed by the thumb-ſcrew *c;* which adjuſtment may be made while the clock is going, provided the ſlit be again cloſed by the thumb-ſcrew, otherwiſe the effective length of the pendulum will be meaſured from the piece *d*, and will remain the ſame whether this piece be raiſed or depreſſed.

The

The bearing piece *d*, *Plate* XII. is the remote end of a tranſverſe lever, *c d*, moveable like the teleſcopic tube of a tranſit-inſtrument on two pivots, *a b*, of an axis ſupported by two ſmall cocks, *a* and *b*, clearly ſeen in the figure, ſo that when the pivots are ſtrong and without play in their holes, it is evident that the end *d* can have no lateral ſhake, provided the bar *c d* has no ſpring, *i.e.* provided it be ſtrong; alſo when the interior end *c* is fixed with reſpect to elevation, the exterior bearing end muſt be alſo fixed; the contrivance therefore for elevating or depreſſing the end *d* in the body of the clock caſe at any time may be as well applied to the end *c* in the front of the frame; for a depreſſion of one end of the lever *c d* always produces a correſponding elevation of the other, and *vice verſâ*, ſuppoſing the axis exactly in the middle; but when the axis is out of the middle of the lever, an elevation or depreſſion of either end is in proportion to its diſtance from the axis of motion directly. At *e* is a circular plate of braſs, better ſeen detached in *fig.* 4, called the *riſe* and *fall*, with an arbor ſquared at the projecting extremity, oppoſite a perforation on the face or dial of the clock, to receive a key of regulation; this circular plate, which is pivoted into the cock *f* attached to the front plate of the frame, has a ſpiral aperture through which a round pin in the end of the lever, *c d*, paſſes ſo as juſt to go in without ſhake; hence, whenever the plate *e* is turned by the key one entire revolution, the pin of the lever, which paſſes through the ſpiral aperture, aſcends or deſcends from one end of the ſpiral to the other, and at the ſame time depreſſes or raiſes the bearing end, *d*, a proportionable quantity and with it the bob of the pendulum, as may now be eaſily apprehended; the condition, therefore, with reſpect to the regulation, is, that one turn of the plate, *e*, ſhall effect as great a change in the length of the pendulum, as ſhall be requiſite to bring the clock to a true rate, or keep it ſo when adjuſted, where the pendulum has a compenſation for temperature, or is made of a ſubſtance that alters its dimenſions but little with the variations of temperature of the atmoſphere. In ordinary clocks of this conſtruction, with a ſimple metallic rod for the pendulum, the ſlit may be ſo nearly cloſed at all times as juſt to allow the thin ſpring of ſuſpenſion to paſs, in which caſe the frequent adjuſtments for rate may be made without opening the caſe; but when the pendulum is of any of the beſt compenſating kinds, it will be better to cloſe the ſlit by the thumb ſcrew *c* in *Plate* XI. after the adjuſtment for rate is completed.

Thoſe parts of *Plate* XI. which are viſible in *Plate* XII. are marked with the ſame letters over again, and therefore need not be again deſcribed, but will be of ſervice to elucidate the relative poſitions of thoſe parts, now that the frame is exhibited as mounted. We will begin our deſcription of the ſtriking and repeating work, with the arbor of the centre wheel, G, the end of which is ſeen within at the projecting end of the ſquared part of the tube, or cannon, of the pinion ſeen in *fig.* 2, called the cannon pinion; the tube of this pinion is put tight on the arbor of the hour wheel, which we have alſo named the centre wheel, and has a ſpring placed on the hour wheel arbor, preſſing its poſterior ſurface ſo as to force it forwards againſt the croſs pin that keeps the hands on; this action of the ſpring occaſions ſo much friction, that though the tube is carried round by the hour arbor, yet it is capable of being moved round, by its hand placed on the ſquare end, independently of this arbor, for the purpoſe of ſetting the hand to the requiſite minute in the divided circle of 60 ſpaces, uſually figured with the Arabic characters: the cannon pinion, as it is called, has 40 teeth, and impels a ſimilar pinion, *g*, round alſo in an hour; this pinion, *g*, which is called the pinion of report, has a pinion of 6 on its arbor, and is pivoted into the cock, *h*, ſo that the ſmall pinion of 6 alſo revolves in an hour; this pinion of 6 again impels the wheel *i* of 72 teeth in $\frac{72}{6}$ of an hour, *i. e.* in 12 hours; this 12-hour wheel has alſo a tube, ſurrounding the tube of the cannon pinion, but in ſuch a way, that a third tube, attached to the bridge, *k*, and ſeen in a detached ſtate in *fig.* 5, is interpoſed between the ſaid two tubes of the cannon pinion and 12 hour wheel; the uſe of which third fixed tube, is to prevent the friction that would neceſſarily take place, if the two revolving tubes had been in contact, and had preſſed on one another, while their velocities are to each other as 12 to 1: on the exterior tube of the 12-hour wheel, the hour hand is inſerted, which indicates the hour among the Roman figures; and it is obvious that whenever the minute hand carries the cannon pinion round, the pinion of report, *g*, alſo moves the ſame quantity, and by means of the ſmall pinion of 6, the wheel *i* at the ſame time muſt move $\frac{1}{12}$ of the ſame ſpace, and conſequently the two hands are ſo connected, that one cannot move without the other, ſuppoſing them both to be faſt to their reſpective tubes; but the hour hand is put on the round part of its tube, and kept to it by mere friction, and therefore may be put to any hour without carrying the minute hand round many revolutions; and yet when once placed right, it preſerves its relative velocity, as though it were more firmly attached to its tube. I is the arbor of the ſeconds-hand, which we have ſeen revolves in a minute, and which meaſures the 120th part in its divided ſmall circle, on the face of the clock, at ſo many vibrations of the pendulum, or at ſo many half-ſeconds. To the 12-hour wheel, *i*, is pinned faſt an indented ſpiral piece of metal, called the ſnail, the ſhell of which it reſembles in ſome meaſure, which ſnail conſequently revolves likewiſe in 12 hours; the indentations appear to the eye to be irregular, as to their relative extents, but each ſubtends an angle of $30°\left(\frac{360°}{12}\right)$ ſo that one indentation, whether near to the centre of motion, or remote from it, is exactly the meaſure of an hour's motion of the 12-hour wheel. The ſteel piece, *m n*, is called a rack from the teeth on the croſs-piece, *m*, the lower croſs-piece of which is called the rack-tail; this rack is moveable on a pin or ſtud at the lower angular point, near which the horſe-ſhoe ſpring, *o*, called the rack-tail-ſpring, preſſes to keep a pin on the remote end of this tail, againſt that indentation of the ſnail, which happens to be contiguous to it; this pin is hid from the ſight, but the place may be ſeen on the extremity of the tail where it is inſerted. On the lever between *m* and *n*, is a bend to prevent its touching the winding arbor F of the fuſee, belonging to the ſtriking part; alſo at *m* is a ſtrong ſteel pin, projecting from the rack. Above the rack is a horizontal ſteel bar, *p q r*, moveable round a ſtud at *r*, which is called the hawk's-bill, from the bill or angular piece at *q*, that catches the teeth of the rack. The piece *s* is fixed to the protruding pivot of wheel Q. *Plate* XI. near its lower extremity, and revolving with it gathers up a tooth of the rack at each revolution, on which account it is called the *gathering pallet*, the catch of the hawk's bill having a contrary ſlope, gives way in the mean time, and comes back again by its own gravity. The pinion of the pin-wheel, P, which has 64 teeth and eight pins, has eight leaves, and therefore revolves once every time that the hammer of the bell is lifted; but we have ſaid that its gathering pallet takes up a tooth of the rack at each revolution of its arbor, conſequently a tooth of the rack is gathered up at every ſtroke of the hammer, when the ſtriking part is in motion. The angular piece, *t u v*, moveable round an arbor, denoted by O, in *Plate* XI. is called the *warning-piece*; its

 lower

lower end, *o*, falls in the way of a pin in the fmall hour-wheel, *g*, and its bent end, *t*, paffes through an aperture, *w*, in the front plate of the frame, and is prefented to the pin in one of the croffes of the wheel R, of the ftriking movement within the plate, fo as to reftrain the motion of this movement when in its quiefcent fituation. The action of the different parts may be thus explained; whenever the hawk's-bill, *q*, is lifted from the teeth of the rack, the fpring, *o*, preffing againft a pin near its tail, makes it fall back till it meets with fome obftacle to arreft its motion, that obftacle would be the pin, *x*, in the front plate, if there were no other interpofed before it had fallen fo far back, but if the fnail is in any other pofition than that, wherein its neareft indentation towards the centre is contiguous to the pin of the rack-tail, the tail-pin of the rack will fall upon the edge of the fnail before the rack has fallen back to the pin, *x*, and all the teeth of the rack will not pafs the catch of the hawk's-bill in this cafe, but juft fo many as there are indentations or fteps counted from the remote angular point of the fnail to the ftep on which the tail-pin refts; in the prefent pofition, in the plate, the tail-pin is refting on the ftep fix of the fnail, which denotes that fix ftrokes will be given by the hammer, or that fix teeth of the rack are to be gathered up by as many revolutions of the hawk's-bill; but we fee that only five teeth remain to be gathered up of the rack; hence we know that the clock has ftruck one out of fix, and is in the act of ftriking; accordingly, we fee that the pin in the hour-wheel, *g*, has juft raifed the warning-piece and permitted it to go again; the clock will therefore now continue to ftrike till the upper end of the gathering pallet, *s*, falls on the projecting pin, *m*, of the rack, which will be as foon as the laft tooth of the rack is drawn up to the hawk's-bill, in which fituation the wheel, Q, cannot revolve any farther till another hour has elapfed. After another hour is paft, the pin of the wheel, *g*, will elevate the warning-piece, *v*, the bent end, *t*, of which will firft be raifed out of the way of the pin of the wheel R, and the fly will run on a revolution or two, with a whiftling noife, *i. e.* the clock will *give warning;* but the end, *p*, of the hawk's-bill has not yet been raifed far enough by the preffure of the end, *t*, of the warning-piece, to make the catch, *q*, clear the teeth of the rack, therefore the rack cannot yet fall back; prefently, however, the hawk's-bill is lifted high enough by the pin of the pinion of report, *g*, which has a flow motion; the rack falls back till its tail-pin refts on ftep feven nearer the centre, which has now arrived at the point of contact, and therefore feven teeth of the rack pafs the catch, *q*, in the fall of the rack, and the hour of feven is now ftruck before the tail of the gathering pallet, *s*, falls again on the pin, *m*, of the rack, and ftops the ftriking; at the fame time the bend of the warning-piece catches the pin of the wheel R, and ftops the fly; and in this way any number of hours will be ftruck by the hammer on the bell that the fnail regulates, which we have faid revolves once in every 12 hours; and if any other caufe than the pin of the hour-wheel, *g*, fhould lift the warning-piece within the hour, counting from warning to warning, the fame number of ftrokes will be repeated, though it fhould be a hundred times or more. To convert this ftriking mechanifm into repeating mechanifm, therefore, it is only neceffary to place a lever, *y*, to revolve round a ftud on the front plate of the frame at the point *y*, with a flender fpring, *z*, over it to bring it back to its original fituation, when the end placed under the warning-piece is elevated by depreffing the exterior end, which may be done by pulling a ftring down which is tied to a hole in this end, as reprefented in the figure; and as often as the ftring is pulled, fo often will the clock repeat the ftrokes of the current hour. There is yet remaining the three-armed piece, 1 2 3, undefcribed, called *ftrike* or *filent*, the ufe of which is explained by its name; this piece is differently made in different clocks; in the clock before us it is moveable on a focket, riveted to the end, 3, of one of its arms, round a ftud in the front plate of the frame, and as the focket has fcarcely any fhake, the other two ends, 1 and 2, move always in the fame plane; at the end marked 1, is a pin projecting above the upper circumference of the face or dial of the clock, fo that it may be moved to the right or left at pleafure, when the glafs-door is open; the end marked 2, has a flope, like a wedge, on that fide which is next to the plane of the frame-plate, and the end of the arbor, O, in *Plate* XI. of the warning-piece, projects fo far as to touch the inclined plane; this arbor of the warning-piece has fome fhake, in the direction of its length, within the frame, and its pofterior pivot paffes between the prongs of a forked fpring, X, which, refting againft the fhoulder of the pivot, pufhes it clofe to the interior fide of the front plate of the frame, where a fimilar fhoulder ftops it; when the pin, at 1, is pufhed to the right, the wedge of the end, 2, pufhes the arbor back, notwithftanding the forked fpring, X, juft defcribed, and the end, *v*, of the warning-piece, being carried with its arbor nearer to the frame than it otherwife would be, falls in the way of the pin of the hour-wheel, *g*, and the clock confequently ftrikes the hour regulated by the fnail; but when the pin, at 1, of the ftrike or filent, is pufhed to the left, the end, 2, is withdrawn from the pivot of the arbor on which the warning-piece is faft, the fpring, X, in the frame pufhes it forward fo far, that the end, *v*, of this warning-piece is clear of the pin of the hour-wheel, *g*, which wheel therefore continues to revolve from hour to hour in a ftate completely detached from the mechanifm of the ftriking part, which we have been defcribing. Sometimes there is a hand moveable in a fmall circle in the dial, which anfwers the fame purpofe as the pin at 1; (See CLOCK-*work*,) but this is generally the cafe when there is no circle for the feconds, or when there is fome other circle to which it is intended to correfpond, for the fake of uniformity, which is generally attended to in the dial-work of every clock.

Laftly, the four holes in the front plate denoted by the letters Z, Z, Z, Z, are the holes in which the pillars of the dial, or face, are inferted and pinned within the frame by metallic pins going acrofs the ends that pafs through the plate of the frame; fo that the face is thus firmly attached to the frame, and then the frame to the cafe, which preferves the wheel-work from duft, and the touch of fuch perfons as might otherwife fatisfy their curiofity at the expence of fome of the more delicate parts of the workmanfhip. Indeed, many of the ornamental portable clocks have cafes of glafs, with various devices, fuch as are calculated to recommend them to the fancy rather than the judgment of their purchafers, who wifh to adorn thereby their chimney-pieces. In the formation of thefe ornamental cafes, fpar of different colours, or moulu, and various other fuperb materials are ufed, agreeably to the tafte of the artift who is employed in fuch manufactory.

A portable clock, fuch as we have here defcribed, is eafily converted into a clock with a long cafe, and a fufpended weight for the maintaining power, by fubftituting a cylindrical barrel for the fufee on the arbor of the great wheel, on which barrel the chain is wound, inftead of being made to furround the fufee; for, as a fufpended load acts at all times with the fame power, it is neceffary that the barrel fhould have the fame diameter at every part of it. Of this conftruction are the eight days houfehold clocks in general ufe.

ufe. Alfo by introducing a wheel and pinion between the great and centre wheels, the clock may be made to go a month or more, and by introducing two fuch wheels and pinions, it may be made to go a year at one winding up, with a maintaining power proportionably great.

Eight-days portable Clock, with Chimes and repeating Mechanifm.

Under our article CHIMES, we gave a defcription of the chime-barrel, hammers, and bells of a chime-clock, in a detached ftate, but referred to our prefent article for their connection with the ftriking part of the clock, which could not there be fufficiently explained; as we have now fhown the manner in which the modern ftriking part is conftructed, and have alfo explained the nature of its action, we propofe to refume the fubject of chimes, for the explaining of which in the beft manner, we have thought it neceffary to defcribe here at full length, a portable chime-clock with a fhort pendulum, for which purpofe we have introduced two feparate views in *Plates* XVI. and XVII. of *Horology*, which, we truft, will enable us to make all the mechanifm, complex as it may appear, fufficiently intelligible to any one who has read and underftands the ftructure of the eight-days clock that has been juft defcribed. *Plate* XVI. is a reprefentation of the three movements of a chime-clock, when the front plate of the frame is taken off, the eye being placed perpendicularly over the mechanifm when viewed; we have retained the fame letters of reference as we ufed in *Plate* XI., as far as they will go, and have fupplied the deficiency from the Greek alphabet, which are ufed for the movement of the chimes, fo that the reader will fee, by a *coup d'oeil*, the arrangement of the movements, even without the help of a defcription, if he recollects the parts of the clock defcribed in *Plate* XI. A, A, A, A, ftand at the four corners of the back plate of the frame, on the interior furface; B, B, B, B, are the places of the four pillars; C the fpring barrel of the going part; D that of the ftriking part, and α that of the chime portion; E is the great wheel and fufee of the going part, F thofe of the ftriking part, and β thofe of the chime part; N in each the catch of the guards refpectively; G, H, and I, are the wheels which, with their refpective pinions, conftitute the train of the going part, of which H is the contrate wheel, and I the balance wheel, with its lower pivot refting on the fmall cock or potence K; the pallets cannot be well feen in this plate, but are vifible in *Plate* XVII.; P, Q, and R, are the wheels of the ftriking movement, having each a pinion on its arbor, and S is the pinion of the fly feen at S in *Plate* XVII. The wheel P has the eight pins for lifting the hammer, the arbor and levers of which are at T; the arbor of Q carries as before the gathering pallet for drawing up the rack of the ftriking part, and R has the pin for catching the bent end of the warning piece; W is the hammer tail fpring, and U the counter fpring, which are fomewhat differently placed here from what they are in *Plate* XI., but act in a fimilar manner. The wheels marked with the Greek characters γ, δ, ε, ζ, are thofe of the chime movement, with their refpective pinions; δ is the wheel on the arbor or axis of the barrel, which has the lifting pins for moving the hammer tails when they ftrike the bells, as has been already explained under our article CHIMES of a *clock*; θ is placed near the hammers, ι on the bell of fmalleft diameter, and κ at the fprings which bring the hammers back after each blow, and hold them in a fituation to be caught by the pins of the revolving barrel.

Let us turn now to *Plate* XVII., where we have a perfpective view of the whole frame and mechanifm before the front plate, with the fame letters of reference, as far as they go, as in *Plate* XII., and alfo as in *Plate* XVI., and where the Arabic figures are put to the mechanifm of the chimes that connects them with both the going and ftriking parts of the clock. In this plate, which is intended to explain the ordinary chime-work of a clock, the parts of the mechanifm which effect this purpofe are fo difpofed as to be nearly all feen in their refpective places of action. A, A, are placed on the back plate of the frame; B, B, B, B, are the four pillars of the frame; C, D, and α, are three ratchet wheels for adjufting and preferving the intenfities of their refpective main-fprings, and are placed before the front plate on the fquares of their barrel arbors, as in any other portable or fpring clock; E, F, and β, are the arbors of the three fufees for the key that winds them up; L and L are the parts of the two guards which are attached to the front plate, on the interior fide, out of fight, the fituation and action of which were explained under *Plate* XI., and the end of the third guard is hid by the fnail. K is the bridge of the dial-work, concealed from fight, fuch as is feen at the bottom of *Plate* XII., in *fig.* 3; I is the crown-wheel and pallets, to the verges of which the crutch of the pendulum is faftened behind the frame; and S is the fly of the ftriking part, held by a fmall fpring croffing the middle of its arbor near S, by fimple friction. The cannon pinion is hid from fight behind the 12-hour wheel *i*, but the pinion that acts with it, of the fame number of teeth, is feen at *g*, which is called the pinion of report, and has a pinion of 6 on its arbor pivoted into the cock *h*, and driving the 12-hour wheel of 72 teeth, as in the preceding clock; the fnail, however, of the ftriking part, is not here on the 12-hour wheel as before, but is attached to a ftar, or wheel with 12 pointed teeth, at *l*, one of which teeth is actuated each hour by a pin in the cannon pinion, and the foot, *f*, with a flender fpring, *e*, preffing againft its leg, the heel of which is placed between the two neareft teeth of the ftar, not only prevents their backward or forward motion during the lapfe of each hour, but alfo yielding to the impulfe given by the pin of the cannon-pinion at the end of each hour, allows the tooth to pafs the heel, and then returns to its original fituation by the force of the fpring *e*, at the fame time pufhing on the ftar the exact fpace of one interval, fo that when the ftar moves at all, it muft neceffarily move juft one twelfth part, and remain in that fituation together with its fnail, for the fpace of a whole hour before it is moved again, which is an effential condition where the ftriking part repeats the hour; otherwife the following inftead of the preceding hour might be ftruck by the clock during each latter half-hour of the day. The tail of the warning piece does not here reach to the pin in the pinion of report as in our former ftriking work, explained in *Plate* XII., but has the chime-mechanifm interpofed; in confequence of which arrangement the chimes play firft, and then fet the clock a ftriking when they have ceafed, at the end of every hour; for this purpofe two feparate racks with their refpective fprings, catches, and gathering pallets, &c. are neceffary in this machine, which may have their refpective offices thus explained: The pinion of report, *g*, goes round in an hour, as we have feen before, and has 4 pins which raife the crofs lever, 1, of the hammer, 1, 2, every quarter of an hour, a little before its completion; at the fame time that the crofs lever, 1, is lowered, the end, 2, of the hammer is depreffed, and its tail-piece in contact with the fpring, 3, is raifed, by reafon of its being placed behind the ftud which is at the centre of motion; the fpring, 3, confequently is raifed alfo from its ftate of reft, fo that when the pin of the pinion of report lets go the lever, 1, this fpring, 3, preffing back the hammer-tail, makes its head, 2, ftrike a projecting

projecting pin near the claw of the detent or catch, 7, after which it falls back again by its own weight; this ftroke of the hammer is powerful enough to drive the claw of catch 7 from the teeth of the chime-rack, 4, which therefore immediately falls back, by the force of its fpring, 6, preffing on its projecting end beyond the ftud, round which its motion is performed, until its tail, 5, falls on the neareft of the four fteps of the fmall fnail of the quarters attached to the pinion of report; in this fall as many teeth of the chime-rack pafs the claw of the catch as there are fteps in the fmall fnail, counted from its remote angular point to the end of the tail-piece of this rack; the greateft fall of the rack is the fpace of four teeth, when its tail refts on the ftep neareft the centre of the fnail for the firft quarter of an hour, and when its lever falls back againft the pin, 11, in the front plate of the frame; in the mean time the fupporting bar, 8, moveable on a ftud near its lower extremity, having been forced back by the long tail of the gathering pallet, 10, of the chime movement, now returns by the action of its fpring, 9, preffing its tail-piece below the centre of motion, and prefents its head to a fecond pin in the catch, 7, nearer its centre of motion, contiguous to the figure 7, fo that when the catch, 7, falls back by its own weight after it has been ftruck by the hammer, 2, its fecond pin falls on the head of the fupport, 8, where it remains until the rack of the chimes has its tail placed on the ftep of the fmall fnail; the chime movement, being at liberty as foon as the tail of the gathering pallet, 10, falls from a pin behind the chime-racks, not feen in the drawing, runs on, and prefently the faid tail of the pallet pufhing againft the inner fide of the fupport, 8, difengages the fecond pin of the catch 7, the claw of which now falls again into the teeth of the rack, and holds the rack while the pallet has drawn it home again, *i. e.* till the tail of this pallet falls on the concealed pin on the back of the chime-rack, the chime-barrel in the mean time revolving and raifing the hammer-tails of the eight concentric bells fixed on a common arbor at 1, and fupported by a bar attached to the front plate of the frame. This procefs is repeated three times at the quarters 1, 2, and 3; but when the hour is completed nearly, the fourth procefs does not ftop here; for as the chime-fnail now permits the rack belonging to it to fall back by its fpring, 6, the fpace of four teeth, the pin in the left hand end of this rack ftrikes the tail, *p*, of the hawk's-bill, *qrp*, moveable on a ftud at *r*, and thus raifes the catch-or bill, *q*, from the teeth of the ftriking rack, which now falls back as far as its fnail, *l*, will allow its tail, *n*, to come towards its centre; when the hawk's-bill is lifted in this manner it pufhes up the bent end, *t*, of the warning-piece from the pin of the wheel, R, in the frame, (*Plate* XVI.) and the fly makes two or three revolutions or more, which motion produces the noife of warning, but the little fpring, 13, immediately pufhes the warning-piece down again into the way of the pin of wheel R, the motion of which is thus prefently ftopped: the chimes of the hour are now going on while the four teeth of its rack are in the act of being drawn home, and when the fourth tooth of this rack is brought back, the chime-pallet falls on the concealed pin of the rack behind it, and nearly at the fame time the pin of the rack in front, which pufhed down the tail, *p*, of the hawk's-bill, now raifes the tail, *v*, of the warning-piece, *tv*, and confequently depreffes the end, *t*, of the fame fo far, that the pin of the wheel R is again free, and the ftriking of the clock goes on as in the ufual ftriking part, till the hour regulated by the 12-hour fnail is counted by the proper number of ftrokes; after which the tail of the gathering pallet, *s*, of the hours falls on its refting pin behind the hour rack, *mn*; it being a matter of no importance on which face of the rack this pin is fixed; nor yet on which fide of the bill, *q*, the gathering pallet is placed, provided there be the proper number of teeth in the rack to be drawn up by the pallet. The four holes, *a*, *b*, *c*, *d*, are to receive the four pillars of the dial; the horizontal lever, *y*, taking the pin of the catch 7, may have its tail, *z*, pulled by a ftring coming through the cafe, to make the chimes repeat, and the hour alfo at any time during the firft quarter after ftriking; and, laftly, the pin of the lever 14 15, or ftrike and filent, by being pufhed to the right or left, will bring the heel of the part 15 to a pin in the middle of the hammer lever, 2, or remove it therefrom accordingly as the chimes and ftriking are to be in ufe, or the contrary. After what we have already faid about trains, we think it not neceffary to particularife the numbers of teeth in the wheel-work.

A Clock without Dial-work, by Dr. Franklin.

When clocks had begun to be common, and a variety of complicated contrivances had been introduced into the different conftructions, it was at length deemed defirable to fimplify the mechanifm; and various attempts have been made to conftruct a clock with as few wheels and pinions as poffible. The late Dr. Franklin, and the late Mr. J. Ferguson both fucceeded in diminifhing the ufual number of wheels and pinions to three of the former and two of the latter, notwithftanding hours, minutes, and feconds, were all indicated by their contrivances; we propofe to defcribe them in fucceffion, and fhall begin with Dr. Franklin's firft, as being prior in point of time. *Fig.* 1, of *Plate* XVIII, of *Horology*, will explain fo much of Dr. Franklin's clock, in queftion, as is neceffary for conveying a fuitable idea of its conftruction. The face, or dial plate, is perfectly reprefented, and the dotted circles denote the wheels and pinions in the frame behind the face, on a fuppofition that the face and front plate of the frame are tranfparent; which mode of reprefentation not only places every wheel and pinion in its own place, but fhows their refpective diameters as well as if a fecond figure had been ufed for this purpofe, as has hitherto been ufual, when this clock has been defcribed by other writers. The great wheel, A, of 160 teeth, goes round, by means of the cord with a fufpended weight furrounding a pulley attached to it, in four hours; this wheel drives a pinion, B, of ten leaves, in $\frac{10}{160}$ of four hours, or one quarter of an hour; on the arbor of this pinion is the fecond wheel, C, of 120 teeth, actuating a fecond pinion, D, of 8, in $\frac{8}{120}$ of a quarter of an hour, which is one minute, and together with it, on the fame arbor, the third wheel, E, of 30 teeth, in the fame time; this third wheel is the ufual fwing-wheel of an ordinary 30-hours clock, and has a feconds pendulum, fufpended in the ufual way from a cock, by a piece of the main-fpring of a watch. Thefe are all the wheels and pinions made ufe of in the clock. The face is occupied by a fpiral line, as feen in the figure, and has the hours denoted by the Roman characters, which count from XII, in the order I, II, III, &c. as the fpiral goes, placed at intervals of a quadrant from each other; thefe hours, as well as the 60 minutes, placed four times over in a furrounding circle and denoted by the Arabic characters, are pointed to by a hand, P, placed by friction on the round part of the protruding arbor of the great wheel, A, that revolves in four hours. In the prefent fituation of this hand, the time indicated is forty minutes paft one of the three hours which it has laft paffed. *viz.* XII, IIII, or VIII, and it is prefumed that a miftake of four hours can hardly happen whenever the clock is examined. The fmall hand at D indicates feconds in the ufual way, and therefore requires no explanation. There can be no doubt but that a clock thus conftructed will meafure time as well as any other clock with a fimilar pendulum, provided it be well made. The objections which have

have been alleged againſt this conſtruction are, that it is poſſible a perſon awaking in the night, and examining ſuch a clock, may miſtake his time four hours very eaſily; and that that it will require being drawn up by its cord once every day, unleſs, indeed, the fall of its ſuſpended weight ſhould be made much greater than is uſual, or even convenient, in the generality of houſes. The writer of the preſent article had ſome years ago a clock, conſtructed nearly ſimilar to the preſent one, but its great wheel revolved once in every three hours, which allowed larger minute ſpaces in the ſurrounding circle, and required an additional ſpiral line; for the Roman figures were four deep at each third part of the circle; its uſe was to keep in motion the ſyſtem of Jupiter and his four ſatellites, which it did very well; an endleſs cord, with a detached ratchet, being applied to produce continual motion during winding as well as at other times. We do not learn, however, that Dr. Franklin's clock has been frequently copied.

A Clock with only three Wheels and two Pinions, by Mr. J. Ferguſon.

We have ſaid that Mr. J. Ferguſon alſo contrived a clock to ſhow hours, minutes, and ſeconds, with only three wheels and two pinions; his principal object was to obviate the objections we have ſtated to Dr. Franklin's clock; *fig.* 2 of *Plate* XVIII. gives a ſimilar repreſentation of Mr. Ferguſon's clock, as *fig.* 1 does of Dr. Franklin's: the great wheel, A, of 120, revolves in 12 hours attached to the pulley, round which the cord of the weight is ſtretched; this wheel drives the pinion, B, of 10 leaves, in $\frac{10}{120}$ of 12 hours, or in an exact hour, which is the value of this fraction; the ſecond wheel C, which revolves alſo in an hour, and which has alſo 120 teeth, drives the ſecond pinion D, of 6 leaves, in $\frac{6}{120}$ of an hour, or in three minutes; conſequently the pallet-wheel of 90 teeth, placed on the arbor of this latter pinion, revolves alſo in three minutes; the pallets of the anchor eſcapement, which act with this wheel, give an impulſe to a ſeconds pendulum at each vibration, as is uſual when the ſwing-wheel has only 30 teeth. The hours are marked in Roman characters, on a circular plate that is inſerted by friction, on the arbor of the 12-hour or great wheel, about three of which hour figures appear always through an aperture cut through the dial below the pinion B; and a fleur-de-lis, deſigned on the dial, ſerves as a hand to point to the hour to be indicated. We ſee no reaſon why a circle, a little ſmaller than the plate of the hours, might not have been marked with the Roman characters on the dial itſelf, within the circle of minutes, with a hand placed on the arbor of the 12-hour wheel to point to it as uſual, which method is certainly more ſimple, inaſmuch as that the circular ſmall plate would have been diſpenſed with and alſo the aperture in the dial. There is alſo another circular plate, divided into 60 three times over, and marked with the Arabic figures, placed by friction on the arbor of the three minute or pallet-wheel F, which is pointed to by another fleur-de-lis on the dial, as ſeen through another aperture of one third of a circle in length; this large plate, borne by the pivots of the pallet-wheel, muſt have been very injurious to the performance of the clock before us, where the power is diminiſhed in the ratio of 720 : 3, ſuppoſing the pallet wheel no bigger in diameter than the pulley, and that independently of friction; the inertia of ſo much matter as the ſeconds plate muſt contain, to be overcome at the return of each vibration, muſt have required a large maintaining power; an objection of which the inventor himſelf acknowledges the exiſtence; this objection, however, might very eaſily have been obviated, for a circle as large as the divided plate might have been drawn on the upper half of the dial, and would have admitted of ſimilar diviſions and figures to which a hand, borne by the pallet-wheel arbor, would have pointed, and indicated the ſeconds as truly as they are at preſent. The inventor, not aware of ſo ſimple a reſource, propoſes to get rid of the force of the objection by omitting the ſeconds altogether, as being of no real ſervice except in aſtronomical clocks; but why attempt the introduction of ſeconds at all, when ſtudying ſimplicity, if they are of no uſe? Mr. Ferguſon has alſo allowed another objection to his conſtruction, from which Dr. Franklin's was exempt, namely, that when the minute hand is at any time adjuſted, it does not alter the hour plate, which muſt have a ſeparate rectification, for effecting which conveniently he had 12 holes drilled in the ſmall 12-hour plate at equal diſtances, to receive the end of a pin whenever the hour-plate required adjuſtment; the ſeconds plate muſt alſo have been ſubject to a ſimilar inconvenience. Mr. Ferguſon has himſelf candidly ſtated one other objection to his clock, that ſeems to afford a better proof of his candour than of his judgment in clock-work, which is, that he ſuppoſes the total arc of vibration of the pendulum muſt be too ſmall with ſuch diminutive teeth as the highly numbered pallet-wheel afforded, and ſeems to have thought that a large arc is better than a ſmall one, by reaſon of the greater momentum of the pendulum, notwithſtanding he mentions a cycloidal arc as that in which all lengths are equally iſochronal: the fact however is, as is now univerſally allowed and proved in practice, that a ſmall arc near the point of the pendulum's quieſcence approximates the neareſt of any other part of a circular vibration to a cycloidal one, and that the momentum, which is not acquired by velocity, may and ought to be made up by the weight of the ball, moving in nearly as ſmall an arc as the eſcapement will admit, to ſuſtain a continuance of the vibration.

A Clock for exhibiting the apparent daily Motions of the Sun and Moon, and State of the Tides, &c. by Mr. J. Ferguſon.

Among the other ingenious contrivances, deſcribed in Mr. Ferguſon's "Select mechanical Exerciſes," is the clock of which we have juſt given the title, the ſimplicity of which has recommended it to the notice of various writers and compilers of dictionaries, and induces us to give it a place in our collection. *Fig.* 3. of *Plate* XVIII. is a reduced copy of the dial of the clock under our preſent conſideration, *fig.* 4. the dial-work, or wheels and pinion connected with the going part of a common 30-hours, or eight days clock, and regulating the motions of the different hands, plates, and contrivances for repreſenting the ebbing and flowing of the tide at any given place. The pinion of 19 leaves, in *fig.* 4, is attached to an arbor which revolves in eight hours, by its connection with the wheel-work within the frame, which revolution may be effected by a wheel of 64, taking into a pinion of eight leaves on the centre wheel arbor; then, as three times eight are 24, three times 19 are 57; the wheel of 57 conſequently revolves in 24 hours, with its centre in the centre of the dial; a ſecond wheel of 59 teeth (*viz.* $29\frac{1}{2} \times 2$) is alſo actuated by the ſame pinion of 19 in $\frac{59}{19}$ of 8 hours, which time is $24^h\ 50^m\ 52^6$, or $24^h\ 50^m\ 31^s.58$; this wheel is of preciſely the ſame diameter as the wheel of 57 teeth; and theſe two wheels, which are all that are neceſſary for producing the relative aſpects on the dial, may be called, that of 57 the *ſolar*, and that of 59 teeth the *lunar* wheel. The lunar wheel, which lies next to the clock frame, has a ſolid arbor, but the ſolar wheel, which covers it, has a tube juſt fitting this arbor, and turning on it without ſhake; the tube is ſhorter than the arbor and carries the ſmall circular plate repreſented by *fig.* 5, that has the 24 hours in Roman characters, and within theſe the $29\frac{1}{2}$ days ſpaces of the moon's age in Arabic figures; all equally divided; the attached piece, S, is the ſun's repreſentative, which thus revolves from the upper XII. in *fig.* 3, to the ſame again in 24 ſolar hours. On the face of this ſmall revolving dial is put a darkened ring, gradually increaſing in breadth half way round, and as gradually decreaſing the other half way. Over this revolving plate of 24 hours and $29\frac{1}{2}$ days is placed another ſmall plate (marked twice with the words *high water*, and as often with the words *low water*, alſo with *moon's age*, &c. as ſeen in *fig.* 3.) upon the arbor of the lunar wheel, ſo as to be adjuſtable on its round part by friction merely; this lunar plate has an aperture near its circumference, through which appear about five hours on the ſolar hour plate, and alſo a ſmall circular hole, within the former, as it regards the centre of motion; this plate has, moreover, the figure of an ellipſe darkened on it, which is nearly covered by a ſmaller circular plate, with concentric circles, borne by a ſupporting wire attached to it at one end, and at the other, when bent a little, to the principal dial-plate near the lower XII., as ſeen in the figure. The uſe of the ſmall circular hole in the lunar plate is to ſhow the phaſe of the moon in any part of its ſynodic revolution; for when in the ſituation F, *fig.* 5, as it regards the ſolar plate, no part of the darkened ring appears; which phænomenon denotes full moon a little ſhort of the fifteenth day's age; when it is at N, the dark part of the ſolar plate or ring covers it entirely at $29\frac{1}{2}$ days age, which poſition denotes new moon, and at 90° from theſe ſituations at both ſides, one half is dark and the other half light, which aſpects denote the quadratures. The ſun, S, attached to the ſolar plate is the index for ſolar time marked on the large hour circle of the principal dial, and the moon, M, pointing to the ſame, indicates the 24th parts of the earth's rotations, as they relate to the moon, which may be called ſo many lunar hours; 24 of which by this mechaniſm, we have ſeen, is upwards of $50\frac{1}{2}$ ſolar minutes longer than 24 ſolar hours. From the ſmall circle of the moon is carried a ſtraight wire over the hours of the ſolar plate, to indicate the mean time of the moon's mean paſſage over the meridian on any day of her age; and at the ſpace of $2\frac{1}{2}$ hours behind is another ſimilar one, to point out the time of a mean high water at London Bridge, which might be ſet to a diſtance, behind or before, correſponding to any other place on the globe, that has two tides in ſomewhat leſs than 25 hours. The ſtationary ſmall plate in the centre is intended to repreſent the earth, and the dot, L, between the fiftieth and ſixtieth parallel of latitude, repreſents London, with reſpect to which place the poſition of the ellipſe or tide-piece is altering its poſition viſibly every hour, agreeably to the words marked on the lunar plate. On the back of this lunar plate is likewiſe another ellipſe of ſolid braſs D, placed concentrically, which, revolving with it, lifts the lever E, moveable on a ſtud at F, when either of the ends comes in contact; but when the ſides are preſented the ſaid lever falls below the horizontal poſition; when this lever, F, is thus raiſed and lowered, twice in each lunar day, it carries with it the attached plate, H, above it, when kept in a perpendicular poſition by the four friction rollers, R, R, R, R, which plate has the ſea painted on it, as ſeen in *fig* 3, over the dial at H. The repreſentation of the phænomena depending on the relative poſitions of the ſun and moon thus ſimply effected, affords a pleaſing object for the eye at a trifling expence; but we are not to expect great accuracy when we conſider how many equations are required to reduce the mean places of thoſe two luminaries, particularly of the latter, to the true apparent places. Mr. Ferguſon made this mechaniſm ſo as to be capable of being actuated by a watch on the fuſee arbor of which a pinion of 20 was placed to drive a wheel of 40 round, on the arbor of which 40 was fixed the pinion of 19 that has been deſcribed as revolving in 8 hours, which it would do in this caſe, provided the fuſee itſelf revolve, as is uſual in common watches, in the ſpace of four hours.

There is, however, an inaccuracy in the numbers of the wheel-work adopted in the dial-work of this clock, which would render it too imperfect to be uſed for a conſiderable length of time without a new rectification, even provided the motions of the ſun and moon, or, more properly ſpeaking, of the earth and moon, were quite equable, as the conſtruction ſuppoſes, which inaccuracy may thus be explained; as the pinion of 19 drives both the wheels of 57 and 59, when the former has performed a revolution in a ſolar day, the latter falls two teeth ſhort of a revolution, which it completes not until two teeth of the ſecond revolution of the wheel 57 have been again impelled, ſo that in every 24 hours the little moon loſes $\frac{2}{59}$ of its revolution, which is a part of a relative retrograde motion, as it regards any point, for inſtance the upper hour XII. in the ſolar plate; ſo that as often as 2 are contained in 59, ſo many day ſpaces muſt there be on the ſolar plate, figured in a retrograde direction, as the figures regard the principal plate; but the value of $\frac{59}{2}$ is $29\frac{1}{2}$ exactly, which number of days meaſures the lunation according to theſe wheels exactly; there is, therefore, a monthly error of $44^m\ 3^s$ almoſt, which will amount to nearly an entire day in the ſhort ſpace of about 32 lunations.

But there is, moreover, a practical objection to the two wheels, 57 and 59, being both driven by the ſame pinion of 19, which is, that being of the ſame diameter, the diſtance between their teeth is not the ſame in both, one being $\frac{1}{57}$, and the other $\frac{1}{59}$ of a ſemicircle, ſuppoſing their teeth and ſpaces to be reſpectively equal to one another, but if both wheels are cut in the cutting-engine by the ſame cutter, the inequality will fall in the teeth entirely; in either caſe the action of one of the wheels muſt be bad if the other is properly proportioned, and periodic jerks will be the conſequence, which, in wheel-work going by a clock or watch movement, ought to be avoided. Whether or not Mr. Ferguſon had the dial of the Hampton-Court clock in his eye when he contrived the ſimple mechaniſm of this clock, we will not undertake to affirm, but we think it extremely probable that he had, particularly as he has copied the poſition of the annual train in another of his clocks, as we ſhall have occaſion to ſhew, under our article DIAL-*work*. Being in the habit of calculating numbers proper for repreſenting given periods of time in clocks, watches, orreries, &c. we have turned our thoughts towards the improvement of this clock, as well as of other pieces of mechaniſm, ſo far as relates to accuracy, and beg leave to lay before the reader the alteration that has occurred to us, for rendering the clock before us more perfect than it is in the ſtate we have deſcribed it.

When deſcribing the Hampton-Court clock we endeavoured to prove that when the moon's age is indicated by the difference of the velocities of the two hands, moving in the ſame direction, and repreſenting the ſun and moon, the latter ought to paſs the XII. o'clock point, on each day $50^m.473$ nearly later than on the preceding day; but by

Mr.

Mr. Ferguson's calculations we see the daily retrogradation is $50^m.526$, and the difference .053 amounts to an entire day's motion in a little more than 952 days, or somewhat upwards of 32 lunations, as we have stated. What therefore we want, in this case, is a couple of divisible numbers that shall be to each other very nearly in the ratio of 24^h to $24^h\ 50^m.473$, which numbers, by a peculiar arithmetical process, become familiar to us by practice, we have determined to be 2368 : 2451. These are the nearest possible numbers that can be got without ascending higher in the scale of continual ratios, and are luckily capable of reduction into composite numbers thus; 2368 taken as a product is equal to 74×32 and $2451 = 57 \times 43$; therefore the train $\frac{43}{74} \times \frac{57}{32}$ will be the wheel-work required; the solar wheel of 74 teeth being made to revolve with a tube as an arbor in 24 hours, by the clock-movement, must impel the wheel of 43 placed on a stud, or otherwise on the front plate of the frame, at one side of it, and this wheel of 43 must have the next driver, 32, pinned to it, to impel the last wheel, 57, or lunar wheel, placed on a solid arbor, concentrically behind the solar wheel, according to Mr. Ferguson's position, and the dials and other designs of the clock face may remain precisely as described; so that instead of the pinion of 19 impelling two unequal wheels at once, we shall have a pair of small wheels pinned together, one impelled by, and the other impelling its fellow, where the motion must be taken from an arbor of 12 hours, carrying a wheel of 37 to actuate the 74 in 24 hours, instead of from one of eight hours, as Mr. Ferguson proposed; which mode is equally practicable. As a proof of the accuracy of our calculation, we have by direct proportion as 2368 : 2451 :: $24^h : 24^h\ 50^m.4729729$, &c.; hence the deviation from the data is here only .0000271 of a minute in each lunar day, which will not amount to an error of an entire day in less than 1,862,472 such days, and therefore may be assumed as no bad substitute for the truth itself; seeing the clock will never be expected to go so long without cleaning or stoppage from some external cause.

Should it occur to the reader that 32 lunations constitute a period long enough for the clock of Mr. Ferguson to go, before a new rectification, we beg leave to suggest to him, that in the space of a lunar day there are two tides and two ebbs, consequently an error of three-quarters of an hour in each lunation will place the tide-plate, H, three hours wrong in the space of about four months, and in nearly eight months an high-water will be changed into low-water, and the reverse in the next eight months, which is certainly an indispensable error.

That the clock-maker may not be at a loss how to apply the remedy we have proposed for the inaccuracy of Mr. Ferguson's solar and lunar wheels, we shall conclude our description of the clock before us with an account of the exact dimensions of the parts proposed to be substituted. If we take the wheel of communication of 37 teeth at 12 per inch, measured at the pitch line, its geometrical diameter will be .98 or $\frac{98}{100}$ of an inch, and its practical diameter, with the addendum for the ends of the teeth, 1.04, as may be seen by inspection in our Table of Diameters, under the article Clock-*making*; the wheel of 74 being double will have its geometrical diameter equal to 1.96, and its practical one 2.02; the fellow of this last or solar wheel has its geometrical diameter by the same proportion, 1.14, and its practical one 1.20; the distance of the stud from the centre of motion of the solar and lunar wheels, must necessarily be the sum of the geometrical radii of these two last wheels, namely $\overline{1.96 + 1.14 \div 2}$ which is $= 1.55$; again the sum of the geometrical radii of the remaining two wheels, 32 and 57, must be also equal to 1.55, in order that the centres of motion of the solar and lunar wheels may exactly coincide; but a wheel of a geometrical diameter, equal to $\overline{1.55 \times 2}$ or 3.10 inches and of $\overline{32 + 57}$ or 89 teeth, will have only about 9 teeth per inch, according to our table, and the practical diameters of wheels 32 and 57, by the same, will be respectively 1.21 and 2.1. The calliper suitable for these proportions and dimensions is given, of half their full size, in *fig.* 2, of *Plate* X. which needs no further explanation, except that the wheels, 43 and 32, are so nearly of a size that one circle represents both, as pinned together, and revolving with a contemporary motion round a stud or screw in their centre, going into the front plate of the clock-frame. The small wheel of 32 acts deeper into the teeth of its fellow than the 43, by reason of having larger teeth than the other, though the wheel is of the same size.

Equation Clock by Enderlin, shewing also the relative Situations of the Sun and Moon, &c.

In our history of the successive improvements in clocks, we have mentioned the names of various ingenious men who contrived mechanism for exhibiting on the dial of a clock both mean and apparent time, and consequently the equation of time, which is the difference between these; to give drawings and descriptions of all the different methods of producing such an effect, would be like throwing ammunition at a dead mark; but to gratify the wishes of the curious, in a certain degree, we propose to select a clock of this kind, made on the continent by Enderlin, which appears to us to merit a description better than most of the others.

Fig. 1, of *Plate* XXIV. of *Horology*, exhibits the plan of the mechanism which constitutes the equation-portion of Enderlin's clock (See Traité de Thiout, p. 252, *Pl.* XXV.) and *fig.* 2, is an exact representation of the dial and hands; we will begin with *fig.* 1, or posterior plane of the dial, where the motion is communicated from the wheels within the frame, and proceed, in the order of the communication of this motion, through the different parts of this figure. The small wheel Q, of 24 teeth, borrows its motion from the movement of the striking part, in order that the going part may not be impeded by cumbersome additions; this wheel is not represented as a contrate wheel in the original, which we have copied on a reduced scale, but would act better if it were; it impels the small wheel R, of 32 teeth, with a vertical arbor, held to its position by a small cock, T, on the front plate of the frame, which arbor has a bend and compound joint, below T, and a second similar cock above, that keeps the lower half of the arbor in its position, while the upper and lower ends, or pivots, bear in their respective cocks, which it is not necessary to insert in the figure; this arbor has a single endless screw, S, on the middle of the inclined half, actuating a large wheel A, of 487 teeth, and also a pinion *a*, of 24 leaves, at the lower extremity, actuating a small wheel V, of 32 teeth, and making this revolve, we are told, in 24 hours. From these data we can now calculate the periods of the other wheels that have been mentioned; if V revolves in 24 hours, *a* revolves in $\frac{24}{32}$ of that time, namely, in 18 hours, and with it the bent arbor R T S *a*; also the small wheel, Q, revolves in $\frac{24}{32}$ of 18 hours, or in $13^h\ 30^m$, by means of its connection with the striking part, which may easily be effected by proper numbers in the teeth of the connecting wheel and pinion; likewise the large wheel A, of 487 teeth, revolves in $\frac{487}{1}$ of 18^h, which time reduced, is 8766 hours, or $365^d.6^h$; this

wheel,

wheel, therefore, is called the *annual* wheel. The wheel X, with 62 inclined teeth, and the wheel Z, with 90 teeth, revolve ſeparately round one common centre 5, and are impelled, the former by a ſingle tooth on the 24 hours arbor of the ſmall wheel V, and the latter by another ſingle endleſs ſcrew Y; this ſcrew Y has a pinion 6, of 21 leaves, on its upper end, impelled by the pinion *a* of 24 in $\frac{2}{2}\frac{1}{4}$ of 18 hours, which period is $59^d\ 1^h\ 30^m$ or the ſum of two lunations, where each is $29^d\ 12^h\ 45^m$; wheel X, we have ſaid, has 62 teeth, one of which is impelled every 24 hours, therefore an entire revolution of this wheel, if the motion were continued, would be performed in 62 days; but it will be ſeen by and bye, that it never is permitted to make more than one half of a revolution, and frequently not ſo much, before it is made to retrograde at one jump to its original ſituation, from which it had been moved by the ſingle tooth or pallet at V. Into the plane of the annual wheel A, are inſerted 12 pins, at ſuch diſtances from each other, in a concentric circle, as are determined by the number of days in the correſponding months, which are ſuppoſed to lie reſpectively between the ſaid pins, ſo as to regulate the interpoſed ſpaces. Theſe pins might be denominated the pins of January, February, &c. in ſucceſſion, as they follow one another at the regulated diſtances of $\frac{31}{365}$, $\frac{28}{365}$, &c. On the centre of the ſaid annual wheel, is alſo fixed the centre of a piece of metal, B, ſhaped by an oblong curve, continually varying its radius of curvature half round, and in a ſimilar way back again, which curve is denominated the *equation* curve, from the office it has to perform, which we will recur to preſently. Round the centre 5 of the two wheels X and Z, is moveable the lever 5 6, with a claw at 6, and a tail 5 3, reſting on a pin in the click 2 7 8, which click is moveable round a point at 7; a ſecond lever 10, has alſo its tail reſting by a pin at 3, on the tail-piece of the click, while its inclined end 10 falls in the way of the pins of the months in the annual wheel, by means of the preſſure of a ſlender ſpring near the centre of its motion. The effect produced by theſe levers and click may be thus deſcribed; the pallet at V continues, by its daily motion, to gather up a tooth of X every 24 hours, and the click 2, ſliding over the inclined ſide of the contiguous tooth, lays hold of it when paſt, and keeps it faſt till the next day, when the ſame operation is repeated, the two levers in the mean time remaining quieſcent; this daily proceſs goes on till one of the month pins of the annual wheel, meeting with the end of the lever 10, depreſſes it; at the time this occurrence takes place, the tail of this lever puſhes, by means of its pin of contact, the tail 3 of the click back, and with it the fang of the click, which now quits the tooth it before held; the wheel X being now detached from the click and pallet V, and having a piece of watch main-ſpring coiled round its centre, one end attached to it and the other to a ſtud, is pulled ſuddenly back to the ſituation in which it was before it was gathered up by the pallet; and the hand D, *fig.* 2, being connected with this wheel, returns with it and recommences its motion from the beginning of the graduations of the double ſemicircle A B C, which is divided on the middle line, but figured alternately above and below, to prevent the figures from being crowded. The ſemicircle at D in *fig.* 2, is divided into $29\frac{1}{2}$ equal ſpaces; and a portion of the dial is cut away, as repreſented by the dark portion, and full phaſe of the moon, which conſtitute a portion of the front face of wheel Z, of 90 teeth, which wheel we have ſaid revolves half round in $29^d\ 12^h\ 45^m$; this dark lunar plate has a mark as an index over the moon, which diſappears the moment that a ſimilar one at the oppoſite point of the plate's circumference begins to appear; alſo another moon, at preſent hid from ſight, begins to appear at one day's ſpace of her age, as ſoon as the preſent one diſappears at the ſpace $29\frac{1}{2}$, and ſo on alternately throughout the year. The graduations of the moon's age are on the principal dial. The anterior plane of the annual wheel has the ſun's place in the ecliptic, the months, with their diviſions into days, and ſun's riſing and ſetting in time correſponding to each day, marked on it, which readings appear through the blackened apertures of the dial, above the double ſemicircle of the days indicated by the hand D, and under the hours and minutes of the large circle, as repreſented in *fig.* 2.

When the wheel X, in *fig.* 1, has returned to its original ſituation, the pallet at V goes on, and when it has made half of a revolution it meets with the end of the lever 6, and preſſing againſt it, diſengages the tail 3 from the click 2, which prevented this click from falling into a notch of the wheel X, while this wheel retrograded; but now the click reſumes its office, and the wheel of the months proceeds by a daily progreſs of one tooth: in the preſent ſituation of the different hands and of the annual wheel, the day is the 5th of November, the moon's age almoſt 15, being full moon, the ſun-riſe, $7^h\ 5^m$, A. M. ſun-ſet, $4^h\ 54^m$, P. M. and the ſun's place 12° in Scorpio. The annual wheel, we have ſeen, revolves in $365\frac{1}{4}$ days exactly, therefore the fractional portion of a day will amount to unity every fourth year, and will require that February ſhould have 29 days in each leap year; a proviſion is here made for effecting this purpoſe which ſhews conſiderable ingenuity; a piece of braſs of the ſhape of 15 16 17 18, repreſented by dots chiefly, becauſe nearly hid behind the annual wheel, as now viewed in *fig.* 1, is moveable on the point 15, and has marked on the concealed flat part the 4 years ſucceſſively, *viz.* leap year, and 1, 2, and 3, after, which are brought annually in ſucceſſion to the aperture of the dial above VI. in *fig.* 2, by the annual wheel: the proceſs is thus; the ſtar 20 with eight angular points, has two of its points carried forwards by pins in the annual wheel, one of which acts on the night of the laſt day of December, and the other in ordinary years at the end of the 28th of February; the ſtar has a metallic leg, called by the French *un ſautoir*, preſſed by a ſlender ſpring in ſuch a way, that the heel falls into a notch of the ſtar, the uſe of which is, to limit the quantity the ſtar ſhall move if it is moved at all; for when any following point of the ſtar forces the leg forwards till it paſſes the heel, the ſpring then puſhes the heel in again, and forces the ſaid point forwards a little farther, till it ſtops under the foot, and till the next following point of the ſtar reſts againſt the back part of the leg where the motion is arreſted: a ſnail with four ſteps is faſt to the ſtar, and regulates the poſition of the piece 15 16 17 18 by ſupporting the end 18 of this piece, which projects towards the ſnail; hence it is eaſy to apprehend, that numbers 1, 2, 3, or leap year, will appear in the aperture of the dial, accordingly as ſtep 1, ſtep 2, &c. of the ſnail is preſented to the reſting point 18 of the plate with the four years marked on its anterior face. This contrivance, however, does not yet account for the 29th of February every fourth year: the additional mechaniſm for this purpoſe is the rack without teeth, marked 11, moveable on the centre of the annual wheel, but under it in *fig.* 1, with a concealed ſpring (dotted) preſſing againſt its concealed and dotted bar, ſo as to make it reſt on a ſecond ſnail behind the ſtar; for as the lever 10 for the monthly pins is carried by this rack, the ſaid lever may be made to meet the pins or recede from them, or any one of them, the quantity correſponding to a day, or more if it were required; thus the concealed ſnail, which has a contrary ſpiral, removes the pin anſwering to the laſt day of February ſo far from the correſponding

ponding pin in the annual wheel, that the hand D, in *fig.* 2, has arrived at the 29th divifion or day-fpace, before the click is detached by the lever 10, and thus February has 29 days every fourth year. It may be proper to obferve here, that the fourth ftep of the fnail feen on the ftar has an inclined plane to allow the end 18 of the four years plate to afcend it without being fet faft, after a revolution has been completed. Hitherto we have faid nothing about the equation portion ufually but improperly defcribed firft, which we have deferred becaufe it depends on the motion of the annual wheel. On the point D, in *fig.* 1, the rack, E, moves, while its tail, C, refts on the circumference of the equation curve, as it makes its annual revolution; at O is a fmall box with a fpring, which keeps the cord 15 always ftretched; this cord furrounds a pulley on the plane of a concealed wheel N, under K, but not attached to it, with which wheel, N, the rack acts and is kept always refting on the equation curve in every fituation of the annual wheel; the pinion, I, revolves by the going part of the clock in 60 minutes, and carries the equable hand of the minutes, now pointing to 46 minutes in *fig.* 2; by the pinion I, which has 30 leaves, the wheel K of 60 teeth is driven; which in its turn drives another pinion, L, of 30 leaves; fo that L revolves in an hour as though it acted immediately with I; to L is attached a wheel, H, of 48 teeth, which actuates a fimilar wheel, F, and this again a third fimilar wheel, G, the tube of which furrounds the arbor of I, and carries the hand with a little fun on its pointing to 30^{m} in the dial, *fig.* 2, which hand moves irregularly, and may be called the *equation hand*; its irregularity is thus produced; the wheel N, below K, is pinned to a bar which is not feen, but which bears the wheel H, and pinion L; and as the teeth of the rack are acting with the wheel N, the concealed bar, here fpoken of, is made to move alternately towards I and 15, as the radius of the equation varies during the year; this motion of the bearing bar makes the pinion L fometimes advance, and fometimes retrograde a few teeth, independently of the motion it receives from the revolution of K, and this additional motion, produced by the rack to the bar on which L refts, is alfo communicated to the wheel H in confequence of its connection with L, and hence to both the fimilar wheels connected with it, F and G, the latter of which gives the faid addition and retrogradation alternately to the equation hand borne by it. The motion conftituting the equation, here fpoken of, would be produced if the wheel K did not revolve at all, and therefore is derived entirely from the equation curve, which curve has its fhape afcertained from a table of the equation of time. In the pofition of the two minute hands the diftance between them is 16^{m}, which is the quantity of equation fubtractive at the end of the firft week of November. The pinion of fix or eight leaves attached to F, drives the hour wheel of 72 or 96 teeth not given here, which bears on its tube the hour hand pointing between II. and III. which hand therefore participates of the equated motion, and correfponds to the equation minute hand; the minute hand for mean time is ufed to fhew, by its diftance from the other hand, the quantity of the equation at any given time. The feconds are for mean time, and therefore correfpond with the minute hand of mean time. Between the fmall wheel Q, which we have feen revolves in 24 hours, and the dial, is a fmall circular dial-plate feen through an aperture below 40^{m} in *fig.* 2, the ufe of which is, to regulate the annual wheel, by a key inferted in the hole at 40^{m} on the fquare of the arbor of the wheel Q, agreeably to the hour of the day indicated by the clock, otherwife the monthly pins might not happen to detach the click of the ratchet-wheel that carries the month hand, D, at the exact termination of the month. Laftly, between the wheels I and G, is placed a fmall fpiral fpring, to prevent a fhake in the hand carried by G, which the play in the teeth might otherwife have occafioned. This mechanifm, as we have faid, is very ingenious, and moft of the other equation clocks have a fimilar equation curve, without an allowance for leap year.

We have recently underftood that Quare, of London, was the firft inventor of an equation clock, but we do not vouch for our authority.

An Aftronomical Clock, by Mr. Thos. Reid of Edinburgh.

It has been ufual, among horological writers, to call all thofe clocks indifcriminately *aftronomical*, which are ufed in obfervatories for afcertaining the right afcenfions of the ftars, whether the time indicated is folar or fidereal, and alfo thofe clocks which reprefent the motions of fome or all of the heavenly bodies, by appropriate wheel-work fuperadded to the ufual train: thefe two kinds of clocks, however, are of a very oppofite nature, one of them rejecting every appendage that does not contribute to accuracy in the meafurement of time, under all the changes of atmofpheric temperature, and the other being loaded with cumberfome additional mechanifm, calculated to produce great irregularities of motion in the going train, befides being deftitute of compenfation mechanifm for the effects of heat and cold. They are, indeed, equally the produce of much ingenuity; but one is contrived and executed with the greateft care for ufe, while the other is defigned, calculated, and conftructed for mere curiofity: one is as fuperior in its meafurement of time to a common clock, as a chronometer is to an ordinary watch; but the other may be confidered, in general, as holding a rank, in this refpect, as much below a common clock: we feel ourfelves, therefore, juftified in giving diftinguifhing names to two machines, which, however they may be claffed in one genus, are of very different fpecies; the clock which is ufed in aftronomical obfervations we fhall call an *aftronomical* clock; but the clock which exhibits the relative pofitions of the heavenly bodies, or of any of them, we fhall call a *planetary* clock; of which kind is that which we have partially defcribed as being in the palace at Hampton Court, and of which kind are moft of Mr. J. Fergufon's more complex clocks. Sometimes clocks are made on the principle to which we propofe to give the appellation of aftronomical, which yet are not ufed in obfervatories, but kept at the clock-makers, with the view of being made a ftandard by which to judge of the rate of other clocks and watches not yet brought to time; thefe are ufually denominated regulators, but when they have good compenfating pendulums and the beft efcapements, they differ not from aftronomical clocks in any thing but the name, which is borrowed from the ufe to which they are put of regulating the time by which the common clocks are to have their rate of going adjufted.

We might have prefented the public with a great variety of aftronomical clocks, each differing from the others in fome particular peculiar to itfelf; but if we give the defcriptions of a couple, all the other conftructions, or the greateft part of them, will be eafily apprehended from our articles ESCAPEMENT and PENDULUM, which are as much the characteriftic parts of an aftronomical clock, as the efcapement and balance are of a chronometer, which machine has been already defcribed.

Plate XXII. of *Horology*, exhibits, in different points of view, an aftronomical clock, made by Mr. Thomas Reid, of Edinburgh, of which his brother, who refides in Clerkenwell, London, obligingly permitted our draftfman to take the requifite drawings for rendering all the effential parts intelligible. *Fig.* 1, is a view of the movement, with the eye placed perpendicularly over it, and the front plate of the frame

frame taken off; and *fig.* 2, is the whole frame detached from the case and viewed laterally, which position may be considered as a section, the other being the plan of the works; these two figures are laid down on a scale of one third of the original size; *fig.* 3, is an enlarged representation of the pallets and escapement wheel, on a scale of one half the real size, and in a detached state; *fig.* 4, shows the back of the great wheel of 120 teeth, together with the small ratchet and auxiliary spring and detent in a detached state, of one third the real size; *fig.* 5, exhibits the gridiron, or compensating pendulum, with its cock and other parts of suspension, of one eighth part of the real size; and *fig.* 6, gives an enlarged view of the part that receives the crutch immediately above the superior end of the gridiron portion of the pendulum; we shall describe these six figures in the order by which we have here enumerated them.

A A, in *figs.* 1 and 2, stand at the superior end of the frame plates, above the pillars, which are omitted in the second figure, that the escapement wheel may not be concealed thereby; B, in both figures, is the cylindrical barrel for the cord C, of the weight, which constitutes the maintaining power; the wheels and pinions are denoted by the numbers of their teeth thus; 120 is the great wheel revolving in twelve hours; 10 is the pinion driven by it on the central arbor; the smallest, 96, is the centre wheel; 12, the pinion on the second wheel-arbor, which arbor has also the second wheel of the train, marked 90, impelling another pinion of 12 on the arbor of the third or escapement wheel of 30 teeth; the other two wheels of 96 each, revolving alike in 12 hours, are instead of dial-work. The small ratchet wheel has 40 teeth, and the large one or perpetual ratchet 200, that the teeth of its circumference may be numerous enough to catch the detent G, *figs.* 1 and 4, without allowing the intensity of the auxiliary spring, H, to be remitted; but the exact number of teeth is of no importance, provided they be both numerous and strong enough to answer the desired purpose; the number is here mentioned to show that the wheel has been cut in an engine which divided it into 200 equal serrated teeth. The spring, which is here a substitute for the maintaining power, during the time of winding up the clock, has its extreme ends brought nearer together than was the case with the auxiliary spring of the chronometers which we have above described, but the office performed is precisely the same, the only difference is in the mode of application; here one end of the auxiliary spring is attached to the great wheel, and the other pressing against the side of one of the crosses of the great ratchet, or perpetual ratchet, and, when the two ends of the spring are brought nearly together, their effort to separate becomes a temporary substitute for the maintaining power. The hour hand revolves in 12 hours along with the arbor of the great wheel of 120 teeth, on which it is placed by simple friction without a square; the long hand E, *fig.* 2, shows minutes from the arbor of the centre wheel, on a circle of 60 surrounding all the hands; and the third hand F, shows seconds on the arbor of the pallet wheel; this hand has a very small recoil on account of the faces of the pallets not being portions of concentric circles. The pallets have each a piece of agate to supersede the use of oil, and are of the isochronal kind, acting with teeth of a peculiar shape, both of which will be more particularly described under the article ESCAPEMENT: the two agates are retained in the hollow parts of the anchor of the pallets by three screws applying to each, two of which are visible in each pallet in the enlarged *fig.* 3; and the two vertical lines below the wheel of this figure represent, one of them, the position of the pendulum corresponding to the present situation of the pallets; and the other, which is drawn a little aside, the situation of the same at the instant of escaping; which with other particulars of this escapement will be more minutely detailed hereafter, according to our recent reference. The frame is mounted by four pillars, situated near the four corners of the two plates, at the small circles in *fig.* 1; and is firmly attached to the bracket or bearing piece L, *fig.* 2, at the superior part of the case, by the connecting cocks, K and K, screwed both to the frame and piece L. Though there is no dial-work in this clock, yet there are four little cocks, *a*, *b*, *c*, and *d*, on the front plate of the frame, the use of which is to make the wheels and pinions fall as nearly as can be at the middle of their respective arbors, so that there may not be more pressure at one pivot than at the opposite one of any arbor; a refinement which few other makers have attended to in their constructions. The French clock-makers have given an appropriate name to each end of the arbor, depending on the distance of the pivot from the wheel it bears, one end being called the *tige*, and the other the *tigeron*. The author of this construction attaches, and very properly, some importance to the position in which the great wheel and cord of suspension are placed, with respect to the centre pinion; it may be seen from an inspection of *fig.* 1, that the weight suspended by the cord C, resting on the acting tooth of the great wheel, does not press so much on the pivots as it would do if placed at the opposite side of the barrel: for in one case the pressure on the pivots may be said to be halved, but in the other doubled; supposing the barrel and wheel to be to each other as 1 : 2; besides, an uniformity of appearance is thus preserved in the three circles of the dial, which, we have not before said, is attached to the frame by pillars, two of which are seen in *fig.* 2, at *e* and *f*.

The pivots of all the arbors are made cylindrical, and act in conical holes in the cocks before, and in the plate behind, which holes have the smaller ends continued through the metal, at the back faces of which is presented the conical point of a piece of steel, to each fixed to the cock in such a way, that a drop of fine oil lies between it and the pivot, detained by its cohesive attraction in the form of a spherule: this contrivance keeps the oil in its place better than countersunk holes would do, and preserves it from impurities longer; the steel pieces at the pivots entering the pillar plate, are seen at *g*, *h*, and *i*.

Fig. 4, requires no further explanation than we have given of the great wheel and perpetual ratchet in *fig.* 1, except that the under-side is here presented to view, and that the same letters of reference apply to both figures. *Fig.* 5, is a representation of the gridiron pendulum, which we have said was invented by J. Harrison; the dimensions in the drawing are on a scale of one eighth of the real size of all the parts. The two obtuse angled cocks, *a* and *a*, braced with cross pieces near the top are firmly attached to the bearing part or bracket, L, in *fig.* 1, which has a portion taken out between L and L, in *fig.* 5, for the pendulum to vibrate in, but which cannot be seen in the position in *fig.* 1, where the cocks are on a larger scale, and where one of them is hid by the other, which appears as a straight vertical piece of metal above M; which M denotes the edge of the pendulum in this figure: on the superior ends of the two cocks is made an angular notch called a Y, the top of which it resembles, in which the pivots of a tranverse piece of steel, *b*, rest; this transverse piece has a central square hole that receives a little rod by which the pendulum is suspended by a slip of watch main-spring, tapered a little downwards; the three pins, one above the transverse piece, *b*, and the other two, at the opposite ends of the slip of spring, may all be seen in the figure without more particular reference; the rod attached to the upper end of the gridiron is of steel, and has an oblong hole

hole with a pair of agates fixed in it, ſo as to form the ſides of the oblong hole, ſeen on a larger ſcale in a detached ſtate in *fig.* 6; the uſe of this long aperture is to receive the pin of the crutch, carried by the arbor of the pallets, as ſeen at N, in *fig.* 2. By this method of ſuſpending a pendulum, the ball has liberty to find its own perpendicular ſituation without twiſting the ſlender ſpring of ſuſpenſion, and the tranſverſe piece, *b*, having ſhoulders to keep it ſteady within the frame compoſed of the cocks, *a* and *a*, prevents any motion or ſhake that might otherwiſe be occaſioned by the vibrations of a heavy pendulum; alſo the pendulum may be put into exact beat, by forming the ſhoulders of the tranſverſe piece, *b*, ſo exactly at each ſide of the centre, that the crutch may not require to be bent; and if a ſlight deviation from true beat be obſerved when the cocks are mounted *pro tempore*, a ſlight lateral motion in them, before final fixing, will rectify the beat; or the poſition of the agates of the crutch pin might be reſorted to for adjuſting a very minute deviation from true beat; or, laſtly, an adjuſtment of the bearing part, L, might be recurred to; but we think it is of importance not to alter the central ſituation of the oblong aperture where the crutch-pin acts, leſt this pin ſhould not impreſs its force to a point perpendicularly over the centre of gravity of the pendulum, and ſhould thereby produce a vacillation in the vibrations, which would be injurious to the exactitude of their obedience to the laws of gravity. The compenſation for heat and cold, and adjuſtments for time and temperature of this gridiron pendulum, may be explained thus; as we have not room to put in the letters of reference between the bars of the gridiron portion we propoſe to call the rod attached to the ſpring of ſuſpenſion 1, which is the middle rod of the nine parallel ones; the two contiguous rods, one at each ſide of N°. 1, we ſhall deſignate by the figures 2 and 2, reſpectively; the two next in order to the right and left from N°. 1, we ſhall call 3 and 3; in like manner the two next at each ſide, correſponding to each other, we ſhall name 4 and 4; and the two extreme ones, 5 and 5, reſpectively; this mode of deſcription will ſhorten the circumlocution we muſt otherwiſe have been obliged to uſe: N°. 1, we have already ſaid, is a ſteel rod proceeding downwards from the point of ſuſpenſion; it has a ſhort croſs-piece, to which are faſtened the inferior ends of the rods, 2 and 2, which are of braſs, and conſequently more expanſible than ſteel; theſe rods, which may be conſidered as one ſo far as relates to their expanſion, proceed upwards to another croſs-piece ſomewhat longer than the firſt, to which their upper ends are faſtened: again another pair of rods, 3 and 3, of ſteel are faſtened to this ſecond croſs-piece near its extremities, ſo as to include the three former rods, and deſcend down paſt the ends of the firſt croſs-piece into a third croſs-piece, below the firſt, to which third piece they are united: this third piece alſo ſupports another pair of braſs rods, 4 and 4, which aſcend to the uppermoſt or fourth croſs-piece of the gridiron, to which they are alſo attached; and, laſtly, another pair of ſteel rods attached to the extreme ends of this fourth croſs-piece, ſo as to include all the former ones, deſcend to the loweſt or fifth croſs-piece, to which they are attached, and by which the gridiron frame is completed. The effect of the expanſion of the deſcending rod, 1, is to lower the firſt croſs-piece; but the effect of the expanſion of the two aſcending rods, 2 and 2, is to elevate the ſecond croſs-piece more; the effect produced by the pair, 3 and 3, as this effect relates to the third croſs-piece, is ſimilar to the effect of rod 1, on the firſt croſs-piece; alſo the effect of the pair of braſs rods, 4 and 4, on the fourth croſs-piece, is ſimilar to that of bar 2 and 2, on the ſecond croſs-piece; and, laſtly, the effect of the pair, 5 and 5, of ſteel bars on the height of the fifth or laſt croſs-piece, is analogous to the effect produced by the pair, 3 and 3, on the third croſs-piece, as above deſcribed: the total effects of the deſcending and aſcending bars are ſtated to be thus in the clock before us: *viz.*

Expanſion downwards by rod 1,	=	1
Do. - - by rods 3 and 3,	=	1
Do. - - by rods 5 and 5,	=	1
Total	=	3
Expanſion upwards by rods 2 and 2,	=	1.56
Do. - - by rods 4 and 4,	=	1.56
Total	=	3.12

Hence the difference between the total elevation and total depreſſion of the ball, which is ſuſpended by the loweſt croſs-piece, is $\frac{12}{100}$ which quantity of exceſs of the aſcending bars is taken as an equivalent for the downward expanſion of the ſpring of ſuſpenſion, and that part of the ſteel rod, 1, which is above the gridiron portion; and if this quantity on trial ſhould be found to be an exact balance, the compenſation of the pendulum will be perfect, and the centre of oſcillation of the whole pendulum will continue unchanged as it regards the point of ſuſpenſion under all the variations of temperature. Whenever it is found that the calculation for the reſpective lengths of the expanſion bars has been erroneous, or the materials imperfect, the adjuſtment for temperature muſt be made by altering the ſituation of the firſt croſs-piece, ſo as to lengthen or ſhorten the pair of braſs bars, 2 and 2, as they regard the ſteel bar 1; it is better to make the braſs ſomewhat too long than too ſhort for the ſteel bars, becauſe this croſs-piece can be raiſed when it cannot be lowered beyond a certain limit depending on the proximity of the third croſs-piece; and if that limit ſhould be found too circumſcribed, the croſs-piece above, which has four bars faſt to it, muſt neceſſarily be lowered, which will be attended with ſome trouble.

The adjuſtment for time is made by means of a nut with a milled head, under the ball of the pendulum, the ſocket of which nut, is tapped to ſuit the thread of the ſhort rod attached to the loweſt croſs-piece of the gridiron frame by which the ball is borne, ſo that the ball reſts not with its inferior edge on the nut itſelf as in ordinary clocks, but has a pin in its centre which reſts on the upper end of the nut's ſocket, and which ſlides in a ſlit made in the bearing rod above and below, as well as at the centre of the ball for the ſake of allowing an aſcending and deſcending motion of the inſerted pin borne by the central hole of the ball. The uſe of this contrivance is to prevent any effect on the pendulum produced in general by the upward expanſion of the ball itſelf, for the ball being circular, will in this caſe expand from the bearing pin in the centre, alike upwards and downwards, as well as in a lateral direction to the right and left. Strictly ſpeaking, however, the bearing pin to have its true ſituation for preventing all effect from the expanſion of a large ball, ought to be placed in ſuch a ſituation as ſhall have a regard to the centre of oſcillation of the pendulum conjointly with the weight and dimenſions of the ball, which, in ſo complex a pendulum, conſtitutes a difficult problem to ſolve in all the variety of caſes likely to ariſe in practice.

This pendulum has a weight of 9 pounds, and the clock goes eight days with a weight of 16½ pounds, falling 4 feet 10 inches in this ſpace of time.

We

We are informed by Mr. Reid's brother, that the arc of escapement in each excursion is two degrees, and that the remaining portion of the said excursion is one degree; hence the whole arc of vibration is just 6° with the weight already mentioned. The peculiar property of this escapement, as will be seen in another place, is, that whatever be the maintaining power, or the total arc of vibration, the partial recoil which it has renders all the vibrations isochronal.

It is scarcely necessary to add, that an astronomical clock, like a chronometer, ought to have all its adjustments made with great care, in order that some dependence may be placed on its rate of going under all circumstances.

An astronomical Clock, with Troughton's Pendulum.

Each astronomical clock has usually some part or parts peculiar to itself; the clock, which we are now about to describe, differs from the preceding one in various particulars, but principally in the structure of its escapement, pendulum, suspension, and pivot holes. *Plate* XXIII. explains the works, and *Plate* XXVII. the pendulum and suspension of the clock before us: we will first describe *Plate* XXIII. *Fig.* 1, is a view of the movement, with the front plate removed, and the eye perpendicularly over the centre, which therefore may be called the plan of the movement; *fig.* 2, is a section, or side view of the frame, containing the movement, a part of the suspension and pendulum, and the dial-work, face, and hands; *fig.* 3, is a front view of the front plate of the frame, as seen with the hands and dial off; and *fig.* 4, is the escapement-wheel and pallets of the dead beat kind. In *figs.* 1 and 2, A is the great wheel with 120 teeth, and a perpetual ratchet of the same number of teeth, similar to the one described in the going fusee of Brockbank's chronometer, which great wheel impels the centre pinion, *a*, of twelve leaves in an hour; the arbor of this pinion carries the minute hand, or large hand seen in *fig.* 3, moving from the centre of the dial; B is the centre-wheel, of 96 teeth, on the same arbor, driving the pinion, *b*, of 12 leaves; and C is the second wheel of the train, with 90 teeth, urging the pinion *c*, of 12 leaves, on the arbor of the pallet or escapement-wheel D, which, as usual, has 30 teeth; this wheel revolves in $\frac{12}{96} \times \frac{12}{90}$ of an hour, which is exactly one minute, and therefore carries the hand of the seconds, or small hand, seen in *fig.* 3, above the minute hand. For a particular account of the dead-beat escapement, we beg leave to refer to the article ESCAPEMENT, a description of which in this place would lead us out of our purpose; *e* and *f* are two cocks on the pillar-plate or back plate, into which the arbor, *g*, of the guard gut, *h*, is pivoted; and *i* is the spring of the guard: below D, in *fig.* 2, is the arbor of the detent of the perpetual ratchet; and below *g* in the same figure is the arbor of the small dial-plate, and its wheel *k*, seen on the frame in *fig.* 3, by a front view; this wheel has 192 teeth, and borrows its motion from a pinion of 8 leaves on the arbor of the centre wheel, these numbers of teeth being to each other as 24 to 1; they are the only dial-work this clock has got; the dial-wheel might have had a hand like the centre wheel, but it would have had its motion retrograde, on which account a light plate, equally poised, and neatly engraved, is put on behind the principal dial instead of it, and the hours are read through a circular aperture made in the large dial, and a fleur-de-lis on the front of this large dial serves as a stationary hand for indicating the hour on the revolving plate, which makes a revolution in a day. At *l*, on the arbor of the centre wheel, behind the dial, is a lever for counterpoising the minute hand, which, being long, of course has a sensible weight, that would act alternately with and against the maintaining power as reduced at this arbor, if there were no such counterpoise. The four black circles, at the four corners of *fig.* 1, are the four pillars pointing to the eye, and near the centre wheel is the fifth; their corresponding outer ends are seen in similar situations, with their fastening screws in *fig.* 3; the other four black circles, within the pillars in *fig.* 3, are the holes for receiving the pillars of the large dial, or face of the clock, two of which pilars, *m* and *n*, are seen in *fig.* 2, and also the edge of the dial *o p*. The frame is screwed to E, the bearing part of the case, by two screws passing through this piece and entering the lower pillars of the frame, one of which screws is seen in *fig.* 2, at F, and the other is on the opposite side. The pivots of the pallet's verge, of the pallet-wheel arbor, and of the second-wheel arbor, all run in jewelled holes in the small cocks, 1, 2, and 3, respectively in *figs.* 2 and 3, and the piece 4, in *fig.* 3, covers the end of the detent-arbor pivot. This clock, which was made by the Brockbanks, and which is the property of Mr. Ed. Troughton, is finished in a very superior stile of workmanship; and though for many years it has been regulated by a pendulum composed of a glass tube and bulb containing quicksilver, which was Troughton's improvement on Graham's quicksilver pendulum, yet its performance has at least equalled, if not surpassed, in accuracy and steadiness of rate, the going of any other clock that has been constructed; if we may credit the report of an individual every way qualified to judge not only of its absolute but comparative excellence: the pendulum, however, which is here made a part of this clock, is of recent invention, and we have great reason to believe will shortly supersede all the other compensation pendulums that have preceded it, on account of its possessing the following properties almost exclusively: first, it has all the advantages of opposite expansions which the common gridiron pendulum possesses; secondly, the arrangement of the metallic parts gives it the simplicity in appearance of a single rod, as well as a diminished resistance of the air in its vibrations; thirdly, it has great strength without much weight above the ball; fourthly, its centre of oscillation, compared with the centre of the ball, or with the centre of gravity, may consequently be very nearly determined; fifthly, the motions of the compensating parts upwards and downwards are not effected by jerks, but are progressive and steady, while yet the parts are sufficiently braced to preserve their relative situations and figures; sixthly, the compensation not only includes the spring of suspension, and is adjustable for temperature, but has its adjustment for temperature made, independently of the rate of going, by a new pyrometer of great sensibility, and free from the usual objections against pyrometers; seventhly, it is capable of being put into beat without altering the shape of the crutch, or of any other part; and, lastly, it is capable of adjustment for rate even while going. It would have been more consistent with our general plan to have given an account of the pendulum in question under the word PENDULUM, but as that article is a remote one, the reader will doubtless have no objection to have the description here. The common gridiron pendulum has already been described in our preceding section, and its mode of compensation explained; therefore we need not use much circumlocution in our present detail. *Fig.* 1, of *Plate* XXVII. of *Horology*, shews a front view of the pendulum as suspended by a board or plug attached to a solid wall; and *fig.* 2 is a side view of the same: *fig.* 3, which represents pretty nearly the common gridiron pendulum, is a representation of the compound part of the pendulum with the rods placed to the right and left of the central one, for the sake of being all seen, instead of being arranged round it, as they ought to be agreeably to the circular section below, which is the true arrangement, but which is not so easy to be apprehended by mere verbal

verbal defcription, as we prefume it will be from the prefent, though unnatural, figure. The middle rod, reaching to the fufpenfion, is of fteel, which, as before, we will call 1, defcending to the crofs-piece *a*, which is a circular plate perforated with a central and four other holes, as feen in the fubjoined fection; under the central hole of this circular piece the rod 1 is pinned; then inftead of a pair of brafs rods afcending contiguous to, and at each fide of, this central fteel rod, as ufual, a brafs tube has its lower extremity faft to the circumference of this fmall plate *a*, and afcends within a fecond larger tube to another fimilar circular plate *b*, with five holes likewife, to which it is attached in like manner; the two brafs tubes are fo clofe together, that they appear as one, except that the inner one is darker than the other in the figure; to the plate *b*, above the afcending inner brafs tube, are faftened two fteel rods, which we will here denominate 2 and 2, appearing contiguous to the rod 1 in the figure, but actually at the fame diftance as the next pair denominated 3 and 3, which pairs ftand at right angles to one another, as feen in the fection under the figure; the pair of fteel rods, 2 and 2, having defcended from *b*, pafs through a pair of the holes of *a*, and are pinned under a tranfverfe circular piece, *c*, that has only four holes in it; to the circumference of this piece, *c*, is made faft the outer brafs tube that hides all the rods and other tube, and that afcends to a fourth circular piece, *d*, which forms an end-piece to the large tube, to which it is fcrewed faft; laftly, the pair of fteel rods, 3 and 3, are made faft to the end-piece *d*, and defcend through the other pair of holes of the three pieces, *b*, *a*, *c*, to a fifth crofs-piece, *e*, under which it is pinned, and by the central hole of which the ball of the pendulum is hung at its centre, or nearly fo. This fubftitution of two brafs tubes for four brafs rods, together with their end-pieces, and the mode of arranging the fine fteel rods within the tubes, afford all the properties united which we have previoufly mentioned; and it is eafy to fee that fhortening either the rods of fteel, or the brafs tubes, will alter the ratio of the refpective aggregates of their length, fo as to increafe or diminifh the effect of the oppofite expanfions, till they exactly balance one another, when a delicate and very exact meafure is taken from the point of fufpenfion to the centre of ofcillation: this meafure is taken with great eafe and accuracy by the new pyrometer which we have feen ufed, but which has not yet been publicly defcribed, and the principle of which we do not feel ourfelves at prefent authorifed to announce; when the author has gone through all his pyrometrical experiments, moft probably he will lay before the public the conftruction and the experiments at prefent in hand, both of which, no doubt, will prove interefting. Indeed, we can now venture to refer the reader to our article PYROMETER for the information we have fpoken of, prefuming upon the experienced liberality of the author. Above the end-piece, *d*, of the outer brafs tube is a fmall tube of the fame metal furrounding the fteel rod that afcends to the fufpenfion fpring, the ufe of which is to receive the fork of the crutch and to guard the fteel from the oxiding influence of the air, as well as, it is probable, the other rods will be preferved, by being enclofed in the brafs tubes. If we can difcover any thing like the fhadow of an objection to this ftructure of the compenfation-pendulum, it is, that the included fteel rods not being expofed to the air, may not be fo foon affected by atmofpheric changes of temperature as the brafs tubes; but we can hardly perfuade ourfelves that the difference will ever be fo fenfible as to affect the rate of going of the clock, feeing all the parts are very contiguous to one another throughout the whole length, and muft confequently affect one another's temperature, by that general contiguity, at all times.

Figs. 4, 5, and 7, are views of detached parts of the fufpenfion, and *figs.* 8 and 9 of the crutch, all which we come now to explain, as effential appendages of the pendulum. We have already faid, that E, in *fig.* 2, *Plate* XXIII. is the bracket on which the frame is fixed; to this bracket the brafs triangular frame G is fcrewed faft, different views of which may be feen in *figs.* 1, 2, and 5, of *Plate* XXVII., as well as in *fig.* 2, *Plate* XXIII., in the laft of which the fcale is larger than the others; at the top of this frame is a milled nut, and under it a micrometer head, *a*, in *figs.* 1, 2, and 3, of *Plate* XXVII., with 22 divifions on the circumference of the micrometer head to fuit the finenefs of the threads of the fcrew, by which the adjuftment for rate is made; below the divided head is a fquare piece of metal, *c*, in *fig.* 4, which in *figs.* 1 and 2 is concealed within a tube having a fquare hole, feen at *c*, in *fig.* 5, feparately; to the fuperior end of this tube with a fquare hole is an index made faft, pointing to the divided head, feen alfo in *fig.* 5, which figure 5, is a detached view, on a larger fcale, of the part *b* in *fig.* 2; at the lower end of the fquare piece *c*, in *fig.* 4, is the piece of watch main-fpring, called the fpring of fufpenfion, to which the middle fteel rod of the pendulum is pinned; that part of the piece *c*, which receives the nut and divided head, being tapped, draws the pendulum up and down when the divided head refts on the frame G, and is turned in a proper direction; the cock *b*, with its tube *c*, is fhewn as detached from the frame G, in *fig.* 5, but the manner in which this cock is attached to the frame, may be feen from the fittings; the dove-tailed piece, *d*, when detached from the cock *b*, enters the dove-tail of the cock G from *d* behind, above G, before the faftening fcrews attach it to the cock *b*, while the cock refts on the fhoulder of the frame G. *Fig.* 7, is a front view of the cock *b*, in *fig.* 5, and is on an enlarged fcale, as compared with the fame in *fig.* 1; *b* in this *fig.* 7, is a thumb-fcrew, which draws the two pieces, *e* and *f*, together, between which the fufpenfion fpring is clamped, after the adjuftment for rate is finifhed; fo that in fact, the lower edge of thefe two clamping pieces, *e* and *f*, may be called the point of fufpenfion, as it relates to the diftance from the centre of ofcillation of the pendulum; and is, therefore, the point from which the effective length of the pendulum ought to be meafured; the fcrew over *e*, which holds this piece to the tube, *c*, is in an oblong hole which allows the piece, *e*, to move without the faid tube. *Fig.* 6, is a front view of the frame top, when *fig.* 7, is taken off, the piece, *d*, in the cock, as fhown in *fig.* 5, is here fhown in its place inferted into the oblong dove-tailed hole of the frame, G, together with the two fcrew-holes at each fide of the letter, *d*; when the cock in *fig.* 7, is placed fo as to cover the fliding-piece, *d*, in *fig.* 6, and is fcrewed faft to this piece by two fcrews from behind the frame, it is evident that, when one of the two lateral preffing-fcrews, *s* and *s*, is turned forward and the other back, the fliding-piece, *d*, and with it the whole cock in *fig.* 7, will move from right to left, or the contrary, as the cafe may be, and confequently the pendulum, borne by this cock, as in *fig.* 1, will have a lateral adjuftment, to bring the clock into exact beat without the rough treatment of bending the crutch, which is often done in other clocks; when the fcrews are all faft, however, the cock is as fteady, as though it had neither the vertical nor horizontal adjuftment; and as the pin at the lower end of the fquare piece, *c*, *fig.* 4, is fmaller than the hole in the upper end of the fufpenfion-fpring, the pendulum finds its own vertical pofition before it is clamped, as well as if it were to be fufpended by Y's and a horizontal axis, that admit not of adjuftment for rate while the clock is going. *Fig.* 8, is the crutch, attached by the fuperior end to the verge of the pallets,

pallets, as feen behind the frame, in *fig.* 2, of *Plate* XXIII. where its fork embraces the brafs tube on the upper end of the pendulum, as feen in *fig.* 9, of our prefent plate; in Mr. Reid's clock, we have feen, a folid part of the fork entered an oblong hole in the pendulum, which hole was formed of two jewels to avoid friction; it may not appear to the reader at firft why there fhould be friction in this hole; but if he confiders that the bent end of the crutch moves in a portion of a circle of fmaller radius than the perforated part of the pendulum does, he will perceive that the crutch-pin muft neceffarily afcend in this hole during the excurfion of the pendulum, and the more, the further the pendulum fwings; to avoid the afcent of the forked part of our prefent crutch, it is not firmly attached, as in general, to the lower end of the vertical part of the crutch, but the crutch is firft bent into the form of an L reverfed, when viewed as in the plate, and then a pin is inferted into the remote end, *a*, of the bent part to become a centre of motion for the lever of the fork, which has a fmall tube moving round the pin fo as to keep it in the fame plane always, as feen in *fig.* 9; in *fig.* 8, the curve of the fork is in the fame plane, facing the eye, as its lever; by means of this contrivance the fork, which holds the top of the pendulum-rod faft, moves in the fame radius of curvature with it, and though it rifes and defcends alternately, as it regards the point, *a*, round which it has an unreftrained motion, yet it continues to clafp the fame part of the rod, and never flides from its hold, thereby avoiding all friction, except the infenfible quantity at the point of motion, *a*. The ball of the pendulum is of the lenticular fhape, but a little flattened at the circumference to allow it to hold more lead, and to be heavier than it otherwife would be. A fmall portion of a circle is attached to the frame below the ball of the pendulum, at 48 inches from the point of fufpenfion, and a pointer demitted from the ball fhows the number of tenths of an inch contained in the whole arc of vibration: this clock will go eight days, with a power of 3lb. $1\frac{1}{2}$ oz. avoirdupois, falling fix inches per day; its arc of efcapement, on the fcale, is 2.2 inches, and the total arc of vibration ufually, when the clock is clean, 3.4 inches; which numbers may be converted into degrees and minutes, thus: the circumference of a circle, of 48 inches radius, is very nearly 151 inches; therefore fay, as 151 in. : 360° :: 3.4 in. : 8° 6′, the whole arc of vibration; alfo as 151 in. : 360° :: 2.2 in. : 5° 14′, the whole arc of efcapement; hence each whole excurfion is 4° 3′, and each arc of action on the pallets in that excurfion 2° 37′, in the clock before us; which quantities are to each other very nearly as 17 to 11. We do not profefs to fay that this is the beft ratio to be adopted in practice between the arc of excurfion and the arc of action on the pallets, becaufe we believe that the experiments hitherto made on dead-beat efcapements, to determine the beft flopes of the faces of the pallets, have not been numerous enough to warrant a conclufion on this point, which, notwithftanding, we think of great importance, and recommend to the notice of clock-makers. Mr. Nicholfon, the author of the "Journal of Philofophy," &c. once informed us, in converfation, that his feconds pendulum clock with a dead-beat efcapement, having its curves nicely formed in a lathe, does not vary its daily rate more than a fecond per day, when the maintaining power is increafed as much as four or five times, or more, which we believe, is not ufually the cafe with dead-beat efcapements; we are not, however, at prefent informed of the exact quantity of the arc of action on the pallets, on which moft probably the ifochronal property chiefly depends, though we were at the fame time informed that the fufpenfion-fpring is tapered from the top downwards, in a way that aids the longer vibration by quickening them. The fubject is worthy of minute inveftigation, particularly as Berthoud has afferted, that in his experiments he always found his clocks retard with the dead-beat efcapement by the addition of weight to the maintaining power, and *vice-verfâ*, which was his reafon for contriving his ifochronal pallets. The experimentalift, however, ought to bear in mind, that both the length of the arc of action and the modification of the impulfe vary with the flope of the pallet's faces in the efcapement in queftion.

Clock-*maker*. If we were to define the word clock-maker agreeably to the derivation of the term, we fhould fimply fay that it means a man who makes clocks, and this definition, at one period of the art, would have been fufficient for our purpofe; but fince clocks have become fo common as to be confidered as articles of houfehold furniture, the art of making them has not been confined, as at firft, to one department of mechanics, but has gradually ramified into various branches, fo diftinct from one another, that the maker of one part is frequently unacquainted with the operations requifite for the manipulations of another, equally effential. Since the time that clocks became an article of our manufactories, requiring various tools and engines for facilitating their conftruction, the fubdivifion of the art into various departments was a natural confequence, which has been found to contribute to expedition, and confequently to cheapnefs; and, for the fame reafon that a tailor has no need to underftand either fpinning, weaving, or dying, a finifher of a clock has now no occafion to caft or cut his wheels himfelf, much lefs to make his fprings or enamel his dial-plate. From cuftom, however, that man is called a clock-maker, who finifhes or puts together the different conftituent parts of a clock when made, and who has his profit from the fale of the machine; though the makers, more properly fpeaking, are the workmen employed in making the frame and contained wheel-work. The different operations may, indeed, be moft of them performed by one workman, when the conftruction is intended to be peculiar, or the works of fuperior accuracy, but in general the different departments of the art may be feparately enumerated, agreeably to the fubjoined order, *viz.*

1. The brafs-founder cafts the wheels, plates, pillars, and faces, according to approved models:
2. The fpring-maker forges, fhapes, and tempers the main-fprings, to any required ftrength or dimenfions:
3. The making of the weights, to be ufed as maintaining powers of the balls, or bobs, and hands, may be confidered as one branch:
4. The man who keeps a cutting-engine and a fufee-engine, cuts the wheels and pinions, and forms the grooves on the fufee or barrel, accordingly as a fpring or fufpended weight is ufed as a maintaining power:
5. The movement-maker mounts the frame, makes the wheels, pinions, detents, &c. and places them in the frame, agreeably to the propofed calliper.
6. The clock fmith forges the fteel pieces for the arbors, pinions, pallets, rack, hammer, detents, &c.:
7. The bell-founder cafts the bell, or bells when the clock has chimes:
8. The enameller prepares the ground of the dial, or face, for receiving the colour of the figures, and gets the painter to lay on the figures, agreeably to the calliper, with or without a circle for the feconds:
9. When the face is not of real enamel, a japanner, or imitative enameller, prepares and finifhes the dial:
10. When the face is brafs filvered, an engraver ufually prepares, and fometimes alfo filvers it:

11. A

11. A jeweller is employed for the pallets and pivot-holes of the beſt aſtronomical clocks and regulators:

12. The gilder is frequently employed for preparing the ornamental parts of the caſe:

13. The glazier is applied to for the door of the ſuperior part of the caſe, when a ſeconds pendulum is uſed, and for the principal door ſometimes, when the clock has a ſhort pendulum:

14. The cabinet maker is reſorted to, uſually, for the caſe of the clock; and ſometimes alſo the carver:

15. The chain or cat-gut maker is indiſpenſably neceſſary:

16. Recently the tubular compenſation-pendulum has been made and adjuſted, by the mathematical inſtrument-maker, as being a portion that requires great preciſion:

17. Laſtly, the finiſher, or, as he is otherwiſe called, the *maker*, poliſhes the teeth and ſteel parts, finiſhes the pivots, verifies the engagement, adjuſts the eſcapement, limits the arc of vibration by adjuſting the maintaining power to the weight of the ball, regulates the adjuſtments for beat and rate, finiſhes the ſtriking and repeating parts, and puts the whole machine into a ſtate ready for ſale.

Clock-*makers, Company of*. See Company.

Clock-*making*. Clock-making, or the art of making clocks, ſeems not to hold that rank among the mechanical arts, which its connection with the ſciences, particularly that of aſtronomy, and alſo which the many ingenious improvements it has undergone by the help of ſcientific men, entitle us to expect. The cuſtom of working by piece-meal from eſtabliſhed models, which, it muſt be allowed, contributes greatly to expedition and cheapneſs, has, no doubt, conduced to exclude calculations and geometrical principles from the work-ſhops of the preſent day: whence it ariſes, that if we wiſh to be introduced to the workman who has had the greateſt ſhare in the conſtruction of our beſt clocks, we muſt often ſubmit to be conducted up ſome narrow paſſage of our metropolis, and to mount into a dirty attic, where we find illiterate ingenuity cloſely employed in earning a mere pittance, compared with the price which is put on the finiſhed machine by the vender of more eaſy circumſtances, though the latter has had little more trouble in the conſtruction than to order his name to be inſerted before it is placed for public notice in his bow-window. The practical departments of this art being thus frequently confined to the obſcurity of a garret, it is no wonder that a dexterity at performing certain manual operations, ſuch as hammering, filing, drilling, turning, ſoldering, tempering, poliſhing, &c. ſhould be conſidered as the perfection of the art, and that the reaſon is frequently not underſtood by the workman himſelf, and ſeldom by his employer, why the numbers of his wheels and pinions, and the ſhape, ſize, and diſpoſition of the different portions of his mechaniſm, are deemed preferable to others which he might have adopted as eaſily, if, in his apprenticeſhip, he had been ſo inſtructed. Indeed we have not in the Engliſh language any regular inſtructions for all the ſucceſſive portions of work to be performed in the conſtruction of a good clock, which want is much to be regretted; for, until the clock-maker by profeſſion can proceed in his work on ſcientific principles, he muſt be content to be a mere ſlave of imitation in an art, which is capable of affording him genuine pleaſure, from the opportunities it affords, of calling in ſcience to his aid in every ſtep that he takes, through an infinite variety of practical conſtructions. It would tranſcend the limits preſcribed to our plan, ſhould we enter into a detail of all the minutiæ of the art, but as none of our predeceſſors have given the mechanician any information on this intereſting ſubject, we will give a ſuccinct account of the principal operations, as they preſent themſelves in ſucceſſion, which, if it may not afford the expert and informed workman much inſtruction, will, we preſume, gratify the curioſity of the inquiſitive mind, as far as the detail goes.

The firſt requiſite to be determined, is the kind of machine to be made choice of; *viz.* whether the clock is to be portable or fixed; how long it ſhall go at one winding up; whether its maintaining power ſhall be a ſuſpended weight, or a ſpring; what kind of a dial plate or face it ſhall have for the indication of time; what ſhall be the nature of its eſcapement; of what materials its pendulum ſhall be compoſed; what ſhall be the time of a vibration; and whether or not it ſhall have the ſtriking work: all theſe and ſimilar determinations, muſt be made before the work is put into hand. We will ſuppoſe that a portable eight-days clock, with a half-ſeconds pendulum, and a ſpring for a maintaining power, is fixed upon as the inſtrument to be made, which we will take as an example, on account of the variety of the parts which ſuch an inſtrument conſiſts of; and, to render the account more complete, we will ſuppoſe that it be required to go whilſt it is wound up, and that, for the ſake of accuracy, it have the dead-beat eſcapement, and a compenſation pendulum; the laſt of which is not very uſual in a portable clock.

The going Part of an Eight-days Clock.

1. *Calculation.* Having determined upon the kind of clock to be made, the firſt thing to be done, and that in which the clock-maker is generally deficient, is, to calculate the movement, or proper number of teeth in the wheels, and of leaves in the pinions, of the going part of the mechaniſm. Dr. Derham, in his "Artificial Clock-maker," has treated this ſubject at conſiderable length, and has laid down rules which have tended more to puzzle than aſſiſt the workman in the choice of his numbers; he propoſes to take at random a certain number of vibrations per hour for a pendulum of an aſſumed length, to repreſent his train, and then to find the factors or numbers, which, uſed as multipliers, ſhall give the requiſite product, or nearly ſo; after which each factor is repreſented by a ratio of two optional numbers, to conſtitute a wheel and its pinion. We will not follow the Doctor through his proceſſes here, but merely obſerve, that, by calculating his whole movement at one operation from an aſſumed number of vibrations, he has introduced a variety of ſuch trains into portable clocks and watches, as make a vibration of the ſhort pendulum, and an oſcillation of the balance, no exact fraction of a ſecond; in ſhort he has begun at the wrong end of the buſineſs; has firſt fixed on the length of his pendulum in inches, without conſidering exactly the number of vibrations it would make, and then calculated a train that would ſo nearly ſuit it, that the adjuſtment for time by the bob, would compenſate the defect of the numbers; the conſequence has been, that the exact value of a vibration in a portable clock, and of an oſcillation in an ordinary watch, has hitherto been diſregarded in the conſtruction. On the contrary, we recommend to the clock-maker, firſt to fix upon his number of vibrations per ſecond, and then to calculate the true length of his pendulum, and exact value of his train agreeably to the number of vibrations per ſecond that he previouſly determined. The moſt ſimple way of calculating the numbers proper for the movement of any clock, intended to ſhow ſeconds, is, by dividing it into three portions, and then by calculating the wheels and pinions for each ſeparate portion, by a ſeparate calculation, beginning at the bottom of the train; thus, we firſt fix upon the pinion of the hour arbor to be, ſuppoſe 8, which is a good practical number; and as our piece is to go eight days, we will make the

the fusee to revolve in 12 hours, which construction will require the great wheel on its arbor to be 8 × 12, or 96, because the pinion of 8 revolves, with the minute hand on its projecting pivot, in one hour; hence if we divide 192, the number of hours in eight days, by 12, the time of one revolution of the great wheel, the quotient 16 will be the number of effective spiral grooves necessary to be cut on the circumference of the fusee, in order that the piece may go just eight days. This portion of the movement is not, however, called a part of the train, but only determines, as has been said, the time that the clock shall continue to go after each winding up of the maintaining power; and it is easy to conceive, that if a fusee or a barrel, with 24 turns of the catgut or chain, were placed on the hour arbor, the clock would go a natural day without the large wheel; and also, that if an intermediate wheel and pinion were placed on the arbor between the hour arbor and the great wheel, the time of going might be prolonged to 10, 12, or even 20 times eight days, but then the maintaining power must be proportionably increased, which circumstance renders such a construction by no means desirable in a regulator, particularly as the auxiliary spring now in use will keep the piece in motion during the act of winding up.

The remaining portion of the movement is properly called the *train*, including those wheels and pinions only, which are used for counting the vibrations made in an hour; the train is most easily ascertained by two calculations, one for the two wheels and two pinions which multiply the minutes into seconds, and the other for that wheel and pinion, or those wheels and pinions, which subdivide the seconds into vibrations; the former of these two portions of the train, like the first portion of the movement, or portion for the period of continuance, is the same for all clocks, let the time of vibration be what it may, a circumstance not usually considered; the ratio of velocity to be gained by the pinion on the arbor of the seconds hand, compared with the wheel on the arbor of the minutes hand, is required to be 60 : 1; which effect might be produced by one wheel of 300 teeth, and a pinion of 5 leaves, as is done in some of the ornamental French pieces; but the size of the wheel is cumbersome, therefore a pair of wheels, with a pair of pinions, one constituting a ratio or vulgar fraction equal in value to 8, and the other equal to $7\frac{1}{2}$, making $8 \times 7\frac{1}{2} = 60$, or any other two numbers making a similar product, will produce the same effect with fewer teeth; for if the pinions be each 8, the wheels, in this case, will be respectively 64 and 60, the compound ratio $\frac{8}{64} \times \frac{8}{60}$ being equal to the simple $\frac{1}{60}$; and, by the same process, if pinions of 10 had been chosen, the wheels would have been $8 \times 10 = 80$, and $10 \times 7\frac{1}{2} = 75$, which numbers would indeed have less friction than the preceding ones, by reason of their teeth acting at less depth, the diameters of the wheels remaining the same, and would moreover be capable of acting more behind than before the line joining the centres of the wheel and pinion; in like manner, pinions of 6 would require wheels of 48 and 45, and pinions of 12 wheels of 96 and 90, as may be seen in *Table* II. in our subsequent article, denominated CLOCK-*movement*.

The last portion of the movement, or second portion of the train, for a half-seconds pendulum, will require only one wheel of 60 teeth on the seconds arbor, properly shaped for the escapement; for as one tooth in the dead-beat and common anchor escapements escapes completely at two vibrations of the pendulum, 60 teeth will escape, that is, a whole revolution of the seconds hand will be made, in 120 vibrations; if, however, the pendulum had been required to vibrate seconds, the wheel in question, called usually the swing wheel, in opposition to the crown-wheel, which requires another escapement, would have demanded only 30 teeth for that purpose; and if three vibrations had been fixed upon, the number to correspond must have been 90, otherwise there must have been a wheel and pinion of the value of 3, like $\frac{8}{24}$, or $\frac{10}{30}$, in addition to the usual swing-wheel of 30; or, which is the same thing, a wheel and pinion of the value of 6, like $\frac{8}{48}$, or $\frac{10}{60}$, must have been introduced between the seconds arbor and a pallet, or swing-wheel of 15. (See *Table* III. under CLOCK-*movement*.) Thus all the variety in the calculation of trains, where seconds are indicated, is confined, as we have intimated, to the last portion of the movement, and the calculation itself is so simple, that the mere altering of the numbers of the pallet-wheel will convert a clock with a seconds pendulum into one with half-seconds, and *vice versâ*.

2. *Notation of the Numbers*. The calculation of numbers suitable for an eight-days clock with a half-seconds pendulum being thus readily obtained by three simple operations, which may be had by mere inspection from the three tables contained in the article CLOCK-*movement*, the whole may be represented, and its value estimated again by a compound fraction thus: *viz.* $\frac{8}{96}$ of $\overline{\frac{8}{64} \times \frac{8}{60}}\big|$ of $\frac{1}{60 \times 2}$ of 12 hours, or, which is the same thing in effect, thus, $\frac{8}{96} \times \frac{8}{64} \times \frac{8}{60} \times \frac{1}{60 \times 2} = \frac{512}{44236800} = \frac{1}{64800}$ of 12^h, or 86400 vibrations in 12 hours, which is the time of a revolution of the fusee, and great wheel, 96, on its arbor, and therefore $\frac{86400}{12}$, or 7200 vibrations, each of half a second in duration, in one hour, constitute the value of this train. This mode of notation gives the value better than any other perhaps that has been adopted; but the position of the wheels and pinions will be better understood from the ordinary mechanical method of writing them down; thus:

Great wheel 96
    Pin. 8—64 hour wheel
        Pin. 8 —60 second wheel
            Pin. 8—60 swing wheel
                2 pallets.

Indeed it is difficult to write down the movement by any one notation that shall express, at the same time, both the value and position of the wheel-work, on which account we recommend to the workman to write down his numbers by both forms, taking care in the method, by compound ratios, to put all the drivers under the line of division, and all the driven ones above; so that when an ascending movement is represented, the wheels may be the denominators, and when a descending one, the pinions. In our mode of calculation from the bottom of the train, the notation must be, as we have made it, ascending.

3. *Proportioning*. The calculation of proper numbers being made and noted down, the next stage of the work is proportioning the diameters of the wheels and of their respective pinions, so as to transmit the maintaining power from the fusee, or barrel in an ordinary 30 hours clock, to the pallets, and thence to the pendulum, to compensate the loss of motion which, when unaided, it would sustain from friction and the resistance of the air. If a wheel and pinion were to be made like two rollers, pressing their edges against one another

other, to produce a communication of rotatory motion, their diameters might and ought to be in geometrical proportion directly as their calculated numbers of teeth; but the force of the maintaining power would be too great to be ſuſtained by mere friction at the points of contact of ſuch rolling wheels and pinions; they have therefore been neceſſarily indented, and their teeth mutually inſerted ſo far into their correſponding ſpaces, as to prevent the revolution of one wheel or pinion without a correſponding motion produced in the next adjoining, which would not be the caſe with rollers, if a conſiderable force impelled them at one end of the train, and at the ſame time a retarding force oppoſed them at the other; for their ſurfaces, at the point of greateſt oppoſition to free motion, would mutually rub without effecting a communication of rotary motion beyond ſuch point. If now we call the points of contact of the two rollers, made in geometrical proportion to each other, or, in other words, the points where they pitch againſt one another, the pitch-line, and conceive a number of projecting little levers, or teeth, fixed at proper intervals from each other at theſe points of contact, in the circular pitch line of each roller, we ſhall have a true idea of two wheels properly proportioned to act together, which, when of unequal diameters, will not now be in geometrical proportion to each other, by reaſon of an equal lever, or length of tooth, being added to each ſeparately, after they were in exact geometrical proportion; moreover, it will be eaſily apprehended that the deviation from their original proportion, in the ſtate of rollers, will be the greater, the greater their diſparity of numbers. Hence it will be readily conceived, that the due proportioning of wheels and pinions is an important object in clock-making, for, ſuppoſing the teeth reſpectively of the true epicycloidal forms, inveſtigated under Clock-*movement*, unleſs the reſpective ſizes be properly adjuſted, the tranſmiſſion of the maintaining power, and communication of motion, will both be unequable, and the mechaniſm ſubject to rapid deſtruction. The uſual mode of proportioning, or ſizing wheels and pinions, as it is often called, is, firſt to make both a little too big for the propoſed calliper, and then, having rounded all the teeth of the pinion and a few of the correſponding wheels, to diminiſh the latter in the lathe, or turning frame, gradually, until, by ſucceſſive trials in the clock frame, they are found to act at a proper depth, when placed in the pivot-holes previouſly made; this vulgar mode we reprobate, as calculated to deſtroy the due practical proportions, and hope to ſee it baniſhed from the workſhops by the general adoption of a better method, which we have now to propoſe.

In proportioning wheels and pinions, after the numbers of their teeth are determined upon, two particulars are to be attended to, the coarſeneſs or ſolidity and the ſhape of the tooth; the former may be expreſſed by the number of teeth per inch in the circumference of the wheel; and the latter by the denomination epicycloidal: if a tooth were rounded in a circular ſhape, which we by no means recommend, the pitch-line would be conſidered as at one half the breadth of the tooth from the extreme edge; but when it is rounded, as we have recommended in our article Clock-*movement*, in an epicycloidal ſhape, or as the workmen call it, the *bay-leaf* form, Hatton has found, from numerous experiments, that the depth, or diſtance of the pitch line from the circumference, will generally be $\frac{3}{4}$ of the breadth of the tooth in any wheel or pinion; and as the epicycloidal is the beſt ſhape for the regular tranſmiſſion of force and velocity, we will adopt it as the beſt for practice. We have juſt ſaid that when an epicycloidal tooth is uſed, the diſtance of the pitch line from the end of the tooth is equal to $\frac{3}{4}$ of its breadth, and if we ſuppoſe the tooth and ſpace cut to be reciprocally equal, we ſhall have the true acting diameter of any wheel or pinion greater than the geometrical diameter, which Camus calls alſo the primitive diameter, by $\frac{3}{4}$ of a tooth or ſpace, on each ſide of the centre, or $1\frac{1}{2}$ in the whole diameter; let now a ſpace or a tooth be called a *meaſure*, and there will be double the number of meaſures as teeth in any wheel; alſo let theſe meaſures of the circumference be reduced into meaſures of the diameter by the uſual ratio, of 3.1416 : 1, and then $1\frac{1}{2}$ added to ſuch geometrical meaſures of the diameter will give the proper acting diameter, which may be expreſſed in inches and parts when the meaſures per inch are known. For inſtance, let our great wheel and its pinion $\frac{8}{96}$ be taken at 12 teeth per inch at the pitch-line, which may in practice be more or leſs, according to the thickneſs of the metal compared with the maintaining power as modified at the wheel's circumference; the number of meaſures of the great wheel is 192, *viz.* 96 teeth and 96 ſpaces, each meaſuring 1-24th of an inch; then as 3.1416 : 1 : 192 : 61.1; therefore, if to the geometrical diameter expreſſed by 61.1 meaſures there be added 1.5, the ſum 62.6 or $62\frac{6}{10}$ will be the acting diameter in the ſame denomination, which are ſo many 24th parts of an inch; but $\frac{62.6}{24}$ gives 2.6 inches for the full acting diameter of the wheel in queſtion:—again, the pinion 8 has 16 ſimilar meaſures in its circumference, and by the ſame proportion the diameter will be 5.09 meaſures; to which if 1.5 be added, the acting diameter will be $5.09 + 1.5 = 6.59$, or with ſufficient accuracy $6\frac{6}{10}$, which divided by 24 as before, will give the ſame $\frac{27}{100}$ of an inch, or ſomewhat more than a quarter for the acting diameter of the pinion. Upon theſe principles Hatton ("Introduction to the Mechanical Part of Clock and Watch-work," page 334,) has conſtructed a table of the ſizes of pinions meaſured diametrically, and compared by a pair of callipers with a given number of teeth and ſpaces, in their correſponding wheels, which many workmen copy in practice; but as his calculations are founded on a ſuppoſition that the ends of the teeth are circular, requiring unity as the ſupplemental portion, we find them differ eſſentially from Berthoud's determination in his "Eſſai ſur l'Horlogerie," p. 172, tome i. and ſhall therefore inſert here a new table calculated on a ſuppoſition that the curve is epicycloidal, and that the circumference is to the diameter as 3 : 1, inſtead of 3.1416 : 1; the reſult of which mode agrees very nearly with Berthoud's experiments on the proper ſizes of wheels and pinions, and therefore we recommend it to the notice of the accurate workman.

Table of the true practical Sizes of Pinions.

Teeth in the Pinions.	Meaſures of the Wheel for a Diameter of the Pinion.
3	3.5
4	4.1
5	4.8
6	5.5
7	6.1
8	6.8
9	7.5

Teeth

Teeth in the Pinions.	Meafures of the Wheel for a Diameter of the Pinion.
10	8.1
11	8.8
12	9.5
13	10.1
14	10.8
15	11.5
16	12.1

The procefs by which this table is calculated is fimply this; multiply the pinion by 2 for the meafures in the circumference; divide by 3 for the diameter, and add thereto $1\frac{1}{2}$ for the acting fize; thus for the diameter of a pinion of 6, it is $\overline{6 \times 2 \div 3 + 1\frac{1}{2}} = 5\frac{1}{2}$ or 5.5; namely, $6 \times 2 = 12$, and $\frac{12}{3} = 4$ and $4 + 1.5 = 5.5$ for the meafures; which laft quantity taken by the callipers acrofs the extreme edge of the wheel will be 3 teeth and $2\frac{1}{2}$ fpaces, or 3 fpaces and $2\frac{1}{2}$ teeth, which are here fuppofed to be cut but not rounded.

The application of this table, it is prefumed, cannot be eafily miftaken by any workman who underftands that the figures in the fecond column, to the left of the decimal point, mean fo many meafures, either teeth or fpaces, and the figure to the right of the faid point, fo many tenth parts more of a meafure to be added to the integral meafures. It may be proper to add here that a proportioned pinion muft be made fomewhat fmaller for a fmall wheel than for a large one, and alfo fmaller when driven than when it is the driver. We now know the numbers of our movement, and alfo that whatever the diameters of the wheels may be, our pinions of 8 muft be turned in the lathe till their diameters are precifely each 6.8 meafures, or three teeth and very nearly four fpaces, (taken by a pair of pinion callipers from their refpective wheels, in a ftraight line acrofs the ends of the teeth) either before or after they are flit, as the operation of dividing and cutting is called by the workmen. The diameters of the wheels are ufually made to diminifh as the train afcends, probably becaufe the force to overcome their inertia diminifhes, and the friction alfo is lefs in fine teeth with flender pivots, than in coarfe ones with thick pivots: indeed there feems to want a ftandard rule for the guide of workmen in this particular. Having taken the great wheel at 12 teeth per inch, meafured at the pitch line, we will take the centre wheel of 64 at 14, and the fecond wheel of 60 at 16, which will make fomething like a regular diminution in the fizes in the afcent of the train, and allow us room enough in our plates for the reprefentation. From thefe data, by the help of the foregoing directions, we readily afcertain the requifites for drawing the calliper as expreffed in the fubjoined table.

Table of Wheels and Pinions.

Wheels.		Teeth per Inch.	Acting Diameters in Inches.	Geometrical Diameters meafured from the Pitch Lines
Great wheel	96	12	2.60	2.55
Its pinion	8	12	0.273	0.21
Centre wheel	64	14	1.514	1.46
Its pinion	8	14	0.234	0.18
Second wheel	60	16	1.24	1.19
Its pinion	8	16	0.207	0.16
Swing wheel	60	16	1.24	

Pennington, of Camberwell, the ingenious mechanift who conftructed Mr. Mudge's time-piece, and gave the drawings in Mr. Mudge's pamphlet, has paid particular attention to the fubject of fizing wheels and pinions, and has publifhed a fmall pamphlet, recommending the ufe of his method of calculation by a fector of a peculiar conftruction; but we do not find that its ufe has become general. On conferring with him, we were informed that his practice is, to add $2\frac{1}{3}$ meafures of the geometrical diameter to the wheel, and $1\frac{1}{2}$ to the pinion, in watch-work, when the wheel is the driver; and $1\frac{8}{10}$ to each, when the pinion is the driver. This rule, it will be feen, differs very little from our theory above laid down, where we obferved that the driver ought to be fomewhat larger than the proportion affigned by the addition of $1\frac{1}{2}$ parts or meafures to each, and more particularly where the wheels have fmall diameters, as is the cafe in watch-work. But, as the good action of wheels and pinions depends upon their being duly proportioned and callipered, as well as on a proper fhape being given to the teeth, the latter of which requifites is defcribed under CLOCK-*movement,* we will not fatisfy ourfelves with having given our own method of fizing, which it is poffible the unlettered portion of the workmen may not underftand, by reafon of its requiring fome knowledge of arithmetical proportion, but we will add moreover the method of fizing pinions practifed and recommended by F. Berthoud, in his "Effai fur l'Horlogerie," which is as follows; *viz.*

No. of Teeth.	*The full or acting Diameter of the Pinion.*
4 =	two full teeth of the wheel, unrounded, and the fpace between.
5 =	three teeth, rounded from point to point.
6 =	three full teeth, unrounded.
7 =	three full teeth and a quarter of a fpace beyond.
8 =	four teeth, rounded, from point to point.
9 =	fomewhat lefs than four full teeth.
10 =	four full teeth.
11	no meafure given.
12 =	five full teeth.
13	no meafure given.
14 =	fix teeth, rounded from point to point.
15 =	fix full teeth.

The pinions in watches, he remarks, muft be fmaller in comparifon with their wheels than in clocks, agreeably to what we have above faid.

Copioufly, however, as we have treated the fubject of proportioning the fizes of wheels and pinions in a train of wheel-work, we think the fubject of fuch importance, that we are unwilling to fatisfy ourfelves with merely having explained, and illuftrated, by an example, the method of converting geometrical into practical dimenfions, but hope to render the bufinefs ftill more familiar by an extenfive table containing the geometrical diameters of all numbers from 4 to 64, and alfo twenty different variations in the ftrength of the tooth, which we have newly calculated on purpofe. The data, on which the table is calculated, have been already explained, and its application, it is prefumed, is fo eafy, that the geometrical diameter, in inches and decimal parts, of any wheel or pinion contained therein, may be feen by infpection, and immediately converted into the proper practical diameter, by the fimple addition of the quantity of engagement of the tooth, at the bottom of the fame vertical column, out of which the geometrical diameter is taken, which addenda are calculated on a fuppofition that the teeth are intended to be rounded of an epicycloidal form.

Table of the Geometrical Diameters of Wheels and Pinions.

Wheels and Pinions.	Teeth per Inch.																				
	4	5	6	7	8	9	10	11	12	13	14	15	16	17	18	19	20	21	22	23	24
4	.32	.25	.21	.18	.16	.14	.13	.12	.105	.098	09	.085	.08	.075	.07	.067	.064	061	.058	055	.053
5	.40	.32	.26	.23	.20	.18	.16	.15	.13	.12	.11	.106	.10	.094	.09	.084	.08	.076	.072	.07	.067
6	.48	.38	.32	.27	.24	.21	.19	.17	.16	.15	.14	.13	.12	.11	105	.10	.096	091	.087	.083	.08
7	.56	.45	.37	.32	.28	.25	.22	.20	.18	.17	.16	.15	.14	.13	.12	.12	.11	.106	101	.097	.093
8	.64	.51	.42	.36	.32	.28	.25	.23	.21	.20	.18	.17	.16	.15	.14	.13	.12	.12	.116	.11	.106
9	.72	.57	.48	.41	.36	.32	.29	.26	.24	.22	.20	.19	.18	.17	.16	.15	.14	.136	.13	.125	.12
10	.80	.64	.53	.45	.40	.35	.32	.29	.26	.24	.22	.21	.20	.19	.17	.17	.16	.15	.14	.14	.13
11	.88	.70	.58	.50	.44	.39	.35	.32	.29	.27	.25	.23	.22	.21	.19	.18	.17	.165	.16	.15	.14
12	.96	.76	.64	.54	.48	.42	.38	.35	.32	.29	.27	.25	.24	.22	.21	.20	.19	.18	.17	.17	.16
13	1.04	.82	.69	.59	.52	.46	.41	.38	.34	.32	.29	.28	.26	.24	.23	.22	.20	.196	.19	.18	.17
14	1.12	.89	.74	.64	.56	.49	.44	.40	.37	.34	.32	.30	.28	.26	.24	.23	.22	.21	20	.19	.18
15	1.20	.99	.80	.68	.60	.53	.48	.43	.40	.37	.34	.32	.30	.28	.26	.25	.24	.22	.21	.21	.20
16	1.28	1.02	.85	.73	.64	.56	.51	.46	.42	.39	.36	.34	.32	.30	.28	.27	.25	.24	.23	.22	.21
17	1.36	1.08	.90	.77	.67	.60	.54	.49	.45	.41	.38	.36	.34	.32	.30	.28	.27	.26	.24	.23	.22
18	1.44	1.15	.96	.82	.71	.64	.57	.52	.48	.44	.41	.38	.35	.34	.32	.30	.28	.27	.26	.25	.24
19	1.52	1.21	1.01	.86	.75	.67	.60	.55	.51	.46	.43	.40	.37	.35	.33	.32	.30	.29	.27	.26	.25
20	1.59	1.27	1.06	.91	.79	.71	.64	.58	.53	.49	.45	.42	.39	.37	.35	.33	.32	.30	.29	.28	.26
21	1.67	1.33	1.11	.96	.83	.74	.67	.61	.55	.51	.48	.44	.41	.39	.37	.35	.33	.32	.30	.29	.27
22	1.75	1.40	1.17	1.00	.87	.78	.70	.64	.58	.54	.50	.46	.43	.41	.39	.37	.35	.33	.32	.30	.29
23	1.83	1.46	1.22	1.04	.91	.81	.73	.67	.61	.56	.52	.48	.45	.43	.40	.38	.36	.34	.33	.32	.30
24	1.91	1.53	1.27	1.09	.96	.85	.76	.69	.64	.59	.54	.51	.47	.45	.42	.40	.38	.36	.34	.33	.32
25	1.99	1.59	1.33	1.14	.99	.88	.79	.72	.66	.61	.57	.53	.49	.47	.44	.42	.39	.38	.36	.35	.33
26	2.07	1.66	1.38	1.18	1.03	.92	.83	.75	.69	.64	.59	.55	.51	.49	.46	.43	.41	.39	.37	.36	.34
27	2.15	1.72	1.43	1.23	1.07	.96	.86	.78	.71	.66	.61	.57	.53	.51	.48	.45	.43	.41	.39	.37	.35
28	2.23	1.78	1.49	1.27	1.11	.99	.89	.81	.74	.68	.64	.59	.55	.52	.49	.47	.44	.42	.40	.39	.37
29	2.31	1.85	1.54	1.32	1.15	1.02	.93	.84	.77	.71	.66	.61	.57	.54	.51	.48	.46	.44	.42	.40	.38
30	2.39	1.91	1.59	1.37	1.19	1.06	.96	.87	.79	.73	.68	.64	.59	.56	.53	.50	.48	.46	.45	.41	.39
31	2.47	1.97	1.64	1.41	1.23	1.09	.99	.90	.82	.76	.70	.66	.61	.58	.55	.52	.49	.47	.45	.43	.41
32	2.55	2.04	1.70	1.45	1.27	1.13	1.02	.93	.85	.78	.72	.68	.64	.60	.57	.53	.51	.48	.46	.44	.42
33	2.63	2.10	1.75	1.50	1.31	1.17	1.05	.96	.87	.81	.75	.70	.66	.62	.59	55	.52	.50	.48	.45	.43
34	2.71	2.16	1.81	1.54	1.35	1.20	1.08	.98	.90	.83	.77	.72	.68	.64	.60	.57	.54	.51	.49	.47	.45
Epicycloidal Addenda	188	.150	.125	.107	.094	.083	075	.068	.063	.058	.054	.05	.047	.044	.042	.04	.038	.036	.034	.033	.031

Table of the Geometrical Diameters of Wheels and Pinions (continued.)

Wheels and Pinions.	Teeth per Inch. 4	5	6	7	8	9	10	11	12	13	14	15	16	17	18	19	20	21	22	23	24
35	2.79	2.22	1.85	1.51	1.39	1.23	1.11	1.01	.93	.86	.79	.74	.69	.65	.62	.59	.55	.53	.50	.48	.46
36	2.87	2.28	1.91	1.63	1.43	1.27	1.14	1.04	.96	.88	.81	.76	.71	.67	.64	.60	.57	.54	.52	.50	.48
37	2.95	2.35	1.96	1.68	1.47	1.31	1.18	1.07	.98	.91	.84	.79	.73	.69	.65	.62	.59	.56	.53	.51	.49
38	3.03	2.42	2.02	1.73	1.51	1.35	1.21	1.10	1.01	.93	.86	.81	.75	.71	.67	.64	.60	.58	.55	.52	.50
39	3.10	2.48	2.07	1.77	1.55	1.38	1.24	1.13	1.03	.96	.88	.83	.77	.73	.69	.65	.62	.59	.56	.54	.51
40	3.18	2.55	2.12	1.8.	1.59	1 41	1.27	1.16	1.06	.98	.91	.85	.79	.75	.70	.67	.64	.61	.58	.55	.53
41	3.27	2.61	2.17	1.87	1.63	1.45	1.30	1.19	1.08	1.01	.93	.87	.81	.77	.72	.69	.65	.62	.59	.57	.54
42	3.35	2.67	2.23	1.91	1.67	1.48	1.34	1.21	1.11	1.03	.96	.89	.83	.79	.74	.70	.67	.64	.60	.58	.55
43	3.43	2.73	2.28	1.96	1.71	1.52	1.37	1.24	1.14	1.05	.98	.91	.85	.80	.76	.72	.68	.65	.62	.59	.57
44	3.50	2.80	2.33	2.01	1.75	1.56	1.40	1.27	1.16	1.08	1.00	.93	.87	.82	.78	.74	.70	.67	.64	.61	.58
45	3.58	2.87	2.39	2.05	1.79	1.59	1.43	1.30	1.19	1.10	1.02	.95	.89	.84	.80	.75	.71	.68	.65	.62	.59
46	3.67	2.92	2.45	2.09	1.83	1.63	1.46	1.33	1.22	1.13	1.04	.97	.91	.86	.81	.77	.73	.70	.66	.64	.61
47	3.75	2.99	2.50	2.14	1.87	1.67	1.49	1.36	1.25	1.15	1.07	.99	.93	.88	.83	.79	.75	.71	.68	.65	.62
48	3.82	3.05	2.55	2.18	1.90	1.70	1.53	1.39	1.27	1.18	1.09	1.01	.95	.90	.85	.80	.76	.73	.69	.66	.64
49	3.90	3.12	2.60	2.23	1.94	1.73	1.57	1.42	1.30	1.20	1.11	1.04	.97	.92	.86	.82	.78	.74	.71	.67	.65
50	3.98	3.18	2.65	2.28	1 98	1.77	1.60	1.45	1.32	1.22	1.14	1.06	.99	.94	.88	.83	.80	.76	.72	.69	.66
51	4.06	3.24	2.70	2.32	2.02	1.81	1.63	1.48	1.35	1.25	1.16	1.08	1.01	.95	.90	.85	.81	.77	.74	.70	.67
52	4.14	3.31	2.76	2.36	2.06	1.84	1.66	1.51	1.38	1.27	1.18	1.10	1.03	.97	.92	.87	.83	.79	.75	.71	.69
53	4.21	3.37	2.81	2.41	2.10	1.88	1.69	1.54	1.40	1.30	1.20	1.12	1.05	.99	.94	.88	.84	.80	.77	.73	.70
54	4.30	3.43	2.87	2.45	2.14	1.91	1.72	1.56	1.43	1.32	1.22	1.14	1.07	1.01	.95	.90	.86	.81	.78	.74	.71
55	4.38	3.49	2.91	2.50	2.18	1.94	1.75	1.59	1.46	1.35	1.25	1.16	1.09	1.03	.97	.92	.89	.83	.79	.75	.73
56	4.46	3.56	2.97	2.54	2.22	1.98	1.78	1.62	1.49	1.37	1.27	1.19	1.11	1.05	.99	.93	.89	.84	.81	.77	.74
57	4.54	3.63	3.02	2.59	2.26	2.02	1.82	1.65	1.51	1.40	1.29	1.21	1.13	1.07	1.01	.95	.91	.86	.82	.78	.75
58	4.62	3.69	3.08	2.64	2.30	2.06	1.85	1.68	1.54	1.42	1.31	1.23	1.15	1.09	1.03	.97	.92	.88	.84	.80	.77
59	4.69	3.75	3.13	2.68	2.34	2.09	1.88	1.71	1 .56	1.45	1.34	1.25	1.17	1.10	1.05	.98	.94	.89	.85	.81	.78
60	4.77	3.82	3.18	2.73	2.38	2.12	1.91	1.74	1.59	1.47	1.36	1.27	1.19	1.12	1.06	1.00	.95	.91	.87	.82	.79
61	4.86	3.88	3.23	2.78	2.42	2.16	1.94	1.77	1.61	1.50	1.38	1.29	1.21	1.14	1.08	1.02	.97	.93	.88	.84	.80
62	4.94	3.94	3.29	2.82	2.46	2.19	1.98	1.79	1.64	1.52	1.41	1.31	1.23	1.16	1.09	1.03	.99	.94	.89	.85	.82
63	5.02	4.00	3.34	2.87	2.50	2.23	2.01	1.82	1.67	1.54	1.43	1.33	1.25	1.18	1.11	1.05	1.00	.96	.91	.87	.83
64	5.10	4.07	3.40	2.92	2.54	2.27	2.04	1.85	1.70	1.57	1.46	1.36	1.27	1.20	1.13	1.07	1.02	.97	.92	.88	.85
Epicycloidal Addenda	.188	1.50	.125	.107	.094	.083	.075	.068	.063	.058	.054	.05	.047	.044	.042	.04	.038	0.36	.034	.033	.031

Use

Use of the Table.

1. In the preceding table, the figures at the top of each column, from 4 to 24, denote the numbers of teeth per inch, and thofe at the left hand, from 4 to 64, denote the numbers of fuch teeth in the refpective wheels and pinions, and the fquares, formed by the vertical and horizontal columns interfecting each other, contain, in inches and decimal parts of an inch, the geometrical diameters of the wheels or pinions fo circumftanced, to which diameters, if the quantity of engagement entitled, "addenda" at the bottom of the vertical column be added, the fum will be the practical fize of the wheel or pinion ftanding at the left hand of the faid horizontal column; for inftance, let us take our centre wheel of 64 of 14 per inch, and we fhall find at the interfection of 64 at the fide and 14 at the top, 1.46 for the diameter, and under it .054, as the quantity to be added, which together make 1.514, agreeably to our former calculation; alfo, if we take its pinion 8, the interfection of 8 at the fide and 14 at the top is 0.18, and the quantity to be added again .054, the fum of which is 0.234, as before. The application of the table to all other numbers, both of teeth in the whole, and teeth per inch, is equally eafy and accurate; and even if the numbers run too high for the vertical columns, like our great wheel 96, the refult may be as well obtained by taking the halves, or other component parts, feparately, and ufing their fum afterwards as a whole; for half of 96 is 48 and the interfection of 48 with 12 per inch is 1.27, which doubled gives 2.54 for 96, to which add 063 and the practical diameter is 2.603 as before determined; or otherwife take out for 50 + 46 and the fum will be 1.32 + 1.22 = 2.54 for the geometrical diameter again as before. This mode of application renders the table unlimited in its extent.

2. But the determination of the practical diameter of a wheel, fuitable for a given number of teeth, of a given ftrength, is not the only ufe of this table; any two of the three things contained being given in clock and watch-work, the third may be found by infpection, of which the preceding is but one variety; as another exemplification, if it were required to cut a wheel of fome given dimenfions into a given number of teeth, the fize of the cutter, which is of great importance, may be immediately afcertained; for, fuppofing that a wheel of 1 inch diameter be required to be cut into 45 teeth, if we follow the column 45 horizontally towards the right hand till we come at .95, which has .05 for its epicycloidal addition, the fum 1.00 or 1 is found in the column 15 per inch, which denotes twice the thicknefs of the cutter; and though the fum may feldom be found, in other inftances, to be fo exact without a fraction, yet the neareft number to the given diameter is always readily found, and if the difference between the teeth per inch, in the columns next above and next below, be taken, the proportional part, properly applied, will give a decimal quantity to be added to the thicker cutter, or lower number per inch, provided very great accuracy be requifite in the thicknefs of the cutter.

3. When the cutter and dimenfions of the wheel are given without the number of teeth, which may be the cafe in fome inftruments with rack adjuftment work, where calculations are out of the queftion, come down the horizontal column from the teeth per inch fuitable for the cutter, till the diameter given is met with; thus in the column of 5 per inch, for a wheel of 2 inches, the addition .15 added to 1.85 makes 2.0 exactly, and this coincidence happens in the horizontal column 29, which number will therefore be that of the wheel, for which the engine for cutting muft be fet to perform the requifite work. Thus the table is not only extremely ferviceable to the clock-maker, but to all inftrument-makers where wheel-work is required.

4. Laftly, when the number of teeth per inch is not afcertained in a wheel, and the diameter of a correfponding pinion of a given number is wanted, the meafured diameter of the wheel, excluding the rounded ends of the teeth, will be at any time fufficient for finding, by the table, the correfponding pinion, and *vice versâ*; for fuppofing the wheel to be 48, and its diameter, exclufively of the rounded ends of the teeth, 1.7 or $1\frac{7}{10}$ inches; from 48 on the left hand vertical column proceed horizontally till 1.7 is found, which will be in column 9 per inch; then in that vertical column afcend till the given pinion, 8 for inftance, ftands oppofite on the left hand, and its geometrical diameter is .28, to which, if its epicycloidal addendum be added, *viz.* .083, the fum .363 will be the practical diameter of the pinion wanted: and in this way any wheel or pinion, previoufly made for a foreign purpofe, may readily and accurately be made a portion of a clock-movement; and the calliper made accordingly.

4. *Callipering.*—We come next to laying down the plan or calliper of the clock-movement on pafteboard to be transferred to the plates of the frame when properly hammered, filed, and fcraped, or planed; the difpofition of the calliper depends not on the acting, but on the geometrical proportions of the wheels and pinions, conjointly with the difpofition of the circles of indication on the face; when the wheels are fmall, fuch as we have chofen for our example of a half-fecond's pendulum, the diftance from the minute to the fecond hands will not be too great for an ordinary face, if the arbors are pivoted in a ftraight line, as in *fig.* 3, of *Plate* X. where the dotted circles, taken from the laft column of our Table of *Wheels* and *Pinions*, reprefent the geometrical proportions, or the places of the pitch-lines, and the complete circles taken from the column of "acting diameters in inches," are fuppofed to coincide with the extreme ends of the teeth; hence the little fpaces contained between the dotted and complete circles, at each fide, reprefent the additional meafure and half of each wheel and pinion, fuch as were determined by calculation, to convert the geometrical into the actual fize, or fuch as are given by the large Table of *geometrical diameters* and *addenda*. It is evident, therefore, to the eye, that the diftance between any two arbors or pivot holes is always equal to the fum of the geometrical radii of the wheel and pinion, which act together; this confideration renders the bufinefs of callipering very fimple; for, the centre wheel of 64 being defcribed from any convenient point in the given plane, a portion of a larger circle, A B, may be defcribed with an extent equal to $\frac{1.46+0.18}{2} = .82$, or $\frac{82}{100}$ of an inch, which is half the fum of the geometrical radii of the wheel 64, and of its pinion 8, and the pivot hole of the pinion may be in any point of this chord line; we have fixed upon a point in a line parallel to the fide of the plate, from which as a centre we defcribe the fecond wheel of 60, and alfo the pinion of 8 on its arbor, to be actuated by the centre wheel of 64; we now take another fweep from this determined point or pivot hole, with the extent $\frac{1.19+.16}{2} = .67$, or $\frac{67}{100}$ of an inch, which is again half the fum of the geometrical radii of the fecond wheel and its pinion, according to our table; and the pivot hole of the feconds arbor, or pallet wheel, may be in any point of the chord C D; but we have faid we propofed to have all the pinions in a ftraight line; another

another point is consequently fixed upon in a line parallel to the edge of the plate, which could not have been the case, if the wheels had been large, like those usually adopted to sustain the great maintaining power of a seconds pendulum. Again with the extent $\frac{2.55+.21}{2} = 1.37$, or $1\frac{37}{100}$ inches, or sum of the geometrical radii of the great wheel and centre pinion, which it actuates, we describe the portion of a circle E F, in any point of which the pivot hole may be placed for the fusee arbor; we have placed it, in our calliper, in a point at right angles to the line of the centres of the train, which is a matter of option, and the spring barrel may be either above, below, or on one side of the fusee, as fancy, or the room left by the other work, may direct. From what has been here said, it will be easy to conceive that there is almost an endless variety in calliper-drawing, the disposition depending on the variable sizes of the wheels and pinions of a movement, compared with the distance from the seconds to the minute-hand arbors; but the particulars we have here detailed, being thoroughly understood, will suffice as a guide in all possible cases; for, supposing a face to be previously given, and the centres for the minute and second hands already made, the pivot hole for the second wheel may easily be determined by intersection from the two given centres with the respective extents as above determined; provided the diameters of the wheels be calculated large enough for the distance of the given centres; that is, provided the aggregate of the geometrical radii of the two interposed wheels and pinions exceed the said distance.

As we intend to give a particular account, at some length, of the various kinds of pallets, under the article ESCAPEMENT, we shall here satisfy ourselves with a brief delineation of the dead-beat pallets, which we have proposed to adopt in our half seconds little clock, and defer, for the present, our account of its peculiar properties. The number of teeth in the pallet-wheel has been determined to be 60, of which any portion less than half may be chosen to be included between the points of the pallets, as will be seen hereafter; we will assume sixteen as a suitable number for a convenient construction; then, as 16 teeth bear the same proportion to 60, the whole number, that 96° do to 360°; from the point *a*, at the extremity of the wheel, in the line of the centres, we set off *a b* and *a c*, each equal to 48°, and from the points *b and c*, draw tangents which will meet at the point *d*, which point will be the required centre of motion for the axis of the pallets: from this point *d*, with the extent *d b*, we draw the curve lines *b e* and *c f*, for the extremities of the pallets, and also parallel thereto, with a smaller extent; two other curves for the interior limits of the pallets, care being taken that the thickness of the pallets, or space between the two concentric curves, be equal to almost one half of a space contained between the extreme ends of two contiguous teeth; the body of the pallets may then be described at option by two curves from the centre of the pallet wheel, reserving a little semicircular portion in addition, for the centre of motion. Had we fixed upon 12 teeth only to be contained between the points of the pallets, the centre of motion would have been, as appears in the figure, nearer the pallet-wheel, as at the interior edge of the body of the pallets; and if more teeth than 16 had been taken, the point *d* would have been more distant than is represented, and the pallets would have required a stem to have brought them down low enough; so that the shape, as well as size of the pallets, depends on the number of teeth included between their acting points, it being advantageous for the action that two lines, drawn from the centres of motion of the pallets and the pallet-wheel respectively, to the acting points, should always form a right angle. With regard to the sloping faces of the pallets, their direction is determined geometrically from the angles that the pendulum is required to pass through in its vibration before the wheel escapes, which is therefore called the angle of escapement: we will not attempt here to ascertain what angle of escapement is best for a given angle of the whole excursion, but assume it, for our present purpose, at 2° on each side of the perpendicular, as a guide for the slope of our intended pallets. From the centre of motion of the pallets, *d*, with any extent, *d g*, therefore, we describe the small chords, *g h*, and *i k*, from the dotted tangent lines prolonged, and set off from *g* towards *h*, also from *i* towards *k*, 2° each, and draw two dotted lines from the point *d*, to the said chords, so as to contain each an angle of 2°; then if a short line be drawn from the upper interior to the lower exterior intersection of the dotted lines with the curved ends of the right hand pallet, it will give the proper slope as seen in the figure; also, if another short line be drawn from the lower interior to the upper exterior intersection of the left hand pallet, it will give its proper slope, and the plan of the pallets will be finished.

The spring box, as we have said, may be placed either above the great wheel of 96, or below it, which is more usual, as at G, in such a situation that the band or chain may wind round both it and the fusee attached to the arbor of the great wheel, without interfering with any part of the movement. We shall have occasion to speak more minutely respecting the requisite size of the spring-box, when we come to treat of the articles MAINTAINING *Power* and MAIN-*spring*.

The calliper of the going part being thus finished on pasteboard, we may proceed to draw the end and side lines to enclose the wheel-work, taking care to leave space enough for the wheel-work of the striking part, and for the ends of the pillars not to interfere with the movement; the space included within the four bounding lines, which may constitute either a parallelogram or a square, limits the size and shape of each of the plates, which, together with the pillars, usually placed near their corners or edges, constitute the frame of the clock-work.

We might here extend our account of the preparation of the plates for receiving the calliper, but as the processes of hammering, filing, and scraping are manipulations familiar to the mechanician, as well as to the labouring workman, we may be allowed to pass over them without any further notice, than that we strongly recommend planing to be substituted for scraping, agreeably to the practice of mathematical instrument-makers, which operation makes an even surface.

The calliper for the striking part of a clock, which is usually and properly called the clock-part, might now be added to the pasteboard; but as we proposed first to pursue the operations attending the construction of the going-part, which has also been called the watch-part, on account of its watching or counting the time as it passes silently along, we will return to the striking-part hereafter, and preserve our account of the *going-part* distinct.

The calliper drawn on pasteboard may now be transferred to one of the plates, either by pricking through the centres upon the brass plate, or by delineating again the same figure according to the directions we have given, unless a deepening tool be made use of for callipering the exact depth at which each wheel and pinion have had their action previously examined; this method is very little known in England, and still less used, though it conduces greatly to perfect

perfect action; whenever this tool is used, which is exhibited in our collection of CLOCK-*tools*, *Plate* XXI. *fig.* 6., of course the ultimate delineation of the calliper on the plate must necessarily be deferred till the wheels and pinions are divided and rounded, and their action duly adjusted in the tool, before the transfer is made by means of the fine points of the parallel arbors, used as dividers, for the description of the requisite circles.

5. *The Pillars and Pivot-holes.* We have said that the pillars, which connect the two plates of the frame, and keep them in their parallel position in a firm manner, are placed at the corners or sides of the plates, as at B, B, B, B, B, *fig.* 1, of *Plates* XI. and XII; the reasons are obvious; the work will thus be more firm than if the pillars were contiguous, and they will be out of the way of the wheel-work and cocks; their number may be three, four, or five, as circumstances may require. Before the holes are drilled for the ends of the pillars, the plates are usually pinned together at each end, and filed to the same dimensions, so that one piercing with the drill and opening with a broach, perforates both plates alike, and ensures the perpendicularity of the pillars when they come to be inserted into their places. It is also usual to pierce with a small drill the pivot holes for the arbors of the different wheels while the plates are pinned together, that the arbors may pass across the frame at right angles to the surfaces of the plates, which is an essential condition in the planting of the wheel-work, and requires the workman to drill in as perpendicular a direction as possible, otherwise the plane of the wheel would not be parallel to the surfaces of the plates, and consequently the communication of motion and transmission of the maintaining power would have an obliquity in their direction, which would produce injurious friction among the teeth.

The strength of the pillars depends chiefly upon the maintaining power and scale upon which the works are constructed; in pieces with a heavy weight they must necessarily be pretty thick, but in spring pieces much metal only gives a heavy appearance, without adding to the utility; the length of the pillars, however, is not so optional; for it depends entirely upon the number of turns of the fusee in spring pieces, and of the barrel in clocks with a suspended weight; in our little eight-days' piece, two inches will be a suitable distance between the interior surfaces of the plates to allow for sixteen turns of the fusee, with the addition of the guard-gut, or stop, at one end of the arbor, and the great wheel and double ratchet, with the auxiliary spring (hereafter described) to produce constant motion, together with space for the rim of the centre wheel at the other. The pillars are generally riveted into the back plate, and pass through the front one to shoulders against which the plate rests, in which situation they are fixed by pins passing through their projecting ends; which mode of fixing the plates admits, indeed, of their being readily dismounted, but is by no means so neat as the method of mounting frames for wheel-work used by the mathematical instrument-makers, which ought to be adopted by clock-makers. The method alluded to is, to screw the pillars into the back plate, even with its exterior surface, and to fix the front plate by screws instead of pins, which screws go into the ends of the pillars, and cover the perforations of the plate by means of intervening collars, as represented in the front of our frame, *Plate* XII. *fig.* 1. The preference of this mode consists in its allowing the plates to preserve their surfaces uninjured and perfectly parallel after being dressed, which the act of riveting would distort; besides, when the clock is at any time cleaned, the plates thus mounted are made handsome again, as they were at first, with little trouble, and the general appearance is thus more workmanlike than in the ordinary way. The shape of the pillars seems to be mere matter of fancy, when they are left strong enough to effect their purpose; but custom has sanctioned the bead, or spheroidical enlargement, at the middle.

Main-spring, Barrel, Arbor, and Ratchet.

The frame being put together and ready to receive the works, a suitable spring must be obtained for a small clock from the spring-maker, whose art is distinct, and will form a separate account in its place; the breadth of the spring we will suppose to be an inch, which will require a box or barrel, C, *Plate* XI. sufficiently deep to hold it, and to be of a diameter sufficient to admit of as many effective coils or spirals as will turn the fusee sixteen times round, before its force is expended in unbending itself. Sometimes it may be necessary to try two or three springs before a good one is met with, that will act with a due degree of regularity; and it has been asserted that a spring will act more regularly and be less liable to have undue friction among the coils, if the breadth be gradually diminished from the exterior to the interior end; but we pledge not ourselves for the existence of the fact, though we conceive that the friction of the sides of the spring against the ends of the barrel, will thereby be greatly diminished. The spring arbor must be strong in proportion to the force of the spring, particularly at the pivots, the front one of which must be thick enough to admit of being squared to hold a ratchet, or small serrated wheel, C, at the outside of the frame, (see *Plate* XII.) the teeth of which ratchet must be strong enough to hold the arbor in any situation to which it is turned, which it does by means of a click attached by a screw to the exterior surface of the front plate of the frame; the spring-arbor has a strong pin inserted into it at the middle, within the barrel, on which pin a hole made near the interior end of the spring hooks, while the exterior end is riveted to the circular side of the box; hence it is not difficult to conceive that when the spring fills the box in its relaxed state, and has its coils most close at the rim of the barrel, it may be coiled up close to the arbor in the centre, or, in other words, it may be wound up, by two different methods; either the barrel may be held fast, and the arbor be turned backward by its ratchet, or by a key fitting its square; or otherwise, which is the general and better practice, the ratchet may be suffered to detain the arbor in its place, and with it the interior end of the spring, and the barrel itself, to which the exterior end is riveted, may be turned forwards by a chain or cat-gut attached to it by a knot at one end and wound round it, as seen at C, in *Plate* XI. We have said the latter method is the better, and the reason is, that, when the greatest and smallest forces of the spring are adjusted to the shape of the fusee, or rather the fusee to them, the ratchet cannot be altered without deranging this adjustment. The arbor is turned in a turning-frame with pivots and shoulders sufficiently remote from each other to reach the interior faces of the plates, but to have just so much play endways as will prevent friction; and the chain or gut must be long enough to fill the spiral grooves of the fusee, and have at least one half-turn on the barrel to spare; also care must be taken that the depth or side of the barrel must be nearly equal to the effective length of the fusee, otherwise the gut will be liable to slip off at the ends of it. The remote end-piece of the barrel is soldered fast, and has a large pivot-hole, against which an inner shoulder of the arbor rests, and the nearer end-piece is turned in the frame so large as to be capable of being forced or sprung into a receptacle turned for it, round the inner part of the edge of the circular

cular rim of the barrel, in which situation it rests against a corresponding inner shoulder of the arbor, and completes the barrel; when this adjustable end is to be taken off for the purpose of examining or taking out the spring, a slight stroke at the remote pivot of the arbor will force it out of its place; some skill is necessary for putting the spring into the barrel, when a tool on purpose is not at hand, which will be described among the WATCH-*tools*, and which seems necessary to be more generally used in clock-work to prevent accidents consequent upon a manual insertion of the spring.

6. *Fusee, Ratchet, and Guard-gut or Stop.*—The spring-barrel and its appendages being finished, a rough estimate of the power of the spring may now be made by first coiling the gut, in a proper direction round it a few times till it is nearly all wound up, the arbor being held by its ratchet, or in a vice, and then by suspending a weight to the spare end, such as will just pull the barrel two or three times round from its relaxed state; this weight will denote the smallest power, which suppose to be one pound; then add such a heavier weight as will uncoil so much more of the gut as may be supposed to fill the fusee, and note it, which we will again suppose to be two pounds and a half for the greatest power of the spring; now this proportion of 1 : 2½ or 2 : 5, may be taken as a guide for the respective diameters of the conical piece of metal, E, *fig.* 1, *Plate* XI, called the fusee, which is introduced to equalize the varying power of the spring, by acting, as it were, with a succession of levers of different lengths, reciprocally proportionate to the power of the spring in any given situation, so that when the power is great it is pulling by a short lever, and *vice versâ.* The piece of solid metal intended for the fusee must be drilled through the centre, and opened with a broach, and then have a steel arbor of considerable strength driven tight into it, by which it is turned into a conical or rather paraboloidal shape, that has its thicker end somewhat smaller generally than the diameter of the barrel, and the other end smaller in the proportion, according to our supposition of 2 : 5, but sometimes in a greater ratio, without the thickness of the gut; the length of the fusee must be shorter than the pillars by as much as will admit the great wheel and two ratchets, with the centre-wheel behind them, to be introduced between it and the plate at one end, and a contrivance for stopping the revolutions when the spirals are filled with the gut, called therefore the guard-gut, at the other end of the arbor, as we have already said. The fusee may now be grooved into sixteen complete spirals by a fusee-engine, the method of doing which will be explained under this term hereafter; after this operation, a pair of strong pivots may be turned on the fusee arbor, the pivot-holes opened by a pivot-broach held perpendicularly with respect to the faces of the plates, and the fusee introduced into the frame, parallel to the spring-barrel arbor; a hole is now drilled at the large end of the spiral, perpendicularly into the metal, the fusee being taken from the frame, and another hole to meet it from the plane of the thick end, about a quarter of an inch from the circumference, which two holes are just large enough to receive the gut; the latter is then enlarged by a chamfering tool to form a bed for a knot to be made at the end of the gut, when inserted into the hole made in the spiral groove. If now a square be made, either on the front or back pivot, which must project through the plate, accordingly as it is intended to be wound up in the face or behind, and if a key be inserted upon it, the spring may be wound up, and it will appear whether or not the gut is too long, and how much, nearly, which may accordingly be altered. Hitherto the work has proceeded, on a supposition that the fusee has been turned of a paraboloidal shape, and that the spring is perfect at the two extremes, as well as at all the intermediate degrees of tension; but it yet remains to be proved, by mechanical adjustment, that these coincidences have been effected, or are even capable of being accurately effected, without material subsequent alterations in the length of the spring and shape of the fusee; for this purpose a long graduated lever, with an adjustable weight, hereafter mentioned under CLOCK-*tools* by the title of an *adjusting-tool*, (*fig.* 10, *Plate* XXI.) is inserted on the square end of the fusee, at N, when the frame is mounted, as in *Plate* XII. and the weight is gradually removed along the bar, until by trial it is found to be an exact counterpoise to the spring previously wound up a few turns by means of the ratchet on the barrel-arbor; such balance being effected, the spring may be wound up by the adjusting tool, used as a key, till the sixteenth spiral at the top, or small end of the fusee, be filled with gut, in which situation, if the weight of the tool still constitute an exact counterpoise to the power of the spring, it is to be presumed that the spring is properly fixed, with respect to its quantum of intensity, by its ratchet; but if, in the latter situation of the tool, it turns out to be more than a counterpoise, either the spring is of too low an intensity in the present situation of its ratchet, or the fusee is too small at the small end, or both may be so circumstanced; on the contrary, if the tool is not a counterpoise for the spring when wound up, either the spring is set too high by the ratchet, or the small end of the fusee is too thick; a few successive trials of similar adjustment for the opposite ends of the fusee, by an increase or decrease of intensity being gradually given to the spring by means of turning its ratchet, will generally determine whether the failure in the adjustment is occasioned by the spring or fusee, and the former may be shortened, or the latter altered, by a detached tool to run in the groove as it revolves in the turning frame, if a fusee-engine is not at hand; though, it must be confessed that some experience in this business will greatly facilitate the determination of the proper means of final adjustment. We will now suppose the spring fixed, and the fusee adjusted by the tool, so as to render the maintaining power precisely the same at the bottom and top of the spiral groove; the adjustments must next be made for all the intermediate turns of the helix successively, by means of the same adjusting tool with the weight unaltered, the spring arbor also resuming at every trial its original position, which we will suppose to have been marked on the holding tooth of the ratchet. When the spring is good, and the fusee approaching to a conical shape, it will be found on trial, that the maintaining power is too great for the tool of adjustment to balance before it is wound up half way; in consequence of which increase in the maintaining power, the fusee must necessarily be again put into the fusee engine and have its groove deepened so as to make a parabolic curve instead of a straight line from the top to the bottom of the fusee; after this alteration the frame must be remounted, the spring coiled up again to its determined position, and the weight of the adjusting tool kept unaltered in its situation; the intermediate grooves in the helix may not yet be found all sufficiently deep to render the maintaining power equal in its effects throughout the whole length of the fusee, but the adjusting tool will detect the particular places where the power predominates; which places when marked may be again altered in the fusee engine, and the parts replaced in the frame, when, after three, four, or perhaps more alterations of the fusee, and adjustments of the spring, at length the effect produced by the power of the spring is the same whateve

whatever part of the fusee be actuated by the gut; the accuracy of this adjustment is of the utmost importance, and should be minutely attended to, otherwise the piece may be made to vary its rate of going on each successive day of the week, by reason of the irregularity of the maintaining power, unless indeed such a consequence be obviated by the nature of the escapement, or other contrivance, which ought not to be depended upon while there is a fundamental remedy. Hence it is evident that, whenever the original main-spring of a clock (or watch) happens to be broken, or by any means altered, another spring, though of the same dimensions ought not to be substituted, as is often injudiciously done, without a corresponding alteration in the fusee, if found necessary, by a trial of the adjusting tool. Were an optician to put a thermometer tube containing mercury, already hermetically sealed, into an old scale previously graduated, the indication of temperature with such an instrument could not be depended upon to any thing like accuracy. Of the same nature is the probability of an imperfect measure and indication of time arising out of an exchange of the main-spring without a corresponding adjustment in the shape of the fusee, more particularly if the crown wheel escapement happen to be adopted, which is almost constantly under the influence of the maintaining power.

During this labour of adjusting the fusee to the spring, it will occur that the gut might be wound up beyond the end of the fusee, if it were turned more that sixteen times round, on which account a snail piece of soft steel, equal in diameter to the small end of the fusee, independently of the claw or projecting piece, is usually screwed against this end of the fusee at N, *Plate* XI., to prevent the gut from slipping off when it comes to the projection of the claw in question, which contrivance it has been said is called the guard-gut; but it may be remarked, that the gut might wind back again by a second course, like the cord of a kitchen jack on its barrel, when it comes to the snail-piece of the guard-gut; to prevent such effect, there are superadded a spring, M, screwed at one end to the inner face of the front plate, which plate is taken off in the drawing, and a lever, L, moveable on a pin as a centre, which pin passes through a stud in the front plate, according to the position given in *Plate* XI.; the remote end of the spring, M, stretches itself towards the back plate of the frame, and carries the lever before it; of which the consequence is, that as the gut approaches the guard in winding, it at length meets with the lever, L, and pressing against it drives it forwards till at length a shoulder near its remote end, seen near the pillar, is presented to the claw, N, of the snail-piece, for which it becomes a stop at the instant that the sixteenth spiral is filled by the gut, and the winding is then necessarily finished.

The contrivance for allowing the fusee to turn in the act of winding up, while the great wheel retains its position unaltered, but which prevents the return of the fusee to take place by the pulling of the spring without the great wheel being actuated, is called the *fusee* ratchet; it usually consists of a serrated wheel with sloping teeth, and a click to catch the teeth, like the mechanism already mentioned as being at the projecting arbor of the spring barrel, but with the addition of a spring to press upon the click to keep it in the teeth; the ratchet wheel may be attached either to the end of the fusee, and the spring and click to the plane of the great wheel, or otherwise the spring and click may be at the end of the fusee, and the ratchet wheel on the plane of the great wheel, but the direction of the slope of the teeth will not be the same in both cases, for in one case the wheel drives the click during the time of going, and in the other the click must drive the wheel; it is usual to turn a circular groove between the edge of the ratchet wheel and the circumference of the fusee large enough to form a bed for the click and its spring, which therefore are hid from sight, when the spring is carried by the great wheel; but when it is carried by the end of the fusee, the said groove must be in the plane of the great wheel to answer the same purpose of concealment; the number of teeth in the ratchet wheel is optional, provided they be strong enough to sustain the maintaining power, and numerous enough to prevent a considerable return of the fusee backward after winding, before the click catches a tooth; the construction will be sufficiently understood without further description, from a reference to *fig.* 5, of *Plate* VIII., to *fig.* 7, of *Plate* XV., and to *fig.* 4, of *Plate* XXII. of *Horology*.

7. *Auxiliary-spring.* But we proposed that our piece should continue to go while it is wound up, which effect requires an additional apparatus to the ratchet we have just described, as being usual in an ordinary spring-clock; the addition usually consists of another ratchet-wheel with teeth inclined in a contrary direction, and of a larger diameter, of a circular or sometimes a horse-shoe spring, and of a lever operating as a detent with one end fastened to the inner face of the back plate, and the other resting upon and sliding over the sloping face of the teeth of the large ratchet when the clock is in motion, but which prevents the ratchet's return when the clock is wound up. One end of the spring is pinned or screwed to the rim of the great wheel, and the other end to the plane of the large ratchet, and in this case the small ratchet before described actuates this large one, instead of the great wheel, exactly as above described; and then the large ratchet contracts the spring connecting it with the great wheel, till its resistance in a contrary direction is equal to the maintaining power, in which situation the great wheel is then impelled in the same manner as if it were immediately connected with the small ratchet; the effect of this beautiful contrivance is, that when the clock is under winding, the spring attached to the great wheel being no longer contracted by the maintaining power, immediately endeavours to extend itself, but the large ratchet-wheel, to which it is riveted, is stopped from going back by the end of the lever or detent, which is always in a position to catch the end of some one of the sloping teeth of this ratchet, the consequence of which is, that as the ratchet will not go back, the great wheel is impelled to go forward at the opposite end of the spring by a force, at the commencement of its action, equal to the maintaining power; and this force will continue for a much longer time than is necessary for the winding-up of any clock. This apparatus may be more clearly apprehended by a careful inspection of *fig.* 4, of *Plate* XXII., and of *figs.* 7 and 8, of Brockbank's chronometer in *Plate* XV., which differ in no other respect, but in their size, from the same parts as used in a clock.

8. *Arbors, Pinions, and Wheels.* Hitherto we have said nothing of the manipulation of the arbors, pinions, and wheels of a clock, but have merely spoken of their diameters, numbers of teeth, and respective situations of their pivot-holes. It will not be necessary to enter into a detail of the manner of forging the arbors, or of casting the brass by models of given dimensions, the ironmongers and tool-sellers having on sale sets of wheels, and arbors with pinions of different numbers ready slit, and also pinion steel-wire drawn into a proper shape for the teeth of small pinions, of all which clock-makers usually avail themselves instead of preparing them; otherwise the brass-founders in Chancery-lane, and in other parts of London, as well as in Lancashire,

ſhire, will caſt wheels to any model at a certain price per pound, which is a great convenience to the workman who has occaſion for unuſual ſizes in his wheel work.

It will not, however, be deemed foreign to our purpoſe to mention, that it is often found requiſite to ſoften the pinions of ſteel, and their arbors, before the graver for turning can cut them with ſufficient eaſe in the turning frame; this ſoftening is uſually effected by putting the ſteel pieces into a wooden fire for ſome hours, and leaving them to cool gradually as the fire goes out. The wheels are generally croſſed in their original caſt ſtate to make them light, in order to avoid the effect of their inertia, and to prevent friction on the pivots of the arbors; but it is neceſſary to hammer the rims and other parts of a wheel well before they are filed flat, to be put on the arbor for turning, otherwiſe the metal would be too ſoft to wear well, and the teeth of a wheel would not ſtand the graver in the operation of being turned, after it is cut by the engine, in caſe the diameter ſhould require to be reduced, as is commonly done, to work well in its calliper with its pinion. When the wheels are hammered, and filed flat to nearly their exact thickneſs, according to the force they are deſtined to tranſmit, they are turned in the turning-frame to their practical diameters, after being previouſly perforated in the centre, and ſcrewed faſt on a ſuitable arbor, ſuch as is repreſented by *fig* 2, of *Plate* XIX.; they are then in a ſtate to be cut by an engine made on purpoſe, which is not always in the poſſeſſion of the clock-maker, particularly in large towns, but which is uſually kept by ſome individual, whoſe buſineſs is chiefly to cut wheels for the clock-makers at a certain price per ſet. An ingenious workman, of the name of Brown, who lived in King-ſtreet, Seven-dials, but who is now dead, was noted in London for his dexterity and accuracy in cutting clock-wheels, and occaſionally pinions, into any number of teeth; his engine, however, is now in the hands of Mr. Fidler, mathematical inſtrument-maker, Oxford-market, London, who has lately begun to uſe it; but the beſt engine for this purpoſe at preſent in conſtant uſe, belongs to Mr. Edward Troughton, of Fleet-ſtreet, and is uſed by James Fayer; at No. 35, White-lion-ſtreet, Pentonville, London.

This engine, which was contrived and made by the late ingenious mechaniſt Rehe, will round the teeth at the ſame time that it cuts them, if required, and is valued at 300*l.* in its preſent ſtate. As we ſhall have occaſion to treat further of the conſtruction and operations of the CUTTING-*engine*, we ſhall here conſider that our wheels have now undergone their operation, and are returned into the hands of the clock-maker for finiſhing; we cannot, however, forbear remarking here, that the clock-makers of the preſent day have greatly the advantage of thoſe who laboured in the art in the infancy of clock-making, in that they have ſome of the moſt difficult operations in theory, ſuch as dividing and cutting the wheels, and forming the ſpiral groove of the fuſee, done by engines not only in leſs than one-hundredth part of the time, but with infinitely more accuracy, than they could be performed by hand with manual tools; and it is much to be deſired that they would, like watch-makers, have their wheels cut and rounded in the engine at the ſame time; that they would make uſe of the deepening tool, *fig.* 6, *Plate* XXI. for pitching their depth of action; and that they would in that ſtate transfer the diſtances of the pivots to the calliper, as we before recommended; for then the equable tranſmiſſion of motion and maintaining power would be enſured, provided the teeth be of a due ſize and form; the expence in cutting would, indeed, be ſomewhat more, but the labour of rounding by a file would be ſuperſeded, and conſequently the work would be more expeditiouſly, as well as more accurately, performed.

But it has been remarked, by a philoſopher well qualified to make the remark (Mr. W. Nicholſon), that the introduction of new inſtruments, and of new operations, requires the ſame ſpace of time that is neceſſary for inſtructing another generation; ſo much does that facility, which ariſes out of habit, militate againſt the adoption of new practices: we muſt, therefore, ſuppoſe our wheels returned from the engine with their teeth not rounded, and proceed with our detail. When a wheel has been cut, in the ordinary way, by the engine, there are uſually ſome filaments of metal, denominated burs, left at the edges of the teeth by the cutter, as well as general roughneſs on the ſides and at the bottom of the ſpaces; theſe are firſt cleared away by a fine file, juſt thin enough to go into the ſpace left by the cutter, which is called an equalling file, (ſuch as is repreſented in *Plate* XXI, *fig.* 14.) from its ſuppoſed property of leaving all the ſpaces equal, when the burs are removed. The croſſes of the wheel and interior edges of the rim are next dreſſed, firſt by a rough file, then by a ſmooth one, and laſtly by a burniſher of poliſhed ſteel, all ſhaped like *fig.* 17, of *Plate* XXI.; the arbor is next turned in a frame, by a well-tempered tool, or ſtrong graver, like that ſeen in *fig.* 9, of *Plate* XXI., to its propoſed thickneſs, and the pinion reduced to its practical diameter; after which its teeth are rounded, hardened, and poliſhed, each of which operations we will ſuppoſe to be underſtood, and the wheel is riveted on a ſhoulder left on the proper end of the pinion, exterior or interior, as the work may require, if it is on the hour arbor; otherwiſe it may be riveted, or ſtill better fixed by two oppoſite ſcrews, on a braſs collet, which is previouſly ſoldered upon ſuch part of the arbor, to which it belongs, as the place of the pinion which it actuates may require. The French call that portion of an arbor, which is between the wheel and the remote pivot, a *tige*; and that portion a *tigeron*, which lies between the wheel and the nearer pivot. Some of the moſt ſkilful workmen contend that the wheel ought always to be placed on its arbor, ſo as to be equally diſtant from both pivots, and we have ſeen clocks conſtructed with cocks, at the back of the plates, to hold the pivots of projecting arbors, in order to effect this purpoſe, but we will not undertake to affirm that this additional work is compenſated by any advantage thus gained; we rather conceive that the grievance complained of in the ordinary method of pivoting in the plates, *viz.* the alleged unequal preſſure and conſequent unequal friction in the oppoſite pivot-holes, when a pinion is at one end of the arbor, is to be attributed to another cauſe principally; *i. e.* the too great aperture of the pivot holes: the workmen have a maxim, that "the pivots muſt have play to avoid friction," but they ſeem not always to underſtand what this play means; it certainly ought not to mean, that the holes ſhould be much larger than the pivots which are to turn in them, for in that caſe the pivots would be driven round the interior circumference of the holes, and cauſe the wheel to act at different depths in the pinion, which would be a great evil; the meaning of the maxim is, that the ſhoulders of the arbors ſhould not preſs againſt the plates when mounted, but that each arbor ſhould have a little play in the direction of its length, or, in other words, be left ſo as to be at liberty to move a little backwards and forwards; which condition ſeems requiſite.

We will now ſuppoſe all the wheels and pinions rounded neatly in the bay-leaf form, and their action tried in the deepening tool, with the correſponding pivot-holes drilled with

with drills refpectively proportioned to the propofed thicknefs of the pivots; the next ftep will be to attach them to their arbors: the great wheel, however, notwithftanding what we have faid about riveting, fcrewing, and foldering, is not fixed by any of thefe operations, but is attached to its arbor by a method which admits of its being taken off at pleafure, thus;—a hole is opened by a broach in the centre of this wheel, large enough to take the arbor of the fufee without play, and, when it is preffed clofe againft the end of the fufee, or, in our prefent inftance, againft the plane of the fecond large ratchet, two marks are made, with a fine file, at oppofite fides of the arbor, clofe to the plane of the wheel; it is then taken off the arbor, and two fquare-fided notches are filed carefully in the faid marks of the arbor, but not deep enough to injure its ftrength; a collet with a circular hole, like that at the centre of the wheel, made at one fide of its centre, and with a ftraight-edged flit acrofs the centre from the faid hole, is then put on the arbor, after the wheel has been firft put on, and is pufhed forcibly along the notches of the arbor till, by means of the oblong aperture, it is concentric with the wheel, againft the plane of which it preffes, when in the notches, and keeps it clofe in its place, and at the fame time allows it to turn on its centre without the arbor. The auxiliary fpring is fcrewed or pinned, in the next place, at one end to the great ratchet, and at the other to the great wheel, as before noticed. When the wheel is thus firmly attached to the fufee arbor, with a power to recede but not to proceed, on account of the click, without carrying the arbor with it, it muft be put into the turning frame and examined, as to its being truly centered, and alfo as to its being in a plane perpendicular to the arbor; which trial may detect fome flight alterations, neceffary to be made, to fulfil thefe two conditions: after which, its pivots may be turned to their exact fize, hardened, and polifhed.

The practice of fome workmen is, to folder their collets on the arbors with hard folder, but we difapprove this practice, and recommend foft folder, particularly if the arbors have been before hardened; it is fcarcely neceffary to add, that if the ends of the collets are opened a little within, the folder will there have beds to contain a quantity fufficient to keep the wheel firm on its arbor. What we have here faid will not be equally applicable to the hour or centre wheel, becaufe it is ufually riveted on the end of its pinion, which will require a riveting punch and clamp, on purpofe (*fig.* 16, of *Plate* XIX.) to prevent any injury being done to the pinion. Before the wheels are all firmly attached to their arbors, they muft be tried in the turning frame, or callipers with a ftraight edge for that purpofe, to fee if they are concentric, and that their planes are perpendicular to their arbors, which conditions the workmen call being "in the round," and being "in the flat;" and when properly adjufted, in thefe refpects, they may be finally fixed as above defcribed, and their arbor-pivots finifhed. Should a wheel, that has had its teeth rounded in the engine, be at any time found a little eccentric in the trial, before it is fixed on its collet, which will not happen if the central hole be enlarged with a good broach and with due care, the remedy in this cafe would be, to mark the fide of the wheel which has the longeft radius, and enlarge the central hole carefully on that fide moft, and then to make a new collet for it, after the hole is again made perfectly round, and found to be concentric on an arbor that fits it; but when the teeth have been rounded by hand, the wheel may have the eccentricity rectified on its own arbor, and be again rounded where the teeth have been touched by the graver, which is the common practice, and which conftitutes the greateft recommendation of manual rounding of the teeth.

The files ufed in rounding the ends of the teeth of a wheel are fmooth, and curved on one fide (fee *fig.* 16, of *Plate* XXI.), and have each a projecting pivot at the remote end, which a dexterous workman holds againft one finger of the left hand as a reft behind the wheel, while the right hand guides the file from one fide of the tooth to the other, alternately, with a degree of rapidity which furprifes the fpectator.

We have hitherto fuppofed our piece not to be jewelled, nor bufhed with bell-metal, which addition enhances the price, but greatly diminifhes the friction in the pivot holes: if the holes are carefully enlarged with a good broach, fmall pieces of metal holding ruby, agate, or bell-metal, may be made to fit them exactly, the holes in which may be refpectively equal to the original pivot holes, and then the good action of the wheel-work will not be altered thereby. The friction in the pivot holes is greatly diminifhed too by the application of fine nut-oil; they are, therefore, generally chamfered, or counter-funk, at the exterior furfaces of the plates, in order that the oil may be retained; but what fhould be the exact depth of the bearing part of the pivot-hole, is a matter not abfolutely decided. Mr. Reid, we have feen, prefers, to a counter-funk hole, a conical point, which holds the oil in a globule by cohefion, and which is held by a fmall cock.

In fome clocks which we have feen well-finifhed, the ends of the pivots are conical, and bear againft holes of nearly the fame fhape, not entirely perforated; but we conceive that, unlefs the pillars and arbors were all of the fame metal, the difference of their expanfibilities muft materially alter the quantity of play, lengthwife, at different feafons of the year, and in cold weather create confiderable friction, by affecting the brafs pillars more than the fteel arbors; an effect which the workman probably does not take into his account, who thinks of avoiding friction by fuch a conftruction.

As the pallet-wheel makes many more revolutions than any other in the movement, it is neceffary that the metal, of which it is made, fhould not be very deftructible, particularly when pallets are ufed which rub againft its teeth; we therefore recommend a tempered fteel wheel to be ufed, which ought to be alfo divided and cut with extraordinary care, becaufe any irregularity in the fhape of the tooth, or diftance between the teeth, would injure the efcapement, and produce befides fuch irregularity in the motion of the feconds hand, placed on this wheel's arbor, as would offend the eye. We referve our obfervations on the fhape of the teeth proper for different efcapements to act with, until we treat this part of our fubject more particularly in its proper place.

9. *Pallets*. There is no part of a clock which requires greater nicety in the execution than the efcapement, or part which limits the intenfity and duration of the impulfe given to the pendulum by the maintaining power, and which keeps up the due quantity of motion, that would otherwife be gradually diminifhed to a ftate of quiefcence, by reafon, as we have faid, of the unavoidable friction at the point of fufpenfion, and of the refiftance which the air affords to the folid parts of the moving pendulum: but for the fame reafon that we have poftponed our particular directions concerning the conftruction of the pallet or fwing-wheel, till we come to the article Escapement, muft we alfo fatisfy ourfelves here with a few general directions and obfervations, which apply exclufively to our dead-beat pallets. We have already detailed, under our fubdivifion, entitled *calliperíng*, the method of laying down the plan of the pallets in queftion; the fhape and dimenfions there afcertained muft be exactly copied, or otherwife projected again, either on one of the plates

plates or on a fmooth fheet of brafs, as a plate of trial for the efcape, which will admit of pivot-holes being drilled, exactly as in the plates of the frame, for the centres of the pallet and pallet-wheel arbors; a piece of good fteel muft then be forged nearly into the fhape of the anchor, compared with the plan on the frame or brafs plate, but fomewhat larger: after the arbor hole is drilled in the anchor, and enlarged to the propofed aperture, the requifite circles may be defcribed, with extents borrowed from the brafs calliper, by means of a pair of fmall bullet compaffes, and the flopes may be copied or re-traced for the faces of the pallets; the excluded metal may then be filed away very nearly, and all the furfaces be fmoothed, firft with fine files, and then with oil-ftone duft and oil. It has been faid that the breadth of each pallet muft be fomewhat lefs than half of a fpace meafured from tooth end to tooth end of the pallet wheel, but the quantity of diminution muft not depend on conjecture; therefore the breadth is left at firft equal to one half as nearly as can be afcertained, fo that the diminution of breadth may be effected by tentative adjuftment, firft partially when the wheel has its teeth finifhed, and when both it and the anchor of the pallets are inferted on pins urged into the pivot-holes of the frame, or trial plate; and again more minutely when they are fixed on arbors, and mounted in the frame. To enfure the perfect portion of a circle at the extremities of the anchor, we recommend that it be put on a motion arbor, that juft fits the central hole, and that it be turned in a frame or lathe, like a wheel, before it is cut; for then it is certain there will be no recoil in the pallet-wheel, and feconds hand, when thofe parts prefs againft the ends of the teeth, during an excurfion of the pendulum; and if the inner circles could be turned alfo after the efcape is nearly adjufted, it would be defireable; however, the point of a graver may trace in the turning frame this inner circle, and then a proper curvi-linear file, made and kept on purpofe, may take off the interior fuperfluity of metal. In adjufting the flopes and breadth of the pallets, it will be ferviceable to infert an index on the pallet's arbor, after it is finifhed and the pivots turned, and to mark on the frame plate the quantity of the efcapement angle, in this inftance 2° at each fide of the perpendicular demitted from the pivot-hole, which will be a good guide for the true final adjuftment of the efcape. Particular care muft now be taken that there be as little drop as poffible, *i. e.* that as foon as one tooth has completely efcaped the face of its pallet, the next acting tooth fhall be clofe to the back of the following pallet, fo as not to ftrike it with a jerk; the place where the firft contact takes place between the end of the tooth and interior or exterior circle of the pallet, accordingly as it is the leading or following pallet, muft be very near the commencement of the flope, but not upon it, nor yet on the angular point of interfection. Indeed it is extremely difficult to give complete verbal directions for this delicate adjuftment, which requires long and attentive practice to do perfectly; for frequently, after the pallets are hardened, which they muft be as much as poffible, the fhape of the anchor is found to be altered, and the adjuftment of the pallets confequently deranged: to remedy this confequence, it is ufual to harden only the pallet parts of the anchor, fo that by certain ftrokes given near its arbor, the pallets may be brought in or fet out a little to rectify them; but after fuch rectification it will always be neceffary to try, in the turning frame, or at leaft by a pair of bullet compaffes, if the circular parts are again perfectly concentric, without which condition the pallets will not be truly what are called dead-beat.

Hitherto we have confidered the back pivot hole of the pallets arbor, as being in the plate of the frame, but it becomes neceffary to cut away that portion of the back plate where the pivot hole falls, by reafon of the crutch, or little rod of fteel which muft be fcrewed to a collet attached behind the frame to the arbor, to form an L; which contrivance impreffes the force that the pallets receive from the maintaining power upon the pendulum; the bent end of the crutch is ufually inferted into a flit made in the verge or rod of the pendulum, but when the bent part is divided and enclofes the pendulum rod, it is denominated the fork: the crutch is moft ufually about one fixth part of the whole length of the pendulum rod, but there feems to be no fixed rule laid down by which its beft length might and ought to be determined in different cafes, which therefore we think deferves further confideration.

But to return to the pivot-hole of the pallet's arbor; this, for the reafon we have juft given, is finally placed in a cock at the back of the pofterior plate, which is generally fo fhaped as to furnifh alfo a point of fufpenfion for the pendulum. (See *a*, in *fig.* 3, of *Plate* XI.) The exact placing of the cock, fo that the arbor pivoted into it fhall be perfectly at right angles to the furface of the plates, is of the greateft importance, and therefore it ought to be placed and its fteady pins fixed, before the original pivot-hole, through which it muft protrude, is cut away in the plate; for in that cafe the protruding end of the arbor, while in its proper pofition after the adjuftment of the efcapement, will be like a fixed arbor on which to flide the cock, and fix its pofition before the fteady pins are applied and the fcrews fitted to their places.

It is, however, the practice of fome workmen to adjuft the efcapement, by moving the cock before the fteady pins are inferted.

We might have noticed, that after the anchor of the pallets is fcrewed to the collet of its arbor, it fhould be fufpended by the pivots of the verge, which is the name given to this arbor, to try if the weight of each pallet exactly balance that of the other, which may be effected by dimifhing the thicknefs of the heavier pallet a little by a fine file before it is finally polifhed; alfo before the crutch is fcrewed it fhould be hung on the verge of the pallets arbor, after the pallets are balanced and fuffered to find their place of quiefcence, in order to find its own perpendicular direction, and then it fhould be fixed in that fituation; for without this care it will require to be bent fo as to offend the eye, for the purpofe of putting the clock into beat, or otherwife will require a flit in it acrofs the centre, to admit of an eccentric adjuftment, or fome fuch contrivance. When all the adjuftments of the efcapement are thus made, the pallet faces, if not jewelled, and alfo the pivots, muft be harden d and finally dreffed by the ufual fucceffive operations of polifhing.

10. *Pendulum and Sufpenfion.* The precautions we have hitherto dictated, in our directions for the fucceffive operations in clock-making, have for their principal object the regular tranfmiffion of a certain quantity of the maintaining power to the pendulum, in order to preferve the arc of vibration unaltered, which is one of the two effential qualities of the going pendulum, on which the exact meafurement of time depends; the other, which is an indifpenfable quality of the pendulum, where great exactnefs is required, is that by which its length is preferved unaltered in all the variations of temperature, as the centre of ofcillation regards the centre of fufpenfion. Thefe two properties of the pendulum, its confta arc, and conftant length, conftitute the excellence to which all the other parts of the mechanifm are fubfervient, and without which no clock will continue invariably

variably to indicate true time at all ſeaſons of the year, however exquiſite the workmanſhip of the movement and other parts. In an ordinary clock, the iron or ſteel rod of the pendulum is liable to conſiderable alternate expanſions and contractions, which render a compenſating contrivance neceſſary for pieces deſtined for aſtronomical purpoſes; or, otherwiſe, a deal or ebony rod is ſubſtituted for the metallic one, which natural ſubſtances, when of a ſtraight grain and gradually dried by age, are found to be much leſs liable to alteration in their length by changes of temperature, than iron or any other metal; their dimenſions, however, are a little altered by moiſture, which alteration renders them in our opinion objectionable.

In our half-ſeconds pendulum we propoſed to make uſe of a compenſation, in which the unequal and oppoſing expanſibilities of two different metals produce the deſired effect. It would, however, extend our preſent article to an improper length, if we entered here into the geometrical theory of the pendulum, or even if we detailed again the arrangement of the bars which conſtitute the mechaniſm of the compenſation propoſed; but under our article PENDULUM, the reader will, we preſume, find the former omiſſion amply ſupplied; and the latter has been anticipated in our deſcriptions of Reid's and Brockbank's aſtronomical clocks. Reſpecting the ſhape and ſize of a requiſite bob or ball for the pendulum, we have already made our remarks under the word BOB, and therefore ſhall not repeat them here, particularly as we ſhall have occaſion to reſume the ſubject under the article MAINTAINING *Power;* it ſeems proper, notwithſtanding, to add a few words on the ſubject of the ſuſpenſion of the pendulum. Berthoud, the juſtly admired author of many French works on clock and watch-making, has affirmed (in conſequence of ſome of his experiments on the length of time that pendulums, differently ſuſpended, take in coming to the ſtate of quieſcence, after being moved the ſpace of a given angle from the line of direction), that what is called a knife-edge ſuſpenſion is preferable to that in which a piece of watch main-ſpring is uſed at the point of ſuſpenſion; we will not undertake to decide to which mode the preference is due, but adopt that which our Engliſh clock-makers, perhaps without a ſufficient compariſon of the two, have brought into general uſe. In ordinary clocks a ſlit is made in the moſt prominent part of the cock, into which the piece of watch ſpring is inſerted, which carries a ſmall piece of braſs riveted to its extremity, by which the weight of the pendulum is ſuſpended on the cock, and a hole drilled through both the clock and braſs piece receives a pin to keep the pendulum in its ſituation; but this mode of ſuſpenſion is liable to two conſiderable objections: firſt, if the pendulum ſpring happen not to coincide with a perpendicular line paſſing through the pivot-hole of the pallet's arbor, one ſemi-arc of vibration will be greater than the other, even after the bending or eccentric adjuſtment of the crutch has brought the clock into beat, for which imperfection this mode of ſuſpenſion affords no remedy; and ſecondly, the adjuſtment for time, as determined by the going of the clock, cannot be made without ſtopping the pendulum to ſcrew up the adjuſting ball at the bottom of the rod; to obviate theſe two evils, we might have ſuſpended our pendulum in a way which admits of both a lateral and longitudinal adjuſtment, without ſtopping the motion of the pendulum, according to the drawings contained in *Plate* XXVII and deſcribed under the ſection of *Aſtronomical Clock*, by Meſſrs. Brockbanks; but for the ſake of variety, we propoſe to introduce a different mode of limiting the effective length of a pendulum, which mode is frequently uſed, but we believe has never before been publiſhed. The method we mean, is that which we have deſcribed under the head of *an eight-days portable clock*, contained in *Plates* XI. and XII.; from which deſcription, it is preſumed, the reader will already have apprehended all that is neceſſary to be underſtood relating to the mechaniſm, and its application to the regulation of the going pendulum.

11. *Dial work.*—In the deſcription of the ſame eight-days clock which we have juſt referred to, has alſo been particularly explained the common dial-work of a clock, the different parts of which are ſo diſtinctly repreſented in *Plate* XII., that no further directions ſeem neceſſary in this place, eſpecially as we propoſe to give an ample account of the different methods of indicating time on the faces of clocks and watches, under the article DIAL-*work*.

12. *Striking part.* As we have dwelt ſome time on the different ſucceſſive parts of the mechaniſm of the going-part of a clock, which are the moſt eſſential to be attended to, we hope to be excuſed, if we treat more generally, and claſs under one head what remains to be ſaid on the ſtriking part, which requires leſs of ſcience, and more of mere mechanical contrivance, than the parts we have hitherto treated of.

Neither do we mean to deſcribe over again here, the offices of the different parts of the ſtriking mechaniſm, but to point out the order in which the various conſtituent pieces ought to be ſucceſſively made, and to give our reaſons for their requiſite diſpoſition; leaving the mechanician and workman in poſſeſſion of their own mechanical reſources, to be applied in their own way.

The directions which we have given for the callipering and manipulation of the train in the going part of the clock, will equally apply to the wheels and pinions of the ſtriking part, and the ſpring box, D. *fig.* 1. *Plate* XI. together with the great wheel, F, and its fuſee, may be made preciſely in the ſame way as thoſe of the going part, C, and E, have been directed to be made; alſo the guard-gut muſt be ſimilar. To ſhew the manner in which the main-ſpring is coiled within the box, the lid or end piece is left out in our drawing. The numbers of teeth proper for the wheels and pinions of the ſtriking train, and alſo the number of pins in the pin-wheel, have been before explained; as alſo the ſtructure and poſition of the fly, hammer, and bell; which explanations therefore, need not be repeated.

When the dial-work is finiſhed, and the clock is meant to ſtrike the hours only, which is our ſuppoſition here, the warning pin may be attached to the wheel which revolves in an hour, which we have elſewhere called the minute wheel, (ſee *fig.* 2, *Plate* XII.) becauſe it carries the minute hand as it revolves; or otherwiſe it may be put, as we have placed it, in the ſecond wheel of a ſimilar number of teeth, marked *g*, in *fig.* 1, of *Plate* XII. where it is better ſeen. The warning-piece, *t u v*, muſt revolve round the angular point, by means of a tube fitting nicely on a fixed ſtud, in ſuch a ſituation on the front plate of the frame, that the piece will not require much force, to be deducted from the maintaining power, to be moved from its ſtationary ſituation at the end of each hour; and that this may be the caſe, the tail-piece, *v*, ſhould lie at right angles with the tangent line, in which the warning pin is moving at the time of their contact; the bent end, *t*, muſt neceſſarily be at ſuch a diſtance from the centre of motion of the warning-piece, as the pin in wheel R, *fig.* 1, *Plate* XI. which it muſt fall in the way of demands. Alſo the hawk's-bill *p q r*, *fig.* 1, *Plate* XII. which is lifted by the bent end of the warning-piece, ſhould lie in a line, *p r*, *p r*, perpendicular to the tangent-line, in which the end of the warning piece moves at the instant of lifting, in order that the least impressed force may detach the bill from the teeth of the subjacent rack; and that the hawk's-bill may have the mechanical advantage of a long lever, its centre of motion must be at the remote end *r*; likewise, in order that it may always move in the same plane, it must have also, like the warning-piece, a tube moving round a stud in the plate of the frame considered as its centre of motion. The weight and strength of the materials of these two pieces, which are usually of steel, should be proportioned to the power of the main-spring which lifts them; for which reason, the lighter they are the better, provided they do not bend with the forces they have to sustain as detents, when at rest. The length of the gathering pallet, *s*, in the last named plate, must be guided by the strength of the main-spring, as exerted at the arbor on which this pallet is placed, compared with that of the counter-spring, *o*, of the rack-tail: for if the latter is comparatively strong, the pallet must necessarily be short, to have power enough to gather up the rack; but the length of the tail of this pallet, depending on the distance that the catching pin, *m*, on the rack, is from the last tooth of the rack, may be optional; only it may be observed, that the longer the tail is, the more power it has to arrest the motion of the striking train, when the hour is struck. It is of little importance, whether the gathering pallet be before or behind the bill of the hawk's-bill, provided they act clear of one another, and provided the rack have 12 notches, to be caught successively by the pallet.

The size and shape of the rack, *m n*, depend on the distance of the centre wheel arbor from the gathering pallet, conjointly with the size of the snail *l*; when the quarters are not struck, the snail is usually attached to the 12-hours wheel, and revolves with it, as in our drawing; but it might with equal propriety be placed on a second wheel, revolving in the same space of time, as is the case in the clock with chimes, represented in *Plate* XVII. of *Horology*. The steps which form the notched spiral outline of the snail have their depth depending partly on the length of the tail-piece, *n*, of the rack, and partly on the size of the teeth of the rack; the distance from the outer to the inner end of the irregular spiral of the snail must be such, and the length of the rack-tail so proportioned to the body of the rack, that the 12 teeth of the rack will only just pass the gathering pallet, while the pin, in the extremity of the said tail, moves down the straight line that connects the two ends of the spiral, formed by the boundary of the snail; hence, supposing the rack previously made, the point for its centre of motion must be found such, that the two conditions will be exactly fulfilled; but the best way is, to fix upon a point for the centre of motion first, and then to proportion the rack and snail to each other, for the assumed length of tail, from the said centre of motion; the steps will ascend by equal additions of height, if the teeth of the rack are equidistant. When the snail is designed, care must be taken, that each step shall subtend exactly the twelfth part of a circle, or 30 degrees; the apparent increase in the size of the steps is not owing to any increase in the angles successively subtended, but to the increase of the successive radii of curvature: for as the wheel that carries the snail round revolves in 12 hours, every step must correspond to an hour's motion, otherwise the same hour might be struck a second time sometimes, when a greater space than the hour has elapsed, as indicated by the minute hand. The rack, like the warning-piece and hawk's-bill, is usually made of steel; but the snail may be made of brass, which is more easily cut by the file than steel. The bar for repetition, and the strike or silent, may be added with little trouble, if deemed desireable, but the account we have given of these parts before, when describing a portable eight-days clock, will be deemed sufficient by the generality of readers, and those, who have been instructed in the art, will need no further direction from us respecting those parts of the workmanship which are merely mechanical; for were we to enter into a detail of all the nick-nacks which have been introduced into the striking part of a clock, we might write a whole quarto volume on the subject. Neither do we think it incumbent on us to enter minutely here into the particulars of the enamelling, silvering, &c. of the dial, or of the manufactory of the hands, and case, which are separate departments: it may, however, be proper to observe, that some care is requisite in making the pillars of the dial to be all of an equal length, that the plane of the dial may be exactly at right angles to the axes of the hands; otherwise they will approach the dial in some points of their respective revolutions more than in others, which will offend the eye of a spectator. It is a matter of very little importance whether the bell of the striking part be fixed vertically or horizontally; this point is generally determined by the size and shape of the case that is fixed upon.

13. *Adjustment for Rate.* Suppose now our proposed clock to be finished, and fixed in its case, and this firmly attached to a solid wall, to avoid all casual motion, that might be derived from the floor of the room in which it is placed, when trodden upon; suppose moreover, its position such, that when the pendulum is put into motion, the two alternate excursions are exactly similar in extent and time, or, in other words, that the clock is in perfect beat; it then only remains to be brought to true mean time, solar or sidereal, as its destination may be; if the latter is fixed upon, the adjustment for the length of the pendulum must be made from night to night successively, till a star, seen through a transit-instrument, shall be found to cross the central thread of the eye-piece, exactly at the same hour, minute, and second, for two or three successive nights, which nicety of adjustment may be effected at any season of the year, provided the compensating mechanism be perfect; indeed an examination of the passage of the same star, whatever it be, will detect the inaccuracy of the compensation if there be any, provided the transit-instrument have its adjustments perfect. It is hardly necessary, perhaps, to add, that when any known star is passing the middle thread of the field of view of a transit-instrument, properly set in the true meridian line, the sidereal clock ought to be indicating the exact hour, minute, and second, denoted by the said star's right ascension for the year and day in question. When solar or common time is intended to be indicated by the clock, which is the case for all civil purposes, the clock may be tried by a comparison with a regulator previously adjusted, or more accurately by successive observations, taken of any of the heavenly bodies, as explained in some of the problems contained in our article CHRONOMETER.

If a transit-instrument is used as the instrument of observation, an allowance of $3^m\ 55^s.9$ must be made for each twenty four sidereal hours, which is the quantity by which a solar day is longer than a sidereal one, as has been explained under the article just referred to. It may save some time in making calculations for this purpose, if we subjoin a table computed for the service of those who wish their clocks to be properly regulated.

Table

Table for regulating Clocks or Watches.

Revolutions of the Stars.	Mean Solar Time corresponding.				Acceleration of the Stars in Solar Time.		
	d.	h.	m.	s.	h.	m.	s.
1	0	23	56	4.1	0	3	55.9
2	1	23	52	8.2	0	7	51.8
3	2	23	48	12.3	0	11	47.7
4	3	23	44	16.4	0	15	43.6
5	4	23	40	20.5	0	19	39.5
6	5	23	36	24.6	0	23	35.4
7	6	23	32	28.7	0	27	31.3
8	7	23	28	32.8	0	31	27.2
9	8	23	24	36.9	0	35	23.1
10	9	23	20	41.0	0	39	19.0
11	10	23	16	45.1	0	43	14.9
12	11	23	12	49.2	0	47	10.8
13	12	23	8	53.3	0	51	6.7
14	13	23	4	57.4	0	55	2.6
15	14	23	1	1.5	0	58	58.5
16	15	22	57	5.6	1	2	54.4
17	16	22	53	9.7	1	6	50.3
18	17	22	49	13.8	1	10	46.2
19	18	22	45	17.9	1	14	42.1
20	19	22	41	22.0	1	18	38.0
21	20	22	37	26.1	1	22	33.9
22	21	22	33	30.2	1	26	29.8
23	22	22	29	34.3	1	30	25.7
24	23	22	25	38.4	1	34	21.6
25	24	22	21	42.5	1	38	17.5
26	25	22	17	46.6	1	42	13.4
27	26	22	13	50.7	1	46	9.3
28	27	22	9	54.8	1	50	5.2
29	28	22	5	58.9	1	54	1.1
30	29	22	2	3.0	1	57	57.0
40	39	21	22	44.0	2	37	16.0
50	49	20	43	25.0	3	16	35.0
60	59	20	4	6.0	3	55	54.0
70	69	19	24	47.0	4	35	13.0
80	79	18	45	28.0	5	14	32.0
90	89	18	6	9.0	5	53	51.0
100	99	17	26	50.0	6	33	10.0
200	199	10	53	40.0	13	6	20.0
300	299	4	20	30.0	19	39	30.0
360	359	0	24	36.0	23	35	24.0
365	364	0	4	56.5	23	55	3.5
366	365	0	1	0.6	23	58	59.4

The application of the preceding table can hardly be miftaken, but, for the fake of illuftration, we will fuppofe an example that fhall include all the difficulties that are likely to occur in practice; let it, for inftance, be required firft to put the clock to mean folar time, when Spica Virginis is paffing the meridian hair of a tranfit-inftrument, on the evening of the 1ft of May, 1807, and that, on the evening of the feventh fucceeding day, the faid clock being obferved to be indicating 10h 15m 29s at the moment, it be required to afcertain the daily lofs or gain on an average of the faid fidereal days? The work may be thus performed; *viz.*

	h.	m.	s.
R. A. of Spica Virginis for 1806 (Tab. III. CHRONOMETER)	13h	14m	59s.29
Annual variation for one year, add			3.15
Ditto for four months, do.			1.05
	13	15	3.49
Star's corrected R.A for May 1, 1807	13h	15m	3s.49
Sun's R. A. for noon of ditto, fubtract	2	30	56.3
Approx. folar time of ftar's paffage	10	44	7.19
Propor. part of 3m 48s.7 (daily dif.) fubtract		1	42.03
True time of the ftar's paffage to which the clock muft be fet	10	42	25.16
Seven days acceleration of the ftar from the Tab. fubtract		27	31.3
True mean folar time	10	14	53.86
Time by clock per fuppofition	10	15	29
Amount of error in feven days	+0	0	35.14

Hence the daily error, in excefs in each fidereal day, is $= \frac{35^s.14}{7} = 5^s.02$, which was to be determined.

For the other methods of afcertaining the rate of a clock or watch, and for the manner of applying the equation of time, the reader is defired to confult the problems under the article CHRONOMETER.

CLOCK-*movement* is a term in *Horology*, which fometimes implies a combination of wheels and pinions employed in the ftriking part of a clock, but moft ufually that fucceffion of wheels and pinions which move one another, from the maintaining power to the pallets in the going part, and which are employed to tranfmit the force of that power in an equable but modified manner to the regulator, at the fame time that they count and indicate the number of its vibrations in a given period of time. That thefe offices of the going part, to which we will confine ourfelves principally, may be performed in a proper manner, it is requifite that the wheels and pinions of the movement fhould have their number of teeth properly calculated; that their diameters fhould be exactly proportioned to act in fuitable pairs; that they fhould be properly callipered to make the pitch-line of each wheel coincide with that of its pinion; and that the teeth fhould be of a proper fize and fhape, to tranfmit the motion and force they have received, to the teeth acting with them in an uniform manner in every poffible fituation of the acting parts.

The three firft of thefe requifites have already occupied a confiderable portion of our attention under the article CLOCK-*making*, to which the reader is referred; and the fourth, which relates to the fize and fhape of the teeth, might have been properly deferred till we come to the word TOOTH, had it not fallen fo far into the alphabet; we fhall therefore introduce the fubftance, appropriated to that word, in this place, that the fubject may appear in as complete a ftate as our arrangement will admit, at an early period of our work.

Before we proceed to examine fcientifically the requifites which ought to guide the practical conftruction of the teeth of wheels and pinions, it feems neceffary that we fhould premife fome obfervations on the principle by which a communication of motion and a tranfmiffion of force in general are effected in wheel-work.

In *Plate* III. *fig.* 1, of *Horology*, let A B be a lever, or rod without weight, moveable on C as a centre, and let W and *w*, fuppofed to be two bodies with weight, have their maffes fo proportioned to each other, that the mafs *w* may be to the diftance C A, as the mafs W is to C B; or, in other words, if the mafs *w*, multiplied by its diftance from the centre of motion,

motion, B C, give a product, or momentum, equal to the product of the mass W multiplied by its distance C A; then the two bodies, *w* and W, will remain in equilibrio in any situation, A B, or *a b* of the free lever; but if we suppose the two bodies to have equal masses, or, which is the same thing, equal weights, it is equally evident, that the body, *w*, would preponderate in consequence of its greater distance from the centre of motion, which we will suppose to be three times as great as that of W, and its velocity would, in this case, be three times that of W; consequently it would require an opposing force, three times as much as W would require, to arrest its motion; but if W has its mass or weight increased three times, the distances remaining unaltered, the opposing forces, becoming similar, would then arrest one another, and produce an equilibrium.

The same effects would follow if the threads of suspension were not left free at the ends of the lever, but were folded round the circumferences of two circles, described round the common centre, C, with their respective radii, C A and C B, *fig.* 2; in this case, also, the smaller the circumference of the dark circle is, the greater must be the suspended weight to preserve the equilibrium of the two circles, and *vice versâ*, the ratio being constantly reciprocal.

Let us suppose, now, *fig.* 1 placed contiguous to *fig.* 2, with the large circle of one converted into the portion of a wheel, and the small one of the other into a pinion, as in *fig.* 3, and let us see what will be the consequence, when the teeth are connected and the weights, W and *w*, applied as before; if we suppose the materials of which the portion of the wheel and the pinion are made, to be without weight, an equilibrium will still take place; but remove the small weight *w*, and what will then be the consequence? Why, the pinion D will be impelled at a mean rate, in a direction contrary to that of wheel B, with a force equal to the weight of the small mass, *w*, or, as we have assumed, equal to one third of the large mass, W, which may now be called the maintaining power. But when a tooth of the pinion moves with only one third of the force of W, applied at the point, A, of the wheel, which is at a distance from the centre, C, equal to the radius of the pinion, it moves with the velocity of point B of the wheel, tooth for tooth, and therefore the pinion makes three revolutions for the wheel's one, supposing their numbers to be respectively 16 and 48, as their radii; hence 3, the acquired angular velocity of the pinion's circumference, multiplied by $\frac{1}{3}$, its diminished force, makes the momentum unity, which will also represent the momentum of the maintaining power in motion, *viz.* force 3 multiplied by $\frac{1}{3}$ comparative velocity. Again, let us suppose the pinion, D, actuated by a body, *w*, equal to $\frac{1}{3}$ of the weight of W, and disregard its velocity for the present, the portion of another wheel, E, attached to it by the lever E D, will have its motion contemporary with it; but, by the same mode of reasoning, the velocity of the point E will be increased to three times that of the pitch-line of the pinion, as before was the case with the wheel B, compared with the point A; but its force will be diminished in the same ratio: that is, the velocity will be $3 \times 3 = 9$, and the force, disregarding friction, &c. will be $\frac{1}{3} \times \frac{1}{3} = \frac{1}{9}$; but $\frac{1}{9}$, the force, multiplied by 9, the velocity, will still be unity as before; accordingly, a small body, F, having one-ninth part of the weight of W, the maintaining power, carried over a small fixed pulley, G, will be sufficient to preserve the equilibrium with respect to the maintaining power, when the second body, *w*, is removed. If an additional wheel and pinion were added to the above, and were supposed to have the same dimensions, the velocity produced at the pitch-line of this third wheel would be $9 \times 3 = 27$, and the diminished force, on our former supposition of no friction or opposing force, would be $\frac{1}{27}$ only of the maintaining power. In this illustration of the principle by which an increase of velocity, and a corresponding decrease of power, accompany a communication of motion in alternate directions in wheel work, the teeth may be considered as the remote ends of so many levers, which have their fulcra at the centres of the wheels and pinions respectively, as in *fig.* 4; and it is easy to see, that an accumulation of velocity, accompanied by a corresponding diminution of force, may be produced in any machine where the wheels drive the pinions, as is the case in clock-movements, to such an extent, even where the ratio of the wheel to the pinion is 3 : 1 only, that the friction of the acting parts, together with the resistance of the air opposed to the moving parts, may become a complete counterpoise to the maintaining power. In higher ratios of 12 : 1, 8 : 1, &c. which are used in clocks, the number of wheels and pinions, to produce such an equilibrium, would not require many pairs beyond what an ordinary clock has for its movement. If, however, the pinions had been the drivers instead of the wheels, the reverse of what we have stated would have taken place; the velocity would have decreased, and the force applied have been augmented in the same ratio we have explained, which is the case in cranes, and other engines for raising heavy bodies. Hence, if the movement of a clock were composed of strong wheels and pinions with thick arbors, and had the power applied at the balance-wheel, which may be called the top of the movement, the barrel of an eight days clock having a thick cord coiled round it, would raise a very ponderous body, and become no mean engine for lifting weights, surpassing the ordinary efforts of man's natural strength.

A slight consideration of what we have here advanced, will suffice to convey an idea to any one, who has not previously considered the subject, how admirably well adapted for the purposes of a clock-movement this property in dynamics is, by which an increase of velocity is always accompanied by a proportionate abatement of force; for by means of it, the maintaining power, however large, may be so economically portioned out in minute quantities, sufficient only to compensate the loss of motion which the pendulum or balance sustains, at each vibration, that it may be made to last for many days, or even weeks, before it is exhausted by the great number of successive minute deductions, which are made at periodical intervals; at the same time, certain wheels and pinions, by their numbers and due arrangement, divide and subdivide, by means of hands and divided circles, the sexagesimal portions of each hour, denominated minutes, and also of each minute by the name of seconds. The various offices of communicating motion, of increasing its velocity as at first produced, of diminishing the original force of the maintaining power, of dividing, subdividing, counting, and indicating the hours, minutes, and seconds, all performed by the simple contrivance of a clock-movement, or watch-movement, in conjunction with the regulator, we hesitate not to affirm, when duly considered, exhibit one of the most striking instances of human ingenuity. The experiments of the philosopher, the calculations of the mathematician, and the persevering skill of the mechanist, have combined to produce ultimately this admirable piece of mechanism, of which we hardly know which most to admire, the simplicity of the construction, or the complexity of the various offices, constantly and most correctly performed.

As a further illustration of this fundamental part of our subject

ſubject, we will here introduce the reader again to the calliper of our propoſed half-ſeconds clock, (ſee CLOCK-*making.*) where we will ſuppoſe the centres of the wheels and pinions arranged in a ſtraight line, as ſeen in *fig.* 4, of *Plate* III. We will ſuppoſe the main-ſpring, or power, pulling with a force, at the point P of the great wheel A, equal to the weight of a body, W, of three pounds; then becauſe this wheel has 96 teeth, and actuates a pinion, *a*, of 8 leaves, their numbers of teeth being directly to each other as their radii, which are repreſented as levers acting together, with their fulcra at the reſpective centres, the velocity at the circumference of the wheel will be communicated to the circumference of its pinion *a*, which therefore will make $\frac{96}{8}$, or 12 revolutions, during the time that the wheel makes one; hence if the great wheel be aſſumed to revolve once in 12 hours, as we propoſed, the centre pinion *a*, together with its wheel B, will revolve in one hour, and its arbor will be proper for the hand to indicate minutes on a circle of 60 on the clock face; but it muſt be obſerved, that the velocity is not increaſed 12 times, as it reſpects the point P, where the maintaining power is applied, nor is the force diminiſhed in the ſame ratio as it regards that point; but, to make the calculation more ſimple, we will conſider the diſtance of P, from the centre of A, equal to $\frac{1}{12}$ of the radius of wheel A, which is the radius of pinion *a*, and correct the concluſion drawn from ſuch a ſuppoſition afterwards: the pinion *a* then, we ſay, has its velocity at the pitch-line, or its number of revolutions equal to 12, and its force equal to $\frac{1}{12}$; in the next place, the radius of a pinion *a*, is to the radius of the concentric and contemporary wheel B, as 8 : 64, and $\frac{64}{8} = 8$; the circumference of this centre wheel, therefore, will move with a velocity equal to 12×8, or 96, and with a force only of $\frac{1}{96}$; both which will be imparted to the pitch-line of the pinion *b*, of 8 leaves; again, the radius of the latter pinion *b*, is to the radius of its concentric and contemporary wheel C, as 8 : 60; but $\frac{60}{8}$ is a fraction, the value of which is $7\frac{1}{2}$; and $96 \times 7\frac{1}{2}$ therefore make 720 to expreſs the velocity, or number of revolutions of pinion *c*, actuated by the wheel C; and $\frac{1}{720}$ will expreſs the force, or fractional part of three pounds, the quantity aſſumed for the maintaining power, which is equal to 24 grains of troy weight. We have now found, that while the wheel A is making one revolution, the laſt pinion *c* will make 720, and that 24 grains troy, ſuſpended by a ſmall pulley over the pitch-line of pinion *c*, and attached to it, will balance three pounds of the ſame ſpecies of weight, ſuſpended at $\frac{1}{12}$ of the radius from the centre of the great wheel, if we diſregard the effects of friction. Let us examine the conſequence of theſe calculations as they regard our arrangement of the calliper: the pinion *c*, we have ſeen, revolves in $\frac{1}{720}$ of 12 hours, according to our original aſſumption of A revolving in 12 hours; but in twelve hours there are juſt 720 minutes; one revolution conſequently of the laſt pinion *c*, is performed in exactly one minute; its arbor will therefore be proper for the axis of the ſeconds hand, which goes round its circle of 60 in this time: but the maintaining power, W, is not actually ſuſpended at $\frac{1}{12}$ of the radius of wheel A, according to our ſuppoſition when we calculated, conſequently the reſult, as it relates to the modified force at the pitch-line of pinion *c*, requires yet to be corrected, or ſomehow compenſated; this is done by making the radius of the ſwing or balance wheel D, equal to the diſtance of P from the centre of A, in which caſe, the calculated velocity and force at any point in the circumference of D will be in the ſame proportion to what thoſe of the maintaining power ſuſpended at P are, as the velocity and force at the pitch-line of pinion *c* would have been, if the ſame maintaining power had been actually ſuſpended $\frac{1}{12}$ of the radius of A, from its centre; the body W of 24 grains therefore attached to the circumference of wheel D, and carried over the ſmall pulley above it, will balance three pounds, ſuſpended at the point P, of the great wheel, if, as before, we diſregard friction; and the puſh made, in any direction, againſt the pallet face, will be equal to a force of 24 grains, acting in the ſame direction: the ſame reſult would alſo have accrued, if the body W had been ſuſpended at the circumference of wheel A, provided the ſwing-wheel had been of the ſame radius; but, ſuppoſing the body W to remain at P, and the ſwing-wheel D made equal to the great wheel, then the velocity of the former would have been increaſed, and its force diminiſhed, in the proportion of the diameter of the ſwing-wheel to the diameter of the barrel, by which we ſuppoſe the body W to be ſuſpended at P: in this caſe, the ſmall dotted body, *x*, ſuſpended at *y*, at a diſtance equal to the radius of wheel A, would have been ſufficient to keep the body W in equilibrio.

The pendulum propoſed to be adopted, being a half ſeconds pendulum, the ſwing-wheel muſt have 60 teeth, in order that one tooth may eſcape the pallets at every ſecond vibration, which therefore becomes a proper regulator for our preſent movement.

When the whole movement is previouſly given in any clock, and we want to aſcertain the relative revolutions of the firſt and laſt wheels, a little conſideration of what we have ſaid will ſhow, that the product of all the wheels, divided by the product of all the pinions at one operation, will give the reſult at once thus, $\frac{96}{8} \times \frac{64}{8} \times \frac{60}{8} = \frac{368640}{512} = 720$, as before.

Alſo, when the relative forces of the maintaining power, at the barrel, and at the end of a tooth of the pallet-wheel are thus aſcertained, this calculated force muſt be altered, inaſmuch as the pallet-wheel has its diameter greater or leſs than that of the barrel, accordingly as we have ſhewn above. When a ſpring and fuſee are uſed, the power of the ſpring may be aſcertained either by the inſtrument of adjuſtment uſed as a lever with a ſliding weight, or, what will be leſs equivocal, a barrel may be attached to the ſquare of the fuſee arbor, when the ſpring is adjuſted, and then a heavy body ſuſpended by it in a ſcale to admit of weights of adjuſtment, which barrel and weights may be ſubſtituted for the ſpring and fuſee in the calculation, firſt of the whole force applied as a maintaining power, and then of its modification at the face of the pallets.

If we were to take the oppoſing force of friction into our calculation, we ſhould find the problem extremely complicated; for there are not only various cauſes of friction in a movement, ſuch as that cauſed by the action of the teeth, and in the pivot-holes, &c.; but the continual variation of the quantum of friction, ariſing from the deſtruction of the rubbing parts, the thickening of the oil uſually applied, and the admiſſion of duſt, is ſuch, that no regular and conſtant data can be obtained whereon to ground ſuch a calculation as would prove ſerviceable. The readieſt, and perhaps beſt, practical way of aſcertaining how much of the maintaining power is expended in friction, is really to ſuſpend ſuch ſmall weight by a tooth of the ſwing-wheel, as will balance the maintaining power, and compare this with the requiſite force obtained by calculation: the difference will give the collective quantum of friction, at the time of the experiment, in the whole movement.

Friction, however, is not the only obſtacle to the maintaining

ing power in the works of a clock in motion; the tranfmiffion of force is not conftant, but effected at fuch equi-diftant intervals of time, as depend on the vibrations; for inftance, in our half-feconds piece, the wheels and pinions are put into motion, and ftopped again alternately 60, and where there is recoil 120 times in each minute; hence the inertia of the matter compofing the wheel-work is as often to be overcome as the arrefted motion is re-produced, and if the wheels are not made very light at the upper end of the train, where the force is greatly reduced, a very confiderable portion of the calculated force will be employed in moving the works from a ftate of reft, at every vibration. We prefume it is on this account, more than on account of friction, that the workmen have found it neceffary to diminifh the wheels, and to reduce their weight, as much as is confiftent with ftrength, accordingly as the train afcends. To overcome the obftacles to the due effect of the maintaining power, arifing from friction, and the inertia of the wheels and pinions, more force is ufually given to it than would otherwife be neceffary, and the requifite addition, over what calculation gives, muft in every inftance depend upon tentative adjuftments of the maintaining power to the momentum of the regulator, fo as to produce the due excurfion of the pendulum or balance beyond the efcapement angle. In fome delicate machines for keeping true time, the pivots of the arbors are made to reft on friction rollers, or otherwife, are bufhed with fome of the precious ftones that take a high polifh, which expenfive additions admit of the maintaining power being fmall in comparifon with the momentum of the regulator; which circumftance, when the regulator is rendered unchangeable by fome compenfating mechanifm, is beft calculated to preferve the angle of vibration alfo unchanged, provided the maintaining power remain uniformly the fame; and even if fome flight alterations of force fhould arife out of an imperfect adjuftment of the fufee, where a fpring is ufed, the great controul of the regulator would, under fuch circumftances, reftrain the effects of the flight irregularities of the impulfes on the pallets.

Hitherto we have confidered the action of the wheel with its pinion, or what the French call *engrenage*, to be fo perfect, that the velocity and force, at the circumference of every wheel, is truly and conftantly imparted to its refpective pinion; which is fuppofing not only the wheel-work to be proportioned and callipered with the utmoft exactnefs, but alfo their teeth fhaped in the beft manner. There are, however, three very common caufes of bad action; firftly, whenever the wheel is too fmall for the pinion, though ever fo well callipered, its teeth will pitch againft the ends of the pinion's leaves, and require more than ordinary force to be confumed in their difengagement; fecondly, when the wheel is too large, it will impart to the pinion too much velocity during the action, and part of the force will be expended in the drop that will take place before the action commences againft each following leaf of the pinion, after it has ceafed to act with the leading leaf; and thirdly, if the curve of the tooth be ill formed, the tranfmitted force will, in fome fituations, exceed, and in others fall fhort of a mean force; in all thefe cafes the varied intenfities of the tranfmitted force will confiderably affect the ifochronifm of the regulator with any of the ordinary efcapements. Of the two firft of thefe caufes of bad action, we have pointed out the remedy, when we treated of the proper method of proportioning and callipering a movement, under the word CLOCK-*making*; the third, which is of the utmoft importance, not only in clock-work, but in wheel work of every defcription, prefents itfelf now for difcuffion.

In *fig.* 5 of *Plate* III., let us fuppofe *b* C A and *b c a* two bent levers, refpectively moveable on the points C and *c*, as their centres, and let us conceive that the parts, C A and *c a*, which are in the direction *c* C of the line of the centres of motion, be unchangeable in length, and be loaded with their refpective weights W and *w*; but that the parts, *b* C and *b c*, be variable in length, fo that in any fituation their extremities may meet and act reciprocally on each other; we affirm that the fame weights W and *w*, which will keep the two variable levers in equilibrio at the point P, in the line of the centres, will alfo keep them in equilibrio when their ends reft againft each other at any other point, *b*, in the circumference of a circle, of which *c* P, or C P, is the diameter.

Demonftration. Let *b d*, the fhort line perpendicular to *b* C, which may reprefent the fhort arc made in an inftant by the lever *b* C, turning on the centre C, exprefs the abfolute force of the weight W, acting at the extremity of the lever; and let this abfolute force be decompofed into two others, *b e* and *b f*, the former of which, *b e*, may be perpendicular to *c b*, in the direction of P *b*, the angle at *b* being a right angle, by reafon of its being formed at the circumference by two chords from oppofite ends of the diameter, and the latter, *b f*, may be parallel to the line *b* C; it is evident that *b e* expreffes that portion of the force *b d*, which is employed in moving the point, *b*, of the lever *c b*, from *b* towards *g*, and, confequently, in making the lever, *c b*, revolve, whilft the other portion, *b f*, expreffes that part of the force, *b d*, which has a tendency to pufh the point *b* towards C; but C is a fixed refifting point, which therefore deftroys this part of the decompofed force: we fay, likewife, that *b d* expreffes the abfolute force exerted at the end, *b*, of the bent lever, *a c b*, by the weight *w*, to oppofe the revolution of the point, *b*, round its centre, *c*; for, becaufe the right line, A W, is perpendicular to C A, and alfo *b d* to *b* C, the weight, W, is to the force, *b d*, as the line, *b* C, is to C A; but in taking C B perpendicularly to the line, *b* P, prolonged, the like triangles, *d b e*, and C *b* B, give *b d* : *b e* : C B : C *b*, therefore, by multiplying, W : *b e* :: C B : C A: now let the abfolute force with which the weight, *w*, urges the point, *b*, in a direction oppofite to *b g*, be called *x*, and we fhall have *w* : *x* :: *c b* : *c a*, or, on account of the fimilarity of the triangles, *c b* P, C B P, as C B to C P; but on the fuppofition that the two weights, *w* and W, will balance each other, when their forces are exerted reciprocally at the point P, on account of the equality of the levers, *c a*, *c* P, the weight, *w*, may be fuppofed to be fufpended at the point P; it will be then as W : *w* :: C P : C A; likewife, by multiplying, W : *x* :: C B : C A; but we have juft proved that W : *e b* :: C B : C A, therefore *e b* = *x*; hence there will be an equilibrium, when the force, *e b*, is equal to the force with which the point, *b*, is urged to revolve by the weight W. It is equally demonftrable, that an equilibrium will take place between the two levers when their point of contact falls in any other part of the circumference of the circle, *c b* P; and alfo, that an equilibrium will not take place if fuch point of contact fall either within or without the circumference of the faid circle. Vide Berthoud's "Effai fur l'Horlogerie," tom. i. p. 41.

Corollary. From the preceding demonftration an inference is deducible, of the utmoft importance in wheel-work; namely, if we fuppofe the point *b* to be always in the circumference of *c b* P, while the lever *b* C A impels the lever *b c a*, the circles H K and *h k* defcribed from the centres of motion C and *c*, and touching one another at P, when attached each to its own lever, will move with the fame force and the fame velocity.

For 1ft. Let F and *f* be the refpective forces with which

which the circumferences of the circles H K and *h k* are urged, and we ſhall have *b d* : F :: C P : C *b*, and *f* : *b e* :: *c h* : *c b* or *c* P ; but we have ſeen above that *b e* : *b d* :: C *b* : C B, therefore *f* : F :: C P × *c b* : *c* P × C B ; but the ſimilar triangles *c b* P and C B P give *c* P : C P :: *c b* : C B ; hence *c* P × C B = C P × *c b*, therefore *f* = F ; conſequently, if the circle H K be moved by any force uniformly, the circle *h k* will alſo be moved uniformly.

2dly. Whatever may be the velocity of the point *b*, of the lever A C *b*, that which it will communicate to the lever *b c a*, in the direction *b e*, perpendicular to the point of contact, will be the ſame as that of point *b* ; that is, if we demit the perpendicular *d g*, the velocity in the direction *b d*, being repreſented by *b d*, the velocity in the direction *b e* will be repreſented by *b g*. If then V be the velocity of the circumference of the circle, or wheel, H K, and *v* that of the ſmaller circle, or pinion *h k*, it is evident that V : *b d* :: C P : C *b*, likewiſe, that *b g* : *v* :: *c b* : *c h* ; alſo, the like triangles, *b d g*, and C *b* B, give *b d* : *b g* :: C *b* : C B ; therefore, V : *v* :: C P × *c b* : *c* P × C B ; but theſe two laſt products we have ſeen are equal, therefore *v* = V.

By a ſimilar reaſoning it is equally demonſtrable, that neither the force nor velocity of the wheel would be equally communicated to the pinion, nor thoſe of the pinion to the wheel, if the point of action were ſituated either within or without the ſemi-circle *c b* P.

To apply the preceding corollary to our preſent purpoſe, of effecting an equable communication of force and velocity in wheel-work, let us conceive the variable levers A C *b*, and *a c b*, in *fig.* 1, *Plate* IV. ſo circumſtanced, that their remote points, *b*, meet at P, and that the points H and *h*, of the wheel H K, and pinion *h k*, be alſo in contact at P ; let the wheel now move through five ſucceſſive portions, H α, α β, β γ, γ δ, and δ P, ſucceſſively, and conceive it to drive the pinion *h k*, by ſimple contact ; in this caſe, the pinion will move in like manner, through ſimilar arcs *h* α′, α′ β′, β′ γ′, γ′ δ′ and δ′ P, and at each of the ſucceſſive points of contact, α α′, β β′, γ γ′, &c. the reſpective lengths of the levers, in order to meet in the ſemi-circle *c b* P, will be A C 1 and *a c* 1, A C 2 and *a c* 2, A C 3 and *a c* 3, A C 4 and *a c* 4 ; and, laſtly, A C *b* and *a c b*, continually varying : hence we may conceive a curve, H *b*, ſuch, that ſome point of the variable lever *a c h*, between *b* and *h*, ſhall always reſt on it in every ſucceſſive ſituation of the wheel and pinion, beginning at *h*, and ending at *b* ; if now we ſuppoſe the curve H, on which the variable end of the lever *c b h* reſts, to be attached to the wheel H K, and to be brought back into its original ſituation P, and alſo *c h* to be coincident with *c* P, it is evident that, though the wheel and pinion were not ſo in contact as to impel one another, yet if the wheel were to be moved uniformly as before, the curve-piece attached to it would drive the lever *c b h* before it in ſuch a manner, as to effect a conſtant variation in its length, but whether ſuch motion of the pinion occaſioned by the attached ſtraight lever, when urged by the curve-piece H *b*, will or will not be uniform, depends entirely on the nature of the curve which we have not yet eſtabliſhed. Let us examine the figure a little more cloſely. The arcs P H, of the wheel, and P *h* of the pinion, we have aſſumed to be equal in length, though one contains double the number of degrees as the other, by reaſon of having only half its radius ; but the ſemicircle, *c b* P, is deſcribed from only half the radius of the pinion ; conſequently the dotted lines, *c* 1, *c* 2, *c* 3, &c. which meaſure degrees, by being prolonged, on the circumference of the pinion, meaſure double degrees on the ſemi-circle, *c b* P, by reaſon of their meeting at the end of the diameter, inſtead of at the centre ; therefore the arc P *b*, is alſo equal in length to either of the others, P H, and P *h*. Theſe conſiderations will enable us to determine the requiſite curve, thus ; transfer the diviſions, H α, H β, &c. from P, back towards K ; and through the interſections 1, 2, 3, 4, *b*, of the ſmall ſemi-circle, *c b* P, draw parallel arcs, 11′, 22′, &c. in dotted curves from C, as a centre ; which extents will be the ſucceſſive acting lengths of the radius of wheel H K, in the ſucceſſive ſituations, as it revolves from P to H ; alſo the ſmall extents P 1, P 2, P 3, &c. will be ſo many radii of curvature applied ſucceſſively from the different points of diviſion ; for inſtance, P 1, applied to the firſt point below P, will interſect the innermoſt dotted arc at 1′ ; P 2, from the ſecond point in P K, will interſect the ſecond dotted curve line at 2′ ; P 3, from the third point, will interſect the third dotted line at 3′ ; P 4 at 4′ ; and, laſtly, P 6, at 6′ ; and if the interſtices of the curve ſo formed be completed, it will have the peculiar property of driving the ſtraight line, *b h*, and its pinion, with an equable force, and an equable velocity, provided the wheel H K, to which it is attached, move equably ; for the acting point will always be found in the ſemi-circle *c b* P, which is fulfilling the condition of the problem ; alſo, reverſing the motion, the ſtraight line, *b h*, will drive the curve, *b* H, and its wheel, back again with an equable force and velocity, as though the pinion drove the wheel by contact. We have here ſuppoſed the curve piece attached to the wheel, and driving the pinion, but the ſame projection, or rather the ſame mode of projecting the curve, will apply when the pinion has the curve, and is the driver according to the repreſentation given in *fig.* 2, of *Plate* IV. which is the reverſe of *fig.* 1. We have alſo hitherto ſuppoſed, that a wheel or pinion moves always in the ſame direction ; but as it is frequently required in wheel-work, that the works ſhould turn both backward and forward, it is neceſſary to have a counter-curve on the following part of the tooth, which in its turn may occaſionally be the leading part ; the ſame geometrical proceſs will give this reverſe curve, which muſt neceſſarily begin at the diſtance from the other of the whole breadth of the tooth ; in our drawing, we have given the tooth equal to four diviſions, in order to ſhow more clearly the proceſs of deſcribing it, but in ordinary works, particularly in clock movements, the ſtrength of the tooth is too ſmall to admit of being deſcribed and demonſtrated in this way ; we have notwithſtanding thought it our indiſpenſable duty to inveſtigate, and lay before the reader, the fundamental principle on which the proper ſhape of a tooth depends. According to the preceding inveſtigation, the curved portion of one tooth, it will be ſeen, muſt always drive a ſtraight edge of the other, and *vice verſâ* ; alſo a line drawn from the point of action to the primitive circles, or pitch line, where it is interſected by a line joining the centres, will be always perpendicular to both the ſtraight line and the curve.

Here the reader will naturally be led to aſk, what is the preciſe nature of the curve we have determined ? and what name ſhall we give it ? A little reflection on its property and delineation will ſoon convince the geometrician that it is an epicycloid, or rather a portion of an epicycloid, for the generation of which the wheel conſtitutes the baſe, and a circle, equal in diameter to the radius of the pinion, the generating circle. Camus in his "Cours de Mathématiques," Liv. x. and xi. has inveſtigated the epicycloid, as it affords a rule for the formation of teeth in wheel-work, which portion of the work has lately been tranſlated into Engliſh, but the tranſlator has added ſome practical directions reſpecting the ſhape of a tooth, taken from "Imiſon's Elements of Science and Art," the principle of which we think it neceſſary here to correct, at the ſame fame time that we avail ourfelves of the elucidation of our fubject which Camus's mafterly treatment of it affords.

In *fig.* 3, of *Plate* IV. let H K reprefent a portion of the fame wheel which is reprefented by the fame letters in *fig.* 1, and let the epicycloidal curve, H 2′ 4′ 6′, be defcribed in the ufual way by a fixed point in the circumference of a fmall circle, with a radius equal to that of the femicircle, *c b* P, in *fig.* 1, coming in contact with the circle H K, and thereby effecting a revolution; then it is evident that the epicycloidal curve in our prefent diagram is precifely the fame as the curve P 1′ 2′ 3′ 4′, &c. in *fig.* 1, which fimilarity proves the latter to be alfo epicycloidal; for the chords 2 2′, 4 4′, &c. in *fig.* 3, are precifely the fame as P 2, P 4, &c. with which the points 2′, 4′, &c. are defcribed in the curve P 1′ 2′ 3′, &c., in *fig.* 1; and in *fig.* 3, the arcs, 2 2′, and 4 4′, &c. are refpectively equal to 2 H, 4 H, &c. of the fame *fig.* in the fame way that P δ', P γ', &c. are refpectively equal to P δ, P γ, &c. in *fig.* 1. This epicycloidal curve, P 1′ 2′ 3′ 4′ *b*′, in *fig.* 1, or H 2′ 4′ 6′, in *fig.* 3, is of the kind called exterior, by reafon of the generating circle revolving round the outfide of the bafe H K; but if it were made to revolve by a fimilar contact with the interior fide of a circle, the generated curve would be of a different fhape, and would be denominated an interior epicycloid. In the particular cafe where the generating circle has its diameter, exactly equal to the radius of the circle which conftitutes the bafe, the line generated by a fixed point in the circumference of this generating circle will not be a curve, but an exact ftraight line paffing through the centre of the circle which is made the bafe: thus in the lower part of *fig.* 3, if we conceive the fixed point, in the generating circle, for defcribing the line required, to be at *a*, a point in the circumference of the bafe, *a d e*, and the generating circle to move forwards to the points *d* and *e* fucceffively, it is demonftrable that the faid fixed point, *a*, will be found fucceffively at *b* and *c*, and will confequently defcribe the ftraight line or radius, *a b c*, in the fame time that the generating circle revolves down one quadrant, *a d e*, of the circle conftituting the bafe; this confequence muft follow from the confideration that the arc, *d b*, is equal to the arc, *d a*, and that the femi-circle, *e c*, is equal to the quadrant, *e d a*, according to our affumption of the diameter of the fmaller circle being exactly equal to the radius of the larger.

With thefe demonftrable truths in our recollection let us now turn to *fig.* 4, and fee how both the exterior and interior epicycloids are concerned in the formation of an epicycloidal tooth, or tooth that has the property of tranfmitting both the force and velocity it has received without alteration to its fellow-tooth, when it is alfo fhaped according to the fame principle: if we fuppofe the portion of a circle, H K, to drive the fmall generating circle, *c b* P, equably by fimple contact at the point, P, the fixed point *b* will defcribe the portion, H *b*, of an exterior epicycloid on the plane of H K, extended, which portion we will fuppofe to be one fide of a tooth attached to H K, to which curve the line P *b* is always perpendicular; again, if we fuppofe not only the faid generating circle *c b* P, but alfo at the fame time the circle *k h*, to revolve, by means of a fimilar contact with H K at the point P, then the inner or generating circle, *c b* P, will make two revolutions, while the outer one, *k h*, makes only one; the confequence refulting from fuch a combination will be the fame, as though the generating circle, *c b* P, moved only once round within the circle, *k h*, confidered as ftationary, that is, the circle *k h* becomes the bafe to the generating circle, and the defcribing point *b* fo circumftanced will trace on the plane of circle *k h*, the ftraight line or interior epicycloid, *h b c*, which is the radius to the circle conftituting the bafe in this cafe. Hence if we conceive the points *b*, *h*, and H, to coincide at H, for the original pofition of the three circles, and alfo of the tracer *b*, it is eafy to comprehend, that if a flit were cut in the plane of the circle *k h*, in the direction of the radius *c h*, fuch as would juft admit the tracer to pafs through it, provided the great circle, H K, communicates its motion equably to the other two by contact at the point P, the tracer will pafs along the flit, confidered as an interior epicycloid, and at the fame time will defcribe an exterior epicycloid on the plane beneath both the circles, *c b* P, and *k h*. A fmall inftrument might readily be made to prove practically that the refult we have here pointed out would follow from this arrangement of three circular metallic plates kept in their places by a fuitable frame. But this is not all; if the curve H *b* be made to form a part of the circle H K, confidered as a wheel, and the point, P, of contact be removed a very fmall diftance from the two circles, fo as not to touch them, the points, *b*, *h*, and H, may be again brought into their original coincident ftate, either by pufhing the tracer, attached to the extremity of the generating fmall circle, againft the edge of the epicycloidal curve, or otherwife by fubftituting a ftraight lever in the place of the flit, as a radius of the circle *k h*, and by preffing it, inftead of the tracer, againft the faid curve; in either cafe an equable motion will be communicated to the great circle or wheel, K H, as well as though it had been actuated at the point of contact, P: nay, moving the tracer back from *h*, while in the flit at P, would give an equable motion to all the three circles: alfo reciprocally, the curve H *b*, by preffing againft the tracer, while the wheel H K is in motion, will give an uniform motion to the generating circle, *c b* P; or otherwife it will uniformly drive the larger circle *k h*, by preffing againft its radial lever; in either cafe the motion will be as equable as though communication were to take place by mere contact, or by teeth infinitely fmall, at the point P.

This view of the fubject, it is prefumed, will afford the reader a full and clear elucidation of the application of the exterior and interior epicycloidal curves to the formation of teeth in wheel-work, and, at the fame time, eftablifh an effential difference, which has been overlooked hitherto in practice, between the curve generated by a circle equal to one of the acting wheels or pinions, and the curve generated by a circle equal in diameter to only *half* of one of the acting wheels or pinions, the correfponding wheel or pinion being in both cafes taken of its full geometrical diameter as the bafe of generation. The tranflator of a part of Camus, and the editor of "Imifon's Elements of Science and Art," have pofitively though erroneoufly afferted that the generating circle fhould in all cafes of wheels and pinions be equal to the fellow of the wheel on which the curve is to be defcribed, in which opinion fome very refpectable mechanicians agree; but others, on the contrary, affert, with equal confidence, and more truth, that the faid generating circle fhould have its diameter equal to only one half of the diameter of the faid fellow. (See Camus, Dr. Young's Syllabus, and Brewfter's Edit. of Fergufon's Select Exer. &c.) A careful examination of Camus's demonftrations would of itfelf have reconciled the difagreeing parties, which we truft a due attention to our elucidation, by means of the tracer and radial lever, will not fail to effect. The fact is, that, where pins like our tracer, or fpindles are ufed for teeth in any wheel or lantern, as is frequently the cafe in large works, the generating circle muft be equal in diameter to the diameter of the acting wheel or lantern which it reprefents, in order to trace the epicycloidal teeth

teeth of its fellow; but in clock-movements, and in all other inftances in wheel-work where both the wheels and pinions have the epicycloidal formation, the generating circle muft be only one half in diameter to what it is required when lanterns are ufed, for in this cafe, which is what is moft frequent, the interior and exterior epicycloids impel each other alternately, the former being a portion of the radial lever, and the latter a portion of the epicycloidal curve: thus, if a pin were attached to the extremity of the generating circle *cb*P, and a radial lever to the plane of the circle, *k b*, of twice its diameter, they would be alike driven by the epicycloidal tooth, H *b*, of the driving great wheel, H K. Hence the fhape of the tooth of the driven wheel or pinion will always determine whether the curve of a tooth in the driver muft be defcribed by a circle of the full or half fize ufed as a generating circle, neither of which, it appears, will apply exclufively in all cafes. It may be remarked, however, here, that the beginning of the curve, or that part which is formed on the end of a tooth, will in practice be, as nearly as may be, the fame whichever of the two generating circles be made ufe of, particularly if the teeth to be formed are fmall; for in *fig.* 1, it will make little difference whether P 1 and P 2, or P δ and P γ, be taken as fucceffive extents for defcribing the curve P 1′, 2′, &c.; but if P*b* were taken inftead of P *b*, the difference at *b*′, the remote end of the curve generated, would be confiderable. Hence the argument in favour of the erroneous principle, adduced from the trial of feven years wear without repair, by the tranflator of Camus, is without weight, and proves only that a convenient approximation to the truth may fometimes be fubftituted for the truth itfelf in practice; as the fmall circular arcs of a pendulum are fubftituted for cycloidal arcs, to which they are very fimilar at the loweft point; but where the truth itfelf is attainable by equally fimple means, approximations ought to be inadmiffible.

Our theory, it will be obferved, hitherto fuppofes that the whole tooth of the driving wheel or pinion be formed by interfecting portions of an exterior epicycloid, and that the whole tooth of the driven wheel or pinion be formed like a ftraight radial lever, which is eafily fhaped; thefe formations will act pretty well together, and when they are adopted, the outer ends of the teeth of the driven one muft be in its primitive circumference; but in the driver the whole tooth muft project beyond its primitive circle; fo that the two primitive circles, geometrically obtained, muft not meet at what is called the pitch-line, but be feparated by a fpace equal to the length of the driving tooth; but in practical works, particularly in horology, the teeth of both the driver and driven wheel or pinion are ufually formed nearly alike; partly by means of the exterior epicycloid, and partly by means of the interior one, or ftraight line; the exterior ends are formed ufually by being curved, and the interior parts are bounded by radial ftraight lines, either of which parts will drive the other, which office they do alternately; this fhape is not only found to be the moft practicable, but admits of a contact of the two primitive circles at a point of each tooth, called the *pitch-line*; the confequence of which formation is, that nearly an equal addition may be made to each curve for the rounded parts of the teeth of both the wheel and pinion; it is this formation that we mean when we fpeak of ordinary teeth, and to which our table is applicable, which is given under the article CLOCK-*making*.

Olaus Roemer, the celebrated aftronomer and mechanift of Denmark, according to Wolfius and Leibnitz, was the firft who pointed out the utility of the epicycloidal curve, when applied to delineate the fhape of a tooth; but De la Hire took up the fubject after him, and demonftrated, that if a tooth of either a wheel or pinion be formed by portions of an exterior epicycloid defcribed by a generating circle of any diameter whatever, the tooth of its fellow will be properly formed by portions of an interior epicycloid, defcribed by the fame generating circle, which curious circumftance allows of an infinite variety in the two correfponding curves that form the teeth of the wheel and pinion, if they were practicable. Nay De la Hire has fhewn, that if the teeth of any wheel be triangular, circular, or of any other regular figure, an uniformity of force and velocity may be mutually imparted, provided the teeth of the correfponding wheel or pinion have its teeth formed by a figure, compounded of the epicycloid and faid figure, which he has further fhewn the method of effecting in a variety of cafes, not however adapted for practice. But whether the mechanift may choofe to ufe his exterior and interior epicycloids jointly in the fame tooth, or feparately, in different wheels acting together, this practical rule ought never to be loft fight of, *viz.* the outer end of his interior, and alfo the inner end of his exterior epicycloid fhould univerfally commence in the primitive or geometrical circle of his wheel or pinion.

The reader is now prepared to be told, what otherwife might have appeared paradoxical, not only that the fame pinion, of eight leaves for inftance, will require the teeth of a wheel of thirty to be fomewhat differently rounded at the ends, from thofe of a wheel of fixty, or any other number, in order to have like action in both cafes, but that, however accurately the teeth of wheels are rounded, all numbers are not equally good to be ufed indifferently for wheels and correfponding pinions. This latter part of our fubject has not been much attended to in practice, but is curious, and may contribute to great utility, particularly in horology, where an equable tranfmiffion of velocity and force is defirable.

The whole of what we have hitherto faid refpecting the action of epicycloidal teeth, has been upon a fuppofition that the impelling force begins at the line which joins the centres of any pair of wheels, or of a wheel and pinion, and is exerted outwards always on one fide of this line until the teeth efcape one another, which mode is allowed to be the beft, when it can be effected; but there are many ratios, and thofe in common ufe between a wheel and its pinion, which will not admit of this kind of action, however good the fhape of the teeth. Indeed, Camus has fhewn, that no pinion lefs than one of eleven leaves, will entirely anfwer the purpofe of acting always on one fide of the line joining the centres, and that confequently the common pinions of fix are very ill calculated to effect an equable tranfmiffion of velocity and force, by reafon of their leaves acting alternately before and behind the line of the centres.

In *fig.* 5, fuppofe H K to be a portion of a wheel of 50 teeth, and *k b* a pinion of feven thin leaves or levers, each like *c b b*; Camus has proved, that this wheel will not impel its pinion in an uniform manner, by acting always behind the centres. His reafoning is to this effect; in order that the leaves of a pinion of feven, may be impelled only behind the line joining the centres, the tooth, P *b* H, of the wheel muft not quit the leaf *c b*, until the next following leaf, *c* I, has reached the point, P, in the line of the centres, in order that it may be impelled in its turn by the next tooth, K I, behind that line. The angle I *c b*, or quantity that one leaf of the pinion muft be moved before the following one comes to the line of the centres, is one feventh of 360°, or 51° 25′ 43″ almoft; confequently the angle C *c b*, or C *c b*, is 51° 25′ 43″. Then taking the radius of the pinion at feven parts, the fide *c b* of the fmall right angled triangle, when folved, will be 4.364

4.364 of fuch parts; alfo in the triangle b C c we have the two fides $c b$ as before, and c C equal 57, with the included angle $b c$ C, from which data, by a folution of the problem, we have the angle b C c equal 3° 35′ 50″ nearly. The angular quantity of the wheel H C I, which one tooth and one fpace occupy, or $\frac{1}{50}$ of 360°, is 7° 12′, from which, if we deduct the angle b C c, or 3° 35′ 50″, the remainder, 3° 36′ 10″, will be the angle H C b. Now, as the two epicycloidal portions of the tooth P b and H b are equal and fimilar, and alfo fimilarly placed with refpect to the full radius C b, and as the angle H C b has been found to be 3° 36′ 10″, the angle H C P, which ought to contain one tooth and fpace, will be 7° 12′ 20″; but we have feen that the angle H C I is only 7° 12′, confequently the angle H C P will be greater than H C I by 20″, which is impoffible, becaufe a part cannot be greater than the whole. It is, therefore, impoffible that a wheel of 50 fhould move in an uniform manner, a proportionate pinion of feven leaves impelling them only behind the line joining the centres. A wheel of fewer teeth than fifty will be ftill lefs proper, and one of a greater number will not leave fpace enough for a fufficient thicknefs of leaf in a pinion. Hence it appears, that when a pinion of feven leaves is ufed, it will be impelled by its wheel, partly before and partly behind the line joining the centres, as may be feen more fully in *fig.* 6, where the pinion is fuppofed to be impelled by the wheel from right to left.

In the fame manner it may be proved, that if a wheel of 57 were made to drive a pinion of eight, the whole arc for both the tooth and fpace would be 6° 18′ 57″, of which 5° 7′ 40″ would be occupied by the tooth of the wheel, and only 1° 11′ 17″ by the fpace, or by the tooth of the pinion, which quantity is not enough for the thicknefs of an acting tooth; therefore if the teeth of the wheel are made nearly equal to the fpaces, they will drive the pinion of eight both before and behind the line of the centres.

Alfo if a wheel of 64 were to drive a pinion of nine leaves in fuch a way, that the impulfe might be only behind the line of the centres, the arc at the pitch-line of the wheel, for both tooth and fpace, will be 5° 37′ 30″, of which the tooth will occupy 3° 45′ 42″, and the fpace only 1° 51′ 48″, which will not leave room for a leaf fufficiently thick for the pinion.

Likewife where a wheel of 72 drives a pinion of ten leaves behind the line of the centres, the arc of the wheel's tooth is 2° 47′ 16″, and of the fpace between two teeth 2° 12′ 44″ only, therefore here the tooth and fpace cannot be equal.

In pinions of 11, 12, &c. the impulfe may take place entirely behind the line of the centres, and the extreme ends of the teeth might be taken away, as in the pinion of *fig.* 6, and in the two teeth of the wheel, which are fuppofed to have paffed the centre. Nay, in thofe cafes where the pinion is always driven by an impulfe made only behind the line of the centres, the addition to its tooth beyond the geometrical diameter, may, as we have faid, be nearly difpenfed with; that is, the acting and the geometrical diameters may be almoft the fame, provided the angular points be a little rounded to prevent their catching and fcraping the teeth of the wheel, though it is fafer to give a little addition for the curves.

Under our article CLOCK-*making*, we faid that the driving wheel or pinion ought to be fomewhat larger than according to its calculated proportion; but we did not give the reafon there, which is, that in thofe cafes where the teeth are actuated both before and behind the line of the centres, the impulfe of the tooth before the line of the centres takes place later than it otherwife would do, as well as occafions a fmaller fhock at the commencement of the impulfe. The defect attending fuch a conftruction is a little more velocity in the driver than in the driven wheel or pinion, but this is confidered as of lefs importance than the fhock otherwife occafioned in movements, with pinions of low numbers.

It is of the utmoft importance, that the mechanift fhould bear in mind another practical rule, which we have now to offer, as it arifes out of the preceding difcuffion of this fubject, which is, that, in all ordinary cafes of wheel-work, the portion of an interior epicycloid of any tooth, whether that portion be a curve or ftraight line, fhould always drive the portion of an exterior epicycloid of the tooth of its fellow, before the line joining their centres; but that, on the contrary, a portion of an exterior epicycloid muft always drive a portion of an interior epicycloid, whether the latter be a curve or ftraight line, after the line of the centres.

We are not, however, to infer from this rule, that it is a matter of indifference, whether a wheel drive its pinion, and a pinion its wheel, before or after the line joining their centres; for in the action that takes place before this line, there is much friction occafioned, by the fliding of the tooth of the driver, along that of the driven wheel or pinion, as well as an accumulation of dirt occafioned at the bottom of the tooth of the latter; whereas, when the action is after the line in queftion, the teeth roll on one another, and the dirt is carried out of the fpace between the teeth; this confideration is the greateft recommendation of that kind of action, where an exterior epicycloid impels an interior one only after the line of the centres, which mode we have feen, cannot take place with low pinions; from which view of the fubject the direct inference is, that pinions of high numbers ought to be adopted in every cafe that admits of fuch an adoption. We may alfo refer to the fame caufe the origin of the common bay-leaf fhape of a tooth, which impels both before and after the line of the centres, and acts at the pitch-line, at the moment of paffing this line.

The epicycloidal teeth, fuch as we have defcribed, are not however the only ones which have the requifite property of tranfmitting the force and velocity uniformly, though we have hitherto confined our obfervations to thefe fhapes, by reafon of their being moft generally ufed. Two teeth which are formed by the evolution of threads from their refpective geometrical circles, as feen in *fig.* 7., and which are fhaped by the interfection of portions of their refpective involutes, though lefs known and confequently lefs ufed, have, according to the late profeffor Robinfon and others, the fame requifite property as the foregoing ones, and may be ufed inftead of them in practice. The curve formed at any point, i, by a thread refting at the point H, of the wheel or point, h, of the pinion, is perpendicular to the thread, and the thread is a tangent to the circumference of the circle from which it is evolved; whenever therefore one tooth drives the other, the point, i, of action lies in the line H h, which is a common tangent to both circles, and the force is conftantly exerted in the direction of the common tangent, in every pofition of the acting teeth; the nature of the action confequently will be precifely the fame as though the points H and h were in contact, and impelled each other at the point of contact, hence the angular velocity of K H, will be to the angular velocity of k h, as the radius, c h, is to the radius C H; that is, the motion and force communicated will be uniform.

It is evident from an infpection of the figure, that when teeth compofed of interfecting involutes are ufed in a pair

5 of

of wheels, or in a wheel and pinion, the geometrical radii muſt have the whole length of their reſpective teeth added to them to conſtitute their acting radii ; and alſo, that the two proportionate circles which come in contact and deſignate the place of the pitch-line in ordinary epicycloidal wheel work, muſt be removed from each other to a diſtance equal at leaſt to the length of the longer tooth. This length will depend upon the geometrical radius of the wheel or pinion, and coarſeneſs of the tooth conjointly. In large wheels the tooth will be long, and in ſmall ones the contrary, where the coarſeneſs is the ſame, the breadth of the tooth being in every inſtance the abſciſs, and its length the ordinate of the involute or curve that forms one ſide of the tooth. The full length of any tooth of a given coarſeneſs, in a wheel of a given diameter, may conſequently be aſcertained by a fluxional proceſs, which we intentionally avoid as being too intricate for ordinary purpoſes.

This formation of a tooth, where the involute is derived from the wheel or pinion's own circumference, is recommended by profeſſor Robiſon, on account of its admitting the ſimultaneous action of many teeth, thereby participating the force among them ; but as he does not profeſs to ſay that they act without friction, we conceive that horology would not be benefited by their adoption, even if the formation were practical. Mr. Brewſter, in his new edition of Ferguſon's Lectures, vol. ii. page 223, has properly obſerved, that the principle of the profeſſor's teeth is not new. De la Hire having long ago conſidered the involute of a circle as the laſt of the exterior epicycloids, which it may be proved to be, if we conſider the generating ſtraight line as a curve with an infinite radius ; and the involute may be deſcribed moſt conveniently by a ruler or other ſtraight edge, bearing a tracing point moved round the circular baſe, while they are kept in contact. Accordingly, our rule for the ſituation of the geometrical or primitive circles holds good. The teeth which project beyond theſe circles are formed by portions of exterior epicycloids; but our rule for the action before and after the line of the centres, cannot, in this caſe, apply, becauſe here the interior epicycloid is not concerned, and the formation of the exterior one has no reference to the diameter of its fellow. (Vide the articles Evolute and Involute.)

After having ſaid ſo much on the requiſite ſhape of a tooth in a wheel or pinion, to tranſmit the motion and force equably through a movement ; we might here add the neceſſary directions for calculating the ſuitable numbers for conſtituting movements of various kinds ; but theſe have been anticipated under the article Clock-*making*, to which the reader is deſired to refer, and in which he will ſee that we have divided the whole movement of the going part into three portions ; any one of which may be altered at pleaſure, without affecting the other two. Upon the plan there laid down, we have calculated three ſeparate tables of the different portions, which we ſubjoin, with a view of giving a great variety of trains, to be had by mere inſpection, and of thus ſaving the clock-makers the trouble of going through tedious calculations, according to the prevailing cuſtom of taking into one calculation the whole movement at once, and of breaking it into portions by a ſubſtitution of ratios equal in value to the different factors of an aſſumed product, which proceſs makes the determination intricate.

TABLE I.

The firſt Portion of a Clock-movement.

Hours.	6	7	8	9	10	11	12	13	14	15	16
3	18	21	24	27	30	33	36	39	42	45	48
4	24	28	32	36	40	44	48	52	56	60	64
5	30	35	40	45	50	55	60	65	70	75	80
6	36	42	48	54	60	66	72	78	84	90	96
7	42	49	56	63	70	77	84	91	98	105	112
8	48	56	64	72	80	88	96	104	112	120	128
9	54	63	72	81	90	99	108	117	126	135	144
10	60	70	80	90	100	110	120	130	140	150	160
11	66	77	88	99	110	121	132	143	154	165	176
12	72	84	96	108	120	132	144	156	168	180	192
13	78	91	104	117	130	143	156	169	182	195	208
14	84	98	112	126	140	154	168	182	196	210	224
15	90	105	120	135	150	165	180	195	210	225	240
16	96	112	128	144	160	176	192	208	224	240	256

TABLE II.

The ſecond Portion of a Clock-movement.

Factors.	6	7	8	9	10	11	12	13	14	15	16
4 15	24 90	28 105	32 120	36 135	40 150	44 165	48 180	52 195	56 210	60 225	64 240
$4\frac{2}{7}$ 14	84	30 98	112	126	140	154	168	182	60 196	210	224
$4\frac{1}{2}$ $13\frac{1}{3}$	27 80		36	120	45		54 160		63	200	72
$4\frac{8}{13}$ 13	78	91	104	117	130	143	156	60 169	182	195	208
5 12	30 72	35 84	40 96	45 108	50 120	55 132	60 144	65 156	70 168	75 180	80 192
$5\frac{5}{11}$ 11	66	77	88	99	110	60 121	132	143	154	165	176
6 10	36 60	42 70	48 80	54 90	60 100	66 110	72 120	78 130	84 140	90 150	96 160
$6\frac{2}{3}$ 9	40 54	63	72	60 81	90	99	80 108	117	126	100 135	144
7 $8\frac{4}{7}$	42	49 60	56	63	70	77	84	91	98 120	105	112
$7\frac{1}{2}$ 8	45 48	56	60 64	72	75 80	88	80 96	104	105 11	120	120 128

TABLE III.

The third Portion of a Clock-movement.

	6	7	8	9	10	11	12	13	14	15	Vibrations per second.	Length of the pendulum.
12	30	35	40	45	50	55	60	65	70	75	2	9.80
15	24	28	32	36	40	44	48	52	56	60		
20	18	21	24	27	30	33	36	39	42	45		
30	12	14	16	18	20	22	24	26	28	30		
12			50				75				$2\frac{1}{2}$	6.27
15	30	35	40	45	50	55	60	65	70	75		
25	18	21	24	27	30	33	36	39	42	45		
30	15		20		25		30		35			
15	36	42	48	54	60	66	72	78	84	90	3	4.35
18	30	35	40	45	50	55	60	65	70	75		
30	18	21	24	27	30	33	36	39	42	45		
45	12	14	16	18	20	22	24	26	28	30		
15	42	49	56	63	70	77	84	91	98	105	$3\frac{1}{2}$	3.20
21	30	35	40	45	50	55	60	65	70	75		
35	18	21	24	27	30	33	36	39	42	45		
15	48	56	64	72	80	88	96	104	112	120	4	2.45
20	36	42	48	54	60	66	72	78	84	90		
24	30	35	40	45	50	55	60	65	70	75		
30	24	28	32	36	40	44	48	52	56	60		
40	18	21	24	27	30	33	36	39	42	45		
15	54	63	72	81	90	99	108	117	126	135	$4\frac{1}{2}$	1.935
27	30	35	40	45	50	55	60	65	70	75		
45	18	21	24	27	30	33	36	39	42	45		
15	60	70	80	90	100	110	120	130	140	150	5	1.57
25	36	42	48	54	60	66	72	78	84	90		
30	30	35	40	45	50	55	60	65	70	75		
50	18	21	24	27	30	3	36	39	42	45		

Explanation of Table I.—Table I. contains numbers suitable for the great wheel, and the pinion on the arbor of the centre wheel, together denominated "the first portion of a clock-movement," which is that on which the duration of motion at one winding up depends: the numbers on the uppermost horizontal column, from 6 to 16 inclusively, represent so many pinions; the numbers in the first vertical column, from 3 to 16 inclusively, are the respective hours in which the barrel or fusee turns once, as the case may be; and the larger numbers beginning with 18, and ending with 256, in the intersections of the horizontal and vertical columns are the great wheels. By way of exemplification, suppose we want a fusee to revolve in twelve hours, with a pinion of 8 on the arbor of the centre wheel, the intersection of 12 hours at the left-hand side, and eight at the top, gives 96 for the number of teeth in the great wheel, to produce the desired effect; or if the pinion had been required for a wheel of 96, to revolve in 12 hours, over 96, in the column of 12 hours, stands the pinion 8. In the same manner for 4 hours, and a wheel of 48, the pinion will be 12, and *vice versâ.*

Explanation of Table II. Table II. contains wheels and pinions suitable for the second portion of a clock-movement or first portion of the train, being that which effects a multiplication of 60, and regulates the velocity of the seconds' hand, by making its arbor revolve in one minute. This might be done by one large wheel of 300, and a small pinion of 5, but such a construction would require a large frame; a composition of two wheels and two pinions is therefore substituted to produce the same effect more conveniently, the first wheel being placed on the centre or hour arbor that carries the minutes' hand, and the last pinion on the arbor of the seconds' hand that revolves in a minute. The uppermost horizontal column contains the pinions from 6 to 16 inclusive, like Table I, and the left-hand vertical column contains couplets of factors, which, multiplied together, produce 60 always as a product. Any pair of factors, which are coupled together by a }, may be taken at pleasure, and the wheels in the spaces made by the intersections of the vertical and horizontal columns, under the pinions made choice of, and opposite the factors selected, will be proper for the second portion of the movement. For instance, when pinions of 8 are used with the factors $\left.{7\frac{1}{2} \atop 8}\right\}$, the wheels, found in the space formed by the intersection of the vertical column under 8 and horizontal column in which $\left.{7\frac{1}{2} \atop 8}\right\}$ stand, are 64 and 60; but if the pinions were to be assumed, one 10 and the other 8, then the wheels would be either 80 and 60, or otherwise 75 and 64, accordingly as the pinion 10 is made a part of the ratio to represent the factor 8, or the factor $7\frac{1}{2}$, which it may be taken to do indifferently; hence the notation may be $\frac{10}{80} \times \frac{8}{60}$, or $\frac{8}{80} \times \frac{10}{60}$, or $\frac{10}{75} \times \frac{8}{64}$, or $\frac{8}{75} \times \frac{10}{64}$, the result in point of accuracy is the same, but in the construction, the wheels of largest diameter and weight are required to be taken first in the movement, because the diameters diminish as the train ascends, by reason of the diminution of the transmitted force, which otherwise would not overcome the inertia of the wheels, which it is required to do at each vibration of the pendulum: any two pair of wheels and pinions properly taken out of the corresponding columns when reduced to a simple ratio by the usual method of multiplying the numerators together for one numerator, and the denominators for one denomination, will be found equal to $\frac{1}{60}$; thus $\frac{10}{80} \times \frac{8}{60} = \frac{80}{4800} = \frac{1}{60}$, and $\frac{10}{75} \times \frac{8}{64} = \frac{80}{4800} = \frac{1}{60}$, as before; therefore, if the first wheel revolve in an hour, the last pinion will, in this and in every other instance, revolve in a minute. The diminution of the diameter of the third wheel of the movement, which is the second wheel of the train, may be effected in two ways, either by taking the seconds' pinion smaller than the first from a column at the bottom of the table, where the factors are nearly of equal value, or by taking the pinions alike, and the wheels from one of the higher horizontal columns where the factors differ considerably in value; thus $\frac{7}{60} \times \frac{6}{42}$ from column $\left.{7 \atop 8\frac{4}{7}}\right\}$ and $\frac{7}{70} \times \frac{7}{42}$ from column $\left.{6 \atop 10}\right\}$ are each equal to $\frac{1}{60}$, but in the first portion of the movement the ratio of the size of the two wheels is 60 : 42, and in the other 70 : 42, from which mode of comparison of the columns it will be seen that a decrease in the diameters of almost any given ratio may be adopted from this table, so comprehensive is its extent, as it relates both to the variety of 11 pinions, and to the choice of 10 pair of factors, which begin with the ratio 4 : 15, or 1 : 4, and end with $7\frac{1}{2}$: 8. The

The principal care to be taken is, that every wheel be taken under its own pinion and oppofite its own factor; for another inftance, in the horizontal column of factor $\left.\begin{matrix}6\frac{2}{3}\\9\end{matrix}\right\}$, the pinions of 9 will do for wheels 81 and 60; or otherwife, wheel 90 may be taken with pinion 10, and wheel 60 with pinion 9; that is, any wheel may be taken oppofite its own factor, provided the pinion in the fame column over it be ufed as its fellow; fo that examples may be given to a very great extent, which will afford abundant means of afcertaining by experiment the beft poffible numbers for this part of the movement, without the trouble of calculation.

Explanation of Table III. Table III. contains the third portion of a clock-movement, or fecond portion of the train, which is that on which the number of vibrations per fecond depends. When a vibration is performed in an exact fecond, one wheel only is neceffary for this purpofe, with 30 teeth, becaufe one tooth completely efcapes the pallets at every fecond vibration, which wheel muft be placed in this cafe on the arbor of the feconds' hand, and is ufually denominated the fwing wheel; alfo a wheel of 60, fimilarly placed, will be proper for half feconds, 75 for $2\frac{1}{2}$ vibrations per fecond, and 90 for 3 per fecond; but thefe laft numbers are found too high for portable clocks, time-pieces, &c. therefore a wheel and pinion are introduced in addition to the efcapement-wheel, in order to diminifh its diameter into a convenient fize for a portable conftruction, as well as to render it light, fo as to have but little inertia to be over come by the diminifhed force, which acts at this part of the movement. In ordinary pieces the two wheels are, one a contrate-wheel, and the other called the crown-wheel, with the pinion on its arbor, but the wheels may all have the ordinary form of the wheels of a clock that fwings feconds, it being not the fhape, but the numbers of the teeth of the wheels and pinions that determine the frequency of the vibrations. The uppermoft horizontal column in this table, like that in the two preceding tables, contains all the variety of pinions from 6 to 15 inclufively; the pinion of 16 being omitted, however, to make room for two additional vertical columns at the right hand of the table, the firft for the number of vibrations per fecond made by the pendulum, and the other for the length of the correfponding pendulum in inches and decimal parts, meafured from the centre of fufpenfion to the centre of ofcillation, which two data once fixed upon determine the horizontal column out of which the wheels muft be taken with a given pinion; the left-hand vertical column is that in which one of the two wheels is found, and the number ftanding on the fame horizontal line, under the given pinion, is the other; for as the product of the two wheels, divided by the pinion under which one of them ftands, is always equal to 60 in the higheft large column, or column of 2 vibrations per fecond, it is of no importance to the accuracy which of the two wheels is made the pallet-wheel; the determination of this point being a matter of practical convenience; in the fame manner in the fecond large horizontal column, the quotient arifing from the product of any two wheels ftanding in the fame line taken horizontally, one in the firft vertical column, and the other under the chofen pinion, divided by the pinion, is always 75; in the third parallel column the quotient fo obtained is 90; in the fourth 105, and fo on; hence any of the combinations adopted will be equal in value refpectively to thofe large fimple numbers ufed as pallet-wheels without fuch combination, which, it has been faid, are objectionable in practice.

Let our firft example be to afcertain the requifites contained in the table for a half-feconds' pendulum, where a pinion of 8 is ufed?

In the firft place, 12 with 40 may be taken as the required numbers for the two wheels; in the fecond, 15 with 32; in the third, 20 with 24; and, laftly, 30 with 16; any one of which wheels may be the pallet-wheel, as the nature of the efcapement may require; if the crown-wheel efcapement is ufed, which requires an *odd* number of teeth for its action to take place at oppofite fides of the wheel, 15 muft neceffarily be the pallet-wheel, and 32 the other, with the given pinion of 8: but whichfoever of thefe couplets of wheels is adopted, the effective length of the pendulum muft, according to the laft vertical column, be 9.80 inches.

Again, for a pendulum that is required to vibrate 3 times per fecond with a pinion of 12: we have, in the firft place, 15 with 72; fecondly, 18 with 60; thirdly, 30 with 36; and, laftly, 45 with 24; fo that either 15 or 45 may be in this inftance the crown-wheel, and for a different efcapement any of the eight numbers mentioned; and the choice is equally extenfive with any other pinion that may be chofen from 6 to 15: the pendulum in this cafe is only 4.35 inches.

In this way, by means of the table before us, the third portion of a clock-movement may be varied almoft at pleafure without affecting the other two portions, and a pendulum of any of the calculated lengths may be applied to works previoufly conftructed; which circumftance will enable the workman to metamorphofe a portable clock into a large one, and the contrary, without difficulty, or without even a knowledge of the operations of arithmetic.

We might have extended this table to take in different intermediate lengths of the pendulum and correfponding number of vibrations per fecond, but our object is to banifh incommenfurate vibrations from clock-movements, and to introduce thofe only which conftitute fome convenient fubdivifion of the fecond, to anfwer the nicer purpofes of philofophy without extra labour and expence. The three or four laft horizontal columns indeed, it may be faid, will not be of much fervice in clock-movements, by reafon of the fhortnefs of the pendulums; but they will be found particularly ufeful in watch-work, as will alfo the other tables, which are fo comprehenfive as to apply with equal propriety to clock and watch-movements.

Clock-*tools*. While clock-making was in its infancy, and the different parts of the machine continued to be made by the fame workman, each clock-maker was obliged to defign and manufacture his own tools agreeably to his own inclination; but as foon as the art branched out into different occupations, tool-making became alfo a bufinefs of itfelf, and is now quite diftinct from clock-making, nay, has itfelf ramified into various branches. In the prefent ftate of this manufactory by far the greateft number of clock-tools are made in Lancafhire, and become an article of commerce among the ironmongers all over his majefty's dominions. In London there are a few houfes that deal chiefly in clock-tools, engines, and rough materials ready prepared, as to fize and fhape, for the workmen; one of the principal of thefe is Fenn's in Newgate Street, No. 105, where tools of all kinds may be had at moderate prices. It is much to be regretted that a Swifs tool and engine-maker, called Petitpierre, who lately worked in an attic of No. 161 Fleet Street, and who made tools and engines in a very fuperior manner, was fome months ago obliged to quit the kingdom for want of regular employ, though his prices were by no means exorbitant.

It

It will not be expected of us, to present the reader with all the variety of tools that individuals have respectively contrived, but to lay before him such a collection as custom has sanctioned in the trade; which we propose to do in three plates, devoted entirely to this purpose; neither will it be expected, that we should particularize all the different uses to which any individual tool may be applied, much less to point out the positions in which they ought to be held in the act of working, which would require almost an unlimited variety of plates; it seems to be sufficient for our general purpose, if we give a list of the different tools at present in common use, with references to the plates, where the reader will comprehend the nature and use of each tool by a survey of its figure, better than from any verbal description without the figures. We shall, therefore, confine the remaining portion of this article to a simple explanation of the plates.

Plate XIX. *of Horology.*

Fig. 1. A motion-arbor, or arbor for turning wheels on, with a fastening nut.
2. Motion-arbor, that screws up to the shoulder, with a small nut to fasten.
3. A plain arbor for collets, or tubes, to hold by friction.
4. A slit arbor for holding and turning small pieces of metal.
5. A pair of cutting bullet-compasses for fitting any central hole, in describing or cutting circles.
6. A cutting-leg for ditto.
7. A marking-leg for ditto.
8. A stake, or small anvil, for hammering, &c. on.
9. Common flat, short-nosed pliers for holding any piece of metal.
10. A pair of callipers, with a straight edge, adjustable by a thumb-screw, of use for trying if a wheel is placed at right angles to its arbor, or what is called in the flat, and also if it is perfectly concentric, or in the round.
11. Common callipers for measuring diameters.
12. A frame-guage, inside and out.
13. A pinion-guage, with spring and screw adjustment.
14. Beam-compasses for cutting out circular pieces of metal from a solid plate, describing large circles, dividing rectilinear and curvilinear lines, &c.
15. A square or rectangular piece of brass.
16. A clamp for holding pieces of metal to be filed or riveted.
17. A drill-arbor and drill in a socket for various drills, to be used with a bow and gut.
18. A drill detached to fit the said arbor.
19. A drill of larger size.
20. A tool or graver for cutting grooves, which may be of various shapes and sizes.

Plate XX. *of Horology.*

Fig. 1. A saw for metal, with a wooden handle.
2. Cutting pliers for shortening pins or for cutting wire.
3. A bench-vice, or vice to be clamped to a bench.
4. A hand vice for holding a small piece fast.
5. Clamping pliers for holding pins, &c. fast in filing.
6. Pendulum-pliers or long-nosed pliers.
7. Pivot-drill with a friction-ferril.
8. Drill-arbor in a drill-frame, to be held in a bench-vice.
9. A drill to fit the socket of the arbor, when it has a square tapering hole.
10. Ditto of larger bore.
11. Screw-head tool, including the arbor and frame with a rest, to be put into a bench-vice.
12. A holding piece of ditto, detached from the end of the arbor, and tapped with a female thread, to hold the screw to be dressed.
13. Ditto with a different thread.
14. A detached ferril for a drill or arbor.
15. Ditto in two halves, with adjusting-screws to fit any arbor or drill.
16. A screw plate with different holes tapped.
17. A tap to be held in a hand-vice, for making a female thread in any hole. *N. B.* When long screws are required to be made, a die, such as the mathematical instrument-makers use, is better than a screw-plate, for preserving the screw from bending; there must be as many taps as the plate has different holes tapped.
18. A screw-driver, of which there are various dimensions.

Plate XXI. *of Horology.*

Fig. 1. A brace for receiving various bits.
2. A chamfering or counter-sinking tool to fit the brace, for which a large drill may be substituted.
3. A pentangular or five-sided broach to fit ditto.
4. A round broach to fit ditto.
5. A square broach to fit ditto.
6. A deepening tool, for adjusting the engagement of wheels with wheels or pinions, not as yet much used in England.
7. Turning-frame, or clock-lathe, of which there are various sizes, and constructions; some going by a bow like the present one, some by a hand-wheel, and some by the foot with a large wheel and crank actuated by a lever which is trodden upon. See LATHE.
8. A graver for cutting the metal in a turning-fame.
9. A large ditto.
10. An adjusting-tool for fusees, with sliding weights, to suit any given maintaining power of a clock or watch.
11. A tool for turning pivots in, when inserted into the end-hole of the turning frame.
12. A triangular tapering-file.
13. A file for slitting or cutting the teeth of pinions.
14. An equalling file, or file for the spaces between the teeth of a wheel, when cut in an engine.
15. A common hand-file with a safe edge for ordinary work.
16. A rounding-off file, for the ends of the teeth in wheels and pinions.
17. A file for crossing-out, or forming the crosses and rim of a wheel.

Besides the above there are various other files differing in size, shape, and coarseness, according to the work they are designed to perform.

CLOCK-*work.* The word *Clock-work*, in its original signification, imported those wheels and pinions, latches, catches, springs, fly, hammer, &c. which constitute the striking, or what was formerly called the clock, part of the works of a large horological machine; but since clocks became common and portable, the term has been applied in a more general sense to the mechanism of the going part, as well as of the striking part, and even sometimes to the works of those machines, which resemble the works of a clock in their appearance and action. Under this head, therefore, we mean to describe such detached pieces of clock-work as are not to be generally met with, but which seem, either from their utility or curiosity, to merit a public notice.

1. *Striking Part, with one Wheel and one Pinion.*—*Fig.* 1, of *Plate* XXV. of *Horology*, is a front view of the ftriking part of a clock, purchafed in the year 1806, at the fale of a gentleman in Suffex, who is now deceafed, but who indulged his fondnefs for mechanical contrivances, at an almoft incredible expence; this ftriking work is fo fimple, that it has only one wheel and one pinion, and no other fly but the hammer, itfelf, and is, notwithftanding, capable of repeating the laft hour at any time, with the addition of a ftring attached to the lifting piece; after our minute defcription of the ftriking mechanifm of an eight-days clock, with the fnail and rack, it will not be neceffary for us to repeat what we have already faid on their action, but merely to confine ourfelves to the movement, bell, and hammer. The frame of this clock is in the fhape of a crofs, confined by the pillars A, B, C, and another below, not fhown; the great wheel of the ftriking part has 300 teeth, with a barrel and ratchet as ufual on its arbor; it is actuated by a weight fufpended by the chain D, and drives a pinion of fix leaves, when at liberty to move; F is a rack placed on a long lever, the tail of which, as ufual, refts on a fnail, on the 12-hour wheel; and G is a lifting piece to raife the click or hawk's-bill H, which it does by means of a pin fliding underneath the piece H, whenever the pin in the one hour wheel moves its tail, as ufual: on the protruding end of the pinion's arbor, a gathering pallet is fixed in the common way, the long tail of which refts on the pin at the right-hand end of the rack, when the rack is gathered up to the laft tooth; and the click, H, is heavy enough to fall by its own weight, after fliding over the inclined part of any tooth; all which actions are fimilar to thofe we have before defcribed in *Plate* XII. The bell is mounted on the uppermoft pillar, as fhown in the figure, and a lever, I, is faft to the middle of the pinion's arbor within the frame; this lever has what may be called a leg, fufpended by a pin at its remote end, forming what we will call the knee-joint of the leg, the foot of which leg forms the hammer: now, it is eafy to perceive, that when the pinion begins to revolve fuddenly, the foot of the leg flies out by its centrifugal force, and the toe moves in the curve *a*, where it meets with the edge of the bell, againft which therefore it ftrikes and rebounds, by permiffion of the knee-joint, to *b*, we will fuppofe, the place where it would have been found if the motion had been fo flow as to have no centrifugal force; but this point is in the interior part of the bell, the foot therefore proceeds within the bell, and the revolution is completed filently, but at the next, and at every fubfequent revolution, this ftroke and confequent rebounding are repeated, till the rack is gathered up, which regulates the number of ftrokes. The ball, K, is only a counterpoife fixed on the tail of the lever I, and affifts with it to form a kind of fly, to regulate the velocity with which the ftrokes are fucceffively made. This mechanifm is very fimple, and appears to have been made many years, from the dirty ftate in which we faw it; but has not, that we know of, been before defcribed.

In the year 1803, the Society of Arts, in the Adelphi, London, voted to Mr. Edward Maffey, of Hanley in Staffordfhire, a premium of twenty guineas, for a new ftriking part of a clock; and in the fame year, another premium of thirty guineas, to Mr. John Prior, of Nesfield in Yorkfhire, for another contrivance to anfwer the fame purpofe: the drawings and accounts of both thefe inventions, were publifhed in the tranfactions of the faid fociety, for the year above fpecified; but the defcriptions, as there given, are fo imperfect, that we have found it neceffary to have new drawings taken from the models themfelves, which are preferved in the repofitory of the fociety, from fuch points of view that the action of the different parts may be rendered as intelligible as poffible. We fhall defcribe Mr. Maffey's contrivance firft, as being fomewhat analogous to the one we have juft defcribed, inafmuch as that it has one wheel and one pinion only, inftead of a train of wheels and pinions, as is ufual.

2. *Maffey's ftriking Part.* *Fig.* 1, of *Plate* XXVI. of *Horology*, is a perfpective view of Mr. Maffey's model, with the front plate of the frame removed, or fuppofed to be tranfparent, to fhow the mechanifm included. A is the great wheel with 78 teeth, like the ordinary count-wheel, impelling the pinion *a* of eight leaves; on the arbor of pinion *a*, but at the oppofite end, is a circular plate, B, with eight pins placed in a concentric circle, at equal diftances, fo that one pin correfponds to one tooth of the pinion; thefe pins being near the circumference of the circular plate act with the pallets *b*, *b*, and form a kind of efcapement; the pallets are connected with a pendulum of about 9 inches long, which therefore vibrates pretty nearly half-feconds, between the efcape of each fucceffive pin refpectively. It is not neceffary to fhow the pendulum which may be hung on a cock, or on the pivot of the pallet's arbor, which is more fimple for this purpofe, and not objectionable, as it would be in the going part. The lever, *d*, is a locking detent, with a claw for locking the pins of the plate B, when it falls in the way of any one of them; on the arbor of the detent, *d*, is another detent with a triangular claw, *e*, which falls in the way of the pins of the great wheel, placed at unequal diftances, like the notches in the locking-plate of a count-wheel; on which account the great wheel may be confidered as a count-wheel alfo; to the fame arbor of *d* and *e*, is alfo attached a third concealed lever, which reaches to the one hour wheel, and is moved by its pin once in each hour, in the ordinary way. On the pofterior face of the locking plate, B, are eight pins in a fmall concentric circle, dotted in the figure, which lift the hammer tail as ufual, to which a counteracting fpring is applied, to give a fmartnefs to the ftroke of the hammer, as it ftrikes the fide of the horizontal bell, mounted over the frame, as fhown in the figure. The action is thus; when the detent, *d*, is raifed, by the pin of the one hour wheel of the dial-work not feen, the weight fufpended by the cord C, draws the barrel on the arbor of the great wheel round, and with it its ratchet, which takes the click of the great wheel, and therefore carries it alfo, the detent *e* being lifted at the fame time with the detent *d*; the wheel now impels the pinion and plate B, which is not only the locking but ftriking plate alfo; the pins begin to lift the hammer in fucceffion, and the ftriking goes on; in the mean time the detent, *d*, falls within the circle of pins, as in the figure, in confequence of the lifting pin of the one hour-wheel letting go the concealed lever, but fo as not to impede the paffage of the pins of the locking plate; prefently the neareft following pin of the great wheel fliding on the inclined face of the detent *e*, raifes it far enough to move the claw of the detent, *d*, into the way of the pins of the locking plate, and then all motion is arrefted, till the dial-work raifes the detents again at the end of another hour, when the fame procefs is repeated, the vibrations of the pendulum regulating at all times the velocity with which the lifting pins of the hammer-tail fhall move fucceffively during the period of ftriking.

3. *Prior's ftriking Part.* *Figs.* 2 and 3 of the fame plate, exhibit in perfpective the mechanifm of Mr. Prior, which we have already mentioned. A is the great wheel of 78 teeth, which impel the double endlefs fcrew, B, cut on the arbor of the fly C C; like *fig.* 1, this figure has the front plate of the frame fuppofed to be tranfparent, to expofe the different levers, &c. to view in their refpective fituations of action; on the arbor of the great wheel is the barrel, as in *fig.* 1;

fig. 1; in a concentric circle, near the edge of the large wheel, are seven pins, at unequal distances, which catch the claw of a detent *a*, placed fast to the arbor *b*, which arbor *b* has a play or motion in the direction of its length, that will allow its pivots to go more or less into the plates of the frame respectively, as circumstances may require, . and a spring pressing on the end of the pivot, which comes to the front plate, pushes it back towards the wheel, whenever this pressure is not taken off; the spring alluded to being on the front plate, of course cannot be seen in the figure; also a small circular plate of metal, carried by the arbor of *b*, just fits into a circular groove turned out on a contiguous and parallel arbor *d*, in such a way that the arbors can revolve independently; but when one moves lengthwise across the frame, the other must necessarily move with it, for which purpose the second arbor, *d*, has also liberty or play lengthwise like the arbor *b*; to this arbor *d*, is fixed a second detent *e*, with a claw turning inwards; *f* is a third arbor, parallel to the other two, but below them in the frame, carrying about its middle the long tail *i* of the hammer K, and at the end next the front plate a smaller tail, to be pressed by the spring, *g*, in the act of striking; on the plane of the great wheel next the eye, are 13 pins, which successively lift the hammer by the tail *i*, while the great wheel revolves; on the face of the great wheel, beyond the inferior circle of pins, and within the superior one, is a spiral groove, with six complete turns, into which the outer end of the detent *e* slides as the great wheel revolves; so that if the groove were equally deep and smooth throughout, the bent end of this detent would be carried gradually from the innermost to the outermost helix without stopping; but the fact is, that at certain intervals in the six-fold spiral, there are deep notches at the bottom of the groove into which the detent inserts its claw by the pressure of the spring, which we have said is not to be seen; and whenever one of these depressions of the detent *e* takes place, not only its arbor *d*, but also the arbor *b* are carried towards the great wheel by their motion lengthwise, so that the claw of the detent *a* comes in the way of the superior pins of the great wheel; the excavated notches in the bottom of the six-fold groove are so arranged, in point of relative distances, that the 12 are placed at such intervals of the spiral as gradually increase, like the intervals between the notches of the locking plate on a count-wheel. The action takes place in this manner; the arbor *b* has a lever connected with the pin of the one-hour wheel of the dial-work, by which it is lifted every hour; this lever is not seen by reason of its being at the other side of the plate B A; at the same time the detent *a* is raised, and sets the pins at liberty, the weight suspended by the cord *h* now turns the wheel and fly; the detent *e* in the mean time advances in the spiral groove, and the 13 pins in the inferior circle raise the hammer, which as often strikes the bell mounted horizontally over the frame; when the detent is at the beginning of the first spiral of the half dozen, one stroke only is made, until the excavation at the bottom of the groove allows the claw to drop into it, which drop presents the claw of the detent *a* to the nearest pin, and stops the striking; after another hour the detent *a* is again raised, and the wheel proceeds to move, so that when the detent drops again it does not fall in the way of the pins, but continues at a little distance from their heads, so long as the bent end of detent *e* does not fall into the next excavation of the groove; when this happens, however, it then falls in the way of the said pins, but in the mean time two strokes have been given, and in the same manner the hours 3, 4, 5, &c. will be struck till the hour of 12 arrives; at this hour the wheel is unlocked as before, and as soon as 10 out of the 12 strokes of the hammer are made, the end of the detent *e* comes at an inclined plane *l*, which it ascends till it is above the plane of the wheel, and at that instant falls by its own gravity to the beginning of the spiral nearest the centre, in which it then begins to move, while the remaining two strokes of the 12 are making, after which it falls into the excavated notch where we first found it, and the whole process has now been gone through. It may have occurred to the reader, that during the fall of the detent *e*, by its own weight, from the outermost to the innermost groove of the six-fold spiral, the spring acting on the nearest pivot of the arbor *b*, would carry it against the plane of the wheel, and make it catch one of the superior grooves before it arrived at the bottom one; this would actually have been the case, if there had been no precaution taken to prevent such accident; but the contrivance shown in *fig.* 3, is introduced as a preventive, which acts thus; the little lever under the slender spring turns on a stud screwed into the interior face of the pillar plate, or that which we have represented as the pillar plate; this lever rests on the shoulder of the projecting pivot, which is brought forwards into the frame, when the end of *e* ascends the inclined plane we have mentioned; at this instant the spring forces the lever down between the shoulder of the pivot and the pivot hole, and prevents the return of the pivot, when the lever *e* is falling; and for a short time after; but while the two last strokes of 12 are making, a pin in the back face of the great wheel seizes the tail of the lever, and raises it from the shoulder and pivot hole again, until the pivot has returned to the situation it occupies when the wheel is locked. This mechanism, we think, is ingeniously contrived, and is very simple; inasmuch as only one wheel is used as great wheel, count-wheel, locking wheel, and striking-wheel, without even so much as a single pinion; there are 78 teeth in the wheel, and 13 pins for the hammer-tail, $\frac{78}{13} = 6$, therefore, are the teeth that pass at each stroke; but the screw is double, and takes two teeth at once, hence the fly makes just three revolutions at each stroke of the hammer; and a small weight is sufficient, not only to maintain the motion of the wheel, but also to overcome all the obstacles it meets with, which recommendations bid fair for bringing this striking mechanism into use in those clocks, where the repetition is not desired.

4. *Strike or silent.* When we described an eight-days clock with the mechanism of repetition, we had occasion to explain one of the methods of causing the clock to strike or be silent by the mere sliding of a pin to the right or left; but there is also another method of doing the same thing, which we propose to describe here as a detached piece of mechanism. *Fig.* 2, of *Plate* XXV. shows all the parts that are necessary for explaining the mechanism we have alluded to. A is a cock on the front plate of the frame, into which a small circular plate *b* is pivoted, on the arbor *a*, of which is inserted the square hole of a hand seen on the dial, pointing to one of the words *strike*, or *silent*, engraved, or otherwise marked, in a small circle near one corner of the dial usually, but sometimes at the top, accordingly as the other works require; B is a lever slit open at the top, and moveable on a stud, at the angular point where the tail *c* is inserted; and *d* is the portion of an ordinary rack with two pins in it, one at the end of the circular part as usual, and the other at *d*, where the tail, *c*, of the slit lever, B, is seen resting; lastly, *e* is the pin in the front plate against which the rack falls as usual, when the snail is off, or when the clock strikes 12. The use of this mechanism may be explained in a few words thus; when the hand on the arbor *a* is turned to *silent*, as in the present position, the tail *c*, of B, falls in the way of pin *d*, and

d, and prevents the fpring of the rack-tail from throwing the rack back; in confequence of which obftacle, the tail of the gathering pallet continues on the other pin, at the end of the curved part of the rack, and the ftriking mechanifm continues to be locked: but when the hand we have fpoken of is turned to the word *ftrike* on the dial, the tail, *c*, of the flit lever is raifed above the pin, *d*, the rack falls without impediment, and the ftriking takes place as though the mechanifm before us had not been in the clock.

5. *Endlefs Cord of Huygens.* The reader has already read a partial defcription of an endlefs cord, for keeping a clock in motion during the act of being wound up, when he perufed our account of a thirty-hours clock; to do juftice to Huygens, the original inventor, and to make the contrivance more clearly underftood, we beg leave to refer the reader to *fig.* 3, of *Plate* XXV. where the contrivance in queftion is diftinctly exhibited in a way that we prefume will make it clearly underftood. A is a metallic pulley, with pins inferted into the bottom of its groove, and placed faft to the centre-wheel without a ratchet; B is a fimilar pulley, with pins and a ratchet wheel faftened to it; the pulley and wheel, together with the click C, and its fpring D, are all placed faft in a detached ftate on the frame, or cafe of the clock, as may be moft convenient; E is a pulley, round which the endlefs cord goes, after this cord has embraced both the fuperior pulleys, and to this pulley is attached the maintaining power, F, one half of the weight of which refts on the pulley, A, and the other half on the pulley, B, fo that this power is double the weight of an ordinary power, fufpended by a fingle cord round a barrel; the cord, G, which goes round all the pullies, as reprefented in the figure, is made fhort in the plate, for the fake of being made an endlefs one, and is kept ftretched by the fmall weight H, and correfponding pulley. Suppofe it to be neceffary now to draw up the power, F, at any time; the hand takes hold of the cord at G, and pulls it down, the pulley, B, turns round in confequence of its pins being caught by the cord; and the click flides over the inclined teeth, and holds the ratchet in whatever fituation it is left when the hand lets go the cord; the pulling down of the cord, G, pulls up the part, I, of the cord, and with it the power, F, while the pulley, F, rolls along the part, K, of the fame as it afcends, which part, K, does not alter its fituation, except by the flow revolution of the pulley, A, *i. e.* by the going of the clock; in the mean time the fmall weight, H, falls towards G, and defcends at the fame time that F afcends, till this power, F, approaches the two pulleys A and B, when the winding concludes by means of a ftop applied over the pulley, E, to prevent the further afcent of the power, F. From this brief account of the mechanifm before us it is evident, that one half of the power, F, continues to actuate the pulley, A, during its afcent, as well as during its defcent, and that the quantity of fall compared with half the circumference of the pulley, A, will determine the continuance of going after each winding up.

6. *Forcing Spring.* We have feen the application of an auxiliary fpring, attached to the fufee of a chronometer, and to the barrel of an aftronomical clock, conftituting the former a going fufee, and the latter a perpetual ratchet, both of which anfwer the fame purpofe as the endlefs cord we have juft defcribed; but there are other applications of a fpring to produce the fame effect, that have not yet been defcribed. *Fig.* 4, of *Plate* XXV. reprefents a fpring of this kind acting in the teeth of the fecond wheel of the train, as we have feen it in fome old Englifh clocks. A and B are portions of the two plates of a clock frame; C, the fecond wheel of the train; D, the arbor of a fliding pallet; H, moveable in the fmall oblong frame; I, attached to the arbor; K, a flender fpring on the arbor preffing againft the ftraight tail-piece of the pallet, which paffes through a hole in the arbor; E, a fpring fcrewed to one of the plates of the frame, and preffing on the tail-piece; F, of the arbor's pivot; and G, a lever with a pin projecting through the dial, with its oppofite end reaching over the hole of winding, which it covers when the clock is going. The mode of application is thus; when the clock is to be wound up, the pin of the lever, G, is pufhed down a circular aperture of the dial to uncover the hole for the key that is ufed in winding up, which key could not otherwife be put into its hole in the dial; the lever, G, being faft to the front pivot, the arbor, D, turns it a little way round during this motion, for uncovering the key-hole; the tail, F, of the arbor, confequently is lifted upwards, and with it the fpring, E, which is the auxiliary fpring that moves the train by its effort to return the tooth of the pallet, H, is fo fhaped, that it will flip over a tooth of the fecond wheel in one direction, but not in the other, without carrying the wheel with it, as may be feen from its fhape in the figure; when, therefore, the pin of G is moved downwards, before winding, a fmall fpace, or portion of a revolution of the arbor, D, which bears the pallet and its frame, it carries the end of the pallet next the wheel over a tooth or fometimes over two teeth; the fmall fpring, K, in the mean time allowing the pallet, H, to recede in its frame, I, till it has paffed the faid tooth or teeth; but as the fpring, K, inftantly pufhes the pallet into the fpace next to the tooth it has juft paffed, and holds it there, the pallet is urged back again by the auxiliary fpring E, and carries with it the wheel fo far that it has room to efcape from the end of the tooth actuated by it, which efcape does not take place till fome time after the clock is wound up: hence the clock continues to go during the period of its being wound up, and when the force of the fpring is an equivalent for the maintaining power, which it may be made to be when ftretched back to a certain point; the contrivance will be a convenient temporary fubftitute for the maintaining power.

7. *The French Forcing-fpring.* The forcing-fpring made ufe of in fome of the French clocks is fomewhat different from the one we have juft defcribed, and we think more fimple in its conftruction, which is this; F A E, in *fig.* 5, of *Plate* XXV., is a rod of metal, moveable on a ftud fcrewed into the front plate at A, where is a fmall tail-piece in contact with a quiefcent fpring, B, alfo fcrewed to the fame plate; at F is a pin in the rod, which comes through a curved aperture in the dial, defcribed from the point, A, with the radius A F; C is as before the fecond wheel of the train, or at leaft a portion of it; D is a pallet moveable on a pin fcrewed into the fuperior part of the lever F A E; and E is a flender fpring preffing againft the pallet to keep it into the teeth of the wheel; the application of this mechanifm is in this wife; the lever is fuppofed to cover the key-hole at fome point near F, to uncover which the pin, F, is carried upwards towards G in the aperture of the dial; the end, E, of the lever is by this motion brought downwards, and the pallet, D, being curved on the back of it, flides over a tooth, or fometimes two teeth, below it, by its motion on the pin at its upper end, the flender fpring, E, during this time receding; but when the pallet, D, has paffed the required tooth or teeth, its fpring, E, pufhes it into the fpace contiguous, and its ftraight interior face lays hold of the tooth next above it; in this fituation the mechanifm is ready for action, but would remain inactive if the fpring, B, or auxiliary fpring, had not at the

ſame time been ſtretched back by the tail-piece at A; this ſpring, B, however, exerts its force to come back to its quieſcent ſtate; that is, to bring the pin back again from G to F, its original ſituation, and conſequently raiſes the end, E, with the pallet and pallet-ſpring, to its original ſituation, which cannot take place without urging the wheel, C, round along with the pallet; this contrivance, therefore, is a ſubſtitute for the maintaining power, ſo long as its action continues; and it is eaſy to perceive, that after a few minutes continuance the pallet will eſcape the tooth which it impels *pro tempore*, and the maintaining power will then reſume the ſole command of the train.

8. *The Bolt and Shutter.*—Another contrivance to anſwer the purpoſe of making a clock go while it is under winding is the bolt and ſhutter, in which a weight is ſubſtituted for a ſpring; *fig.* 6, of *Plate* XXV., repreſents the great wheel and barrel, together with the temporary apparatus, to be ſubſtituted for the maintaining power thus; the arbor, A, is pivoted into two collars, *a* and *b*, within the frame of the works, in which it is at liberty to revolve; this arbor carries three levers or arms, B, C, and D, all of which are made faſt to it, and conſequently revolve with it, whenever it is turned round; the lever, B, is juſt long enough to extend to the hole in the dial for winding, oppoſite the ſquare end of the barrel-arbor, which hole it juſt covers when in its ſtationary ſituation, by means of its circular extremity; the lever, C, carries a ſmall bolt at its extremity, which is puſhed out by a cylindrical ſpring of moderate ſtrength, while it is confined to move in the direction of the length of this lever; and D, the third lever, has the weight appended, which is uſed *pro tempore* as a ſubſtitute for the maintaining power. From this ſhort deſcription it is eaſy to perceive, that, when the clock is to be wound up, the lever, B, muſt neceſſarily be removed up or down, to allow the key to enter the hole in the dial's face; it is prevented from aſcending, and therefore any perſon unacquainted with the nature of the mechaniſm, who is going to wind up the clock, finds that he muſt puſh down the covering end of B, to gain admiſſion for his key; this motion of the lever, B, downwards, pulls down at the ſame time the ſecond lever C, and alſo elevates the third lever D, which is behind the centre of motion; but the lever, C, cannot deſcend till the bolt, which meets with the teeth of the wheel, is puſhed in by them, after which it will paſs; the cylindrical ſpring however inſtantly puſhes it out again, and makes it fit a ſpace between two teeth; in this ſituation the weight, D, acting on the third lever, now urges both the levers, B and C, back again to their original ſituations, and conſequently urges the wheel forwards by means of the bolt, till they have attained thoſe ſituations; which will require ſome minutes when the bolt is applied to the great wheel, according to our figure; but we are of opinion that it would be better to apply the bolt to the centre, or even the ſecond wheel, where a ſmaller weight would ſuffice, and where it would continue to act a ſhorter time; for the objection to the bolt and ſhutter, as well as to the forcing-ſprings juſt deſcribed, is, that their action continues for ſome time after the act of winding in addition to the maintaining power, which therefore it doubles for a time, ſuppoſing the temporary power to be exactly equal to the permanent one, agreeably to the intention of the contrivance.

CLOCK-MOVEMENT.

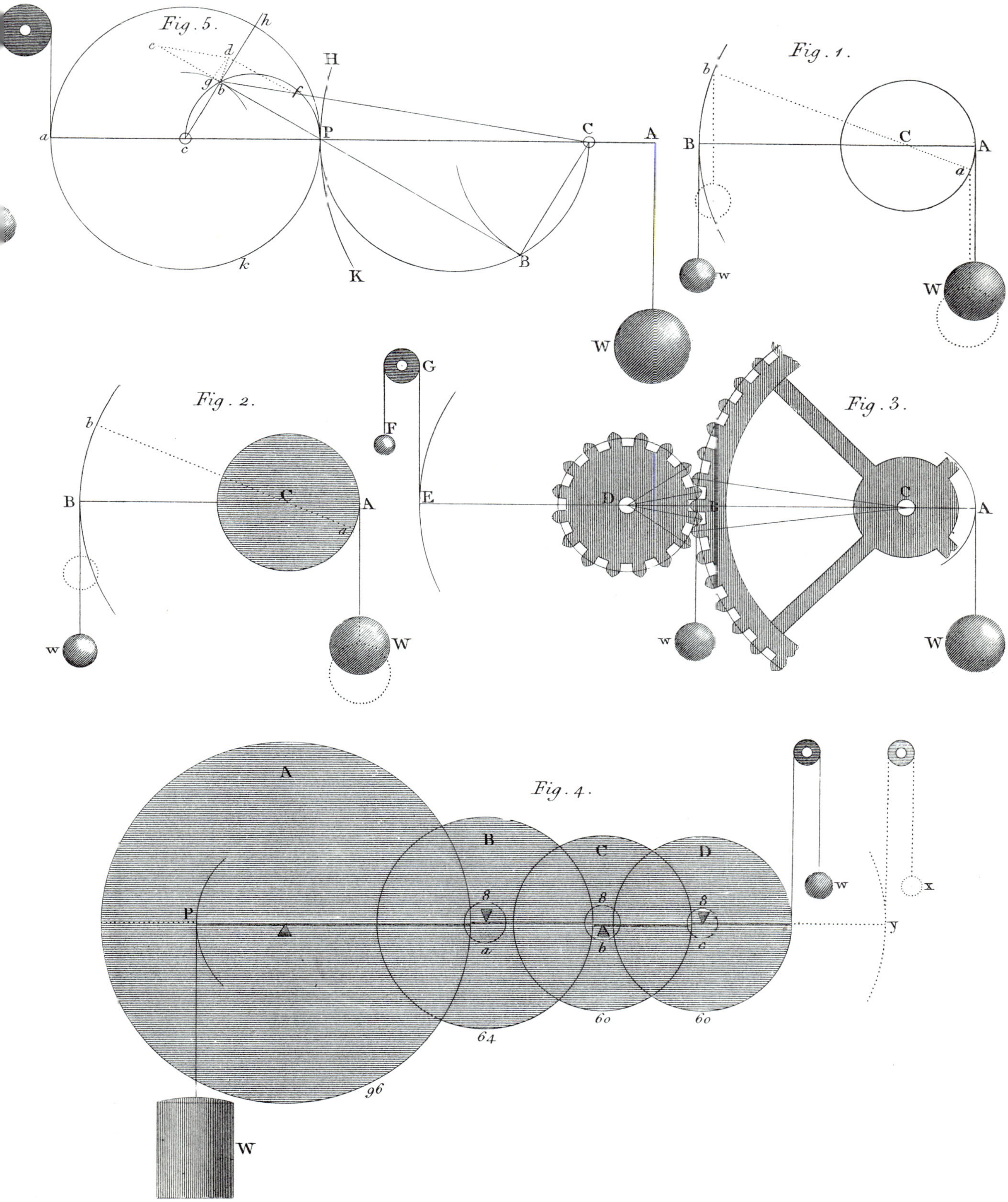

Published as the Act directs, June 2^d 1806, by Longman, Hurst, Rees & Orme, Paternoster Row.

Engraved by Wilson Lowry.

HOROLOGY.

CLOCK-MOVEMENT.

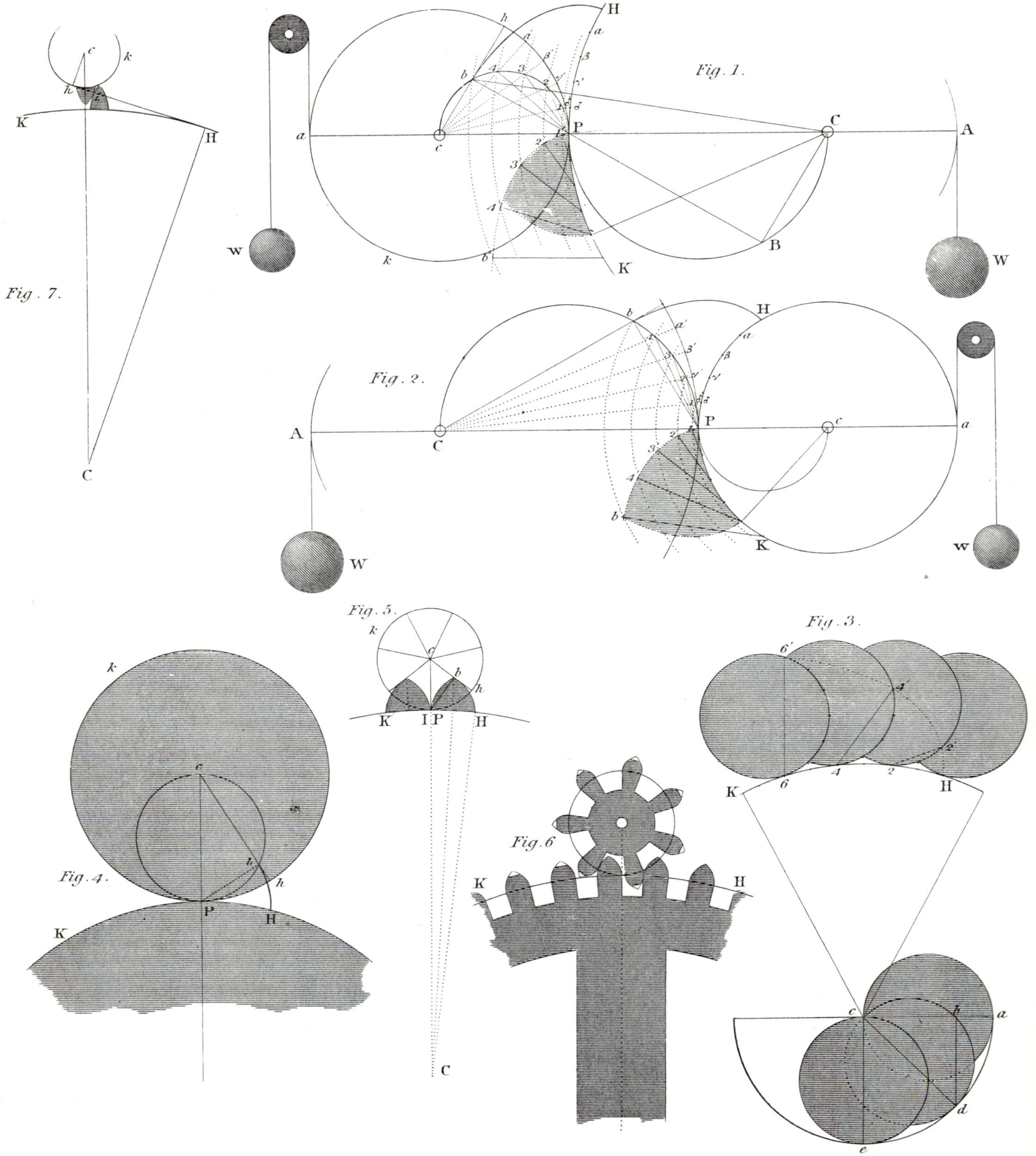

Published as the Act directs, June 2d 1806; by Longman, Hurst, Rees & Orme, Paternoster Row.

Engraved by Wils

ANCIENT CLOCK BY HENRY DE WICK, IN 1370.

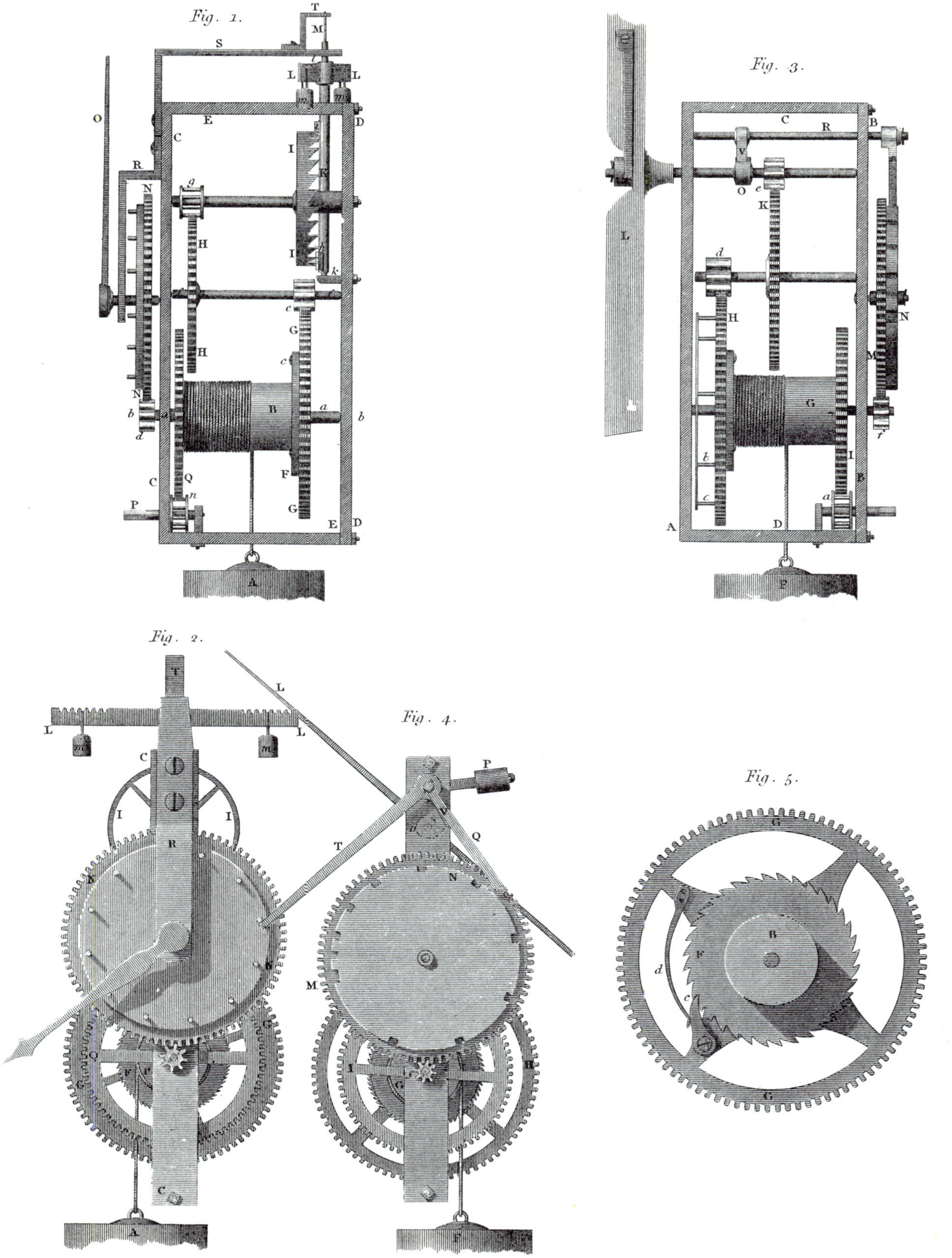

Published as the Act directs, 1808, by Longman, Hurst, Rees & Orme, Paternoster Row.

Engraved by Wilson Lowry.

A Thirty hours Clock with "Alarum and Count Wheel Striking Work.

Fig. 1.

Fig. 3.

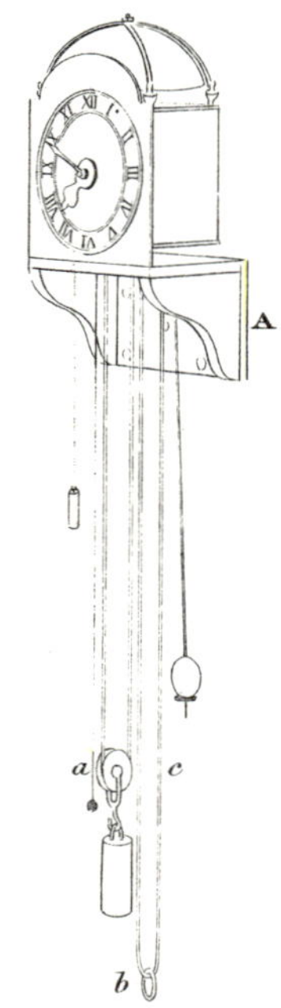

Fig. 2.

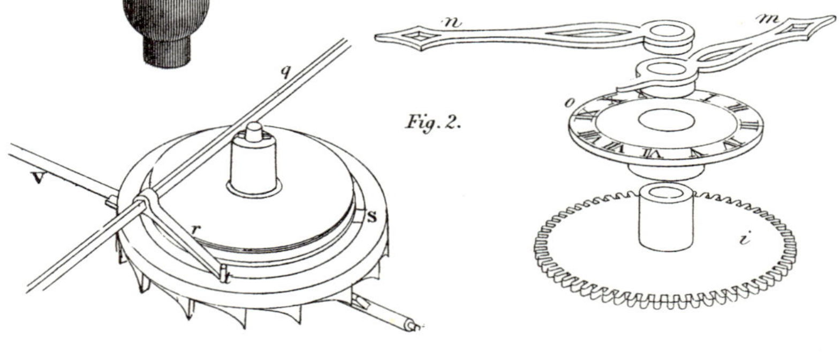

J. Farey Jun. del. Published as the Act directs, 1816, by Longman, Hurst, Rees, Orme, & Brown, Paternoster Row, London. Lowry sc.

CALLIPERING.

(ART. CLOCK)

Fig. 1.

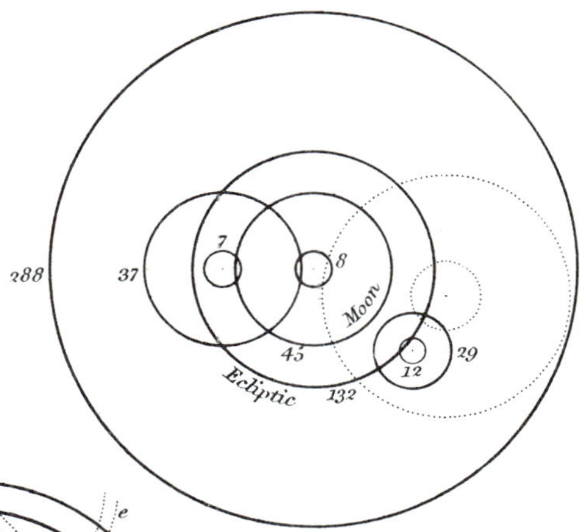

Fig. 2.

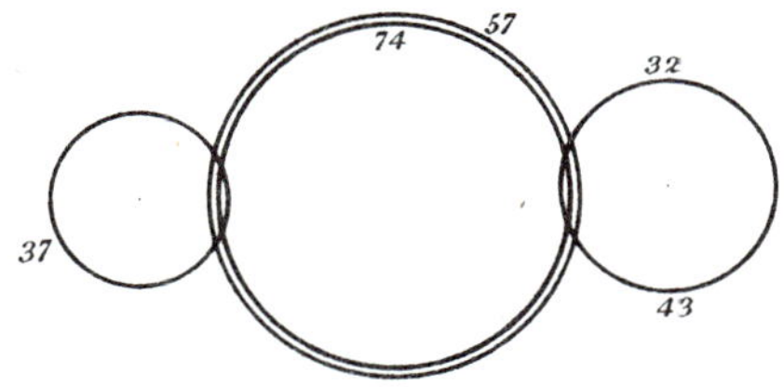

Fig. 3.

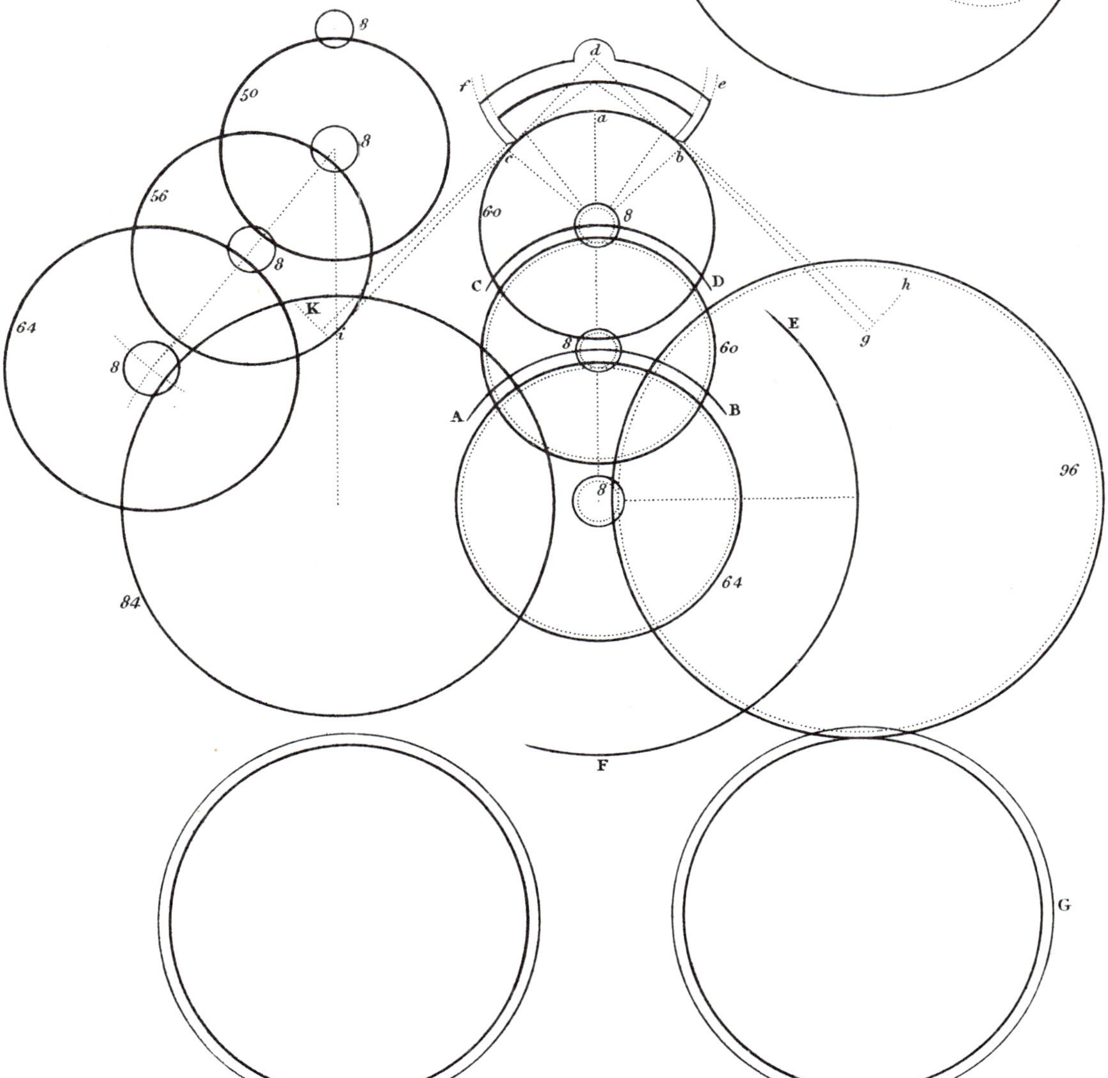

Published as the Act directs, 1808, by Longman, Hurst, Rees & Orme, Paternoster Row.

Engraved by Wilson Lowry.

Portable eight days Clock.

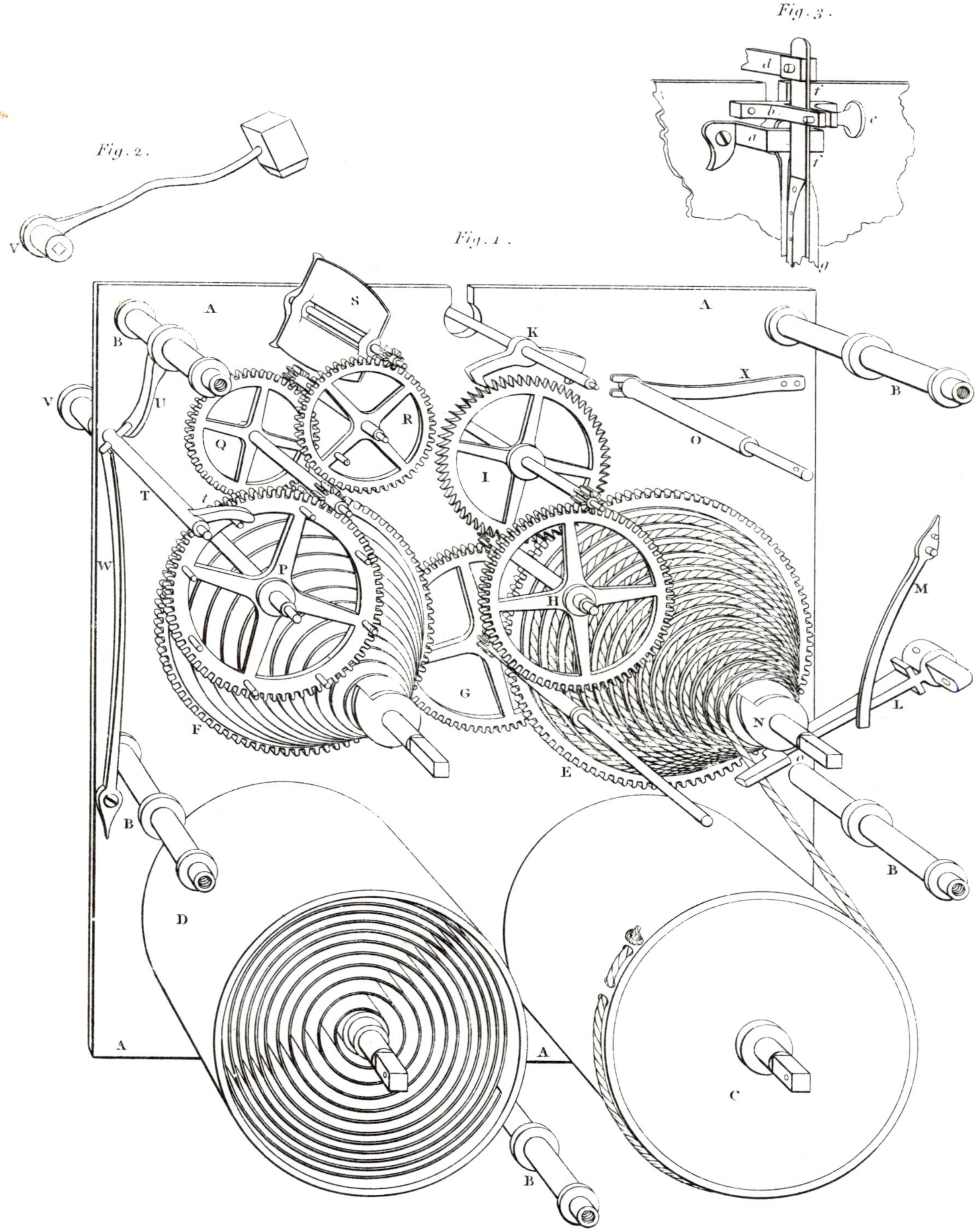

*J. Farey Jun*r. *delin.* *Published as the Act directs, 1812, by Longman, Hurst, Rees, Orme & Brown, Paternoster Row.* *Engraved by W. Lowr*

DIAL WORK AND STRIKING PART OF AN EIGHT DAYS CLOCK.

Fig. 1.

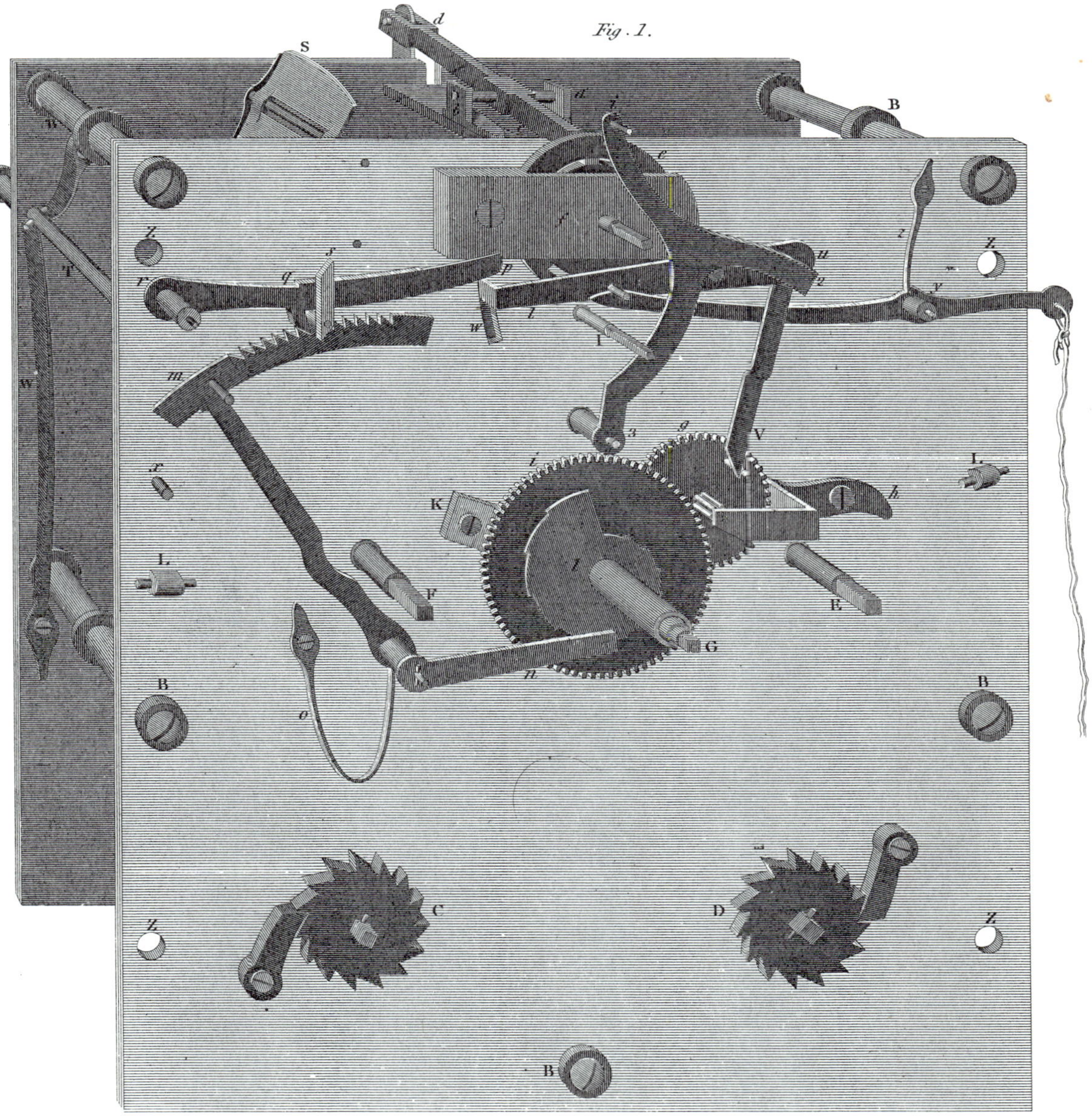

Fig. 2.

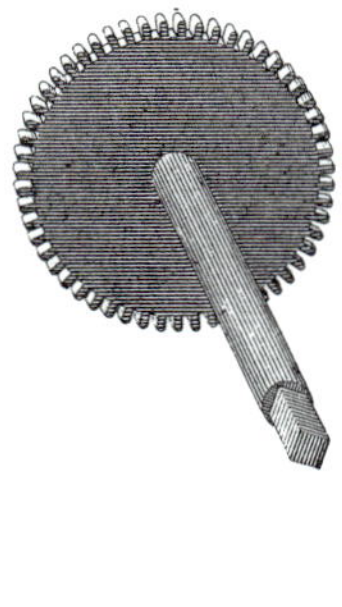

Fig. 3.

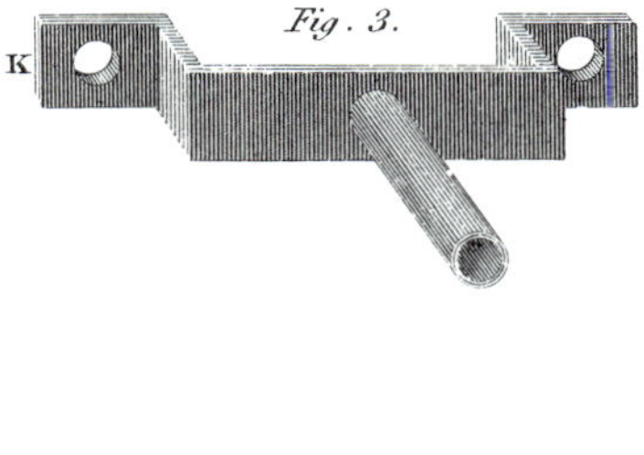

Fig. 4.

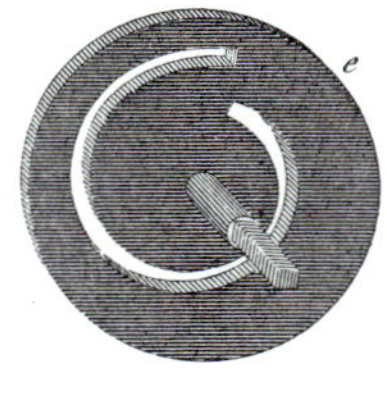

…awn by J. Farey Jun.r *Published as the Act directs, 1808, by Longman, Hurst, Rees & Orme, Paternoster Row.* *Engraved by Wilson Lowry.*

CLOCK WITH CHIMES.

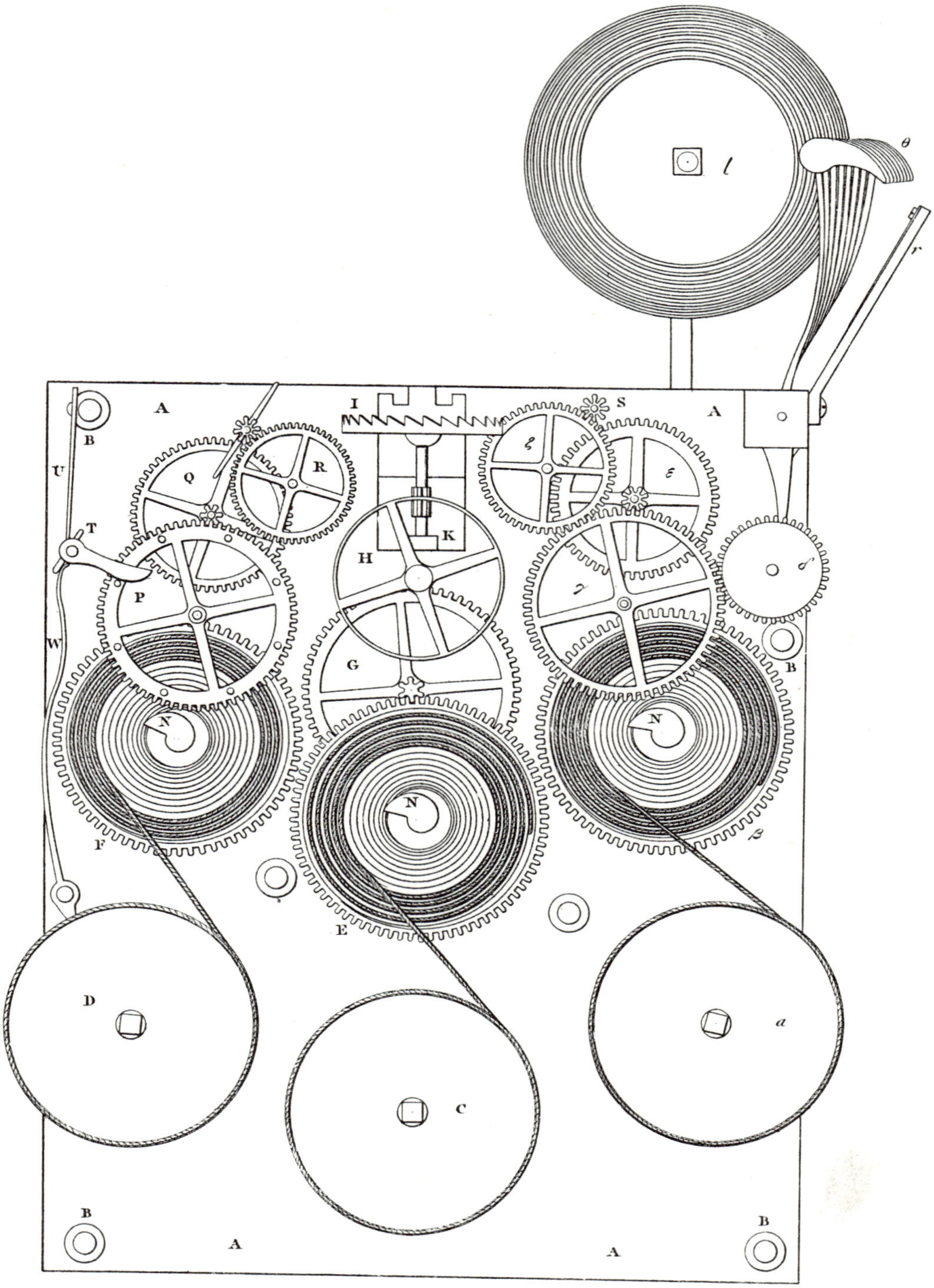

J. Farey Jun.r delin. Published as the Act directs, 1808, by Longman, Hurst, Rees & Orme, Paternoster Row. Engraved by Wilson L

CLOCK WITH CHIMES.

rey Junr. delin. Published as the Act directs, 1808, by Longman, Hurst, Rees & Orme, Paternoster Row. Engraved by Wilson Lowry.

2nd Clock by Mr. Ferguson

Fig. 3.

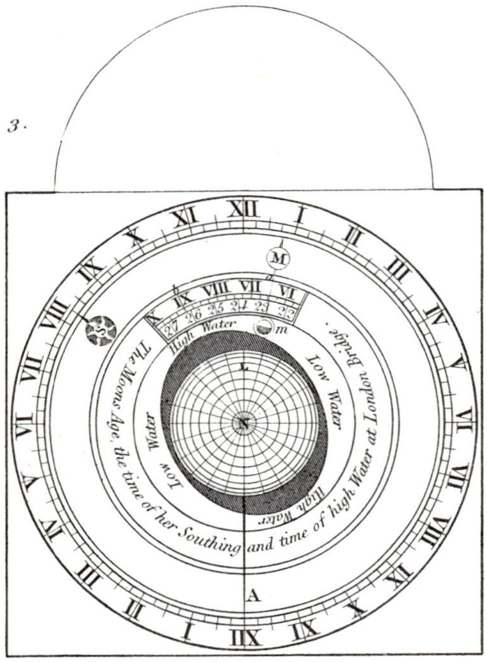

Clock by Dr. Franklin

Fig. 1.

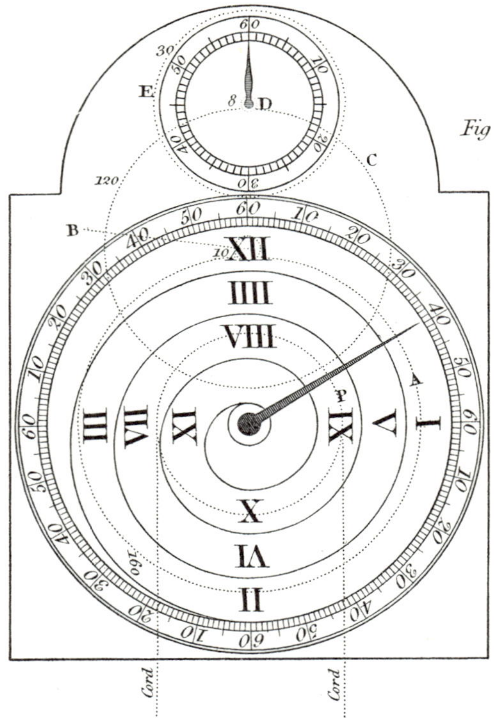

Fig. 4.

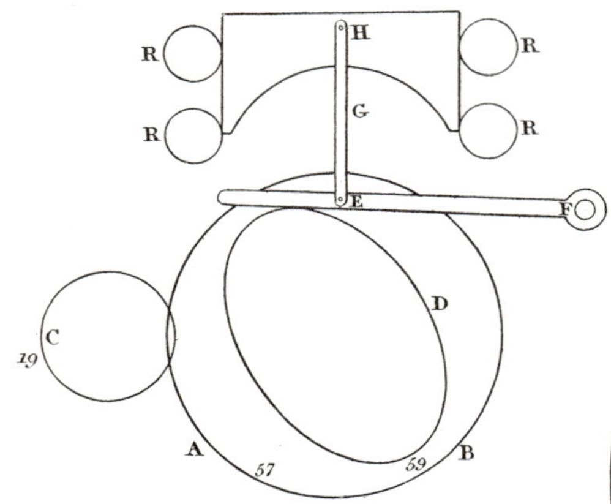

Clock by Mr. J. Ferguson

Fig. 2.

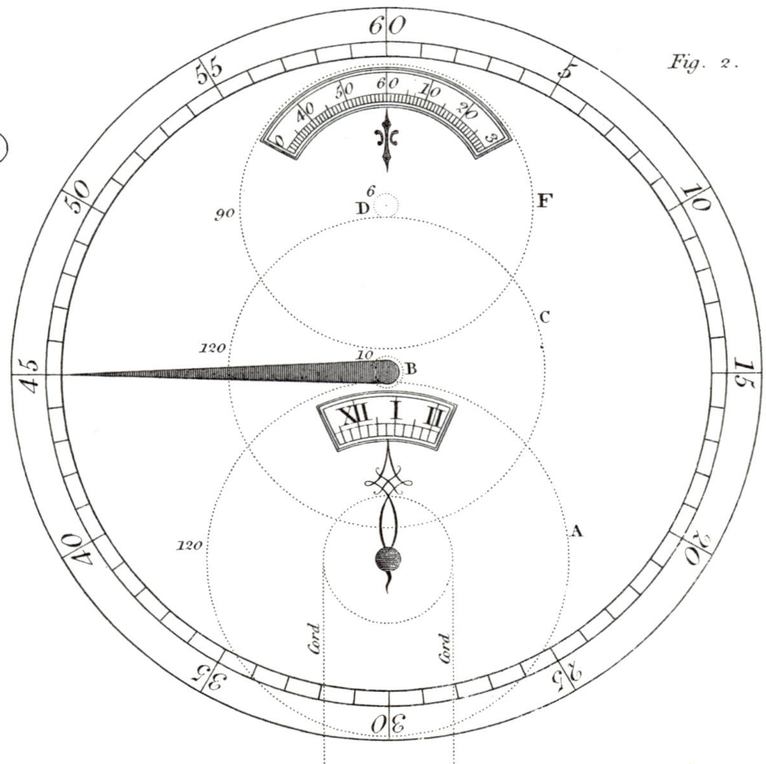

Fig. 5.

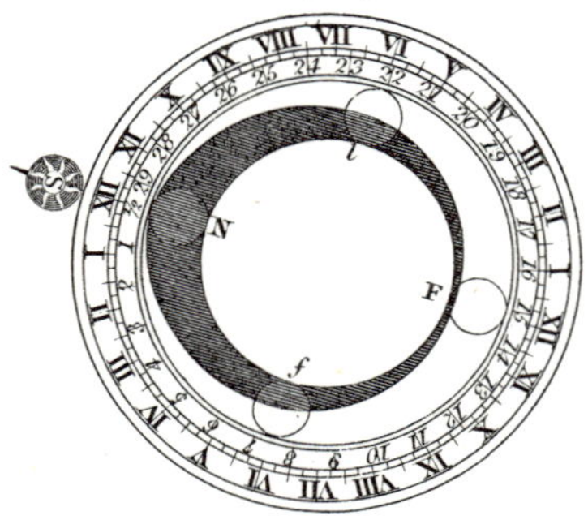

Published as the Act directs, 1808, by Longman, Hurst, Rees & Orme, Paternoster Row.

Engraved by Wilson Lowry

HOROLOGY.

CLOCK TOOLS.

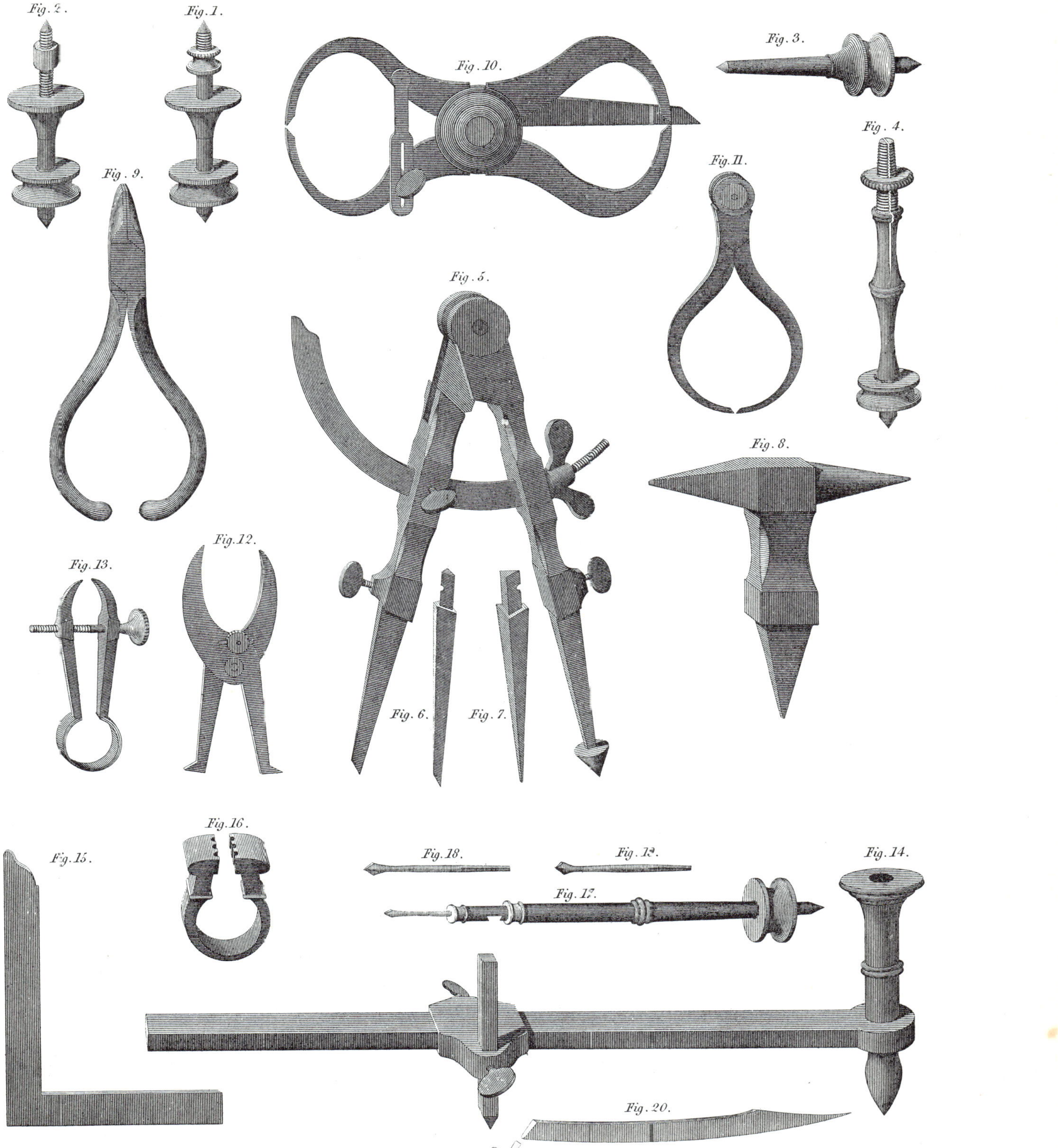

Published as the Act directs, 1809, by Longman, Hurst, Rees and Orme, Paternoster Row. Engraved by Wilson Lowry.

CLOCK TOOLS.

Published by Longman, Hurst, Rees & Orme, Paternoster Row. March 1st 1809. Engraved by Wilson Low

CLOCK TOOLS.

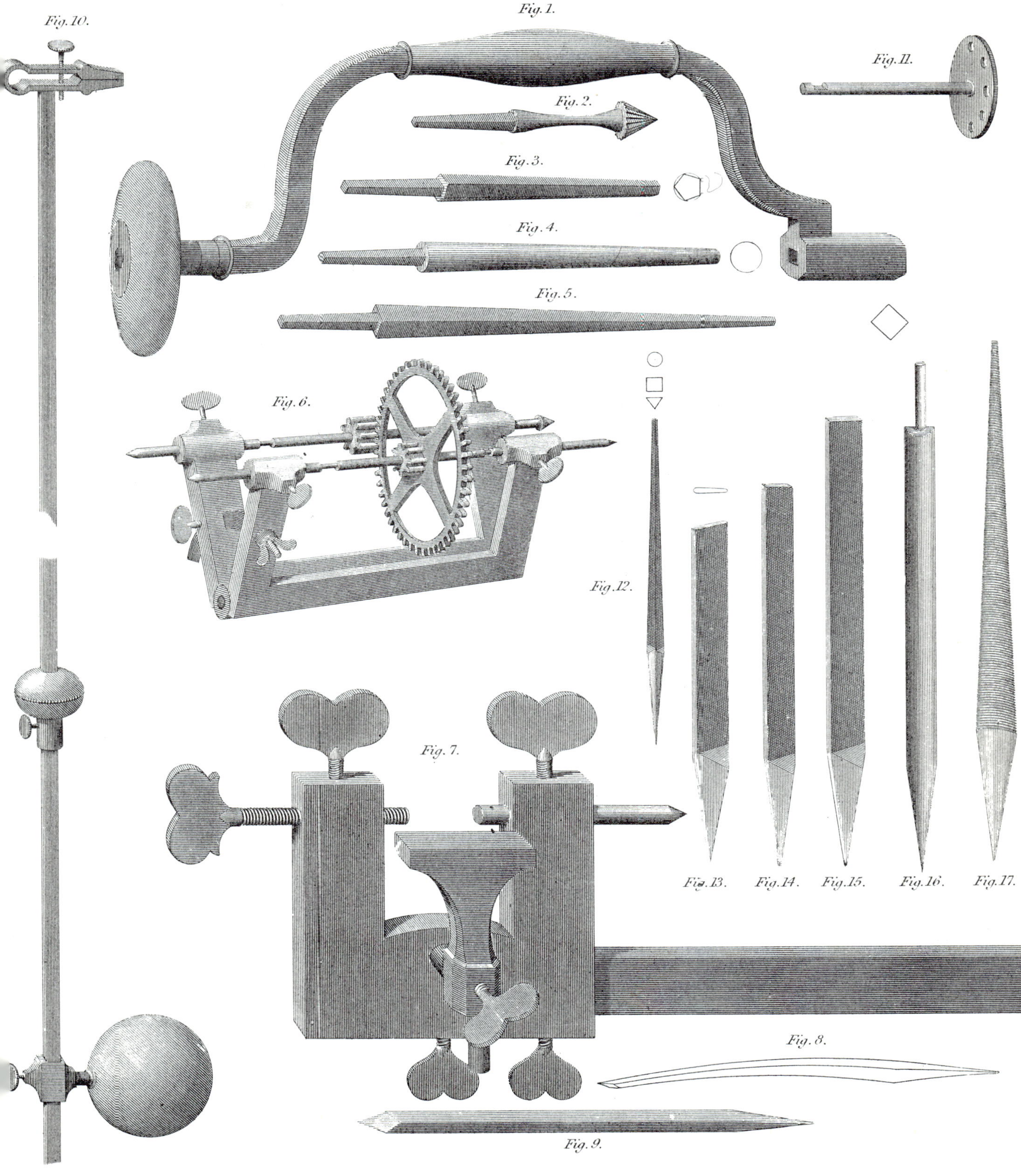

Published as the Act directs, 1815, by Longman, Hurst, Rees, Orme, & Brown, Paternoster Row, London. Lowry sculp.

ASTRONOMICAL CLOCK BY REID.

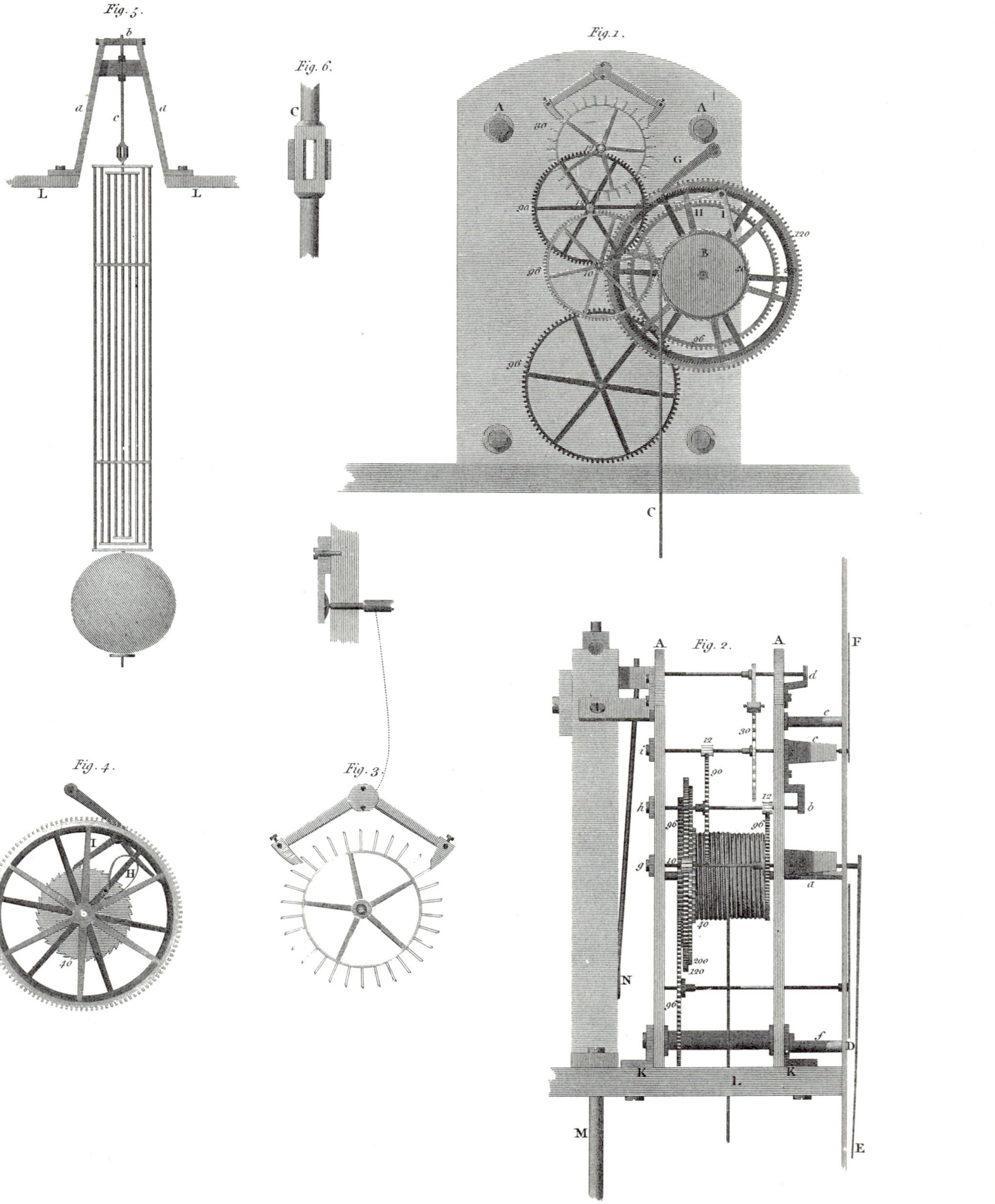

Published as the Act directs, 1810, by Longman, Hurst, Rees & Orme, Paternoster Row. *Engraved by Wilson Lowry.*

Astronomical Clock by Mess.rs Brockbanks.

Fig. 1.

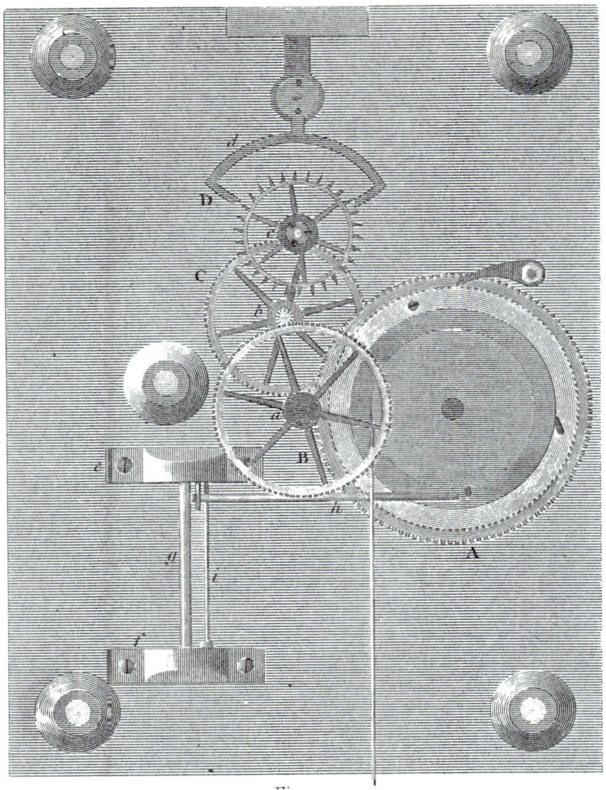

Fig. 2.

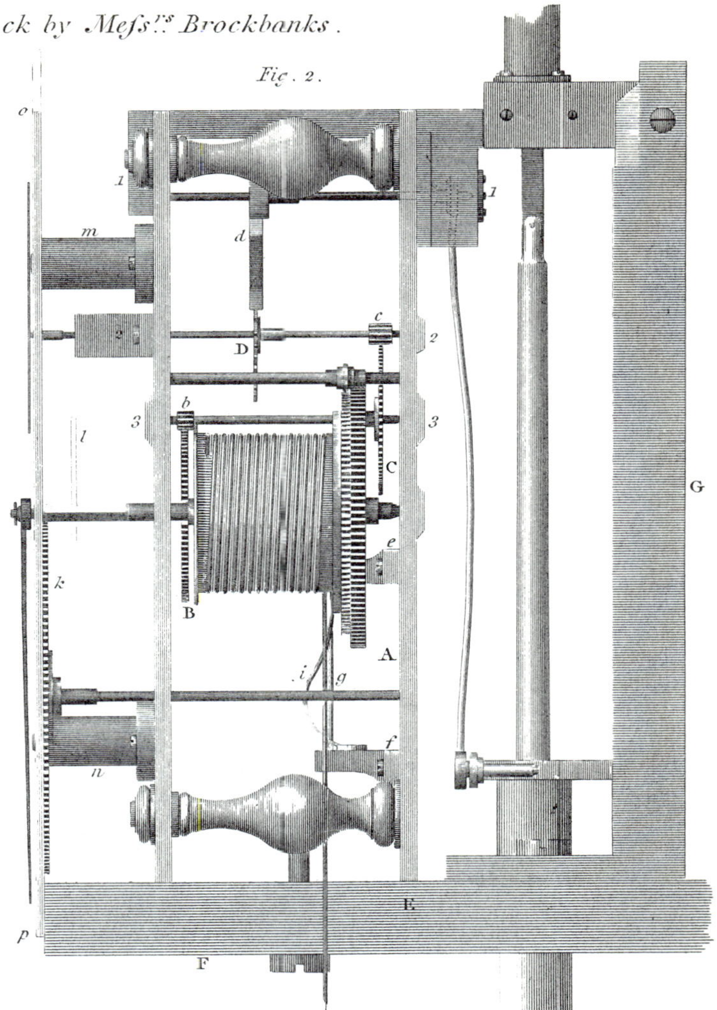

Fig. 3.

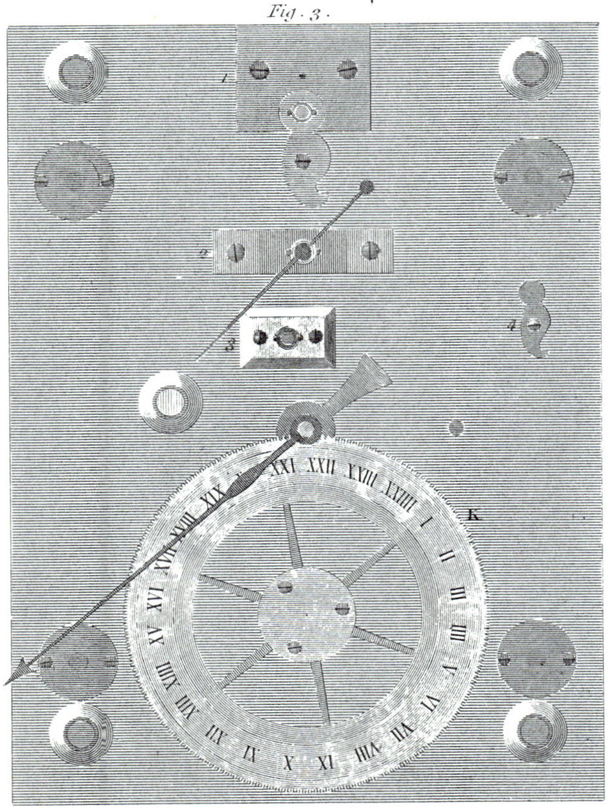

Fig. 4.

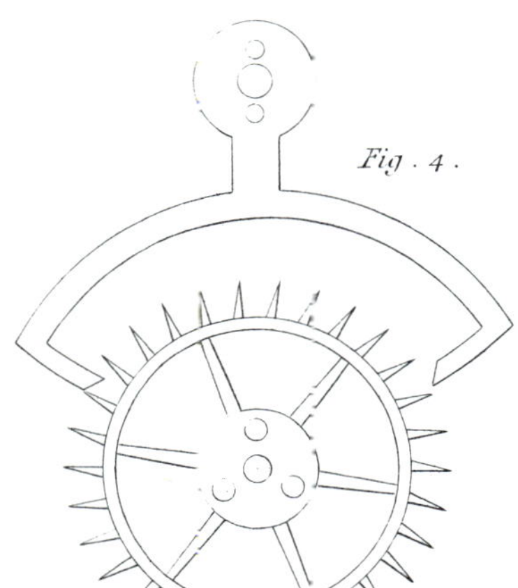

Farey delin. Published as the Act directs, 1811, by Longman, Hurst, Rees, Orme & Brown, Paternoster Row. Engraved by W. Lowry.

HOROLOGY.

EQUATION CLOCK BY ENDERLIN.

Fig. 1.

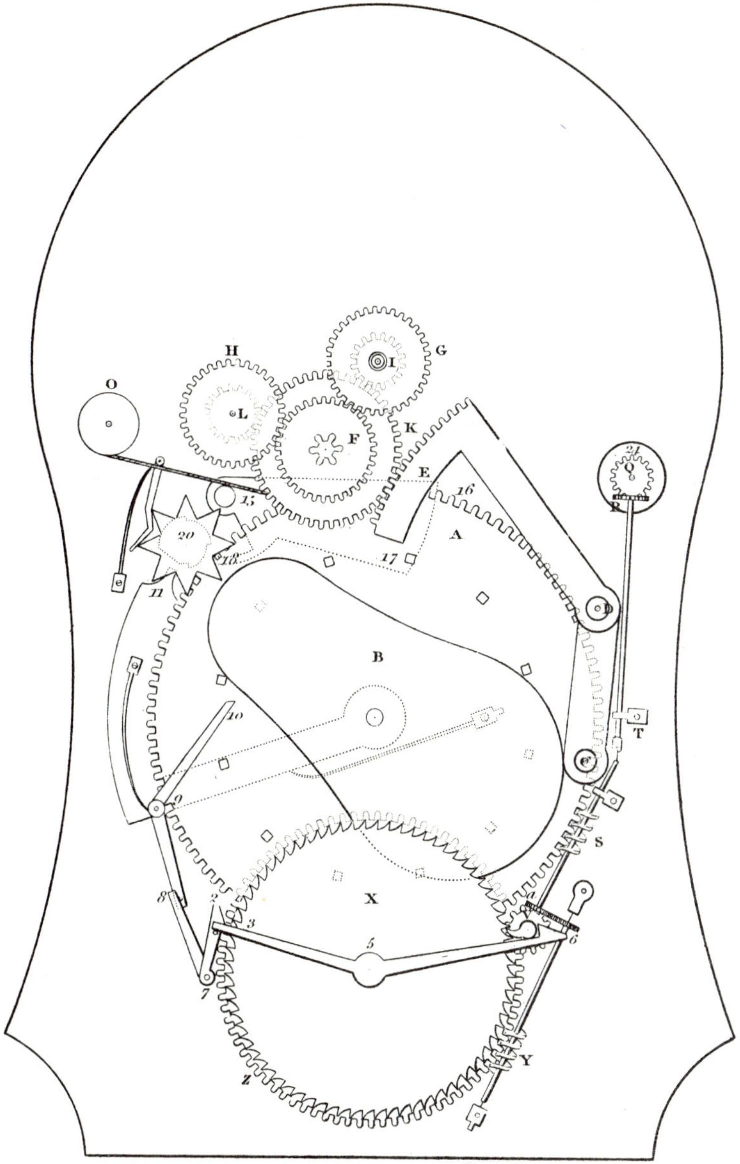

Fig. 2.

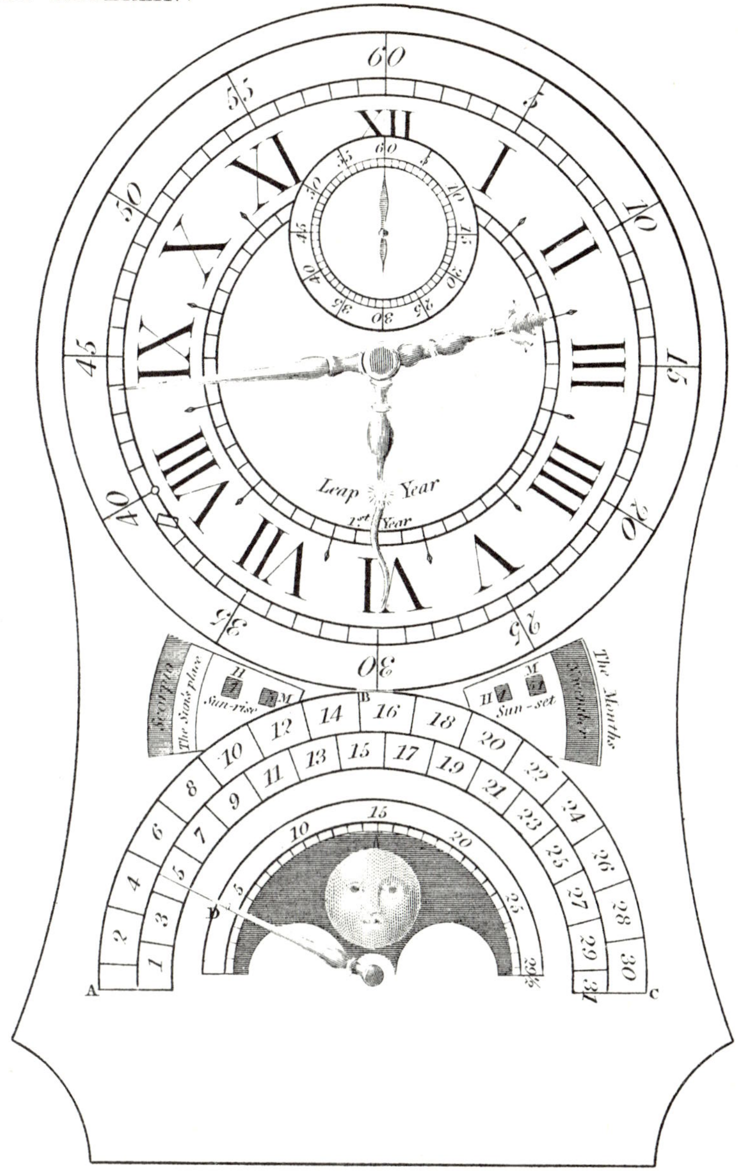

Published as the Act directs 1808 by Longman, Hurst, Rees and Orme, Paternoster Row.

CLOCK WORK.

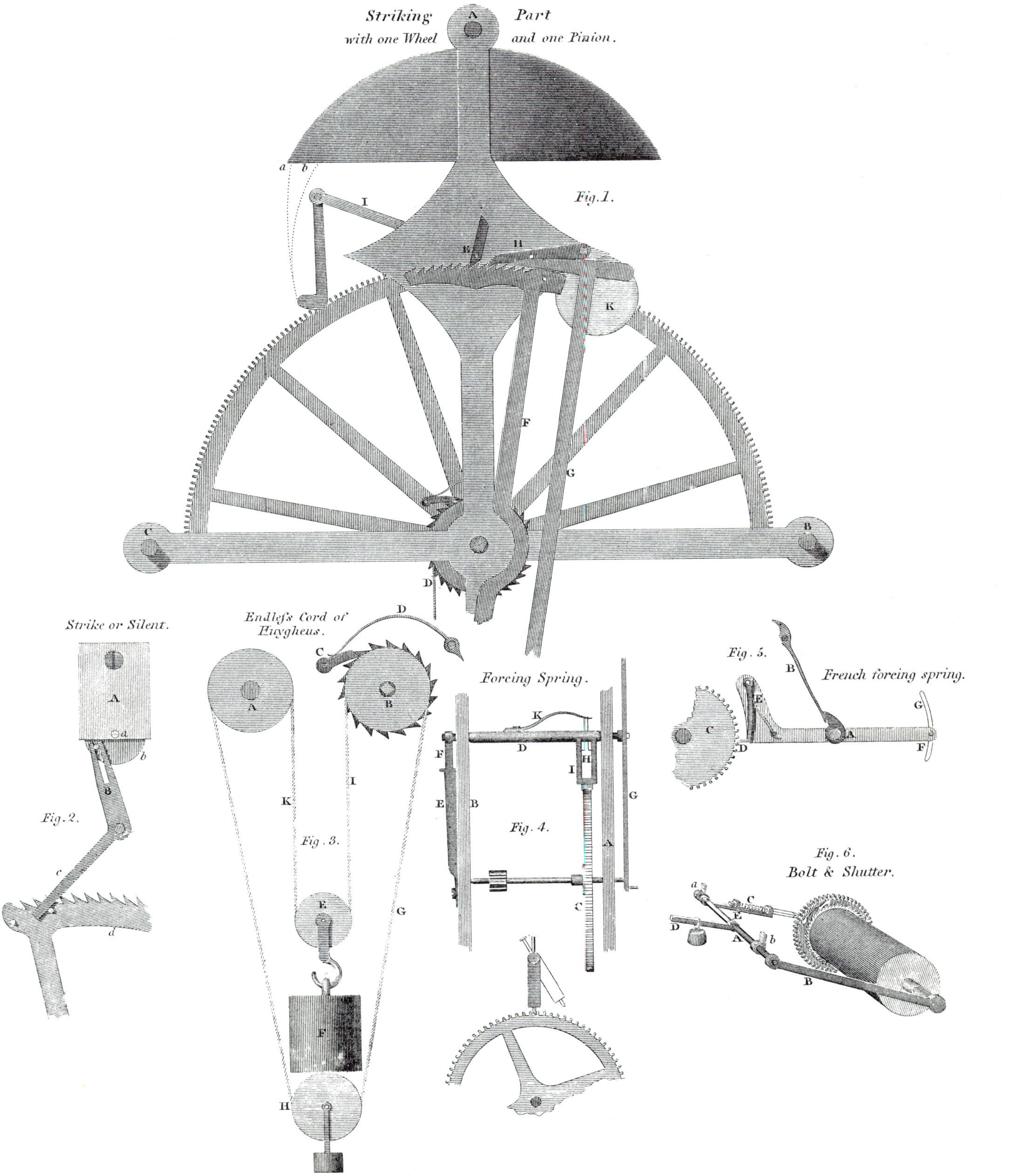

Published as the Act directs, 1809, by Longman, Hurst, Rees & Orme, Paternoster Row.

Engraved by Wilson Lowry.

Fig. 1.
Massey's striking part.

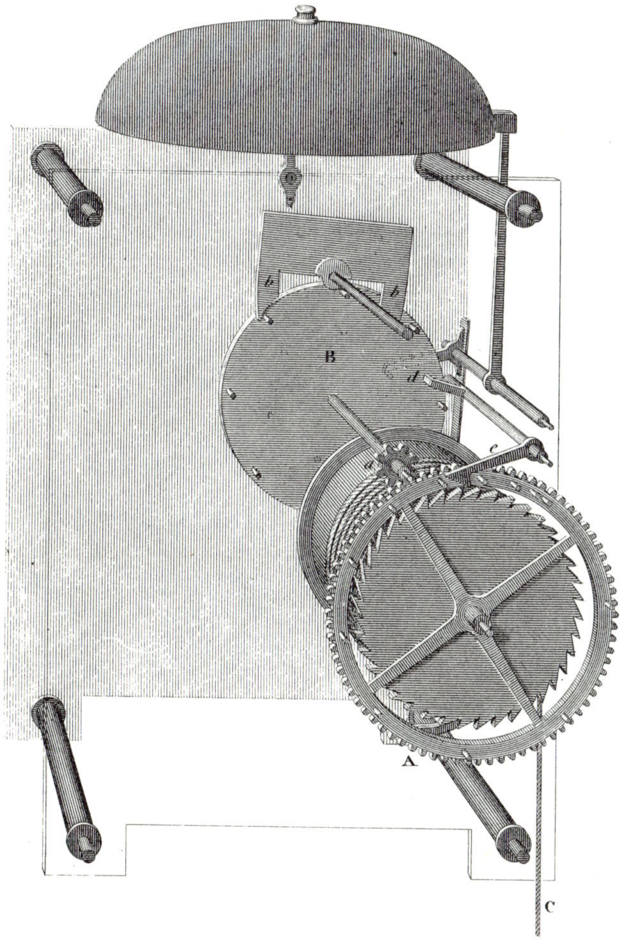

Fig. 2.
Prior's striking part.

Fig. 3.

J. Farey Jun. del.

Published as the Act directs, 1812, by Longman, Hurst, Rees, Orme and Brown, Paternoster Row.

Lowry sculp.

Compensation

Compensation-*Balance*, in *Horology*, is the balance of a chronometer, so ingeniously contrived, that two opposite actions counteract each other's natural effects at all times upon it, and equalise its momentum under all the changes of temperature experienced in different climates and seasons. The opposition of two natural effects, produced by an artificial arrangement of the acting parts, is very properly called the *compensation*, which has been effected in various ways. The most simple compensation that has been produced in a machine for measuring time, is that which depends upon the variable fluidity of the oil applied to the pivot holes of the balance arbor, and of the other arbors of a watch; oil, it is well known, however good, is more fluid in hot weather than in cold, and therefore diminishes the friction at the pivots more in summer than in winter, the consequence of which variation in the friction of a watch with the ordinary crown-wheel escapement, would have its rate accelerated in the former season, and retarded in the latter, by this simple cause, if there were no compensating property in the structure of some part of the mechanism; this property naturally exists in the balance and balance-spring, both which are subject to have their dimensions enlarged by heat, and contracted by cold; and these changes alternately occasion retardations and accelerations in the rate of going; but in such a way, that when the watch is disposed to have its rate accelerated by the most fluid state of the oil, it is also disposed to have the same retarded by the contemporary enlargement of the balance and balance-spring; hence the quantity that one of these dispositions prevails over the other in any watch, when regulated to mean time at a mean temperature, constitutes its liability to vary its rate of going in different months of the year. Berthoud, who attempted to proportion the thickness of his pivots to suit the oil with a given balance, so as to produce an approximation towards a good compensation, has called this the *natural* compensation, which is a very proper appellation, as it exists in a certain degree in all chronometers and watches that have oil applied to them. This natural compensation, however, is found to be a very imperfect one, and unfortunately interferes with the effects produced by the better compensations, which, in opposition to this, may be called the *artificial* compensations. To banish the interference of the variable effects produced by oil, is one of the difficulties, and indeed now the principal one, that opposes the efforts of the chronometer-maker, to make the action of his mechanism permanently uniform.

The mechanism of the artificial compensations may be arranged under three divisions, *viz.* 1st, That which applies exclusively to the balance-spring; 2dly, That which applies exclusively to the balance itself; and, 3dly, That which applies partly to the spring and partly to the balance: for the sake of order, we propose to describe the different compensation-balances agreeably to these divisions, which are also nearly agreeable to the order of time in which they were invented.

Compensations at the Spring.

1. *Harrison's Compensation Curb.*—We have said, under our article Chronometer, that Mr. James Harrison of Barton, in Lincolnshire, was the first person who applied a self-compensating piece of mechanism to the balance of a watch, which contrivance, according to his provincial dialect, he called a *kirb*, and which we have taken the liberty to name a *curb*. *Fig.* 1. of *Plate* XXVIII. of *Horology*, represents the balance, balance-spring, small cocks, and curb, as they would be seen by an eye placed directly over the back of the pillar-plate, when the large cock is taken off, that bears the verge-pivot: AA is the plate on which the mechanism is placed; B the balance of one uniform metal; C the spiral spring of regulation; DE the thermometrical curb, so contrived as to lengthen the effective length of the spring in the different degrees of temperature, without manual adjustment; Gg a double cock to which the curb DE is attached; and F the small cock, or stud, to which the exterior end of the spring is pinned. The thermometrical curb, DE, is composed of two long slender slips, one of brass and the other of steel, pinned

pinned together in different places, and is attached to the fmaller half g of the double cock $G g$, fo that it may be brought nearer to G, or removed farther from it at pleafure, before g is fixed to G, in order to place it at the requifite point between a and b of the fpiral; the double cock is reprefented in the figure as though it were fixed faft by a fcrew and fteady pin to the plate; but in the original time-keepers, it was adjuftable between two parallel-pieces, that allowed it to approach the ftud F, or recede from it, for the purpofe of afcertaining particularly the beft length of the curb, for effecting the requifite compenfation: the fide of the curb next to D was fteel, and the fide next to E brafs, and the effect produced by changes of temperature was thus: when the time-keeper was expofed to heat, the balance, and balance-fpring, had their dimenfions enlarged, fo that the momentum of the balance was increafed, fuppofing its arc of vibration unaltered, and the force of the fpring was diminifhed, or at leaft would have been by its elongation, if the curb had not been applied; and the natural confequence would have been for the piece to retard; but the fame heat, which naturally produces thefe changes in the balance and balance-fpring, produces fuch a change in the fhape of the curb, as counteracts thefe natural effects; in the cafe of heat, the brafs part, E, of the curb becomes more lengthened than the fteel part D, by reafon of its fuperior expanfibility; the confequence is, that the compound piece, D E, becomes convex on the brafs fide E, and concave on the fteel fide D; that is, the remote or free end of the curb, confidered as an index, moves from a towards b, and as this end carries two pins between which the ftraight end of the fpring paffes, thefe pins, and not the ftud, form the limit of length at all times of the regulating fpring, which confequently is curbed as to its effective length, under all the variety of elongations, that it experiences from different degrees of heat; fo that it may be faid to be always both lengthening and fhortening in nearly the fame proportion: on the contrary, when the piece is expofed to cold, or to fpeak more fcientifically, is expofed to a low temperature, the fteel fide, D, becomes convex, and the brafs fide, E, concave from its greater contraction, and the free end of the curb moves from b towards a, thereby mechanically lengthening the fpring, as it naturally contracts, in a degree fufficient to check the velocity which the balance would acquire from its diminifhed dimenfions. Thus, whether it is fummer or winter, in hot or in cold climates, the compenfation-curb is calculated to counterbalance the natural effects of changes of temperature on the dimenfions of the balance and of its regulating fpring.

It has been afferted, that the *elafticity* of the balance-fpring is affected by changes of temperature; but we conceive it to be its *force*, or that power which depends on its *length* that is naturally altered, and not, at leaft in any fenfible degree, that property which depends upon its *temper*. This contrivance, we have faid, in another place, laid the foundation of modern chronometry, though it was afterwards found to be a better application of the compound bar, that the balance itfelf fhould bear it, in order that the dimenfions of the balance and of the compenfation pieces might be affected in a contemporary manner by the variation of atmofpheric temperature, fo that the artificial changes going on in the balance itfelf might compenfate the changes naturally produced in it and in the fpring, inftead of the artificial changes in the fpring being made to compenfate the natural changes in its dimenfions, and in thofe of the balance. We have before faid, that this mechanifm was invented and ufed fo long ago as the year 1726, and that it is incapable of the ufual adjuftment for rate, except at the fixed ftud, from which the faftening pin muft neceffarily be taken, to admit of the fpring's being lengthened or fhortened.

Compenfation by F. Berthoud.

In the year 1760, which was about feven years before the Board of Longitude publifhed the principles of Harrifon's Time-keeper, Mr. F. Berthoud, the famous clock and watch-maker of Paris, contrived and introduced, into No. 1, of his marine clocks, a compenfation-curb, acting on the fpiral fpring of the balance in a manner fimilar to the action of Harrifon's curb, except that the backward and forward motion of his *pince-fpiral*, or curb, was produced by the difference of two direct expanfions agreeably to the difpofition of the grid-iron pendulum, which Harrifon had previoufly invented, (for the defcription of which fee the fection, "Aftronomical clock, by Mr. Reid of Edinburgh," under the article CLOCK.) The mechanifm by Berthoud, is reprefented in *fig.* 2. of *Plate* XXVIII., the eye being, as before fuppofed, to be directly over the pillar plate of the frame; the circle A A denotes the plate juft mentioned; B B, the grid-iron frame, attached to the faid plate, and compofed of alternate bars of fteel and brafs, joined together by end-pieces like the compenfation-pendulum alluded to, which therefore need not be again minutely defcribed; the pair of brafs bars in the middle of the frame, or grid-iron, are thofe which act by protruding more or lefs in different degrees of heat againft the tail-piece D, of the double lever D E, moveable on an arbor with pivots borne by the plate and cock G, or by G made as a double cock; H is the principal cock, in which the balance pivot runs in the triangular hole, made by three friction-rollers, not fhewn in the figure; the *pince-fpiral b*, and arm a are faft to an arbor which is pivoted into the cock H, and a fmall cock under it, which is attached to it; this arbor is parallel to the balance verge, and is fo contiguous to it, as to be in one of the eccentric points, from which the fpiral of the fpring which it holds is geometrically defcribed; that is, at half the diftance from the verge, that the two neareft fpirals are from each other; on each of thefe two levers, a and b, is a fliding piece of metal; that on a for the lever E to reft againft, and that on b to have a flit for receiving the thread of the fpiral fpring; they are both fixed by preffing fcrews to their refpective levers, when adjufted to their proper places thereon; but b is prolonged to d, which is attached to the graduated L piece by the fcrew e, and this L piece has a pointer, ufed as an index for the graduations on the bridge I, that lies acrofs the grid-iron frame, when attached to the plate A A. K is a fpring, preffing againft a tail-piece of the lever a to keep it always in contact with the lever E; C is the fpiral regulating fpring, as in an ordinary watch, and F an adjuftable ftud, for the exterior end of the fpring, with an oblong perforation for the preffing fcrew; the interior end of the fpring being faft to a collet on the verge, which is adjuftable by friction. The mechanical action is thus produced; the difference of the expanfibilities of the fteel and brafs rods of the grid-iron frame becomes fenfible at the pin D, in the tail-piece of the bent lever D E, moveable in a point of bridge G, between thofe two letters; the part E, being longer from the centre of motion than the part D, which is acted on, moves fafter and carries before it the lever a, and together with it the *pince-fpiral b*, and L piece with the fcrew e; and it is obvious, that, as the fliding piece, borne by lever a, limits its effective length from its place of contact with lever F, the relative lengths of thefe two levers may be varied at pleafure by adjuftment, till the effect produced

duced on the fpring by the grid-iron frame, through the medium of thefe levers, is according to its due quantity; and the pince-fpiral will go and come in a manner fimilar to the motion of Harrifon's curb, though the law of their motion is different, one depending on direct expanfion, and the other on lateral curvature. The graduations of the L piece are for bringing the machine to time, the end, *d*, being the index, and the fcrew, *e*, the inftrument of adjuftment; and the index of the L piece fhows, on the graduated cock I, the fituation of the pince-fpiral under all the variations of temperature. Befides No. 1, Nos. 3, and 8 of the marine-pieces of this author had the compenfation we have here defcribed, which, though complex, would no doubt anfwer its purpofe as well as Harrifon's more fimple contrivance; and though the idea did not originate with the contriver, yet it will be acknowledged, that there is evident ingenuity in the arrangement and adaptation of the acting parts of the mechanifm; the effect alfo produced by the mechanifm before us muft be regular and permanent, when the materials are good, and the work well executed. (See "Traité des Horloges Marines" par F. Berthoud. Paris 1773.)

Harrifon's Compenfation improved by F. Berthoud.

After the Board of Longitude had publifhed the principles of Mr. Harrifon's Time-keeper, Mr. F. Berthoud perceived that Harrifon's compenfation-balance might be improved by introducing the action of a lever, capable of adjuftment for any given quantity of effect, without moving the cock that holds the compenfation bar; the conftruction he adopted was thus; A, in *fig.* 3; of the plate already named, is the cock for the balance pivot, ("Hiftoire de la Mefure du Temps") which balance is not neceffary to be fhewn, B is the curb lever, detached from the compenfation bar, and turning on a fhort arbor pivoted into the cock A, and an under cock borne by this not feen, the centre of its motion being at one fide of the balance verge, as in Berthoud's former balance; the fpring under the cock, that urges it towards the fcrew of C, is concealed from fight; C is the compound or compenfation bar, fixed in the chops of a faft bridge D, and having, at its free end, a fcrew with its point bearing on the tail-piece H of the curb-lever B, which curb-lever has a fliding piece with a narrow groove to admit the thread of the fpring, and is fixed by a preffing fcrew in any given fituation; E is what the French call a *piton*, or adjuftable frame of the ftud, ftanding on four fcrews, as feet, which allow it to be brought into the plane of the fpiral-fpring, fo as not to force it from its natural pofition; the ftud F, that holds the exterior end of the fpring flides in the *piton*, when under adjuftment for diftance, and is then fixed by a preffing fcrew, paffing through the contiguous end of the fpring G, which turns on a ftud or fcrew, at its remote end when loofe. The manner in which this mechanifm acts will be readily apprehended from what we have faid of Harrifon's, and the preceding compenfation curb, of which it conftitutes an union, thus; when the temperature is high, the compound bar, C, is affected exactly in the way we have juft defcribed Harrifon's to be affected, and the chops of the bridge, D, allow the lengths of the curb lever B, and of its tail-piece, preffed by the fcrew, to be adjufted according to any given ratio as they refpect each other; and as the quantity of effect to be produced on the fpiral fpring depends upon this ratio, with a given compound bar balance, efcapement, and maintaining power, it is obvious that the adjuftment for temperature may be brought to the requifite degree of exactnefs by a few fucceffive trials in the oppofite extremes of temperature.

The Application of Harrifon's Curb to a common Watch.

After Berthoud had made the alterations, which we have juft defcribed, in Harrifon's curb, he turned his mind towards fimplifying it fo as to become ufeful in a common watch; the manner in which he fucceeded in doing this is fhewn in *fig.* 4 of our prefent plate, where it will appear by infpection how the action takes place, particularly as we have placed the fame letters of reference to the fame parts, as in *fig.* 3; the difference of the two contrivances confifts chiefly in the ftud, F, being here fixed faft to the circular plate of the frame; in the curb-lever being bent and having its centre of motion on a feparate fmall bridge, H; and in the oppofing fpring, K, being attached to the face of the plate itfelf; otherwife the adjuftment for temperature is performed as in *fig.* 3, and is equally capable of taking in the extreme limits of temperature. (For another difpofition fomewhat fimilar to the prefent one, fee Berthoud's "Supplement au Traité des Horloges Marines." *Plate* III. *fig.* 8.)

Compenfation by Mr. Alexander Cumming.

Mr. Cumming publifhed his book on "The Elements of Clock and Watch-work," in the year 1766, in which he has given an ingenious though complex method of compenfating the effects of heat and cold on the balance and balance-fpring, different from the foregoing ones, and depending on the difference of the direct expanfions of two feparate and different metals, aided, like Berthoud's, by the introduction of adjuftable levers; whether the invention preceded or followed Berthoud's grid-iron frame and lever, is a confideration which does not, perhaps, affect either of their two claims to originality, as Harrifon's curb was moft probably heard of, if not previoufly known, by both the mechanifts in queftion; nor does it appear that either of the two complex methods here fpoken of has been copied by others in practice. *Fig* 1. of *Plate* XXX. of *Horology*, reprefents a view of Cumming's mechanifm of compenfation, without the balance, as viewed by the eye placed directly above it, when fixed to the back of the pillar-plate of the frame. The outermoft circular rim, AAA, is of fteel, and has eleven fmall rollers revolving on as many pins or ftuds inferted into the plane of the rim, and the interior rim, BBB, which refts againft the rollers, is of brafs, but is not an entire circle: one end of this interior, or brafs rim, is fcrewed faft to the plane of the fteel rim, by means of its enlarged extremity being made to lie upon the plane of the fteel rim, as fhewn in the figure; in this fituation it is clear that the excefs of the brafs rim's expanfibility will make its oppofite ends approach each other in a high temperature, and recede in a low one, while the friction-rollers will prevent jerks; hence any object oppofed to the free end of the brafs rim will be pufhed forwards by it, as the temperature is raifed, which effect affords the means of alternately lengthening and fhortening the fpiral fpring of the balance in the following manner; the free end of the brafs rim preffes againft the end, *a*, of a lever moveable on a ftud about the middle of it, and the oblong flit made in the oppofite end of the fame lever receives a pin inferted into the contiguous end of a fecond lever *b*, moveable about its remote end round a focket concentric with the balance-verge, or nearly fo; the lever, *b*, which may be called the curb-lever, has two pins in underneath, not feen in the drawing, between which the outermoft thread of the fpiral fpring paffes; hence it is plain, from an infpection of the figure, that when the end, *a*, of the flit lever is pufhed outwards, its oppofite end draws inwards the pin of the curb-lever, and with it the curb, fo as to fhorten the effective length of the

the ſpring in this caſe; and on the contrary, when the braſs rim ſhortens by a low temperature, the contrary effect takes place: for the end, *a*, of the ſlit lever follows the free end of the ſaid rim as it retires in contracting, in conſequence of a ſpring, not ſeen, impelling it into contact with this end at all times; the ratio of the two acting ends of the two levers from their reſpective centres of motion is adjuſted by the horizontal arbor, *d*, ſquared at the end to receive a key; for as the interior end of this arbor is formed into a ſcrew that paſſes into a tapped hole in a ſquare nut on which is the centre of motion of the ſlit lever, the ſaid ſquare nut is made to ſlide in an oblong hole of the plate below the lever; and thus the different parts of the ſlit may be made, ſucceſſively, to impel the pin of the curb-lever, as the compenſation may be found on trial to require. The balance, of courſe, is placed above this mechaniſm, and vibrates free from it. We obſerve that in the original drawing given by Cumming, (*fig.* 5 of his *Plate* XIV.) the ſpiral is made to turn the wrong way about, ſo that the effect of heat will be to elongate the ſpring, which conſtruction would double the natural effect of variations of temperature inſtead of compenſating them, as the contrivance is intended to do, and would do, if rightly conſtructed; but we ſee no reaſon to doubt this being as effectual and permanent as any of the foregoing modes of producing a compenſation, when the ſpiral is bent the right way round, as we have placed it.

Berthoud's Compenſation by a Braſs Arch.

Another method of producing a compenſation, different from any of the preceding ones, in its mode of application, comes next under our conſideration, as another production of the French clock and watch maker, whoſe name we have had frequent occaſion to mention with reſpect.

Fig. 5. of *Plate* XXVIII. exhibits the plan of all the eſſential parts of this contrivance, in an intelligible manner; A, as uſual, is the principal cock over the pillar plate; B, the balance of one metal; C, the ſpiral ſpring; D, an arch of braſs with its ends reſting in two ſloping notches filed into the ſteel bar, E, which is fixed by a ſcrew, near E, in the middle, to the plate, ſo that it may be at liberty to yield both ways to the action of increaſed or diminiſhed temperature on its total length; F is, as before, the ſtud to hold the exterior end of the ſpring, while the interior one is faſt to a collet on the balance-verge; G*g*, is a bent lever, with a pin in its ſhort end, *g*, and its long end pointed as an index, to reach to the diviſions, H, made on the plane of the plate, which bears the mechaniſm; the effect is thus produced; the braſs arch, D, is more elongated than the ſtraight ſteel bar, E, with the ſame increaſe of heat; this arch carries a ſmall ſcrew, the head of which is juſt oppoſite the pin, *g*, already mentioned, between which ſcrew and pin the outermoſt coil of the ſpiral ſpring paſſes; now if the ſcrew and pin were made to approach, till they both ſhould touch the ſpring at oppoſite ſides, they would become in place of a curb, and there limit the length of the ſpring that comes into action; on the contrary, if they were made to recede from each other, till the ſpring acted up to the ſtud, F, placed on the plate, they would be ſuperfluous; but when they are placed at ſuch an intermediate diſtance from each other that the ſcrew limits the opening of the ſpiral, and the pin its cloſing, before the part of the ſpring near the ſtud is moved by the motion of the balance, then the natural effective length of the ſpring is controlled, and the pin and ſcrew together become a ſpecies of curb, that limits the effective length of the ſpring from an undiſtinguiſhed point, lying ſomewhere between the ſtud and the ſaid pin and ſcrew; which indiſcriminate point will always be nearer to the ſtud the more the pin and ſcrew are removed from each other, and *vice verſâ*; hence it is eaſy to ſee, that moving the index, G, forwards or backwards, on the graduated portion of a circle, H, will bring the pin, *g*, nearer to, or remove it farther from, the ſcrew head, than a mean diſtance, and will thence affect the rate depending on the effective length of the ſpiral ſpring, with a given balance, maintaining power, and eſcapement; but it is the ſame thing if the index be ſuffered to remain, and the ſcrew be brought nearer to the pin; for the one limits the right hand and the other the left hand excurſion of the balance; this is the preciſe operation of the contrivance; the ſuperior elongation of the braſs arch over the ſteel bar ſhortens the radius of curvature of the arch, and makes the ſcrew-head approach the ſpring, more or leſs, agreeably to the change of temperature, to which the acting parts are expoſed. The law by which the approach is effected is ſomewhat different, however, from that by which an expanſion-rim is guided, with one of its ends at liberty; and the manner alſo in which the effective length of the ſpring is limited, allows a long ſcale of lengths, in a correſponding ſhort ſcale of diſtances between the ſcrew-head and pin; the latter of which laws is evidently diſadvantageous to the nicety of the adjuſtment for temperature; beſides, when the braſs arch and the ſteel bar are found on trial not to be in their due proportions to each other, the remedy cannot be applied without varying the original proportions, by lengthening or ſhortening the verſed ſine, as the caſe may be, which alteration will require the mechaniſm to be diſplaced.

Improvement on the laſt Compenſation.

Berthoud, having diſcovered the inconvenience and objections to the laſt deſcribed compenſation curb, propoſes another in his "Hiſtoire du Temps," tom. ii. p. 101 and 102, thus: A, *fig.* 6, *Plate* XXVIII, is the principal cock, mounted as uſual on the pillar plate behind; B is again the balance of one metal; C, the uſual ſpiral ſpring; D *d b*, a rack moveable on a ſocket concentric with the balance verge, with ſome additional apparatus carried by it; E is a ſecond cock; and F a third, bearing the *piton*, or ſliding ſtud, with its frame and ſecuring ſpring as in *fig.* 3 of the ſame plate already deſcribed; G is a pinion with an arbor ſquared at the projecting end, to receive a key, which pinion has its teeth formed ſuitable to impel the teeth of the rack D *d*; *a* and *b* are two compound bars of braſs and ſteel, ſoldered together with hard ſolder, inſerted at *d* into a bearing piece, and kept faſt by a preſſing ſcrew *e*, urging them together with the looſe piece, *d*, interpoſed, and kept parallel by a clamping piece *c*; the rack has a circular perforated groove, which allows it to paſs backward and forward by the action of the pinion, while the heads of the two ſcrews, *f* and *g*, confine it to the ſame plane on the back of the plate; and, laſtly, the compound bars, *a* and *b*, have each a pin at their free extremities, which contain between them the laſt coil of the ſpiral ſpring, the action of which is limited preciſely in the manner deſcribed in our laſt compenſation mechaniſm; here, however, the law of flexure is again the ſame as in Harriſon's, though the law of limitation in the effective length of the ſpring is different; the latter, as we have ſeen, depends on the quantum of play between the pins, whereas, in the former, there was no play at all, and the curb itſelf was the limit; ſeeing the effective length of the ſpring diminiſhes as the pins approach each other, the exterior laminæ of the compound bars, *a* and *b*, muſt neceſſarily be braſs, and the interior ones ſteel, to ſhorten the ſpring in high temperatures, and the contrary; when the effect produced is too great, the clamping piece, *c*, is carried outwards towards the pins, in order to ſhorten the bars, which have their lengths meaſured from this clamp; but when the effect

fect produced is too little, the said clamp must be brought inwards towards *d*, to lengthen the bars; so that the scale of adjustments here runs along the length of the two parallel compound bars, as it relates to the variations of temperature; while the adjustment for a mean rate is made by the index *b*, which shews the quantum on the contiguous portion of a divided circle. Still, we think, that while a small variation in the play, between the curbing pins, produces a great variation in the effective length of the spring; this contrivance can have no pretensions to rank in utility with some of the more recent inventions, that place the compensating parts on the balance itself.

Compensation by Breguet.

A compensation similar in effect to the two foregoing ones by Berthoud, but more simple than the latter, is that which Breguet of Paris has applied with success to his common watches; this compensation is shewn in *fig.* 7 of *Plate* XXVIII. where the bounding circle denotes the back plate of the frame; A the principal cock; B the balance of one metal; C the common spiral spring; D*dc* a three armed piece, called by the French a *rateau*, to which is attached the mechanism of compensation thus: the curvilinear fork, *a b*, is composed of two metals: each prong consisting of a slip of steel and a slip of brass, so attached together, that the steel is the exterior, and the brass the interior part in each; the end *a* is screwed to the lever D, and the end *b* is loose, consequently at liberty to approach or recede from the spiral spring, as the changes of temperature affect the fork; the interior end, *a*, of the fork bears a pin near the letter *b*, not seen by reason of the lever D, which lies over it, and a second pin similar to the other is carried by the lever D, between the two largest coils of the spiral: these two pins limit the effective length of the spring exactly in the way that the two preceding compensating contrivances have been explained to do, for which reason we do not repeat what we have already said on this part of the subject; the alternate inward and outward motions of the pin, *b*, regulate the distance between the two pins, on which the play depends, and consequently the length of the spring also, as it has regard to the action; the prong *a* has a tendency to become straight by heat, in consequence of the concave part being brass, but the prong, *b*, has a tendency to become still more convex in consequence of the brass being the convex part; the joint effect of which alterations of shape and position will be, that, as the balance and its spring become enlarged, the space between the curbing pins will become diminished, and the contrary. The watch is regulated for mean time by the index part *d* of the three-armed *rateau*, and the third arm, *c*, contains a pin about the middle of the spiral spring, which we conceive to be the banking pin, though no mention is made of it in the original description. "Histoire de la Mesure du Temps, by Berthoud."

Compensation by Mr. James Scott.

In May of the year 1805, Mr. James Scott, of Grafton Street, Dublin, published an account in Mr. Nicholson's Journal, (vol. XI. p. 19—21., 8vo. series) of a compensation curb which acts in a manner somewhat analogous to that which we have described as the invention of Cumming; but it is evident he had not previously seen the one alluded to, nor yet Berthoud's, otherwise he would not have said in his letter to Mr. Nicholson, that "artists have not been able to invent a compensation-curb adjustable to the exact expansion required." The contrivance now before us is ingenious, and we presume might be applied to an ordinary watch with considerable advantage; though we cannot persuade ourselves that any method of curbing the regulating spring, after it has been adjusted for isochronism, affords so good a compensation, as the mechanism of compensation now usually applied to the balance itself. *Fig.* 2. of *Plate* XXX. of *Horology*, shews all the parts of Mr. Scott's compensation-curb, as given by himself in the original account; but as his description did not seem very intelligible to us at the first reading, we hope to be excused the liberty we take of giving our own account of the contrivance: A A is an index of steel, with a circular part embracing a dove-tailed groove that surrounds the balance verge in the upper plate of the frame; to this steel index is attached, by a screw and steady pin, one end of a compound circular bar, B B, so formed, that the steel part is the exterior, and the brass part the interior; to this compound bar is fastened a second compound bar C C, also circular, by means of a clamping piece and screw at D; a loose piece of metal, interposed between the two circular bars, being pressed by the screw, the outer one keeps the inner one close to the shoulder of the clamping piece, which cannot be seen in the figure; but it is easy to conceive that such a clamping piece will apply at any part of the circular bars; the bar, C C, has the brass part exterior, and the steel part interior; E is the stud in the plate, that holds the outward end of the spiral regulating spring, intended to be curbed; and the interior compound bar, C C, passes between two steady pins, inserted in the plate at *a*, a short distance from *b*, where the curbing pins are fixed, that embrace the last coil of the spring: whenever the index, A A, is moved, in regulating for mean time, it carries with it the two concentric compound bars, and the curbing pins slide along the coil of the spring, so as to lengthen or shorten it as the case may be; suppose, now, that the regulation for rate be made, and that the balance has its temperature elevated, the tendency of the outer circular bar will be to open, or enlarge its radius of curvature, which action will bring the interior compound bar and its curbing pins nearer to the index, and shorten the balance spring a little; but this action is not sufficient to produce the whole requisite effect; the tendency of the interior circular bar C C, on the contrary, is to close, that is, to approach the index, in consequence of its radius of curvature becoming shorter, while its end at D is clamped to the free end of B B; and it is the amount of these two contemporary motions, produced in the interior compound bar, C C, that constitutes the motion of the curbing pins at *b*, along the spiral spring, in a direction towards the index A A. In a low temperature, just the reverse takes place; the outer compound bar, B B, has its radius of curvature shortened, and the inner one, C C, has its radius of curvature, on the contrary, lengthened in this case, so that the joint effect is to make the curbing pins retire from the index to lengthen the regulating spring; which alternate lengthening and shortening of the balance-spring's effective length in heat and cold, form the compensation for the corresponding variations in the momentum of the balance and force of the spring. The excellence of the present contrivance consists in its affording a scale of adjustments; for moving the clamping piece, D, forwards, towards the middle of the concentric circular bars of compensation, lessens the effect, and removing them back, towards the open ends, increases the effect to be produced on the balance-spring, by giving a greater or smaller range of motion to the curbing pins. The inventor's rule for the thickness of the compound bars is, that each lamina be of the thickness of an ordinary main spring of a common watch, so that the compound pieces may be each of twice the said thickness.

COMPENSATIONS *on the Balance.—Thermometer of Peter le Roy.*

We have said under our article CHRONOMETER, that Harrison was the first person who suggested the idea of the balance carrying its own compensation, and Peter le Roy was the first who in 1766 succeeded in such a construction; in our notice of this invention, however, there is a typographical omission which we beg leave to correct here before we proceed to our description; it is said under the article just referred to, that the compensation was effected "by means of two thermometers, one of mercury and the other of alcohol, attached to, and carried by, the balance itself;" instead of, "one end of each bearing mercury, and the other end alcohol, &c." as will now be more clearly understood. *Figs.* 1 and 2 of *Plate* XIX. of *Horology*, exhibit so much of the balance of Peter le Roy as is necessary for explaining its construction; A A, in *fig.* 1, is a portion of the arbor or verge to which the balance B is attached by screws; C is a small ring attached to the inferior part of the verge which holds the two screws of regulation for mean time, and which also clamps the horizontal parts of the two glass thermometrical bent tubes, D and E, placed diametrically opposite to each other, one of which is seen in a detached state in *fig.* 2; these tubes are also held fast by two other clamps, to the middle of the verge, as may be seen in *fig.* 1; the superior end of the straight vertical part of each tube is open to admit the pressure of the atmosphere on the mercury, which is now seen standing at the same level in both the upright portions of each thermometer, after the inferior or horizontal connecting part is filled; but the bulbs and parts above the letters D, and E, are filled at a mean temperature with alcohol, respectively resting on the columns of mercury: when the balance and balance-spring are enlarged by heat, the alcohol has also its bulk increased, and descends below the letters D and E, for instance, in very high temperatures as low as the horizontal connecting part, in which case the mercury pressed upon by it descends under the alcohol, but ascends in the same proportion at the other end, till it reaches nearly the open superior end of the vertical straight part of the tubes respectively; the mercury which was before in a column parallel to the verge, but at a distance from it, is now also parallel but contiguous, and, as it is a ponderous fluid, the momentum of the balance is considerably diminished by the approach of the mercury to the centre of motion, which change of position of the mercury constitutes the compensation required to balance the effect of an enlargement of the balance itself, and of an elongation of its regulating spring taken jointly: in a low temperature the contrary change takes place in the situation of the mercury, for as the bulk of the alcohol contracts and retires round the acute angular point into the bulb, the exterior vertical branch becomes filled by the mercury. Peter le Roy, we have before said, was disposed to prefer this balance to the balances he made entirely of metal, but the frangibility of the bent tubes, and the liability of the fluids to be agitated, if not to be mixed, sometimes constituted objections to its portability, that prevented its being copied by other chronometer makers. The adjustment for the extremes of temperature must have been made by varying the quantity of mercury in a given tube; and probably tubes of various dimensions were tried successively before the exact effect was produced, both at the extremes, and at all the intermediate degrees of temperature. The balance being heavy, was suspended by a thread above the superior end, A and the slender cylindrical part, underneath the same A, rotated in a triangular hole, made by three surrounding friction rollers, not shewn in our figure; lastly, two regulating springs, wound in contrary directions, were placed under the ring, C, near the lower pivot of the verge, which springs are purposely left out, that the reader may not have to view more parts than are necessary for explaining the compensation balance itself.

Compound metallic Balance of P. le Roy.

Peter le Roy, however, was not content with trying his thermometrical tubes, only to effect a compensation on the balance, but he so far adopted Harrison's idea on the requisite construction of a balance, carrying its own compensation in motion, that he tried moreover his compound bars of steel and brass, not, indeed, as curbs, applied to the springs according to Harrison's method, but as rims to the balance itself: *fig.* 3 of *Plate* XXIX. shews a balance of this contrivance, where A is the diametrical bar fixed to the verge, and B B and C C two separate semicircular portions of the compound rim, which were each composed of laminæ of steel and brass respectively rivetted together in different places, in such a way that the brass pieces were the exterior or concave ones, and the steel pieces the interior or convex ones; the loads are not given in the drawing, but the inventor says "Memoire sur la meillure maniere, &c." that his contrivance made a considerable portion of the circumference to approach the centre of the balance in a high temperature; so that either he must have had loads on the ends of the semicircular parts of the rim, or otherwise these parts themselves must have been thick, which would prevent their due obedience to the variations of temperature. *Fig.* 4 is a kind of register contrived to examine the law by which the flexure of a compound bar was guided, as compared with a mercurial thermometer, thus; A is the crosses and rim of a balance of one metal, steel, we will suppose, and the compound piece B, screwed at the end *a*, to the plane of the rim, has its end, *b*, at liberty to obey the changes of temperature; near *b* is a small cock into which a roller, bearing the index C, is pivoted above, against which roller the free end, *b*, of the compound bar, B, acts, whilst the interior end of the index points at the divisions made on the portion of a circle, D, near the centre of the balance: the experiments that were made with this instrument, according to P. le Roy's account, shewed that the rim of the index, so circumstanced, corresponded very well with the rim of a mercurial thermometer, in all the different parts of the scale; and if this was found to be really the case, we cannot help thinking it extraordinary that the contriver of the compound metallic balance, which has stability and portability to recommend it, did not prefer it, in actual practice, to the mercurial balance, which was liable to the objections we have stated.

Arnold's Balances.

Some pains have been taken to prove that Mr. Arnold, senior, did not invent his compensation balances himself, but borrowed them from Peter le Roy; we confess that we have not met with any conclusive argument in favour of such a supposition, but, on the contrary, have asserted under our article CHRONOMETER, that, "a variety of different shapes were given by Arnold to his balances, and actually tried in practice, before he adopted the one in present use, some of which balances are yet in existence;" and we have it now in our power to say further, that many of the balances here alluded to are yet in use, as ourselves have witnessed very lately; which balances have been proved to be excellent regulators, and some of which have public testimonies in their favour. We propose therefore now to describe them in succession, or at least such of them as have fallen in our way.

No.

No. 1. In his preface to his printed certificates, Mr. Arnold, fenior, fays, that his firft attempt to improve clocks and watches commenced in the year 1764; but we find it was not till the year 1767, that he turned his mind ferioufly to the conftruction of chronometers; the firft compenfation balance that he actually brought into ufe, was the one reprefented in *fig.* 5. of *Plate* XXIX. which was taken from one of the original balances while in our poffeffion. The rim is 2.4 inches in diameter, and is, together with the three radial arms, of flender brafs: the compenfation bar is of brafs and fteel coiled round the balance verge, at a little diftance from the mechanifm which lies between it and the plane of the balance; the exterior part of this compound fpiral bar is brafs, and the interior part fteel; *a b* is a piece of fteel moveable round a tube furrounding the balance verge, and having a longitudinal aperture or flit, each way from the centre to nearly each extremity, *a* and *b* refpectively; the end, *a*, is turned up a little, and has the exterior end of the compound fpiral bar of compenfation attached to it, fo as to partake of its motion under all the changes of temperature; *c d* and *c d* are two bracket pieces of metal carrying each a pin under them, that exactly fit the longitudinal flit of the piece, *a b*; confequently thefe bracket pieces alfo partake of the motion produced by the compound fpiral bar; the ends, *c* and *c*, of the bracket pieces are attached to the balance rims at the points, *e* and *e*, by the ftraight flender fprings, *e c*, and *e c*, refpectively; to the ends *d* and *d*, of the faid brackets are attached the long pins, *d f* and *d f*, which pafs through the fmall fteady bridges on the rim, and carry the fectoral loads of temperature, *f* and *f*; the action produced on this mechanifm by the changes of temperature may be thus explained; when the temperature is elevated, the exterior end of the compenfation fpiral bar, moves in a direction from *a* towards *c*, and carries with it the end, *a*, of the crofs-piece, *a b*, at the fame time the oppofite end of this piece moves from *b* towards the *c* which is contiguous to it; thefe oppofite motions of the ends, *a* and *b*, of the crofs piece carry the two brackets in fuch direction that the loads of temperature, *f* and *f*, are both made to approach the balance, fo as to leffen its momentum when in motion; which is the effect required in the above-named temperature; on the contrary, when the temperature is low, the motion of the ends, *a* and *b*, is towards *d* and *d* refpectively, and the loads of temperature are thrown outwards, to increafe the momentum of the moving balance. The quantum of effect is regulated by enlarging or diminifhing the loads, *f*, *f*, of temperature, as compared with the fcrews, *g* and *g*, or weights of adjuftment for mean time; or otherwife, by fcrewing them into a new fituation, out or in, as the cafe may be; the fprings, *e c* and *e c*, do not interfere with the inward and outward motions of the loads of temperature, but only confine the brackets to the plane of their action, and limit the direction of their motion. This compenfation-balance was applied to ten or twelve different chronometers, between the years 1775 and 1778, but we have not been able to afcertain its comparative merit; its principle is fimple, in that there is but one compound bar, to move two loads of temperature, but its mechanifm is complex, which was probably the principal reafon why it came into difufe.

No. 2. The next balance, which Mr. Arnold contrived, was that which is reprefented in *fig.* 6. of *Plate* XXIX, in which the mechanifm is much more fimple than in its predeceffor; the rim and diametrical bar, by which it is attached to the verge-collet, are both fteel, the parallel bars, *a b* and *a b*, are compofed of brafs and fteel refpectively, the brafs being here interior and the fteel exterior, and are inferted into pivot-holes made in the rim of the balance; in fome of the chronometers, the end, *b*, was faft to the balance, and the end, *a*, of the fame bar loofe, while the end, *a*, of the other bar was faft, and the end, *b*, of the fame was loofe, but in other chronometers both the ends of each bar were formed into pivots with fhoulders, and were inferted loofely by bending the bars about the middle; the crofs pieces, *c* and *c*, which carry the loads of temperature, are of fteel, and the rim itfelf is fo notched as to admit of the loads being contiguous, that they may meet with very little refiftance from the air, when in motion; the mode of action is obvious from a view of the figure, and from what has been previoufly faid of a ftraight compound bar, being liable to become convex on the brafs fide by an elevated temperature, and the reverfe in an oppofite temperature; the bars, *a b* and *a b*, becoming convex, with increafed heat, towards the centre of the balance, draw the loads, *d* and *d*, nearer to the rim, than they are when the bars are ftraight at a mean temperature, and, on the contrary, as the heat diminifhes, the fame parallel bars become convex on the fides next the rim, and carry the loads from the balance; and thefe alternations of motion, keeping pace with the changes made in the dimenfions of the balance itfelf and of the balance-fpring, conftitute the compenfation for the effects of temperature: the exact quantum of effect is adjufted as before, by altering the magnitude of the loads, *d* and *d*, as compared with the weights of the adjuftment-fcrews, *e* and *e*, for mean time; or otherwife have their relative pofitions altered in regard to their diftances from the centre. About twenty of thefe balances were made and applied as regulators, all of which performed their office, we are told, very well; the famous gold chronometer, No. 36, tried for 18 months by Dr. Mafkelyne, at the Royal Obfervatory, while worn in the pocket, under the different degrees of temperature, agitations of the body, and changes of pofition, had, and ftill has, one of thefe balances; and its rate, which has been publifhed, eftablifhed its credit fo highly, that it was fold for 1000 *l.* or guineas, notwithftanding the detent of its efcapement, after the fafhion adopted by the French, is fixed to an arbor revolving on pivots. The balances of the prefent conftruction were made from the year 1778 to the year 1780, and the only objection we have heard alleged againft them, was the play they neceffarily had in their pivot-holes, which we think muft have been fcarcely perceptible when the pivots were well made and fitted tight into their holes.

No. 3. The objection to the motion at the pivot holes in the laft balance, (No. 2.) was attempted to be remedied by the conftruction exhibited in *fig.* 7. of the fame plate, where the expanfion pieces are of brafs and fteel, arranged in the form of a long S, which form was afterwards adopted by Emery, the Swifs watch-maker, who fettled in London, and who has been the reputed inventor of this arrangement; the two halves of the S are joined at the middle, and the brafs, as before, is interior throughout; the rim and diametrical bar of fteel, in the balance we examined, which was 1.2 inches only in diameter, by reafon of its having been ufed in a pocket-chronometer; the exterior end, *a*, of the S is attached to one of the fhortened croffes of the balance, and the interior end of the fame bears the long pins, *c* and *c*, that carry the loads of compenfation *d* and *d*; the long pins, *c* and *c*, do not pafs through the rim, but are held to their proper pofitions by the two flender fprings, *e c* and *e c*, as in No. 1, or *fig.* 5. as before, the relative weights of the loads of compenfation, *d*, *d*, and *f*, *f*, the loads of adjuftment for mean time, conftitute the limit of effect to be produced by the expanfion bars, fupſuppoſing their diſtances from the centre fixed; but if the loads remain unaltered, the effect to be produced may be adjuſted by varying the diſtances of the ſcrews reſpectively as they regard the centre of the balance. Of theſe balances about forty were made and applied to uſe between the years 1779 and 1782, and have been found to perform well.

No. 4. The length of the S pieces in the laſt balance having been found difficult to execute of uniform thickneſs and ſhape, by reaſon of their great length, the conſtruction repreſented by *fig.* 8. was ſubſtituted, which is very nearly allied to its predeceſſor, and indeed differs from it only as the bends are united by an interpoſed piece of ſolid metal, thereby dividing what was before one half of the S, into two parallel compound bars; in all other reſpects, the deſcription juſt given of No. 3, together with the letters of reference, will equally apply to our preſent balance; which circumſtance renders a further detail ſuperfluous.

Of theſe balances not more than ſix or eight were conſtructed, by reaſon of the more ſimple form, which is now in general uſe, having been adopted in 1782, when a patent was taken out for it. The ſpecification, however, contains the drawings of Nos. 2 and 3 (*figs.* 6 and 7), as well as of the common balance, which we have fully deſcribed under our article Chronometer, to which the reader is reſpectfully referred, and which was known to the workmen by the name of the Z balance, by way of diſtinction from the S balance. Beſides the above-deſcribed conſtructions of the compenſation balance, Arnold, we are aſſured, contrived various others which never came into uſe, one of which had the expanſion pieces parallel to the verge of the balance, and borne by it; and others had platina in the expanſion pieces, inſtead of ſteel; but we conceive it to be unneceſſary to purſue the ſubject of his inventions further, than merely to obſerve, that ſo regular a ſucceſſion of different balances actually made and uſed, where ſo many were conſtructed, and where each was evidently meant to be an improvement on the laſt preceding, may be taken, if not as a poſitive, at leaſt as a preſumptive proof, that the imputation of plagiariſm, unleſs otherwiſe fairly made out againſt him, is in contradiction to the facts we have here adduced; which are all facts open to ſtill further inveſtigation.

For the deſcription of the compenſation-balance as made by the Brockbanks and Earnſhaw, the reader is deſired to turn to the article Chronometer, where this part of our preſent article has been anticipated. The deſcription of Earnſhaw's balance and eſcapement, however, which we have quoted as the production of this chronometer-maker, it may be proper to ſtate, was not his, but the production of Mr. Firminger, the late aſſiſtant of Dr. Maſkelyne at the Royal Obſervatory, as we have been aſſured by the author himſelf, ſince the publication of the pamphlet in which it appeared oſtenſibly as Mr. Earnſhaw's own account. And now that we are on the ſubject of Mr. Earnſhaw's balance and eſcapement, we beg leave to add further here, an acknowledgment of our miſinformation reſpecting the patent, which we were credibly informed, and ſaid, was never taken out; but which we now find, on more minute examination, was actually taken out and regiſtered. How far his pretenſions to originality of invention are warranted by his conſtruction, is not the object of our preſent enquiry.

A Compenſation balance by F. Berthoud.

Berthoud informs us, "Hiſt. de la Meſure, &c." p. 104 tom. ii., that he had propoſed a means of compenſation by the balance itſelf, ſo long ago as in the year 1754; but that was not until 1787 that he effected his purpoſe. The reader is led to ſuppoſe, that he was not indebted to P. le Roy, or any other maker, for the contrivance, but that it was the reſult of his own inventive genius; and that the contrivance was his own is rendered ſomewhat probable from this conſideration, that in his "Traité des Horologes Marines," two balls only were propoſed to be uſed with two compound bars; but after a calculation of the effect to be produced, contained in page 195 of the Supplement, four balls were found to be neceſſary with as many compenſating bars. The propoſed arrangement is ſhewn in *fig.* 9. of plate XXIX, and is certainly original; A, A, A, and A are four radial bars of metal attached to the verge collet of the balance by two ſcrews, as ſeen in the figure; the ſlender bars, *a*, *a*, *a*, and *a*, attached reſpectively to the exterior ends of the four radial bars by ſcrews and fixing pieces, are four compound bars of ſteel and braſs, the exterior laminæ being of braſs; and each of thoſe compound bars has at its free end a ſtem formed into a ſcrew, upon each of which a ball, B, with a long ſocket, ſcrews towards or from the verge, to adjuſt the momentum to the maintaining power, and regulating ſpring, and alſo to effect the equilibrium, when the machine is brought to time. From what we have repeatedly ſaid of the flexure of a compound bar, it is eaſy to ſee, that the braſs portions being exterior, will cauſe the balls to approach the centre of motion in hot weather, and the reverſe in cold; which effect conſtitutes the compenſation. The author acknowledges that the exact effect to be produced depends on the relative length and thickneſs of each compound bar, and that many trials are neceſſary to come at the due proportion; conſequently, the want of an adjuſtment for temperature, after the balance is tried, is in this conſtruction a fundamental objection to its general adoption; otherwiſe we ſee no reaſon to doubt its accuracy of performance.

This balance was uſed in No. 8. of Berthoud's marine pieces.

Another Compenſation-balance by F. Berthoud.

The compenſation-balance which we have juſt deſcribed being found liable to objection, the ſame author contrived and uſed, in his No. 63, a different arrangement of the compenſation bars, which he again reduced to two, and introduced two additional ſcrews for adjuſtment of mean time, placed at right angles to the former; the contrivance is ſhewn in perſpective in *fig.* 10. of *Plate* XXIX., where all the eſſential parts may be viewed at once: A A is the balance, made light and of the beſt braſs, having four croſſes; B and B are the two compenſation bars fixed to the rim of the balance at one end, and bearing the ſcrews or weights of temperature, C and C, at the other reſpectively; and D and D are the ſcrews or weights of adjuſtment for mean time, borne by two of the oppoſite croſſes, as ſhewn diſtinctly in the figure. The balance as uſual is attached to the verge collet by a couple of ſmall ſcrews; the exterior laminæ of the two compound bars, B and B, are of braſs, and the interior ones of ſteel, ſo that a high temperature brings them towards the centre of motion, as the balance itſelf enlarges, and as its ſpring elongates and becomes weaker. The adjuſtment for temperature may be made in two ways; either by altering the relative ſizes of the ſcrews of temperature and ſcrews of mean time; or otherwiſe, which is more readily done, by ſcrewing one pair in and the other out, ſo as to produce a ſimilar effect without taking either of the pairs entirely out; it is hardly neceſſary to explain that, in all the conſtructions, when the momentum of the balance conſiſts more of the moving weights of temperature, than of the moving weights of mean time, the effect of the compenſation

compenſation bars will be the greater in the ſame proportion, and the contrary; ſtill, however, the quantity of flexure of the compound bars in the preſent balance remains unaltered, and conſequently the velocity of their inward and outward motions is not capable of adjuſtment in any other way, than by altering the ratio of their length to their thickneſs reſpectively, which may be called an alteration rather than an adjuſtment.

Compenſation balance by Joſias Emery.

We have ſaid, under our article CHRONOMETER, that Emery's balance is one of the varieties previouſly conſtructed by the ſenior Arnold, (ſee his N° 3.) and the reader will now judge for himſelf how far our aſſertion was founded in fact. We take our drawing from Berthoud's figure of Emery's balance, which Sarron took over into France, and which was much admired for the accuracy and permanency of its regulation. *Fig.* 11 of *Plate* XXIX, preſents the plane of Emery's balance to the eye placed over it; A A is the balance made with four croſſes, the rim of which is braſs; B B is a ſteel flat ring ſcrewed to two of the croſſes; C and C are the two compenſation bars, each bent twice over into the ſhape of a long narrow S; D and D are the weights for mean time, formed into adjuſting ſcrews to ſuit the female ſcrews of the protuberances made on the plane of the balance; the ſcrews, E and E, are the adjuſtable weights for temperature, but are tapped only at the parts neareſt their heads; the points being faſt to the exterior ends of the compound bars, and the intermediate pins being without threads and moving ſteadily in the two ſockets reſpectively fixed to the rims, or rather formed out of them in the ſolid; the S pieces have the braſs laminæ as uſual on the exterior ſides, ſo that at the middle of the S of each compound bar, the poſition begins to be reverſed, in order that the condition may be fulfilled; the interior end of each compound bar preſſes againſt the circumference of the ſteel ring B B, which we have ſaid is faſt to two of the croſſes; conſequently, whenever a change takes place in the temperature of the parts, the whole motion takes place at the foot of the ſcrew of temperature, at each oppoſite ſide of the balance. The ſmall interior circle F is the upper end of the cylindrical ſpring of the balance, together with the pin that holds one end of it, and the part that forms the banking, which have nothing to do with the compenſation. In this balance, as in Berthoud's latter one, the adjuſtment for temperature muſt neceſſarily be made, by altering the relative momenta of the ſcrews of mean time, and of the ſcrews of temperature.

Mr. W. Hardy's Compenſation-balance.

Among the moſt recent improvements in chronometry, may juſtly be reckoned the new compenſation-balance of Mr. W. Hardy of Clerkenwell, communicated to the Society of Arts in the Adelphi, in March 1805, and renumerated with thirty guineas. The balance in common uſe, that brings the weights of temperature alternately towards and from the verge, by ſimple flexure of the compound rim-pieces, is objectionable, in the opinion of Mr. Hardy and others, on account of theſe conſiderations: 1ſt, The two metals of the compound pieces of the rim, when united by fuſion of the braſs, are liable to partial ſeparations, which affect the permanency of the adjuſtment for temperature; 2dly, The variation of the centrifugal force of the weights of temperature in large and ſmall arcs of vibration overpower the elaſticity of the expanſion pieces, and affect the momentum of the balance, as adjuſted for mean arcs of vibration; the weights being thrown from the centre more in the long arcs than in the ſhort ones; and, 3dly, The adjuſtments for rate and temperature, however carefully made, will affect the equilibrium of the balance, and the rate of the machine placed in different poſitions, without ſubſequent rectifications; to remove theſe objections, Mr. Hardy has availed himſelf of the direct expanſion of two different metals, where their difference of expanſion produces a mechanical action on the loads of temperature, ſuch as bids fair for general introduction in practice.

Fig. 3. of *Plate* XXX. of *Horology*, exhibits a lateral view of the balance before us, and *fig.* 4. the balance inverted; A A is the balance arbor or verge of the ordinary conſtruction; B B is a ſteel diametrical bar, borne by the collet of the verge, 1.6 in. long, 0.232 broad, and 0.032 thick; this ſteel bar is made uniformly except at the two extremities, which are made narrow, and in each of which a notch is cut in the inferior ſurface, ſo far, that the remaining thin metal conſtitutes a ſpring above each notch; beyond theſe ſprings, which we have ſaid form a part of each end of the diametrical bar of the balance, are attached two ſtems at right angles, ſo as to be parallel to the verge when the ſmall ſprings are not acted on by any force; the upper parts of the ſtems, in the original drawing, are formed into ſcrews with threads adapted to the tapped holes of the two balls C, C, which they ſupport, and which are weights of adjuſtment for temperature, but in our more recent drawing, the weights ſlide and are fixed by ſide ſcrews; D, D, are two ſimilar ſcrews for rate which ſcrew reſpectively into the inferior or thicker parts of the ſtems, ſo as to be capable of increaſing or diminiſhing the effective momentum of the balance according to circumſtances; and E, E, are ſcrews of adjuſtment for equilibrium; each end of the ſteel diametrical bar has, moreover, a ſtud or ſhoulder, pointing downwards under the ſtem juſt deſcribed, which ſtud cannot be ſeen in the figure in a ſeparate ſtate; underneath the ſteel bar, and exactly parallel to it and to each other, are placed two bars of braſs, ſhorter but broader than the ſteel bar; the length of each is 1.47 in, the breadth 0.078, and the thickneſs 0.032, like that of the ſteel; each of the two braſs bars has one of its ends ſecured to one of the ſtuds deſcribed, and preſſing with its oppoſite end againſt the other, but in ſuch a way, that their poſitions are reverſed with reſpect to the fixing of the ends; the fixed end of the one being contiguous to the free end of the other reciprocally; a ſecond view is given of theſe bars in *fig.* 4, where they are repreſented in an inverted ſtate. The action of the mechaniſm it is not difficult to underſtand, though the author has rendered his account ſomewhat confuſed by his mention of ſhort levers, which the reader does not ſee in the figure. When the balance is expoſed to heat, the ſteel bar elongates a little both ways from the verge in the centre, but the two braſs bars ſeparately elongate more, and conſequently each detached end, having liberty to move a quantity equal to the difference of the two expanſions, puſhes againſt the ſtud or ſhoulder that is contiguous, and forces it outwards; the ſlender ſpring-parts, above the notches of the ſteel bars, now yield to the forces impreſſed reſpectively on the ſtuds, and the parts of the two ſprings which yield moſt may be conſidered as centres of motion of a pair of levers; the diſtance from the yielding point of each ſpring to the point urged by the detached end of its braſs bar, may be called the length of the ſhort end of each lever, and the diſtance from the ſaid point of each ſpring to the centre of the load carried by the neareſt ſtem, may be taken as the length of the long end of the lever, which two ends ſuppoſe as 1 to 20 reſpectively; then if the difference of of the expanſions of the ſteel and one of the braſs bars be called 1 outwards, the quantity of inward motion in the ſame denomination will be 20; conſequently, the ſame heat which elongates the ſteel bar, and carries the balls from the centre of motion, will, at the ſame time, by the exceſs of expanſibility of the braſs bars, bring the ſame balls inwards towards the ſame centre, and the two contrary actions may be adjuſted, by moving the loads up or down their proper ſtems, till theſe oppoſite effects exactly balance one another; for as thoſe diſtances that have been denominated the ſhort ends of the levers are variable in length, or very nearly ſo, varying the longer ends only, *viz.* the diſtances of the balls from the yielding points of the ſprings, conſidered as centres of motion, will alter the ratios, which may be done in the requiſite proportion, till the compenſation is found in the tentative adjuſtments to be perfect. When once the exact compenſation is effected, it is obvious, that the law by which it is preſerved is uniform and conſtant, as that according to which the grid-iron pendulum is conſtructed, which experience has proved to be permanently uſeful in practice. When the balance has its momentum properly adjuſted for rate, with a given maintaining power, eſcapement, and balance-ſpring, and when the balance is put into an exact ſtate of equilibrium, the ſtems being at all times either exactly parallel to the verge, or, if inclined a little by the extremes of temperature, alike inclined, it is ſelf-evident, that, under theſe reſtrictions, the adjuſtment of the loads up or down for temperature will not deſtroy the adjuſtments of rate, and of equilibrium for the poſitions, previouſly made, provided the two loads be of equal weight, and placed in ſimilar poſitions on their reſpective ſtems; and even if one happen to be left a little higher on the ſtem than the other, no ſenſible bad effect will be produced thereby, which circumſtance is one of the greateſt recommendations of the preſent balance. The effects produced by variations of temperature, are ſaid to be produced not by jerks, but regularly, notwithſtanding the united ſtructure of the parts is ſufficiently firm; and, laſtly, the conſtruction is ſo ſimple, that a common workman may ſucceed in making and putting together the conſtituent parts of this balance without difficulty.

Compenſation on the Balance and alſo at the Spring.

Double Compenſation by Berthoud.—F. Berthoud having experienced much inconvenience, and loſt much time in making the adjuſtments of his balances for rate, temperature, and poſition, at length determined to avail himſelf of the curb, as a ſupplemental aid, in addition to the compenſation carried by the balance itſelf; the union of the two modes of effecting the due compenſation, is repreſented in *fig.* 5 of *Plate* XXX. of *Horology*, in ſuch a way that all the acting parts may be ſeen and underſtood: A is the balance-rim of braſs, and bears four croſſes, beſides two projecting bars, B, B; theſe bars carry each a compound or compenſation bar, C, of the uſual conſtruction, the free ends of which are perforated and attached to the ſcrews or loads of temperature, D and D, or, more properly ſpeaking, to the ſliding ſockets on which thoſe loads ſcrew in and out, in making the adjuſtment; E and E are the ſcrews or loads of adjuſtment for mean time: FF is the cock, and G a ſmall cock, bearing a jewel mounted on it, to take the pivot of the balance. The action of the compenſation bars, C and C, needs no explanation, after what we have ſaid about the effect of ſuch pieces borne by other balances: I, is the compenſation bar of the curb for the ſpiral ſpring, ſhewn in the figure, and acts on the bent ſteel bar, K, which holds the curbing pins, and which has its centre of motion under the cock, at a ſhort diſtance from the balance-verge; the quantity of action of I on K is regulated by the holding piece, *a*, ſliding in a dove-tailed groove, between the parallel bars, *b* and *b*, ſcrewed to the piece N, which piece, N, is moveable round a pin under it, near the part, *a*, in ſuch a way as to admit of a little circular motion, together with the compound bar, I, to regulate for mean time without altering the length of I; and a ſlow motion may be given to the piece, N, by the graduated piece, M, which is moveable on the pivots of its arbor, in the plate and ſmall cock, L, while the tooth, or rounded part, lies in a concave part at the end of N, and impels it almoſt imperceptibly as the end, M, moves; the graduations of the piece, M, are ſeen through a hole made in the interior caſe of the watch; and the adjuſtment for rate may be made there without trouble, and without deranging the previous adjuſtments of the balance itſelf; but then, if there is an iſochronal point in the acting length of the ſpiral ſpring, the adjuſtment made by means of the curb will derange its iſochroniſm, which will become ſenſible as the arc of vibration varies in uſe; otherwiſe the adjuſtment for temperature might be completed by varying the length of the compound-bar, I, after this adjuſtment has been nearly effected on the balance itſelf, and after the balance has been brought into a ſtate of equilibrium for the poſitions; but whether the effect produced by the joint agency of two cauſes, will be uniform in all degrees of temperature, has not been ſaid by the author.

Compenſation for the Spring's want of Iſochroniſm.

That property of a balance's regulating ſpring which makes the balance perform all its vibrations, long, ſhort, and intermediate, preciſely in the ſame time, is called its *iſochronal* or *iſochronous* property, from the two Greek words *ισος*, *equal*, and *χρονος*, *time*: it has been aſſerted that there is a certain proportion between the length and thickneſs, or between the length and the breadth and thickneſs taken conjointly, which only aſcertains the iſochronal point in any uniform ſpring, of whatever ſhape; this doctrine has been maintained by Dr. Hooke and Peter le Roy ſeparately, and has been ſubſcribed to by Arnold the ſenior, Berthoud, and others, though Pennington of Camberwell ſays there are different points in the ſame ſpring equally iſochronal; now allowing that there is only one point in the length of every ſpring that determines its iſochroniſm, and that conſequently all other lengths of the uniform ſpring regulate the balance uniformly, only while the arcs of vibration remain unchanged, to determine the iſochronal effective length of ſuch ſpring would be to render all compenſations, for want of iſochroniſm, unneceſſary; and this point may in moſt ſprings be aſcertained by repeated trials with long and ſhort arcs ſucceſſively, and by marking down the reſpective reſults; for it is found from experiments, that if ſhortening the regulating ſpring makes the long arcs of vibration to be performed in leſs time than the ſhort ones, and *vice verſâ*, the ſpring under trial is capable of having its total length reduced gradually till the iſochronal length is determined, but, generally ſpeaking, not otherwiſe. Another method of managing the ſpring, which may be ſaid to be effecting a ſpecies of compenſation for want of iſochroniſm, in the total length of an uniform ſpring, is tapering it from the outer to the inner end, till, with a given length, the long and ſhort arcs are performed in equal times; this method was ſtrongly recommended by Cumming in his Eſſay towards the improvement of watch work, and after him, Earnſhaw has laid much ſtreſs on this particular, ſo much indeed, that he conſiders his ſkill in the adjuſtment of of the iſochroniſm, by a tapering ſpring, as conſtituting one of his chief claims to public encouragement; nay, he profeſſes to manage the buſineſs ſo nicely, that he can make ſuch a ſlight deviation from perfect iſochroniſm as will compenſate the *laſſitude* alſo, in the intenſity of his ſprings in uſe!

The firſt notice that we have of a regular compenſation, for want of perfect iſochroniſm in a regulating ſpring, is Harriſon's, when he propoſed to introduce the cycloid-pin which we have already confeſſed we do not underſtand, from the want of ſhading in the drawing of it, as attached to his compenſation curb, in the publication by the Board of Longitude; we may, however, ſtate, that Harriſon, not knowing, probably, that there is naturally an iſochronal length in moſt uniform ſprings, concluded, that the long arcs are naturally performed in leſs time than the ſhort ones, reaſoning from the circumſtance of one of his pieces having been obſerved to go faſter in a horizontal poſition, than in a vertical one, though the arc of vibration was ſhorter in the latter than in the former poſition; he therefore applied his cycloid-pin, as Mudge did after him, as may be ſeen in the drawing of his eſcapement, in ſuch a way that the ſhort vibrations were accelerated by its puſh more than the long ones, that were not ſo long acted on by auxiliary means; but the fact is, the uſe of ſuch a contrivance would neceſſarily have an effect juſt oppoſite to the intended one, if the balance-ſpring happened to be pinned to the ſtud at the contrary ſide of the iſochronal point; that is, if the long vibrations were ſlower than the ſhort ones, the cycloidal contrivance would make the difference ſtill more ſenſible.

Berthoud's Compenſation for want of Iſochroniſm.

Berthoud has given an ingenious method of compenſating the want of iſochroniſm in a balance-ſpring in his "Supplement au Traité des Horloges Marines," which contrivance he calls an *iſochronal compenſator.* (Compenſateur Iſochrone.) The contrivance is repreſented in *fig.* 6. of *Plate* XXX, where A repreſents the balance with three radii; B an eccentric collet of poliſhed ſteel, placed on the balance-verge, and preſſing againſt the large roller C, which is carried by and pivoted into a ſmall frame at the end of the ſlender ſpring D, attached to the cock, E, on the plate of the watch frame, whenever the long vibrations are found by trial to be performed in a longer time than the ſhort ones, the poſition of B and C are reſpectively as in the drawing, at the time when the balance-ſpring is quieſcent; for then the preſſure of the eccentric piece, B, againſt the roller urged by the ſpring, is greater in the ſmaller arcs than in the long ones; but when the long arcs are performed in the leaſt time, the poſition of B muſt be at *b*, and then C will be found by the force of its ſpring at *c*, where it is evident, that the preſſure will be leſs in the ſhort arcs than in the long ones comparatively; and the quantity of eccentricity of B muſt depend on the greateſt difference of the two extreme vibrations. The principal reaſon why Berthoud adopted this iſochronal compenſator in five clocks, and in his portable watch for the longitude, was, that whenever the cock was diſmounted the iſochroniſm of his ſpring, however well adjuſted, was generally found to be deranged when re-mounted. The eccentric collet was of courſe capable of being put to any given eccentricity as well as into any given poſition for accelerating long or ſhort arcs at pleaſure. The Engliſh watch-makers have not, that we know of, ever uſed this contrivance, but depend on their own patience and ſkill, in aſcertaining the beſt length and moſt deſirable ſhape of their regulating ſprings for effecting their own purpoſe.

Mr. W. Hardy's Compenſation for want of Iſochroniſm.

Mr. W. Hardy publiſhed a ſhort account of a new method of making the balance of a chronometer perform its long and ſhort vibration, in the ſame time, in Mr. Nicholſon's Journal for September, of the year 1806, (Supplement of vol. xiv. 8vo. Series) which account, we propoſe, to copy with ſome ſlight verbal deviations, that are more adapted to our figures, as drawn in perſpective. The ſpring *a b*, *fig.* 8 of *Plate* XXX, is ſcrewed to the inferior ſide of the cock A, *fig.* 7. and lies over the upper part of the regulating-ſpring, proceeding in a right line to the axis of the balance B, and having a bend to clear the verge, from which it paſſes on to the end of the ſpring which it holds. This ſtraight ſpring is made of ſuch a ſmall ſtrength, that it comes into action before the regulating ſpring. The other piece *c*, *fig.* 8, is attached to the cock, in a line with the ſlender one, but on the oppoſite face of the cock; the part of the piece, *c*, which is bent down, is divided into a fork and admits the part, *b*, of the ſlender ſpring, to paſs and play between its prongs; at each ſide of the prongs is a lateral ſcrew, preſenting its point to the ſpring, in ſuch a way, that the play between the prongs, or rather between the points of the two equidiſtant ſcrews, may be limited according to circumſtances, provided they are not farther aſunder than the quantity of the eſcapement angle. The other end of the cylindrical regulating ſpring, as uſual, is attached to a piece of metal carried by the verge of the balance. The effect produced by the compenſating ſpring, *a b*, is thus; when the balance is firſt moved from its place of reſt, the ſlender ſpring, *a b*, moves firſt, till it meets with the point of one of the two ſcrews of the fork, then the cylindrical-ſpring begins to be acted on, as in the chronometers in general; this motion of the ſtud, at the end of the ſlender ſpring, retards the motion of the balance, which begins to be accelerated only, when the ſtud is at reſt; conſequently the contrivance before us ſuppoſes that the ſhort arcs are in general performed in leſs time than the long ones; and require on that account, to be checked in their velocity at ſtarting; and the quantity of play in the fork limits the quantum of retardation; but as we have ſeen the long arcs may be, and often are, performed in leſs time than the ſhort ones, with ſprings not adjuſted for iſochroniſm; we conceive, therefore, that, like Harriſon's cycloid pin, which owed its origin to a contrary ſuppoſition, the application of the invention we are deſcribing may happen to be productive of an effect, juſt oppoſite to that which it is meant to produce: for it can be uſed as a compenſation for want of iſochroniſm, only in the particular caſe, when the ſhort arcs of vibration are performed in leſs time than the long ones, which we have ſeen is juſt the oppoſite application of that adopted by Harriſon, and founded, as we have ſaid, on an oppoſite ſuppoſition; whereas the fact is, that ſometimes one of the two contrivances may prove to be neceſſary, and ſometimes the other, with a limited length of ſpring, accordingly as the long, or the ſhort arcs of vibration, may turn out on trial to be performed in the leſs time; and the trial neceſſary for aſcertaining, whether the ſpring is too long or too ſhort, will afford the means of aſcertaining very nearly, if not quite, the exact iſochronal length, ſuch as ſhall require no compenſation at all; to attain which object, though with ſome additional trouble, we are diſpoſed to think the preferable practice.

COMPENSATION CURBS.

Fig. 1.

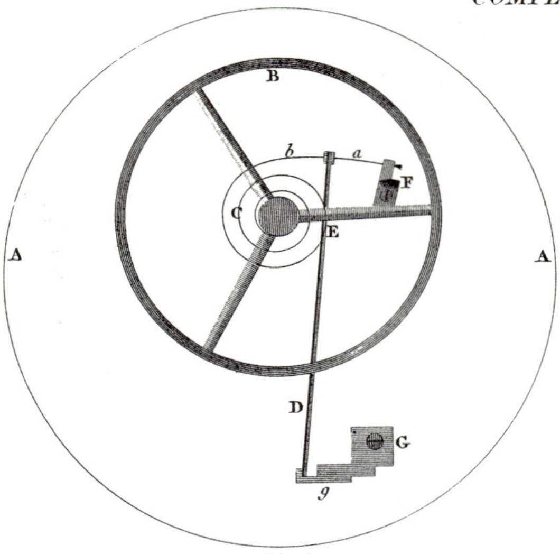

Fig. 3.

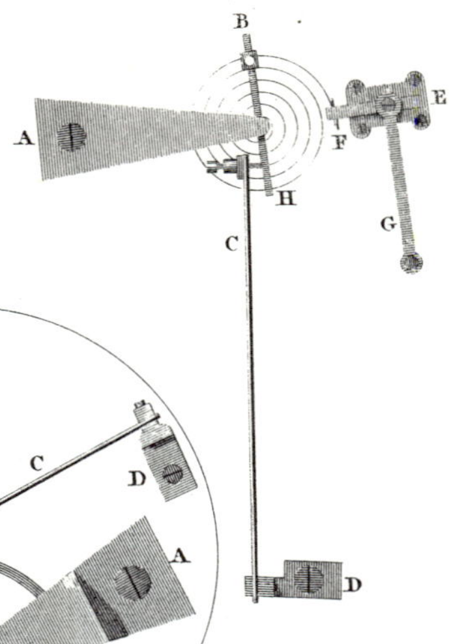

Fig. 4.

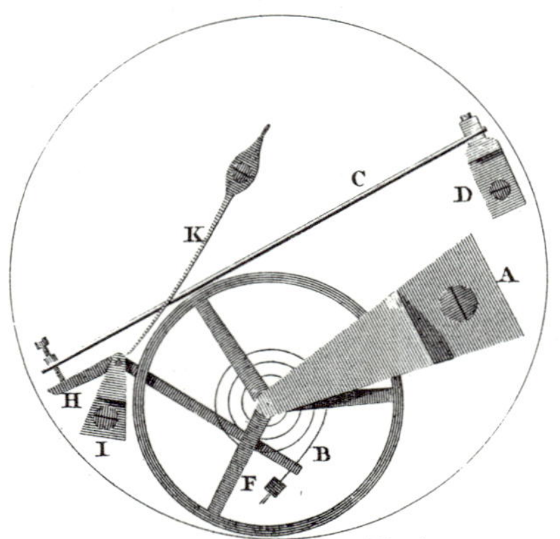

Fig. 2.

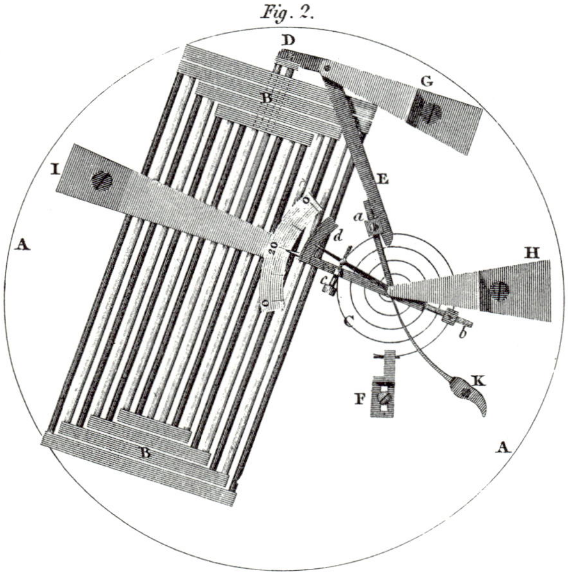

Fig. 5.

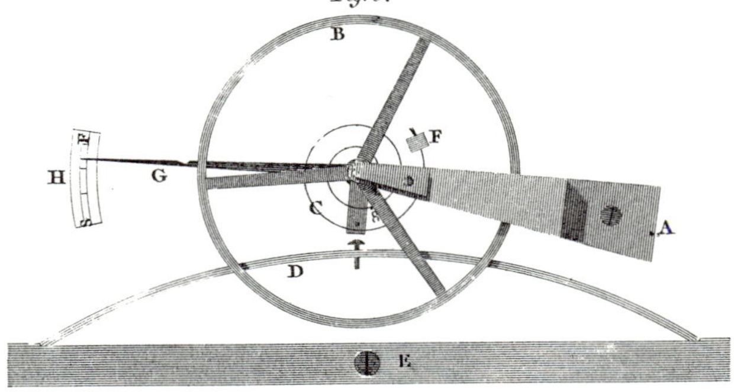

Fig. 7.

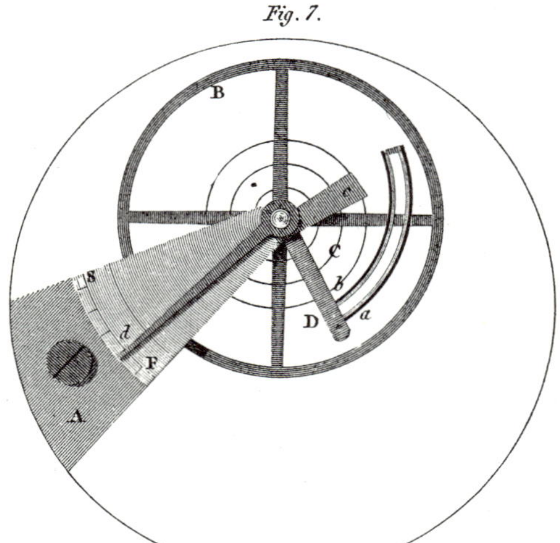

Fig. 6.

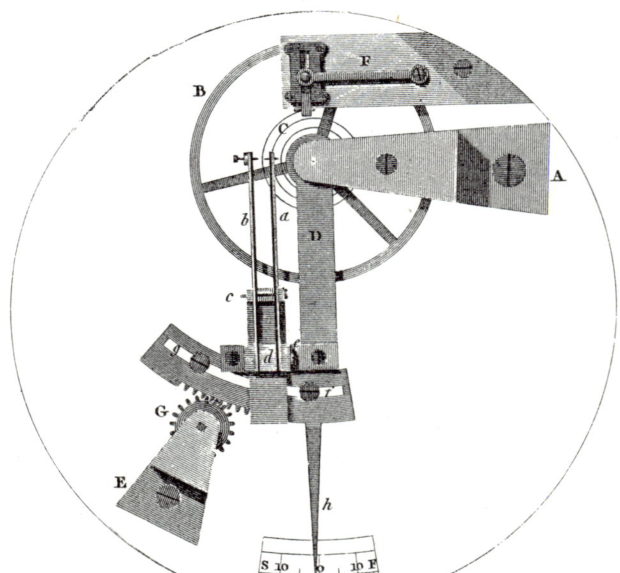

Published as the Act directs, 1809, by Longman, Hurst, Rees & Orme, Paternoster Row. Engraved by Wilson L

COMPENSATION BALANCES.

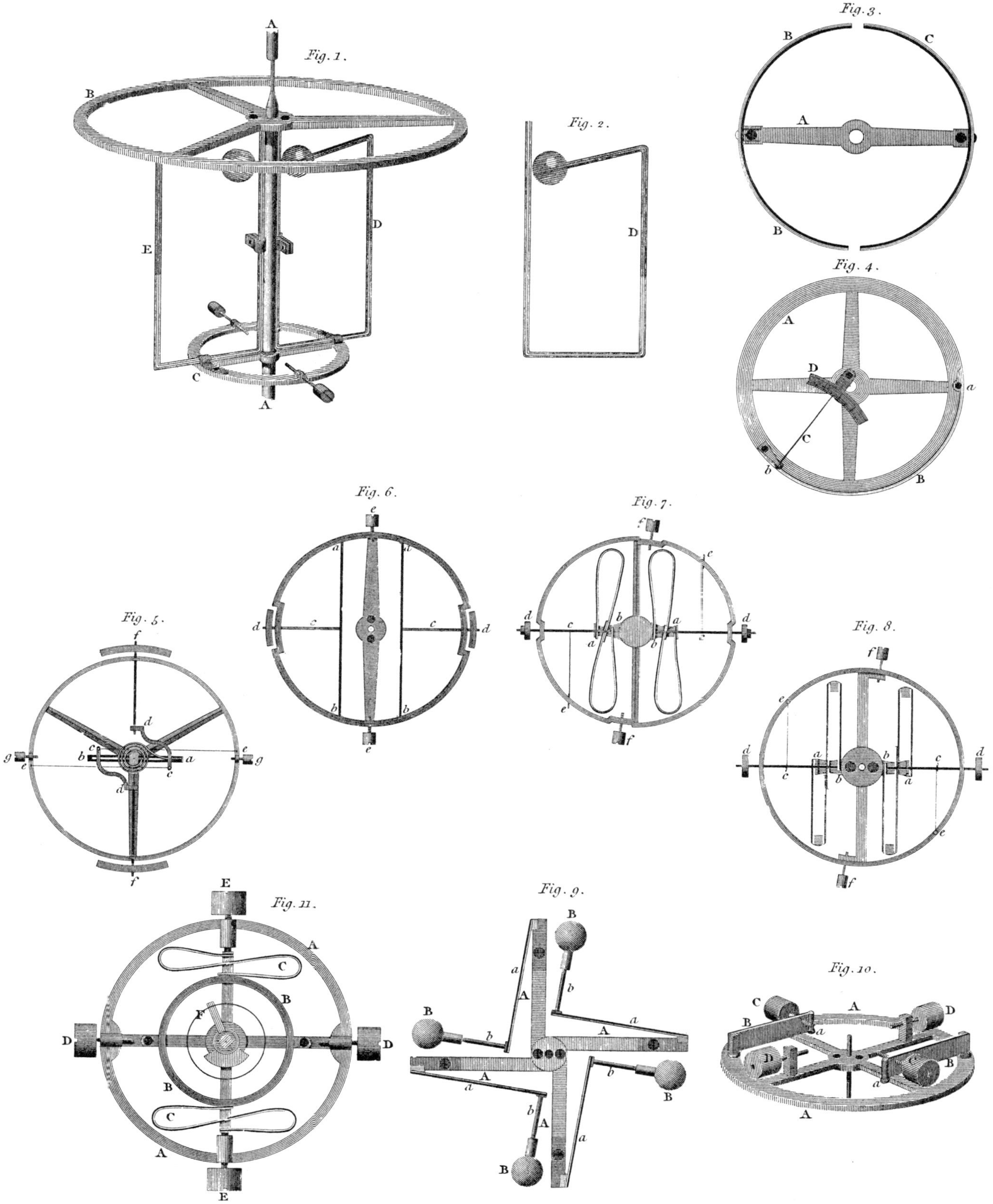

Published as the Act directs, 1811, by Longman, Hurst, Rees, Orme & Co Peternoster Row.

Engraved by W. Lowry.

HOROLOGY.

COMPENSATION CURBS AND BALANCES.

Mr. Scott's,

Fig. 2.

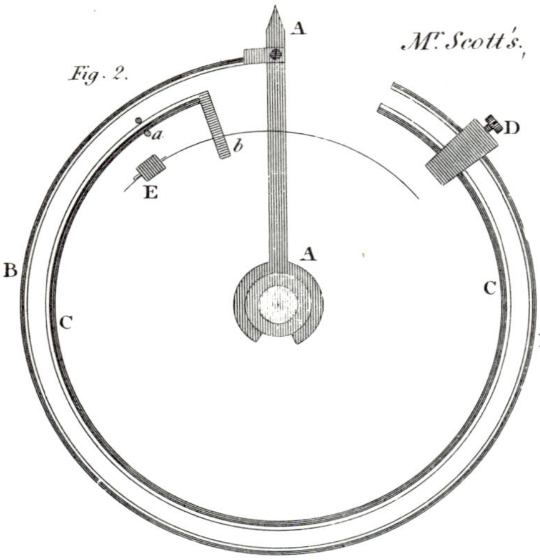

Mr. Cumming's.

Fig. 1.

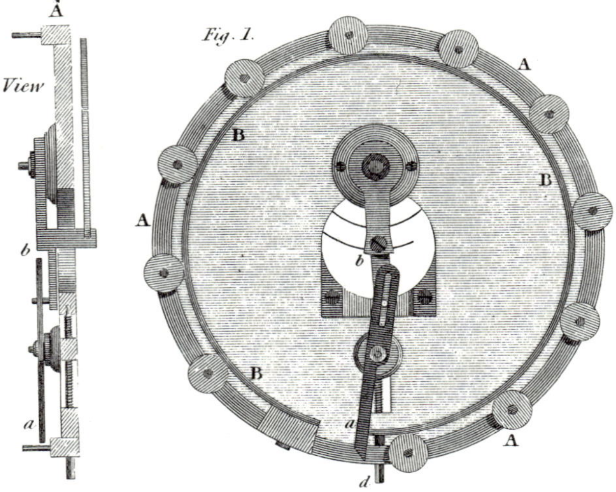

Fig. 5.

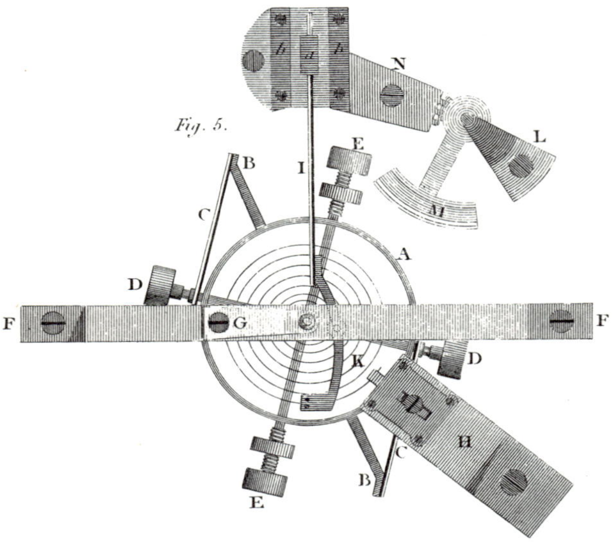

Mr. Hardy's balance.

Fig. 3.

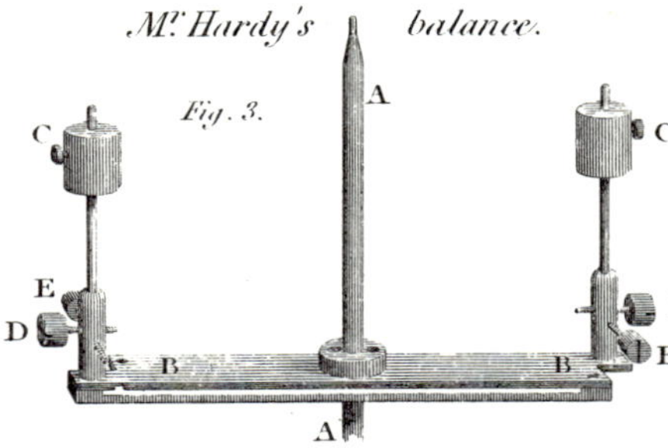

Fig. 4.

Berthoud's.

Fig. 6.

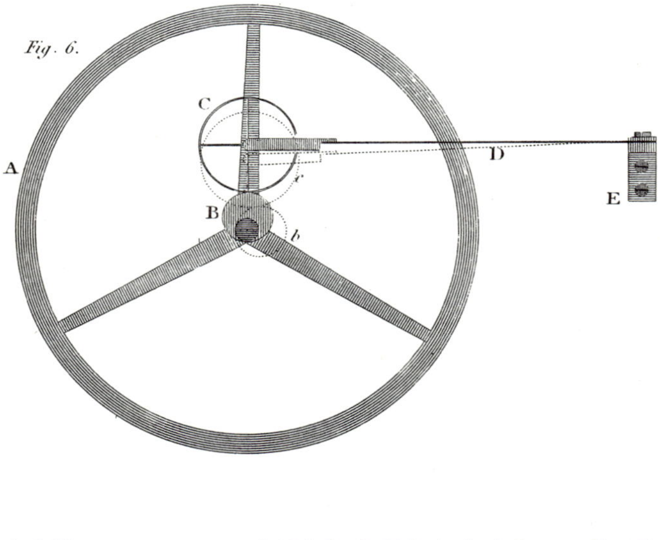

Mr. Hardy's, Isochronal compensation.

Fig. 7.

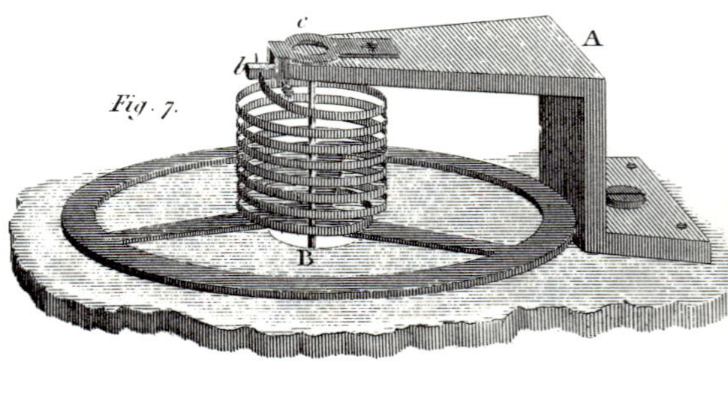

Fig. 8.

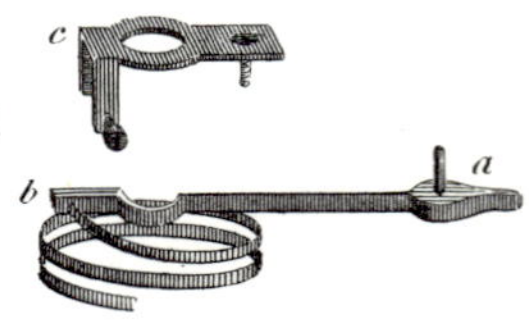

J. Farey Junr. deliv. Published as the Act directs, 1817, by Longman, Hurst, Rees, Orme & Brown, Paternoster Row. Lowry Sc.

Dial and Dialling

DIAL, or Sun-Dial, an instrument serving to measure time by means of the shadow of the sun.

The word is formed from *dies*, day, because indicating the hour of the day. See Hour.

The ancients also call it sciathericum, from its doing it by the shadow.

Dial is more accurately defined a draught, or description, of certain lines on a plane, or surface of a body given, so contrived as that the shadow of a style, or ray of the sun passed through a hole therein, shall touch certain points at certain hours.

The antiquity of dials is beyond doubt. Some attribute their invention to Anaximander, or his disciple Anaximenes Milesius, who flourished in the sixth century, B. C.: and others to Thales. Vitruvius mentions one made by the ancient Chaldee historian Berosus, on a reclining plane, almost parallel to the equinoctial. Aristarchus Samius invented the hemispherical dial. And there were at the same time some spherical ones, with a needle for a gnomon. The discus of Aristarchus was an horizontal dial, with its limb raised up all around to prevent the shadow stretching too far.

But it was late before the Romans became acquainted with dials. The first sun-dial at Rome was set up by Papirius Cursor, the Roman general, about the year of the city 461, B. C. 293; before which time, says Pliny, there is no mention of any account of time but by the sun's rising and setting: this first dial was set up at or near the temple of Quirinus, but it was imperfect; about thirty years after, as Pliny informs us, on the authority of Varro, M. Valerius Messala being consul, brought from Catania in Sicily another dial, which he set up on a pillar near the rostrum; but for want of its being made for that latitude, it could not point out the hours with exactness. They, however, made use of it ninety-nine years; till the censor Marcius Philippus set up another more accurate, and constructed for the latitude of Rome, near that of Valerius. In that interval dials became common at Rome, as appears from a fragment of Plautus, preserved by Aulus Gellius. It is an hungry parasite that speaks.

"Ut illum dii perdant, qui primus *horas* reperit,
Quique adeo primus hic statuit solarium,
Qui mihi comminuit misero articulatim *diem!*
Nam me puero uterus hic erat solarium,
Multò omnium istorum optumum et verissimum,
Ubi iste monebat esse, nisi cum nihil erat.
Nunc etiam quod est, non est, nisi soli lubet.
Itaque adeo jam oppletum est oppidum solariis,
Major pars populi aridi reptant fame."

However, as a sun dial did not serve in cloudy weather, Scipio Nasica, five years after, remedied this defect, by introducing a method of dividing the night as well as the day into hours, by means of a water-machine, or *clepsydra* (which see); called by Pliny an *horologium*.

But there seem to have been dials among the Jews much earlier than any of these. Witness the dial of Ahaz, who began to reign four hundred years before Alexander, and within twelve years of the building of Rome, 741 years B. C.: mentioned by Isaiah, chap. xxxviii. ver. 8.

The ancient Bramins seem to have been acquainted with this instrument, and to have applied it to astronomical purposes. The Chinese and Egyptians were also acquainted with it at a very early period. Learned men have generally ascribed the invention of it to the Babylonians, from whom the Jews derived it before the time of Ahaz. Herodotus traces the knowledge of the art of dialling among the Greeks to the same source.

The diversity of sun dials arises from the different situation of the planes, and the different figure of the surfaces whereon they are described; whence they become denominated equinoctial, horizontal, vertical, polar, direct, erect, declining, inclining, reclining, cylindrical, &c. For the general principles of their construction, see Dialling. See also *Gnomonic* Projection.

Dials are sometimes distinguished into primary and secondary.

Dials, *primary*, are those either drawn on the plane of the horizon, called horizontal dials; or, perpendicular thereto, on the planes either of the meridian, or prime vertical, called vertical dials: to which number are also usually added, those drawn on the polar and equinoctial planes, though neither horizontal, nor vertical.

Dial, *equinoctial*, is that described on an equinoctial plane, or a plane representing that of the equinoctial.

A plane oblique to the horizon, either hangs over towards it, and makes an acute angle with the plane of the horizon, or it falls off backwards from it, and makes an obtuse angle therewith. This latter is called a reclining plane; which, if it recline back at an angle equal to the complement of the latitude of the place, then it lies in the plane of the equinoctial; and a dial drawn thereon, is denominated an equinoctial dial.

Equinoctial dials are usually distinguished into upper, which look towards the zenith; and lower, which respect the nadir.

Now, as the sun only illumines the upper surface of an equinoctial plane, while he is in our hemisphere, or on the northern side of the equator, an upper equinoctial dial will only shew the hour, during the spring and summer season.

And again, as the sun only illumines the lower surface of an equinoctial plane, while he is in the southern hemisphere,

or

or on the other fide of the equator; a lower equinoctial dial will only fhew the hour in autumn and winter.

To have an equinoctial dial, therefore, that fhall ferve all the year round, the upper and lower muft be joined together; that is, it muft be drawn on each fide of the plane.

And, fince the fun fhines on one fide or on the other of an equinoctial plane the whole day, fuch a dial will fhew all the hours of an artificial day.

To defcribe an equinoctial Dial *geometrically.*—The equinoctial is the firft, eafieft, and moft natural, of all dials; but the neceffity of drawing it double, prevents its being much in ufe. However, as its ftructure fhews the reafon of the other kinds, and as it even furnifhes a good mechanical method of drawing all the other kinds of dials, it fhall be here laid down.

Firft, then, to defcribe an upper equinoctial dial: from a centre C (*Plate* I. *Dialling fig.* 4.) defcribe a circle A B D E, and by two diameters A D, and B E, interfecting each other at right angles, divide it into quadrants A B, B D, D E, and E A. Subdivide each quadrant into fix equal parts, by the right-lines C I, C II, C III, &c. which lines will be hour lines. Through the centre C drive a ftyle, or pin, perpendicular to the plane A B D E.

The dial thus defcribed being raifed fo as to be in the plane of the equator, the line C XII, in the plane of the meridian, and the point A looking towards the fouth, the fhadow of the ftyle will fhew the hours both of the forenoon and afternoon.

For horary circles include arcs of the equator of fifteen degrees each; confequently, the plane A B D E being fuppofed in the plane of the equator, the horary circles will likewife include arcs of 15 degrees of the circle A B D E. Wherefore, fince the angles XII C XI, XI C X, X C IX, &c. are each here fuppofed 15 degrees, the lines C XII, C XI, C X, C IX, &c. are interfections of horary circles with the plane of the equinoctial.

Again, fince the ftyle paffing through the centre C reprefents the axis of the world; its diftance from the centre of the earth being inconfiderable, and it being the common diameter of the horary circles, its fhadow will cover the hour-line C XII, when the fun is in the meridian, or circle of 12 o'clock; C XI, when in the circle of 11 o'clock; C X, when in the circle of 10 o'clock, &c.

Secondly, to defcribe a lower equinoctial dial: the method is the fame as that for the upper dial, already defcribed; except that no hour-lines are to be drawn beyond that of fix o'clock.

Thirdly, to defcribe an univerfal equinoctial dial: join two metal, or ivory planes, A B C D, and C D E F, (*fig.* 5.) fo as to be moveable at the joint. On the upper furface of the plane A B C D, defcribe an upper equinoctial dial, and upon the lower a lower, as already directed; and through the centre I drive a ftyle. In the plane D E F C cut a box, and put a magnetic needle G therein; fit on the fame plane a brafs quadrant nicely graduated, and paffing through a hole cut in the plane A B C D. Now, fince this may be fo placed, by means of the needle, as that the line I 12 fhall be in the plane of the meridian; and, by means of the quadrant, may be fo raifed, as that the angle B C F fhall be equal to the elevation of the equator; it will ferve as a dial in any part of the world. When the fun is in the equator, thefe dials can be of no ufe.

The beft univerfal dial is that of Mr. Pardie, of which the following defcription is given by Mr. Fergufon. This dial confifts of three principal parts; the firft is called the horizontal plane A (*fig.* 6.) becaufe in practice it muft be parallel to the horizon; in this plane is fixed an upright pin, which enters into the edge of the fecond part B D, called the meridional plane, which is made of two pieces, the loweft whereof, B, is called the quadrant, becaufe it contains a quarter of a circle divided into 90 degrees; in which part, near B, the pin enters. The other piece is a femicircle D, adjufted to the quadrant, and turning in it by a groove, for raifing or depreffing the diameter E F of the femicircle, which diameter is called the axis of the inftrument. The third piece is a circle G, divided on both fides into 24 equal parts, which are the hours. This circle is put on the meridional plane, fo that the axis E F may be perpendicular to the circle; and the point C be the common centre of the circle, femi-circle, and quadrant. The ftraight edge of the femicircle is chamfered on both fides to a fharp edge, which paffes through the centre of the circle. On one fide of the chamfered part, the firft fix months of the year are laid down, according to the fun's declination for their refpective days, and on the other fide the laft fix months. And againft the days on which the fun enters the figns, there are ftraight lines drawn upon the femicircle, with the characters of the figns marked upon them. There is a black line drawn along the middle of the upright edge of the quadrant, over which hangs a thread H, with its plummet I, for levelling the inftrument. From the 23d of September to the 20th of March, the upper furface of the circle muft touch both the centre C of the femicircle, and the line of ♈ and ♎; and from the 20th of March to the 23d of September, the lower furface of the circle muft touch that centre and line. In order to find the time of the day by this dial, it muft be fet on a level place in fun-fhine, and adjufted by the levelling fcrews *k* and *l*, till the plumb-line hangs over the black line upon the edge of the quadrant, and parallel to the faid edge; then move the femicircle in the quadrant, till the line of ♈ and ♎ (where the circle touches) comes to the latitude of the place in the quadrant; and turn the whole meridional plane B D, with its circle G, upon the horizontal plane A, till the edge of the fhadow of the circle falls precifely on the day of the month in the femicircle; and then the meridional plane will be due north and fouth, the axis E F will be parallel to the axis of the world, and will caft a fhadow upon the true time of the day, among the hours in the circle, which is now in the plane of the equinoctial. The upper furface of the circle will ferve from the 20th of March to the 23d of September; and the lower for the reft of the year.

The months and figns are laid down on the femicircle in the following method. Draw the right line A C B (*fig.* 7.) equal to the diameter of the femicircle A D B, and crofs it in the middle at right angles with a line E C D, equal in length to A D B; then E C will be the radius of the circle F C G, which is the fame as that of the femicircle. Upon E, as a centre, defcribe the circle F C G, on which fet off the arcs C *h* and C *i*, each equal to 23½ degrees, and graduate them accordingly, for the fun's declination. Then, laying the edge of a ruler over the centre E, and alfo over the fun's declination for every fifth day of each month, as found in tables of declination, mark the points on the diameter A B of the femicircle from *a* to *g*, which are cut by the ruler; and here place the days of the months, anfwering to the fun's declination. Set one foot of the compaffes in C, and extending the other to *a* or *g*, defcribe the femicircle *a b c d e f g*; divide this into fix equal parts, and through the points of divifion draw right lines parallel to C D, for the beginning of the fines (of which one-half are on one fide of the femicircle, and the other half on the other) and ſet the characters of the ſigns to their proper lines. Ferguſon's Lectures, &c. Lect. 10.

DIAL, *Horizontal*, is that deſcribed on a horizontal plane, or a plane parallel to the horizon.

Since the ſun may illuminate an horizontal plane at all times of the year, while he is above the horizon; an horizontal dial may ſhew all the hours of the artificial day throughout the year: ſo that a more perfect dial than this kind cannot be required.

To deſcribe an horizontal DIAL *geometrically*.—Draw a meridian line A B (*fig.* 8.) on the given immoveable plane; or aſſume it at pleaſure, on a moveable one. See MERIDIAN *line*.

From a point taken at pleaſure, as C, erect a perpendicular C D, and make the angle C A D equal to the elevation of the pole. In D make another angle C D E, equal likewiſe to the elevation of the pole, and draw the right line D E, meeting A B in E. Then make E B equal to E D, and from the centre B, with the radius E B, deſcribe a quadrant, E B F, which divide into ſix equal parts. Through E draw the right line G H, cutting A B at right angles. From the centre B, through the ſeveral ſubdiviſions of the quadrant E F, draw right lines B *a*, B *b*, B *c*, B *d*, B H, meeting the line G H, in the points *a*, *b*, *c*, *d*, H. From E, upon the right line E G, ſet off the intervals E *a*, E *b*, &c. *viz.* E *a*, from E to *e*, E *b*, from E to *f*, E *c* from E to *g*, &c. from the centre A deſcribe a little circle, and applying a little ruler to A, and the little points of diviſion *a*. *b*, *c*, *d*, H, and *e*, *f*, *g*, *h*, G, draw the lines A XI, A X, A IX, A VIII, A VII, and A I, A II, A III, A IV, A V. Through A draw a right line VI, VI, perpendicular to A B. Continue the right line A VII, beyond the little circle to VII, A VIII to VIII, A V to V, and A IV to IV. Round the whole ſcheme draw a ſquare, circle, or oval figure. And laſtly, in A fix an index, making an angle D A C, with the meridian A B, equal to the elevation of the pole; or in C erect a perpendicular ſtyle equal to C D; or at A E fix a triangular plate A D E perpendicular to the plane of the dial. In this caſe, the lines, A XI, A X, A IX, &c. are the hour-lines of the forenoon; and A I, A II, A III, &c. thoſe of the afternoon; and the ſhadows of any of the gnomons, or ſtyles abovementioned, at the ſeveral hours, will fall on the reſpective hour-lines.

To deſcribe an horizontal DIAL *trigonometrically*.—In large dials, where the utmoſt accuracy is required, geometrical lines are beſt ſet aſide; and in lieu thereof, the lines of the dial are to be determined by trigonometrical calculation. M. Clapies, in the Memoires de l'Academie Royal des Sciences, anno 1707, has done the world good ſervice in this reſpect; having rendered the calculation of the hour-lines, which before had been operoſe enough, exceedingly eaſy and expeditious: his canons, or analogies, we ſhall lay down under the reſpective kinds of dials. And, firſt, for an horizontal dial; the elevation of the pole of the place being given to find the angles which the hour-lines make with the meridian in the centre of the dial.

The analogy, or canon, is this: as the whole ſine or radius is to the ſine of the elevation of the pole of the place, ſo is the tangent of the ſun's diſtance from the meridian, for the hour required, to the tangent of the angle required: that is, as the ſide A C (*fig.* 9.) is to D C, ſo is the tangent of F D C to the tangent F C, of the angle F A C. See TANGENT, &c.

To deſcribe an horizontal DIAL *by the terreſtrial globe*.—Let a common globe of twelve inches diameter, upon which are uſually drawn twenty-four meridian ſemi-circles, be elevated to the latitude of any given place, and turned about until any one of theſe meridians cuts the horizon in the north point, where the hour of XII is ſuppoſed to be marked; the reſt of the meridians will cut the horizon at the reſpective diſtances of all the other hours from XII. Let theſe points of diſtance be marked on the horizon, the globe be removed, and a flat board or plate, even with the ſurface of the horizon, put into its place, then ſtraight lines drawn from the centre of the board to theſe points will be the hour-lines of an horizontal dial for that latitude; and the gnomon will be in the ſituation of the axis of the globe, or make an angle with the plane of the dial equal to the latitude of the place. If the globe be elevated to the complement of the latitude of the given place, and you proceed as above, you will have a direct ſouth dial: obſerving, that the gnomon makes an angle with the dial plane equal to the co-latitude of the place, and that the hours on the dial muſt be ſet the contrary way to that on the horizontal. If the globe has more than twenty-four meridian ſemi-circles, the proceſs is ſomewhat different. Elevate the pole to the latitude $51°\frac{1}{2}$ N. *e. gr.* of London, and bring the firſt meridian, or that of London on the Engliſh globe, to the north: ſet the hour-index to the uppermoſt XII on the horary circle; then turn the globe weſtward, till the index points ſucceſſively to I, II, III, IV, V, and VI in the afternoon; or until 15, 30, 45, 60, 75, and 90 degrees of the equator paſs under the brazen meridian; and the firſt meridian will mark the following degrees on the horizon from the north towards the eaſt, *viz.* $11\frac{2}{3}$, $24\frac{1}{4}$, $38\frac{1}{12}$, $53\frac{1}{2}$, $71\frac{1}{15}$, and 90; which are the reſpective diſtances of the above hours from XII on the plane of the horizon.

In order to transfer theſe and the reſt of the hours to an horizontal plane, draw the parallel right lines *a c* and *b d* (*fig* 10) upon that plane, at a diſtance equal to the thickneſs of the gnomon; and the ſpace between them will be the meridian or twelve o'clock line of the dial. At right angles to this meridian, draw *g h*, and on *a* and *b* as centres, with the ſame radius of any length, deſcribe the quadrants *g e*, and *f h*, and graduate them. The centre of the plane C, and the centres of theſe quadrants *a* and *b*, are at a little diſtance from each other, in order to enlarge the hour diſtances about noon. Lay a ruler on the centre *b*, and on the ſeveral diſtances already found, *viz.* $11\frac{2}{3}$, $24\frac{1}{4}$, &c. in the quadrant *f h*, and draw the reſpective afternoon hour-lines to I, II, &c. As the ſun riſes about four in the morning, on the longeſt days in London, continue the hour-lines of III and V in the afternoon through the centre *b* to the oppoſite ſide of the dial. Then lay the ruler on the centre *a*, through the ſame diſtances, reſpectively in the quadrant *e g*, draw the forenoon hour-lines of XI, X, &c. and as the ſun does not ſet before eight in the evening on the longeſt days, continue the hour-lines of VII and VIII through the centre *a*, to the other ſide. Through $51\frac{1}{2}$ degrees of either quadrant from its centre draw the right line *a g*, which will be the hypothenuſe or axis of the gnomon *a g i*; and from *g* let fall a perpendicular to the meridian *a i*; and a plate ſimilar to the triangle *a g i*, ſet up between *a c* and *b d*, will have its hypothenuſe *a g* parallel to the axis of the world, when the dial is truly ſet, and will caſt a ſhadow on the hour of the day. A line of cords will ſuperſede the neceſſity of graduating the quadrants. An erect ſouth dial (*Plate* II. *fig.* 15.) is conſtructed in the ſame manner, obſerving the rule already given with reſpect to the elevation of the pole, and of the gnomon, and the poſition of the hours. Ferguſon's Lect. lect. 10.

To deſcribe an horizontal DIAL *mechanically, by the dialling ſcales*.—Draw a double meridian line *a b*, *c d*, (*fig.* 11.) on the horizontal plane, and croſs it at right angles by the ſix o'clock o'clock line *f e*; take the latitude of the place with the compaſſes, in the ſcale of latitudes, and ſet that extent from *c* to *e*, and from *a* to *f*, on the ſix o'clock line; then taking the whole ſix hours between the points of the compaſſes in the ſcale of hours; with that extent ſet one foot in the point *e*, and let the other foot fall where it will upon the meridian line *c d*, and at *d*. Do the ſame from *f* to *b*, and draw the right lines *e d* and *f b*, each of which will be equal in length to the whole ſcale of hours. When this is done, ſet one foot of the compaſſes in the beginning of the ſcale at XII, and extending the other to each hour on the ſcale, lay off theſe extents from *d* to *e* for the afternoon hours, and from *b* to *f* for thoſe of the forenoon; and thus the lines *d e* and *b f* will be divided in the ſame manner as the hour ſcale is divided, at 1, 2, 3, 4, 5, and 6; on which the quarters may alſo be laid down, if required. Then, laying a ruler on the point *c*, draw the firſt five hours in the afternoon from that point through the dots at the numeral figures 1, 2, 3. 4, 5, on the line *d e*, and continue the lines of IIII and V through the centre *c* to the other ſide of the dial for the like hours of the morning; then lay the ruler on the point *a*, and draw the laſt five hours in the forenoon through the dots 5, 4, 3, 2, 1, on the line *f b*; and continue the hour-lines of VII and VIII through the centre *a*, to the other ſide, for the like hours of the evening; ſet the hours to the reſpective lines, and make the gnomon in the manner already directed for the horizontal dial.

An erect ſouth dial may be made by taking the co-latitude of the place from the ſcale of latitudes, and proceeding in all reſpects for the hour-lines, as in the horizontal dial; only reverſing the hours, as in *fig.* 15. and making the angle of the ſtyle's height equal to the co-latitude.

An horizontal dial may be eaſily deſcribed with only a line of cords, or a graduated quadrant of a circle. With any opening of the compaſſes, as Z L, (*fig.* 12.) deſcribe the two ſemicircles L F *k* and L Q *k*, on the centres Z and *z*, where the ſix o'clock hour line croſſes the double meridian line, and divide each ſemicircle into twelve equal parts, beginning at L; then connect the diviſions, which are equidiſtant from L, by the parallel lines K M, I N, H O, G P, and F Q. Draw V Z for the hypothenuſe of the ſtyle, making the angle V Z E equal to the latitude of the place; and continue the line V Z to R. Draw the line R *r* parallel to the ſix o'clock line, and ſet off the diſtance *a* K from Z to Y, the diſtance *b* I from Z to X, *c* H from Z to W, *d* G from *z* to T, and *e* F from Z to S. Then draw the lines S *s*, T *t*, W *w*, X *x*, and Y *y* each parallel to R *r*. Set off the diſtance *y* Y from *a* to 11, and from *f* to 1; the diſtance *x* X, from *h* to 10, and from *g* to 2; *w* W from *c* to 9, and from *h* to 3; *t* T from *d* to 8, and from *i* to 4; *s* S from *e* to 7, and from *n* to 5. Then laying a ruler to the centre *z*, draw the forenoon hour-lines through the points 11, 10, 9, 8, 7: and laying it to the centre *z*, draw the afternoon lines through the points 1, 2, 3, 4, 5; continuing the forenoon lines of VII and VIII through Z, to the oppoſite ſide of the dial, for the like afternoon hours; and the afternoon lines IIII and V through *z* for the like morning hours. Set the hours to the lines, and erect the ſtyle, and the dial is complete. A ſouth dial may be conſtructed in the ſame manner, by drawing the line V Z to make an angle with the meridian Z L equal to the co-latitude of the place, &c. reverſing the hours, as in *fig.* 15. Ferguſon ubi ſupra.

Dial, *Vertical*, is that drawn on the plane of a vertical circle. See Vertical.

Of theſe there are ſeveral varieties according to the vertical pitched upon. The verticals chiefly uſed, are the prime vertical, and the meridian; from which reſpectively ariſe, ſouth. north, eaſt and weſt dials.

Dials which reſpect the cardinal points of the horizon, are particularly called direct dials. See Direct.

If any other vertical be choſen, the dial is ſaid to decline. Farther, if the circle, whoſe plane is uſed, be perpendicular to the horizon, as is ſuppoſed to be the caſe in all thoſe now mentioned, the dials are particularly denominated erect. *E. gr.* erect ſouth, erect north, &c.

Otherwiſe the plane, being oblique to the horizon, they are ſaid either to incline or recline.

Dial, *South*, or, more particularly an *erect direct ſouth* Dial, is that deſcribed on the ſurface of the prime vertical circle looking towards the ſouth.

Since the ſun then illumines the plane of the prime vertical looking to the ſouth, when, in its progreſs, he paſſes from the prime vertical to the meridian, or returns back, from this to that; in which he is employed ſix hours before, and ſix after noon; a ſouth dial ſhews the hours from ſix in the morning to ſix at night.

Dial, *To draw a vertical ſouth.*—On the plane of the prime vertical looking ſouthwards, draw a meridian line A B (*fig.* 13.) and taking the interval A C, at pleaſure, for the magnitude of the future dial, in C erect a perpendicular, of an indefinite length C D, and making an angle C A D equal to the elevation of the equator; draw a right line A D, meeting the perpendicular C D in D. Then in the point D make the angle C D E likewiſe equal to the elevation of the equator, and draw the right line D E cutting the meridian in E. Through E draw the right line G H, cutting the meridian A B at right angles. Take E B equal to E D, and with this radius deſcribe a quadrant E F. The reſt is performed as in an horizontal dial, except that the hours of the afternoon are to be written on the right hand, and thoſe of the forenoon on the left; as may be conceived from the figure. Laſtly, in the point A fix an oblique ſtyle, in an angle equal to the elevation of the equator; or in C erect a perpendicular ſtyle, equal to C D; or, laſtly, a triangular plate A D E, upon A E, ſo as to be perpendicular to the plane of the dial.

Then will the ſhadow of any of theſe indexes touch the ſeveral hour-lines, at their reſpective hours.

Dial, *North*, or *erect direct north* Dial, is that deſcribed on the ſurface of the prime vertical looking northward.

Since the ſun only illumines this ſurface while he advances from the eaſt to the prime vertical, and proceeds from the ſame vertical to the weſt; and ſince he is in the prime vertical at ſix o'clock in the morning, and at ſix in the evening; a north dial ſhews the hours before ſix in the morning, and thoſe after ſix in the evening. And hence, as in autumn and winter-time, the ſun does not riſe before ſix, and ſets before ſix in the evening, a north dial is of no uſe all that time; but this being joined with a ſouth dial, ſupplies the defects thereof.

Dial, *To deſcribe a vertical north.* Draw a meridian line E B, (*Plate* II. *fig.* 14.) and from A deſcribe a little circle at pleaſure. At A make the angle D A C equal to the elevation of the equator, and from the point C, taken at pleaſure, erect a perpendicular C D, meeting A D in D. Make another angle, C D E, likewiſe equal to the elevation of the equator, and draw likewiſe a line D E, meeting A E in E; then take I B equal to E D, and through I draw G H; cutting A B at right angles; and from the centre B, with the radius I B, deſcribe a quadrant, which divide into ſix equal parts: through the two extreme diviſions draw lines from the centre B, *viz.* B *h*, and B G, meeting G H in *h* and G, and make I *d* equal to I *h*, and I H equal to to I G; then applying a ruler to A and *b*, and G, and again to A and *d*, and H, draw the right lines A V, A IV, A VII, and A VIII. Laftly, in A fix an oblique index A D, making an angle D A E, with a meridian line in the plane of the meridian, equal to the elevation of the equator; or the perpendicular index in C, equal to C D; or, inftead of an index, a triangular plate E D A, on the meridian line E A, perpendicular to the plane of the dial.

Then will A IV, A V, A VI, be the hours of the forenoon; and A VI, A VII, and A VIII, thofe of the afternoon; and accordingly will be pointed out by the fhadow of the feveral indexes. Or thus, in a fouth dial (*fig.* 13.) if the hour lines IV and V, as alfo VII and VIII, be continued beyond the line VI A VI; and the triangle A D E turned about its pole A, till A E fall directly on A XII produced; it is evident, a north dial is hereby had, only obferving what has been faid about writing the hours.

To draw a vertical north or *fouth* DIAL *trigonometrically*.

Thefe only differ from the horizontal dial, in that the angle C A B is equal to the complement of the elevation of the pole of the place; fo that the fame analogy ferves as for the horizontal one; only making the fecond term the complement of the elevation of the pole of the place.

For the method of defcribing this dial by a globe, and by *dialling fcales*, fee *Horizontal* DIAL.

DIAL, *Eaft*, or *erect direct eaft* DIAL, is that drawn on the plane of the meridian, looking to the eaft.

Since the fun only illumines the plane of the meridian looking eaftward, before noon, an eaft dial can only fhew the hours till twelve o'clock.

DIAL, *To defcribe an eaft*. On the eaftern fide of the plane of the meridian draw a right line A B (*Plate* II. *fig.* 16.) parallel to the horizon, and to this join A K, making with it an angle K A B, equal to the elevation of the equator: then, with the radius D E, defcribe a circle, and through the centre D draw E C perpendicular to A K, by which means the circle will be divided into four quadrants. Each of thefe quadrants fubdivide into fix equal parts. And from the centre D through the feveral divifions draw the right lines D IV, D V, D VI, D VII, D VIII, D IX, D X, D XI. Laftly, in D erect a ftyle equal to the radius D E, perpendicular to the plane; or, on two little pieces perpendicularly fixed in E C, and equal to the fame radius D E, fit an iron rod, parallel to E C.

Thus will each index at the feveral hours project a fhadow to the refpective hour-lines IV IV, V V, VI VI, &c.

DIAL, *Weft*, or *erect direct weft* DIAL, is that defcribed on the weftern fide of the meridian.

As the fun only illumines that fide of the plane of the meridian looking to the weft after noon, a weft dial can only fhew the hours from noon to night.

This, therefore, joined with an eaft dial, fhews all hours of the day.

DIAL, *To draw a weft*. The conftruction is perfectly the fame as that of an eaft dial, only that its fituation is inverted, and the hours are written accordingly.

DIAL, *Polar*, is that defcribed on a plane paffing through the poles of the world, and the eaft and weft points of the horizon. It is of two kinds, the firft looking up towards the zenith, and called upper; the latter down to the nadir, called lower.

The polar dial, therefore, is inclined to the horizon in an angle equal to the elevation of the pole.

Since the polar plane P O Q S (*fig.* 17.) paffes through the eaft and weft points O and S, a quadrant of the equator is intercepted between it and the meridian; confequently the upper furface is illumined by the fun from fix in the morning to fix at night; and the lower from the fun's rife to fix in the morning, and from fix in the evening to fun-fet. A lower polar dial, therefore, fhews the hour of the morning from fun-rife to fix o'clock, and that of the evening, from fix to fun-fet; and an upper, the hours from fix in the morning to fix in the evening.

DIAL, *To draw an upper polar*. Draw a right line A B (*fig.* 18.) parallel to the horizon; and if the plane be immoveable, find the meridian line C E. Divide C E into two equal parts, and through C draw a right line F G parallel to A B. Then from the centre D, with the interval D E, defcribe a quadrant, which divide into fix equal parts; and from the fame centre D, through the feveral points of divifion, draw right lines D I, D II, D III, D IV, D V, and the intervals E I, E II, E III, E IV, E V, fet off the contrary way, *viz.* E XI, X, IX, VIII, and VII. From the points V, IV, III, II, I, &c. raife perpendiculars meeting the line F G in the correfpondent points. Laftly, in D erect a perpendicular ftyle equal to D E, or on two equal pieces, E and C, fix a crofs iron rod.

Then will XII XII, I I, II II, III III, &c. be hour-lines, to be pointed out at the proper times by the fhadow of the indices.

DIAL, *an upper polar*, only differs in fituation, and the manner of writing the hours, from eaft and weft dials, joined together in the line of fix o'clock.

DIAL, *a lower polar*, is had by putting out the hours of the forenoon, IX, X, and XI, and thofe of the afternoon, I, II, and III, with the noon-hour XII, itfelf; and only leaving the hours VII and VIII of the morning, and IV and V in the evening.

To make three DIALS *on three different planes, fo that they may all fhew the time of the day by one gnomon.*—On the flat board A B C (*fig.* 19.) defcribe an horizontal dial, with its gnomon, F G H, the edge of the fhadow of which fhews the time of the day. To this horizontal board join the upright board E D C, touching the edge G H of the gnomon: then making the top of the gnomon at G the centre of the vertical fouth dial, defcribe it on the board E D C. Befides, on a circular plate I K defcribe an equinoctial dial, and by a flit *c d* in the XII o'clock line, from the edge to the centre, put it on the gnomon F G, as far as the flit will admit. The fame gnomon will fhew the fame hour on each of thofe dials.

To draw all the primary DIALS *on the fame block or poft.*—
1. Let the plane A B C D, (*fig.* 20.) in the proper pofition of the block, be fuppofed horizontal; and thereon defcribe a horizontal dial. See *Horizontal* DIAL.

2. Draw the right lines E M and F L parallel to D C, which, accordingly, in the proper pofition of the block, will be parallel to the horizon: then let the plane B N M C make an angle with E M, equal to the elevation of the pole C M E; and thereon defcribe an upper polar dial.

3. Let the oppofite plane, A D E, make with E M an angle D E M, equal to the elevation of the equator; and on this draw an upper equinoctial dial.

4. Let the plane, K L H I, make with F L an angle H L F, equal to the elevation of the equator; and on this infcribe a lower equinoctial dial.

5. Let the oppofite plane, F G, make with F L an angle G F L, equal to the elevation of the pole; and here draw a lower polar dial.

6. Let the plane M N K L, and the oppofite one E F, be perpendicular to F L; and on that draw a fouth dial, and on this a north dial.

7. On the plane E M L F defcribe a weft dial; and on the oppofite plane an eaft dial.

If,

If, then, the block be fo placed, as that the plane M N K L looks to the fouth, and the plane of the meridian bifect it in the line of 12 o'clock in the horizontal dial A B C D, and fouth dial M N K L, all the hours of the day will be indicated by feveral planes at once.

Dials, *Secondary*, are all thofe drawn on the planes of other circles, befides the horizon, prime vertical, equinoctial, and polar circles; or thofe which either decline, incline, recline, or deincline.

Dials, *Declining*, are erect or vertical dials, which decline from any of the cardinal points; or they are fuch as cut either the plane of the prime vertical, or of the horizon, at oblique angles.

The ufe of declining dials is very frequent; as the walls of houfes, whereon dials are ufually drawn, commonly deviate from the cardinal points.

Of declining dials there are feveral kinds, which are denominated from the cardinal points; which they feem moft to refpect, but from which they have a real declination: decliners from the fouth and from the north, as a fouth-eaft decliner, a fouth-weft decliner, a north-eaft decliner, and a north-weft decliner, and even decliners from the zenith.

To draw a vertical declining Dial *trigonometrically.*

1. The declination of the plane, and the elevation of the pole of the place, being given; to find the angle formed in the centre of the dial, by the meridian and fubftyle.

Canon. As the whole fine is to the tangent of the complement of the height of the pole of the place G F (*fig.* 21.) fo is the fine of the angle of the declination of the plane G F D, to the tangent G D, of the angle required G A D.

2. The declination of the plane being given, and the elevation of the pole of the place, to find the angle formed in the centre of a vertical declining dial, by the fubftyle and axis.

Canon. As the whole fine is to the fine of the complement of the elevation of the pole G F; fo is the fine of the complement of the declination of the plane D G F, to the fide D F, the fine of the angle D A B required.

3. The declination of the plane, and the elevation of the pole, given; to find the difference of longitudes, that is, the arc of the equator, comprehended between the meridian of the place and the meridian of the plane.

Canon. As the whole fine is to the fine of the height of the pole of the place, fo is the tangent of the complement of the declination of the plane, to the tangent of the complement of the difference of longitudes.

4. The angle of the difference of longitudes, and that of the axis with the fubftyle, being given; to find the angles formed in the centre of a vertical declining dial, between the fubftyle and hour-lines.

This problem admits of three cafes; for the hour-lines, whofe angles are fought, may be either, 1. Between the meridian and fubftyle; or, 2. Beyond the fubftyle; or, 3. On that fide of the meridian where the fubftyle is not. In the two firft cafes, the difference is to be taken between the fun's diftance from the meridian that hour, and the angle of the difference of longitudes found by the laft problem; and in the third cafe, the fum of thofe two angles is to be taken, and the following canon ufed.

Canon. As the whole fine is to the fine of the angle between the axis and fubftyle, fo is the tangent of the difference of the fun's diftance from the meridian, and the difference of longitudes, or of the fum of thofe two angles, to the tangent of the angle required.

5. The angle formed by the fubftyle, with the hour-lines, and that of the fubftyle with the meridian, given; to find the angles formed between the meridian and hour-lines in the centre of vertical declining dials.

1. The angles of the hour-lines between the meridian and fubftyle are found by fubtracting the angle formed by the fubftyle with the hour-line from the angle formed by the fubftyle with the meridian.

2. The angles beyond the fubftyle, and on the fide oppofite to that of the meridian, are found by adding thofe two angles.

3. Thofe on the other fide of the meridian are found by taking their difference.

To defcribe a vertical Dial, *declining from the fouth to the eaft or weft geometrically.*—Find the declination of the plane, as already taught, under the article Declinator. Then draw upon a paper an horizontal dial, fuppofing the line of contingency of the horizontal with the equinoctial plane to be G H (*fig.* 22.) through the point E, wherein the meridian line A E cuts the fame, draw a right line I K, making with G H an angle H E K, equal to the declination of the given plane. Thus, as G H reprefents the interfection of the prime vertical with the horizon, I K will be the interfection of the declining plane and the horizon; whence we alfo conceive, that the part I E muft be raifed above G E, in cafe the given plane decline to the weft; or it muft be depreffed below the fame G E, in cafe it decline to the eaft. Draw a right line parallel to the horizon on the given plane, or wall, to anfwer to I K; and affuming a point therein anfwering to E, fet off from the right line I K on the paper the feveral horary diftances E 1, E 2, E 3, &c. Then from the point E erect a perpendicular E C, equal to the diftance of the centre of the horizontal dial, from its line of contingency. Draw lines thence to the feveral points of the hour-lines E 1, E 2, E 3, &c. let fall a perpendicular A D, from the centre of the horizontal dial A, to the line of contingency I K, and transfer the diftance E D, from the point E upon the wall: then will C D be the fubftylar line.

Wherefore, joining A D and D C at right angles, the hypothenufe A C will be an oblique index, to be faftened on the wall in the point C, according to the angle D C A.

To conftruct a vertical Dial, *declining from the fouth towards the eaft or weft, by the globe.*—Elevate the pole to the latitude of the place, and fcrew the quadrant of altitude to the zenith. If the dial declines towards the eaft, count the degrees of declination in the horizon from the eaft point towards the north, and bring the lower end of the quadrant to that degree of declination, at which the reckoning ends. Then bring any meridian (*e. gr.* the firft meridian) directly under the graduated edge of the upper part of the brazen meridian, and fet the hour index to XII at noon. Turn the globe eaftward on its axis, the quadrant of altitude being kept to the degree of declination in the horizon, and obferve the degrees cut by the firft meridian in the quadrant of altitude, counted from the zenith, as the hour index comes to XI, X, IX, &c. in the forenoon; and the degrees, then cut in the quadrant by the firft meridian, are the refpective diftances of the forenoon hours from XII on the plane of the dial. For the afternoon hours, turn the quadrant of altitude round the zenith till it comes to the degree in the horizon oppofite to that where it was placed before, or as far from the weft point towards the fouth as it was at firft from the eaft towards the north; and turn the globe weftward on its axis, till the firft meridian comes to the brazen meridian, and the hour index to XII; then turning the globe weftward to the afternoon hours I, II, III, &c. the firft meridian will cut the quadrant of altitude in the refpective number of degrees from the zenith, that each of thefe hours is from XII on the dial; and the limit of this dial will be determined by the hour-index,

index, when the firſt meridian goes off the quadrant at the horizon, both in the forenoon and afternoon. Lay down theſe hour-diſtances on the dial plane, either by dividing a ſemicircle into two quadrants, beginning at the hour line of XII, or by the line of cords, as directed under *horizontal* DIAL. As the ſubſtyle makes an angle with the meridian line in this kind of dials, falling among the forenoon hour-lines in an eaſt decliner, and among thoſe in the afternoon in a dial declining towards the weſt, its diſtance may be found in the former, by counting the degrees of the declination of the dial in the horizon from the eaſt point towards the north, and bringing the lower end of the quadrant of altitude to that degree of declination where the reckoning ends; then turn the globe till the firſt meridian cuts the horizon in the like number of degrees, counted from the ſouth point toward the eaſt; and the quadrant and firſt meridian will croſs one another at right angles, and the number of degrees of the quadrant, intercepted between the firſt meridian and the zenith, is equal to the diſtance of the ſubſtyle line from the 12 o'clock line; and the number of degrees of the firſt meridian, which are intercepted between the quadrant and the north pole, is equal to the elevation of the ſtyle above the plane of the dial. If the dial declines weſtward from the ſouth, count that declination from the eaſt point of the horizon towards the ſouth, and bring the quadrant of altitude to the degree in the horizon at which the reckoning ends; both for finding the forenoon hours, and diſtance of the ſubſtyle from the meridian; and for the afternoon hours bring the quadrant to the oppoſite degree in the horizon, as far from the weſt towards the north; and then proceed as above. The north dial declining eaſt or weſt by the ſame number of degrees may be eaſily had from the former, by only extending the hour-lines, ſtyle, and ſubſtyle, quite through the centre; for thus the ſouth-eaſt decliner will produce the north-weſt decliner, and the ſouth-weſt decliner the north-eaſt decliner. Ferguſon, lib. cit.

To draw a vertical DIAL, *declining from the north towards the eaſt and weſt.*—Take the declination of the planes, as already taught; then, as north dials are only ſouth dials inverted, draw a vertical dial declining from the ſouth, and invert it in ſuch a manner, as that the centre C looks to the horizon, and the point E to the zenith; and the hours on the right hand ſet off towards the left, and contrarywiſe, omitting all hour-lines, which in ſuch a plane cannot be ſhewn.

For the practice, the beſt way is, after drawing a ſouth decliner upon paper, to prick the ſeveral points thereof through with a pin; then applying the face of the paper to the wall, the back ſide thereof will ſhew you all the points neceſſary for the north declining dial.

DIALS, *Inclined*, are thoſe drawn on planes not erect, but inclining, or leaning forward towards the ſouth, or ſouthern ſide of the horizon, in an angle, either greater or leſs than the equinoctial plane.

Such an inclined plane may be conceived by ſuppoſing one part of the plane of the equator lifted up towards the zenith, and the other depreſſed towards the nadir; and thus to revolve upon a line drawn from the eaſt to the weſt point of the horizon.

DIAL, *To draw an inclined.* 1. The inclination of the plane as DC (*fig.* 23.) being found by a declinator, as taught under DECLINATOR; if it fall between the equinoctial plane C E, and the vertical one C B, in ſuch manner as that the angle of inclination D C A is greater than the elevation of the equator E C A; on the upper ſide draw a north dial, and on the lower a ſouth dial, to an elevation of the equator, which is equal to the aggregate of the elevation of the equator of the given place, and the complement of the inclination to a quadrant.

2. If the inclined plane C F fall between the horizontal one C A, and the equinoctial C E, ſo as that the angle of inclination F C A, is leſs than the elevation of the equator E C A, deſcribe an horizontal dial to an elevation of the pole, equal to the aggregate of the elevation of the pole of the given place, and the inclination of the plane. Dials, thus inclined, are drawn after the ſame manner as primary dials; except, that the index in the former caſe muſt be fitted under the angle A D C; and in the latter, under the angle D F C; and that the diſtance of the centre of the dial from the line of contingency, in the former caſe, is D C, and in the latter F C.

DIALS, *Reclining*, are thoſe drawn on planes not erect, but reclined, or leaning backwards from the zenith towards the north, in an angle greater or leſs than the polar plane. A reclined plane may be conceived by ſuppoſing one part of the polar plane raiſed towards the zenith, and the other depreſſed towards the nadir; and thus revolving about a line drawn from eaſt to weſt. To find the reclination of a plane, ſee RECLINATION.

DIAL, *To draw a reclining.* 1. If the reclined plane H C fall between the vertical plane B C, and the polar plane I C, ſo as that the angle of reclination B C H is leſs than the diſtance of the pole from the zenith B C I, deſcribe two vertical ſouth and north dials to an elevation of the equator equal to the difference between the elevation of the equator of the given place, and the angle of reclination. 2. If the reclined plane, as K C, fall between the polar plane I C, and the horizontal one C L, ſo as that the angle of reclination B C K is greater than the diſtance of the pole from the zenith I C B, deſcribe an horizontal dial thereon to an elevation of the pole, equal to the difference between the angle of reclination, and the elevation of the equator of the given place.

To draw inclining and reclining DIALS *trigonometrically*.—The inclination and reclination of the plane, and the elevation of the pole, being known, to find the angles made in the centre of an inclining or reclining dial, by the meridian and hour lines.

Such dial is properly an horizontal dial, for a latitude equal to the particular elevation of the pole on the plane of the dial. Its angles, therefore, are found by the canon laid down for horizontal dials.

As to the elevation of the pole on the dial plane, it is thus found: the plane being inclined, either its inclination is greater than the elevation of the pole of the place, or leſs; or it is equal thereto. In the two firſt caſes, for upper ſouth, or lower north dials, the particular elevation of the pole on the plane is had by taking the difference between the elevation of the pole of the place, and the inclination of the plane; and in the latter caſe, the dial is a polar dial, wherein the hour-lines will be parallel, by reaſon that the plane being placed on the axis of the world, neither of the poles can be repreſented thereon.

For upper north, and lower ſouth dials. 1. If the inclination be greater than the complement of the elevation, the complement of the inclination muſt be added to the complement of the elevation. 2. If it be leſs, the inclination muſt be added to the elevation. 3. If it be equal, the dial will be an equinoctial dial, wherein the angles at the centre will be equal to the ſun's diſtance from the meridian.

DIALS, *Deinclined*, are thoſe which both decline and incline, or recline. See DECLINERS.

The uſe of inclined, reclined, and eſpecially deinclined dials,

dials, is very rare; the geometrical and trigonometrical construction of these last, therefore, as being somewhat intricate, we here choose to omit, and refer such as may have a fancy for such a dial, to an universal mechanical method of drawing all kinds of dials here subjoined

An easy method of describing a DIAL *on any kind of plane, by means of an equinoctial* DIAL, *or circle*.—Suppose, *e. gr.* a dial required on an horizontal plane: if the plane be immoveable as A B D C (*Plate* II. *fig.* 24.) find a meridian line G F; or, if moveable, assume the meridian at pleasure. Then, by means of the triangle E K F, whose base is applied on the meridian line, raise the equinoctial dial H, till the index G I becomes parallel to the axis of the world (which is had if the angle K F F be equal to the elevation of the pole), and the 12 o'clock line on the dial hang over the meridian line of the plane, or the base of the triangle. If then, in the night-time, a lighted candle be successively applied to the axis G I, so as the shadow of the index, or style G I, fall upon one hour line after another; the same shade will mark out the several hour-lines on the plane A B C D.

Noting the points, therefore, on the shadow, draw lines through them to G; then an index being fixed in G, according to the angle I G F, its shadow will point out the several hours by the light of the sun.

If a dial were required on a vertical plane, having raised the equinoctial circle, as above directed, push forward the index G I, till the tip thereof, I, touch the plane.

If the plane be inclined to the horizon, the elevation of the pole should be formed on the same; and the angle of the triangle K E F should be made equal thereto.

See a new method of constructing sun-dials, for any given latitude, without the assistance of dialling scales, or logarithmic calculations, by Mr. Ferguson, in the Phil. Trans. vol. lvii. for 1767, art. 36, and in his Select Exercises, p. 95.

Note. Beside the several species of dials above-mentioned, with are said to be with centres, there are others, called dials without centres. See CENTER *of a Dial*.

DIALS *without centres*, are those whose hour-lines do really converge, but so slowly, that the centre towards which they converge cannot be expressed in the given plane.

DIALS, *Horizontal, without centres*, are to be made for places, the elevation of whose pole is either very small, or very great.

DIALS, *Vertical, without centres*, are for places, the elevation of whose pole is very great.

DIALS, *For the furniture of*. See FURNITURE.

DIAL, *Quadrantal*. See *Horodictical* QUADRANT.

DIAL, *Reflecting*. See REFLECTING *Dial*.

DIAL, *Cylindric*, is represented in *Plate* III. *fig.* 27. This dial may be delineated on paper, and then pasted round a cylinder of wood, and it will shew the time of the day, the sun's place in the ecliptic, and his altitude at any time of observation. Draw the right line *a* A B (*fig.* 28.) parallel to the top of the paper, and with any opening of the compasses on the centre *a* describe the quadrant A E, and graduate it. Draw the right line A C at right angles to *a* A B, and touching the quadrant A E at the point A. From the centre *a* draw right lines through as many degrees of the quadrant as are equal to the sun's altitude at noon, on the longest day of the year, at the place for which the dial is to serve; which altitude at London is nearly 62 degrees; continue these right lines till they meet the tangent line A C; and from the points of concourse, draw straight lines across the paper, parallel to the first right line A B, and these will be the parallels of the sun's altitude in whole degrees, from sun-rise to sun-set, on all the days of the year. These parallels of altitude must be drawn out to the right line B D, which must be parallel to A C, and as far from it as is equal to the intended circumference of the cylinder on which the paper is to be pasted. Divide the space between the right lines A C and B D, at top and bottom, into twelve equal parts, for the twelve signs of the ecliptic; and from one mark to the opposite one, draw right lines parallel to A C and B D; and place the characters of the twelve signs in the twelve spaces at the bottom, beginning with ♑ Capricorn, and ending with ♓ Pisces. These spaces should likewise be divided by parallel lines into halves and quarters. At the top of the dial make a scale of the months and days of the year, so that the days may stand over the sun's place, found in an Ephemeris, for each of them in the signs of the ecliptic. Compute the sun's altitude for every hour, in the latitude of your place, or take it from a table, when he is in the beginning, middle, and end of each sign of the ecliptic; and in the upright parallel lines, at the beginning and middle of each sign, make marks for these altitudes among the horizontal parallels, reckoning downward in the order of the numeral figures at the right-hand, answering to the like divisions of the quadrant at the left. Through these marks draw the curve hour-lines, and set the hours to them, as in the figure, reckoning the forenoon hours downward, and the afternoon hours upward. The sun's altitude may also be taken and laid down for the half hours and quarters. Then cut off the paper at the left-hand on which the quadrant was drawn, close by the right line A C, and all the paper at the right-hand, close by the right line B D, and cut it also close by the top and bottom horizontal lines, and it will be fit for pasting round the cylinder. This cylinder, (*fig.* 27.) is hollow for holding the style D E, when it is not used. The style must stand out, perpendicular to the side of the cylinder, just over the right line A B, (*fig.* 28.) where the parallels of altitude begin, and its length, or distance of the point *e* from the cylinder, must be equal to the radius *a* A of the quadrant A E (*fig.* 28.) When this dial is used, the horizontal foot B C is placed on a level table, or hung by the ring F, where the sun shines, and the horizontal top A D, which is made moveable, is turned till the style stands just over the day of the month. Then the cylinder is turned round, till the shadow of the style falls upon it, parallel to the upright lines which divide the signs; or till the shadow is parallel to a supposed axis in the middle of the cylinder; and then the point, or lowest end of the shadow, will fall upon the time of the day, as it is before or after noon, among the curve hour-lines, and will shew the sun's altitude at that time among the cross parallels of altitude, encompassing round the cylinder; and it will also shew in what sign of the ecliptic the sun is at that time, and the degree may be nearly estimated by the eye. Ferguson.

When the sun is in the equinoctial, and has no declination, his altitude may be easily found by the following proportion. As radius is to the cosine of the latitude, so is the cosine of the hour from noon to the sine of the altitude: but if he has north or south declination, say, as radius is to the sine of the declination, so is the sine of the latitude to the sine of the sun's altitude at six o'clock. For the sun's altitude at other hours the two following proportions must be used, *viz.* 1. As the cosine of the hour from the meridian to radius, so is the tangent of the latitude to the tangent of a fourth arc, from which the declination is to be subtracted, when north, and to which it is to be added, when south, for a fifth arc. The supplement of the sun's declination, added to the fourth arc, must be used for this fifth arc, when they exceed a quadrant, in finding the altitudes before and after six.

2. As the fine of the fourth arc is to the cofine of the fifth arc, fo is the fine of the latitude of the place to the fign of the altitude for the given hour and declination.

The altitude in the middle of the fign may either be had by taking the arithmetic mean between the next greateft and leaft, or by computation in the manner above directed. Leybourn's Dialling, Tract vi. p. 23, &c.

To find the altitude by the globe, fee GLOBE.

DIAL, *Portable, on a card*, is reprefented in *fig.* 29. and may be eafily conftructed. Draw the occult line A B (*fig.* 30.) parallel to the top of the card, and crofs it at right angles with the fix o'clock line E C D: then, upon C, as a centre, with the radius C A, defcribe the femicircle A E L, and divide it into twelve equal parts, beginning at A, as A *r*, A *s*, &c. and from thefe points of divifion draw the hour-lines *r*, *s*, *t*, *u*, *v*, *w*, and *x*, all parallel to the fix o'clock line E C. If each part of the femicircle be fubdivided into four equal parts, they will give the half hour lines and quarters, as in *fig.* 29. Draw the right lines A S D *o*, making the angle S A B equal to the latitude of the place. Upon the centre A defcribe the arc R S T, and fet off upon it the arcs S R and S T, each equal to 23½ degrees, for the fun's greateft declination, and divide them into 23½ equal parts, as in *fig.* 29. Through the interfection D of the lines E C D and A D *o* draw the right line F D G at right angles to A D *o*. Lay a ruler to the points A and R, and draw the line A R F through 23½ degrees of fouth declination in the arc S R; and then laying the ruler to the points A and T, draw the line A T G through 23½ degrees of north declination in the arc S T; fo fhall the lines A R F and A T G cut the line F D G in the proper length for the fcale of months. Upon the centre D, with the radius D F, defcribe the femicircle F *o* G, which divide into fix equal parts F *m*, *m n*, *n o*, &c. and from thefe points of divifion draw the right lines *m h*, *n i pk*, and *q l*, each parallel to *o* D. Then fetting one foot of the compaffes in the point F, extend the other to A, and defcribe the arc A *z* H for the tropic of ♑; with the fame extent, fetting one foot in G, defcribe the arc A E O for the tropic of ♋. Next fetting one foot in the point *h*, and extending the other to A, defcribe the arc A C I for the beginnings of the figns ♒ and ♐; and with the fame extent, fetting one foot in the point *l*, defcribe the arc A N for the beginnings of the figns ♊ and ♌. Set one foot in the point *i*, and having extended the other to A, defcribe the arc A K for the beginnings of the figns ♓ and ♏; and with the fame extent fet one foot in *k*, and defcribe the arc A M for the beginnings of the figns ♉ and ♍. Then fetting one foot in the point D, and extending the other to A; defcribe the curve A L for the beginnings of ♈ and ♎; and the figns will be finifhed. This done, lay a ruler from the point A over the fun's declination in the arc R S T, found by a table, for every fifth day of the year; and where the ruler cuts the line F D G, make marks, and place the days of the months right againft thefe marks, as in *fig.* 29. Laftly, draw the fhadow line P Q parallel to the occult line A B; make the gnomon, and fet the hours to their refpective lines, as *fig.* 29. and the dial will be finifhed.

The lines *a d*, *a b*, and *b c*, of the gnomon, muft be cut quite through the card; and as the end *a b* of the gnomon is raifed occafionally above the plane of the dial, it turns upon the uncut line *c d* as on an hinge. The dotted line A B muft be flit quite through the card, and the thread C muft be put through the flit, and have a knot tied behind, to keep it from being eafily drawn out. On the other end of this thread is a fmall plummet D, and on the middle of it a fmall bead for fhewing the hour of the day.

To rectify this dial; fet the thread in the flit right againft the day of the month, and ftretch the thread from the day of the month over the angular point, where the curve lines meet at XII; then fhift the bead to that point on the thread; and the dial will be rectified.

To find the hour of the day; raife the gnomon, and hold the edge of the dial next the gnomon towards the fun, fo that the uppermoft edge of the fhadow of the gnomon may juft cover the fhadow-line; and the bead then playing freely on the face of the dial, by the weight of the plummet, will fhew the time of the day among the hour-lines, in the forenoon or afternoon.

To find the time of fun-rifing and fetting; move the thread among the hour-lines, until it either covers fome one of them, or lies parallel betwixt any two; and then it will cut the time of fun-rifing among the forenoon hours, and of fun-fetting among the afternoon hours for that day of the year to which the thread is fet in the fcale of months.

To find the fun's declination; ftretch the thread from the day of the month over the angular point at XII, and it will cut the fun's declination, as it is north or fouth, for that day, in the proper fcale.

To find on what day the fun enters the figns; when the bead, as above rectified, moves along any of the curve lines which have the figns of the zodiac marked upon them, the fun enters thofe figns on the days pointed out by the thread in the fcale of months.

DIAL, *Univerfal, on a plain crofs*, is reprefented by *fig.* 31. and is moveable on a joint, C, for elevating it to any given latitude on the quadrant C *o* 90, as it ftands upon the horizontal board A. The arms of the crofs ftand at right angles to the middle part; and the top of it, from *a* to *n*, is of equal length with any of the arms *n e* or *m k*.

This dial is rectified, by fetting the middle line *t u* to the latitude of the place on the quadrant, the board A level, and the point N northward by the needle: thus, the plane of the crofs will be parallel to the plane of the equator. Then from III o'clock in the morning till VI, the upper edge, *k l*, of the arm, *i o*, will caft a fhadow on the time of the day on the fide of the arm *e m*; from VI till IX, the lower edge, *i*, of the arm, *i o*, will caft a fhadow on the hours, on the fide *o q*. From IX in the morning to XII at noon, the edge, *a b*, of the top part, *a n*, will caft a fhadow on the hours on the arm *n e f*; from XII to III in the afternoon, the edge *c d* of the top part will caft a fhadow on the hours of the arm *k l m*; from III to VI in the evening, the edge *g h* will caft a fhadow on the hours on the part *p q*; and from VI till IX, the fhadow of the edge, *e f*, will fhew the time on the top part *a n*. The breadth of each part, *a b*, *e f*, &c. muft be fo great, as never to let the fhadow fall quite without the part or arm on which the hours are marked, when the fun is at his greateft declination from the equator.

To determine the breadth of the fides of the arms which contain the hours, fo as to be in juft proportion to their length; make an angle, A B C, (*fig.* 32.) of 23½ degrees, which is equal to the fun's greateft declination; and fuppofe the length of each arm, from the fide of the long middle part, and alfo the length of the top part above the arms, to be equal to B *d*. Then, as the edges of the fhadow, from each of the arms, will be parallel to B *e*, making an angle of 23½ degrees with the fide B *d*, of the arm, when the fun's declination is 23½°: it is plain, that if the length of the arm be B *d*, the leaft breadth that it can have, to keep the edge B *e* of the fhadow, B *e g d*, from going off the fide of the arm, *d e*, before it comes to the end of it, *e d*, muft be equal to *e d* or *d* B. But, in order to keep the fhadow within the quarter divifions of the hours, when it it comes near the end of the arm, the breadth of it ſhould be ſtill greater, ſo as to be almoſt doubled, on account of the diſtance between the tips of the arms.

The hours may be placed on the arms, by laying down the croſs, *abcd (fig. 33.)* on a ſheet of paper; and with a black lead pencil, held cloſe to it, drawing its ſhape and ſize on the paper. Then take the length *ae* in the compaſſes, and with one foot in the corner *a*, deſcribe with the other the quadrant *ef*. Divide this arc into ſix equal parts, and through the points of diviſion draw right lines *ag*, *ah*, &c. continuing three of them to the arm *ce*, which are all that can fall upon it; and they will meet the arm in thoſe points through which the lines that divide the hours from each other, as in *fig. 31*, are to be drawn right acroſs it. Divide each arm, for the three hours contained in it, in the ſame manner; and ſet the hours to their proper places, on the ſides of the arms as they are marked in *fig. 33*. Each of the hour ſpaces ſhould be divided into four equal parts, for the half hours and quarters, in the quadrant *ef;* and right lines ſhould be drawn through theſe diviſion-marks in the quadrant, to the arms of the croſs, in order to determine the places thereon, where the ſubdiviſions of the hours muſt be marked.

This is a very ſimple kind of univerſal dial; it is eaſily made, and has a pretty uncommon appearance in a garden. Ferguſon.

Dials, *Refracted*, are ſuch as ſhew the hour by means of ſome refracting tranſparent fluid.

If a pin or ſtick be ſet up, or any point be aſſigned in a concave bowl or diſh for the centre of the dial, let an horizontal dial be applied over the ſame, aſſigning the meridian line on the edges of the bowl, and marking out the reſt of the hour-lines alſo on the edges of the bowl; take away the horizontal dial, and elevate a ſtring or thread from the end of the ſaid pin over the meridian line, as much as is the latitude or elevation of the pole of the place: then, by bringing the thread to caſt a ſhadow on any hour point formerly marked out on the edges of the bowl, by a candle or the like, that ſhade in the bowl is the true hour-line; and if the bowl be full of water, &c. when this is done, it will never ſhew the true hour by the ſhadow of the top of the pin but when it is filled again with the ſame liquor.

Dial, *Ring*, is a kind of dial, uſually ſmall and portable, conſiſting of a braſs ring or rim, ſeldom exceeding two inches in diameter, and one-third of an inch in breadth. In a point of this rim there is a hole, through which the ſunbeams being received, make a lucid ſpeck on the concavity of the oppoſite ſemicircle, which gives the hour of the day in the diviſions marked therein.

Theſe diviſions are made by deſcribing a circle, *(Plate* IV. *fig. 34.)* to repreſent the ring, and drawing an horizontal chord E F; with this as radius deſcribe the quadrant FD, and graduate it; and through the ſeveral degrees which mark the ſun's altitude for every hour at the time of the equinox, draw lines from E to the oppoſite part of the circle, and there ſet the correſponding figures. Thus the XII o'clock line will paſs through 38° 38′, the co-latitude of the place; and the XI and I line through 36° 56′; the X and II through 32° 36′; the IX and III through 26° 5′; the VIII and IV through 18° 8′; and the VII and V through 9° 17′. This may be eaſily calculated by the globe.

But it only holds good about the times of the equinox. To have the dial perform throughout the whole year, the hole is made moveable, and ſigns of the zodiac, or the days of the month, are marked on the convex ſide of the ring, by taking, *e.g.* E T and E *t (fig. 34)* on each ſide of E equal to double the ſun's declination when he enters any particular ſign, as Taurus or Piſces; and there marking the character of the ſaid ſign, or correſponding month, and ſo for all the others; and by means of theſe the dial is rectified for the time. It is evident, from a view of the figure, that E XII ♉ is = F E XI I = the altitude of the ſun in the equinox; but T XII E is equal to the ſun's declination in Taurus, becauſe it is an angle in the circumference ſtanding upon an arc, which is double the declination; and therefore T XII ♉ is equal to the meridian altitude when the ſun enters Taurus; and a ray paſſing through the hole at T will mark the XII o'clock hour of that day; but this dial will not point out the other hours exactly: becauſe T III *d* exceeds E III *d*, the equinoctial hour by the angle of declination; and ſo of the reſt. Some have contrived to remedy this inconvenience, by making the concave ſurface of the ring wider, and deſcribing upon it ſeven circles, the middle one to repreſent the equinoctial, and the extremes the tropics; and in theſe circles they have marked the forenoon and afternoon hours from a table of altitudes.

To uſe it, put the moveable hole to the day of the month, or the degree of the zodiac the ſun is in; then ſuſpending it by the little ring, turn it towards the ſun, till his rays, as before, point out the hour among the diviſions on the inſide.

Univerſal or *aſtronomical Ring* Dial, is a ring-dial, which ſerves to find the hour of the day in any part of the earth; whereas the former is confined to a certain latitude. Its figure ſee repreſented in *Plate* IV. *fig. 35*.

It conſiſts of two rings or flat circles, from two to ſix inches in diameter, and their breadth, &c. proportionable. The outward ring, A, repreſents the meridian of any place you are at, and contains two diviſions of 90° each, diametrically oppoſite to one another, ſerving the one from the equator to the north, the other to the ſouth pole. The inner ring repreſents the equator, and turns exactly within the outer, by means of two pivots in each ring at the hour XII.

Acroſs the two circles goes a thin reglet or bridge, with a curſor C, that ſlides along the middle of the bridge. In the curſor is a little hole for the ſun to ſhine through. The middle of this bridge is conceived as the axis of the world, and the extremities as the poles; and on the one ſide are drawn the ſigns of the zodiac, and on the other the days of the month. On the edge of the meridian ſlides a piece, to which is fitted a ring to ſuſpend the inſtrument by.

In this dial, the diviſions on the axis are the tangents of the angles of the ſun's declination, adapted to the ſemidiameter of the equator as radius, and placed on either ſide of the centre: but inſtead of laying them down from a line of tangents, a ſcale of equal parts may be made, whereof 1000 ſhould anſwer exactly to the length of the ſemi-axis, from the centre to the inſide of the equinoctial ring; and then 434 of theſe parts may be laid down toward each end from the centre, which would limit all diviſions on the axis, becauſe 434 is the natural tangent of 23° 29′. And thus, by a nonius fixed to the ſliding-piece, and taking the ſun's declination from an ephemeris, and the tangent of that declination from the table of natural tangents, the ſlider might be always ſet true within two minutes of a degree. And this ſcale of 434 equal parts might be placed right againſt the 23½° of the ſun's declination, on the axis, inſtead of the ſun's place, which is there of little uſe. For then the ſlider might be ſet in the uſual way, to the day of the month, for common uſe; but to the natural tangent of the declination, when great accuracy is required.

Uſe of the Univerſal Ring-Dial.—Place the line *a* (on the

the middle of the fliding-piece) over the degree of latitude of the place (*e. gr.* 51½° for London) put the line which croffes the whole of the curfor to the degree of the fign, or day of the month. Open the inftrument fo as that the two rings be at right angles to each other, and fufpend it by the ring H, that the axis of the dial, reprefented by the middle of the bridge, may be parallel to the axis of the world. Then turn the flat fide of the bridge towards the fun, fo that his rays, ftriking through the little hole in the middle of the curfor, may fall exactly on a line drawn round the middle of the concave furface of the inner ring; in which cafe the bright fpot fhews the hour of the day in the faid concave furface of the ring. Note, the hour of XII is not fhewn by this dial, becaufe the outer circle being then in the plane of the meridian, hinders the fun's rays from falling on the inner; nor will this dial fhew the hour when the fun is in the equinoctial, becaufe his rays then fall parallel to the plane of the inner circle.

DIAL. *Univerfal Meridian*, fo called by Mr. F. Wollafton, F. R. S., who has fhewn how to conftruct it, fo as to admit of being fet to any latitude, and adapted for fhewing the mean folar time of noon by infpection, without any calculation. Although this dial will not give the time at any other part of the day, it will always fhew the inftant when it is 12 o'clock without any regard to the equation of time; as well the inftant of apparent noon, and confequently the equation of time alfo, which is always the difference between the two. This method is defcribed in a 4to. pamphlet, entitled " Directions for making an Univerfal meridian Dial," &c. and publifhed for Wilkie in 1793.

Nocturnal, or *Night* DIAL, is that which fhews the hours of the night.

Of this there are two kinds; lunar and fiderial.

Moon-DIAL, or *Lunar*-DIAL, is that which fhews the hour of the night by means of the light, or fhadow, of the moon, projected thereon from an index.

To defcribe a Moon DIAL.—Suppofe, *e. gr.* an horizontal moon-dial required: draw, firft, an horizontal fun-dial; then erect two perpendiculars, A B and C D (*Plate* II. *fig.* 25.) to the line of XII o'clock; and, dividing the interval G F into twelve equal parts, through the feveral points of divifion draw lines parallel thereto: now, appropriating the firft line CD to the day of the new moon, and the fecond to the day when the moon comes an hour later to the meridian than the fun; their interfections with the hour-lines will give points, through which to draw a curve line XII XII for the meridian-line of the moon. After the like manner determine the other hour-lines, I I, II II, III III, &c. which the fhadow of the moon, projected from the ftyle of the dial, interfects at the refpective hour; then blot out the hour-lines of the fun dial, together with the perpendiculars whereby the lunar hours were drawn, and divide the interval G F, by other parallel lines, into fifteen equal parts, anfwering to the fifteen days between new and full moon; laftly, to thefe lines write the feveral days of the moon's age.

Now the moon's age being learnt from a calendar, the interfection of the line of the moon's age with the lunar hour-lines, will give the hour of the night.

After the fame manner may any other fun-dial be converted into a moon-dial.

To draw a portable Moon DIAL.—On a plane, that may be raifed according to the elevation of the equator, defcribe a circle A B (*fig.* 26) and divide its circumference into 29½ equal parts: from the fame centre C defcribe another moveable circle D E, which divide into twenty-four equal parts, or hours; in the centre C erect an index, as for an equinoctial dial.

This dial being duly placed, after the manner of an equinoctial dial, and the XII o'clock line brought to the day of the moon's age; the fhadow of the index will give the hour.

To ufe a Solar as a Lunar-DIAL. *i. e. to find the hour of the night by a* fun-dial.—Obferve the hour which the fhadow of the index points at by moon-light, find the moon's age in the calendar, and multiply the number of days by ⅘, the product is the number of hours to be added to the hour fhewn by the fhadow, throwing out 12, if it exceed, to give the hour required: the reafon of which is, that the moon comes to the fame horary circle later than the fun by about four-fifths of an hour every day; and at the time of new moon the folar and lunar hour coincide. Or thus: If you know the time of the moon's fouthing, count how many hours and minutes the fhadow on the dial falls fhort of 12 o'clock; fubtract thefe from the time of her fouthing, for the hour of the night. But if the fhadow fall beyond 12, add thefe hours and minutes (on the dial) to the time of her fouthing, rejecting 12, if it exceed: and you have the hour of the night.

DIALS, *Placing of.* Every dial muft be fo placed, that the upper edge of the ftyle may point directly to the pole, and that the horizontal line be perfectly level, and that it have its proper declination and inclination; and in an upright dial, that the 12 o'clock line be perpendicular to the horizon. Having prepared the plane on which the dial is to reft, the hour may be found with tolerable accuracy by a large equinoctial ring-dial, to which a watch is to be fet; and then the dial may be fixed by the watch at leifure. But in order to have the time more exactly, the fun's altitude fhould be taken by a good quadrant, and the precife time of obfervation noted by a clock or watch: then the time fhould be computed for the obferved altitude, and the watch fet to agree with that time. Otherwife, take two equal altitudes of the fun in the fame day in fummer; one between feven and ten in the morning, the other between two and five in the afternoon; mark the moment of the two obfervations by a clock or watch; and if the watch fhews the obfervations to be at equal diftances from noon, it agrees exactly with the fun; if not, the watch muft be corrected by half the difference of the forenoon and afternoon intervals; and then the dial may be fet true by the watch. To find a meridian line, fee MERIDIAN.

It is neceffary to obferve, that if a dial be made according to the ftrict rules of calculation, and truly fet at the inftant, when the fun is in the meridian, it will be a minute too faft in the forenoon, and as much too flow in the afternoon, by the fhadow of the ftyle; becaufe the edge of the fhadow that fhews the time is even with the foremoft edge of the fun in the forenoon, and with his hindmoft edge in the afternoon, on the dial, whereas the fun's centre determines the time in the (fuppofed) hour circles of the heaven; and as the fun is half a degree in breadth, he takes two minutes to move through a fpace equal to his breadth; and there will be two minutes at noon in which the fhadow will have no motion at all on the dial. It likewife appears, that if the dial be fet true by the fun in the forenoon, it will be two minutes too flow in the afternoon; and *vice verfâ*. The way of remedying this error, is to fet every hour and minute divifion on the dial one minute nearer XII than the calculation requires. Every dial is alfo too faft in the forenoon and too flow in the afternoon, on account of the refraction of the fun-beams, by which the fun is raifed higher, and the fhadow brought nearer the fubftyle. Ferguson's Tables and Tracts, p. 73, &c.

DIAL-*planes*, plain fuperficies, upon which the hour-lines of dials are drawn. See DIAL.

DIAL,

DIAL, *Siderial.* See NOCTURNAL.

DIAL, *Tide.* See TIDE-*Dial.*

DIAL-*Bird*, in *Ornithology* the *dial-grekle* of Latham, and *Saularis Gracula* of Gmelin, which fee.

DIAL-*Plate*, is the face of an inftrument which fhews the time or hour of the day either by means of a moving hand or index, or by the fhadow of a gnomon that falls upon it; hence we hear of clock-dials, watch-dials, and fun-dials.

The clock makers have alfo given the name of dials to a particular fort of time pieces; *viz* fuch as are generally ufed in kitchens, fhop , &c. They are eight day pendulum movements, which do not ftrike, but only fhew the hours and minutes.

The dial-plates of clocks and watches are almoft univerfally divided into twelve hours, which are denoted by Roman numerals, and each of thofe hours is fubdivided into five minutes, fo as to divide the whole circumference into fixty equal parts or minutes. Inftead of Roman numerals, the hours of fome clocks and watches are often indicated by the common numerical or Arabic figures; and fometimes they are even fancifully indicated by the letters of fome particular word, or name. We have fometimes, though feldom, feen the dial-plate of a clock divided into 24 hours, or twice 12; and in them the hour hand of courfe goes round once only in 24 hours. In certain particular time-pieces, the hours on the dial plate are difpofed in a fpiral manner; fo that four or fix of them fill up the whole circumference, and the reft are marked under thofe, and nearer to the centre (as in Dr. Franklin's clock, and others); the hour hand then going round the whole circumference in lefs time than 12 hours. See the next article.

DIAL-*Plate* of a *Clock* or *Watch*, called alfo the *face*, is that fixed plate which contains the divided circles of hours, minutes, and feconds, pointed to by the refpective hands, and which is pinned to the frame by the dial-pillars. This plate, in the ancient machines, was made of brafs, engraved into various devices, and filvered, fometimes only partially; but the modern makers prefer enamel to filvering, and that either real or imitative, according to the price intended to be charged for the workmanfhip. The maker's name is alfo ufually put on fome confpicuous part of the dial-plate; and if the machine is contrived fo as to have the repeating mechanifm, the days of the month, the moon's age, the days of the week, or fun's place in the ecliptic, &c. The divided circles containing thefe, or fome of thefe, are introduced, and perforations to fhew the divifions of other fmaller plates, revolving within or behind the principal plate, are frequently fuperadded.

DIAL-*Work* of a *Clock* or *Watch*, properly fpeaking, is that work which relates to the dial-plate, hands, dial-pillars, and fmall revolving plates, that are fometimes calculated to perform their revolutions in certain given periods of time; but, becaufe the refpective periods, and relative fituations of the moving parts depend on the motion-work, or wheel-work, contained between the dial-plate and the frame, the term dial-work may be faid to include alfo the motion-work, on which depend the figure, dimenfions, and circles of indication on the dial-plate, as well as the number of indicating hands. In the infancy of clock-work, there was but one hand, indicating the hours and fubdivifions of the hour, and inferted on the protruding arbor of the fufee, or barrel, made to revolve in twelve hours; but when the motion-work was introduced to fubdivide the hour more fenfibly into minute portions, it was found that any modified period of time might be reprefented by wheel-work, borrowing its motion from the going-part of the machine, without materially altering its rate of going, particularly when the motions produced are flow. Hence arofe the practice of reprefenting planetary phenomena, and of introducing various devices, either for ufe or ornament, which enhance the value, or rather the price of many clocks and watches, of which a great portion of the mechanifm is fuperfluous, and frequently renders the rate of their going unfteady. It might be confidered as unneceffary now, that clocks and watches are in every perfon's poffeffion, to give a minute detail of the circles and hands on an ordinary dial-plate; but as many of our readers may wifh to know the connection between the hour and minute hands, in regard to the mechanifm behind the dial, we beg leave to refer to *Plate* XII. of *Horology*, and to the defcriptions of an eight-days' portable clock with repeating mechanifm, under the head CLOCK, where the requifite information has been already given. The fame defcription is equally applicable to the dial-work of an ordinary watch: and when the feconds are fhewn by a long hand moving from the centre of the face, the calliper is fo laid down, that a cannon wheel revolves in a minute, (over a bridge that has got a tube furrounding the interior cannon wheel of the hours) in confequence of its connection with a fimilar wheel on the arbor of the wheel that ufually carries the feconds' hand out of the centre. The dials fhewn in *Plate* XXIII. already defcribed, as contrived by Dr Franklin and Mr. Fergufon, have no motion-work, but fimply the hands, or revolving plates attached to the arbors of the going part of the clock, which we mention here with a reference to the word CLOCK, left the reader fhould confult our prefent article for them in vain. We have alfo defcribed the dial-work of Enderlin's equation clock, as feen in *Plate* XXIV., in *fig.* 2, of which is feen a dial with the hours, minutes, and feconds out of the centre, together with other appendages, with the account of which the curious reader will be interefted in the article already referred to. We have, however, referved for our prefent article three or four different fpecimens of dial-work, that have not been before defcribed, and that, we truft, will therefore be acceptable to the public.

Plate XXXI. of *Horology* prefents two figures, the firft of which exhibits the motion-work under the dial of a clock, that indicates the days of the week and month, together with that of the moon's age, and at the fame time reprefents the moon's phafes; and *fig.* 2, is the reprefentation of the correfponding dial, with the parts feen through the perforated parts. In *fig.* 1, the letter *a* is placed on the wheel of 72 teeth that revolves, as in common, in twelve hours, by means of its connection with the cannon pinion of fix leaves; to this wheel is attached another fmaller one, *b*, of 30 teeth, or any other convenient number, which actuates a third of fixty, *i. e.* double its number, in 24 hours exactly, but in a contrary direction, which is feen denoted by *c*. This wheel of 60 teeth has two pins projecting from its plane, feen near the fmall letter *c*, one of which only we will regard at prefent. The pin nearer to the extremity of the wheel comes in contact with a tooth *d*, projecting from the edge of the vertical bar *de*, once in every 24 hours, and lifts this bar, which is kept in a true vertical pofition by a little bridge, *e*, below, and a pin within the flit at the fuperior end, which pin is held by a fmall cock, over the plane of the twelve-hours' wheel: the confequence of this diurnal lifting is, that a jointed tooth, *f*, attached to the faid vertical bar, takes one of the feven points of the ftar *g*, and turns the ftar till the heel of the fautoir, *h*, paffes the point of the ftar at prefent refting againft its leg, and pufhes it one-feventh part round by the action of the fmall fpring, *i*; the vertical bar then defcends by its weight, its tooth being clear of the pin in wheel *c*, till its jointed tooth, *f*, falls on the next fucceeding point of the ftar, on which

which it would rest if it had no joint; but the weight of the bar is sufficient to make this tooth, *f*, turn on its pivot or joint, till it has passed the said point of the star, after which the little spring underneath the back part of the tooth restores its horizontal position, below the point of the star that is to be caught the succeeding day; and in this way, the star is made to revolve in seven days, by as many sudden daily leaps, and it presents an attached plate with seven days of the week on it, as they occur, to an aperture at one of the inferior corners of the dial where the word *Monday* is now seen exhibited in *fig.* 2.

Again, the inner pin, projecting from the plane of the 24 hour wheel *c*, meeting with the curved triangular part of K*l*, which is a piece of metal moveable on a stud at the point *l*, slides along it, and at the same time raises it upwards, while the jointed tooth, attached to its opposite end, is depressed, so as to urge a tooth of the serrated rim, kept in its place by the three rollers, *m*, *n*, and *a*; presently, however, the pin of *c* comes in contact with the extreme corner of K, that points downwards, and passing it, allows it to return from its elevated position, till its tail-piece meets with a pin in the small cock below *l*; but its weight is sufficient to make it fall with velocity enough to bend the joint of the tooth that impels the rim; therefore, a spring is attached to the small cock, and acts on the back portion of the tail-piece, to give the requisite force in the fall. The tooth here has also a re-acting spring, like the jointed tooth *f*, and finds its horizontal position by means of its action on the back part of the tooth seen resting on a pin, that prevents its falling beyond the horizontal line. Now, as the rim in question has 31 teeth on its interior edge, one of which is moved thus every day, it is evident, that if its plane be divided into 31 equal spaces, they will be so many day spaces, to indicate the day of the month, as seen through some convenient opening in the dial, which, for the sake of uniformity, may be just above the point VI of the dial, as we have given it in *fig.* 2. This rim, it will be observed, will require a monthly adjustment by hand, that No. 1. may begin with the first day of each month. We have, indeed, seen a contrivance, different from Enderlin's already described, for making the month-plate adjust itself at the end of each month, in a clock made by a Scotchman of the name of Smith, of Pittenween, in Fifeshire, and exhibited in 1808, for public inspection, at No. 27, Leicester square; but as some part of the value of the superb machine consisted of this contrivance, which was shewn us in confidence, we feel not at liberty to make it public, without the proprietor's previous consent. On the rim we have been describing is a pin, which takes hold of one of the 12 points of the star *p*, the action of which is precisely the same as that of star *g*, already described; on a plate connected with this star, and revolving with it in 12 months, are the months designated and shewn through another opening at the inferior corner of the dial, in *fig.* 2. opposite to that of the days. The two engraved plates of the days and months are designated in *fig.* 1. by dotted circles only. The wheel *a*, which we have said revolves in the space of 12 hours, has also its pin for the moon's wheel, independently of its pin for making the clock strike, which pin takes the interior end of the lever *q*, that carries a third jointed tooth, in every respect like those described, so far as relates to the action; and whenever the interior end of the said lever is depressed by its pin of wheel 72, which happens once in each 12 hours, the jointed tooth pushes a tooth of the moon's wheel, which consist of 118; hence two of these are actuated in the space of each 24 hours, and the sautoir *r* keeps this wheel steadily to its place after each push; thus the moon's wheel, denoted by *s*, revolves once in 59 days, or in a period somewhat more than two lunations. On the plane of the moon's wheel, which is a solid plate, are painted two moons diametrically opposite each other, which appear to put on all the natural aspects successively, by the help of a triangular aperture in the dial-plate, two sides of which figure are semi-circles to the radius of the moon's curvature, as represented in *fig.* 2, which is the situation for Monday, July 4th 1808, when the moon's age is 12. The edge of the moon's plate, beyond the painted part, is figured twice over into 29½ spaces, which are the spaces of the moon's age seen through the curved opening over the moon herself, at any given time. The hour and minute hands, and also the circles of the dial, being of the ordinary kind, require no explanation. The practical objections to the dial-work here described, are, that the regularity of the maintaining power is greatly interrupted by the hourly, daily, and monthly opposing forces to be overcome by only short intervals of action, particularly when the weights and springs that constitute the opposing forces are considerable, which they must be to answer their purpose; secondly, the means used, though they profess to be simple, in as much as they avoid solar and lunar calculations for appropriate wheel-work, yet are actually more complex and expensive than proper trains would be, without re-acting springs, and other mechanical appendages; and, lastly, manual adjustment of the monthly plate is requisite many times within the year, as well as occasionally a rectification of the moon's age necessary, when the deficiency of 44 minutes 3 seconds amounts to an entire day. We might here proceed to shew how Ferguson, Jenkins, and others, have obviated the objections we have stated, by introducing the respective calculations of trains into their astronomical, or rather planetary, clocks; but as the accounts of them have been published by the authors themselves, in Ferguson's Mechanical Exercises, and in a description of several astronomical and geographical clocks, by Henry Jenkins, London, 1778, we will give the dial-work of a small portable clock, contrived by the writer of the present article in the year 1800, which indicates all the periods of the preceding clock, with the addition of the moon's place in the ecliptic, and of her nodes, latitude, and apogee, in a simple and unobjectionable manner, and which is the only clock that has been made of the kind.

Fig. 1. of *Plate* XXXII. of *Horology*, shews the motion-work of the little clock alluded to, and *fig.* 2. its dial, forming together what may be called its dial-wheel. Partly on account of the simplicity of the wheel-work, which stands in place of the ordinary motion-work, and partly for the sake of novelty, in the mode of reading the time indicated, the hours and minutes are both pointed to, at all times, by the same hand; the minutes are indicated in the outer large circle of Arabic figures, and the hours in the inner one, or circle of Roman numerals, which, at first sight, will appear a paradox, but when the reader is told that the inner circle moves as well as the common hand, the wonder will cease; their respective velocities are 60 : 55; so that supposing the point XII and the hand to start together from the point 60, in the fixed circle, when the hand has gone round, which it does in an hour, and indicates minutes on the fixed circle, the hour-plate has its point XII at 55, and its point I at 60, under the end of the hand; and at every successive revolution of the hand, the hour-plate falls back the twelfth part of a circle, and in the same proportion for any smaller quantity, by which means both the hour and minute, at any period of the day, are both immediately under their common hand. At present the indication is half past one o'clock, as seen in *fig.* 2; and puzzling as this mode may seem at first, it is actually more simple for a child to learn, than the ordinary mode by two hands, where one may be mistaken for the other; the only objection

objection is, that the hour cannot fo well be known at a diftance. The moveable plate being fmall, and equally poifed, by being circular, conftitutes no obftacle to free motion. The relative velocities are thus effected: the hand is carried by the arbor of the hour-wheel in the ufual way, and a tubed or cannon pinion, *a*, of 22 leaves, is put on tight, in the common manner, which gives its velocity to a pinion, *b*, of a fimilar number, and this pinion being made long and conical, turns on a ftud in the front of the frame, and actuates a third pinion of 24, concentric with the firft and covering it, round the tube of the little bridge that takes off the friction as ufual, fo that, as the velocities of the firft and third pinions are to each other inverfely as their refpective numbers of leaves, the revolving hand and hour-plate move with the velocities 24 and 22, that is, 60 and 55, as before ftated. In this mode of indicating time, three pinions without wheels are all that are neceffary for the motion-work, whereas three wheels and one pinion, or two wheels and two pinions, are ufually wanted to make the velocities of two hands 12 : 1.

But this fimple mode of indicating time has no connection with the other motions in the upper part of the dial-work, and the indication might be effected in the ordinary way without interfering with what follows, provided the ftriking work fhould be introduced and require it. It may be proper to obferve here, that the long fide pinion will act better when made not quite conical, but like two pinions, varying a little in diameter, placed clofe in contact. The fmall wheel *c* in *fig.* 1, revolves in feven days by means of its interior connection with the fufee-arbor, from which the moving power is taken, and is always detached, by pufhing in the key, during the act of winding, which is performed on the pofterior part; it is not neceffary to ftate more particularly how the multiplication from 12 hours to feven days arifes, as the operation is of the fimpleft nature; being only as 1 to 14; the arbor of this wheel *c*, which has 32 teeth, carries the fmall hand that points out the feven days of the week in one of the two correfponding fmall divided circles in *fig.* 2; by this wheel of 32 teeth is a large one of 135 driven, which is concentric with the lunar plate, that is carried round by its arbor in $\frac{135}{32}$ of 7 days, or in $29^{d}\ 12^{h}\ 45^{m}$. The arbor of this large wheel, as well as that of wheel 32, paffes through both plates of the clock frame, and is therefore fteady. On the plane of the moon's plate is painted a black fpiral on a white ground, tapering both ways from the middle in fuch a way, that the greateft breadth will juft cover the round hole made at the top of the dial, through which the moon's prefent afpect may always be feen, as it correfponds to her age. The arbor of the lunar wheel projects and receives three hands; that with a lune on it points out the moon's age in the outermoft circle of $29\frac{1}{2}$ fpaces of Arabic characters, and alfo the time of fouthing in the fecond circle of Roman numerals; the ftraight hand indicates the time of high water at each tide in the faid fecond circle, when fet for any particular place at new moon; and the third hand, that has a crofs near its extremity, fhews the day of the month in the outermoft circle, but requires adjuftment at the end of the month. To the lunar arbor is attached a pinion of 24 leaves, not feen, driving a wheel *d*, of 107 teeth round a ftud in the front plate, and another pinion of 31 leaves, faft to it, actuates a fecond wheel of 86 teeth round, in $\frac{107}{24}$ of $\frac{86}{31}$ of $29^{d}\ 12^{h}\ 45^{m}$, which period reduced is $365^{d}\ 6^{h}\ 1^{m}\ 5^{s}$, and a hand, inferted on its arbor, points out the month and fun's place in the fmall ecliptic, correfponding in fize to the week circle, in which the days are indicated. Upon the annual arbor of wheel 86 is alfo put a fmall circular plate divided into $29\frac{1}{2}$ fpaces, which fits into a round hole cut in the dial, the arbor being pivoted acrofs the frame, like the lunar and weekly arbors; the ufe of this fmall plate is to fhew the moon's place in the contiguous fmall ecliptic; for whatever be the moon's age indicated in its proper circle, the fame age looked to, in this little revolving circle of $29\frac{1}{2}$, will be oppofite the moon's mean place in the ecliptic at that time. Juft behind the fmall lunar plate which revolves in a year, and in the fame plane with the principal dial-plate, are fixed faft to the annual arbor two other wheels, one of 98 teeth, and the other of 55; the former clofe behind the dial, and the other contiguous to that; thefe drive each a correfponding wheel round the lunar arbor, confidered as a common ftem; the 98 drives a fellow of 93 teeth, and the 55 a fellow of 62; the former pair of wheels regulate the plate of the moon's nodes and latitude, marked on the plane of the wheel itfelf, which is not croffed, and the other pair carries a double hand, having A at one end, and P at the other, to fhew when the moon's hand comes in conjunction with the mean apogeal or perigeal point of her variable orbit. At the time the contriver of this clock adopted the numbers for the retrograde motion of the nodes, and progreffive motion of the apogee and perigee, he was not aware that they had been ufed in orreries, but he has fince found that the ratios he calculated for thefe purpofes have been chofen for the like ufes in fome other inftruments of illuftration. According to thefe numbers, the periods produced by the wheel-work are refpectively $18\frac{2}{3}$, and $8\frac{6}{7}$ civil years: and wherever the lunar hand is feen at any time, the latitude is indicated thereby on the innermoft circle, and the moon's pofition, with refpect to the apogee or perigee, may be feen, and confequently her apparent diameter and horizontal parallax, depending on her diftance from the earth, may be judged of at the time of infpection, and might even be indicated if another ftill fmaller plate were marked therewith, and fubftituted for the double hand. The pofitions of the refpective hands, and fmall plates, are for Monday July 4th, of the year 1808, when the moon's age is 12, her fituation at perigee, and alfo a little to the north fide of the afcending node, the fun's place, a little fhort of the middle of Cancer, and the moon's, a few degrees fhort of the end of Scorpio. In our article CHRONOMETER, we mentioned that the late Margetts contrived the dial-work of his chronometers fo ingenioufly, that both folar and fidereal time, in the three different denominations of hours, minutes, and feconds, are conftantly indicated at the fame time, and confequently, the fun's mean right afcenfion alfo, which is the difference between mean folar and fidereal time, at any given inftant; and as we promifed to defcribe the contrivance under the prefent head, we propofe to introduce it next.

Fig. 1. of *Plate* XXXIII. is a plan of the wheels and pinions employed in the motion-work, and *fig.* 2. is the reprefentation of the dials, of which there are four, one large and three fmall, and of the three hands; both figures being of the real fize. We will explain, *fig.* 1. firft; after premifing that we have inferted the letters of reference *a*, *b*, *c*, *d*, and *e*, on the margin of the face to avoid confufion, and have referred by dotted ftraight lines to the centres to which they refpectively belong, and where the reader is defired to conceive them placed: at the centre *a*, the hour-wheel or centre-wheel arbor projects, as ufual, from the frame of the chronometer, and takes the minutes-hand round the large circle, beyond the fmall face to the right in *fig.* 2, in a folar hour on a tube of fteel that fits tight by friction on the faid arbor, which tube is fquared as ufual to receive its hand; to the lower extremity of this tube, or cannon, as the workmen call it, are attached two cannon pinions, the inferior one of 80 leaves,

leaves, and the superior one of 15, both which consequently revolve together, in the same period of one hour; the under pinion of 80, being of small diameter, has its teeth so diminutive, that the naked eye can scarcely perceive that there are any, and drives the large wheel of 487 teeth round the centre *c* where a tube surrounds the solar seconds' hand, and forms a stud for this wheel, in order that the arbor of the seconds' hand, projecting as usual from the frame, may have no friction from contact, which it would have if the said fixed tube were not interposed; the period of this large wheel is $\frac{487}{80}$ of a solar hour, or $6^h\ 5^m\ 15^s$, and it carries round with it the lowest small face in *fig.* 2, in a direction contrary to that of the motion of the seconds' hand, which points to its divisions, and also to the fixed divisions of the circle surrounding this small face on the large dial itself. Again, the upper cannon pinion, of 15 leaves, actuates one of the two wheels of 75 teeth that revolve on the fusee's arbor, as on a stud, at the point *d*, both of which wheels, being of the same diameter and construction, and lying over one another, appear to an eye placed over them in the plan as one wheel; the wheel, however, in question, is the under one, to which a pinion of 15, not seen, is attached, that drives the under one of two similar wheels of 72 teeth each, round a stud at the point *b*, in the period $\frac{75}{15} \times \frac{72}{15}$ of a solar hour, that is, in 24 hours of solar time; and upon a tube on this under wheel of 72 is the solar hour hand placed, that revolves in an exact solar day, and indicates the 24 hours of mean solar time on the large circle of Roman characters, that surrounds the left-hand small face. Thus, the three hands indicate the solar hours, minutes, and seconds on the three fixed circles of the large plate or dial. We have already seen, that the great wheel of 487 teeth revolves, by its connection with the cannon pinion, of 80, in $\frac{487}{80}$ of an hour, or in $6^h\ 5^m\ 15^s$, which period is exactly $365\frac{1}{4}$ minutes of solar time; to this wheel a pinion of 12 leaves is fast, that impels a wheel of 72 teeth round a stud, or pin, on the cock *f*, that lies over a part of wheel 487, at the point *e*; this wheel has again a pinion fast to it of 8 leaves, that impels the wheel of 80 teeth round the cannon at *a*, in $\frac{72}{12} \times \frac{80}{8} = \frac{60}{1}$ of $365\frac{1}{4}$ minutes, which period is $15^d\ 5^h\ 15^m$, or $365\frac{1}{4}$ hours; and to the tube of this wheel of 80 the retrograde right-hand small plate, or dial, is attached by friction, and, consequently, revolves with it, as the small dial on the inferior part of the face does with the large wheel of 487, to which it is also attached by the friction of their tubes; and, lastly, a train of $\frac{75}{15} \times \frac{72}{15}$, in every respect similar to the one of this denomination, already described, converts the $365\frac{1}{4}$ hours into $365\frac{1}{4}$ days, for the period of the third, or left-hand small dial's retrograde revolution; that is, the pinion of 15, fast to the wheel of 80 teeth, drives the upper 75 round the point *d*, and its pinion of 15 again drives the upper 72 in the manner before described, the train being, in both cases, equal to $\frac{24}{1}$. Now, the manner in which sidereal time is indicated is, by means of the solar hands pointing to the three small dials that revolve in a retrograde direction, as they regard these hands, in $365\frac{1}{4}$ days, $365\frac{1}{4}$ hours, and $365\frac{1}{4}$ minutes respectively; the sidereal hours, minutes, and seconds being shorter in duration than solar ones, by such a minute quantity, as amounts, in the aggregate, to an entire sidereal day in a year, constitute that species of time which is used as the measure of the right ascension of any of the heavenly bodies, and which is itself measured by the earth's absolute rotations on its axis, as they have a reference to a star, or any fixed point in the heavens; whereas, solar, or common time, is measured by the said rotations, as they respect the sun, in apparent motion; which apparent motion is the consequence of the earth's annual progress in the ecliptic. The difference between a solar and sidereal day, in solar time, is somewhat less than four minutes, or $3^m\ 56^s.55$, so that a difference, or acceleration of sidereal, on solar time of one second, takes place in about six minutes of solar time, namely, in $\frac{365^m.25638}{60} = 6^m.0876$; hence, supposing the figures of the small dials to be put, at any time, to coincide with the same figures on the circles surrounding them on the fixed face, after about six minutes, or entire revolutions of the solar seconds'-hand, the revolving seconds'-face or small dial, ought to fall back *one second space*, but the other two small dials will appear unaltered, by reason of the slowness of their motion; in the next six minutes *two seconds* of difference will be indicated by the seconds'-hand, *i. e* the solar seconds will be pointed to on the fixed face, and sidereal seconds on the moveable dial, which, therefore, we call the sidereal seconds'-dial, and the difference of the two indications will be two seconds; in like manner, after sixty such periods of six minutes nearly, that is, after upwards of six hours, one minute-space of difference will be indicated by the minutes'-hand, when the time, shewn on the moveable dial, is compared with the time on the fixed circle of minutes, and the seconds'-hand will now, as at first, indicate both solar and sidereal seconds alike, by reason of the first minute of difference being completed; and thus, in the course of the whole 24 solar hours, there will be a difference of nearly four minute spaces shewn by the minute hand; but as yet there is scarcely any alteration perceptible in the hour-circles, nor will there be a difference of one hour's space till upwards of 15 days have transpired. Strictly speaking, the period in which the acceleration will amount to a second, by the wheel-work, will be $\frac{487}{80}$ of a minute, or 6.0875 minutes, and as the annual period of the retrograde motion of the sidereal hour-dial is $365^d\ 6^h\ 0^m$, instead of $365^d\ 6^h\ 9^m\ 11^s.5$, the whole error in the indication of sidereal time, by the solar hands pointing to the sidereal, or moveable dials, will be little more than one second and a half in a sidereal year, which, it will be allowed, is hardly worth naming. The adjustment of the dials is performed by the sun's mean right ascension, which is determined from the apparent right ascension, given for noon of each day, in the Nautical Almanac, and in White's Ephemeris, by applying the equation of time, together with the acceleration added thereto, at the rate of one second in each six minutes of equation, to convert it into sidereal time; but care must be taken to apply the equation with a contrary sign, to what is applied when apparent solar time is converted into mean. When the sun's mean right ascension is thus obtained in hours, minutes, and seconds, at the noon of any day, the hour on the sidereal hour-dial must be put to coincide with XII on the solar dial, the minutes on the sidereal minute-dial must be put to coincide with 60 on the solar minute-dial, and the seconds so determined must be put as counted on the sidereal seconds'-dial to 60 on the solar seconds'-plate, in which situation of things, the chronometer will indicate not only solar and sidereal time, but also the sun's mean right ascension at mean noon, from which the apparent

parent right afcenfion may be obtained, by the application of the equation of time, increafed by the acceleration, with the fign fet down in the almanac. In *fig.* 2, the rectification of the dials and hands is for the 3d of Sept. 1808, when the folar time is noon or 24^h, the fidereal is $10^h\ 49^m\ 46^s$, the fun's mean right afcenfion alfo $10^h\ 49^m\ 46^s$, and the equation — 50″, which makes the apparent right afcenfion $10^h\ 48^m\ 56^s$, agreeably to the almanac.

From this view of the dial-work of Margett's chronometers, it is eafy to fee, that when the folar time is accurately indicated, the fidereal time correfponding muft neceffarily be indicated, accurately too, within a fecond in the firft eight months after rectification, and within two feconds for the next eight months, and fo on for any length of time; and when the time of rectification has been noted, the error may be applied, as a rate, if neceffary, at any fubfequent period, which confideration annihilates the objection to a want of complete accuracy. The motions produced by the wheels and pinions in this complex motion-work, being all very flow, deduct very little from the maintaining power, fo little, indeed, as to be abfolutely imperceptible in practice, particularly when a good detached efcapement is ufed. The ufe of fuch dial-work in a chronometer, that is regulated by a tranfit inftrument, muft be evident to every perfon fkilled in aftronomy, navigation, or that branch of furveying which has to do with long lines, the bearings of which are afcertained by the greateft elongations of the pole ftar. The fmall dials will tell, at all periods of day or night, throughout the year, what heavenly bodies, the right afcenfions of which are previoufly known, are on or near the meridian of the place for which the chronometer has its time adjufted.

We might enlarge this article by the addition of various other contrivances, introduced by ingenious workmen into the dial-work of clocks and watches, but we will fatisfy ourfelves with a defcription of a curious method of indicating time that has hardly yet gained a name, but which ranks among the other modes of pointing out time, though it has not yet been made public. A German mechanift happened to be at Copenhagen, felling a valuable clock which he had contrived and made, when the Englifh caufed that city to capitulate, and the artizan was brought over to England, where he remains in effect a prifoner; being thus feparated from his family, and having no means of fubfiftence but his fkill in mechanics, he invented the contrivance we are going to give a brief fketch of, and the plan of which is given in *fig.* 3, of *Plate* XXXIII. A B is a circle of brafs divided into twelve hours, and fub-divided into five minute fpaces, which circle is fupported, as feen in the figure, in a vertical pofition on a wooden bafe; C and D are two weights, apparently balanced on the oppofite ends of a bar, that turns on an axis, placed horizontally at right-angles, and fupported by a bearer in the fhape of a fifh refting on its head, the counterpoife C is ornamented with a ftar for the fake of a deception, that the mind may not be induced to fuppofe that any mechanifm is included. The curiofity, as well as utility of the contrivance, confift in this apparently felf-acting principle; move the bar, that now refts at $7^h\ 55^m$, forcibly round, fo as to give it confiderable velocity, and it will continue to fpin round for fome time, but whenever it comes to reft, it will come to a ftate of quiefcence, at any hour and minute of the day or night, with its index at the very point of the large divided circle that the prefent hour and minute require, though a fuperficial view of the thing will not difcover any caufe for fuch tendency to reft always at the prefent hour and minute, any more than the polarity of a magnetic needle can be difcovered by ocular examination. The fact, however, is, that a watch is contained in the box C, which makes a wheel revolve in twelve hours, that is loaded with a weight heavy enough to alter the common centre of gravity of the loads C and D, as the faid wheel revolves, thereby throwing the centre of gravity to the right and left of the axis of motion alternately, and caufing the ends of the bar, with its loads, to preponderate alternately, and, confequently, to revolve gradually from XII round to XII again, as the watch governs the rate. It will be feen, however, that the velocity of the bar will not be uniform in every part of its revolution, becaufe the cofines of the angles, denoted by equal times, do not vary by equal quantities, and the ingenious remedy for the natural inequalities of motion among the equally divided fpaces of the circle, conftitutes the fecret on which the accuracy of the thing, as a meafure of time, depends, when the watch is good, and which, therefore, we feel not at liberty to difclofe under the prefent circumftances of the manufacturer. We underftand, that their royal highneffes the dukes of York and Kent have purchafed each one of thefe horofcopes, as they may be not improperly called, as has alfo general Campbell; and Meffrs. Rundell and Co. of Ludgate-hill, have kindly undertaken to vend this ingenious article for the benefit of the inventor, who puts no higher price on it than about feven guineas.

DIALLING, the art of drawing fun, moon, and ftar-dials, on any given plane, or on the furface of any given body. See DIAL.

The Greeks and Latins call this art gnomonica, and fciatherica, becaufe it diftinguifhes the hours by the fhadow of a gnomon. Some call it photo-fciatherica, becaufe the hours are fometimes fhewn by the light of the fun. Laftly, others call it horologiography.

Dialling is wholly founded on the firft motion of the heavenly bodies; and chiefly the fun; or rather on the diurnal rotation of the earth; fo that the elements of fpherics, and the fpherical aftronomy, fhould be maftered, before a perfon advances to the doctrine of dialling: the doctrine, or theory, we fay, for as to the practice, or the operations themfelves diftinct from the demonftrations, nothing is more eafy and obvious.

The principles of dialling may be eafily deduced from the method, already illuftrated, of conftructing dials by the globe. Or, they may be farther explained, by fuppofing the whole earth P *c p* (*Plate* IV. *fig.* 36.) to be tranfparent and hollow, like a fphere of glafs, and its equator to be divided into twenty-four equal parts by fo many meridian femicircles *a*, *b*, *c*, *d*, *e*, *f*, *g*, &c. one of which is the geographical meridian of any given place, as London, which is fuppofed to be at the point *a*; and if the hour of XII were marked at the equator, both upon that meridian and the oppofite one, and all the reft of the hours in order on the other meridians, thofe meridians would be the hour-circles of London: becaufe, as the fun appears to move round the earth, which is in the centre of the vifible heavens, in twenty-four hours, he will pafs from one meridian to another in an hour. Then, if the fphere had an opake axis, as P E *p*, terminating in the poles P and *p*, the fhadow of the axis, which is in the fame plane with the fun, and with each meridian, would fall upon every particular meridian and hour, when the fun came to the plane of the oppofite meridian, and would confequently fhew the time at London, and at all other places on the fame meridian. If this fphere was cut through the middle by a folid plane A B C D in the rational horizon of London, one half of the axis E P would be above the plane, and the other half below it; and if ftraight lines were drawn from the centre of the plane to thofe points where its circumference is cut by the hour circles of the ſphere, thoſe lines would be the hour-lines of the horizontal dial for London; for the ſhadow of the axis would fall on each particular hour-line of the dial, when it fell upon the like hour-circle of the ſphere. If the plane which cuts the ſphere be upright as A F C G (*fig.* 37.) touching the given place, *e. g.* London, at F, and directly facing the meridian of London, it will then become the plane of an erect direct ſouth dial; and if right lines be drawn from its centre E, to thoſe points of its circumference where the hour-circles of the ſphere cut it, theſe will be the hour-lines of a vertical or direct ſouth dial for London, to which the hours are to be ſet in the figure, contrary to thoſe on a horizontal dial; and the lower half E *p* of the axis will caſt a ſhadow on the hour of the day in this dial, at the ſame time that it would fall upon the like hour-circle of the ſphere, if the dial plane was not in the way. If the plane, ſtill facing the meridian, be made to incline, or recline, any given number of degrees, the hour-circles of the ſphere will ſtill cut the edges of the plane in thoſe points to which the hour-lines muſt be drawn ſtraight from the centre; and the axis of the ſphere will caſt a ſhadow on theſe lines at the reſpective hours. The like will ſtill hold, if the plane be made to decline by any given number of degrees from the meridian towards the eaſt or weſt; provided the declination be leſs than 90 degrees, or the reclination be leſs than the co-latitude of the place; and the axis of the ſphere will be the gnomon; otherwiſe, the axis will have no elevation above the plane or the dial, and cannot be a gnomon.

Thus it appears, that the plane of every dial repreſents the plane of ſome great circle on the earth, and the gnomon the earth's axis; the vertex of a right gnomon the centre of the earth or viſible heavens; and the plane of the dial is juſt as far from this centre as from the vertex of this ſtyle. The earth itſelf, compared with its diſtance from the ſun, is conſidered as a point; and therefore, if a ſmall ſphere of glaſs be placed upon any part of the earth's ſurface, ſo that its axis be parallel to the axis of the earth, and the ſphere hath ſuch lines upon it, and ſuch planes within it, as above deſcribed, it will ſhew the hours of the day as truly as if it were placed at the earth's centre, and the ſhell of the earth were as tranſparent as glaſs. Ferguſon, ubi ſupra. See Gnomonic *Projection.*

The principal writers on dialling are, Vitruvius, Sebaſtian Munſter, John Dryander, Conrade Geſner, Andrew Schoner, Fred. Commandine, Joan. Bapt. Benedictus, J. G. Schomberg, Solomon de Caus, J. B. Trolta, Deſargues, Kircher, Hallum, J. Mark, and Clavius, who demonſtrates all, both the theory and the operations, after the rigid manner of the ancient mathematicians; but ſo intricately, that probably few ever read them: Dechales, Ozanam, and Schottus, give much eaſier in their Courſes, and Wolfius in his Elements.

M. Picard has given a new method of making large dials by calculating the hour-lines; and M. de la Hire, in his Dialling, printed in 1683, a geometrical method of drawing hour-lines from certain points, determined by obſervation. Eberhardus Walperus, in 1625, publiſhed his Dialling, wherein he lays down a method of drawing the primary dials on a very eaſy foundation. The ſame foundation is deſcribed at length by Sebaſtian Munſter, in his Rudimenta Mathematica, publiſhed in 1651. Sturmius, in 1672, publiſhed a new edition of Walperus's Dialling, with the addition of a whole ſecond part, about inclining and declining dials, &c. In 1708, the ſame work, with Sturmius's additions, was republiſhed with the addition of a fourth part, containing Picard's and De la Hire's methods of drawing large dials, which makes much the beſt, and fulleſt book on the ſubject. Peterſon, Michael, and Muller, have each written on dialling, in the German tongue; Coetſius, in his Horologiographia Plana, printed in 1689; Gauppenius, in his Gnomonica Mechanica; Leybourn, in his Dialling, fol.; Bion, in his Uſe of Mathematical Inſtruments; and Wells, in his Art of Shadows. We have a treatiſe by M. Deparcieux, in 1740. Mr. Ferguſon has alſo written on this ſubject in his Lectures on Mechanics: Mr. Emerſon, in the 9th volume of his works; and Mr. W. Jones, in his Inſtrumental Dialling.

Dialling *Cylinder, Univerſal,* is repreſented by *fig.* 38. A B C D is a glaſs cylindrical tube, cloſed at both ends with braſs plates, in the centres of which a wire or axis E F G is fixed. The tube is either fixed to an horizontal board H, ſo that its axis may make an angle with the board equal to that which the earth's axis makes with the horizon of any given place, and be parallel to the axis of the world; or it may be made to move on a joint, and elevated for any particular latitude. The twenty-four hour-lines are drawn with a diamond on the outſide of the glaſs, equidiſtant from each other, and parallel to the axis. The XII next B ſtands for midnight, and the XII next the board H for noon. When the axis of this inſtrument is elevated according to the latitude, and the board ſet level with the line H N in the plane of the meridian, and the end N towards the north, the axis E F G will ſerve as a gnomon or ſtyle, and caſt a ſhadow on the hour of the day among the parallel hour-lines, when the ſun ſhines on the inſtrument. As the plate A D at the top is parallel to the equator, and the axis E F G perpendicular to it, right lines drawn from the centre to the extremes of the parallels will be the hour-lines of an equinoctial dial, and the axis will be the ſtyle. An horizontal plate *e f* put down into the tube, with lines drawn from the centre to the ſeveral parallels cutting its edge, will be an horizontal dial for the given latitude; and a vertical plate *g c*, fronting the meridian and touching the tube with its edge, with lines drawn from its centre to the parallels, will be a vertical ſouth dial: the axis of the inſtrument ſerving in both caſes for the ſtyle of the dial; and if a plate be placed within the tube, ſo as to decline, incline, or recline, by any given number of degrees, and lines being drawn, as above, a declining, inclining, or reclining dial will be formed for the given latitude. If the axis with the ſeveral plates fixed to it be drawn out of the tube, and ſet up in ſunſhine in the ſame poſition they were in the tube, A D will be an equinoctial dial, *e f* an horizontal dial, and *g c* a vertical ſouth dial, and the time of the day will be ſhewn by the axis E F G. If the cylinder were wood, inſtead of glaſs, and the parallel lines drawn upon it in the ſame manner, it would ſerve to facilitate the operation of making theſe ſeveral dials. The upper plate with lines drawn to the ſeveral interſections of the parallels, which appears obliquely in *fig.* 38, would be an equinoctial dial, as in *fig.* 39, and the axis perpendicular to it be its ſtyle. An horizontal dial for the latitude of the elevation of the axis might be made, by drawing out the axis and cutting the cylinder, as at *e f g h*, parallel to the horizontal board H; the ſection would be elliptic, as in *fig.* 40. A circle might be deſcribed on the centre, and lines drawn to the diviſions of the ellipſe would be the hour lines; and the wire put in its place again, as E, would be the ſtyle. If this cylinder were cut by a plane perpendicular to the horizontal board H, or to the line S H N, beginning at *g*, the plane of the ſection would be elliptic, as in *fig.* 41, and lines drawn to the points of interſection of the parallels on its edge would be the hour-lines of a vertical direct ſouth dial, which might be made of any ſhape, either circular or ſquare; and F the axis of the cylinder would be its ftyle. Thus alfo inclining, declining, or reclining dials might be eafily conftructed, for any given latitude. Fergufon, ubi fupra.

Dialling *Globe*, is an inftrument made of brafs, or wood, with a plane fitted to the horizon, and an index; particularly contrived to draw all forts of dials, and to give a clear exhibition of the principles of that art.

Dialling *Lines*, or *Scales*, are graduated lines, placed on rules, or the edges of quadrants, and other inftruments, to expedite the conftruction of dials.

The principal of thefe lines are, 1. A fcale of fix hours, which is only a double tangent, or two lines of tangents, each of 45 degrees, joined together in the middle, and equal to the whole line of fines, with the declination fet againft the meridian altitudes in the latitude of London, fuppofe, or any place for which it is made: the radius of which line of fines is equal to the dialling fcale of fix hours. 2. A line of latitudes, which is fitted to the hour-fcale, and is made by this canon: as the radius is to the chord of 90 degrees; fo are the tangents of each refpective degree of the line of latitudes to the tangents of other arches: and then the natural fines of thofe arches are the numbers, which, taken from a diagonal fcale of equal parts, will graduate the divifions of the line of latitude to any radius. The line of hours and latitudes is generally for pricking down all dials with centres. For the method of conftructing thefe fcales, fee Scale. There are feveral other lines put upon fcales for dialling, adapted to particular latitudes. Such are, 1. A line of chords: 2. A line of the fubftyle's diftance from the meridian: 3. A line for the height of the ftyle above the place: 4. A line of the angle of the hours of 12 and 6: 5. A line of the plane's difference of longitude, or inclination of meridians. All thefe are calculated for every degree of declination, for fome particular latitude, *e. g.* that of London, by the following problem. Let the latitude of the place and the declination of the plane be given; and let it be propofed to find the height of the pole above the dial-plane, the diftance of the fubftyle from 12 o'clock, and the plane's difference of longitude. Let N E S W (*fig.* 42.) be the horizon, C its centre, NS the meridian, P the pole, and Z the zenith; H A the declining vertical plane; P 1, P 2, P3, &c. meridians or femi circles, interfecting the circle H A in *t*, *v*, &c. Let the meridian P *v* 2 be perpendicular to H A, then the plane of the meridian P *v* 2 will be perpendicular to the plane of the dial; and therefore the angle Z P*v* is the plane's difference of longitude, or the angle on the fphere contained between the meridian of the place and the meridian of the plane, called alfo the "inclination of meridians," and the arc P *v* is the height of the pole above the plane; and the arc Z*v* is the diftance of the fubftyle from the meridian. Therefore in the right-angled fpherical triangle PZ *v* all the three requifites will be found. For we have P Z the complement of the latitude, and the angle P Z A the complement of the declination A E; whence will be found P *v*, Z *v*, and the angle Z P*v*. Thus 1. As radius is to the fine of declination, fo is the cotangent of latitude to the tangent of the fubftyle's diftance from 12 o'clock; for by a cafe of right-angled fpherical triangles, rad. : cof. P Z *v* :: tang. PZ : tang. Z *v*. 2. Radius is to the cofine of declination as the cofine of latitude is to the fine of the ftyle's height; for by right-angled fpherical triangles, rad. : S . PZ :: S. P Z *v* : S . P*v*. 3. Sine of latitude is to radius as the tangent of declination is to the tangent of the plane's difference of longitude; for, rad. : cof. PZ :: tang. PZ*v* : cotang. Z P *v* :: tang. Z P*v*. cotang. PZ *v*. 4. For the hour angles, it will be, as radius is to the fine of the pole's height, fo is the tangent of the hour-arc from the meridian of the plane to the tangent of the hour angle from the fubftyle. For let P *t* 1 be any hour-circle; and in the triangle P *t v*, we have the pole's height P *v*, and the hour-arc, equal to the angle *t* P *v*; and therefore to find *t v*, fay, rad. : S . P *v* :: tang. *t* P *v* : tang. *t v* = the angle at the centre, or hour-angle. That thefe feveral conclufions are rightly deduced will appear thus. The planes of the two meridians PZ, P*v*, interfect one another on the fphere in an angle Z P *v*, equal to the plane's difference of longitude. And the fame meridians interfect the plane of the dial in an angle equal to Z V, as it fhould be. For they interfect one another in the centre of the fphere, which is the centre of the dial. And they interfect the dial-plane H A in Z and *v*, and Z *v* meafures that angle at the centre. For the fame reafon, *t v* meafures the hour-angle from the fubftyle. Again, P *v* meafures the angle of the ftyle above the fubftyle. For P *v* is perpendicular to H A, and the arc P *v* meafures the angle formed at the centre of the fphere, which is the centre of the dial; one fide being the axis drawn from P, the other the fubftyle drawn from V. Moreover, Z P *v* has been fhewn to be the plane's difference of longitude. It appears farther, that the plane's differencce of longitude on the fphere is converted into the fubftyle's diftance from 12, by the interfection of the two meridians (of the place and of the plane) with the dial-plane: For the angle Z P *v* expreffes the one, and the arc Z *v* the other. Again, every hour-angle on the fphere is converted into the hour-angle in the dial, by the interfection of the fame two hour-circles, with the plane of the dial; for the angle *t* P *v* expreffes the hour-angle on the fphere, and *t p* the angle at the centre. Similar refults may be obtained from other data by different proportions. Thus, 1. For finding the fubftyle's diftance: cof. of pole's height above the plane P*v* : S. lat. or cof. P Z :: radius : cof. of the diftance of the fubftyle from 12. Or, radius : S. declination or cof. P Z *v* :: tang. of the pole's height above the plane P*v* : S. of the fubftyle's diftance Z*v*. 2. For finding the ftyle's height: cof. of fubftyle's diftance, Z *v* : S. lat. or cof. P Z :: radius : cof. ftyle's height P *v* : or, rad. : S. fubftyle's diftance, Z *v* :: cotang. declination, or tang. PZ*v* : tang. of ftyle's height, P*v*. 3. For the plane's difference of longitude; cof. lat. or S. PZ : rad. :: S. fubftyle's diftance, Z*v* : S. plane's difference of longitude, Z P *v*. Or, rad. : tang. lat. or cotang. P Z :: tang. ftyle's height, P *v* : cof. of plane's difference of longitude, Z P*v*. Or, rad. : cof. declination, S . P Z *v* :: cof. fubftyle's diftance Z*v*: cof. of plane's difference of longitude, ZP*v*. Or, cof. ftyle's height, P*v* : S. declination, or cof. P Z *v* :: rad. : S. of plane's diff. of longitude, Z P *v*. Or, S. of ftyle's height P*v* : rad. :: tang. of fubftyle's diftance, Z *v* : tang. of plane's diff. of longitude, Z P*v*. On a fcale the feveral requifites for any declining dial are had by infpection. In ufing it, count the plane's declination on the line of chords, and a line drawn directly acrofs, will interfect all the other lines in their proper points; which the numbers of graduation will indicate, and thefe give all the requifites without calculation.

When a fcale of this fort is not to be had, the requifites may be found by Gunter's fcale, extending upon the feveral lines, according to the rules and proportions laid down for that purpofe.

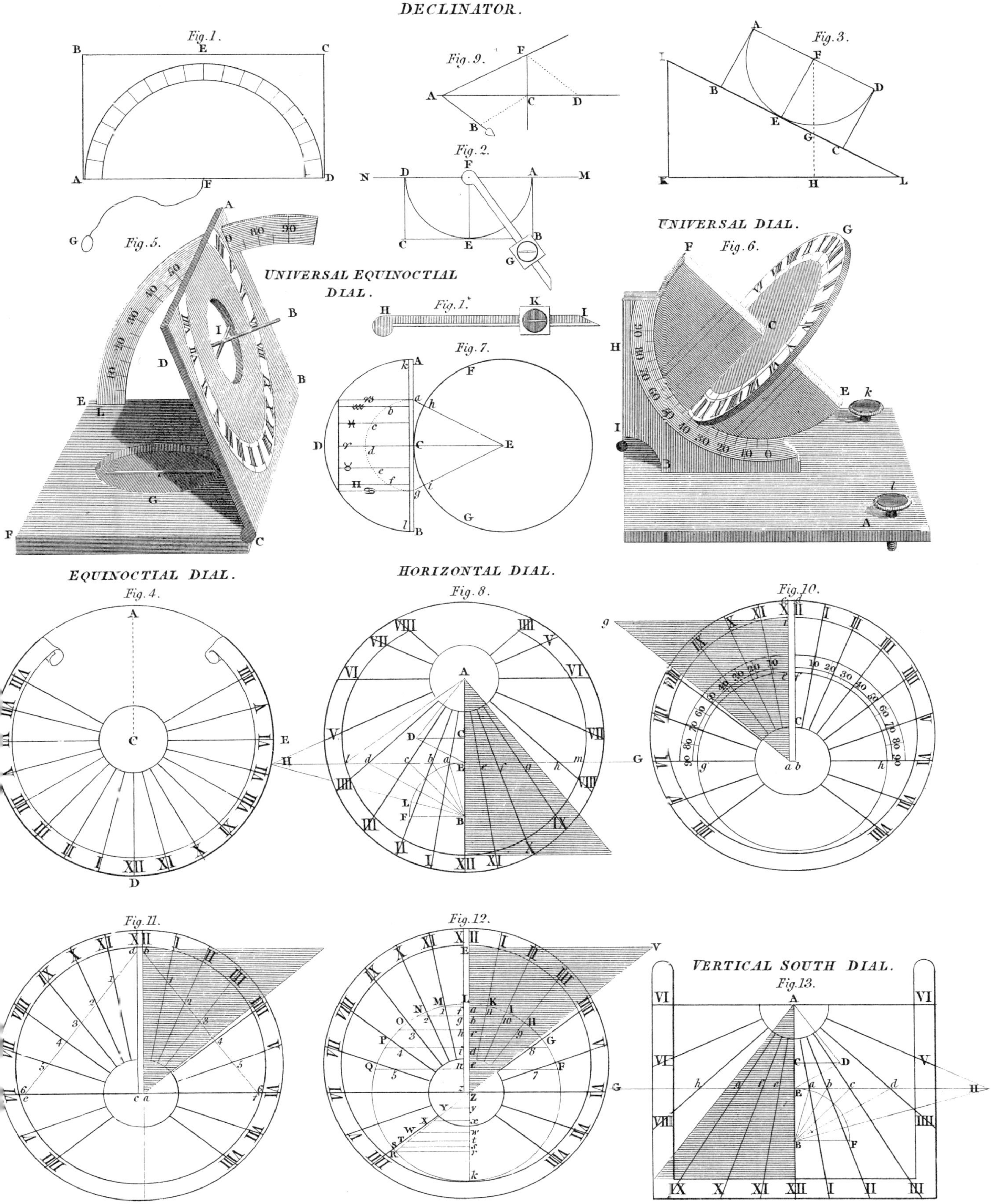

Published as the Act directs, 1809, by Longman, Hurst, Rees & Orme, Paternoster Row.

Engraved by Wilson Lowry.

Fig. 14.
VERTICAL NORTH DIAL

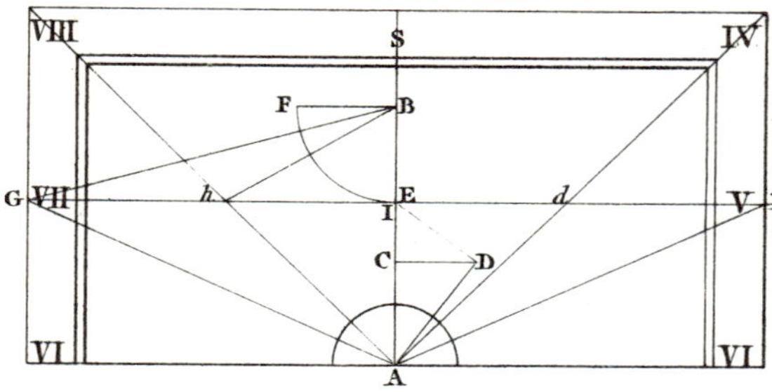

POLAR DIAL
Fig. 17.

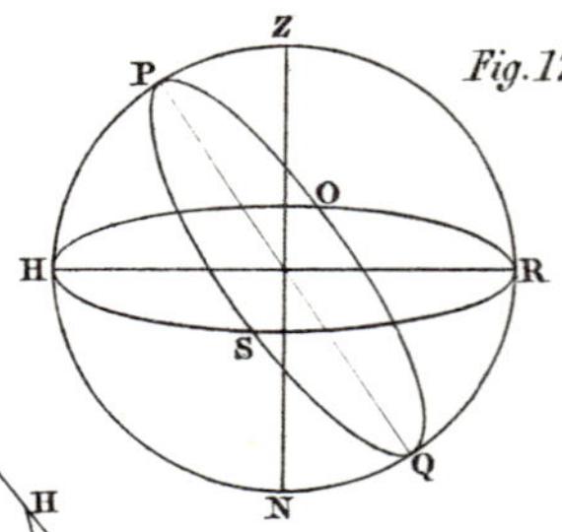

Fig. 15.
SOUTH DIAL

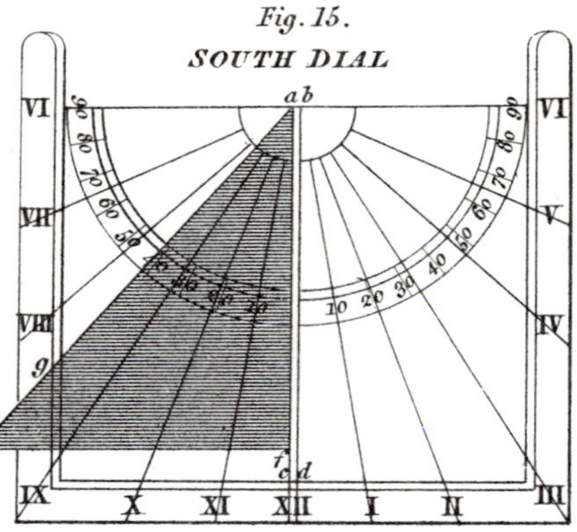

Fig. 16.
EAST DIAL

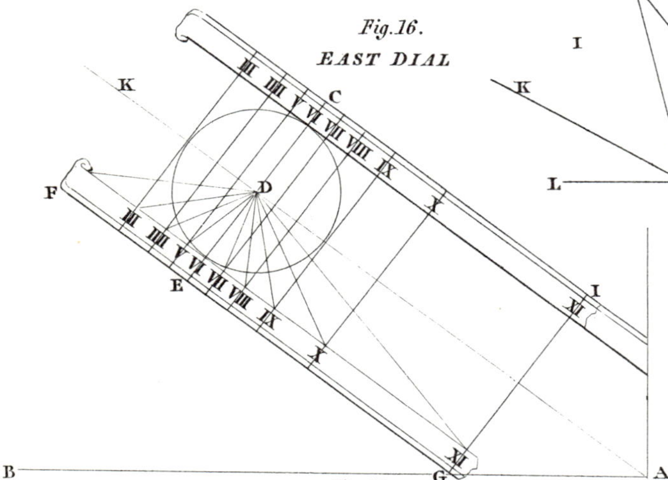

Fig. 23.
INCLINED DIAL

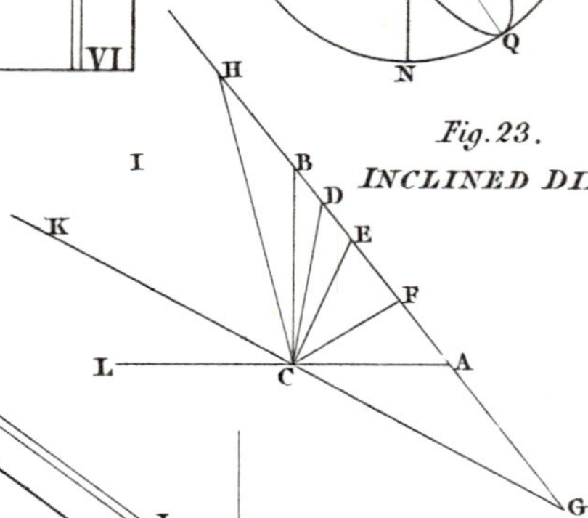

Fig. 22.
VERTICAL DECLINING DIAL

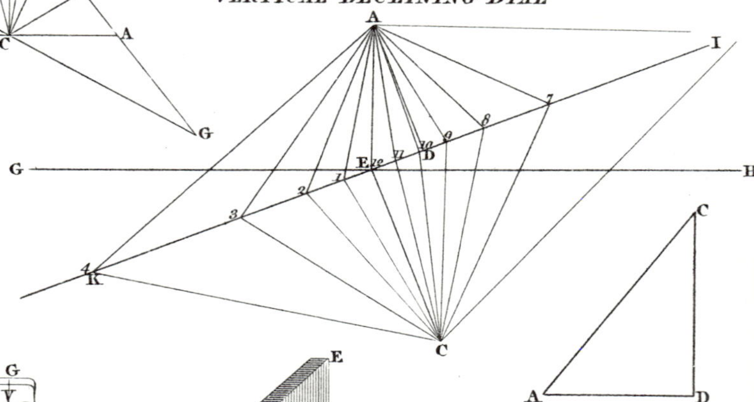

Fig. 18.

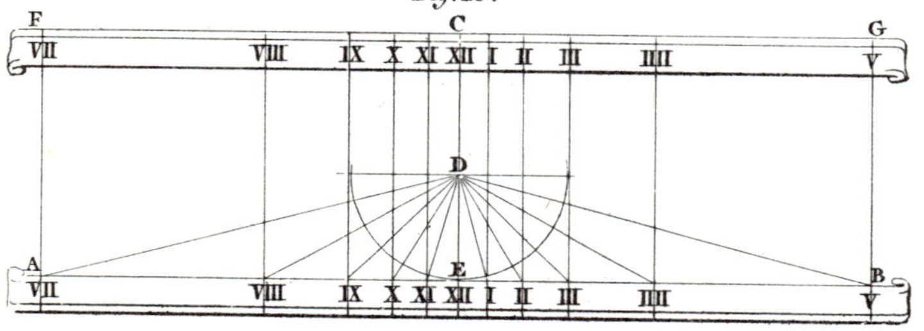

Fig. 19.
DIAL ON THREE PLANES

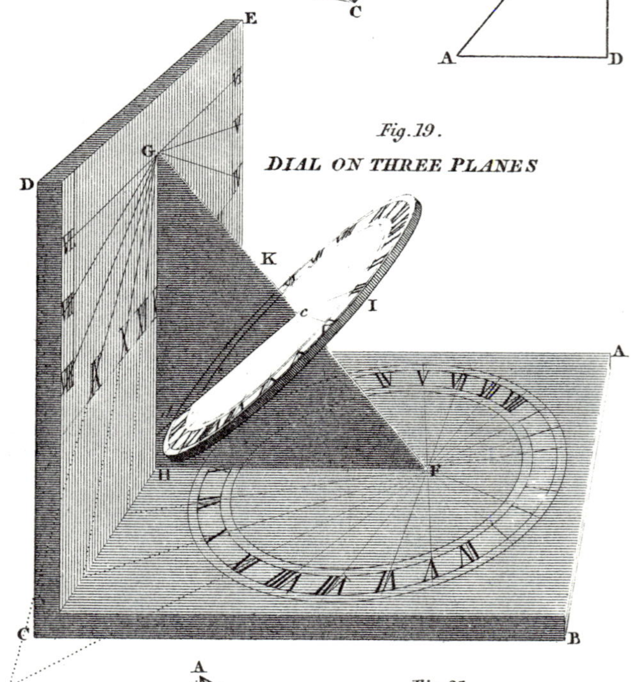

DIALS
Fig. 20.

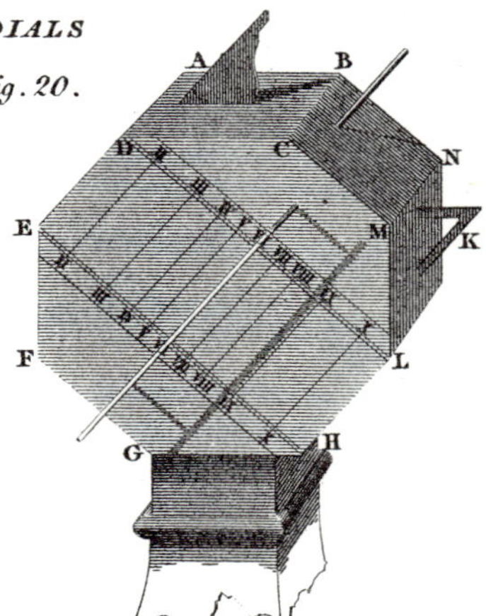

Fig. 24.
UNIVERSAL MECHANICAL DIAL

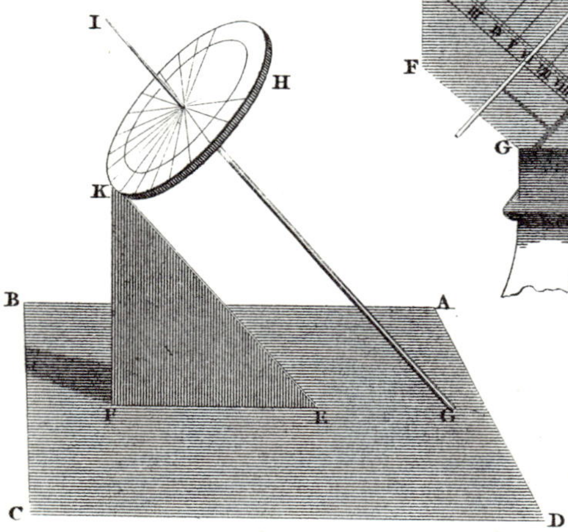

Fig. 21.
VERTICAL DECLINING DIAL

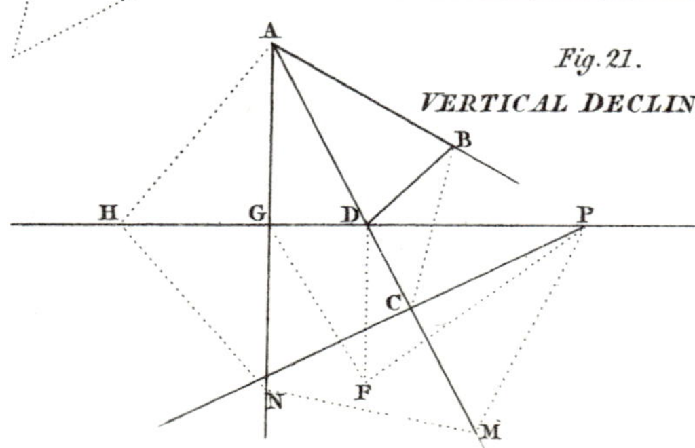

Published as the Act directs, 1809, by Longman, Hurst, Rees & Orme, Paternoster Row.

Engraved by Wilson Lo

Cylindrical DIAL, Fig. 28.

Fig. 27.

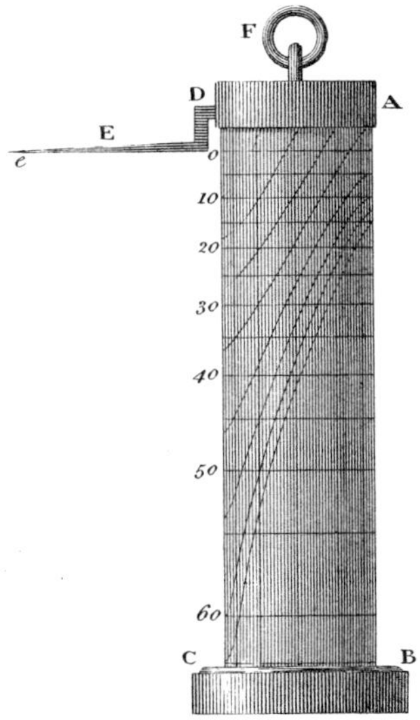

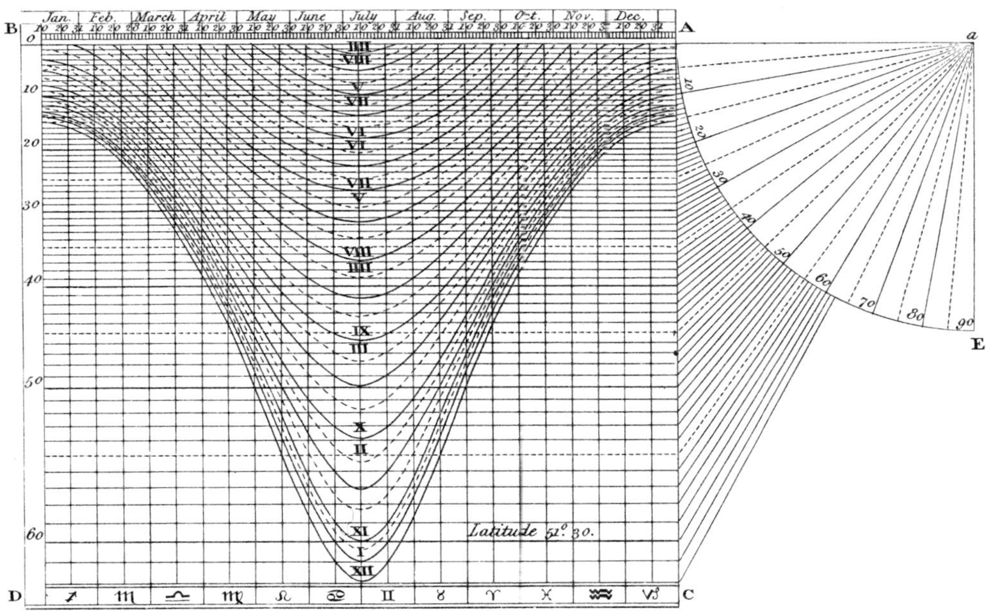

Portable DIAL.

Fig. 29.

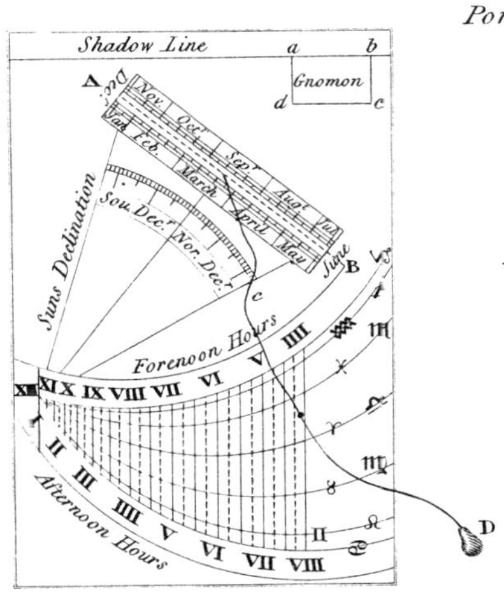

Fig. 30.

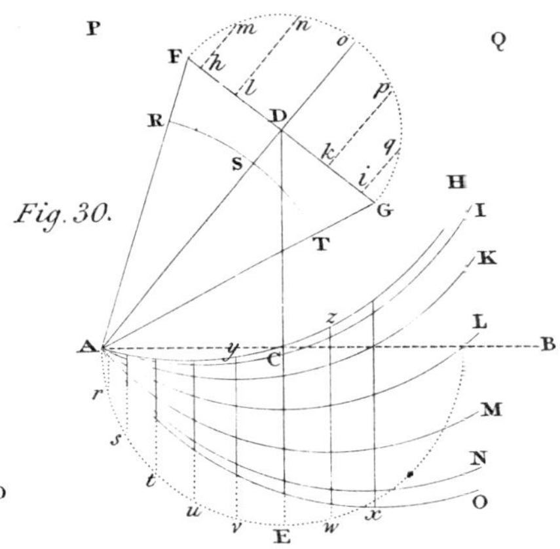

Fig. 32.

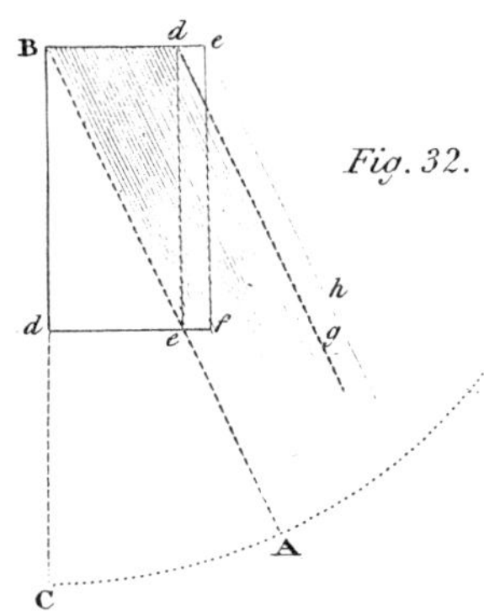

Fig. 31.

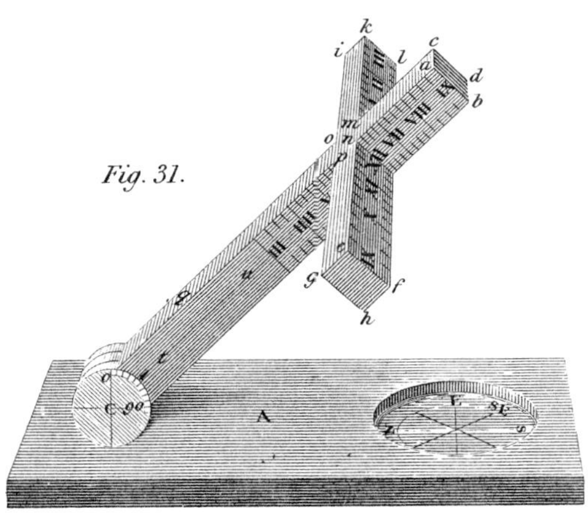

Universal DIAL on a Cross.

Fig. 33.

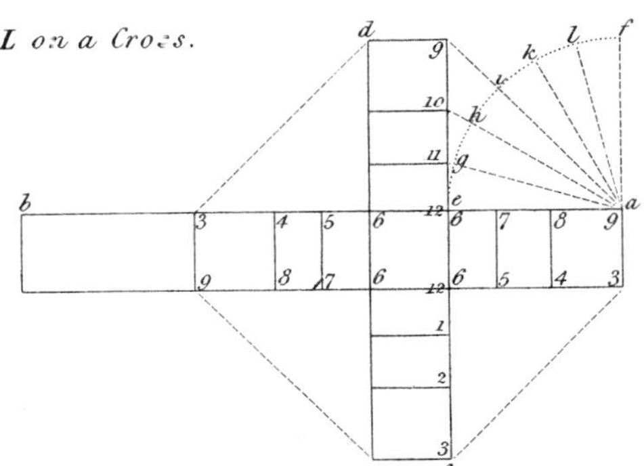

Published as the Act directs, 1813, by Longman, Hurst, Rees, Orme & Brown, Paternoster Row, London.

Lowry sculp.

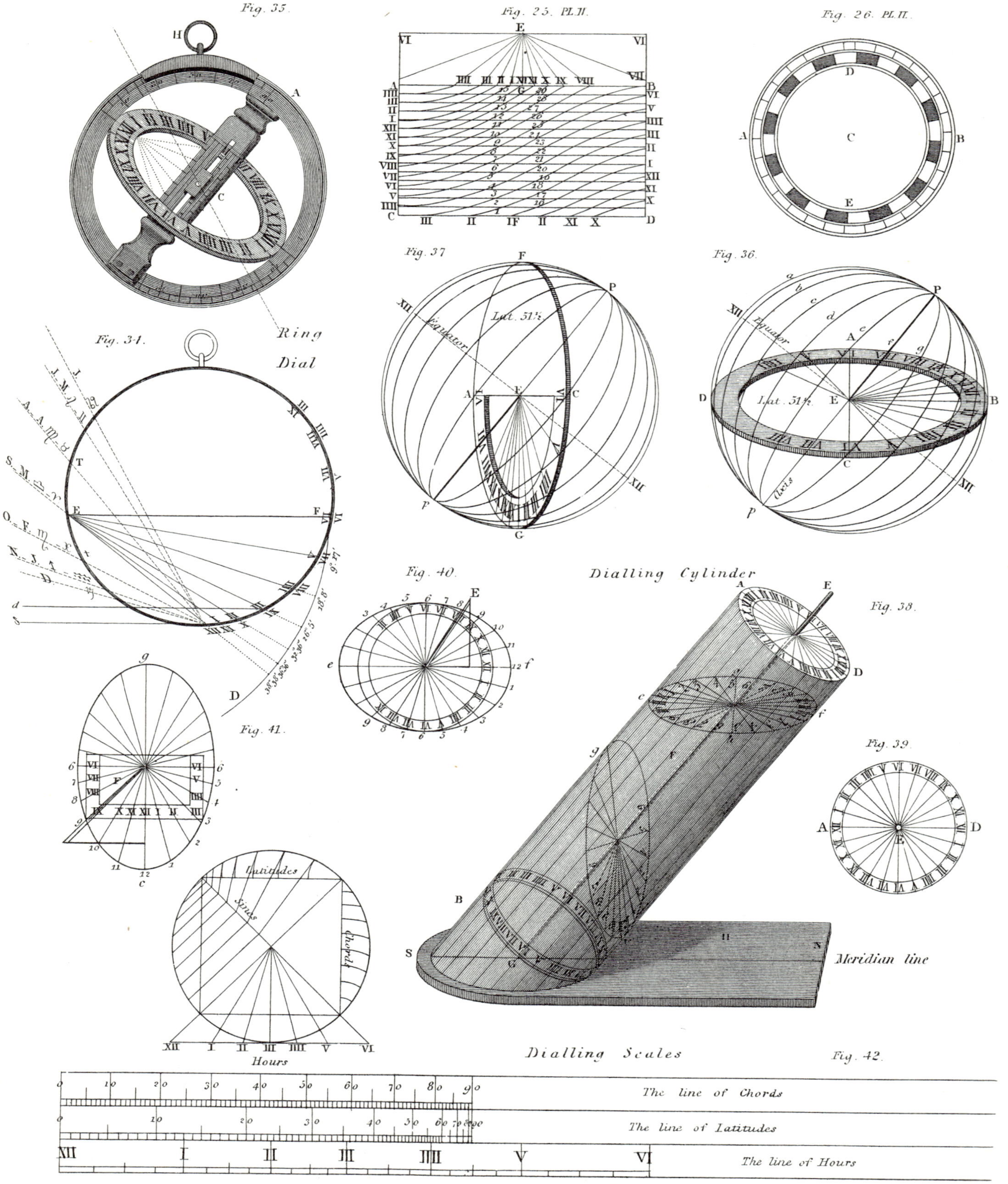

J. Farey. Junr. delin.

Published as the Act directs, 1816, by Longman, Hurst, Rees, Orme, & Brown, Paternoster Row

Lowry Sculp

DIAL WORK OF A CLOCK SHEWING THE MOON'S AGE, PHASES &c.

Fig. 1.

Fig. 2.

J. Farey Jun.r delin.

Published as the Act directs, 1809, by Longman, Hurst, Rees & Orme, Paternoster Row.

Engraved by Wilson Lowry.

page 182

HOROLOGY.

New DIAL work of a small spring Clock shewing the Phenomena of the Moon &c.

Fig. 2.

Fig. 1.

J. Farey Jun.r delin.

Published as the Act directs, 1811, by Longman, Hurst, Rees, Orme & Brown, Paternoster Row.

Engraved by Wilson Lowry.

Dial Work.

Fig. 1.

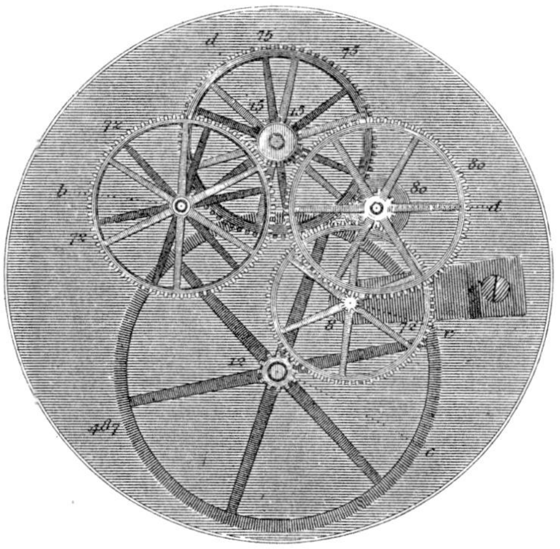

Fig. 2.

Fig. 3.

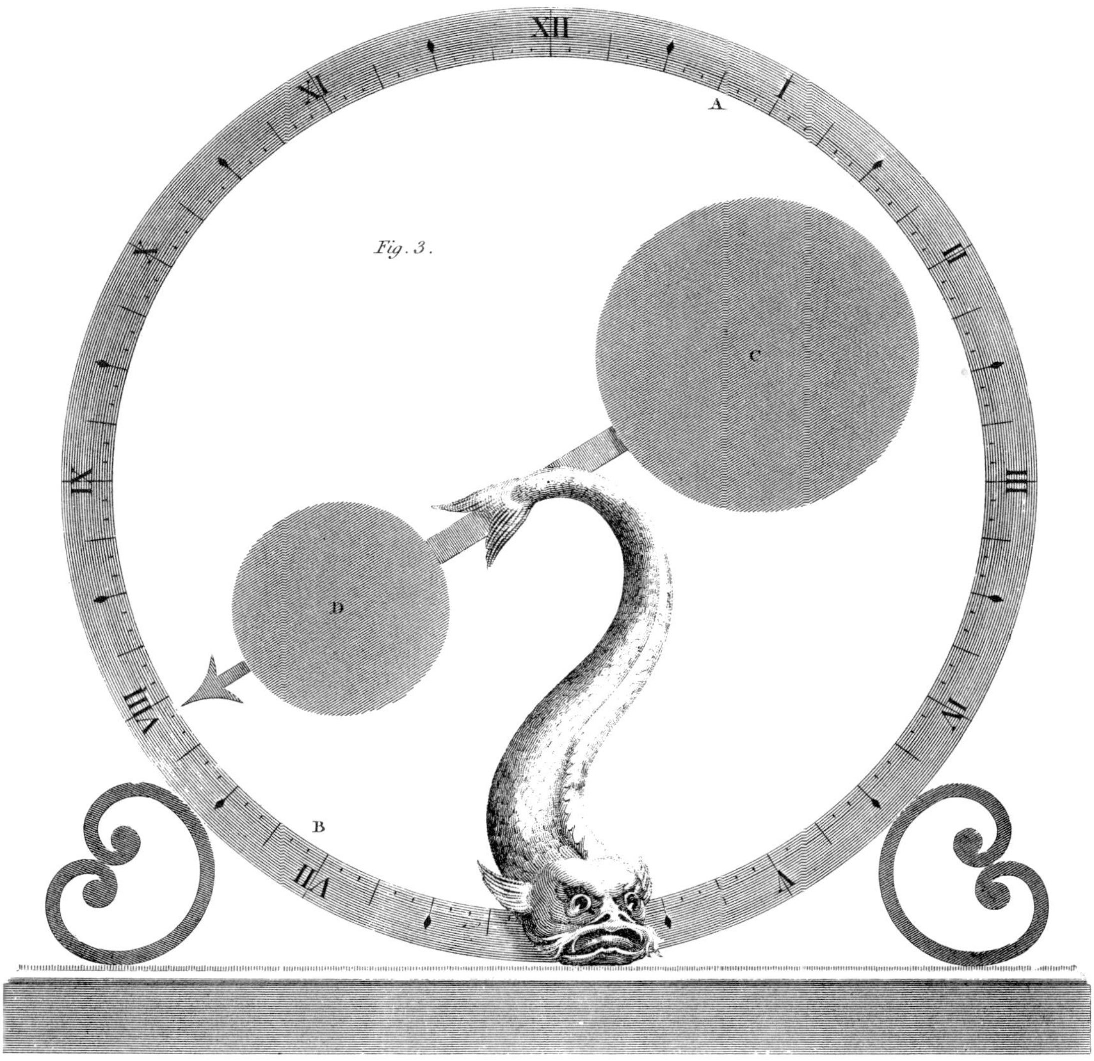

by J. Farey Jun.r *Published as the Act directs 1811, by Longman, Hurst, Rees, & Orme, Paternoster Row.* *Engraved by W. Lowry.*

Escapement

ESCAPEMENT, in *Horology*, is an ingenious mechanical contrivance for tranſmitting, in a modified way, and at equidiſtant intervals of time, the maintaining power of a clock or watch to the regulator, whether balance or pendulum, in order to reſtore that loſs of motion in every vibration or oſcillation, which neceſſarily ariſes from the friction of the acting parts, and the reſiſtance of the air in every machine. It has been ſaid, under the article CLOCK, that the maintaining power, whether conſiſting of a weight or ſpring, when ſuffered to expend itſelf unreſtrained, would make the wheel-work run on with an increaſing velocity, until the obſtacles to motion, ſuch as the friction in the teeth and at the pivots, the reſiſtance of the air, &c. fixed the maximum of velocity with which it could expend itſelf; at which period the friction and reſiſtance would jointly operate as a regulator, whilſt they and the maintaining power remained unaltered, and the machine ſo circumſtanced would greatly reſemble a common kitchen-jack; and if an index were inſerted on the arbor of the firſt or ſlowly moving wheel, it would indicate ſuch portions of time as the ratio between the maintaining power and the regulating friction and reſiſtance of the air taken together would produce; but it is obvious, that every alteration produced in the friction by wear, foulneſs, &c. or in the denſity of the air by the ſtate of the atmoſphere, would affect the indication ſo much, as to render it extremely irregular; hence the balance and the pendulum were ſucceſſively invented and applied to regulate the indication of time by their iſochronous vibrations; but then as both theſe regulators have alternate motions backwards and forwards, they could neither of them be applied *immediately* to check a wheel that moves continually in one direction, without the intervention of ſome contrivance, connected both with the wheel and regulator, which ſhould convert a direct circular motion into a motion changing its direction by perpetual alternations. To deviſe and execute a piece of mechaniſm that ſhould perform ſuch an office ſteadily and accurately, required much ſcience as well as mechanical ſkill, and the invention once executed was denominated an *eſcapement*, or by contraction among the workmen *'ſcapement*, becauſe it ſuffers a tooth of the wheel, with which it acts, to *eſcape*, or paſs on, at ſuch intervals of time as are meaſured by the regulator, which wheel is therefore alſo called the *eſcapement wheel*.

From this ſhort account of the origin and office of an eſcapement, the reader will perceive, that the accuracy of its conſtruction conſtitutes one of the moſt eſſential conſiderations in an horometrical machine, it being an indiſpenſable condition, that the ſucceſſive impulſes taken from the maintaining power through the medium of the wheel-work, and given to the regulator, ſhould be ſo modified, as to diſturb as little as poſſible the natural iſochroniſm of the regulator, or, if it do diſturb it, that it ſhould ſomehow compenſate that diſturbance. Hence it becomes neceſſary, that the workman, or mechanician, ſhould be fully acquainted with the theory of the natural iſochroniſm of his propoſed regulator, when he undertakes to deſign and execute his eſcapement; he muſt not only know how much of the maintaining power will be fufficient to balance the lofs fuftained in the natural length of a vibration, by reafon of friction and the refiftance of a ftandard medium, but he muft confider and indeed afcertain, by theory or experiment, or both, at what part of the arc of vibration the power muft be applied; likewife how long the impulfion fhall continue to act, and whether that action ought to be uniform, increafing, or decreafing in its intenfity. Thefe and fimilar confiderations have produced, as might be expected, a great variety of conftructions of an efcapement, both with regard to their external fhape and *modus operandi*. As a good efcapement is of the utmoft importance in the ufeful art of clock-making, and as we have no author in the Englifh language who has given us the different conftructions in hiftorical order, together with fuch remarks as may enable the reader to judge of their refpective excellence, we will devote fome time to the detail of both the theory and practical conftruction of fuch different efcapements as feem to merit a general notice.

But before we enter upon our defcription and examination of different efcapements of clocks and watches, as the reader has not yet had an opportunity of perufing the article PENDULUM, and as all efcapements derive their fhape and manner of action from the relation they bear either to the laws of its motion, or to thofe of a moving balance, it will contribute greatly to his better underftanding the fubject, if we premife firft fome of the chief circumftances which relate to the theory of a free pendulum, and to the influence that any external impulfe will have on it when applied to become the regulator of a clock; and fecondly, if we explain the theory of a balance's motion, which we have purpofely referved for this place by reafon of its intimate connection with the efcapement.

1. In *fig.* 1. of *Plate* XXXI. of *Horology*, let C D reprefent a *free pendulum*, or pendulum that has no weight but what is collected at the centre of ofcillation or point D, and let the arc D B reprefent a femi-vibration round the point C, taken as the centre of motion or point of fufpenfion, at which we will fuppofe there is no friction, then if the refiftance of the air and rigidity of the ftring or rod were not to interfere, when the pendulum was once removed from its natural or perpendicular fituation C D to B, and fuffered to defcend again by the fole force of gravity, it would acquire fuch a velocity at its loweft point D, by means of the accelerating force of gravity, as would, under thefe circumftances, carry it up an equal femi-vibration or arc D G, the increments of accelerated motion being equal in the arc B D to the decrements of retarded motion in the equal arc D G, the fame reafoning would apply in the returning vibration from G to B; the velocity acquired at D by the defcent from G to D would carry the pendulum again up to B, and the confequence would be, under fuch *unnatural* circumftances, if they could be effected, that there would be a perpetual alternation of vibrations of the fame length, which would confequently be ifochronal with refpect to each other, fo long as the pendulum remained of the fame length; and fuch a pendulum would of itfelf be a perfect time-meafurer, and would only want a regifter to indicate the number of its vibrations, to make it a complete regulator or chronometer: but the fact is, the friction and rigidity at the point of fufpenfion cannot by human art be entirely annihilated, though ingenuity can greatly diminifh it; nor can the obftacle to free motion arifing from the refiftance of the air be removed, or even rendered uniform; hence it is found that when a pendulum is drawn afide from its point of reft, every fucceeding vibration, after motion has commenced, is fomewhat fhorter than the preceding one, until after fome hours have been fpent in alternate vibrations continually fhortening, the pendulum finds its original point of reft; the time expended in doing this is longer in proportion to the diminution of friction, the air and other circumftances remaining unchanged. Ferdinand Berthoud, the famous French clock-maker, made fome experiments to afcertain this fact, and conftructed the knife-edge fufpenfion of a fecond's pendulum, with a heavy lenticular ball, fo free from friction, that an arc of vibration of 10° was not reduced to 15′ till 29h 46″ had elapfed; but notwithftanding every attempt to reduce the friction, the pendulum came to reft at laft. From this fhort view of the tendency of a pendulum to come to reft after it has had a motion communicated to it, we clearly perceive the neceffity of making fome auxiliary additions to the natural force of gravity, in order that the pendulum may keep up the length of its original vibration in every fucceeding one, notwithftanding the obftructions which continually oppofe the operation of gravity; thefe auxiliary additions to the motion, produced by gravity only, conftitute the operation of the maintaining power, modified by the wheel-work and efcapement, in every pendulum clock; and hence arifes a *difturbance* in the *theory* of the free pendulum, which makes fuch a theory no farther of ufe in clock-work, than as it affords certain laws and a fcale of forces, fuch as the mechanician may labour to imitate, by modifying the joint agency of his *auxiliary additions* and of the operation of gravity, fo that they fhall act in concert; which imitation is to be effected principally, as to its *mode*, by the efcapement, the maintaining power being concerned only as to the *quantity* of any additional power to be applied in aid of gravity.— Now, if we fuppofe the pendulum at the fituation B, or extremity of its arc, and draw A B perpendicular to the vertical line C D, and B E = C B perpendicular to the line A B, or, which is the fame thing, parallel to the line C D; alfo, if we draw B F a tangent to the arc at the point B, and from the point E demit the line E F upon it, to form a right angle at F, in which cafe the line F E will be parallel to the line C B, B E will reprefent the force of gravity, which, by the refolution of forces, is equivalent to B F and F E; but F E is perpendicular to the tangent, or is in the direction C B of the radius, and therefore can neither accelerate nor retard the motion, whilft B F, being in the direction of the tangent, or perpendicular to the radius, is wholly employed in accelerating or retarding the pendulum's motion: hence, the line B E : B F : : the force of gravity : the accelerating force; but from the fimilar triangles B E F, C B A, B E is = C B, and B F = B A, therefore, by fubftitution, C B : B A : : force of gravity : accelerating force, and confequently the accelerating force $= \frac{\text{gravity} \times BA,}{CB}$ in which expreffion gravity and the radius C B are invariable, therefore the accelerating force varies as the fine B A of the arc of vibration. From this law of variation of the accelerating force of a body moving in a circular arc, it follows, that the longer arcs require longer times than the fhorter arcs in the proportion of 34 : 29, as Huygens has demonftrated; for the fine does not increafe fo faft as the arc, which it ought to do, to make the vibrations of the fame pendulum in different arcs ifochronal: if, however, the body could be made to move along the chord or ftraight line B D, inftead of the arc B D, then the vibrations of different lengths would be ifochronal, becaufe two fimilar bodies will run down each a feparate chord of different lengths to the loweft point of the fame circle in the fame time, namely, in the time that each would take in falling through

through the whole diameter of the circle, and the acquired velocity of each, at the loweſt point, would be in proportion to the length of its chord. But the mechanician has not attempted to make his pendulum vibrate in chord lines, which probably is impracticable, but has invented another curve, called the cycloid, generated by a point in a circle revolving on a ſtraight line, which curve poſſeſſes this peculiar property, that the accelerating forces, which a body acquires by moving along it, are exactly proportionable to the *arcs*, (ſee CYCLOID,) ſo that a pendulum vibrating in this kind of curve, will perform its long and ſhort vibrations in the ſame time. This property of the cycloid was firſt diſcovered by the celebrated Huygens, who owed much of his celebrity to it, and adopted a contrivance for making a pendulum move in ſuch a curve, which will be deſcribed under our article PENDULUM, and alſo the reaſons will be given why it was laid aſide in the practice of clock-making. It may be proper, however, to remark here, that with a pendulum vibrating in the arc of a cycloid, with all its weight collected in the ball, and its length rendered invariable, if ſuch a pendulum could be made, it would, notwithſtanding, be of conſequence what the nature of the eſcapement were, becauſe though the variations in the *arcs*, ariſing from varied gravity alone, would not affect the *times* of the ſucceſſive vibrations, yet, when *additional force* is applied in aid of gravity, by means of the maintaining power and eſcapement, the uniform law of acceleration juſt deſcribed would be partially diſturbed by ſuch addition, though the diſturbance would be leſs perceptible than in circular arcs of conſiderable extent.

When, however, a pendulum vibrates in a ſmall arc, it is demonſtrable that ſuch an arc differs very little from a cycloidal arc near its loweſt point, nay almoſt coincides with it; it is alſo found from experience, that a heavy ball is leſs diſturbed by the requiſite additional force, communicated by the maintaining power, than a light one, as might be inferred from theory; on which two accounts the prevailing practice among clock-makers is, to make ſuch eſcapements for a pendulum as require but ſhort vibrations, and, in order to gain momentum, to append a heavy ball; and it has been diſcovered, and is now indeed an acknowledged fact, that a regulator with an exact compenſation for the effects of variable temperature in its pendulum, and that vibrates by means of a good eſcapement in ſmall arcs, with a heavy lenticular ball, goes with a degree of accuracy beyond what Huygens durſt even hope for from his cycloidal contrivance; in effecting which, theory and practice were found to be greatly at variance.

But the cycloidal theory, which, conſidered as a theory, is not only ingenious, but illuſtrative of the principal circumſtances which relate to the motion of a pendulum, will furniſh us with a conciſe and clear notion how the natural arc and time of a vibration will be altered by an acceſſion of external force to the force of gravity, without ſome knowledge of which alteration no mechanician can proceed on ſcientific principles in the conſtruction of an eſcapement.

Suppoſing now the pendulum in *fig.* 2, ſimilar to that in the preceding figure, and the arc to be ſo ſmall as nearly to coincide with the cycloidal curve, let the arc G B be unbent into the ſtraight line *h* H, and with the ſemi-arc D H, as a radius, deſcribe the ſemi-circle H L *h*; ſet the line A B from H to M, and from *h* to *m* perpendicular to the line *h* H, and join *m* M, which line will paſs through the centre D; from any other points I and K in the ſemi-arc or radius, draw the lines I N and K O parallel to H M, and alſo the correſponding ones *i n* and *k o* parallel to *h m*, and we ſhall obtain the following relations of the properties of the pendulum's motion, *viz.*

1ſt. The forces of gravity urging the pendulum towards D from B, at the points I and K of the ſemi-vibration, will be repreſented by, and proportional to, the ordinates I N and K O of the ſtraight line M D *m*; and the correſponding ordinates *i n* and *k o* will repreſent, and be proportional to, the retarding forces of the ſame gravity, at the points *i* and *k* of the other ſemi-vibration, and *vice verſâ* in each half of the returning vibration.

2dly. The velocities acquired by the uniform action of gravity at the points I K and *i k*, will be reſpectively proportional to the ordinates I P, K Q, and *i p*, *k q*, of the ſemi-circle H L *h*; hence the velocity of the pendulum at the loweſt point D, is to its velocity at any other point of the arc I, as D L (= D H, or whole ſemi-arc) is to I P.

3dly. The times of the pendulums vibrating through the portions H I, I K, K D, &c. of its arc, will be repreſented by, and proportional to their correſponding arcs H P, P Q, Q L, &c. reſpectively in the ſemi-circle H L *h*; which proportions will be equally applicable in the aſcending portions of the arc D *k*, *k i*, and *i h*, which will alſo be moved through in times proportional to the arcs L *q*, *q p*, and *p h*, reſpectively; if the pendulum is projected from the loweſt point D with the initial velocity D L, this velocity will be reduced at the point *k* to *k q*, at *i* to *i p*, and at *h* to nothing, whence the pendulum will return, and the operation of gravity will begin to be reverſed.

4thly. If one pendulum deſcribe the arc repreſented by H D *h*, and another deſcribe the arc K D *k*, (ſtill ſuppoſing them to be cycloidal or very nearly ſo,) they will deſcribe them in equal times, but their velocities at D, or greateſt velocities, will be reſpectively proportional to H D and K D, or length of their ſemi-arcs; that is, while a pendulum, projected with the initial velocity D L, from the loweſt point, aſcends to the point *h*, or extremity of its ſemi-vibration, another ſimilar pendulum, projected with the initial velocity R D, will aſcend to the point *k*, or extremity of its arc, D R being = D *k*, and the ſame will be true in the deſcent, where the increments of motion exactly balance the aſcending decrements.

5thly. The areas M H I N, M H K O, &c. are proportional to the ſquares of the velocities acquired in moving down the arcs H I, H K, &c. or to the diminution of the ſquares of the velocities effected by the retardation of gravity in the returning or aſcending ſemi-vibration.

From a due examination of the relations exhibited in the preceding illuſtration of the principles of a pendulum's motion, the reader will readily perceive that there is a certain length of the arc of every vibration, whether performed in a circle or cycloidal curve, which correſponds to its maximum of velocity, or velocity at the loweſt point; which length, in large circular arcs, is determined by the greateſt chord line, but in cycloidal curves, or *ſmall* circular arcs, is the whole arc itſelf: he will alſo underſtand how, when the weight of a pendulum is given, its accelerating action in any given point of the arc may be appreciated; from which and ſimilar conſiderations he will be able to form ſomething like an eſtimate of the effect that will be produced in the above natural relations by the application of an external force of an aſcertained quantity, acting in a given manner: for inſtance, let us ſuppoſe that a maintaining power is applied to keep up the vibrations of a pendulum, as in our laſt figure; that the quantum of power to be added to gravity be to its accelerating force at H or *h*, as *h m* is to *m r*, which we will ſuppoſe to be a number of grains to ſome number of ounces; and that this additional power

power is applied through the agency of an efcapement, which fo modifies it, that its action is exerted but a fhort time, and at the extremity of the arc: the rough conjecture which the reader would be apt to form would be, that as the two forces began to act in concert, a pendulum actuated by them would obey a compound fcale of forces, differing in quantity but not in quality from the fimple fcale of gravity, and that therefore little or no alteration would take place in the time of its vibration; let us fee what would be the refult according to our theory; from the point *m* of the line *h m*, of the greateft acceleration or retardation, as the cafe may be, fet off *m r* equal to its reprefentative, for fo many grains of augmenting or retarding force, and then *h r* will be the whole force urging the pendulum towards D: do the fame for every point of *h* D, and we get a new fcale of forces reprefented by the dotted line *r s t* D, and the fpace *m* D *r*, contained between the two fcales D *m* and D *r*, will reprefent the addition made to the fquare of the velocity in moving over *h* D by the joint action of gravity and of the maintaining power. Then if we make the area D *u v* equal to the area D *h m*, by leaving out as much fpace parallel to the perpendicular *h m*, as we take in beyond the hypothenufe *m* D, the point *u* will limit the excurfion or extent of the returning, as well as of every fucceeding vibration, becaufe the area of the triangle D *u v* is proportional to the fquare of the velocity, and *u* is therefore the point where the initial velocity D L will be extinguifhed. Laftly, if the arc *h w* of the femi-circle be bifected, its half will very nearly reprefent the contraction effected in the time of the outward femi-vibration, agreeably both to theory and practice; and in a fimilar manner might be eftimated the effect to be produced on a natural vibration, or vibration from fimple gravity, by any acceffion of external force differently applied: in the inftance we have given, a temporary impulfion would evidently make a greater derangement when given at the extremity of the arc, than if it had been given at I, and greater at I than at K, and confequently the leaft derangement would have arifen if the impulfion had been given at the loweft point D, which is worthy of recollection; though the fact is, that moft of the efcapements, that are not of the detached kind, continue to act for a fenfible time, and, after all, the thing principally aimed at in modern practice is, to conftruct all the parts of an horological machine fo, that the arc of vibration may vary as little as poffible under the different influences that affect the pendulum's motion; for then, whatever the length of the arc be, whilft there is no variation in it, the vibrations will be ifochronal, whatever be the nature of the curve; but completely to effect fo defirable a purpofe can hardly be expected under all the combinations of variable refiftances arifing from friction, oil, air, &c., however detached the efcapements may be, *i.e.* however fmall the continuance of each of their impulfions. Hence, certain ingenious artifts have attempted fuch a compound fcale of forces, confifting of gravity and maintaining power conjointly, as fhall render even circular vibrations of different lengths ifochronal among themfelves, when they fhall happen to vary in length, which plan would certainly make the beft compenfation for the effects of variable refiftances, if fuch plan could be completely executed at pleafure; and the compound fcale of forces to do this, muft be to the fcale of gravity alone, as the *arc* of femi-vibration is to the *fine* of the fame in every point of the excurfions. For in *fig.* 1, if B F be fuppofed equal to the femi-arc B D, then B E, the whole compound force at the point B, will be greater than C B or force of gravity alone, and will be to it, by fimilar triangles, as B F (= B D) is to the fine A B, as has been demonftrated; therefore the whole compound force at any point B of either femi-vibration fhould be to the force of gravity alone in that point, as the arc D B is to the fine A B, which compound fcale confequently will demand a certain limited variation, or fcale, in the auxiliary force derived from the maintaining power through the modifying medium of the efcapement.

2. When a balance was firft applied as a regulating power, it poffeffed no natural property, like the pendulum, of perpetuating its own ofcillations, but had its motion produced by an artificial force acting alternately on its oppofite pallets, and derived entirely from the maintaining power at certain intervals; fo that to perform all the ofcillations in equal times, the action of the force applied was required to be the fame in mode, quantity, and duration, in each ofcillation, which conditions could not be fulfilled by reafon of the conftant changes that took place in the friction of the works, and in the denfity and confequent refiftance of the air; a fimple balance therefore was no otherwife better than a continued fly, except that its alternate ofcillations prevented the acceleration of motion that would enfue, if there were no fuch frequent checks as the balance experiences at each alternate impulfe of the pallets. Dr. Hooke's invention of a regulating fpring, to be to the natural balance what gravity is to a pendulum, was an important one, to which all the excellence of an ordinary watch may have its origin imputed, and without which the compenfations for different temperatures in the beft chronometers would not be of any utility. The difcovery of a *fpring's force* being *as its tenfion*, or diftance moved from the point of reft, (*ut tenfio fic vis,*) was the firft ftep towards a theory of motion by which the balance and its regulating fpring, taken conjointly, may have their effect eftimated, and by which fuch a balance with the beft compenfation and efcapement now vies in accuracy even with the pendulum itfelf.—George Atwood, efq. of Cambridge, has publifhed an excellent memoir in the Philofophical Tranfactions of London, vol. 84, part i. 1794, intitled "Inveftigations founded on the Theory of motion, for determining the times of Vibration of Watch Balances," from which we beg leave to extract fo much as falls within our purpofe in this place.

"Let P M N S, *fig.* 3. of our laft plate, reprefent the circumference of a watch balance, which vibrates by the action of a fpiral (or helical) fpring, on an axis paffing through the centre C. Let O D B E be the circumference of a concentric circle, confidered as fixed, to which the motion of the balance may be referred. In the circumference of this circle let any point O be affumed, and when the balance is in its quiefcent pofition, fuppofe a line to be drawn through C and O, interfecting the circumference of the balance in the point A; the radius C A will be an index, by which the pofition of the balance, and its motion through any different arcs of vibration, will be truly defined. In the fequel, the motion of the balance, and the motion of the index C A, will be ufed indifferently, as terms conveying the fame meaning. Since the balance is in its quiefcent pofition when the index C A is directed to the fixed point O, on this account O is called the point of quiefcence of the balance, or balance fpring, indicating the pofition when the balance is not impelled by the fpring's elaftic force either in one direction or the other. If the balance fhould be turned through any angle O C B, the fpiral fpring being wound through the fame angle, endeavours by its elaftic force to reftore itfelf, and when at liberty, impels the balance through the arc B O with an accelerated velocity, till it arrives at the pofition O, where the force of acceleration ceafes; with the velocity

city acquired at O, the balance proceeds in its vibration, defcribing the arc O E with a retarded motion. The elaftic forces of the fpring at equal diftances, on the oppofite fides of the point O, are affumed to be equal; it is alfo affumed that the effects of friction, and other irregular refiftances, which retard the motion of the balance, are compenfated by the maintaining power, fo that the time of defcribing the firft arc of vibration B O, by an accelerated motion, fhall be equal to the time of defcribing the latter arc O E, by a retarded motion, and that the entire arc of vibration B O E is bifected by the point O.

To render the conftruction of *fig.* 3. more diftinct, the fixed circle O D B E is reprefented to be at a fmall diftance from the circumference of the balance, but is to be confidered as coincident with it, fo that the arc B O, fubtending the angle B C O, may be of the fame length with an arc of the circumference of the balance, which fubtends the fame angle B C O: on this principle C O or C A may be taken indifferently as the radius of the balance.

The determination of the time in which the balance vibrates, from the theory of motion, requires the following particulars to be known, *viz.*

Firft. The fpring's elaftic force, which impels the circumference of the balance, when it is at a given angular diftance O D, (*fig.* 3.) from the quiefcent point O:

Secondly. The law or ratio obferved in the variation of the fpring's force, while the balance is impelled from the extremity of the femi-arc B, to the point of quiefcence O, where all acceleration ceafes.

Thirdly. The weight of the balance, including the parts which vibrate with it.

Fourthly. The radius of the balance C O, and the diftance of the centre of gyration from the axis of motion C G.

Fifthly. The length of the femi-arc B O.

Suppofe the plane of the balance to be placed vertically, and let a weight P, (*fig.* 4.) be applied by means of a line fufpended freely from the circumference at T, to counterpoife the elaftic force of the fpring when the balance is wound through an angle from quiefcence O C D. This weight P (the weight of the line being allowed for,) will be the force of the fpiral (or other) fpring, which impels the circumference of the balance, either at reft or in motion, when at the angular diftance O D from the quiefcent pofition. It appears from many experiments (of Berthoud) that the weights neceffary to counterpoife a fmall fpring's elaftic force, when the balance is wound to the feveral diftances from the quiefcent point, reprefented by the arcs O G, O H, O I, &c. *fig.* 4, are nearly in the ratio of thofe feveral arcs. It alfo appears, that the fhape, the length, and number of turns of the fpiral, may be fo adjufted to each other, that the forces of elafticity fhall be counterpoifed by weights, which are in the precife ratio of the angular diftances from the quiefcent pofition, or, as it is fometimes expreffed, in the ratio of the fpring's tenfions; at leaft as nearly as can be afcertained by experiment: this law of elaftic force is affumed in the fubfequent inveftigation.

The pofition of the centre of gyration may be always determined when the figure of the vibrating body is regular, by calculating the fum of the products, which arife from multiplying each particle into the fquare of its diftance from the axis of motion, and dividing the fum by the weight of the vibrating body; the fquare root of the refult will be the diftance of the centre of gyration from the axis of motion. When the figure of a vibrating body is irregular, recourfe muft be had to experimental methods, in order to determine the pofition of the centre of gyration. See Atwood's treatife "On the Rectilinear Motion and Rotation of Bodies."

Let the radius of the balance C A or C O $= r$, *fig.* 3, the femi-arc B O $= b$; let the fpring's elaftic force, acting on the circumference of the balance, when wound to any given angle O C D, from the quiefcent pofition, be P, and let the arc O D $= a$; the weight of the balance and the parts which vibrate with it $=$ W; the diftance of the centre of gyration from the axis of motion C G $= g$. Thefe notations being premifed, the refiftance of inertia, by which the mafs contained in the balance oppofes the communication of motion to the circumference, is $\frac{Wg^2}{r^2}$, and confequently the force which accelerates the circumference at the angular diftance O C D from the quiefcent pofition is $\frac{Pr^2}{Wg^2}$. This quantity remaining invariably the fame, while the balance defcribes the arc of vibration B O E, may be denoted by the letter F, fo that $F = \frac{Pr^2}{Wg^2}$; fuppofe the radius C A commencing a vibration from the point B to have defcribed the arc B H, and let O H $= x$, fince the force which accelerates the circumference at the angular diftance from quiefcence O D $=$ F, and the forces of acceleration are fuppofed to vary in the proportion of the angular diftances from the quiefcent point O, the force which accelerates the circumference of the balance at the point H will be $= \frac{Fx}{a}$; let u be the fpace through which a body falls freely from reft by the acceleration of gravity, to acquire the velocity of the circumference at the point H; the principles of acceleration give this equation $\dot{u} = \frac{-Fx\dot{x}}{a}$; (Newtonii Princip. vol. i. prop. xxxix.) and taking the fluents while x decreafes from b to x, $u = \frac{F \times \overline{b^2 - x^2}}{2a}$: if therefore l is made $=$ 193 inches, being the fpace which bodies falling freely from reft by the force of gravity near the earth's furface defcribe in one fecond of time, the velocity of the circumference, when the extremity A of the index C A has arrived at the point H, will be $= \sqrt{\frac{2lF}{a}} \times \sqrt{b^2 - x^2}$. Let t reprefent the time in which the circumference defcribes the arc B H; then will $\dot{t} = \sqrt{\frac{a}{2lF}} \times \frac{-\dot{x}}{\sqrt{b^2 - x^2}}$; and $t = \sqrt{\frac{a}{2lF}} \times$ a circular arc, of which the cofine $= \frac{x}{b}$ to radius $=$ 1, which is the time of defcribing the arc B H expreffed in parts of a fecond; when $x = 0$, that is, when the circumference has defcribed the entire femi-arc B O, the circular arc of which the cofine $= \frac{x}{b}$ is a quadrant of a circle to radius $=$ 1. Let $p = 3.14159$, &c. the time t of defcribing the femi-arc B O $= \sqrt{\frac{a}{2lF}} \times \frac{p}{2} = \sqrt{\frac{p^2 a}{8lF}}$.

In this expreffion for the time of a femi-vibration, the letter a denotes the length of the arc O D (*fig.* 3.); if this arc fhould be expreffed by a number of degrees $c°$, a will then $= \frac{prc°}{180°}$; and this quantity being fubftituted for a, the

a, the time of a ſemi-vibration will be $t=\sqrt{\frac{p^2 r c^2}{8lF\times 180^2}}$; if inſtead of F, its value $\frac{Pr^2}{Wg^2}$ is ſubſtituted in the equation $t=\sqrt{\frac{p^2 r c^2}{8lF\times 180^2}}$, the time of a ſemi-vibration will be $t=\sqrt{\frac{Wp^2 g^2 c^2}{8Prl\times 180^2}}$.

Let the given arc c° be $=90^\circ$; in this caſe $t=\sqrt{\frac{Wp^2 g^2}{16\,Prl}}$. Theſe are expreſſions for the time of a ſemi-vibration, whatever may be the figure of the balance, the other conditions remaining the ſame as they have been above ſtated. If the balance ſhould be a cylindrical plate, it is known that the diſtance of the centre of gyration from the axis is to the radius as 1 to $\sqrt{2}$; wherefore in this caſe $g^2=\frac{r^2}{2}$; and the time of a ſemi-vibration, or $t=\sqrt{\frac{Wp^2 r}{32\,Pl}}$."

Agreeably to this expreſſion, Mr. Atwood wiſhed to try the practical reſults of a chronometer made by Kendal, to prove how far his theory agrees with practice, and aſſuming the centre of gyration to be the ſame as if the balance were a cylindrical plate, he obtained from Mr. Earnſhaw the following data, *viz.*

W, the weight of the balance and vibrating parts $=42$ grains.

P, the force at the circumference of the balance, which counterpoiſes the force of the ſpring when wound to the diſtance of $90^\circ = 24$ grains.

r, the radius of the balance $=1.125$ inches.

l, the ſpace deſcribed in one ſecond of time by bodies which deſcend freely from reſt by the acceleration of gravity $=193$ inches.

p, the circumference of a circle to radius $1=3.14159$, &c.

The balance, when adjuſted to mean time, made juſt five vibrations in a ſecond; the actual time of a ſemi-vibration is therefore 0.1000.

Then *t*, the time of a ſemi-vibration by the theory, will be $=\sqrt{\frac{42\times 3.14159^2\times 1.125}{32\times 24\times 193}}=0.0994$ parts of a ſecond; hence $0.1000-0.0994)=0.0006$ is the difference between the actual time and the time by calculation from the theory; and the near coincidence of the two reſults proves that the ſuppoſition of the centre of gyration being nearly as in a cylindrical plate, is near the truth.

"It is obſervable," ſays our author, "that the ſemi-arc of vibration $BO=b$, does not enter into theſe expreſſions for the time of a ſemi-vibration; if, therefore, inſtead of the ſemi-arc BO, an arc of any other length LO, terminating in the point of quieſcence, (*fig.* 3.) ſhould be ſubſtituted in the preceding inveſtigation, the time of deſcribing LO would be ſtill $=\sqrt{\frac{ap^2}{8lF}}$, or, $\sqrt{\frac{p^2 r c^2}{8lF\times 180^2}}$, equal to the time of deſcribing the other ſemi-arc BO; conſequently, whether the balance vibrates in the largeſt or ſmalleſt arcs, the times of vibration will be the ſame. From the preceding inveſtigations it appears, continues our author, that when the force by which the circumference of the balance is accelerated at the given angular diſtance c° from the quieſcent poſition is $=$ F, the time of a ſemi-vibration $t=\sqrt{\frac{p^2 r c^2}{8lF\times 180^2}}$; and converſely, when the time of a ſemi-vibration is $=t$, the force which accelerates the circumference at the given angular diſtance c° from the quieſcent poſition, that is, $F=\sqrt{\frac{p^2 r c^2}{8lt^2\times 180^2}}$.

"Since time-keepers are uſually adjuſted to mean time when the balance makes five vibrations in a ſecond, the time of a ſemi-vibration will, in this caſe, be $=\frac{1}{10}$th part of a ſecond; the ſubſtitution of $\frac{1}{10}$th for t being made in the preceding equation, the force which accelerates the circumference of the balance, when at any given angular diſtance c° from the quieſcent poſition, will be determined for all time-keepers adjuſted to mean time, in which the balances make five vibrations in a ſecond. Suppoſe the given angle $c^\circ=90^\circ$, then making $c^\circ=90^\circ$, $p=3.14159$, &c. $l=193$, $t=\frac{1}{10}$, the accelerative force at the angular diſtance from quieſcence 90° or $F=\frac{p^2 r\,90^2}{8lt^2\times 180^2}=r\times 1.00408926$. We have therefore arrived at the following concluſion; if the radius of the balance is equal to one inch, and the time-keeper is adjuſted to mean time when the balance makes juſt five vibrations in a ſecond, the force which accelerates the circumference of the balance at the diſtance of 90° from the quieſcent poſition is $=1.00408926$, the accelerative force of gravity being $=1$. And if the radius of the balance is greater or leſs than one inch, the force by which the circumference is accelerated at the diſtance of 90° from quieſcence will be greater or leſs than 1.00408926 in proportion to the radii.

"According to the principles aſſumed in the preceding ſolution, the ſpring's elaſtic force is ſuppoſed to vary in the proportion of the angular diſtances from the quieſcent poſition, and on this condition the vibrations are ſhewn to be iſochronous, whether they are performed in longer or ſhorter arcs; but if the ſpring's elaſtic force at different diſtances from quieſcence ſhould not be preciſely in the ratio here aſſumed, the longer and ſhorter arcs may be deſcribed in times differing in any proportions of inequality. If, for inſtance, the ſpring's force, inſtead of varying in the ratio of the aforeſaid diſtances, ſhould vary in the $\frac{999}{1000}$dth power or $\frac{1001}{1000}$dth power of the diſtances, it does not appear from the preceding ſolution what alteration in the daily rate would be cauſed by this change in the law of the force's variation, when the ſemi-arc of vibration is increaſed or diminiſhed by a given arc. To aſcertain this point fully, other reſearches will be neceſſary, by which it may be known, what alteration of the daily rate of a time-keeper is occaſioned by a given increaſe or diminution of the arc of vibration, when the ſpring's elaſtic force varies in a ratio of the diſtances from the quieſcent poſition, the general index or exponent of which is any number or fraction *n*.

"The force which accelerates the balance being aſſumed in that power of the diſtances, the exponent of which is *n*, let $BO=b$, (*fig.* 5.) be the ſemi-arc of vibration when the time-keeper is adjuſted to mean time; let $DO=a$; the accelerating force on the circumference at the diſtance from quieſcence $OD=F$; ſuppoſe the circumference to have deſcribed the arc BH from the extremity of the arc B; and let $HO=x$: then the force by which the circumference is accelerated when at the angular diſtance from the quieſcent poſition $OH=\frac{Fx^n}{a^n}$; let *u* be the ſpace through which a body falls freely from reſt by the acceleration of gravity, to acquire the velocity of the circumference when it has deſcribed the arc BH: the principles of acceleration gave this equation $\dot{u}=\frac{-Fx^n\dot{x}}{a^n}$; taking the flu-

ents

ents while x decreaſes from b to x, $u = \frac{Fb^{n+1} - Fx^{n+1}}{\overline{n+1} \times a^n}$, and l being 193 inches, the velocity acquired by the circumference, after deſcribing B H, will be $= \sqrt{\frac{4lF}{\overline{n+1} \times a^n}} \times \sqrt{b^{n+1} - x^{n+1}}$; let T be the time of deſcribing the arc BH; wherefore $\dot{T} = \sqrt{\frac{\overline{n+1} \times a^n}{4lF}} \times \frac{-\dot{x}}{\sqrt{b^{n+1} - x^{n+1}}}$. The time of deſcribing the arc B H will be the fluent of this fluxion, while x decreaſes from b to x, and the time of deſcribing the ſemi-arc B O will be the entire fluent of $\sqrt{\frac{\overline{n+1} \times a^n}{4lF}} \times \frac{-\dot{x}}{\sqrt{b^{n+1} - x^{n+1}}}$, while x decreaſes from b to o. Now let the balance commence its vibration from any other point I, (*fig.* 5.) and let I O $= c$; ſuppoſe the circumference to have deſcribed the arc I K, and make O K $= y$; let t be the time of deſcribing the arc I K; then by proceeding in the ſame manner as in the former caſe, it is found that $\dot{t} = \sqrt{\frac{\overline{n+1} \times a^n}{4lF}} \times \frac{-\dot{y}}{\sqrt{c^{n+1} - y^{n+1}}}$; and the time of deſcribing the ſemi arc I O will be the entire fluent of this fluxion, while y decreaſes from c to o. Although the fluents of the fluxions $\frac{-\dot{x}}{\sqrt{b^{n+1} - x^{n+1}}}$, and $\frac{-\dot{y}}{\sqrt{c^{n+1} - y^{n+1}}}$, cannot be expreſſed in general terms, yet the exact proportion of the ſaid fluents may be aſſigned, which will be the proportion of the times in which the balance vibrates in the two ſemi-arcs B O, I O; the multiplying quantity $\frac{\sqrt{\overline{n+1} \times a^n}}{4lF}$ being common to both fluxions; and ſince the entire fluent of $\frac{-\dot{x}}{\sqrt{b^{n+1} - x^{n+1}}}$ is to the entire fluent of $\frac{-\dot{y}}{\sqrt{c^{n+1} - y^{n+1}}}$ as $b^{\frac{1-n}{2}}$ is to $c^{\frac{1-n}{2}}$, it follows that the time of a ſemi-vibration in the arc B O is to the time of a ſemi-vibration in the arc I O as $b^{\frac{1-n}{2}}$ to $c^{\frac{1-n}{2}}$, or as 1 to $\overline{\frac{IO}{BO}}\Big|^{\frac{1-n}{2}}$.

Suppoſe a watch to be adjuſted to mean time when the ſemi-arc of the balance's vibration is = B O, (*fig.* 5.) and let this ſemi-arc be afterwards diminiſhed to I O; the time ſhewn by the watch in any given portion of mean time t, when the ſemi-arc of vibration is I O, will be $= t \times \overline{\frac{BO}{IO}}\Big|^{\frac{1-n}{2}}$; and if t is put $= 24^h$, the alteration of the daily rate, in conſequence of the diminution of the ſemi-arc of vibration from BO to IO, will be $24^h \times \overline{\frac{BO}{IO}}\Big|^{\frac{1-n}{2}} - 1$. To apply this propoſition, let a caſe be aſſumed; ſuppoſe a watch to be regulated to mean time, when the ſemi-arc of vibration is 135°, and let this ſemi-arc be diminiſhed 8°, ſo as to become 127°; let the ratio of the ſpring's elaſtic force deviate from that of the diſtances from the quieſcent poſition by a ſmall difference of $\frac{1}{1000}$dth power of the diſtances, inſtead of in the entire ratio of the ſaid diſtances from the quieſcent poſition. The alteration in the daily rate of the watch will be obtained from the preceding theorem, by making the following ſubſtitutions; B O = 135°, I O = 127°, $n = \frac{999}{1000}$, the alteration of the daily rate $= 24^h \times \overline{\frac{135}{127}}\Big|^{\frac{1}{2000}} - 1 = + 2''.62$."

From this theory of the balance and balance ſpring it is evident, that when there is a very minute alteration in the law of the force's variation, amounting to no more than $\frac{1}{1000}$dth part of the entire ratio of the diſtances, an acceleration is cauſed in the daily rate of more than $2\frac{1}{2}''$, when the diminution of the ſemi-arc is only 8°, and as this alteration in the force is leſs, probably, than can be detected by actual meaſurement of the force by mechanical means, it is very probable that a much greater acceleration or retardation of the rate, than in the example before us, may occur from the ſame cauſe, which can only be corrected by a correſponding alteration either in the length or ſtrength of the balance ſpring, or both, as the caſe may require, and the beſt way of detecting the law of force by which the ſpring acts at different diſtances from the quieſcent poſition, is by increaſing or diminiſhing the maintaining power, ſo as to render the arc of vibration unequal in ſucceſſive trials for a limited time. The neceſſary concluſion, from ſuch trials with different maintaining powers, will be, that when an acceleration of the daily rate accompanies a diminution of the arc of vibration, the elaſtic force of the ſpring uſed varies in a leſs ratio than that of the diſtances from the quieſcent poſition; but on the contrary, when a retardation takes place under the ſame circumſtances, the ſaid elaſtic force varies in a higher ratio than that of the diſtances from the point of quieſcence. We here ſuppoſe the ſame eſcapement to be uſed in all the trials, and that the impulſe given to the balance is no more than ſufficient to overcome the inertia of the balance and the reſiſtance it meets with from friction and the denſity of the air; but as there is generally an overplus of power communicated to the balance to guard againſt ſtopping by the acceſſion of dirt or thickening of the oil, this ſuperfluous power has ſuch an influence on the balance and its ſpring when in motion, as to produce a compound law of varying forces, ſomewhat different from the ſimple law derived ſolely from the elaſticity of the ſpring, and the more this compound law varies from the ſimple law derived from the ſpring's elaſticity, the greater will be the diſcrepancy between the theory and performance of a watch balance. The buſineſs of a good eſcapement of a watch or balance clock is to keep the balance in continual motion with as little interference as poſſible with the ſimple law of its natural varying forces, which it has been ſhewn are in proportion to the diſtances from the point of quieſcence.

After theſe remarks on the laws by which a pendulum and a balance together with its ſpring become the regulators of an horological machine, the reader will be able to form an opinion, founded on ſcientific principles, on the merits or demerits of the different eſcapements that are ſucceſſively deſcribed, not only in this article, but in the articles Chronometer and Clock, which have been already publiſhed.

Henry Sully, who was an ingenious clock and watch maker about the beginning of the 18th century, wrote a hiſtory of the eſcapements of his time down to the year 1727, which Julien le Roy added to Sully's "Regle Artificielle du Temps," in the ſecond edition, publiſhed at Paris, 1737, from which Berthoud has extracted his account of ſeveral eſcapements in his "Hiſtoire de la Meſure du Temps," and which we mean to avail ourſelves of in a certain degree in our ſubjoined deſcriptions.

The

The different efcapements of clocks and watches may be divided into four claffes; *viz;*

1. Thofe with a *recoil*, or which produce a backward motion of the efcapement wheel and fecond's hand in fome parts of the vibration; (echappement a recul.)

2dly. Thofe that have the *dead-beat*, or that allow the efcapement wheel and hand to reft while the whole vibration is made; (echappement a repos.)

3dly. *Detached*, or *free* efcapements, which act on the balance or pendulum an almoft infenfible time, and fuffer the vibration to be performed almoft without connection with the pallets; (echappement a vibrations libres.)

4thly. *Detached*, or *free* efcapements with a *remontoir*, or auxiliary fpring frequently wound up by the maintaining power, to equalife the arcs of vibration; (echappement a vibrations libres et a remontoir d'égalité d'arcs.)

The moft ancient efcapement in ufe at this time, and probably the firft that was ever contrived, is the *crown-wheel efcapement* at prefent employed in our table-clocks and common watches; we know not who was the original inventor, but find mention made of it by Leonard de Vinci, who died in the year 1513; and Venturi thinks that the horological machines of Wallingford and Dondi, early in the fourteenth century, had the fame, as we have feen, under our article CLOCK, that Henry de Wyck's clock had in the year 1364. The idea of converting a rotatory into a vibratory motion, and the means adapted to effect this purpofe mechanically are equally ingenious; as an effort of genius this thought is much to be admired. We have already given a general defcription of the connection that the efcapement-wheel has with a train of wheel work on one fide of it, and with the balance on the other, under our article juft referred to, but as the balance had no rim, and was without the *regulating*, or, as it is fometimes called, the *pendulum fpring*, we think it right to give a feparate drawing and defcription of this efcapement, as made at prefent, together with fome remarks on its properties.

1. *Crown-wheel efcapement.*—In *Plate* XXXI. of *Horology*, *fig.* 6. reprefents the crown-wheel and balance in connection, which may be in either a horizontal or vertical pofition, and is in fact fometimes in one pofition and fometimes in the other in a common watch agreeably to the different pofitions of the body of the wearer; and it is eafy to conceive that a fhort pendulum might be fubftituted for the balance when the crown-wheel moves horizontally in a fixed pofition, and the axis of the balance is placed horizontally, which is the cafe in moft of the portable clocks. The name of crown-wheel is taken, no doubt, from the refemblance it has to an old fafhioned *royal diadem*; but when applied to a watch it is frequently called fimply the *balance-wheel*, from its connection with the *balance*: this wheel and the balance, including the pallets and regulating fpring, may be confidered as conftituting the *efcapement*, the action of all which may be underftood from the following fhort account.

Suppofe the pinion *a*, on the arbor *a b* of the balance-wheel or crown wheel *c d*, to be actuated by the main-fpring, or weight which forms the maintaining power, by means of the train, in the direction *c f d e*, while the pallets *e* and *f*, attached to the axis of the balance *g h*, and ftanding at right angles to each other, or very nearly fo, are long enough to fall in the way of the ends of the floped teeth of the wheel, when turned round an angle of 45°, fo as to point in oppofite directions, as in the figure; then a tooth of the wheel below, for inftance, meets with the pallet *f*, fuppofed to be at reft, and drives it before it a certain fpace, till the end of the tooth *efcapes*; in the mean time the balance attached to the axis of the pallets moves in the direction *h k g*, and winds up the fmall fpiral fpring *i*, one end of which is faft to the axis and the other to a ftud on the plate of the frame, in this operation the fpring oppofes the momentum given to the balance by this pufh of the pallet *f*, and prevents the balance from going quite round before the pallet *e* meets with another tooth at the oppofite end of the wheel's diameter; here this pallet receives a pufh which carries the balance back again, having as yet but fmall momentum, in the direction *h l g*, and aids the fpring, which now unbends itfelf till it comes to its quiefcent pofition, but beyond that point is unwound, partly by the pufh from the maintaining power on the pallet *e*, and partly by the acquired momentum of the moving balance, particularly when this pallet has efcaped; at length pallet *f* again meets with the fucceeding tooth, and is carried *backward* by it, in the direction the balance is now moving in, till the maintaining power and force of the unwound fpring together overcome the momentum of the balance; during which time the *recoil* of the fecond's hand is apparent, when put on the pivot of the arbor *a b*; at length the wheel brings the pallet *f* back again till it efcapes, and the fame procefs takes place with pallet *e* that has been defcribed with refpect to pallet *f*, and thus two contrary excurfions, or ofcillations of the balance, take place before one tooth has completely efcaped, which is the reafon why there muft be always an odd number of teeth in this wheel, that a fpace at one fide of the wheel may always be oppofite to a tooth at the other, in order that one pallet may be out of action while the other is in action. The fame account will apply to the vibrations of a fhort pendulum fubftituted for the balance, when this efcapement is ufed. This efcapement, which is of the firft clafs, is evidently calculated to derange the natural law of the fpring's varying force, which, we have feen, is in theory in proportion to the diftances from the quiefcent pofition; for as the action of the maintaining power continues to prefs on the pallets, either with or againft the balance's motion, during a large portion of each vibration, this extraneous force acting along with the elaftic force of the fpring makes a new law compounded of the two forces, which may or may not be a regular law, as modified by circumftances, that are to be taken into confideration; for inftance, the relative forces of the maintaining power and of the regulating fpring, the uniformity of each force feparately confidered, the ratio between the radius of the pallet-wheel and length of the pallets, the weight and diameter of the balance, the angle of action in the efcapement relatively to the whole arc of vibration, the adjuftment of the length and ftrength of the fpring for ifochronifm, and of its quiefcent place to the pofition of the pallets, together with the friction and inertia of the acting parts, as well as refiftance of the air to the moving balance muft all enter into the account, which complex data render all calculations founded in theory very inadequate to their purpofe of determining the practical refults independently of experimental trials, which are neceffarily adopted by the workmen in finifhing the adjuftments. The bad effect of an almoft conftant force on the balance, as derived from the maintaining power, is evident by the application of the key to the fufee-arbor of a watch, and by urging it in a direction contrary to that of winding up, for the additional force thus given to the maintaining power, will very fenfibly accelerate the beats of the watch, and leffen the arc of vibration; hence any inequality occafioned in the maintaining power, by the acceffion of dirt, or by want of perfect compenfation in the fufee of the varying force of the main-fpring, &c. will alter the daily rate very materially, though the mechanical

chanical adjustments at the balance may, in the first instance, have produced jointly a law of varying forces in the regulating mechanism that shall approximate near enough to the isochronous law of the balance spring, while the arcs of vibration remain unaltered. It is of the utmost importance, therefore, when a crown-wheel escapement is used in any watch, that, whenever a new main-spring is put in, the fusee should also be altered, so as to become as perfect a compensator as possible for that *identical spring*, which, we believe, is seldom, if ever, attended to in practice. This escapement usually has a pin in the rim of the balance that may catch a notch on each side of the cock beyond the extremities of the longest vibrations, called the *banking*, to prevent the derangement of the balance and of its regulating spring by any shake or accident.

2. *Escapement a pirouette by Huygens.*—About the year 1675, when Huygens was contriving the best construction he could devise of a marine time-piece, he naturally dwelt upon the portability of the balance, as affording the likeliest plan of accomplishing his purpose, but he was aware that the balance vibrating in short arcs was not comparable in point of steadiness to a pendulum, as a regulating principle, seeing it is too much under the controul of the maintaining power, when used with the crown-wheel escapement, the only one at that time known. The balance under such strong influence, and that a variable influence, had the property of putting itself into motion when in a state of rest: Huygens, therefore, made such additions to the balance and escapement then in use, as produced much longer vibrations than had been before witnessed, and introduced an additional wheel and pinion between the pallet-wheel and the balance to effect this object; thus, in *fig.* 7. of our last plate, is a side view of a small frame containing the mechanism of Huygens' escapement, consisting of two wheels, one pinion, a pair of pallets, and a balance with the spiral spring, together with the requisite cocks to bear the pivots of the arbors; *a b* is the crown-wheel or pallet-wheel of the usual shape and number of teeth, placed in a vertical position, like the contrate wheel of a common watch or table clock; *c d* is a contrate wheel placed horizontally, like the pallet-wheel of a common watch, having two pallets *e*, *f*, as usual on its arbor, at right angles to each other, and acting exactly as we have explained above; the contrate wheel *c d* drives a pinion *g* on the axis ot the balance *h*, under which, as usual, is the spiral spring *i*; the action of this escapement differs from that of its predecessor only, inasmuch as the contrate wheel impels the pinion several times round at each impulse of one of the pallets, thereby occasioning several revolutions of the balance itself at each oscillation; and Sully says, that when the spring is not used he has seen the oscillations performed in the space of two seconds, and with a spring in one, in imitation of a seconds' pendulum clock, to which this escapement is not applicable. The influence, however, of the maintaining power must have been exerted on the balance in this construction also, during at least one of its revolutions, and the extraneous force interfering with the law of the spring's varying forces would produce a compound scale of forces, in which, as in the former instance, one would be nearly constant while it acted, and the other varying; besides, the friction in the teeth of the wheel and pinion, however well executed of the respective sizes and shapes, would produce checks that would be unfavourable to the steady oscillations of the balance, we therefore are not surprised to find that this escapement, which was of the first class, was soon discontinued.

3. *Escapement with two balances by Dr. Hooke, and by Du Tertre.*—Dr. Hooke, says Sully, produced a balance in the year 1675, which he had invented in 1658, with two balances engaging each other by means of teeth cut in their edges, and one of them having a spiral spring on its verge or axis; but Berthoud attributes the invention of this escapement with two balances, without the spring, to a German artist on the authority of Thiout (Traité d'Horlogerie de Thiout) where there is a drawing of the escapement in question; J. B. Du Tertre afterwards made some additions, which we have given in *fig.* 8. of *Plate* XXXI., and of which the following is a description, *viz.*; A and B are two balances acting together, like wheels with equal numbers of teeth, and having on one of the crosses of each a pallet, at *a* and *b* respectively; at equal distances from the axes of these balances is the arbor of the last wheel of the train, bearing two stars with each five radii, one numbered 1, 2, 3, &c. which is smaller than the other, lying the lower in position, and the other marked *d e f*, &c. with radii long enough to reach almost to the centres of motion of the balances, which they can pass only when the semi-circular parts that are cut away are presented to them; but at other times they rest against the central semi-circular pieces that remain uncut: the radii of the small star reach far enough to fall on the pallets *a* and *b* successively, which they impel alternately, and give motion to both balances at the same time; but the escape takes place by means of the radii of the large star, that rest successively against the semi-circular portions of metal not cut away, above the centres of the balances respectively. The three axes stand in the angles of an isosceles triangle, and have their upper pivots supported by three separate cocks, placed over the upper plate of the frame, the lower pivots of the balances resting on the upper plate, but the arbor of the two stars passing across the frame to take the last wheel. The action is thus; radius 1 of the small star has parted with the pallet *a* of balance B, and the long radius *d* is escaping; presently radius 4 will fall on the pallet *b* of balance A, and impel it in a direction from 4 to 5, the balance B at the same time moving in a contrary direction; this impulse will continue till the radius 4 falls from the pallet, soon after which the long radius *g* will fall on the semicircular central piece of balance A, and stop the further motion of the two stars, which move together on the same arbor, and the balances will go on in their vibrations till the spring under balance B brings them back again, so as to present the cut half of the central piece of balance A to the point *g* of the long radius, which, being impelled by the train, now escapes; the stars go on till radius 4 of the small star meets with the pallet *a* of balance B, the motion of which it now opposes, and is itself brought back, so as to make a recoil of the hands till it has stopped the motion of the balances, which now begin each another oscillation: this radius 4 escapes the pallet *a*, and the long radius *g* falls on the semi-circular piece, at the centre of balance B, as it did before on that of balance A; the oscillation continues forward again till the spring brings it back, when another following radius of the small star falls again on the pallet *b* of balance A, as before, and drives it; soon after which a succeeding long radius falls on its semi-circular central piece, and rests till at the return of the oscillation it can escape; thus an alternation of oscillations is produced by the short radii acting alternately on the two balances, while the long radii suspend the impulses by resting on the semi-circular parts of the centres of the balances, during the latter portion of each excursion. This escapement was evidently of the first class, like its predecessors, and though it avoided the jerk usually given to one balance by the stroke of the pallet, yet was subject to the friction arising from the action of the teeth, like the preced-

 ing

ing one; it had, however, a capability of making the angle of action and angle of the balance's motion before the escape in any ratio to each other, by altering the lengths or number of the radii respectively, as well as the distance of the pallets from the centres of motion; it had also the good property of preventing an increase in the maintaining power from affecting the balances in the same proportion, by the long lever pressing against the semi-circular central part of the balances alternately, during the latter part of the excursion of the balances, so as to check any accession of momentum occasioned by an accession of maintaining power: the law of this pressure, however, might be modified by changing the figure of the semi-circular piece into a spiral shape, so that the pressure might bear a proportion to the angular distances of the balances from the spring's quiescent position. The friction in both cases, it must be admitted, would be detrimental to the performance. The banking, we presume, in this escapement, was the same as in an ordinary watch, as no remark is made thereon. The invention, it appears, must be attributed to three successive artists; first the German, whoever he was, invented the escapement without the balance spring, and with the small star only, in which state it would act under the controul of the main-spring, without the regulating spring: secondly, the regulating spring was added by Dr. Hooke, and, thirdly, the large star was introduced by Jean Baptiste Du Tertre, about the year 1724, which addition completed the escapement as given in our drawing, which in its present state will not act without a regulating spring. Of course this last addition followed Tompion's and De Baufre's escapements hereafter described, from which probably the idea of the large star pressing on the circular part of the axis of motion was derived. Each addition was evidently an improvement on the original German escapement as described by Thiout; but the number of pivots in motion, and the increased friction occasioned by the pressure of the semi-cylindrical pieces at the centres of the balances, were obstacles too powerful to be overcome in this construction.

4. *First dead beat escapement by Tompion.*—The ingenious Tompion of London was probably the first watch-maker who contrived an escapement without recoil, or escapement of the second class, called a *dead-beat* escapement from the circumstance of the second's hand remaining motionless, or, as it were, dead, during the motion of the balance after the escape. Sully says that he succeeded in effecting this about the year 1695. In this escapement, of which *fig.* 7. of *Plate* XXXV. is a plan, the balance verge carried a cylindrical piece of metal *a b c*, terminating with a pallet *c*, with a notch cut between *a* and *c* to allow the escape of each successive tooth of the balance wheel. This wheel had its plane parallel to the planes of the plates of the frame, and its teeth bent at the end, and so distant, that the cylindrical piece and pallet could revolve between any two of them. When the cylindrical part of the pallet presented its notch to a contiguous tooth of the escapement wheel, it received a push therefrom, which put the balance in motion, and the next succeeding tooth falling on the circular or cylindrical portion of the pallet, now in motion in the direction *c b a*, rested against it without recoil till the balance spring had brought the notch back again, when it escaped in its turn by giving its push, and a third tooth fell on the cylindrical part of the pallet, and in like manner remained motionless till the notch came back again and allowed the escape: thus one escape took place at every alternate vibration, and the contrivance was admirably calculated to prevent the bad effect on the balance of any change in the impulses derived from the main-spring; but the friction on the back of the pallets when large, and at the balance pivots occasioned by the pressure of the resting teeth, constituted the same objection in practice as applied to Du Tertre's modification of the escapement with two balances. This escapement, however, was the archetype of all the succeeding dead-beat escapements, and therefore is deserving of the notice we have given it.

5. *Cylinder escapement in diamond by De Baufre.*—A native of Geneva of the name of Facio had applied rubies to the pivot holes of a watch about the year 1700, and went into partnership with De Baufre, a French watchmaker, established in London; the latter artist finding the advantage of applying the polished surfaces of the precious stones for diminishing friction at the parts of action, contrived a new escapement about 1704, of which the pallets were of diamond. Sully says that a watch of De Baufre, with the diamond pallets, was put into his hands by sir Isaac Newton, who had found its accuracy of performance admirable. The following account will be intelligible, we presume, to any artist who understands the mechanism of a watch. The balance had two pallets on its verge consisting of a semi-cylindrical diamond, and the escapement wheels, of which there were two, were vertical, with their common arbor at right angles to the verge of the balance; *i. e.* with their planes parallel to the verge: these wheels, which were similar, were placed one opposite the other on contrary sides of the verge in such way, that a tooth of one fell in the middle of the space of the other, and *vice versâ*; the pallets were $\frac{1}{24}$th of an inch thick, and two and a half twelfths in diameter, and the distance between the ends of the teeth of the double escapement wheel about $\frac{1}{12}$th of an inch, or twice as much as the thickness of the pallets; the pallets had their planes parallel to the plane of the balance, and had their semi-circle sloped in an angle of about 45° at the ends for the pallets to slide over, after pressing on their planes alternately during the respective oscillations. We will endeavour to explain the action, as we comprehend it to have been by a verbal description. Let the leading pallet be called *a*, and the following pallet *b*; and let the balance be supposed in motion with a tooth of the escapement wheel resting on the plane of the diamond; when the sloped end of pallet *a* arrives at the said tooth, the tooth falls off, accelerating the balance by its sliding motion, and goes on till an opposite tooth of the other wheel falls on the plane of the diamond in motion; this tooth rests without recoil, as did the former one, till the sloped end of the pallet *b* comes to it, when it falls off in its turn, and gives motion to the balance, which has thus its spring wound up by the momentum of the balance sufficiently far to bring the balance back again after every impulse; and thus an alternation of vibrations is continually maintained while the force of the mainspring lasts. This escapement is also of the second class, or of the dead-beat kind, but is subject to the constant friction of its immediate predecessors, though this friction is greatly diminished by the substitution of the diamond for metallic pallets, and by the bushing of the pivot holes with rubies; it possessed all the good properties of Tompion's escapement just described. (See Machines and Inventions approuveés, &c. vol. vii. p. 137.) Larcum Kendal used two crown wheels in a similar manner, which W. Howell improved by adding a forked detent to make the escapement detached, which was rewarded by the Adelphi Society, in 1792. See their Transactions, vol. x. p. 216.

6. *Sully's modification of De Baufre's escapement.*—The praise which sir Isaac Newton bestowed on De Baufre's new watch induced H. Sully to bestow much attention on it, and in the year 1721 he adopted an alteration in it, which made only one wheel necessary for the escape at both oscillations

lations; but Berthoud has expreffed an opinion that his alteration conftituted no real improvement, but made the adjuftment of the pallets more difficult. Sully having applied his modification of the efcapement in queftion both to a pendulum clock and a marine time-piece, prefented two memoirs on its conftruction to the Royal Academy of Sciences in the years 1723 and 1724; which are alfo printed in his defcription of marine time-pieces, publifhed in 1726. *Fig.* 1. of *Plate* XXXII. of *Horology*, reprefents a fide view of this efcapement, where A is the efcapement wheel, B the contrate wheel between it and the balance C, which has a pinion only on its verge, that is actuated by the contrate wheel; the pallets 1, 2, are two circular pieces of agate, placed at the diftance from one another of one fpace of the efcapement wheel nearly; thefe pallets have each an inclined or floped end, like De Baufre's, the flopes alfo being at the contrary ends of each; the intention of this modification is not only to make one wheel A perform the whole bufinefs of the efcapes, but to produce long vibrations of the balance, even beyond a complete circle. The action was thus; on giving motion to the balance C, fuppofe the tooth *b* to reft on the plane of the pallet or agate 1, it foon meets with the end flope, and falls off till it meets with the pallet 2 below, on which it refts, after having given a pufh to the pallet during the act of fliding off; the balance goes on, till the fpring, now wound up, brings it back, there being in the mean time no recoil, by reafon of the arbor of the contrate wheel being alfo the axis of the pallets, and at right angles thereto; the tooth *b* now finds, after the return of the balance, the floped end of pallet 2, and efcapes entirely; but at the fame inftant tooth *c*, which follows, falls on the plane of pallet 1, and is then in the fame fituation we at firft found *b* in, therefore the fame action is repeated with this, and every fucceeding tooth, till the power of the main-fpring is exhaufted. This modification of De Baufre's efcapement is confequently of the fecond clafs alfo, and enables us to comprehend more clearly the action of the original one, which had its double efcapement-wheel placed with refpect to the balance verge, as the prefent one has its fingle one, with refpect to the arbor of the contrate wheel. The effect produced muft have been nearly the fame on the pallets, though the vibrations of the original balance muft neceffarily have been fhorter than of the latter, but then it had not the friction in the wheel-work between the pallets and balances.

7. *Efcapement with two cylindrical pallets.*—In the year 1727 Julien le Roy fhewed Sully fome watches juft brought from England into France, which had each two cylindrical pallets with notches cut in them like Tompion's fingle pallet, which kind of efcapement feems to have differed from his, only in having an efcape at every vibration, and in having the old crown-wheel as an efcapement wheel; but as we have not been able to trace the inventor of this efcapement, we muft confider it as a lineal defcendant from Tompion's, and therefore as a variety of Englifh extraction. Its utility has been, however, evinced by the circumftance of Julien le Roy's having adopted it as a favourite, and ufed it in fome of his productions: it had the property of converting the recoil of the crown-wheel into a dead-beat.

8. *Cylinder or horizontal efcapement of Graham.*—We are not informed in what particular year Mr. Geo. Graham, the celebrated watch and clock-maker, invented the efcapement with a cylinder, called alfo the horizontal efcapement, from its wheel being in an horizontal pofition, but it appears that Julien le Roy introduced this conftruction into France in the year 1728, when wonderful properties were attributed to it, but particularly the property of compenfating all the inequalities of the maintaining power, which at that time was the great defideratum among watch-makers. The contrivance bears a great analogy to Tompion's efcapement, and was probably originally derived from his, which, however, does not leffen the merit of the contrivance, as it requires as much ingenuity to improve an exifting inftrument, frequently, as to invent a new one. *Fig.* 2. of *Plate* XXXII. is a perfpective view of Graham's horizontal efcapement, in which A is the efcapement-wheel, B the balance, and C the hollow cylinder fixed on the verge of the balance; each tooth of the wheel, which may have any convenient number of teeth, is formed of a triangular inclined plane or wedge, fupported by a ftem projecting perpendicularly from the plane of the wheel at its circumference; the length of the wedge of each tooth is juft a diameter of the internal part of the cylinder, fo that the cylinder may revolve round the wedge when the balance is in motion; in this fituation the anterior point of the wedge juft touches the cylinder's interior furface, but the pofterior point of the fame is free, fo as not to touch and ftick faft in the cylinder, but to give it a pufh on the edge when the contact ceafes; the diftance between the teeth is alfo great enough to allow the exterior portion of the cylinder to come between them, when the efcape has taken place, and while the balance continues its ofcillation. The action is performed in the following manner; fuppofe the wedge to be juft efcaping from the hollow part of the cylinder, and to be giving its impulfe to the edge of the cylinder which it quits, the balance goes on and winds up the fpring, as ufual, at the fame time prefents its exterior face, or convex furface to the point of the fucceeding tooth or wedge; this point falls on the faid convex furface and refts there, without recoil, till the ofcillation is finifhed, at the return of the balance, by means of the fpring now wound up, the hollow or concave part of the cylinder comes to the point of this following wedge, and being rounded at this end, admits it eafily into its concavity, at which inftant the other edge of the cylinder receives a fecond impulfe in a contrary direction; the balance now makes another ofcillation, and the point of the wedge refts againft the interior face of the cylinder, as it did before on the exterior, without recoil, and in both fituations of reft the wedge has no other effect on the balance than that of retarding it by its preffure againft the circular furface; the fecond tooth is now in the fame fituation as we found the firft, and prefently in its turn gives another pufh at leaving the cylinder altogether, and leaves the third wedge to perform the fame office by firft refting on the back of the cylinder, then entering it, and laftly giving its parting impulfe to the fecond edge of the faid cylinder, as the two wedges that preceded had fucceffively done before. At one period this efcapement was in high eftimation in England, as well as on the continent, and even now the common watches that are conftructed with this efcapement will fetch greater prices than thofe with the ordinary crown wheel efcapement, but the detached efcapements of the beft chronometers are now held in much higher eftimation than either. We have feen that the hollow cylinder, though conftituting but one pallet, in effect performs the office of two pallets, and permits but one tooth or wedge to efcape entirely at two ofcillations, during one of which the point refts at the outfide, and during the other at the infide of the cylinder, while the two edges of the cylinder perform the part of feparate pallets on receiving feparate impulfes in contrary directions, it is, therefore, of importance in this, as indeed in all efcapements with a regulating fpring, that the quiefcent pofition be duly adjufted to render both ofcillations precifely fimilar in extent and duration at each fide, otherwife the performance will be a fpecies of hobbling, and the compound law of motion, derived

derived from the ſpring's action and impulſes on the pallets conjointly, will be far from regular; conſequently not likely to be an iſochronal law. Berthoud has deſcribed a cylinder eſcapement with two balances connected by two pinions, but it did not prove ſucceſsful.

9. *French eſcapement a Virgule.*– The French eſcapement known by the appellation of *a virgule*, from the pallet's reſemblance of a *comma*, is repreſented by *fig.* 6 of *Plate* XXXV. A, B, C, are three teeth of the balance wheel, placed very obliquely compared to a true radial line, and carry each a pin at their extreme ends, ſtanding vertically like the ſupports of Graham's wedges for his cylindrical eſcapement; the pallet B A has its centre of motion at *c*, the centre of both the concave and convex portion, exactly over the verge of the balance; this centre is in the circumference of the circle that paſſes through the points or pins of the teeth; the pallet, which is of hardened and poliſhed ſteel, has its plane parallel to that of the wheel, and is ſupported between *f* and *e* by a crank piece put over the verge, which cannot be admitted as high as the pallet; this end *a o*, which connects the convex and concave portions, is a portion of an equi-angular ſpiral, and the part *b d* of the pallet is called the *horn*, probably from its reſemblance of the beak of a bird; the inferior face *b* A may be ſtraight like the dotted line, or curved, at the option of the maker, and the ſuperior part *f* A may be of any ſhape: the convex part *a e f* contains about 100°. The length of the pallet from *a* to *d* is ſuch, that when the pin at B preſſes againſt the curved part at *a*, the point *d* of the horn will juſt miſs the pin at A of the preceding tooth, and the height at which the pallet ſtands is ſuch as to make it paſs over the rim of the wheel, but not over the pins at the ends of the teeth. When the balance is at reſt, and has its ſpring in equilibrio, the point *d* of the horn is at the point *i*, about 30° from *d*; but the preſent poſition is that which the pallet has when the tooth A has juſt eſcaped from the point *d* of the horn. To explain the action, on a ſuppoſition that the tooth A has juſt eſcaped, the ſucceeding pin B has juſt fallen on the point *a*, joining the ſpiral and convex portions of the pallet, and the balance has juſt received its impulſe from tooth A as it paſſed from *b* to *d* along the inferior face of the horn; it, therefore, proceeds in its motion in the direction *d g h* of the dotted circle, while the point B ſlides on the convex ſide *e f*, and is detained without recoil; when the point *d* of the horn has arrived as far as *h*, more than 90° from A, the momentum of the balance is exhauſted, and the balance returns; the point *d* comes to about the point *i* before the pin of tooth B comes to *o*, and enters the concave part of the pallet, where it ſlides till the point arrives about *k*, where the momentum of the balance ſpring is again exhauſted; the balance returns again, and when the point *d* comes to *i*, the pin of tooth B eſcapes the concave part at *b*, and ſlides along the inferior face of the horn, at the ſame time giving it a puſh till it eſcapes entirely, and is in the ſituation of tooth A, where the proceſs began; in the ſame manner the following tooth C, and every ſucceeding tooth in its turn, will fall on the convex ſide of the pallet, return and enter the concave ſide, and then finally eſcape. We muſt obſerve, however, that the pallet oppoſes the motion of the balance as it paſſes along the ſmall ſpiral part from *a* to *o*, which may be ſo proportioned as to be a counterbalance to the impulſe previouſly given to the balance by the leading tooth, in caſe the friction on the back of the cylindrical part is not competent to retard the balance's impulſe ſufficiently. The effect produced by this eſcapement muſt be very ſimilar to that produced by Graham's cylindrical eſcapement, but has this advantage, that the inclined plane formed on every wedge of Graham's conſtruction is transferred to the pallet, and ſerves for all the pins to ſlide on in ſucceſſion, thereby removing the difficulty of forming all the wedges alike in ſize and ſhape. Beſides, the convex and concave ſides of the pallet may be reduced to any dimenſions, provided the acting tooth may not ſlide at once from the ſpiral to the horn, and allow the train to run down, which may be the caſe if the parts of action be ſmall, and there be much play in the pivot holes of the wheel and balance. The conſtruction is a compound of Tompion's and Graham's watch eſcapements, to both of which it has certain points of reſemblance. In 1752, Le Paute altered this eſcapement, by putting pins for teeth in both planes of the wheel, as he did in Amant's eſcapement *a cheville*, which addition gave it ſome advantages.

10. *Eſcapement with one pallet and one detent by Thiout.*— Thiout, a French clock-maker of conſiderable eminence, publiſhed a work in 1741, entitled "Traité de l'Horlogerie Mechanique et Practique, approuvé par l'Academie Royale des Sciences," 2 vols. 4to, in which he has deſcribed an eſcapement of a watch that has been miſtaken ſometimes for a real detached eſcapement, and conſidered as the firſt that has been made of ſuch a conſtruction, though the inventor himſelf did not conſider it as ſuch, as is evident from his own deſcription. This eſcapement is given in *fig.* 8. of *Plate* XXXII., and to uſe the author's own account, "it is an eſcapement of a watch in which half of the vibrations *ſeem* independent of the train of wheels, while they are performed. The detent B ſtops the eſcapement wheel, the balance bringing back the pallet A, the detent recedes, to leave the wheel free to ſtrike the pallet; and ſo on. This eſcapement could not perform without a ſpiral ſpring." - It hence appears, that the pallet A is on the verge of the balance, but that the detent B is on a ſeparate arbor, while they are connected by the tail of the pallet entering the ſlit in the remote end of the detent, ſo that the balance is never diſengaged from the detent, and therefore cannot be conſidered as of the detached claſs. When the balance moves in a direction from A to B the pallet gets out of the teeth, but the detent falls in, and *vice verſâ*. The curves of the claw of the detent appear to have been drawn from the centre of its motion, and therefore we are diſpoſed to think that there would be no recoil in the eſcapement wheel.

11. *The anchor eſcapement by Clement, or Dr. Hooke.*—While theſe improvements were going on in watch eſcapements, an equal attention was paid to the contrivance of new clock-eſcapements. Huygens had placed the pallets of the crown-wheel eſcapement at an angle of about 60° inſtead of 90°, as they were made for a watch, in order that they might be accommodated to the ſhort vibrations of a pendulum, or rather, perhaps, to the vibrations of ſome length performed in a cycloidal curve, by means of his cycloidal cheeks near the point of ſuſpenſion; but ſtill the arc of action on the pallets was too long to allow practice and theory to coincide, as to the performance of long and ſhort arcs of vibration in the ſame time. The preference that ſeemed due to ſhort arcs, when circular, ſuggeſted the propriety of contriving an eſcapement that would admit of very ſhort arcs of vibration. This was firſt effected about the year 1680 according to Smith's account, by Clement, who was a clock-maker in London; but Dr. Hooke has diſputed the priority of the invention, and has affirmed that he exhibited a pendulum with what is now known by the name of the *anchor* eſcapement, to the Royal Society of London, in the year 1666, ſoon after the great fire of that year. We will not endeavour to ſettle the point that relates to priority of invention, but proceed to deſcribe the contrivance itſelf, which has continued, and is likely to continue, to be of permanent uſe

use in ordinary clocks. *Fig.* 3. of our last named plate is the representation of the anchor escapement, as it is commonly constructed for a pendulum clock that swings seconds, which pendulum, in conjunction with the anchor escapement, gained the appellation of the *Royal pendulum.* The escapement, or swing-wheel A, has usually thirty teeth, when a vibration is performed exactly in a second, in which case a second's hand is usually inserted on its elongated pivot, to indicate seconds; but if no attention is intended to be shewn to the indication of seconds, the train may be made to admit of any other convenient number of teeth; B C is the anchor and pallets, taking its name from a resemblance to a ship's anchor, that may be observed in its shape; the distance from its centre of motion *a* to that of the wheel *b* is determined by the number of teeth that the pallets take in between them, it being essential to the freedom of action that two lines, drawn from the two centres of motion *a* and *b*, should constitute a right angle at each of the two points of action, as seen included by the dotted lines, meeting at the ends of the acting teeth *d* and *e*: in the pallets which we have made to include nine teeth, the centre of the pallet's motion falls at the point *a*, but if they were made to include eleven teeth the said point would fall above at *h*; and if they included fewer than nine the point for the centre of motion would fall below *a*: in general ten or twelve are contained between the pallets of a second's pendulum, but this must depend on the arc that the pendulum is intended to vibrate; for the more teeth are taken in, the greater is the distance of the two centres of motion, and consequently the smaller the angle or arc of action on the pallets. If a circle *f g* be described by a radius equal to half the distance of the centres of motion *a b*, round the point *a*, the two faces of the pallets B and C usually lie in the tangent lines *e i* and *d g* of that circle, which lines are the guides for making the slopes. The action of the escapement before us may be explained thus; suppose the tooth *d*, seen resting on the slope of pallet C, to be urging this pallet outwards from the centre of the wheel, by which a motive force is applied to the moving pendulum through the medium of the crutch; presently it escapes, and pallet B is brought inwards towards *b*, so as just to avoid the tooth *e*; the wheel is now at liberty to move, and goes on, by the force of the maintaining power, till the succeeding tooth *k* falls on the sloped face of pallet B, which is now urged outwards, and endeavours to give a contrary motion to the pendulum; but the momentum of the moving pendulum is too powerful for its impulse, and carries the tooth *k* back again till the momentum is expended, at which instant a recoil of the wheel and second's hand placed on its arbor take place; presently the pendulum returns, and the tooth *k* now pushes it on, and escapes; the pallet C is now brought back, so that the tooth following *d* falls on its sloped face, and in its turn experiences a recoil till the pendulum returns, when it pushes on and escapes; then another tooth next after *k* falls on the face of B, and experiences the same recoil and subsequent power of escapeing that the others did; and thus the alternate vibrations of the pendulum are perpetuated as long as the maintaining power continues. The effect of this escapement on the pendulum is similar to that of the crown-wheel escapement on the balance; the pendulum is either accelerated or retarded in every part of its vibration, except during the almost insensible instant of the drop of the succeeding tooth, when one has escaped; but as the momentum of a heavy pendulum in motion is greatly exceeding that of a balance in motion, the bad effect is smaller in degree. When an additional weight is added to the weight constituting the maintaining power, the arc of vibration is increased, whence one might be led to expect a retardation in the time of a vibration, the arcs being circular; but the fact is, that the vibration is accelerated very sensibly by means of the pendulum being both accelerated in its descent and opposed in its ascent, and the pendulum comes sooner back than gravity alone would bring it. Here then the compound law of motion differs from the simple law of gravity, but not in the ratio of the arcs to the sines of the angle of vibration; for if the compound law were to the simple law of gravity as the *arcs* are to their *sines*, the vibrations would be isochronal, whatever their extent might be. It is hardly necessary to add that this escapement is one of the first class.

12. *Graham's dead-beat.*—The dead-beat escapement, as originally contrived by Graham, is a modification of the anchor escapement we have just described, from which it differs only by taking off a part of the slope of each pallet, and by making the pads, that is, the back of one pallet and interior part of the other, portions of circles described from the anchor's centre of motion, so that when the point of a tooth rests on those parts, while the pendulum is finishing a vibration, no recoil takes place, but the wheel continues motionless, or, as it were, *dead*. *Fig.* 4. of the same plate gives the figure of the dead-beat escapement-wheel and pallets, where we have represented the wheel, as we did the wheel of recoil, with only four instead of six radii, that there might be more space for our geometrical lines. In this construction, as in the preceding one, the distance of the centres of motion *a*, *b*, is determined by the tangent lines meeting the radii at the points of the acting teeth; when this distance is an exact diameter of the escapement wheel, we find that the pallets take in just ten teeth out of thirty, which is the case in the figure before us, but when twelve teeth are taken in, the centre of the anchor's motion falls at *h*, just a diameter and a half from the centre of the wheel; and this is Berthoud's rule for giving the distance *a b* without any reference to the tangent and radius; for he has found that the arcs of action on the slopes of the pallets *a m n*, and *a o p*, are each only one degree, but that if fewer teeth than twelve were included between the pallets, the anchor would be shorter, and the arcs of action proportionally longer. This rule might be a good one with a given maintaining power, that is, where the whole arc of vibration is of a given extent, but it does not follow that one degree for the arc of action before the escape is the best in all cases where the motive powers vary. We are of opinion that experiments might be instituted to ascertain this important point, by applying various weights successively, first with pallets that have twelve teeth included, and then with such as have eleven, ten, &c. but the experiment could not go beyond 14 teeth, because when half the wheel is included the tangent becomes a vertical line, parallel to the line of distance *a b*, and therefore this distance becomes *infinite*. There is probably a certain maintaining power that is best for a certain angle of action, with a pendulum of a given weight, but that does not seem to have been ascertained either by calculation or experiment hitherto. The thing to be aimed at is, to make the compound law of the pendulum's motion such that the varying forces may accelerate or retard it in proportion to the arcs of distance from the lowest point, which is the case with the simple law of gravity, as it regards the *sines* of the said arcs. We have put the same letters of reference to our present figure as to *fig.* 3, that the reader may the more easily comprehend the correspondence between them, and to prevent the necessity of further minute description of the parts; the slopes of the pallets being determined by the same tangent lines *e i*, and *d g*, as before. We shall, therefore, confine the remainder of

of our defcription to the features that differ, and to the mode of action depending thereon. The circle B C, paffing the points of the firft and eleventh teeth, including the ten fpaces between the pallets, forms the interior faces of the faid pallets B and C, and the exterior portions B *e* and C *d* are formed by a concentric circle of larger radius. Thefe circles are defcribed from the point *a*, or centre of the anchor's motion, therefore a tooth refting at any time againft any one of thofe circular portions remains without recoil; the diftance between the faid circles is fomewhat lefs than one-half of the fpace contained between the extreme ends of any two contiguous teeth, and the nearer it is to one-half of a fpace, the fmaller will be the drop after an efcape, or as the workmen fay, the clofer the pallets will be laid. The teeth are fomewhat varied in fhape from the inclined teeth of the recoil efcapement, a portion of the flope at the bottom being cut away, that the end of the pallet may enter the fpace, when the pendulum vibrates beyond the arc of action, without coming in contact with the metal of the wheel, which would make the pendulum rebound, and difturb the regularity of the vibration. Thefe obfervations being premifed, the reader will now be prepared to comprehend the action. The tooth at *e* has juft paffed the flope of pallet B, and is efcaping, after having given its ftroke or impulfe gradually, in fliding along the flope from *e* to its prefent fituation; the pallet B is now going from the centre of the wheel's motion *b*, and the pallet C is come back from its excurfion, in time to allow the tooth *b* to fall on its inner circular part, near the angular point of the flope; the pendulum goes on in its vibration till its gravity overcomes its momentum, during which time, the pallet B is no longer in contact with the wheel, but the pallet C enters the fpace between *b* and *d*, and the point of the tooth *b* refts on the inner circle of the pallet; prefently the pendulum returns, the point of tooth *b* flides back again till the flope *b d* prefents itfelf, when urged by the maintaining power, it efcapes after having made its pufh to the pallet C, which now continues its motion outwards, and confequently pallet B falls in the way of the tooth *k*, which follows, but which refts on the exterior circle B *e* of this pallet, till the pendulum returns as before; it then meets the flope at *e*, and, giving its pufh to the pallet, efcapes, as its preceding tooth had done; and thus the vibrations are alternately aided in future, while, as has been feen, one tooth only completely efcapes at the interval of two vibrations: which is the reafon why a wheel of 30 teeth is proper for the indication of 60 vibrations of a fecond's pendulum. When the wheel is made of hardened fteel, and the pallets of ruby or agate, the friction is greatly diminifhed, and will admit a fhort angle of action, or diminution of power.

13. *Modification of the dead beat by Grignon.*—When we gave the "Hiftory of the fucceffive Improvements in Clocks," under our article CLOCK, we faid that the fenior Thomas Grignon, of Covent Garden, London, improved the dead-beat efcapement of Graham, and prefented a clock to the Society of Arts in 1759, which remains in one of their rooms to this day, that has got this improved efcapement, the peculiar property of which is faid to be, that it performed alike with *four* and with *twelve* pounds fufpended as a maintaining power; if this is a fact, the efcapement muft have the property, fo much defired in all efcapements, of rendering all the arcs of vibration ifochronal. We have been favoured with a manufcript account of this efcapement, and of its principle of conftruction. It will be doing but juftice to the memory of departed genius, if we give the author's own figures, demonftration, and account of his improvement, as he has left it in his own hand-writing. *Fig.* 5. *of Plate* XXXII. is the *fcheme*, and *fig.* 6. the *demonftration*, as the author himfelf has intitled the figures, the latter of which has no letters of reference, but is fufficiently intelligible in conjunction with the other. "It appears from the fcheme," fays the author, "that the diftance of the centre of the pallets from the centre of the fwing wheel in that machine (at the Adelphi) is *one whole diameter* of the wheel, which has this peculiar excellence, that the triangles formed at the ends of the pallets, by drawing the tangents for the flopements are *equal*; (nor indeed can they ever be equal, continues he, where the diftance is more or lefs than *one whole* diameter of the fwing wheel,) and, on the contrary, where that diftance is duly obferved, all the triangles will ever be equal; for then all the interfections, made by drawing the two tangents to the circle of the pallets, will fall exactly alike, (*fig.* 6.) interfecting the intermediate circle, as may be feen by the figure, confequently the tangent lines, together with the chord of the fwing wheel form a rectangled triangle, the upper angle of which, by altering the flopement only, changes its place in the circumference of the circle, the fame chord always continuing its bafe. For a farther demonftration of which, fee 21ft propofition of the the 3d book of Euclid.

"Again, when the diftance is duly obferved, we fhall find that one of the lines for the flopement of the pallets will be the hypothenufe, and the other the perpendicular of rectangled triangles, whofe bafes are equal to 120°, the halves of which muft be 60°. That they are rectangled triangles may be eafily proved; for if the bafe be 120°, and perpendicular 60°, the hypothenufe muft be 180°. See 1ft Prop. of the 12th book of Euclid."

In another manufcript paper, Mr. Grignon fays, "that the teeth of the fwing wheel have an equal effort, when they act upon the flopements or faces of the two levers or pallets, may be thus demonftrated.

"Draw the lines A, A, upon the faces or flopements of the pallets or levers *a*, *a*, now thefe are the lines of direction; let fall the perpendiculars B, B, from the centre of motion of the pallets C; now, I fay, that if you meafure the lengths of thefe perpendiculars, they give you the efforts of the arms of the levers or pallets D C, D C; for by the 9th prop. of Mr. De la Hire's Treatife of Mechanics, the efforts of a weight or power are not to be meafured by the length of the arms of the levers, but by perpendiculars drawn from the hypomochlion upon the lines of direction of the weight or power. That thefe levers are equal is thus proved; for, taking C for your centre, defcribe the circle E E E; now if the perpendiculars B, B, are both radii of the circle E E E, they muft, of confequence, be equal, which is what was to be proved. Again, that the faces or flopements of the pallets or levers are equal may alfo be thus proved: *a* A, *a* A, are the lines of direction of the action of the teeth, and being both tangents of the fame circle E E E are confequently equal."

The remarks that have occurred to us on contemplating Mr. Grignon's *fcheme*, as he has called it, are, firft, that his diftance between the centres of motion require *ten teeth* out of thirty to be taken in by his pallets, as we have fhewn in our laft article; fecondly, the circle E E E is defcribed with lefs than half the diftance of the centres, and is yet made the circle to which the tangents are drawn, that form the flopes, the reafon of which he has explained, the radius C B of this circle, it appears, muft be determined by a line *a* A, drawn parallel to the line of diftance, which muft be a tangent to the faid circle, and which affords an eafy and practicable mode of drawing the flopes agreeably to this conftruction; thirdly, if the flopes thus formed by tangents drawn

drawn to a diminished circle have the isochronal property, they are worthy of future imitation, under the circumstance particularly of ten teeth being included between the pallets; and lastly, we observe, that the *same circle* forms the *exterior* curve of one pallet, and the *interior* one of the other, which is not the case with Graham's pallets, but which certainly makes the friction alike on both pallets, seeing the circular surfaces rubbed by the points of the wheel's teeth are at equal distances from their common centre of motion. These considerations are all in favour of Mr. Grignon's modification of the dead-beat escapement, as each excursion of the pendulum must be influenced by similar circumstances, and if moreover it possesses the isochronal property attributed to it, it is somewhat remarkable, that the Adelphi society have been labouring to reward inferior escapements in great variety, while one of the best *dead-beat* escapements is beating *dead* to the society, having *escaped* their notice, while performing its functions within their own walls.

14. *Mr. Bennett's modification of the dead-beat.*—Mr. J. Bennett of Norwich has published an account of what he considers an improved method of describing the curves and faces of Graham's dead-beat pallets (see Nicholson's Jour. vol. xv. 8vo.) which account we think deserves a place here in common with Grignon's method.

Draw the line A B (*fig.* 7. *Plate* XXXII.) says the author, on which describe the circle B, the size of the intended swing wheel; then, according to the number of teeth the pallets are intended to escape over say, as 60 (the double of the number of the wheel's teeth) is to 360°, so is double the number intended and one more to that proportion; thus, suppose the number intended to escape over nine, the double of which is 18, to which add one, and there are 19; then say,

As 60 : 360° :: 19 : 114°; the half of 114°, which is the whole arc included, is 57°; then on the circle already drawn set off on each side of the line A B 57°, from which points draw lines to the centre of the circle; then on these points where the circle is crossed erect perpendiculars, which will intersect at the centre of the pallet's motion in the line A: from this point as a centre draw the arc C C, to cut the points where the radial lines cut the circle; the arc thus drawn describes the receiving and leaving pads of the pallets. The inclination or inclined plane of each pallet must form an angle of 60° with the said radial lines; thus from the points of intersection, near E and G, draw the small semicircles D and G with any radius, and set off 60° from *f* to *g* on the semi-circle D, and from *h* to *m* on the semi-circle G, then will lines drawn from the centres of the semi-circles, through the points *g* and *m* respectively, give the proper direction of the slopes of the pallets' faces.

This construction is very similar to Grignon's, though differently projected: the centre of motion A is determined in the usual way, and the odd unit given to double the number of teeth included is an allowance for the breadth of the pallet, to have as little drop as possible; the slope of the pallet G is parallel to the line A B, in Grignon's construction, but is not quite so here, for the difference of the angles at the opposite sides of the radial line is $\overline{60°-57°}=3°$, which is what these lines want of being parallel, which quantity of deviation seems to constitute the principal difference between the two constructions; if the radial line of pallet G had been drawn to the lower point of the pallet, so as to have made the angle at the centre 60°, and the teeth included had been exactly ten, then the two methods would have coincided, and the point A of the pallets' motion would have been an exact diameter of the wheel from its centre. We conceive that the effects of the two constructions may be so nearly alike, that their difference in practice may be almost imperceptible. If there is any difference, varying the maintaining power considerably in different trials will detect it, by rendering the alterations in the rates perceptible from the comparisons. Otherwise the alternate angles might each have been 57°, instead of one of them being 60°.

15. *Isochronal escapement for a pendulum.*—So long ago as about the year 1720 Saurin, Julien le Roy, and Enderlin, wrote memoirs on the nature of the curve that was necessary to give to the pallets of an anchor, instead of the concentric circles, in order that the swing wheel may have just so much recoil only, as will render all vibrations, performed in arcs of different extent, isochronal; and since their time Berthoud in France, Reid in Edinburgh, Melville and others in London, have put in practice a method of constructing pallets founded upon the same principle. We think it not improbable, however, but that the Scotch and English artists have borrowed their contrivance from the French author, who published his construction in his "Essai sur l'Horlogerie" in the year 1786, before which time we do not find any traces of an isochronal escapement except Grignon's just described, and Smeaton's, which follows. The construction of the escapement in question differs from the dead-beat only in one particular, which is, that the circular portions of the pallets, which in the dead beat are concentric, described from the centre of the pallet's motion, in the isochronal escapement are *excentric*, with respect to each other, those curves on which the ends of the teeth of the swing wheel rest being described from points out of the centre of motion. Berthoud had made several experiments with different escapements, and different maintaining powers, from which he found that the same addition to the power which accelerated the vibrations of the pendulum, when a recoil escapement was used, retarded the same when a dead-beat escapement was substituted; hence he concluded that there must be a medium between much recoil and no recoil, which would make the pendulum vibrate in the same time in all arcs, or, which is the same thing, with any addition or diminution of the maintaining power; he therefore fixed, after various trials, on a construction producing but a *small recoil*, which answered his most sanguine expectations, and which we shall be happy to describe here, particularly as it promises to answer a better purpose than the common dead-beat, which has been long in high estimation; and as it has not, we believe, been described by any English author. *Fig.* 1. of *Plate* XXXIII. is a plan of Berthoud's isochronal escapement, where A, as before, is the swing-wheel, B and C the pallets of the anchor, and *a* the centre of their motion; the distance of the centres *a*, *b* also, as before, depending on the number of teeth taken in by the pallets; the angles *m a n* and *o a p*, are the angles of action by which Berthoud has given his slopes of the pallets without the tangent lines *e i*, and *d g*, but we have thought it right to shew how they may be given both ways. Berthoud's method of sloping to suit any angle previously given is more extensive in its application, but the quantity and direction of the impulse will vary as the angles *m a n*, and *o a p*, vary; whereas the method of sloping by the tangent lines can only vary by varying the circle *f g*, as Grignon has proposed: in our plan we have taken the circle's diameter equal to the distance *a b* of the centres for the sake of uniformity of plan, that the reader may see how the pallets, so formed, differ in shape from the dead-beat pallets. The exterior portion of the circle *d* C, of pallet C, is described with the extent *a d* from the centre *a*, but the interior portion *b* 4 is described with the same extent from a point 5 below

below the centre; the difficulty confifts in determining the true fituation of this excentric point 5; Berthoud's rule, founded on the refult of experiments, is this; having determined the breadth of the pallets, equal to fomewhat lefs than half a fpace of the fwing-wheel, defcribe from *b*, in the line *a p*, the concentric dotted circle 1, 2, 3, &c. fet the diftance between the two concentric circles, or breadth of the pallet, *three times* over along the inner curve to 1, 2, and 3, then in a direction from 3 towards *a* fet the fame *once* to 4; in the next place, with the extent *a b* find the point 5 by interfection from the points *b* and 4, and this point 5 is the excentric point defired, from which the curve *b* 4 is defcribed with the extent *a b* as before, which is the interior curve to be fubftituted for the concentric curve 1, 2, 3 of the dead-beat efcapement; and it is evident that while the pendulum is in motion inwards, with a tooth refting on this excentric curve, the wheel muft have a certain recoil. Again, the exterior curve of excentricity, of pallet B, is thus defcribed; the curve 1, 2, 3, &c. is defcribed, from the point of the tooth that is in the line *a m*, and the points therein are taken as before at the diftance from each other of the breadth of the pallet, or at fomewhat lefs than half a fpace of the wheel; the point 4 is here fet outwards in the direction from *a* to 3; then with the extent *a e* the excentric point 6 is determined by interfection from *e* and 4, which point is above the centre; laftly, with the extent *a e*, as before, defcribe the excentric curve *e* 4 from the point 6, which being fubftituted for the concentric circle 1, 2, and 3, will produce a fimilar recoil in the pallet B, as the curve *b* 4 produces in pallet C. The inner curve of pallet B having no action, like the outer curve of pallet C, may be defcribed from *a*, or be a ftraight line. The writer of this article has a half-fecond's pendulum clock with this efcapement, and Troughton's tubular pendulum, which goes admirably with any power that will keep it in motion.

16. *Smeaton's efcapement of the turret-clock at Greenwich.*—The efcapement which Mr. Smeaton contrived for the turret clock at Greenwich refembles the ifochronal efcapement, which we have juft defcribed, in a certain degree, and was, no doubt, intended to anfwer the fame purpofe. *Fig.* 2. of *Plate* XXXIII contains the plan of this efcapement, in which A B is the radius of the efcapement wheel; B C a tangent line at $3\frac{3}{4}$ teeth from the vertex, drawn from C, the centre of the pallet's motion; B D E is a circle defcribed from the centre C through the points B and E, which are each $3\frac{3}{4}$ teeth from the vertex; F G is a line drawn parallel to the vertical line A C at the diftance from the vertex of $3\frac{1}{2}$ teeth, and cutting the circle B D E in the point G; then G C B is half of the angle of vibration, and fetting off the arc B H = B G, G C H will be the whole angle of vibration. Draw the line C H cutting F G in I, and through the point G of this pallet from the centre A defcribe a circle that may cut the radius A K in the point K; this will be the point of the other pallet; the line E C being alfo a tangent to the wheel at $3\frac{3}{4}$ teeth from the vertex. Now fetting off K L on the circular arc = G H, the angle of vibration K C L will be equal to the other angle of vibration G C H, and if through the point L the line C L be produced, and L M be fet off = I H, then will the wheel move equally forward, by a fimilar motion of each pallet. Again, with the radius K C, upon K as a centre, defcribe the arc C N, and alfo with the fame radius from the point M defcribe the interfecting arc O P; then from Q, the point of interfection with the radius unaltered, defcribe the curve through the two points K and M to R, which curve will give the working face of the pallet that will produce a fmall recoil nearly equal to that produced by the plane G I; in this conftruction the depth of the teeth is equal to a fpace contained between two contiguous teeth, and the acting fide of the tooth is a portion of a circle defcribed from the bottom of the floped fide of the *fifth tooth* inclufive. The circle to which the flope of the tooth is a tangent is $\frac{1}{6}$th of the diameter of the wheel. When the fwing-wheel has a greater number of teeth than 30, $7\frac{1}{2}$ teeth are ftill contained between the pallets, and the centre of their motion is determined, as in the former conftructions, by a tangent line drawn from the point of the acting tooth. The two ends of the anchor may be made more apparently equal in length by letting the pallet K efcape over $3\frac{1}{4}$ teeth and the pallet G over $4\frac{1}{4}$, in which cafe the arc B D will be of a greater radius than the arc N D. The fteel pallets or end-pieces are fixed by fcrews that allow an adjuftment for diftance or depth of the pallets, fo that they may be detached, newly formed, and replaced, at any time after they have been in ufe.

17. *Efcapement with two levers by De Bethune.*—Thiout fays that De Bethune was the firft contriver of the clock efcapement with two detached levers, and that the origin of this conftruction was derived from the efcapement of Dr. Hooke, or of a German artift, with two balances, which we have defcribed. Thiout made the firft of his in the year 1727, and fays that from that time the efcapement with two levers became common among clock makers in France. *Fig.* 3. of our laft plate is a reprefentation of this efcapement as made for a pendulum; where A and B are the two levers placed faft on each a feparate arbor included within the frame; to the lever B is attached the crutch or fork that communicates with the pendulum not feen, and the fork C which is vifible in the figure; the lever A has a bar of metal D faft to it, that carries the roller E, of a diameter equal to the aperture of the fork C; and the ends of thefe levers conftitute the pallets placed at the diftance from each other of five out of thirty teeth of the fwing wheel. During the action of this efcapement, when one of the levers is acted on by a tooth, the roller and fork are raifed up, and the other lever becomes a detent to the wheel; then when at the next vibration the other lever is acted upon, the roller and fork are depreffed, and the firft lever in its turn becomes a detent; and thus the vibrations are alternately affifted while the maintaining power continues to actuate the fwing wheel. This efcapement has a recoil, and Julien le Roy difcovered (Memoire fur l'Horlogerie par M. le Roy, fils ainé de Julien le Roy, 1750) that when the levers are each the exact length of the radius of the fwing wheel, it poffeffes the ifochronal property, that is, all arcs of vibration will be performed with it in the fame time. We are not aware that this efcapement was ever made in England.

18. *Dead-beat efcapement for a pendulum by Amant, (a cheville)*—*Fig.* 4. of the fame plate fhews the plan of an efcapement by Amant, formerly a clock-maker at Paris, that has an efcapement wheel without teeth, but with pins or pegs inferted in a concentric circle in the plane of the wheel, as appears in the figure; the pallets A and B are both faft to the verge or arbor of the crutch, in the ufual way, and are brought fo nearly together that the fame pin that flides along the floped face of one pallet B immediately falls on the inner circle of the other pallet A, and continues without recoil till the pendulum has made its vibration; at the return of the pendulum the pin meets with the floped face of pallet A and efcapes; then the next pin falls on the interior circle of pallet B, and remains without recoil till the pendulum returns from its vibration, when it flides down the flope and gives a pufh to the pallet as the firft one did before; thus the two pallets are alternately actuated by the fame pin before its

its final escape, and sixty vibrations are made during one revolution of the plate with thirty pins. The slopes appear to stand at right angles to each other, but it is not said, in the account we have seen, how either of the slopes ought to be drawn geometrically. This is the kind of escapement at present used in the large turret clock at Hampton Court, and we have seen many other clocks in England with a similar escapement, which, we have no doubt, may perform as well as with the ordinary dead-beat escapement as originally constructed by Graham, provided the drop from the first to the second pallet, which must necessarily be equal to the thickness of the pin, do not produce a jerk that may affect the vibrating pendulum.

19. *Dead-beat escapement (a cheville) by J. A. le Paute.*—In Le Paute's "Traité d'Horlogerie," published at Paris in 1755, is an account of a dead beat escapement as made by him for a clock, which differs from our last by Amant only in its mode of position, and which therefore requires no figure for its description; the levers of the two pallets open like a pair of dividers to any extent, but are not in the same plane, one of the levers being over the anterior plane of the pin wheel, and the other over the posterior plane; hence two sets of pins are necessary, one on each plane; but the action is the same as we have described it in Amant's escapement, which it probably excels inasmuch as the drop may here be made as little as possible, because the same pin is not required to pass between the ends of the pallets in escaping; but one pin acts only with one pallet before it escapes; there are consequently sixty pins in all, thirty on each plane of the wheel.

The same author also published an account of a watch escapement with pins (a cheville) which the reader will find described at page 198 of the same treatise.

It is easy to conceive, after what we have said about the isochronal pallets, that both Amant's and Le Paute's escapements, and indeed all the dead-beat escapements, may be made to have a little recoil, so as to become of the isochronal kind, provided the acting concentric circles of the pallets were made in a certain degree excentric; that is, if they were described from points out of the centre of motion of the pallets themselves. The only difficulty is in determining the situation of those points under different circumstances, which we think is a fair object for future experiments.

20. *Harrison's clock escapement.*—The escapements which we have hitherto described require oil on the pallets, to diminish the friction of the parts that rub against one another in action; but the escapement we are now going to describe as the invention of the ingenious Harrison requires no oil, on account of his construction being free from rubbing at the time of the pallets' action. *Fig.* 5. of *Plate* XXXIII. exhibits the shape of Harrison's escapement. G D O is a portion of the swing wheel, moving round M as a centre: A is the verge or arbor of the crutch that communicates with the rod of the pendulum; to this are attached the pallets, each consisting of a long and a short lever with a joint at the place of union; the short lever A B is fast to the arbor or axis of motion, and carries the longer lever B C which has a claw or hook *a* C, and a slender spring is so applied as to keep these two levers nearly at right angles to each other; at right angles to A B is also the other short lever A E fast likewise to the axis of motion, and bearing the longer lever E F, as A B bears the lever B C; the joints are at B and E, and a similar small spring holds the levers A E and E F in like manner at right angles to each other, when not otherwise acted on. The lever E F has not only a claw, like the other long lever, but a heel to it, or projecting part *f*. The action of this contrivance may be thus explained; suppose the tooth D urging the pallet or claw C, while the pendulum is moving towards the right hand in its excursion; this pressure of the tooth produces a tendency in the lever B C to move round, but the joint at B yields, and opens the angle A B C a little; during this motion round A, the short lever A E is lowered a little, and with it the long lever E F, which we will call the detent; its claw F therefore falls in the way of the tooth G, which now drops on the back part of the claw, and is inserted into a notch under the projecting heel *f*; the further motion of the pendulum, and consequent depression of the short lever A E, carry the end of the tooth G back again till the vibration is finished, and produce a recoil. At the moment when the tooth begins to have a recoil, the tooth D at the pallet or claw C is withdrawn and escapes; the slender spring of the lever B C, which had been bent from its quiescent state, now resumes its original situation, and throws out the lever B C, till its angle formed at B is again nearly a right angle; in the mean time the wheel is detained by the heel of the detent E F, in which situation it opposes the vibration through the medium of the two levers E F and E A, which are attached to the crutch of the pendulum. The vibration being finished, the pendulum returns towards the left, and the wheel is at liberty to advance again; the pressure of the tooth G on the claw of the detent now assists the vibration through the arc of escapement or arc of action, and causes the pallet C to approach the wheel, till the tooth following D falls on it and experiences a recoil in its turn. This recoil sets the detent F at liberty from the pressure of the tooth G, which now flies back by means of its slender spring to its rectangular situation, as it regards the small lever A E: presently this vibration terminates and the pendulum returns, while the tooth following D aids its vibration by pulling forwards the pallet C, when the process of making two successive vibrations is gone through, and the same operation is repeated. The reader will have observed in this account that the tooth and pallet, or tooth and detent, do not *slide* on one another, but give each other a *direct push*, or pull, the business of the small springs being to remove them from contact by one sudden leap, where rubbing is altogether avoided, which contrivance supersedes the necessity of having oil applied to the pallets, and is certainly an ingenious contrivance, as well as useful in practice. It must be observed, however, that the recoil, such as it is, takes place at the extremities of the vibration, when the momentum of the pendulum is almost exhausted, and is therefore the most unfavourable part of the vibration to occur in; notwithstanding which disadvantage the escapement has performed as well as any other that had preceded, perhaps as well as any that has followed it, as the regulator to which it was applied has seldom if ever been excelled in steadiness and accuracy of performance for fourteen years of uninterrupted succession. The imperfection arising from the recoil taking place at the extremities of the vibrations seems to have been compensated by the immense momentum of the pendulum that this escapement admitted of with a comparatively small maintaining power, while the absence of friction and of oil gave it the advantage of permanently uniform action. Of course the pendulum was of the grid-iron form, that compensated the effects of varying temperature, and allowed the escapement to have its merit appreciated. The report of the performance of Harrison's regulator with the escapement in question was, that it did not vary *one second* per day from any preceding or following day for the space of fourteen years, and that the aggregate of the variations from true mean time did not amount to half a minute.

Haley also constructed clock pallets that required no oil, but

but as none of them have fallen under our notice, we pretend not to ſay how they differed from Harriſon's.

21. *Eſcapement by Alex. Cumming for a clock.*—Mr. Cumming publiſhed his elements of clock and watch-making in the year 1766, in which, among other matters, he deſcribes a conſtruction of an eſcapement of his contrivance, that removes the friction and influence of oil not only during the time of reſt of the pallets, but alſo during the motion of the ſwing wheel, ſo far as they can influence the vibrations of the pendulum; and alſo obviates all the irregularities that can poſſibly ariſe from the maintaining power as tranſmitted through the train. *Fig.* 6. of *Plate* XXXIII. exhibits as much of this eſcapement as is neceſſary for explaining its general action, but thoſe readers who wiſh to ſee all the minutiæ of the conſtruction, as repreſented by different plates of ſections, muſt have recourſe to the book itſelf, as the complexity of the parts repreſented in various points of view demands more minute repreſentation than our plan admits of. Our figure is a copy of plate 5 of the author, in which A B C is the eſcapement or ſwing-wheel; D E the pallets ſeparately moveable on the axis of motion, and having each a little ball or weight H and I reſpectively ſupported by horizontal arms; each pallet with its attached weight is ſeparately moveable from the other round its reſpective axis of motion, ſo that when one is raiſed by the wheel the other remains at reſt; F and G are two detents firmly attached together, like the ordinary pallets of a common dead-beat or recoil eſcapement, and are fixed to a cylinder of braſs P P, that ſurrounds the two axes of motion of the pallets, which are in the ſame ſtraight line but ſeparated from each other at the middle of the cylinder, where is a partition that bears the interior pivot of one axis and the exterior one of the other; the crutch of the pendulum is alſo faſt to the cylinder; ſo that the two detents and crutch move together along with the cylinder, without the motion of either pallet; and each pallet can move ſeparately without the detents and crutch. M is a load to counterpoiſe the detents and crutch in any poſition given them; and *n o* are two pins carried by the pendulum that project far enough to fall alternately in the way of the arms that carry the balls H, I, as the pendulum vibrates.

To explain the action of this eſcapement, let us ſuppoſe the pallets D and E, the detents G and F, the crutch L, and pins *n* and *o*, together with the balls H and I to be exactly as repreſented in the figure, where the lever of the detent G is hid behind the lever of the pallet E, but may be conceived to be parallel to it, the pallet D having been previouſly lifted from its quieſcent poſition, where its point reſted on the rim near the foot of tooth A; then if we conceive the pendulum moving from right to left with the ball I urging the pin *o* attached to an arc fixed to the pendulum, or crutch of the pendulum, the vibration is aided by the gravity alone of this ball. The arc, that bears the pins *n* and *o*, has a contrivance not ſeen in the figure for unlocking the detents at the moment when either of the pins bears againſt its ball, which contrivance it will be ſufficient for our purpoſe to call Z. Conceive now the pin *n* brought into contact with the arm of ball D by the moving pendulum, the detent F is unlocked, and the tooth A of the wheel quits the end of pallet D; the ball H now acts by its gravity on pin *n*, and deſcends, at the ſame time aiding the returning vibration, while the pallet D, united to the arm of ball H, deſcends alſo towards the wheel; but as ſoon as the pendulum firſt began its original vibration, the pin *o* was withdrawn from the preſſure of the ball I, by means of the attached pallet E falling on the tooth C of the wheel, ſo that all the latter part of the vibration was free; this preſſure of pallet E on the tooth C diminiſhed the preſſure of tooth A on the detent F at the inſtant of the unlocking, which we have mentioned; the wheel now in motion by the action of the maintaining power, (through the medium of the train,) goes on till the tooth C, acting on the pallet E, raiſes it outwards, together with its attached ball I, till the detent G ſtops the wheel's motion, when all the parts are at reſt till the pendulum nearly finiſhes its returning vibration, and then its pin *o*, meeting with the elevated arm of ball I, preſſes againſt it and concludes the vibration; at the moment this preſſure is felt the contrivance Z unlocks the detent G, and allows the tooth C to eſcape; in the mean time the ball I by its gravity urges the pendulum for a ſhort time, while the tooth following A, meeting with pallet D, raiſes it and the ball H a ſecond time, the tooth following C comes in the way of pallet E, and ſtops the further action of the ball I on the pendulum, which now vibrates, as before, free, till its pin *n* again meets with the arm of ball H, when the ſame proceſs is repeated; and thus the weights H and I are alternately raiſed by the wheel, and then alternately act for a ſhort time on the pendulum, at the beginning of each vibration, ſo that gravity alone is properly ſpeaking the maintaining power of the pendulum, and whatever may be the maintaining power of the train, it never affects the motion of the pendulum; but only raiſes the balls, that afterwards act on the pendulum by their gravity alternately exerted from a ſtate of reſt. Thus, whatever may be the effects of oil, duſt, friction, wear, &c. on the action of the train of wheel-work, as it has no immediate connection with the pendulum, except during the ſmall inſtant of unlocking the detents, theſe effects are almoſt entirely avoided, as they regard the pendulum or regulating part of the motions. The greateſt objection in theory to this conſtruction is, that the additional gravity commences inſtantaneouſly, and alſo ceaſes inſtantaneouſly, and at a part of the vibration where the inſtantaneous exertion of extraneous force of any ſort is the leaſt favourable for aiding the ſimple gravity of the pendulum's ball, which conſideration is analogous to the obſervation we had occaſion to make on Harriſon's eſcapement: but, like his, the eſcapement before us has ſtood the teſt of practical experience, notwithſtanding its complexity, in his majeſty's private obſervatory. If, however, this eſcapement is not *better* than any of the other eſcapements, the number of its adjuſtments, and variety of its parts, conſtitute a practical objection to its general adoption; and we believe no proof has been given of its being preferable to Harriſon's, or even to the common dead-beat or iſochronal eſcapements, except that the works of the clock that has this eſcapement may not require to be ſo often cleaned. This eſcapement leaving the pendulum free or detached from the wheel-work during the greateſt part of each vibration, is of the third claſs of eſcapements.

22. *Eſcapement by Mudge for a clock.*—The clock eſcapement contrived by Mr. Mudge to detach the pendulum from the train, ſo as to allow it to have nearly a free vibration maintained by gravity alone, may be conſidered as a ſimplification of Cumming's eſcapement juſt deſcribed, from which its principal difference is, that each of his two ſeparate pallets act alſo as detents; the mode of action and effect produced are very ſimilar in both. Mr. Mudge's eſcapement is repreſented by *fig.* 7. of our laſt plate, in which L M is a portion of the ſwing wheel; G A and G B are two levers ſeparately moveable round one common axis of motion, and have each a pallet at the lower extremities; theſe pallets are formed like the common recoil pallets, except that each has a claw at the loweſt point, that prevents the eſcape of the ſliding tooth when acting thereon: the pallet G A has a lever, which

which may be called its tail-piece G O, attached to it, and alfo a fmall metallic ball *u* furmounted on a pin over the pallet, which may be eafily changed for one of a different weight; the pallet G B has a fimilar tail-piece I O, with a fimilar ball *v* over its pallet B; and the dark circular fpot at N is the end of a ftiff pin projecting from the pendulum, or its crutch, far enough to fall in the way of the tail-pieces alternately, as the pendulum vibrates. After having perufed our account of the action of the preceding efcapement, the reader will anticipate the explanation we have to offer of this mechanifm. Suppofe the wheel to be urged by the train in its proper direction from I towards M, and an impulfe given to the pendulum, to produce a motion in it in a contrary direction; the pin N keeps pace with the pendulum towards L, while the tooth at L is locked in the claw of pallet A, (which it is fuppofed to have raifed fo as to bring the tail-piece G O out of its vertical direction,) and at length meets with this tail-piece, to which it gives a pufh, at the extremity of the vibration, fufficiently ftrong to drive the pallet A from its locked tooth L, when the wheel is at liberty to proceed, till the tooth near pallet B raifes it and its furmounted ball *v* until the claw detains the further motion; in the mean time the returning vibration commences, and the ball *u* urges the pendulum, by means of its tail-piece G O being in contact with the pin N; when, however, the tail-piece has gained its vertical fituation, the pallet A meeting with the rim of the wheel ftops; the pendulum now finifhes its vibration in a detached ftate, having gained an acceffion of force by the impulfe from the ball *u* downwards, which has been greater than the refiftance it met with in moving upwards at the inftant of unlocking: the tail-piece I O is now in the fituation of I P, by reafon of the ball *v* being elevated by the wheel, therefore when the pin N of the pendulum arrives at I P, which is at the other extremity nearly of the returning vibration, its ftroke drives the pallet B in its turn from the locked tooth near M, which now efcapes; the wheel is again at liberty to proceed till the tooth behind L has raifed the pallet A a fecond time, when it is again locked; the pendulum begins another vibration, aided by the ball *v*, till its tail-piece refumes its vertical pofition, and then the pendulum is again detached, after being a fecond time urged down a longer arc than it was oppofed up; and thus the vibrations are maintained by the excefs of the downward pufh over the refiftance upwards at every new vibration, which excefs, derived from gravity alone, is in fact the maintaining power of the pendulum. This efcapement is alfo of the third clafs, and being more fimple in its conftruction than its predeceffor and parent, perhaps we may add, is better adapted for practice. The external force derived from the gravity of the balls in aid of the pendulum's gravity is here alfo both applied and withdrawn fuddenly, as well as at the moft unfavourable parts of the vibration, as was the cafe with Cumming's efcapement; but it poffeffes the fame advantages of detachment from the train.

23. *Detached efcapement by Peter le Roy in* 1748.—We have deviated a little from the regular order of time, in having noticed the detached clock-efcapements before thofe that were previoufly applied to the balance, in order that the fucceffion of improvements made in clock-efcapements might not be interrupted by an intermixture with watch-efcapements. We now go back again to the year 1748, when Peter, the eldeft fon of Julien le Roy, invented an efcapement for a balance that vibrated a confiderable portion of its total arc without any connection whatever with the efcapement wheel, and which therefore may be confidered as the father of the third clafs of efcapements, and the origin of the fubfequent more recent improvements in the efcapement of chronometers. *Fig.* 8. *of Plate* XXXIII. will afford us the means of explaining this efcapement, a model of which was prefented to the Academy of Sciences at Paris in the year 1748, previoufly to which year, no detached efcapement had been made public, though Du Tertre is faid to have invented one, and even to have made it at the time; but as the invention, or at leaft the conftruction, was kept a fecret, it did not derogate from the merit of Peter le Roy's contrivance, which muft be confidered as original. G H is the efcapement wheel of the contrate kind, actuated by a pinion on its arbor, receiving the force of the maintaining power; its profile is feen contiguous to it, at *g h*; T V is the balance, the verge of which has a fpiral fpring attached to it at the interior end in the ufual way; a curved pallet A E is faft to the verge under the balance, and the femi-circular piece C I is affixed to the fame above the balance, but under the fpiral fpring; the end C of this piece is rounded, and fo fituated, that a line drawn from it to the centre of the balance would form with the curve A E an angle of about 80°. An angular lever Q P X, turning on pivots under a little cock on the plate at C, is fo proportioned, that the branch P Q will reach the pallet, while the other branch P X with a concave end will reach to a tooth G in the balance wheel, when not refting on a pin K, to which in its quiefcent ftate it is kept by the fpring R M, made faft to the plate on the other fide at R. Suppofe now, the regulating fpring moved from its ftate of natural reft, and the balance to move with it through a certain arc of vibration, more or lefs, till the rounded end C of the femi-circular piece falls on the branch P Q of the angular lever near Q, and pufhes it from its fituation, at the fame time bringing the concave end of branch P X into the way of a tooth G of the wheel; when the acting tooth, which we will call D, efcapes the pallet A E; the wheel is then at liberty till its tooth G falls on the concave end of the branch P X, which may be called the detent, becaufe it now *detains* the wheel from moving further under exifting circumftances. The balance is now perfectly free from all contact of the acting part of the branch P Q, and continues its vibration, after having received a pufh from tooth D on its pallet: prefently its velocity is deftroyed by the tenfion of the regulating fpring, the re-action of which now makes it return with an accelerated motion, when the pallet comes into contact with the efcapement wheel at the tooth following D, which now experiences a recoil for about half of a fpace in the wheel: the branch P X, or detent, is relieved by this partial recoil, and obeys its fpring R M, which takes it back again to its ftate of reft at the pin K; when the wheel, having ftopped the balance, fends it off again with a new impulfe, by ftriking the pallet in a contrary direction, when the fame operation of the angular lever is repeated, and the vibrations are maintained as in other efcapements. It muft be remarked of this efcapement, that it does not allow the balance to be detached for an entire vibration, but oppofes the balance back from the point where it gave its forward pufh, thereby producing a certain recoil, which, fo far as this is a leading confideration, would entitle us to rank the efcapement among thofe of the firft clafs, but inafmuch as the balance vibrates in a detached or free ftate in the remaining portion of its vibration, we may place this efcapement with propriety at the head, in point of time of the third clafs.

24. *Peter le Roy's fecond or improved detached efcapement.*—Some time after Peter le Roy had completed his firft detached efcapement, and had confidered its principal defects, he fet about contriving another, lefs liable to objections, in which he fucceeded in a very confiderable degree; when in

the year 1766 he presented a time-piece to the Academy of Sciences, which was afterwards tried at sea by the French government, with his improved escapement, which we shall now describe. *Figs.* 1, 2, and 3, of *Plate* XXXIV. exhibit sufficient of Peter le Roy's improved escapement to enable our readers to form a complete idea of its construction; the leading feature of which is to give the balance one temporary push that shall maintain its motion for two successive vibrations, one forwards and the other back, and that this push may be given at a moment of the balance's vibration that is most favourable for avoiding a derangement of the law of the spring's forces. *Fig* 1. is a view of the balance-wheel, which has six radial levers for teeth, but no rim, that it may be as light as possible, the ends of which levers are bent downwards, that is, in an opposite direction to those on the horizontal wheel with Graham's cylinder escapement; the connection of this horizontal balance-wheel with the train is by means of the pinion on the opposite end of its arbor; *fig.* 2. is the arbor of the detents, on which there are three levers, one double one for locking, and two single ones for unlocking, all clearly shewn in the figure; *fig.* 3. shews the wheel, detents, unlocking levers, and balance in their relative situations, and in one of the acting positions. A is a portion of the balance, on the plane of which is screwed a pallet *p*, so as to fall in the way of one of the levers of the balance-wheel. The same letters denote the same parts in *fig.* 3. as in *figs.* 1 and 2, which therefore explain themselves. The action may be explained thus: suppose one of the levers of the balance-wheel to be detained by the claw of the detent D, while the balance put in motion is vibrating in the direction from *i* to A; when the force of the balance spring, not seen, overcomes the momentum of the balance, it returns by means of the spring's force, which is now at its greatest tension, or distance from the point of quiescence; the direction of the balance, we have said, is now from A towards *i*; in this return a pin at *i*, situated in the superior plane of the balance, strikes the end of the lever H F of the detent arbor, unlocks by this stroke the detent D from the claw of the wheel's radial bar or tooth, and at the same instant directs the other end H *e* of the double detent into the way of a succeeding radial bar of the wheel K *r*, which now rests on this end of the detent; this part of the action is called by the inventor the *preparation*, the vibration goes on to finish, and the spring brings the balance back; at a certain part of this vibration the balance receives its stroke from the radial bar K, which impels the pallet *p*, screwed on the balance itself, in this manner; a pin, situated like the former pin at *i*, but placed on the inferior plane of the balance, out of sight, and a little farther from the balance's centre than the former pin, strikes the end of the short lever H *c*, and unlocks the radial bar of the wheel, that has been resting against the opposite end of the detent D, and that now runs on to D, which is in the way to catch it; during this short period the wheel is free, the pallet *p* of the balance being at that instant in the situation F, the point K of the radial bar K *r* of the wheel pursues the pallet and makes its stroke on it, about the time when the balance-spring, as we suppose, is at its quiescent point, and therefore when the balance is moving with its most accelerated velocity; the stroke is of short duration on that account, as well as on account of the short distance the wheel has to run before it is again detained by the detent D: the backward vibration is made as before, and the same process is repeated of two lockings and unlockings at each stroke of the wheel on the balance. The stroke, however, which the balance receives, has more force than both the strokes that its pins give in the acts of unlocking, and therefore the difference of these contrary strokes constitutes the real maintaining power of the balance, which was of the thermometrical kind, with alcohol and quicksilver, as described under our article Compensation-*balance*. The step made towards perfection, when this escapement was devised, was very long, and the principal objection to it in practice is, that it has too many pivots that require oil, as well other parts that come successively in contact during the action, which therefore are not sufficiently free from friction and consequent wear. It may be necessary to add, that to prevent the detent's being displaced by any jerk, a long tail-piece behind it is contiguous to a circular appendage of the balance, seen from *i* to F, against which it falls, but only when any derangement of the detent accidentally takes place, and thus reinstates it. The stroke is given to the balance not only when its velocity is greatest and least liable to be deranged, but the pallet that receives it, being removed from the centre of the balance, requires but a comparatively small impulse in this situation; besides, the spring, which the inventor considered as not desirable, for holding the angular lever to its place of locking, is here entirely dispensed with. This was considered as an improvement at the time, but subsequent experience has proved the utility of spring detents, which are now in universal use in the best English chronometers.

25. *Detached escapement by F. Berthoud.*—Ferdinand Berthoud, one of the members of the National Institute of France, and the author of various works on horology, has contrived different escapements for clocks and watches at different times, and has written more on the subject than any other author, but it will not come within our plan to detail here all the alterations that have presented themselves to him; we shall satisfy ourselves with describing a few of those which he has himself selected as most worthy of public notice in his last publication; "Histoire de la Mesure du Temps." Others may be seen in his "Traité des Horloges Marines;" or in the "Supplement" to that work. The first that we mean to describe is contained in *fig.* 4. of *Plate* XXXIV., which is a perspective view of all the parts necessary to be seen: A represents the verge of the balance, to which is affixed by two screws the circle of escapement B as a pallet; C is the escapement wheel, *a b c* the detent, which has at *b* a claw to catch a tooth of the wheel, when it suspends its motion, during the period of the balance's free vibration; the portion *a b* of the detent is a flexible spring, that yields first at the point *a*; this, therefore, is the *spring of the detent*; the other part *b c* properly forms the detent, which has a claw rounded at the end *c*; another slender spring *b e* lies cross-wise near the said end, and is held fast to the circle B by a screw and fixing-piece *f*; this second spring carries a pin *d* near its extreme end, which acting on the end of the detent, disengages the locked tooth of the wheel, when the balance vibrates in a direction from B towards A; at that instant the succeeding tooth of the wheel acts on the notch or pallet part of the circular piece B at *h*, and transmits an accession of force to the balance, which now vibrates in free space; at the balance's return, the pin *d* of the small spring *d e*, borne by the balance's circular piece B, applies itself to the end of the detent, so as to slide past it without disturbing its position with regard to the claw, against which the next tooth is now resting; when the balance has finished this second vibration it returns, the pin *d* again pushes aside the detent, and unlocks the wheel, which now gives another push at *h*, and the same operation is resumed that we have been describing. Thus two vibrations are performed at once unlocking, and at one stroke of the wheel on the notch of the circular pallet piece. The small spring *b e* may be

be called the lifting ſpring, in oppoſition to the detent ſpring, or ſpring detent; and to give a clear idea how it detaches the claw of the detent in paſſing forwards, but not in paſſing back again, it may be neceſſary to obſerve, that this lifting-ſpring has a pin in the circular piece B, againſt which it reſts, to prevent its laying too much hold on the back ſloping part of the detent's extremity; but it is at liberty to yield in the contrary direction towards the verge of the balance; and as the ſpring is very ſlender, and has its yielding point or centre of motion at the remote end *e*, it gives way to the detent in the direction that puſhes the detent againſt the tooth of the wheel, which becomes a prop to ſteady it; but when the impulſe is in a contrary direction, the detent, having no ſuch prop, and its end being at that ſide ſuited to receive the blow of the pin without its ſliding, itſelf yields and withdraws its claw from the wheel, which is then unlocked. The ſpring of the detent is alſo confined from rebounding too far by a pin *g* fixed in the plate. Here we have an inſtance of a retrograde ſtep in the improvement of eſcapements, the ſpring which Peter le Roy laboured to reject in his ſecond conſtruction is here reſumed with an additional ſpring, and the adoption of the *ſpring-detent* has become general in the ſubſequent conſtructions. Berthoud began his detached eſcapements in the year 1754.

26. *Another modification of a detached eſcapement by Berthoud.*—The eſcapement which Berthoud employed in ſome of his watches, particularly N° 60, was conſidered by him to be more ſure in its action than the preceding one, as well as more eaſy of conſtruction; *fig.* 5. of the ſame plate repreſents the plan of one of theſe eſcapements, where L is the eſcapement wheel; *m* the circle of eſcapement carried on the verge of the balance; *p o r* the detent, the claw of which at *r* ſuſpends the motion of the wheel; the verge carries above the circle *m* a ſmall tube made faſt by friction, and to this tube is made faſt a projecting forked claw, in which a pin is adjuſted to act on the detent and raiſe it. The pin, which we will call *s*, in turning in one direction, acts on the part *p* of the detent, and withdraws the claw *r* from the wheel; this is now at liberty to impel the circle *m* by its notch and with it the balance; the ſlender ſpring *u* brings back the claw of the detent, and the vibration goes on; at its return the pin meets with the end of the detent-ſpring fixed on an arm at *t*, this ſpring yields to the retrograde motion of the balance. The balance at its return makes its pin (of the fork) urge the ſpring *p* againſt a pin *s* in the arm *n* of the detent; this arm turning on its centre of motion, withdraws the claw of the detent from the wheel, which now acts again on the notch of the circle *m*, as before. The pin *q* ſerves as a ſtop to the detent, and the arm *k* of *precaution* is to lock the wheel when the balance is diſmounted.

27. *Another eſcapement without a ſpring by Berthoud.*—*Fig.* 6. of the laſt plate is a view of an eſcapement by Berthoud, in which he has diſpenſed with ſprings altogether. The action of the wheel A is tranſmitted to the balance, as before, by the notch of the circular piece B on the verge; the preſent poſition of the parts is for the moment when the detent allows the wheel to begin its action on the notch of B. The detent C has two arms, with each a claw *a* and *b*, the faces of which are circular; the claw *a* ſerves to ſuſpend the action of the wheel after its ſtroke is made; but the claw *b* receives the tooth of the wheel when the balance in its return makes the tooth at *a* to eſcape. Theſe effects are produced by two levers C *d*, C *e*, forming the fork that moves the detent C: this fork is fixed on the centre C of the detent, and its levers lie one above the other. The lever C *d*, which is contiguous to the detent, correſponds to the ſmall ſemi-circular piece *d e*, placed on the circle of eſcapement B; this ſemi-circle is notched by lines, one of which tends to the centre; this ſide of the notch unlocks the wheel by its action on the arm C *d* whenever the balance retrogrades from the firſt vibration of the pair; during this effect the wheel advances but a ſhort way, juſt ſufficient to unlock the claw *a*, and to ſuffer the balance to proceed in its retrogradation. The balance, having finiſhed in a detached ſtate this ſecond vibration, returns, when the ſemi-circle *f g*, placed over the other ſemi-circle, and notched in like manner, preſents its acting ſide to the upper arm *f* C of the fork, and unlocks the tooth at claw *b*; at this moment another tooth, catching the notch at *c* of the circular piece B, gives it another impulſe, and conſequently puts the balance in a ſtate of renewed force, and the proceſs already deſcribed is renewed, while the ſemi-circular pieces *d e* and *f g* on the verge alternately ſerve to keep the arms C *d* and C *e* of the fork in a firm poſition to enſure the deſired effects of the eſcapement. It is eaſy to perceive that there muſt be much friction in this eſcapement, and that the ſemicircular pieces on the verge of the balance, if not ſomehow counterpoiſed, muſt affect the rate of going in different poſitions of the time-piece.

28. *Detached eſcapement with a detent and ſpring by Berthoud.*—*Fig.* 7. of the ſame plate is the plan of another of Berthoud's eſcapements with a detent and ſpring, and that which we underſtand has been moſt copied by others. It was uſed in the author's marine time-piece N° 9. A is the circle of eſcapement; C the wheel; *a b d* the detent with three arms; the arm *a* ſuſpends the motion of the wheel while the balance oſcillates in a free ſtate; the ſpring *d* ſerves to bring the detent *a* back again as ſoon as the pallet *c* has ceaſed to act on the arm *b*; at that moment one of the teeth of the wheel impels the circle by acting on the roller *h*, and communicates its force to the balance; this having finiſhed its oſcillation returns, and in the return the pallet *c* meets with the end *b* of the detent, which turns the pallet back towards B, the centre of the circle, till it has paſſed without unlocking the wheel; its ſpring *b* brings the pallet back again to its original ſtate of reſt on the pin at *c*, the oſcillation ceaſes, and in the return of the balance the pallet *c* again meets with the arm *b* of the detent, which it turns aſide, the tooth being no prop to it in this direction; the unlocking therefore takes place; the wheel's motion is reſumed, and a new impulſe is given to the moving balance. Here the detents are arms or levers acted on by a ſeparate ſpring, as is alſo the pallet, conſequently there ſtill remain pivots that require oil, though we are diſpoſed to conſider this as the beſt of Berthoud's watch eſcapements; perhaps equal to any of the Engliſh conſtructions.

29. *Iſochronal eſcapement of a watch by F. Berthoud.*—Among the other contrivances of Berthoud is an eſcapement for a watch with iſochronal pallets, ſimilar to thoſe applied to a pendulum which we have deſcribed above, but the mode of applying them is different. In *fig.* 8. of our laſt plate, A is the eſcapement wheel; B the anchor with the iſochronal pallets producing a ſmall recoil, and formed in the manner before deſcribed, (ſee *fig.* 1. of *Plate* XXXIII.) C D the balance; *a* a pinion on the verge above the wheel, that is actuated by a ſectoral toothed piece *b*, attached to the anchor of the pallets, and moving with it round the common centre of motion of it and the pallets at E; the intention of this conſtruction is to produce a long oſcillation of the balance compared to the motion of the wheel and toothed ſectoral piece, and to render all the oſcillations, however long or ſhort, of equal duration, the arcs of action being large in proportion to the whole oſcillations. This conſtruction

may

may have given riſe to that which is now known by the name of the Liverpool eſcapement, which ſo greatly reſembles it, that they may be conſidered as the ſame thing. We obtained a watch by Litherland of Liverpool, with a view of giving a ſeparate deſcription and drawing, but we found that the only difference is, that the pallets are placed a little on one ſide of the wheel, inſtead of the three pivots being in one ſtraight line, and that the centre of the pallet's motion is between the verge and pallets requiring a counterpoiſe behind: a further deſcription is therefore unneceſſary.

30. *Modern eſcapements of Engliſh chronometers.*—If we had not already anticipated our deſcriptions of the modern eſcapements for chronometers, as they are now made in England, we ſhould have given them a place here; but as we have been very minute in our account of the eſcapements of Meſſrs. Mudge, Arnold, the Brockbanks, and Earnſhaw, under our article Chronometer, and have given drawings of them in our *Plates* XIII. XIV. and XV., it would be ſuperfluous to enter into a ſecond detail of the ſame here, and therefore we paſs over them by a ſimple reference to the article already named.

It may not be improper to mention here, that Mr. John Watkin, of Weſt Smithfield, has propoſed an alteration in the ſituation of the pallets in the modern ſpring detent eſcapement, for which he received thirty guineas as a premium from the Adelphi Society, in the year 1804. His alteration conſiſts in bringing the pallets into a ſituation between the centre and circumference of the wheel, and making the wheel a contrate one with the teeth upwards. (See the Tranſactions of the Society, vol. xxiii. p. 375.)

31. *Mudge's forked eſcapement.*—Mr. Mudge contrived another detached eſcapement beſides the one we have referred to as already deſcribed, which we ſhall now notice, though he did not ſeem to place ſo much value on it as Emery, Margetts, and others have done after him; we ſhall not be able to make the reader comprehend the conſtruction without two figures: *fig.* 1. of *Plate* XXXV. is the plan of this eſcapement, and *fig.* 2. a profile or ſide view of the frame that contains the mechaniſm: A is the eſcapement wheel, B the anchor of the pallets, which are of Graham's dead-beat kind adapted to a watch, and which give the neceſſary force to the balance for maintaining its motion; *a* is the centre of motion of the anchor, nearer to the wheel than to the balance; *b* is the forked part of the anchor, the two prongs of which appear in *fig.* 1. to be cloſe together in the ſame plane, but *fig.* 2. ſhews that they are one above the other. The pallets are on the verge of the balance, and are of the cylindrical ſort, one above the other, ſo as to fall in the way of the two prongs alternately, pallet 1 is urged by *b*, and pallet 2 by the other prong in its turn; the time of action of the prongs on the two pallets is very ſhort, and the blows are made alternately one in each vibration; the remainder of the vibration is perfectly free, the wheel being in the intervals reſting on the circular portions or pads of the anchor's pallets, which may in this caſe be conſidered as detents and pallets alternately. A timepiece by Emery with this eſcapement performed with wonderful accuracy, though the detachment continues only for a portion of one vibration before the balance has its force renewed, which conſequently might be ſuppoſed to have its motion more diſturbed than if the impulſe had been given at each alternate vibration. The queen's watch by Mudge with this eſcapement had a ſlight recoil, which kept one of the prongs preſſing gently againſt the cylinder of the verge to ſteady the fork. The performance was excellent.

32. *Margetts' modification of Mudge's forked eſcapement.*—The late ingenious Margetts adopted Mr. Mudge's eſcapement, which we have juſt deſcribed, for his chronometers that indicated both ſolar and ſiderial time, but made ſome alterations with the fork. *Fig.* 3. of the ſame plate ſhews the plan and profile of Margetts' modification, where A is the wheel; *a b* the anchor moveable at *a*; *b* the fork where the prongs are in the ſame plane; *c* a piece of ſteel carrying a pin *e* under its extreme end that ſtrikes the prongs of the fork *b* alternately, one at each oſcillation of the balance *g*, which ſtrokes unlock the wheel; *h* and *i* are two banking ſcrews, the points of which act as ſtops to the anchor at oppoſite ſides of the centre of motion, and keep the fork in its place to be caught by the pin *e*, one or other of them at each vibration; *d* is a circular piece of metal, under the pallet piece *c*, that has a notch in it bounded by ſtraight lines going towards its centre; and *f* is a pin carried under the anchor, as ſeen in the figure, the end of which gives the circular piece *d* a puſh, by acting on its notched ſide at each unlocking of the wheel, thereby aiding the vibration: the pin *f* is always at the notch when the pin *e* is in the fork, therefore the notch in *d* and the pin in *c* are always placed in the ſame vertical line over one another, and as they move together on the verge will always remain ſo, while the pin *f* and fork *b* are alſo together on the anchor. The writer of the preſent article has a chronometer by Margetts with this eſcapement, which acts very well when ſtationary, but when carried in the pocket the pin is apt to catch the outſide of the prong, when any jerk has ſent the anchor back during the balance's excurſion.

33. *Modification of Mudge's forked eſcapement by R. Robin of Paris.*—R. Robin, a watch-maker of Paris, publiſhed a memoir in 1794, containing an account of a free eſcapement, which he conſidered as new, but which we conſider as another modification of Mr. Mudge's forked eſcapement, though we allow that in theory it appears to be an improvement. *Fig.* 4. of the ſame plate is a plan of this eſcapement, in which the wheel reſtores the exhauſted force of the balance once only in two vibrations, which circumſtance conſtitutes the improvement; to effect this, pallets G *a*, and G *b*, acting as detents, are made of unequal length from G, the centre of their motion, and the fork, which is ſeen ſeparately at *e*, as well as the tooth or pallet *d* and lever *f*, are concealed under B, the circular piece carried by the balance; the fork *e* is faſt to the end of one of the pallets *a*, inſtead of forming an anchor with the pallets, but this ſhape makes no difference in the action; the lever *f* is alſo faſt to the ſame underneath the fork; but the tooth *d*, which acts with the fork, is faſt to the verge under B, as we have already ſaid. We have given three poſitions of the circular part B and pallets, to ſhew more clearly the nature of the action, as ſome of the acting parts are concealed. One of the claws *a* of the pallets being removed, the wheel A eſcapes, and acts on the notch of the piece B, thereby aiding the balance; the claw *b* receives another tooth and holds it as a detent, and the balance finiſhes its oſcillation; at its return it unlocks the tooth at *b*, which is the longer pallet, and the wheel runs on a very ſhort way till a leading tooth falls on the pallet *a*, and the balance vibrates without a new puſh; at its return the tooth at *a* is unlocked, and its eſcaping tooth gives the notch of B another puſh, which ſerves for other two vibrations; but if the two ends of the pallets G *b*, and G *a*, had been of equal lengths, the eſcapement would have acted in the uſual way, and would have given its puſh at each oſcillation. The manner in which this effect is produced, however, does not yet appear evident; the lever *f*, which is faſt to the end *a* of the pallet or detent G *a*, reſts in a groove made in a circular collet on the verge, and holds the fork *e* from going on one ſide, like as the banking ſcrews

fcrews do in Margetts' efcapement, then the tooth *d*, faft to the balance, ftrikes the two prongs of the forks alternately, one at each vibration, which ftrokes make two unlockings during the time that one pufh is made to the balance by a tooth; becaufe the faid tooth previoufly made two fteps forward, one long and one fhort, before it efcaped to make its pufh. We have not been informed how this modification performed, but we fee no reafon to doubt its competency, while we fee much reafon to admire the ingenuity of the application of two detents of unequal length, formed like the pallets of the anchor efcapement, which detents we have named pallets on that account.

34. *Prior's modification of Mudge's forked efcapement.*—In the year 1798 Mr. John Prior received 30 guineas as a premium from the Society for the Encouragement of Arts at the Adelphi, for his invention of what he confidered a new detached efcapement; but its fork differs from Mudge's principally in its having a roller for a pallet borne by an arbor carried by the balance itfelf; the wheel of efcapement, together with its acting pallets, were fimilar to Le Paute's improvement on Amant's wheel, and therefore had nothing original in it. The reader who wifhes to fee the account and drawings of this efcapement, compofed as it is of Mudge's and Le Paute's contrivances, which notwithftanding the author might not have feen or known, may find it in the 16th vol. of the Tranfactions of the Society juft named, and alfo in the 2d vol. of Nicholfon's Journal, 4to. ed. p. 363 and 364, where the editor has remarked that Mr. Crofthwaite of Dublin had made a pendulum clock with fimilar pallets in 1788, which was defcribed in the Memoirs of the Irifh Acad. vol. ii.

Prior junior has very recently been rewarded by the fame fociety for another efcapement contrived by him for a clock, but the account of it is not yet publifhed in their Tranfactions, and therefore cannot with propriety be given here.

35. *Duplex efcapement.*—We are not informed who was the original contriver of the efcapement known by the name of *Duplex* efcapement, a name probably taken from the circumftance of its having a wheel with *two* fets of teeth neceffary for the efcaping; neither are we aware of its date; but the efteem in which it is held by thofe who have tried it, entitle it to a place in our collection. *Fig.* 5. of *Plate* XXXV. exhibits on a large fcale enough of the acting parts to enable us to convey an idea of its action. A B is a part of the efcapement wheel, having teeth *a*, *b*, *c*, &c. at the periphery, intended for detaining the wheel while it ofcillates beyond the angle of action. An agate cylinder *d e f*, placed on the verge of the balance, acts with thefe teeth to produce the defired effect. This cylindrical agate has a notch *d*, which paffes by the tooth at D, that is fuppofed to reft on the cylindrical furface, when the motion is in the direction *d e f*; but when the motion of the cylinder is in the contrary direction *f e d*, the tooth at D falls into the notch and accompanies it, preffing on one of its fides till the notch comes into the fituation *d*: the tooth, then at the fituation of *b*, efcapes from the notch, and a fucceeding tooth falls on the convex furface of the cylinder at D. The other fet of teeth are erect on the plane of the wheel's rim, and are ftrong and fquared at the fides; they ftand a little fhort of the middle of the fpaces between the pointed teeth on the edge of the wheel, but confift of the fame number; they are to be feen at *h*, *i*, &c. denoted by darkened parallelograms: above the fmall cylinder we have named as being made of agate, is a larger one E F G without a notch, and placed high enough on the verge to go over the edge of the wheel's rim; this bears a pallet of ruby or fapphire G, long enough to fall in the way of the fquare vertical teeth, when they happen to be near the line joining the centres of motion of the balance and wheel, but fhort enough to avoid touching the teeth when not near that line. The larger cylinder is fo placed with refpect to the fmall one, that when the tooth *b* has juft efcaped from the notch, the pallet G has juft paffed the fquare tooth *h*, which was at A, while D refted on the fmall cylinder, but is moved from A to *h*, while D moves to *b*. The wheel is now free, and the tooth *h* exerts its force, derived from the main-fpring, in the moft direct way on the pallet G, and urges it forward till another tooth on the edge falls on the convex part of the fmall cylinder *d e f*, and the fame action that has been defcribed is refumed. The angle of action in this efcapement depends partly on the length of the pallet G, and partly on the diameter of the fmall agate cylinder, and advance of tooth D into the notch; for when a horizontal tooth on the edge falls on the faid fmall cylinder, the motion of the wheel is arrefted, and the vertical fquare tooth ceafes to impel the pallet G fo far as it otherwife would have done. This efcapement requires very little oil on the fmall agate cylinder, and none at all on the pallet G, which receives a direct pufh without friction, and its excellence is confidered as at leaft equal to that of Graham's horizontal or cylindrical efcapement.

36. *Free efcapement under the pendulum.*—The clock efcapements which we have before defcribed fuppofe a connection with the upper end of the pendulum, which is the moft common connection, but there are efcapements that give their impulfe at the inferior end, one of which is reprefented by *fig.* 8. of *Plate* XXXV. which we fhall now defcribe, though we know not who was its original contriver. The part A B D C is made faft to the rod of the pendulum, and vibrates therewith; at B is a joint on which the lever B C is moveable; fo that when the end C meets with any refiftance the lever mounts upwards at this end, but cannot fall below a horizontal line; it terminates in a claw with both an interior and exterior flope or inclined plane; D is a pallet of agate or other polifhed ftone, with a flat face next the end C of the lever, and made faft into the vertical part A D; F G H I is a two-armed detent, moveable round H as a centre, but kept in its prefent pofition againft the ftop or pin K, by means of the weight G, when not otherwife acted on. In the prefent pofition the arm H I, which forms the detent, detains the tooth I of a vertical wheel of efcapement, while the pendulum is fuppofed to be vibrating clear of all the teeth. Let us fuppofe the end C of the horizontal lever, moving with the pendulum, to be approaching the end F of the fecond arm of the detent; prefently it gives the end F a blow, the lever C flies up, flides over the inclined face of F, and returns into its horizontal pofition; when the vibration terminates, the pendulum returns by virtue of its gravity, and the interior face of claw C catches the exterior one of the arm F, and gives it a pull fufficient to draw the end of H I from the tooth I; the wheel is then at liberty to run on till the tooth E, moving fafter than the pendulum, overtakes the pallet D, and gives it a direct pufh forwards; this pufh aids the claw C to get over the claw of F, and the weight G replaces the arm H I to catch the following tooth M, which prevents the further impulfe of the wheel on the pallet D, and fuffers the pendulum to finifh its vibration in free fpace; at the return of the pendulum the fame operation is repeated, and the pendulum continues to receive one pufh at every two vibrations, and that when the pendulum is in its vertical pofition, or moving with its greateft velocity, which we have faid is the moft favourable point of the arc. The arc of action, before the pallet is quitted, depends greatly on the

the radius of the wheel compared to the length of the pendulum, when the number of teeth is limited. The pendulum, however, ought to be of the compenfating kind, otherwife an increafe in the length of the rod will derange the action of the claw C, with the claw F, as well as of the tooth on the pallet D, in all the extremes of atmofpheric temperature. The maintaining power of the train that actuates the wheel E M may be either a fpring or weight, as the mechanift may think proper. Berthoud has propofed this conftruction of a free efcapement for a half-fecond's pendulum in his "Hiftoire de la Mefure du Temps," tome i. p. 214; but he does not clearly fay that it was of his own contrivance; on examining his drawings, fince our obfervations were written on this efcapement, we find that this author propofed a compenfation pendulum to be ufed with it; and he propofes fuch a combination of parts as a proper model for an aftronomical clock to vibrate half-feconds. Mr. Nicholfon, who has alfo defcribed the fame efcapement as one he has noticed in England, does not venture to afcribe its invention to any particular artift.

37. *Detached efcapement of a clock by Berthoud.*—Among the numerous contrivances of F. Berthoud in clock-work, we frequently find this author applying the fame contrivance to anfwer different purpofes; for inftance, when he had hit upon an ifochronal efcapement for a clock, he adapted it foon after, as we have feen, to a watch; and in the inftance before us, he has converted a free or detached efcapement of a watch into a clock efcapement, which we fhall here defcribe, not knowing the exact date of the application. *Fig.* 1. of *Plate* XXXVI. is a reprefentation of an efcapement of the third clafs adapted to a half fecond's pendulum, as defcribed by Berthoud himfelf. F is the efcapement wheel, and *a*, *b*, the detents of efcapement, (formed like the pallets of the dead-beat efcapement of Graham,) the action of the wheel F is tranfmitted to the pendulum by means of the notch made in the circular portion of the efcapement piece A B, thus; the detent has two claws, *a* and *b*, formed by portions of a circle, at the holding parts. The claw *a* ferves to fufpend the action of the wheel, when it has communicated its motion to the notch of the efcapement piece, which this does again to the pendulum, by means of the lever C, which it carries, and of the roller D, correfponding to the pendulum. The claw *b* ferves to receive, and to detain the wheel, when the pendulum returns. Thefe effects are produced by two levers *d c*, *e c*, fixed on the centre *c* of the detent; but one of them more diftant from the plane than the other. Thefe two levers form a fork that correfponds to the circle of efcapement *b g*. The lever *d e*, which is neareft to the detent, anfwers to one portion of the circle *g*, carried by the circle of the efcapement D. This part *g* prefents a ftraight line to the fork *d*, which being acted on, detaches the claw *a* of the detent, and, confequently, fets the wheel at liberty. The notch *f* prefents itfelf, and the tooth of the wheel acts on the notch, and reftores to the pendulum the force that it had loft. The wheel having finifhed its action on the notch *f*, a tooth prefents itfelf to the claw *b*, and the pendulum finifhes its vibration in a free manner, at its return, the portion of the circle *g*, which is moft elevated, meets with the arm *e* of the fork, and difengages the claw *d*; the tooth that was here detained now efcapes, and runs on a fhort fpace before another tooth drops on the claw *a*, and the procefs is repeated. We have not feen this efcapement in England.

38. *Efcapement by Mr. Nicholfon.*—Some time in the year 1784, Mr. Nicholfon fays (Phil. Jour. vol. ii. 4to. p. 59.) that he contrived a clock efcapement to go without oil, and which is of the dead-beat kind; it was conftructed in the year 1798, and when the account was written had been going freely for a year or more. His account is this, *fig.* 2. of *Plate* XXXVI. "is a fketch, in which G H reprefents a fteel wheel; D and E are pallets of agate, with flat polifhed faces. The pallet D is fixed to the lever D C, which is confined to its prefent fituation by the loaded branch or arm C B refting on the ftop *k*. The lever E C is alfo kept in its fituation by the loaded arm C A refting upon I. A pin N proceeds from the pendulum rod to its fituation between the levers. The centre of motion of the pendulum is in the continuation of the axis of the pallets. Suppofe the vibrations to be begun; the pin N in its progrefs will lift the pallet D towards L, while part of the weight B will be fupported by the tooth of the wheel which will follow, and at laft efcape and fuffer the oppofite tooth F to fall on the pallet E. But in its return, or defcent, the pendulum will be acted on by the whole weight B, and, confequently, its vibration will be kept up. When the pin N raifes the pallet E, a fimilar effect will take place on the fide of the perpendicular towards M, and thus the procefs will go on as long as the wheel G H retains any force." In this efcapement, the two weights are not raifed entirely by the action of the wheel, but are chiefly lifted by the pendulum itfelf; therefore though the vibration is made during the repofe of the wheel, in a ftate detached from the train, yet the pendulum is not detached from the weights, and, confequently, the compound law of forces by which the pendulum is maintained confifts of the action of the pendulum's gravity, of the action of the ball's gravity, which acts in concert with the other, and of a fhort action of the wheel on the agate pallet; this compound action differs from Mudge's and Cumming's in this refpect, that their weights are raifed by the wheel entirely without oppofing the pendulum, otherwife than by the fhort ftroke made at unlocking.

39. *Crank efcapement by Simon Goodrich.*—In the year 1799 the Adelphi Society rewarded Mr. Goodrich for his contrivance of a crank efcapement, which is exhibited in *figs.* 3. and 4. of *Plate* XXXVI. of which *fig.* 3. is a front view, and *fig.* 4. a fide view of the fame, where the fame letters of reference apply to both. A A A A is the back plate of a clock frame; B the crank fixed to the end of C the arbor of the laft pinion of the train; D, D, are two fmall wire chains, compofed of two links only attached to the crank B, by means of two collets, with great liberty of play. Thefe chains are faftened to E, E, two fmall fprings fcrewed to F F F, a bar that goes acrofs, and is faftened in the middle to G G, the crutch of the pendulum. I, I, are two fcrews going through the bar F; the ends of which being made to act againft the fprings E, E, ferve to adjuft them to a proper diftance and degree of ftrength, and by that means eafily to put the pendulum into proper beat. In *fig.* 3. the pendulum is partly reprefented by dotted lines, in order to fhew the crank, &c. The advantages attributed to this conftruction are, fimplicity of conftruction and confequent cheapnefs; total filence during the efcape; uniformity of action on the pendulum; durability and certainty of continuance; and the power of acting without oil. Thefe properties may recommend the efcapement before us, but as its influence on the pendulum is conftant, it is but ill calculated to counteract any irregularities in the maintaining power as tranfmitted and varied by the train under different circumftances of wear, dirt, and friction. The originality of the contrivance is that of a rotatory motion converted into a vibratory one, which Sounille had done very nearly in the fame way in the year 1746. (See Machines et Inventions approuvées par l'Academie Royal des Sciences; tome xiii. p. 325.) As there is no efcapement wheel, a

3 T 2 further

further explanation is unneceſſary to every one who has ſeen a treadle in action; but it may be neceſſary to remark that an additional wheel and pinion muſt be requiſite in the train where the laſt pinion revolves more frequently than in other eſcapements.

40. *Eſcapement without a verge on the pallets by Ed. Maſſey.*—Mr. Maſſey's eſcapement without a verge was one of two eſcapements which together gained him twenty guineas from the Adelphi Society in 1803; the other, being of the 4th claſs, will be deſcribed in another place. *Fig.* 5. of *Plate* XXXVI. is a plan of this eſcapement, which is of a novel conſtruction, borrowed probably from the ſpring detent eſcapement of the modern chronometers, and adapted to a clock, where the pendulum is free, during a large portion of its vibration. The pendulum rod is ſeen at *a*, ſuſpended by a piece of watch ſpring in the uſual way, but without a crutch and verge; *b* is the ſwing wheel with its teeth ſhaped like the dead-beat kind, except that they are rounded at the end; this wheel is not within the frame as uſual, but on the outſide of the back-plate; *c* and *d* are the arms of the pallets *f* and *g* reſpectively; theſe arms are attached to the frame, and are adjuſtable for the included angle by the ſcrews *h* and *i*, which preſs againſt the upper ends of the arms, that open like a pair of tongs, and are kept open by a ſlender ſpring *e* forced into notches at the ends. A croſs bar *k*, carried by the pendulum, gives alternate ſtrokes to the two pins fixed in the pallets *f* and *g*, at the time of unlocking. In the figure the pendulum is ſuppoſed to be in motion with the pin of pallet *g* reſting againſt the croſs bar *k*, which bar it impels on the returning vibration, and aids the pendulum, while the inclined plane of pallet *f* is raiſed by the wheel to give its ſtroke, when unlocked, in a contrary direction. Thus the pallets are raiſed partly by the wheel and partly by the pendulum, as is the caſe with Mr. Nicholſon's pallets; for the ſpring in this caſe performs the ſame office that the weight does in that; but in this eſcapement the inclined planes muſt require oil, which is not neceſſary in the other, where the wheel acts by a direct puſh without any ſliding motion. The inventor tells us, that a more than uſually large maintaining power is neceſſary when this eſcapement is adopted, which indeed muſt be the caſe with all detached eſcapements, where the duration of the impulſe is ſhort.

The reader may have obſerved that we have not yet noticed in this article the fourth claſs of eſcapements which act by the aid of *remontoirs*; this omiſſion has been intentional, not only for the purpoſe of avoiding the further enlargement of the article, but that we may give the hiſtory of the invention of the *remontoir*, and of its modifications by the different ingenious men, who have preferred its uſe to that of a detached eſcapement without ſuch additional aid. We have already had occaſion to ſpeak of Harriſon's and Mudge's auxiliary ſprings, or remontoirs, under our article Chronometer, and it will ſuffice, for our preſent purpoſe, to name here that Huygens, Leibnitz, Gaudron, Haley, Breguet, De Lafons, Maſſey, Mendam, Antis, and others have taken a part in bringing the remontoir into notice, as we ſhall have occaſion to relate more particularly under the article Remontoir.

ESCAPEMENTS.

Fig. 1.

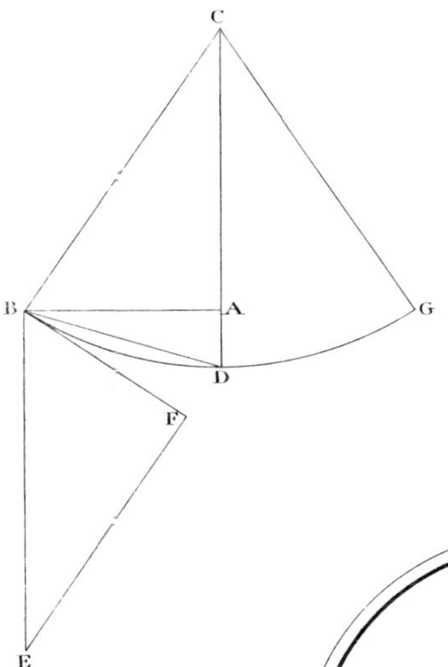

Fig. 2.

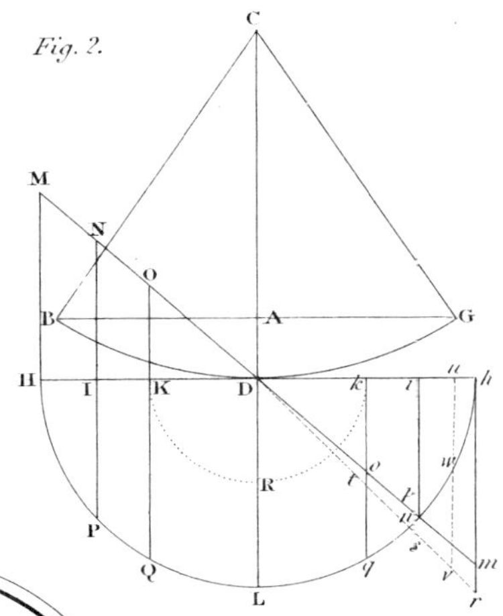

Fig. 3.

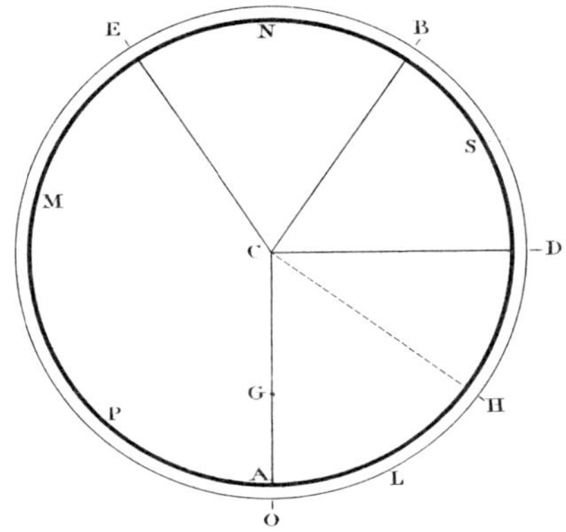

Fig. 4.

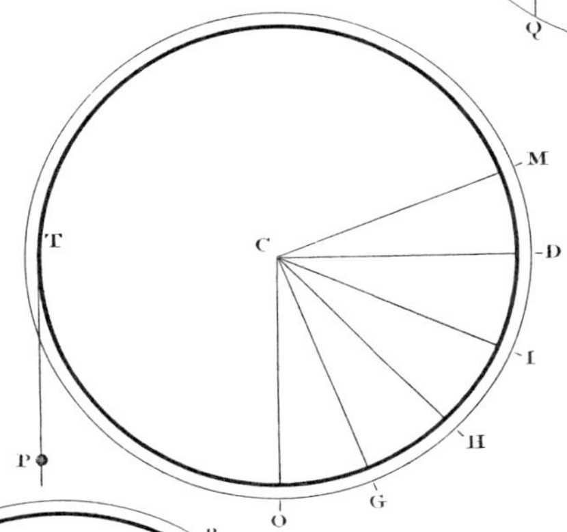

Mr. Atwood's Theory of the Balance.

Fig. 5.

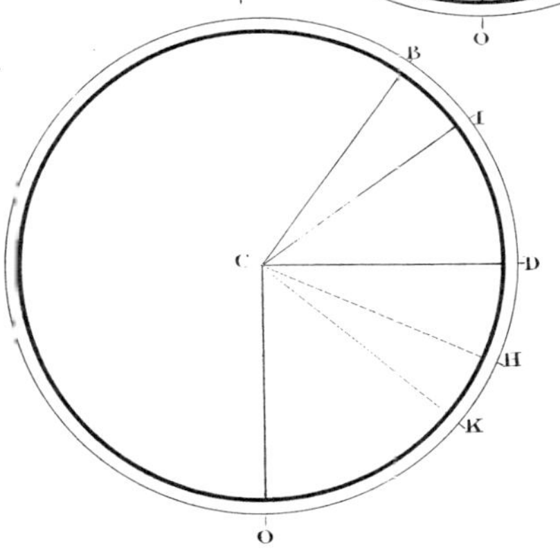

Crown Wheel.

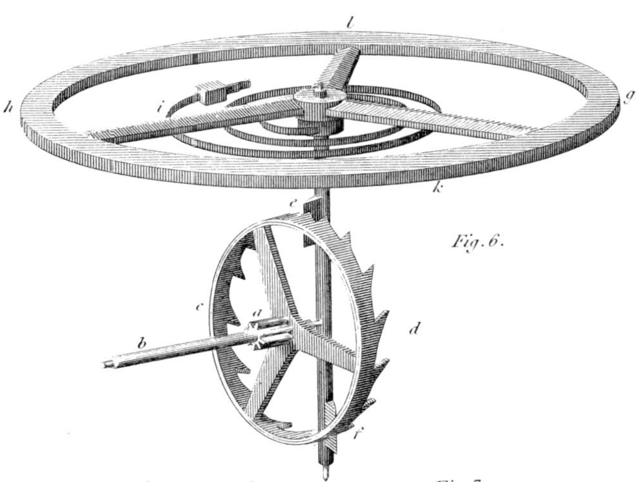

Fig. 6.

Dr. Hook's. & Du Tertre's, Fig. 8.

Huygens's. *Fig. 7.*

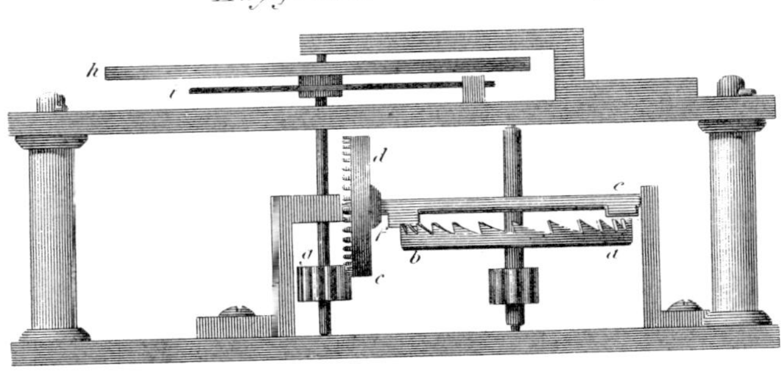

HOROLOGY.

ESCAPEMENTS.

Fig. 1.
Sully's.

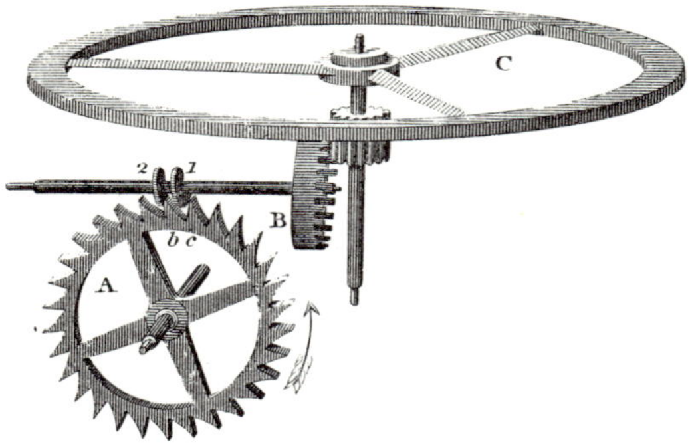

Fig. 2.
Graham's Horizontal.

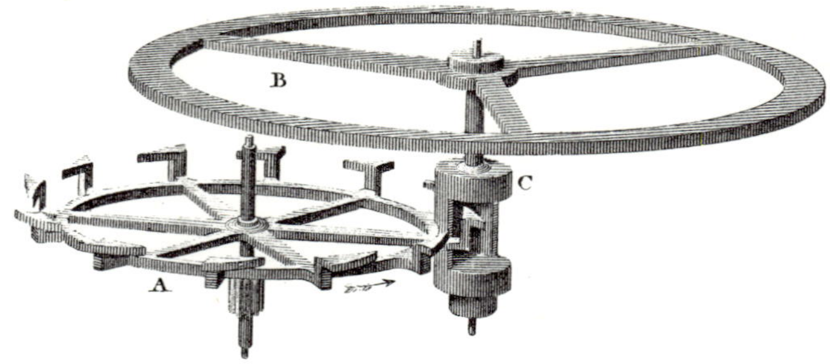

Fig. 3.
Anchor Escapement.

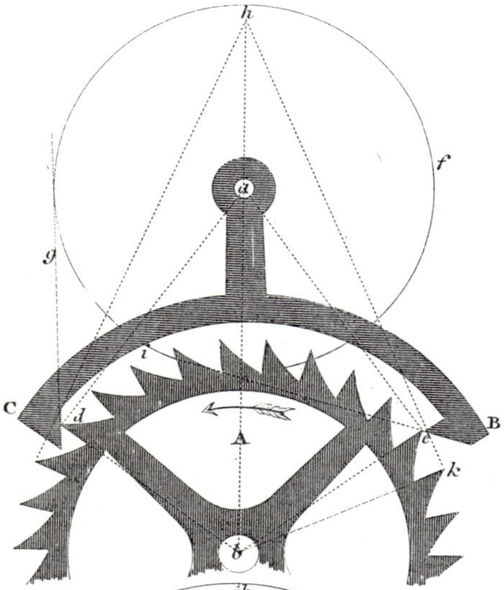

Fig. 5.

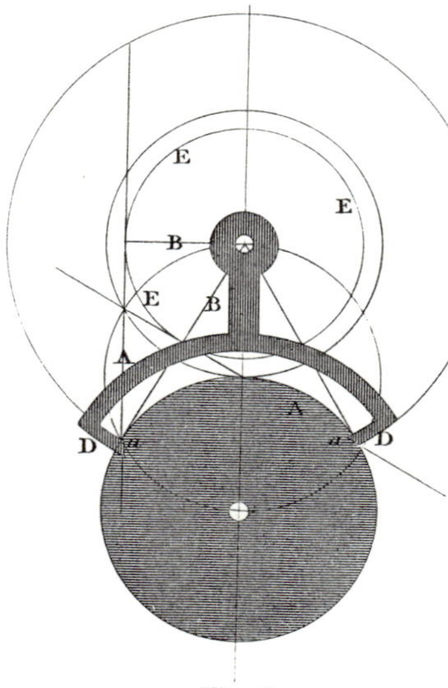

Fig. 7.
Bennet's Dead Beat.

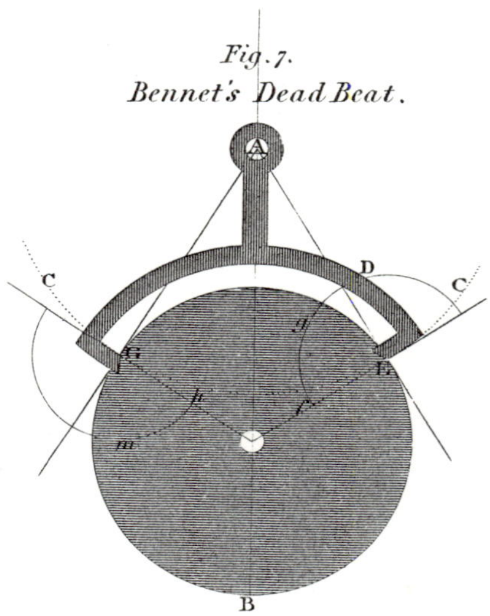

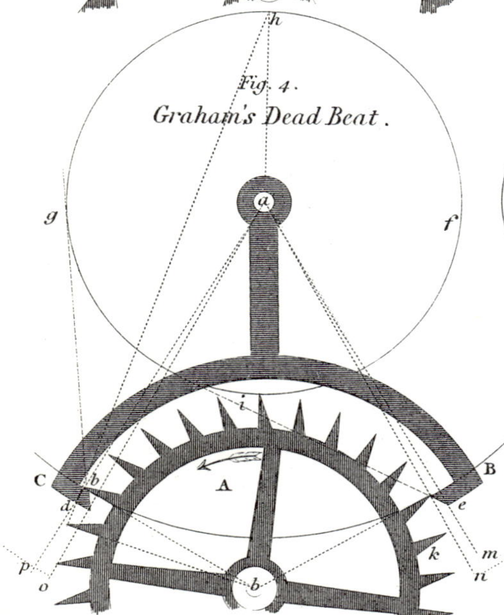

Fig. 6.
Grignon's Dead Beat.

Fig. 8.
Thiout's.

J. Farey delin.   Published as the Act directs, 1813, by Longman, Hurst, Rees, Orme & Brown, Paternoster Row.   Engraved by W. Lowry

HOROLOGY.

ESCAPEMENTS.

Berthoud's.

Fig. 1.

Smeaton's.

Fig. 2.

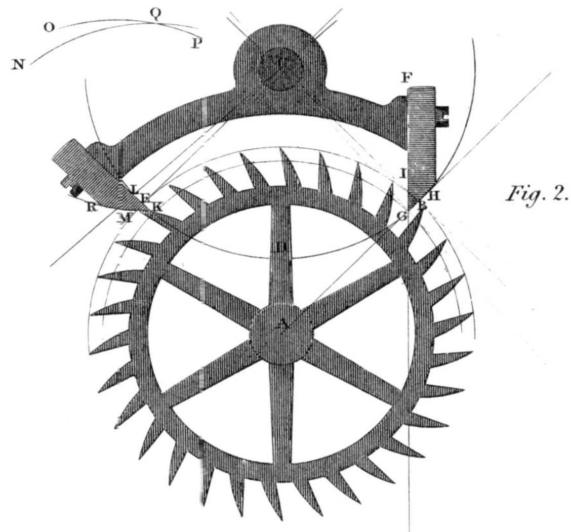

De Bethune's.

Fig. 3.

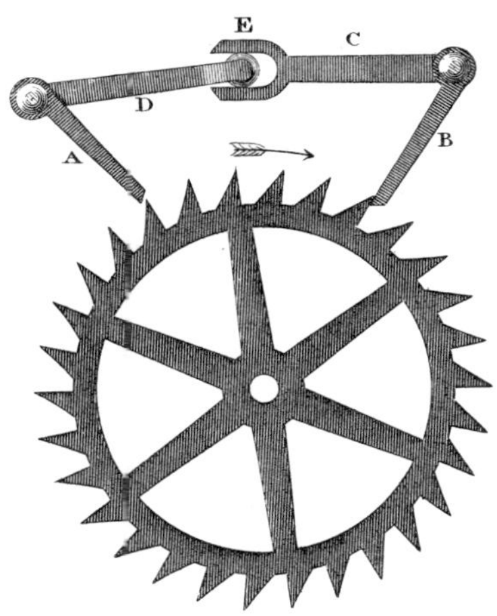

Amant's.

Fig. 4.

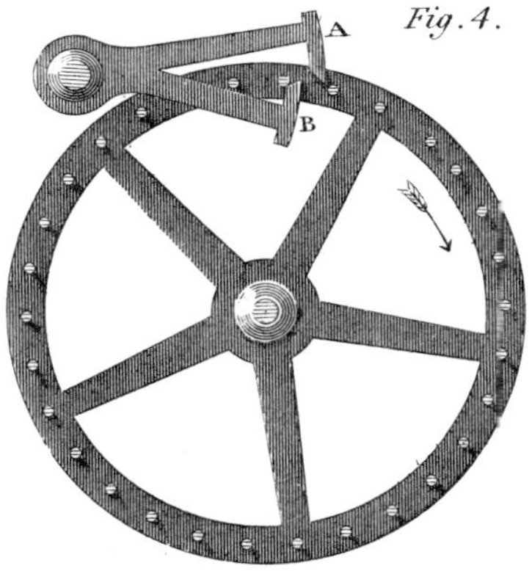

Harrison's.

Fig. 5.

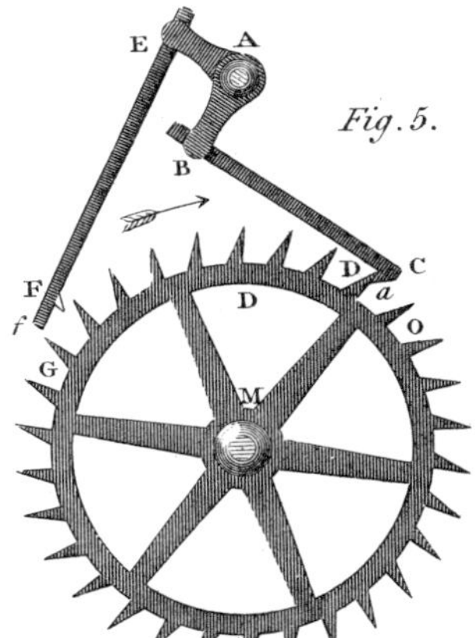

Fig. 6. Cumming's.

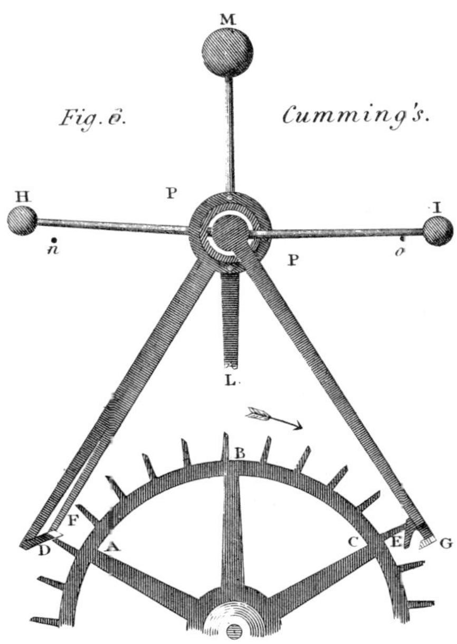

Mudge's.

Fig. 7.

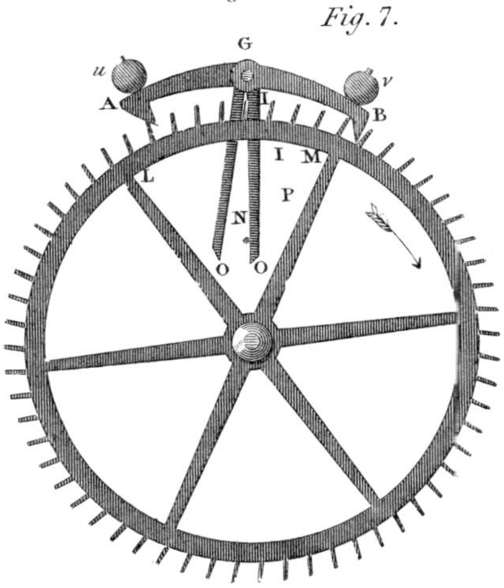

Peter Le Roy's.

Fig. 8.

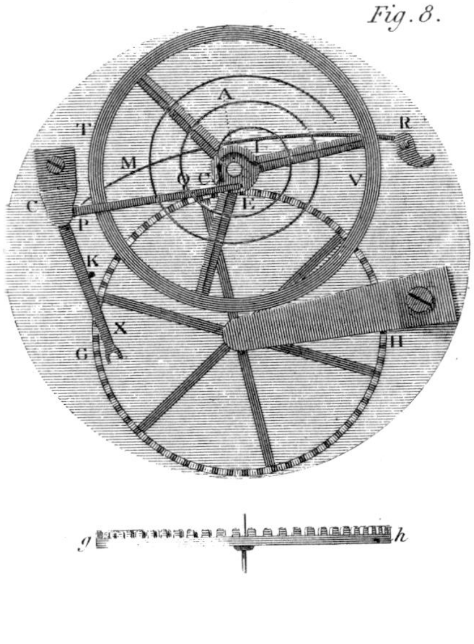

Published as the Act directs, 1813, by Longman, Hurst, Rees, Orme & Brown, Paternoster Row, London.

Lowry sculp.

HOROLOGY.

FRENCH ESCAPEMENTS. for Chronometers

Fig. 1.

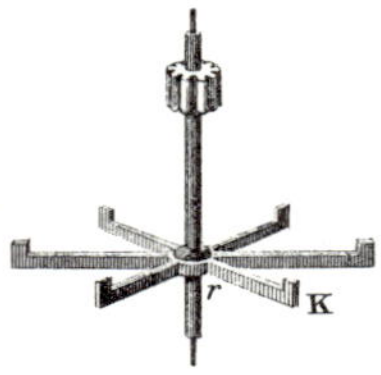

Peter le Roy's Improved

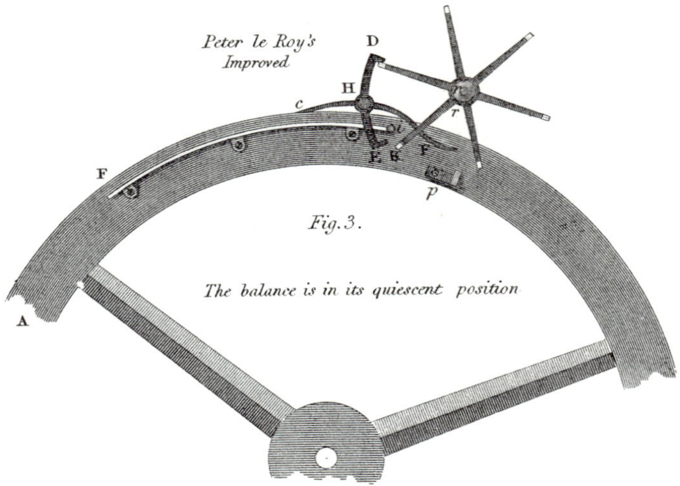

Fig. 3.

Fig. 2.

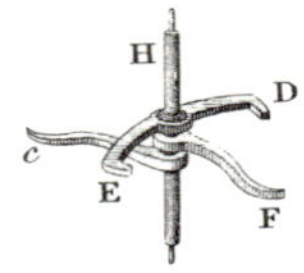

Berthoud's Detached

Fig. 4.

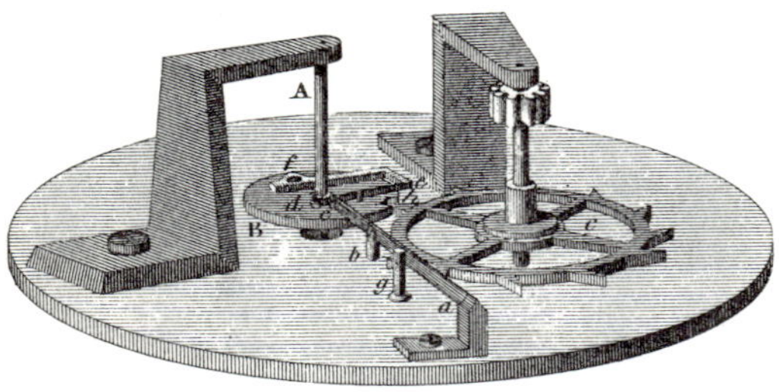

Fig. 6. Berthoud's without a Spring

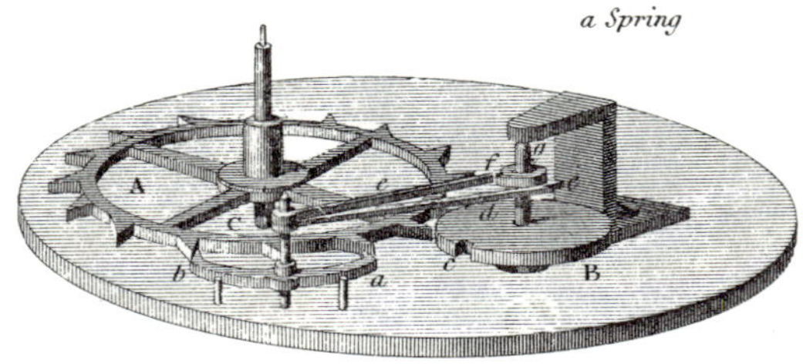

Fig. 5.

Berthoud's Improved Nº. 60.

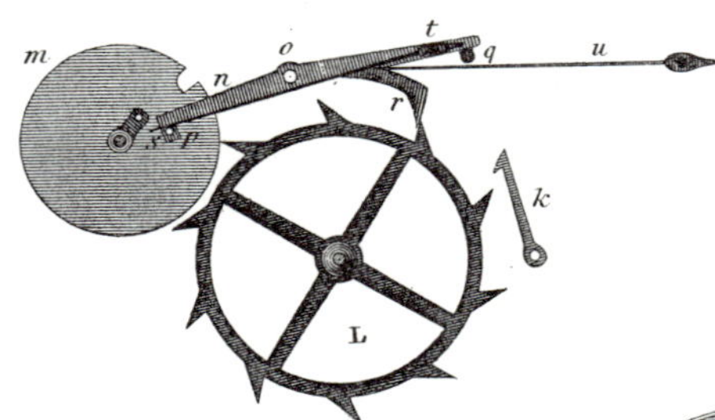

Fig. 7.

Berthoud's with a Spring and Detent in Nº. 9.

Fig. 8. Berthoud's Isochronal

J. Farey del. Published as the Act directs, 1816, by Longman, Hurst, Rees, Orme, and Brown, Paternoster Row, London. Lowry sc

ESCAPEMENTS.

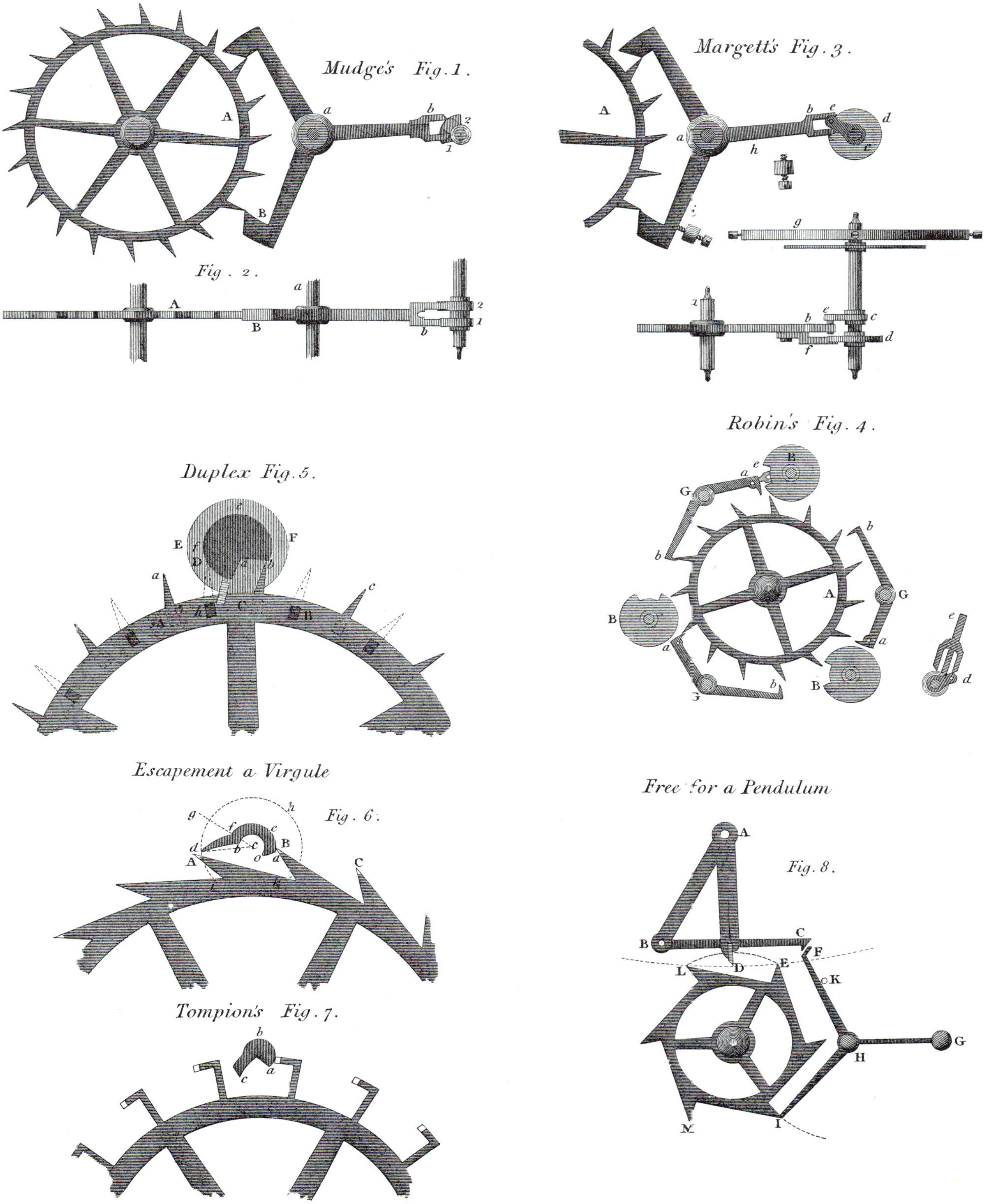

J. Farey delin. Published as the Act directs, 1813, by Longman, Hurst, Rees, Orme & Brown, Paternoster Row. Engraved by W. Lowry.

ESCAPEMENTS &c.

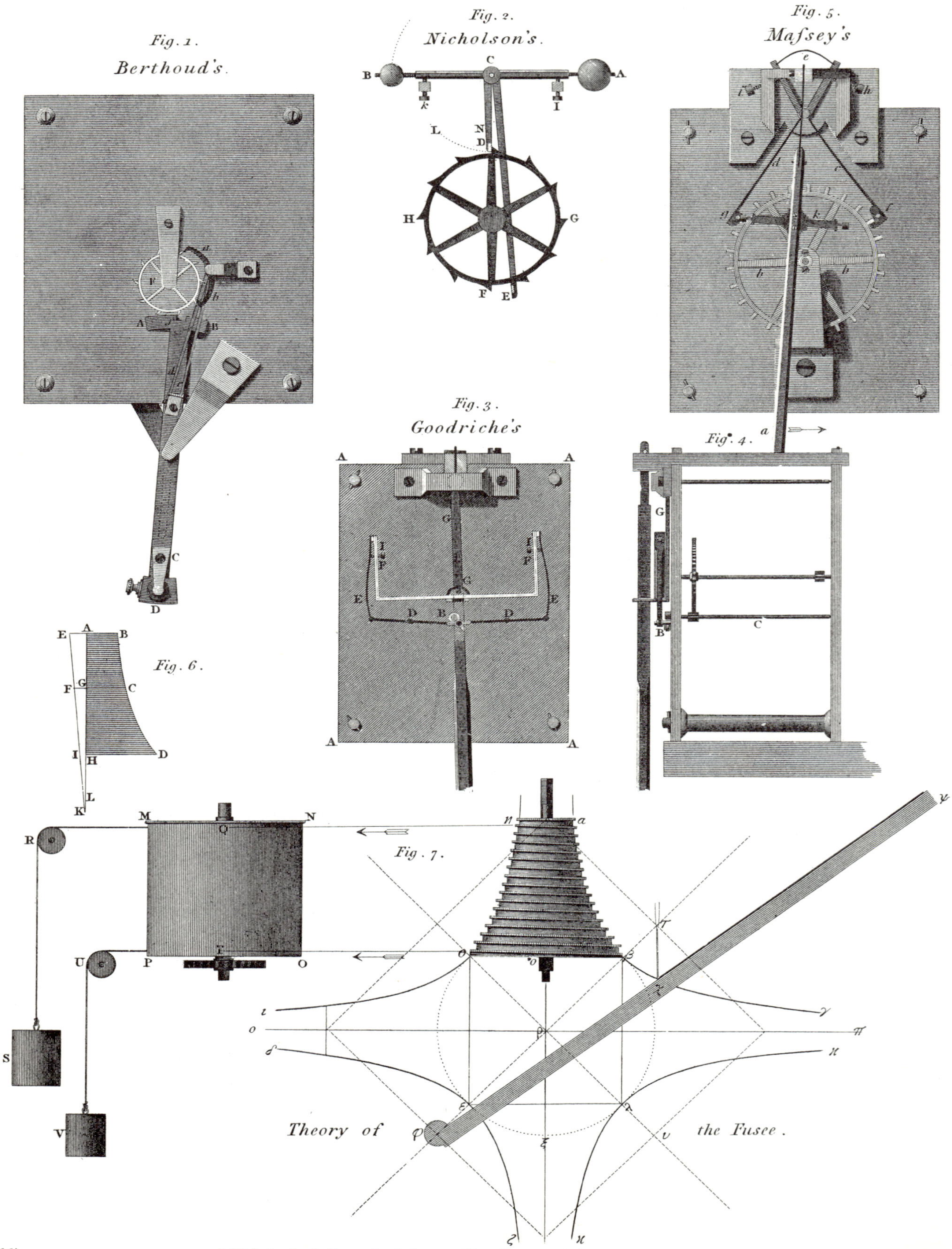

J. Farey delin. Published as the Act directs, 1812, by Longman, Hurst, Rees, Orme & Brown, Paternoster Row. Engraved by W. Lowr

Fusee

FUSEE, in *Horology*, is a mechanical contrivance for equalizing the power of the main-fpring of a watch, chronometer, or portable clock, in all its different ftates of tenfion. When a fufpended weight is ufed as the maintaining power of a clock, the cord, gut, or chain, by which it is fufpended, is wound round a cylindrical barrel, that has not the leaft fenfible variation in its diameter; on which account the fufpended power never varies in its intenfity, as it is wound off in its defcent; in this cafe no contrivance is neceffary for equalizing that power, which is already unvarying from its own nature. But a clock with fuch a power is not portable, or at leaft not conveniently fo; a fpring has therefore been adopted to fupply the place of a fufpended weight in portable machines for meafuring time, and as the power of all fprings varies with its diftance from the quiefcent pofition, the power derived from the force of a fpring requires to be modified according to circumftances, before it can become a proper fubftitute for a fufpended weight acting conftantly, by means of its gravitation, with a flow and controlled velocity, fo as to gain no acceffion by acceleration. The firft contrivance, we have faid in another place, was that which was denominated *ftack-freed*, confifting of two fprings, fomehow oppofing each other's action alternately, in a way that has never been well defcribed; but the contrivance called the fufee has been found of permanent ufe in practice, and confequently merits a particular defcription: we cannot, however, convey to our readers a clear idea of the nature and due conftruction of the fufee, without taking into confideration, along with it, the main-fpring, its box, and the chain that connects them. Thefe indeed

indeed, have been generally defcribed under the articles CHRONOMETER and CLOCK; but as the theory, on which the fhape of the fufee depends, has not been adverted to, it is neceffary that we fhould treat the fubject more particularly in this place.

The fteel arbor, that paffes through the fpring-box, has a fquare on its protruding end, on which a ftrong ratchet-wheel is inferted; and a catch, with a fpring preffing on it, fcrewed to one of the plates of the frame of the works, always holds it faft in any pofition given it; fo that the box will turn freely, when urged round, without any motion in its arbor; upon this fixed arbor is a ftud within the box, on which the interior end of the fpring, intended for the maintaining power, is hooked, by a hole made near the extremity of the faid fpring; while the exterior end of the fame fpring is pinned to the fide of the box, the convex part of one fuiting the concave part of the other, and the breadth of the fpring being a gauge for the depth of the box. The diameter of the box depends upon the length and thicknefs of the fpring conjointly; the ftrength of the fpring depends upon its breadth and thicknefs taken together, when of a given length; and the time it will continue to act at once coiling up, or winding, as it is called, will depend partly on its length, and partly on its ftrength, for as the time in which the fufee is made to revolve once muft be guided by the whole time that the machine is intended to go at once winding, divided by the number of effective fpirals or turns on the fufee, when the fpring is either too fhort to admit of the requifite number of turns to correfpond to the fufee, or too weak to actuate the works for the whole time of going, the dimenfions of that fpring will not anfwer the requifite conditions, and muft be changed for one of more fuitable dimenfions. The relative forces of the fpring when coiled up, and when nearly relaxed, muft alfo be afcertained by weights fufpended by the gut or chain, that is, wound round the exterior furface of the barrel; for on the ratio of thefe extreme weights, or powers, depend not only the relative diameters of the oppofite ends of the fufee, but alfo its actual length; the latter of which meafures, we believe, is ufually taken at random, or agreeably to fome propofed model that has no dependence on, or connection with, the ratio of the extreme powers of the identical fpring that is to be adopted; whereas an attention to the ratio in queftion is a condition in the conftruction, that it will appear prefently ought not to be difpenfed with, particularly when the workmanfhip profeffes to be of a fuperior kind.

Conceiving now the main-fpring coiled up in the box, (which operation requires a tool to force it properly in,) with the interior end faft to the fixed arbor, and the exterior end pinned to the fide of the box; let us fuppofe the gut, or chain, faft alfo to the fide of the box, near one of its ends, and wound a few times round the box, the outer end being at liberty; then, fuppofing the fpring that is coiled to be in a comparatively relaxed ftate, preffing againft the concavity of the box, with an effort calculated to unwind itfelf, in this cafe the central part of the box is not filled by the fpring; pull now the free end of the gut or chain, fo as to wind it from the box, while its faft end prevents its flipping off without carrying the box round along with it, as many times as there are folds thereon, and, as the box is faft to the exterior end of the included fpring, this fpring is thus brought round the fixed arbor that holds its interior end, and its folds are gradually brought up to fill the central part of the box, in which fituation its intenfity is increafed, in confequence of its being moved from its quiefcent pofition: remit now the gut or chain, and the barrel will retrograde and wind on the folds again, while the included fpring is relaxing itfelf gradually by a correfponding diminution of power, till it is again in its quiefcent pofition. Thus the power of the fpring, contained in the box, is always difpofed to pull back the box itfelf whenever the hand, or any external force applied to the chain, has pulled it round in a forward direction.

Again, let us fuppofe the free end of the chain, that is fpoken of as having been pulled, to be wound round a cylinder, placed in a frame parallel to the arbor of the box, and made faft to its end, in fuch a manner, that turning the cylinder round on the pivots of its arbor, will wind the chain round it while it is unwound from the box; in this cafe the inclofed fpring is wound up as before, by being drawn towards the centre of its box; and provided its force were the fame in all fituations, or diftances from the quiefcent point, the cylinder in queftion would be a proper fufee for the machine to be actuated by, after it had the great wheel attached to it, and the ufual ratchet for winding applied; but we have faid that the force of the main-fpring is always varying as it relaxes itfelf, after it has been wound up to the centre of the box; therefore the cylinder muft neceffarily be fo changed into another figure, that the place of action of the chain on its furface fhall be more or lefs diftant from its axis of motion, as the law of variation in the force of the main-fpring fhall demand, in order that the force tranfmitted to the train of wheel-work, and through it to the balance or pendulum, may be invariable, or very nearly fo. It is obvious, from this view of the fubject, that the figure of the body, fubftituted for the cylinder in queftion, as a fufee, muft taper according to fome certain law, either by ftraight lines forming the fruftum of a cone, or by curves of fome order, bounding the fides of the faid figure; and that the point of the chain's action fhall recede from the axis of the fufee's motion, as it revolves, in fuch a way, while the force of the main-fpring is diminifhed by its relaxation, or departure from the centre of its box, that the product arifing from the force of the fpring, multiplied by the acting radius of the fufee, fhall be in every fituation the fame; for then the effectual power of the main-fpring, as it regards the train, will be invariable. It can hardly be expected that the workman can determine the exact figure, and proportional diameters of his fufee, by any mathematical rule that may be laid down from the meafurement of the fpring's extreme powers; but for the fake of illuftrating the theory of a fufee's action more particularly, we will prefent the reader with fome rules and calculations on this fubject, propofed by the late Benj. Martin, which are fufficiently comprehenfive to take in all cafes. The figure of the curve, which will form the fuperficies of the fufee, by a revolution about its axis, may be inveftigated as follows: Let B C D, in *fig.* 6. of *Plate* XXXVI. of *Horology*, be the curve, and A L the axis of the fufee produced; let D be the point where the end of the chain is fixed on the fufee, when the clock or watch is down, or the main-fpring relaxed, and B the point where it lies when the fpring is wound up: from the points B and D, let fall the perpendiculars B A and D H refpectively to the axis A L, in which perpendiculars produced, let there be taken A E and H I proportional to the refpective forces of the main-fpring, when acting by the chain at the points of extreme power B and D; through the points E and I, let the right line E I K be drawn, till it interfects the axis produced at K, and from any point C on the curve, let C F be drawn perpendicular to the axis in the point G, then will F G reprefent the acting force of the fpring when the chain is pulling it round on the point C. Now fince the force acting on the firft wheel of the train ought always to be uniformly the fame, and as this force is always

always as the product or rectangle of the lines $CF \times GF$, this given quantity may be made $= ab$, in which caſe we ſhall have $FG = \frac{ab}{GC}$. To determine, therefore, the equation of the curve BCD, let $KH = a$, $HI = b$, $HG = x$, and $GC = y$; then becauſe of the ſimilar triangles HKI and GKF, we have $HK : HI :: GK : FG = \frac{ab}{y}$; that is, $a : b :: a + x : \frac{ab}{y}$; whence we have $aa = ay + xy$, which is the equation of the curve, and which ſhews it to be that of the *hyperbola* with reſpect to the ſpace between the curve and its aſymptotes. Hence when $x = 0$, then $a = y$, or $HK = HD$; alſo when the point G arrives at A, then $y = AB$; and becauſe $EA \times AB = IH \times HD$, we have $EA : IH :: HD : AB :: a : y ::$ ſo is the greateſt force of the ſpring upon the fuſee to its leaſt force. Again, becauſe the ordinates HD, AB, &c. are at right angles to AL, the curve BCD is that called an *equilateral hyperbola*, by the revolution of which about its axis AK, or common aſymptote, the true form of the ſolid or fuſee is generated, as ſeen in *fig.* 7.

Our *fig.* 7. contains the ſubſtance of three figures, as given by Martin, and the letters of reference are altered, ſo as to prevent the recurrence of the ſame letters, which in the original account rendered the references very ambiguous to the reader; $\alpha\beta\gamma$, $\delta\varepsilon\zeta$, $\eta\theta\iota$, and $\kappa\lambda\mu$, are ſo many oppoſite equilateral hyperbolas deſcribed about the aſymptotes $\nu\xi$, and $o\pi$ interſecting one another in the centre ϱ. Put $\varrho\sigma\ [= \sigma\beta] = 1$, then $\varrho\beta = \sqrt{2} =$ radius of the circle $\beta\theta\varepsilon\lambda$ touching the four equal hyperbolas in their reſpective vertices: again, $\varrho\tau = \varrho\upsilon = \sqrt{\varrho\xi^2 + \xi\upsilon^2} = 2$ is the diſtance of the focus of each hyperbola from the centre; and laſtly, the parameter $ab = \theta\lambda$ the diameter of the circle, agreeably to the properties of the hyperbola. It is therefore plain, that the ſection of any given fuſee, $\alpha\beta\theta\eta$, through its axis $\nu\xi$, is determined by two equal arcs $\alpha\beta$ and $\eta\theta$ of two equal and adjacent hyperbolas, beginning from their vertices β and θ. It is alſo evident, that, ſince $\sigma\beta$ and $\nu\alpha$ do repreſent the two extreme forces of the main-ſpring, in every fuſee properly made, $\theta\beta$ and $\eta\alpha$ of the greateſt and leaſt ends thereof, muſt be exactly proportioned to the leaſt and greateſt force of the ſpring reciprocally.

It alſo follows, that when the proportion of the greateſt and leaſt force of the ſpring is known, or the ratio of $\sigma\beta$ to $\nu\alpha$ is given, then alſo the length of the fuſee $\nu\sigma$ is a given quantity of a determinate length. Laſtly, it is evident, that when the length of the fuſee and one of the two forces $\sigma\beta$ or $\nu\alpha$ are given, then the other correſponding force is determinable, and not to be aſſumed at pleaſure.

Having thus determined the geometrical form of the fuſee, it may not be unacceptable to the reader to give an illuſtration of this theory by examples of the different caſes. Let us therefore put $\sigma\beta = a$, $\nu\alpha = y$, and $\nu\sigma = x$, and then the equation $aa = ay + xy$ will appear in its proper form; whence if a and y be given to find x, we have $\frac{aa}{y} - a = x$; when x is given, or $x = 1$, we have given the ratio of a to y; for then $aa = ay + y$; conſequently $a : y :: a + 1 : a$. When a and x are given, then $\frac{a^2}{a + y} = y$; and laſtly, when a and y are given, we have $a^2 - ay = xy$, and, completing the ſquare, $a = \sqrt{xy + \frac{1}{4}y^2} + y$.

The proper denominations for expreſſing the reſpective forces, a and y, of the main-ſpring, may be ounces and fractional parts, which we will ſuppoſe to be repreſented by ſo many tenth parts of an inch in the meaſures $\sigma\beta$, $\sigma\nu$, and $\nu\alpha$ of the fuſee. Theſe forces may be determined thus; let $MNOP$ be the barrel containing the ſpring, and let $QN\eta\alpha$ be the poſition or direction of the chain or gut upon the barrel and fuſee when the ſpring is wound up; then, ſuppoſing the chain diſengaged from the fuſee, and carried under or behind the barrel, in the direction NQM to the pulley R, ſo as to paſs over and ſupport the weight S in equilibrio, this weight will expreſs the force of the ſpring in this ſituation. In like manner, if $T\theta\beta$ be the chain when the ſpring is down or relaxed, and it be detached and carried under the barrel, in the direction OTP, to the pulley U, ſo as to paſs over it, and to ſupport the weight V, which is juſt large enough to balance the ſmalleſt force of the ſpring, then $\sigma\beta = a : \nu\alpha = y :: S : V$.

Caſe 1.—Suppoſe the weight S to be 63 ounces, and the weight V to be 21; then $a = 63$ and $y = 21$; or, which is the ſame thing, $a = 3$ and $y = 1$; hence we find $x = \frac{aa}{y} - a = 6 = 2a$, that is, when the extreme forces are $3 : 1$, the length of the fuſee $\nu\sigma$ is equal to the diameter of its baſe or thick end $\beta\theta$. When $S : V :: 2 : 1 :: \sigma\beta : \nu\alpha$, then $x = \alpha$, or $\sigma\nu = \sigma\beta$; alſo when $S : V :: 3 : 2$, then $x = \frac{1}{2}a$; and univerſally if $S : V :: m : n :: a : y :: a + x : a$, then it will be $\frac{m - n}{n}a = x$.

Caſe 2.—Given the length of the fuſee $\sigma\nu = 6$, to determine the ratio of the forces or weights S, V, which will give the diameters $\theta\beta$ and $\eta\alpha$ of the oppoſite ends of the fuſee. Since $a^2 = ay + 6y$ agreeably to the laws of the hyperbola, we have $a : y :: a + 6 : a$; then by aſſuming the value of a we have that of y thus; ſuppoſe $a = 3$, then $a : y :: 3 + 6 : 3 :: 3 : 1$; and in this caſe the diameter $\theta\beta$ is three times the diameter $\eta\alpha$; and becauſe $a : y :: m : n$, therefore for any aſſumed ratio, we have $\frac{n}{m - n}x = a$; thus if $m : n :: 2 : 1$, then $x = a$; or if $m : n :: 3 : 2$, then $2x = a$; conſequently if $x = 6$, we have $a = 12$, and $y = 8$.

Caſe 3.—Given the length of the fuſee, and the greateſt force of the main-ſpring to find the leaſt force of the ſame. Suppoſe $x = \sigma\nu = 6$, and $a = \sigma\beta = 3$, then $y = \frac{a^2}{a + x} = \frac{9}{9} = 1$; ſo that if $S = 63$ ounces, we have $V = 21$; and if $x = a$, then $y = \frac{1}{2}a$, or $\theta\beta = 2\eta\alpha$.

Caſe 4.—Given the length of the fuſee, and the leaſt force of the main-ſpring, to find the greateſt. Let $x = 6$, and $y = 1$, then $\sqrt{xy + \frac{1}{4}y^2} + \frac{1}{2}y = \sqrt{6.25} + 0.5 = 3 = a$, ſo that if $y = 21$ ounces, the greateſt force will be $a = 63$: therefore, in every caſe, the form and dimenſions of the fuſee are, or may be, geometrically determined.

But we have yet to ſhew how the requiſite hyperbola $\alpha\beta\gamma$ may be deſcribed, as a pattern for the curve of the fuſee. Suppoſe the extreme forces of the main-ſpring to be, as we have before aſſumed, as 63 to 21; then having drawn two lines $\nu\xi$ and $o\pi$ to interſect at ϱ at right angles for aſymptotes, let the angles $\nu\varrho\pi$ and $o\varrho\xi$ be biſected by the right line $\tau\varrho\varepsilon$, continued each way indefinitely. Then having determined the diameter of the baſe $\theta\beta$ of the fuſee, take that extent in the dividers, and ſet it off each way from the centre ϱ in the line $\tau\varepsilon$ to τ and φ, which points will be the foci of the two oppoſite hyperbolas $\alpha\beta\gamma$ and $\delta\varepsilon\zeta$. In the focus φ fix a ruler $\omega\chi\psi$, ſo as to be moveable round a pin as a centre at φ; then take a thread, as much ſhorter than

than the ruler as is equal to the diameter $\beta\ \iota$ of the circle $\beta\ \theta\ \iota\ \lambda$, and make one end of it faſt to the end, ψ, of the ruler, and the other to a pin fixed in the other focus τ; then a tracer ſtretching the ſaid thread, at the edge of the ruler at any point χ, will trace the required hyperbola, while the ruler is made to turn on its centre of motion φ; for in every caſe the difference of the two lines $\psi\ \chi$ and $\tau\ \chi$, taken from any point x in the deſcribed curve, will always be equal to the diameter $\beta\ \iota$, agreeably to the laws of the hyperbolic curve. When the proper curve is thus formed on paper, for any particular main-ſpring, it may be transferred to card-paper, or ſheet-braſs; as a pattern or gauge to be applied to the fuſee-engine, as will be hereafter deſcribed.

FUSEE-*engine* is an engine contrived for cutting the ſpiral groove, that winds round the ſolid part of the fuſee, which is formed, as we have ſhewn above, by the revolution of an hyperbolic curve. Various conſtructions of this uſeful engine have been deviſed by different ingenious workmen, but it would be difficult to trace each modification to its reſpective inventor, nor indeed is it a matter of much importance, as the general principle is the ſame in all, namely, the application of a ſliding cutting tool to the ſurface of a revolving fuſee; the different methods of adjuſting and limiting the velocities of the moving parts of the engine, and a proviſion for the ſteadineſs of theſe motions, have given riſe to various applications of the mechanical powers to produce the ſame ultimate effect. We ſhall lay before our readers a few varieties of the engine in queſtion, ſuch as we deem the moſt uſeful in practice, from an examination of which, it will be obvious to the mechaniſt, that other combinations, to anſwer the ſame purpoſe, might be deviſed without difficulty.

1. *Fuſee-engine acting by means of an inclined Plane.*—The firſt fuſee-engine that we propoſe to deſcribe, and that which, in our eſtimation, is the ſteadieſt in its action, is repreſented by *fig.* 1. of *Plate* XXXVII. of *Horology*, and is that which was formerly uſed by the late Brower, whoſe cutting engine we deſcribed as made by Hindley of York; and we have reaſon to believe that the fuſee-engine before us was contrived by the ſame ingenious mechaniſt. A A is a ſolid block of hard wood, that may lie on a table, or have feet of its own, to ſupport it: B B are two firm iron ſupports ſcrewed to the ſaid block, and bearing the ſtrong triangular bridges C, C; through theſe triangular bridges paſſes a long priſm of ſmooth iron, the end of which is ſeen at D; upon this ſtrong priſm ſtand three upright iron, or ſteel, pieces E, F, and G, formed like the puppets of a lathe, the laſt of which is moveable at pleaſure, and the whole taken together conſtitute a frame very much reſembling that of a lathe: H we will call the mandril, or arbor, that paſſes through circular holes in the puppets E and F, and has a winch, I, to be turned by; the cylindrical piece of ſteel, K, paſſes through the puppet G, and is fixed at any depth in the hole by the lever and ſcrew L, over it, which ſcrew enters the head of the puppet, and preſſes againſt the convex ſide of K, the inner end of which is pointed by a cone, and its outer end has a conical hole drilled in it, by which means the piece, K, may be made to bear either againſt a hole, or a point in any arbor that a fuſee may happen to have; M is a lever with a projecting pin ſcrewed to the interior end of the mandril, ſo as to revolve with it, and the arbor of the fuſee, O, which bears againſt a cylindrical hole at this end of the mandril with one end, and againſt the cylindrical point of piece K with the other, carries a ſecond lever N, ſeen ſeparately in *fig.* 2, which correſponds to what is called a *dog* in a turning frame; this dog is fixed to the arbor of the fuſee, by a ſmall thumb-ſcrew at its lower end, as ſeen in *fig.* 2, and the oblong hole at its upper end receives the projecting pin of lever M, ſo that when this lever revolves with the mandril, the dog and fuſee are carried round at the ſame time. On the projecting parts, P, P, of the puppets, F, G, which have each a triangular hole, is ſupported a ſecond, but a ſmaller priſm, Q, of iron, or ſteel, that may be fixed in any poſition, to the right or left, by ſliding without ſhake in the two triangular and diſtant holes above-named; to this priſm, Q, is fixed a ſliding perforated piece, R, by means of a thumb-ſcrew ſeen under Q, and at right angles to this ſliding piece, R, is fixed faſt a third ſtill ſmaller priſm, S, which conſequently may be fixed over any part of priſm Q, ſo as to be oppoſite the fuſee O, or any part of it; along this third priſm, S, ſlides the upright bearer T, that may alſo be fixed at any diſtance from the end of S by another thumb-ſcrew, which appears under it; the bearer, T, therefore ſlides alſo along the ſecond priſm, Q, whenever the piece, R, is moved in making the adjuſtment for poſition; at the top of this bearer is a ſquare hole of conſiderable depth, in which the cutting tool, U, ſlides without ſhake, and any ſlight preſſure on the handle at U of the cutting tool will apply its cutting point to the ſurface of the revolving fuſee, the winch, I, being turned at the ſame time; but the tool U, or the fuſee, muſt one of them have a lateral motion, otherwiſe the groove cut by the ſaid point would form a circle only on the ſurface of the fuſee; to render the groove ſpiral, therefore, is the buſineſs of the remaining portion of the apparatus.

W X Y Z is a long metallic frame with its ſides made perfectly parallel, ſo as to ſlide without ſhake within the two fixed parallel bars of metal, *a a* and *b b*, which are ſcrewed to the puppets E and F; this frame is counterbalanced by a weight ſuſpended by a ſtring, not ſeen in the figure, that is faſtened to its ſuperior end, before it paſſes over a fixed friction-wheel, not neceſſary to be exhibited; and on the mandril, behind this frame, is a pinion inſerted that acts with the rack, or toothed bar, that is viſible at the middle of the lower part of the frame, to which it is made faſt; hence, whenever the winch, I, is turned, not only the fuſee revolves, but the frame, W X Y Z, aſcends or deſcends, accordingly as the turns are made direct or retrograde: *c d e* is a metallic bar, ſimilar to one of the bars that conſtitutes the ſides of the frame, and is moveable round a pin at *d*, as a centre of motion; the extreme croſs-pieces or end pieces of the frame have each a circular perforation quite through, large enough to admit each a ſcrew through them, of which the milled nuts are viſible near W and Z reſpectively; it is obvious, therefore, that this bar, *c d e*, may be fixed either parallel to the ſides of the frame, or at any given angle of inclination, within the limits of the circular apertures, the edges of which are, or may be, graduated; for when the bar is turned round its centre from its parallel poſition, as in the figure, the thumb-ſcrews will act on its extreme ends, and ſet it faſt. Laſtly a weight, V, is ſuſpended by a cord, going round a fixed friction-wheel on the cheek *a a* of puppet E, and made faſt to the ſecond priſm, Q, at the end near R, ſo as to pull this priſm, and ſurmounted cutting tool, U, towards the inclined bar of the frame, till an elaſtic croſs-bar, *f f*, attached to the end of the ſaid priſm, comes in contact with the ſtraight edge of the inclined bar, in which ſituation the priſm in queſtion will remain at reſt while the frame is at reſt; but when the frame is made to aſcend by the pinion of the winch, the weight, V, preſerves the contact of the elaſtic bar, *f f*, with the ſide of the inclined bar, and deſcending, pulls after it the priſm Q, with its cutting-tool, the fusee, in the mean time, being made to revolve; it is evident, therefore, that if one hand of the workman turn the winch, I, while the other is pressing gently against the handle, U, of the cutting-tool, a groove will be made on the surface of the fusee of some determined number of revolutions, which shall depend on the quantity of the bar's inclination, that regulates the motion of the prism; and as this inclination is adjustable, a given number of spiral turns may be assigned by adjustment, to a given length of the fusee to be cut. At the same time, as the fusee is not cylindrical, the regulation of the pressure by hand only must be very nice, to make the tool cut the groove of equal depth in every part of the successive revolutions, where the diameter is continually varying its dimensions. To obviate any inaccuracy that is liable to arise from this operation, the French have applied a piece of metal formed into the proper hyperbolic curve, as a guide, or gauge, for the run of the cutting tool, such as is represented by *fig.* 3, which we will describe as adapted to our present engine, though it was applied, according to F. Berthoud, by M. Lelievre to an engine, contrived by him, of a construction nearly similar to the one we are describing, except that the sliding frame moved horizontally, whereas our's moves vertically.

In the figure last named, A B is a brass frame, carrying the cutting tool U, and its supporter T, adapted to the prism Q, in *fig.* 1; but at C, where the cutting point is connected with A B, is a piece of metal formed into an hyperbolic curve, as seen detached at D, near the said figure; this curve, or gauge, is formed by the rules laid down in our preceding article Fusee, and is fixed by a clamping piece, E, with a thumb-screw at the required position, as to depth, while the screws at the ends of A B adjust the lateral position, for the length of the fusee, by means of the oblong holes that allow the whole frame of this figure to slide to the right and left, as occasion may require, before it is finally fixed for action. In this situation of the apparatus, a pin is inserted fast into the stem of the cutting tool near C, so as to be in contact with the edge of the curve; the consequence of which contrivance is, that while the cutting tool goes along with the prism Q, in *fig.* 1, as before described, the pin, resting against the edge of the fixed curve, is pushed outwards at the same time that the tool advances in a lateral direction, along the side of the fusee, and the bottom of the spiral groove thus cut is gauged to a given depth, depending not on the shape of the fusee itself, but of the hyperbolic curve at the edge of the small plate, C, which in fact forms the fusee, at the same time that it is grooved by the cutting point. This contrivance has not, that we know of, been used in England, notwithstanding its ingenuity and practicability, and the reason may be, that, as we have observed before, every particular spring, used as a main-spring, requires its own appropriate curve, and it would be endless to have a curved piece made for every separate fusee that is to be cut by the engine: besides, a spring is rarely found uniform in its variation of force with the variation of tension, throughout its whole time of unwinding; so that, after all, tentative adjustments of the depth of certain spirals, to correspond to the actual force of the spring, at the several states of relaxation, become necessary, to ensure an uniformity of the spring's influence on the train of wheel-work: the tool which our English watch and clock-makers use, to ascertain the proper dimensions of the fusee, when formed and grooved, has been already described among the other Clock *tools*, and may be seen at *fig.* 10. of *Plate* XXI. of *Horology*; and when any particular part of the fusee requires to be diminished, or to have its groove deepened, the fusee is put either into a common turning frame, or into the Fusee *frame*, hereafter described, (*fig.* 3. of *Plate* XXXVIII.) and the cutting point is suffered to run by the guidance of the groove already formed, till the given part is found deep enough, after a repetition of trials, made with the spring properly fixed, as to tension, and wound up to the given part of the fusee that has been altered. Thus every point in the spiral groove of a fusee may be adjusted to the acting force of the spring with which it is to be used; and we beg leave to repeat, that these adjustments of the fusee, by trial of the fusee-tool, ought never to be omitted when a new main-spring is first used in any watch or clock, however old; for as soon might we expect that the expansion of mercury, in a new tube, would correspond to an old scale previously made, as that a new main-spring shall have its extreme and intermediate forces exactly compensated by an old fusee without new adjustments.

2. Fusee-*engine acting by the union of a screw, a double lever, and a pair of wheels.*—The fusee-engine which we now come to describe unites the combined powers of wheel-work, the screws, and the double lever, to modify the respective motions of the fusee and of the cutting tool, in the following manner; A B, in *fig.* 4. of the last plate, is a plate of brass fixed, or otherwise placed on a table, during the act of using the engine now before us; C C and D D are two pair of puppets, respectively screwed to the plate A B, and united respectively by the edge-bars lying on the said plate; the little frame under *fig.* 5, seen there detached, is screwed on the same plate A B, between the two edge-bars, to which it will slide parallel, before it is fixed by its thumb-screw beneath; this small frame receives the fusee between its cylindrical holding pieces E and F, and the piece E is moreover used as a stud for a loose wheel to revolve round; a pin projecting from the plane of this wheel impels a dog, as in our last engine, fixed on the fusee-arbor, to which it communicates motion whenever the wheel revolves; this wheel is impelled by another, placed fast on the tapped arbor, of the winch I, that fits in a hole made in the puppet C, near the winch, and is supported at the other end by a cylindrical piece with a conical point, entering the hole of the other puppet C; whenever therefore the winch is turned, both wheels and the fusee will all revolve, though not all in the same direction, nor yet in the same time, unless the two wheels have like dimensions and numbers of teeth. The perforated cylindrical piece G, which if tapped to receive the screw of the arbor, is prevented from turning round, as the arbor revolves, by a projecting piece below it, that surrounds a fixed strong wire H K, which slides, or is at liberty to slide, in two cocks screwed to the puppets C and C, as seen in the figure; therefore, when the arbor containing the male screw revolves, by means of its winch, the piece G, with the corresponding female screw, is drawn towards the winch gradually, and the regularity of this motion will depend on the due inclination of the threads of the screw, which therefore ought not to be *drunk*, as the workmen call the imperfection in this respect; but the piece G would be at liberty to travel along the whole length of the arbor, if there were no stops at the limited distances: to obviate this, the wheel on the tapped arbor has a pin also projecting from its plane, and the two sliding pieces L and M will clamp at any given points of the wire just described; near the last named wheel the wire is cranked so as to permit the wheel to pass through its crank in its present position, but when the piece under G slides along the wire till it meets with the clamp L, it pushes the wire altogether before it, till the angular part of the crank falls in the way of the wheel's pin, when all further motion is arrested by the locking of the wheel; a contrary motion will also lock the wheel

wheel by a pin behind the wheel, when the clamp M is acted on, and the diftance of the clamps muft depend upon the length of the fufee to be cut, as we fhall fee more clearly prefently. Above and below the moveable piece G are fixed two ftrong wires N and O, which running parallel to each other, are fo fixed, by a pin paffing along focket P, that they have thereon one common centre of motion, in the manner of a long joint; this focket is attached to the double lever, feen detached at *fig.* 6, with five holes near its centre, and with a fimilar focket at its oppofite end; the cock, which is feen in *fig.* 7, is fixed by a fcrew and fteady pin to the bottom plate A B, beyond the fmall frame E F, and fupports the double lever P Q, by a pin placed in fome one of its five holes, accordingly as the oppofite ends of the lever are to be proportioned to each other, that is, accordingly as the central pin is inferted into the firft, fecond, or third, &c. hole of the double lever. R is a long bar, of equal breadth throughout, that fupports the cutting tool S, and that is at liberty to flide within the puppets D and D without fhake; another pair of ftrong wires, like N and O, are made faft to this bar above and below, which wires have alfo their common centre of motion round a long pin, paffing through the focket Q, at the oppofite end of the double lever.

From this arrangement it is evident, that whenever the piece G is drawn towards the winch, by the revolution of the latter, the wires N and O will make their end of the double lever follow; but as the oppofite end of the faid lever will move in a contrary direction, its wires will draw back the long bar R, and with it the cutting tool, the fufee in the mean time being revolving; hence it is obvious that a little preffure on the handle S of the cutting tool will form a fpiral groove, fuch as was defcribed when we detailed the operations of the laft engine. The number of turns on the fufee may be limited either by altering the central pin between the oppofite levers, or by changing the wheels for others of different ratios, with refpect to their diameters and numbers of teeth, or by both jointly when great nicety is required. We are difpofed to think that the variety of fittings in this conftruction, and the union of different powers, are not calculated to render the motions to be produced equally fteady, particularly when the parts of action have been a little loofened by ufe.

It may not be improper to fubjoin here another method of applying the cutting point to the fufee, which feems calculated to prevent parallax or lateral motion of the point; R, in *fig.* 8, as in *fig.* 4, is the edge-bar bearing the cutter D fixed faft to the upright lever B; this lever is faft on the arbor A, which turns on two pivots at the cocks *a* and *a* fixed to the bar R; then fuppofing D to have a handle like S in *fig.* 4, the motion of the arbor A round its pivots will allow the cutting point to approach, or recede from the fufee, according to the quantity of preffure applied to it in the act of cutting. This conftruction is not fo common as that in *fig.* 4, but muft be more fteady, and would probably cut better if there were a joint where D is made faft to B, to preferve the horizontal pofition of the cutting-tool as it recedes or advances.

3. *Fufee-engine* (*common*) *acting by the union of one fcrew and two fingle levers.*—The fufee-engine reprefented in *fig.* 1. of *Plate* XXXVIII. of *Horology*, is the common engine taken from one of the Lancafhire pattern-books, and is the moft fimple of any; A B C D is a plate of brafs, on which the dove-tailed plate E F flides without fhake, and with but little friction, the fides being perfectly parallel. The axis of the handle is fupported by the bearing pieces A and G faft to the under plate, and has a fcrew cut on it throughout its length between the faid bearing pieces; the thick end H of the lever H I is tapped fo as to receive the fcrew of the axis without fhake, to which it can be adapted by the fcrew at H; this lever has an oblong aperture through it, in the direction of its length from H nearly to I, and is moveable round the pin at K, that enters the lower plate through a fecond aperture at the extreme end of the lever; L M is a fecond lever, turning on the pin M fixed to the fliding plate, and fet faft by the milled nut L, over any point of the circular aperture N, as the inclination may require; O L is an endlefs fcrew of flow motion, for regulating the accuracy of the adjuftment to the point where the nut fixes the lever, and T is a pin fixed in the lever L M, that paffes through the oblong apertures of lever H I, and gives motion to the fliding plate E F, whenever the handle turns the axis of the fcrew, and as the cutting point P Q is attached to this plate, by the cocks R and S, it alfo moves with the fame velocity; this velocity depends on the diftance of the pin T from K the centre of its motion, and may therefore be regulated by the nut L and fcrew O L, to the requifite angular velocity, that may fuit each turn of the handle, or fufee attached to it by a clamping piece between G and D.

The operation of cutting the fufee by this engine, it is prefumed, can require no further explanation.

4. *Fufee-engine* (*old*), *acting by the union of a wheel and pinion, a fcrew, a fimple lever, and a fliding frame.*—*Fig.* 2, of the laft plate, exhibits a perfpective view of the old fufee engine as given in the Lancafhire pattern books; A B C D, as before, is the inferior frame, and E F the fliding frame, but inftead of dove-tailed edges this upper frame flides by means of four cylindrical pieces paffing through correfponding holes in the edges of the lower frame, as feen in the figure; there is alfo an axis formed into a fcrew, turned by the handle, acting in the part H of the invifible lever H I, as before, which lever lies here under the frames; but a wheel G on a feparate axis, that carries the fufee, is actuated by a pinion on the axis of the handle. The fecond lever L M of the preceding engine is here wanting; but the nut T is in place of its pin T, and penetrates one of the circular holes in the fuperincumbent piece R T, fo as to lay hold of the lever H I through a concealed aperture in the fliding frame E F; the diftance of the hole ufed from the centre of the lever's motion K regulates the velocity, as before, of the fliding frame, and confequently of the cutting point P Q, fcrewed to it by a pair of embracing cocks, while the fcrew, taking into the nut R, adjufts the exact fituation of the hole occupied by the milled nut T. In all other refpects the operation of cutting the fufee is the fame as in the other engines. Sometimes we have feen a fcrew acting againft the fliding frame and placed between the cylindrical pieces on the inner edge of the under frame to prevent jerks, but when the fcrew and other acting parts are well fitted, this is unneceffary. From thefe four engines it is eafy to perceive, as we have obferved, that various other modifications may be adopted to anfwer the fame purpofe.

Fusee *Frame.*—When a fufee has been cut by an engine, and is found on trial to have fome place or places of too large a diameter for the main-fpring to act equally with, a little frame, like that in *fig.* 3. of the laft plate, is fometimes ufed, where the cutting tool A has its point put in the fpiral groove, and is made to reft on a cylindrical arbor B, that refts on the equal rollers C and D, fixed in the refpective ends of the frame, by means of a grooved pulley, fo that as the fufee revolves, the faid arbor B moves laterally, and keeps the point of the cutting tool in the groove, at right angles to the arbor of the fufee, and thus any particular part only of the fufee may be reduced, by partially turning the handle, as the circumftances of the cafe may require.

Fusee

FUSEE ENGINE acting by an inclined Plane.

Fig. 1.

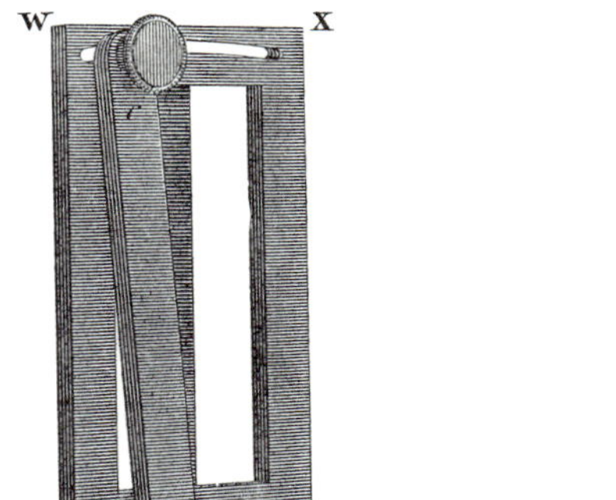
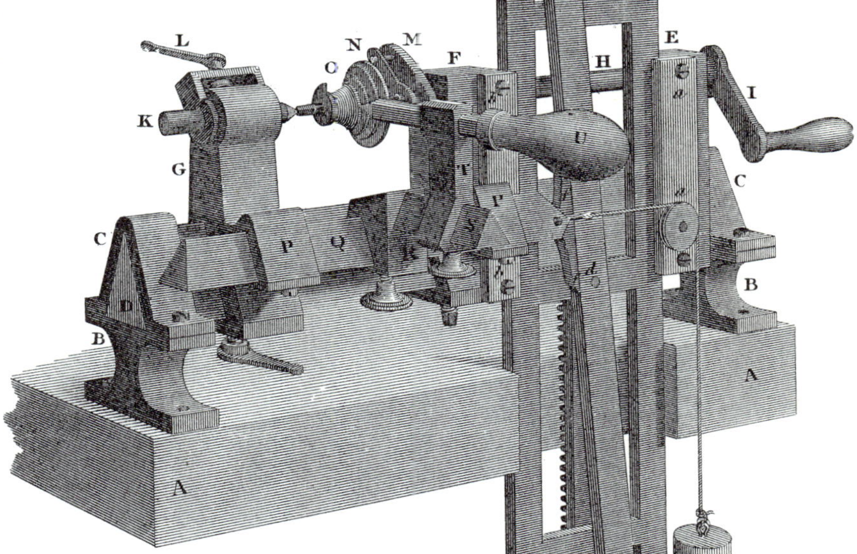
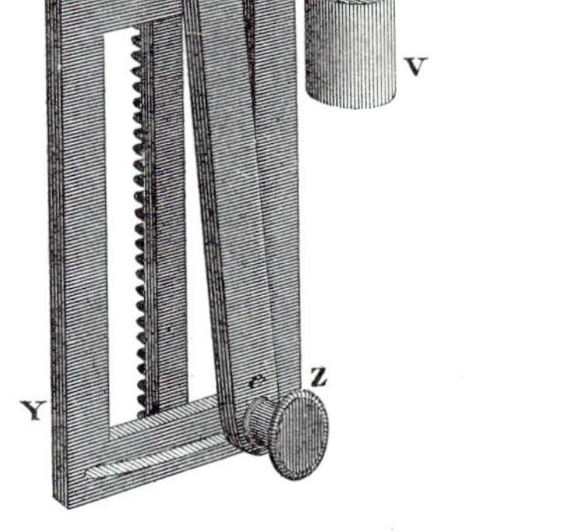

Fig. 2.

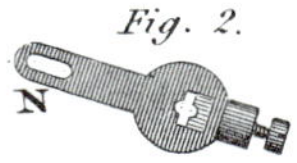

Fig. 3.

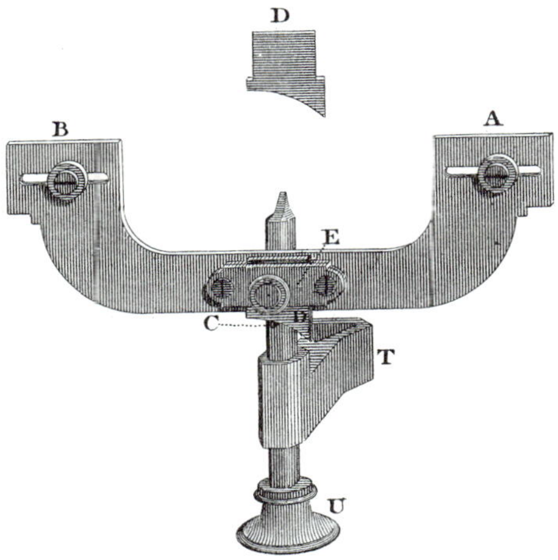

Fig. 7.

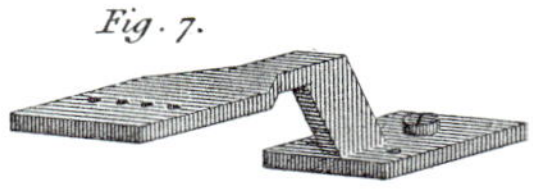

Fig. 6.

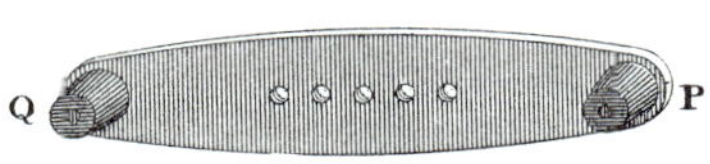

Fig 5.

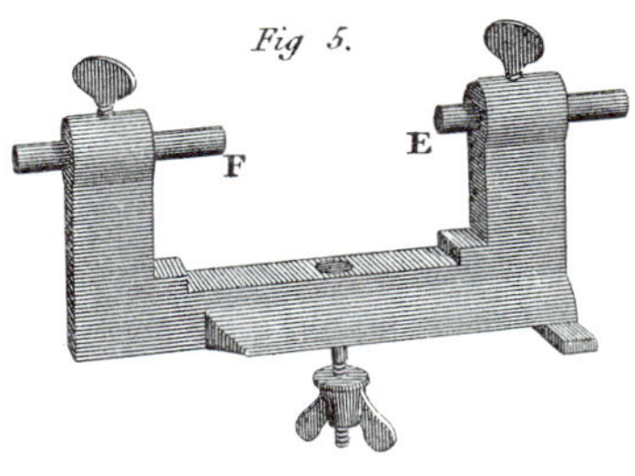

Fig. 8.

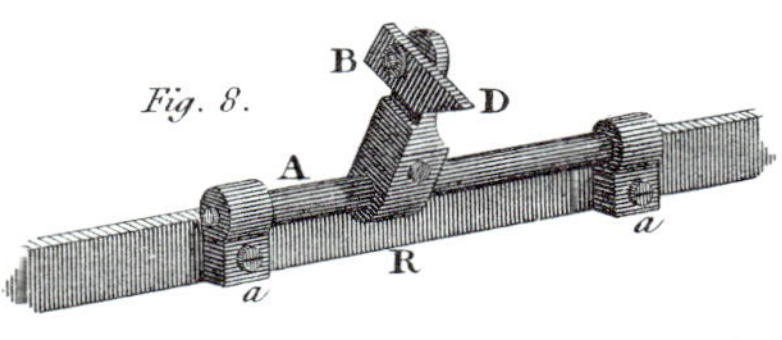

Fusee Engine with a Screw and Lever.

Fig. 4.

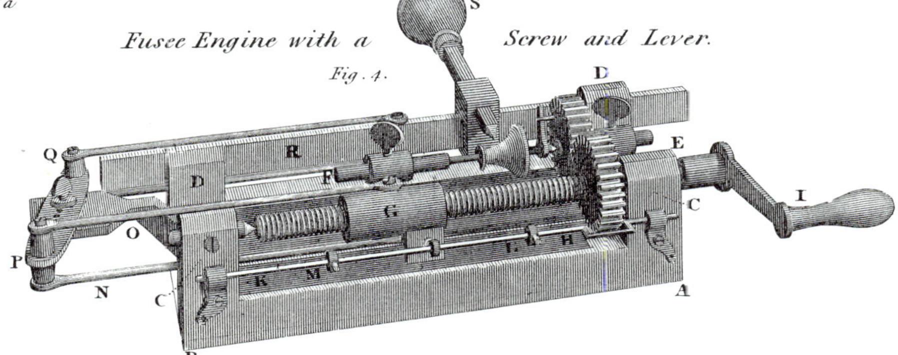

J. Farey delin. *Published as the Act directs, 1817, by Longman, Hurst, Rees, Orme & Brown, Paternoster Row.* *Lowry Sc.*

HOROLOGY.

FUSEE ENGINES and FUSEE FRAME.

Common Fusee Engine.

Fig.1.

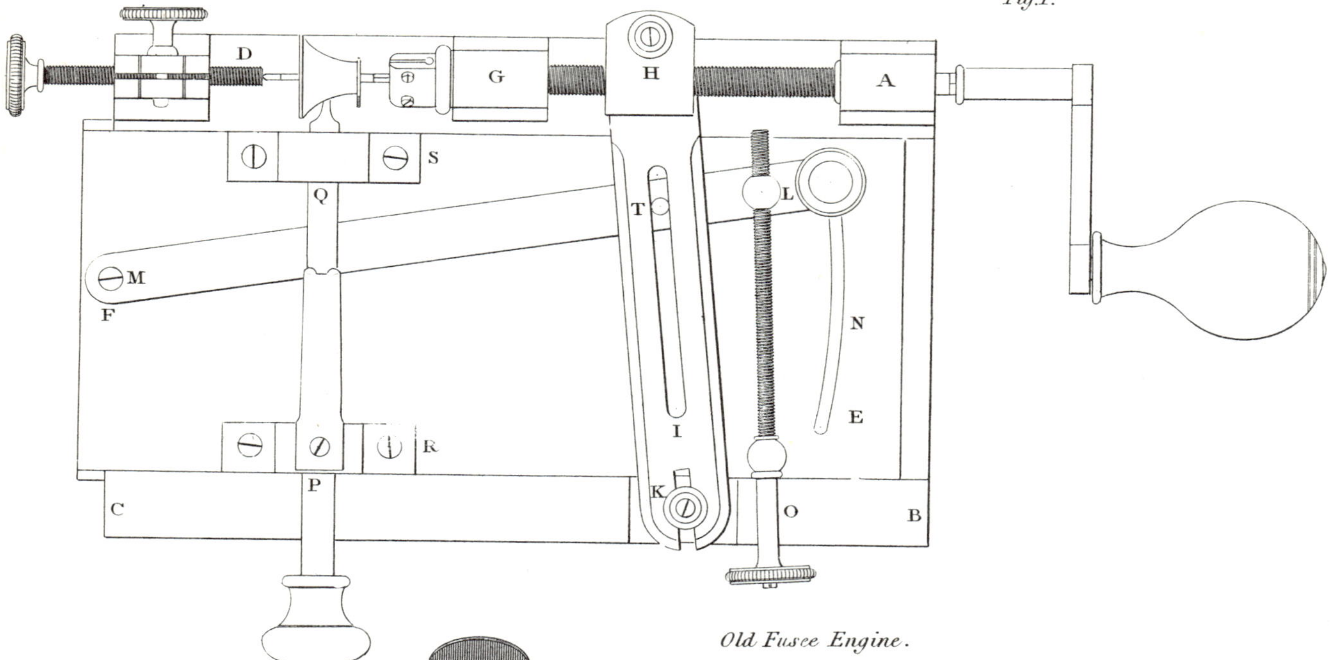

Old Fusee Engine.

Fig. 2.

Fusee Frame

Fig. 3.

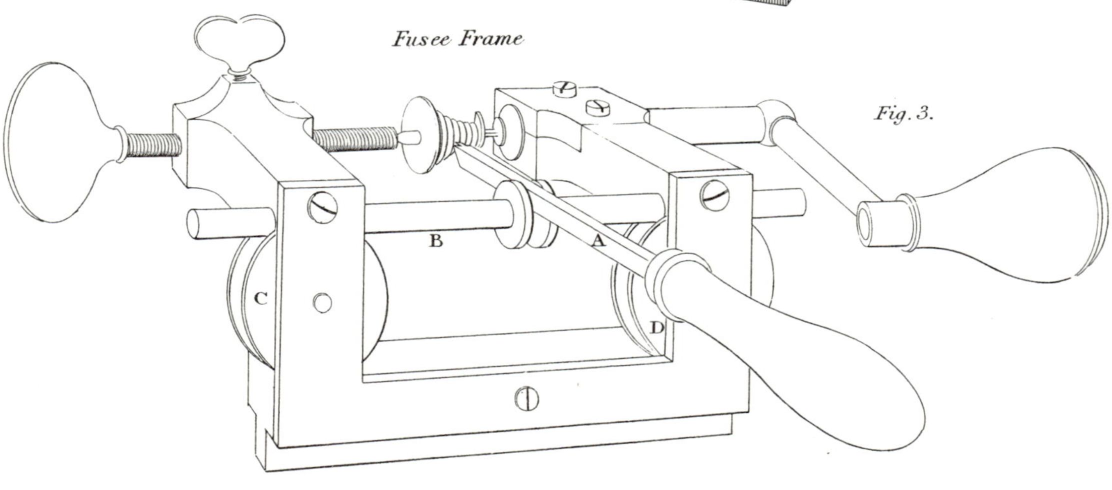

Pendulum,
Bob and Maintaining Power

PENDULUM, in *Dynamics*, is a fimple ponderous body, fo fufpended by a flexible cord from an axis of fufpenfion, that it is at liberty to vibrate by the action of its own gravity alone, when it is once raifed, by any external force, to the right or left of its quiefcent pofition; and in demonftrating the theory of its motion, mathematicians are obliged to affume, that there is no rigidity in the cord; no friction at the axis of fufpenfion; no refiftance to motion oppofed by the air; and no variation in the total length of the cord, arifing from the variable temperature or moifture of the atmofphere; and if thefe affumptions were ftrictly correct, a pendulum, once put into motion, would continue to move *ad infinitum*, without the further acceffion of any external force; but when the pendulum is applied as the regulator of a clock, for which purpofe it is admirably adapted, the affumptions we have named require an equal number of mechanical corrections, of which the theory, fimply confidered, takes no notice. In *Horology*, therefore, the pendulum muft be confidered, not fimply as a felf-moving pendulous body, without any tendency to come to a ftate of reft, or to have its dimenfions altered by the circumftances in which it is placed; but as a body, whofe motion is perpetuated by repeated acceffions of force, in aid of its own gravity, and whofe vibrations are rendered ifochronal by a nice adaptation of mechanical contrivances, that prevent or remedy the prejudicial influences of all natural impediments to uniform and uninterrupted motion. It feemed neceffary, in the firft place, to eftablifh the diftinction that thus exifts between the free or *detached* and *attached* pendulum; and having done this, which we muft beg the reader to bear in mind, we propofe to divide our article into three heads, namely, 1ft, the hiftory of the pendulum; 2dly, its theory in its *detached* ftate; and 3dly, its various conftructions, when ufed in its *attached* ftate, as the regulator of an horological machine.

1. We have already given an hiftorical account, under our article Clock, (fee *Hiftory of the fucceffive Improvements in Clocks*,) of the firft notice that was taken by Galileo of the vibrations of bodies fufpended by long chains, and alfo of the application that was made foon after, by various perfons, of the ifochronal property of the detached pendulum, particularly in aftronomy, as well as of its introduction, in its attached ftate, into horological machines as a regulator of their motions, after the balance had been applied for the fame purpofe: we beg leave, therefore, to refer the reader to the article already named, for the particulars that would have fallen under this head, had not this part of the fubject been anticipated.

2. The fecond head of this article which we propofed to confider, was the theory of the fimple pendulum vibrating in its detached ftate. When Galileo difcovered that ponderous bodies, whether large or fmall, fufpended by cords of equal length, perform their vibrations apparently in equal times, he was led into the contemplation of the laws of uniformly accelerated motion, produced by the conftant force of gravity, and foon found that the theory of the detached pendulum was derivable from the fame laws by which bodies, left to defcend in free fpace, were found to be accelerated, and that gravity was the fole caufe of fuch acceleration. We have already treated of thefe laws at fome length, under our articles Acceleration, Descent, and Motion, which render unneceffary a repetition of this part of our fubject, except fo far as the vibration of the pendulum is concerned. It was, however, foon perceived, that a pendulum, vibrating in long arcs of a circle, requires a fomewhat longer time for the performance of each of thofe vibrations, than when it vibrates in fmall arcs, all other circumftances remaining the fame; it thence became a queftion among mathematicians what the nature of that curve might be, in which, fubftituted for the arc of a circle, both long and fhort vibrations would be performed in the fame time. Though Galileo is acknowledged to have been the firft who difcovered and defcribed the curve firft called by him the cycloid, yet it remained for the celebrated Huygens to explain not only the ifochronal property peculiar to this curve, but alfo that the involute of a cycloid is itfelf a cycloid, which latter difcovery enabled him to make the attached pendulum actually vibrate in a cycloidal curve. Under our article Cycloid we have given at confiderable length both the geometrical and mechanical properties of this remarkable curve, which we recommend to the reader's notice, as preparatory to our doctrine of vibrations performed in cycloidal, and alfo in circular curves; for he will there fee that fhort arcs of vibration, at each fide of the loweft or quiefcent point of the curve, are very nearly fimilar, both as to coincidence in pofition, and alfo to the confequent ifochronal property. He will fee, not only that arcs of all lengths, when cycloidal, are performed in equal times, but that fmall circular arcs approximate fo nearly to their correfpondents in the cycloid, that they may be fubftituted in practice for each other, the former for the latter, without fenfible deviation.

Prefuming, now, that our reader has made himfelf acquainted with the laws of uniformily accelerated motion, agreeably to our references, and that he has perufed with attention alfo the article Cycloid, we will proceed to lay before him the fluxional method of determining the actual times of a detached vibration both in *cycloidal* and in *circular arcs*. This part of our fubject has been recently treated fo ably by profeffor Bridge of the Eaft India college, that we beg leave to avail ourfelves of his labours, at the fame time that we recommend to general notice his "Introduction to the Study of the Mathematical Principles of Natural Philofophy, 1813"

On the Times of Vibration in Cycloidal Arcs.—In *fig.* 1. of Plate XXXIX. of *Horology*, "let A V B be a cycloid, in which a body, P, fufpended from the point S, performs its vibrations between the two *femi-cycloids* S A, S B, and let it begin its vibrations from the point E; alfo let P p be any very fmall arc, defcribed in the defcent from E to the loweft point V; draw E F, P N, p n, parallel to the bafe, A B, of the cycloid, and join C M, C m, M V, m V. Let C V, the diameter of the generating circle, $= d$, F V $= a$, N V $= x$; then F N $= a - x$, M V $= \sqrt{\overline{CV \times NV}} = \sqrt{dx}$. Now the cycloidal arc P V being = twice the *chord* M V $= 2\sqrt{dx}$; let E P $= z$, then E V $= z +$ P V; and fince E V is a *conftant* quantity, $\dot{z} +$ P V $= 0$, ∴ $\dot{z} = -$ the *fluxion* of P V $= -\frac{\sqrt{d.\dot{x}}}{\sqrt{x}}$. The velocity down E P (v) = the velocity down F N $= \sqrt{4m \times FN}$ $= \sqrt{}$

$= \sqrt{4m \cdot \overline{a - x}}$; if therefore t = the *time* down E P, we ſhall have $\dot{t} = \frac{\dot{z}}{v} = \frac{\sqrt{d} \cdot \dot{x}}{\sqrt{x}} \times \frac{1}{\sqrt{4m \cdot \overline{a - x}}} = -\sqrt{\frac{d}{4m}} \times \frac{\dot{x}}{\sqrt{ax - x^2}}$; and $t = -\sqrt{\frac{d}{4m}} \times$ circular arc, rad. = 1, ver. ſin. $\frac{x}{\frac{1}{2}a}$ + cor.; but when $t = 0$, $x = a$, ∴ $0 = -\sqrt{\frac{d}{4m}} \times$ cir. arc, rad. = 1, ver. ſin. = 2, + cor. = $-\sqrt{\frac{d}{4m}} \times$ ſemicircle, rad. = 1, + cor. = (if $\pi = 3.1416$, &c.) $-\sqrt{\frac{d}{4m}} \times \pi + C$, or $C = \sqrt{\frac{d}{4m}} \times \pi$.

The fluent properly corrected therefore is $t = \sqrt{\frac{d}{4m}} \times \overline{\pi - \text{cir. arc, radius 1, verſed ſine } \frac{x}{\frac{1}{2}a}}$ = (when $x = 0$) $\sqrt{\frac{d}{4m}} \times \pi$ = the time of deſcent through the cycloidal arc E V; conſequently the *whole time of vibration* = $\sqrt{\frac{d}{4m}} \times 2\pi = \sqrt{\frac{d}{m}} \times \pi$. Since the quantity a does not enter into this value of t, it is evident that the times of vibration will be the *ſame*, from whatever point of the cy cloid the body P begins to vibrate. Moreover, ſince the time down the axis C V $= \sqrt{\frac{d}{m}}$, we have the *time of a vibration* : the *time down the axis* :: $\sqrt{\frac{d}{m}} \times \pi : \sqrt{\frac{d}{m}} :: \pi : 1$ :: the *circumference* of a circle : its *diameter*;" which concluſion coincides with our former determination under the article CYCLOID.

On the Times of Vibration in circular Arcs.—In *fig.* 2. of the ſame plate, "let the ſmall body P, vibrating in the circular arc E V G, begin its vibrations from the point E; let P V = z, N V = x, F V = a, P p = $\dot{z}$, S P or S V = r, then $\dot{z} = \frac{r\dot{x}}{\sqrt{2rx - x^2}}$. Let t = the time through E P, v = the velocity in the point P, then $v = \sqrt{4m \times FN} = \sqrt{4m \cdot \overline{a - x}}$, and $\dot{t} = \frac{-\dot{z}}{v} = \frac{-r\dot{x}}{\sqrt{2rx - x^2}} \times \frac{1}{\sqrt{4m \cdot \overline{a - x}}} = -\frac{r}{\sqrt{4m}} \times \frac{\dot{x}}{\sqrt{ax - x^2}} \times \frac{1}{\sqrt{2r - x}}$ (for $\sqrt{2rx - x^2} = x^{\frac{1}{2}} \sqrt{2r - x}$), hence $\dot{t} = -\frac{r}{\sqrt{4m}} \times \frac{\dot{x}}{\sqrt{ax - x^2}} \times \frac{1}{\sqrt{2r}} \times \overline{1 + \frac{x}{4r} + \frac{3x^2}{32r^2}}$ &c. by expanding $\frac{1}{\sqrt{2r - x}}$ into a ſeries.

Again, let $\frac{\dot{x}}{\sqrt{ax - x^2}} = \dot{A}$; then ſince $\sqrt{\frac{r}{4m}} \times \frac{1}{\sqrt{2r}} = \sqrt{\frac{r}{8m}}$, we have $\dot{t} = -\sqrt{\frac{r}{8m}} \times \overline{\dot{A} + \frac{x\dot{A}}{4r} + \frac{3x^2\dot{A}}{32r^2}}$ + &c. Therefore, $t = -\sqrt{\frac{r}{8m}} \times \overline{A + \frac{aA - 2P}{8r} + \frac{9a^2A - 18aP - 12Q}{256r^2}}$ + &c. + cor. (where $P = \sqrt{ax - x^2}$, $Q = x\sqrt{ax - x^2}$, &c. &c.)

Now when $t = 0$, $x = a$, ∴ P = O, Q = O, &c.; and at the ſame time A $\left(= f' \frac{\dot{x}}{\sqrt{ax - x^2}} = \text{cir. arc, whoſe rad. is 1, ver. ſin. } \frac{x}{\frac{1}{2}a}\right)$ = a *ſemi-circle* whoſe radius is $1 = \pi$; hence $O = -\sqrt{\frac{r}{8m}} \times \overline{\pi + \frac{a\pi}{8r} + \frac{9a^2\pi}{256r^2}}$ + &c. + cor. The fluent, therefore, when properly corrected, is $t = \sqrt{\frac{r}{8m}} \times \pi \times \overline{1 + \frac{a}{8r} + \frac{9a^2}{256r^2}}$ + &c. $-\sqrt{\frac{r}{8m}} \times \overline{A + \frac{aA - 2P}{8r}}$ + &c. = $\sqrt{\frac{r}{8m}} \times \pi \times \overline{1 + \frac{a}{8r} + \frac{9a^2}{256r^2}}$ + &c. (when $x = 0$, and conſequently, A, P, Q, &c. each = 0) for the time of deſcent down the arc E V; ∴ *the whole time of one vibration* $= \sqrt{\frac{r}{8m}} \times 2\pi \times \overline{1 + \frac{a}{8r} + \frac{9a^2}{256r^2}}$ + &c.

If the arcs be ſo very ſmall, that the verſed ſine, a, may be entirely rejected with reſpect to the radius r, then the time of a *vibration* $= \sqrt{\frac{r}{8m}} \times 2\pi$ = (if $r = 2d$) $\sqrt{\frac{d}{4m}} \times 2\pi = \sqrt{\frac{d}{m}} \times \pi$ = the time of a vibration in a *cycloid* whoſe axis is d or $\frac{1}{2}r$. The time of oſcillation, therefore, in a very ſmall circular arc, is the ſame as the time of oſcillation in a cycloid, whoſe axis is equal to half the length of the cord by which the vibrating body is ſuſpended," as was otherwiſe ſhewn to be the caſe under our article CYCLOID: for in theſe very ſmall arcs, the time of deſcent down E V $= \sqrt{\frac{r}{8m}} \times \pi$. Now the time down the *chord* = the time down the *diameter* of a circle $= \sqrt{\frac{2r}{m}}$. Hence the time down *a very ſmall circular arc* : the time down the correſponding *chord* :: $\sqrt{8m} \times \pi : \sqrt{\frac{2r}{m}} :: \pi \sqrt{\frac{1}{8}} : \sqrt{2} :: \pi : 4$ :: the circumference of a circle : four of its diameters.

If the length of the arc through which the pendulum vibrates be ſuch, that all the terms of the ſeries except the

two

two firft may be rejected, then the time of a vibration = $\sqrt{\frac{r}{8m}} \times 2\pi \times \overline{1+\frac{a}{8r}}$; and, confequently, the time of a vibration in a *circular arc* (whofe radius is r) : the time of a vibration in a *cycloid* (whofe axis is $\frac{1}{2}r$) :: $1+\frac{a}{8r}$: 1. For inftance, let the arc of femi-vibration be 2° at each fide of the loweft point, then a = *verfed fine of* 2° = $\frac{6r}{10000}$ nearly; ∴ $\frac{a}{8r} = \frac{1}{13333}$. In arcs which are lefs than 2°, this difference will be ftill lefs; fo that if the whole arc through which a pendulum vibrates does not exceed 4°, it may, without any material error, be confidered as ofcillating in a *cycloid*, and confequently all its ofcillations will be performed in the *fame time*, whether it vibrate in a greater or a leffer arc *within this limit*.

Laftly, let A = the arc, whofe verfed fine is a, then (fuppofing the arc to be *very fmall*) $a = \frac{A^2}{2r}$; and if n = the number of degrees in this arc, then A : $2\pi r$ :: n : 360, or $A = \frac{2\pi r n}{360}$ ∴ $a = \frac{A^2}{2r} = \frac{2\pi^2 r n^2}{\overline{360}|^2}$ = (for π = 3.1416, &c.) $\frac{r n^2}{6565}$ very nearly; hence $\frac{a}{8r} = \frac{n^2}{52520}$; confequently the variation for N − n degrees is $\frac{N^2 - n^2}{52520}$; if therefore the pendulum keeps *true time*, when it ofcillates through n degrees, the error arifing from its vibrating through N degrees will be $\frac{N^2 - n^2}{52520}$ of *the time of one vibration*. For inftance, if a pendulum which vibrates *feconds*, keeps *true time* when it vibrates through an arc of 2°, on each fide of the vertical line; then if it vibrate through an arc of 3°, it will *lofe* $\frac{3^2 - 2^2}{52520} = \frac{5}{52520} = \frac{1}{10504}$ every fecond, or $\frac{60 \times 60 \times 24}{10504}$ = 8″.2 in a whole day. If the fame pendulum be made to vibrate through an arc of 2° 10′ on each fide of the vertical line, then it will lofe only $\frac{\overline{2.16,\&c.}|^2 - 2^2}{52520} = \frac{.665\ \&c.}{52520}$ in a fecond, or $\frac{60 \times 60 \times 24 \times .665\ \&c.}{52520}$, which is very nearly *one fecond per day*."

Nearly the fame refults will be the confequence of this ftill more fimple procefs; *viz.* "Take $\frac{3}{2}$ of the difference of the fquares of the arcs for the *feconds*, gained or loft by a decreafed or increafed arc;" thus $\overline{3^{\circ} \times 3^{\circ}} - \overline{2^{\circ} \times 2^{\circ}} \times \frac{3}{2}$ = 7.″5 nearly as before.

When two pendulums of different lengths are made to vibrate in a detached ftate, at the fame time, the times of their refpective vibrations will be in the fubduplicate ratio of their lengths: and the refpective lengths of different pendulums, vibrating at the fame time in fimilar arcs, will be in the duplicate ratio of the times of a vibration directly; or in the reciprocal duplicate ratio of the number of vibrations fo made in any one and the fame time. Let the length of a fecond's pendulum be taken in a given latitude at 39⅛ inches = r, let g be $16\frac{1}{12}$ feet, or the fpace that a heavy body falls in free fpace in the firft fecond of time, and let p = 3.1416, &c. the diameter of a circle, whofe radius is 1; then t, the time of vibration in a circular arc, being as $p\sqrt{\frac{r}{2g}}$, where p and g are conftant or given, t will be as $\sqrt{r}$, and r as t^2. Hence the length of a half-feconds pendulum will be $\frac{1}{4}r$, or $\frac{39\frac{1}{8}}{4}$ = 9.781 inches; and the length of a quarter feconds pendulum will be $\frac{1}{16}r = \frac{39\frac{1}{8}}{16}$ = 2.445 inches: and in this manner a table of fquare roots may be readily conftructed of the various lengths of a pendulum, that are required to correfpond to any number of feconds, or portions of a fecond, in each vibration.

The action of gravity, however, is not the fame in all latitudes; nor yet at all heights above the earth's furface in the fame latitude; for inftance, the vibrations are flower at the equator than at the poles, and increafe in quicknefs with the latitudes, north or fouth, as the centrifugal force decreafes, and *vice verfâ*, the force of gravity being counteracted, in a certain degree, by the centrifugal force, which increafes with the diftance from the earth's axis of rotation, and the times of a vibration being reciprocally proportional to $\sqrt{g}$.

By an experiment made on the pendulum that vibrated feconds at Paris, Richter found that at the ifland Cayenna, 4° from the equator, it required to be fhortened a line and a quarter, to perform a vibration in a fecond; and Mr. Campbell, in 173[illegible], by an experiment made in Jamaica in 18° north latitude, determined that if a feconds pendulum is 39.126 Englifh inches long, agreeably to Graham's determination, the length at the equator would be required to be 39.00, and at the pole 39.206 inches. (Phil. Tranf. N° 432.) Thefe experiments led Mr. Emerfon to compute a table for the length of the pendulum to fwing feconds for every five degrees of latitude, together with the length of a degree of latitude in each of thofe parallels, in Englifh miles, which table we fhall here fubjoin.

Degrees of Lat.	Length of a Pendulum.	Length of a Degree.
	Inches.	Miles.
0	39.027	68.723
5	39.029	68.730
10	39.032	68.750
15	39.036	68.783
20	39.044	68.830
25	39.057	68.882
30	39.070	68.950
35	38.084	69.020
40	39.097	69.097
45	39.111	69.176
50	39.126	69.256
55	39.142	69.330
60	39.158	69.401
65	39.168	69.467
70	39.177	69.522
75	39.185	69.568
80	39.191	69.601
85	39.195	69.620
90	39.197	69.628

The

The numbers in this table of courſe ſuppoſe the pendulum vibrating cloſe to the Earth's ſurface in each parallel of latitude.

Profeſſor Bridge has alſo calculated a little table, on a ſuppoſition that the length of a ſeconds pendulum is 39.2 inches in latitude 52°, which is taken as his ſtandard, and he has given the lengths correſponding not only to ſix different latitudes, but to four different heights, ſix at the Earth's ſurface, and at one, two, and three miles of elevation reſpectively. We ſhall take the liberty of giving this table alſo.

Latitude.		Length of Pendulum at the Earth's Surface.	One Mile above the Earth.	Two Miles above the Earth.	Three Miles above the Earth.
At the Equator		39.060	39.036	39.013	38.997
Lat.	30°	39.115	39.091	39.068	39.052
	45	39.170	39.146	39.122	39.107
	52	39.200	39.176	39.152	39.137
	60	39.225	39.201	39.177	39.161
At the Pole		39.281	39.257	39.233	39.218

The greateſt variation in the length of a pendulum vibrating at three miles above the equator, and on the Earth's ſurface at the pole, is only 0.284 of an inch, which affords but a ſmall ſcale for determining either altitudes or the figure of the Earth, to both which purpoſes the obſerved variations in the lengths of the pendulum have been propoſed as a meaſure.

When a pendulum of a given length has been obſerved to gain or loſe a certain quantity daily, it is convenient, in making the adjuſtment for exact time, to have ſome conciſe theorem as a guide to bring it to the point deſired, at one trial in all caſes, which theorem profeſſor Bridge has given in theſe words: "*Multiply* twice the length of the pendulum by the number of ſeconds gained or loſt, and *divide* the reſult by the number of ſeconds in a day; the *quotient* will give the number of inches, or parts of an inch, by which the pendulum is to be lengthened or ſhortened." Suppoſe the *gain* of a ſeconds pendulum to be three minutes, or 180″ in a ſolar day, then $\frac{39.2 \times 2 \times 180''}{86400''} = .163$ parts of an inch, is the quantity, in this caſe, by which the pendulum muſt be *lengthened* to meaſure mean time; but if the three minutes had been *loſs* with a half-ſeconds pendulum, then $\frac{9.8 \times 2 \times 180}{86400} = .041$ of an inch, or the fourth part of the former gain, will be the quantity by which the ſaid pendulum will require to be *ſhortened.* This rule is not only ſhort in its application, but eaſy to be remembered.

Under the title Cycloid, we have ſhewn that if l be the length of a ſeconds pendulum, and π the periphery of a circle, or 3.1416, &c. the expreſſion $\frac{\pi l}{2}$ will be the meaſure of the velocity of a falling body acquired in one ſecond of time, and alſo that the expreſſion $\frac{\pi^2 l}{2}$ will denote the ſpace fallen through in a vertical line in the ſame time; if we would apply theſe expreſſions to experiments in any given latitude, inſtead of the length of the pendulum determined in our own latitude, we muſt ſubſtitute the length given in one of the preceding tables, agreeably to the latitude propoſed, and then the reſult ariſing from ſuch length, as determined by the ſaid expreſſions, will be correſponding; and the ſcale of acceleration will be accommodated to the increaſed or decreaſed force of gravity in that particular latitude; for inſtance, agreeably to Mr. Bridge's Table, the ſpaces fallen through in one ſecond of time at the equator and at the pole will be reſpectively 192.663 and 193.753, the difference between which is only 1.09 inches; which is the *greateſt difference* that exiſts between the ſpaces fallen through in one ſecond on any two points of the Earth's ſurface.

If it ſhould be required to know what the height of a mountain is, on which a ſeconds pendulum, that meaſured mean time truly at its baſe, loſt a given quantity; let $r =$ radius of the Earth, $x =$ the height of the mountain, $T =$ the number of ſeconds in an hour, or the number of vibrations in an hour at the baſe, and $t =$ the ſeconds loſt on the mountain; then $T - t$ will be the number of vibrations in an hour at the top of the mountain. When the length of the pendulum is given, it is known that the number of vibrations in a given time are directly as the *ſquare roots of the forces,* which act upon the pendulum; hence $T : T - t :: \sqrt{\frac{1}{r^2}} : \sqrt{\frac{1}{\overline{r+x}|^2}} :: \frac{1}{r} : \frac{1}{r+x} :: r + x : r, \therefore t : T - t :: x : r$, or $x = \frac{tr}{T - t}$.

If t be *very ſmall* with reſpect to T, then x may be conſidered as equal to $\frac{tr}{T}$. Let T now be taken equal 3600″, and let t be taken ſucceſſively at 1, 2, 3, &c. ſeconds, then x will be equal to $\frac{r}{3000}, \frac{2r}{3000}, \frac{3r}{3000}$, &c.; and ſuppoſing r to be equal to 4000 miles, the heights of mountains upon which a pendulum, that vibrates ſeconds at their baſes, loſes 1, 2, 3, &c. ſeconds in an hour, will be $1\frac{1}{3}$, $2\frac{2}{3}$, $3\frac{1}{3}$, &c. miles reſpectively. Hence, if the pendulum ſhould gain t ſeconds in an hour before it is carried up the mountain, and it ſhould be required to find that point in the aſcent, where it would keep *true time,* we have $T + t : T :: r + x : r, \therefore t : T :: x : r$, or $x = \frac{tr}{T}$; that is, if we multiply the *gain* by the Earth's radius, and divide the product by the number of vibrations performed in the hour, the quotient will be the height of that part of the mountain, where the pendulum muſt be fixed to keep *true time.*

Hitherto we have conſidered the matter that conſtitutes the whole pendulum as collected into one point, and that the centres of *gravity* and of *oſcillation* coincide in that point, which cannot be ſtrictly the caſe with any ſhape of the bob, however ſlender the cord, wire, or rod of ſuſpenſion. For inſtance, if a ſmall ball of lead, or other ponderous body, were ſuſpended by a ſilken thread; ſuppoſe the ball to be divided into two equal ſemi-balls by a horizontal line, as it is ſuſpended by the vertical thread, then the lower ſemi-ball would move with a greater velocity than the upper one, though vibrating together, in conſequence of its greater radius of motion, conſequently the momentum of the lower

ſemi-ball

ſemi-ball would exceed that of the upper one, and conſidering the two ſemi-balls as again united, it is clear that the common point of *oſcillation* will be ſomewhere in the lower ſemi-ball, though the centre of gravity is in the middle of the line of diviſion, even if the ſilk thread be aſſumed as having no ſenſible weight. As we have given the methods of determining the centres both of *gravity* and of *oſcillation* of bodies of different figures, with much minuteneſs, under our article CENTER, it is not neceſſary to purſue this part of our ſubject here, but to beg of the reader to recollect that the centre of *gravity* of any ſingle pendulum is always *above* the centre of its *oſcillation*, and that the diſtance from the point of *ſuſpenſion* to the centre of *oſcillation* is always to be taken as the *true length* of the pendulum, however ſuſpended, or whatever may be the ſhape or ſize of the bob or ball. This obſervation is indeed equally true, however complex the conſtruction of the pendulum may be, for we have ſhewn, (ſee CENTER,) that whatever be the number of particles of matter compoſing a pendulum, the products of the particles lying at one ſide of the centre of oſcillation, when multiplied both by their reſpective diſtances from the point of ſuſpenſion, and alſo from the centre of oſcillation, will make an amount, always equal to the amount of the products obtained from the particles ſo multiplied, lying at the other ſide of the centre of oſcillation, without regard to ſhape or ſize. In a common ſeconds pendulum of 39.2 inches length, with a heavy lenticular bob, and a ſlender verge or rod of ſuſpenſion, the centre of *gravity* is very nearly $\frac{1}{20}$th of an inch above the centre of *oſcillation*; and in all our preceding deductions the length of the pendulum muſt be conſidered as meaſured from the point of ſuſpenſion at the vertex, to the centre of oſcillation.

But we have ſeen a pendulum ſo applied for the regulation of a portable clock, that the point of ſuſpenſion was not at the vertex, but in a point ſomewhere between the vertex and the centre of gravity; by which application, the centres of gravity and of oſcillation were farther removed from each other, than they would have been with the point of ſuſpenſion at the vertex. A pendulum of this deſcription, of a ſhort length, may be made to vibrate in times of different lengths, which will depend on the reſpective diſtances between the centres of gravity and of oſcillation; but the ſcale of adjuſtment for time being very ſhort, and liable to be altered by variable temperature, this conſtruction may be conſidered as an object of curioſity rather than of utility. If it were required, however, to find the diſtance between the centre of oſcillation and point of ſuſpenſion in any pendulum of this ſort, where the *time* of *vibration* is *given*, it may be thus determined: Let T be the time given, L = length of a ſimple or common pendulum that will vibrate in that time, $\pi = 3.1416$, $m = 16\frac{1}{12}$ feet, and F = the force of gravity conſtantly acting; then ſince $T = \sqrt{\frac{\pi^2 L}{2\, m\, F}}$, we ſhall have $L = \frac{2\, m\, F\, T^2}{\pi^2}$ for the diſtance from the point of ſuſpenſion to the centre of oſcillation. Again, in *fig.* 3, *Plate* XXXIX. of *Horology*, let A *a* be a ſolid uniform rod, and G its centre of gravity; then if A be taken as the point of ſuſpenſion in the vertex, the centre of oſcillation will be at O, at two-thirds of its length from A. But let S be put as the point of ſuſpenſion below A, then $L = SG + \frac{AG \times GO}{SG}$. Let $SG = x$, $AG = D$, $GO = d$; then $L = x + \frac{Dd}{x}$, or $x^2 - Lx + Dd = 0$, from which $x = \frac{L \pm \sqrt{L^2 - 4Dd}}{2}$; ſubſtitute $\frac{2\, m\, F\, T^2}{\pi^2}$ for L in this equation, and it determines the diſtance of the point (S) of ſuſpenſion to (G) the centre of gravity of the rod; through which point (S), if the axis of ſuſpenſion be made to paſs, the rod will vibrate in the *given time*. But as x has two values, it ſeems that there are two points at different diſtances from G, the centre of gravity; from which if the rod be ſuſpended, it will oſcillate in the given time. By ſimilar reaſoning it may be proved, that the rod will vibrate in the *leaſt poſſible time*, when S G is a mean proportional between A G and G O; likewiſe, that if *o* be the centre of oſcillation to the point of ſuſpenſion S, the point, G, of the centre of gravity will be in the middle between S and *o*; and, laſtly, that a rod ſo circumſtanced will vibrate in the leaſt time poſſible, when its point of ſuſpenſion, S, is $\frac{21}{100}$ of the whole length of the rod from its vertex A.

3. We come, in the third and laſt place, to treat of the various conſtructions of the attached pendulum, when uſed as the regulator of an horological machine, as we at firſt propoſed. It has been uſual, when treating of the pendulum, to call what we have denominated the free or *detached* pendulum, vibrating by the action of its own gravity alone, the *ſimple* pendulum; and what we call the *attached* pendulum, where external force is introduced in aid of gravity, the *compound* pendulum. But as the attached pendulum is conſtructed ſometimes with a *ſimple* rod or verge, and ſometimes with a *compound* one, we propoſe to apply theſe two terms, *ſimple* and *compound*, excluſively to the different conſtructions of the attached pendulum, agreeably to the ſignifications in which they are uſed by clockmakers.

In writing the article ESCAPEMENT, which is that portion of a clock that tranſmits the maintaining power, in a modified way, to the pendulum, we found it neceſſary to explain at ſome length, how the natural ſcale of forces, ariſing from gravity alone, is diſturbed by the acceſſion of any external force, ſuch as is derived from the maintaining power; and by a reference to *figs.* 1. and 2. of *Plate* XXXI. of *Eſcapements*, we proved that this ſcale is moſt deranged, when the external force is applied at the extremity of the arc of excurſion; and leaſt, at the loweſt point of the ſaid arc. We alſo ſhewed how to eſtimate, in any point of the arc, both the force applied at that point, and the velocity with which the pendulum is moving there; as well as determined the total arc correſponding to the maximum of velocity, and other particulars of eſſential importance both in the theory and practice of clock-making. We muſt, therefore, requeſt the reader to conſult our introduction to the article ESCAPEMENT, ſo far as it relates to the *pendulum*, before he proceeds to peruſe the remainder of our preſent article. Indeed, the different parts of a clock are ſo intimately and mutually connected, that the action of one frequently cannot be clearly explained, without at leaſt a partial explanation of the properties of the other; which is the reaſon why we have anticipated ſeveral particulars relating to the motion of the pendulum, before we arrived at our preſent article.

Of the ſimple Pendulum in Horology.—When a pendulum was firſt applied to be the regulator of a clock, the crown-wheel eſcapement was uſed with it by Huygens, and the pendulum was neceſſarily made to vibrate in long arcs; but it was ſoon found that it had at leaſt three defects: 1ſt, the maintaining power acted on the wheel, and conſequently on the pendulum, by means of the fork, almoſt conſtantly, thereby producing an alternate acceleration and retardation of motion beyond what gravity alone would produce; 2dly, the times of vibration in the longeſt arcs were found to be ſlower than in the ſhorteſt arcs, in the proportion of about 34 to 29; and, 3dly, the verge of the pendulum itſelf was found to have its total length increaſed or decreaſed by variations of atmoſpheric temperature. Hence aroſe the neceſſity of new eſcapements, and of different modifications of the pendulum, to make it act as an uniform regulator of mean time under all circumſtances. As we have had occaſion to deſcribe different pendulums, and alſo the parts connected with them, ſuch as the fork that gives it motion, the cock by which it is ſuſpended, the ſpring of ſuſpenſion, the verge, the bob, and ſcrew of adjuſtment, &c. (ſee CLOCK, and CLOCK-*Making*,) a repetition of what we have before ſaid would be ſuperfluous. We will, therefore, take it for granted that every one, that has ſeen an ordinary clock with a ſeconds pendulum, which, with Dr. Hooke's anchor eſcapement, is called the *royal pendulum*, knows its parts and conſtruction. The true length between the centre of oſcillation and axis of ſuſpenſion of any pendulum is beſt aſcertained by adjuſtment for rate, firſt, from hour to hour roughly, and then from day to day, either by the going of a good clock already adjuſted, or by a tranſit of one of the heavenly bodies. (See CHRONOMETER, prob. vii, viii, and ix.) But from what we have ſaid under our article ESCAPEMENT, it will be ſeen that the length of any pendulum will require to be a trifle more or leſs, accordingly as the *external impulſe*, derived from the maintaining power, is great or ſmall, permanent or temporary, gradual or inſtantaneous; or, which is to the ſame effect, accordingly as the eſcapement has a *recoil*, is of the *dead-beat* ſort, is *detached*, or has a *remontoire*. When, therefore, we ſpeak of the acting length of a pendulum, we muſt always have a reference to both the quantity and quality of the auxiliary force that perpetuates its motion. In ordinary caſes, an additional force applied in aid of gravity accelerates the pendulum, and requires a compenſation by adjuſtment, as will be ſeen preſently, for additional length. This adjuſtment was made by Huygens, by a ſmall ball moving along the verge above the bob, but is generally effected by a tapped nut at the inferior end. This method is, however, inconvenient, inaſmuch as that it requires the pendulum to be ſtopped; for which reaſon, in many regulators, or aſtronomical clocks, the adjuſtment for rate is made at the ſuperior end, above the cock. For the ſhape and ſize of the bob of a given pendulum, we muſt refer to our term BOB, in its proper place, where we have entered into a minute detail of the particulars relating to its ſhape and ſize, with a given maintaining power and eſcapement, both which muſt be conſidered at the ſame time; but further experiments are perhaps wanting, in this department of horology, to determine the beſt poſſible combination of all the requiſite agents in a moving pendulum, including the ſuſpenſion. The metal of which the verge is uſually compoſed is iron or ſteel; but their expanſion by heat, and contraction in cold, render them improper ſubſtances for an accurate pendulum: for, in the extremes of temperature, the rate will alter by an iron pendulum more than 20″ in the 24 hours. Straight-grained deal or ſapadillo, well ſeaſoned, is better for the verge than metal; though theſe, and indeed all ſorts of wood, are ſubject to changes, from moiſture, if not from heat, and therefore cannot be entirely depended on, in all ſeaſons of the year, where great accuracy is required.

From what we have here had occaſion to ſtate reſpecting the motion of an attached pendulum, it will be evident to our readers, that our deductions with reſpect to the exact length of the *detached* one, where the *accelerative force* was conſidered conſtant and uniform, will not be ſtrictly accurate when that force is *variable*, which it is more or leſs in its *attached* ſtate: we will therefore conclude our remarks on the *ſimple* attached pendulum, by ſhewing in what degree the length of ſuch pendulum is affected by any accelerative force. Whatever this force (F) may be, the time of a body's falling down half the length of the pendulum, by the laws of acceleration, will be $\left(= \sqrt{\frac{S}{m\,F}}\right) = \sqrt{\frac{L}{2\,m\,F}}$, where $m = 16\frac{1}{2}$ feet, and L = the length of the pendulum, and $\pi = 3.1416$; therefore T . $\sqrt{\frac{L}{2\,m\,F}}$ :: π : 1, or T $= \sqrt{\frac{\pi^2 L}{2\,m\,F}}$, which *varies* as $\sqrt{\frac{L}{F}}$. Conſequently, the times of vibration of pendulums of *different lengths*, acted upon by *different accelerative forces*, will vary as the *ſquare roots* of the *lengths* directly, and the *ſquare roots* of the *forces* inverſely. If L be given, then $T \propto \frac{1}{\sqrt{F}}$; that is, the times of vibration of pendulums of the *ſame length* vary inverſely as the *ſquare roots* of the *accelerative forces;* and if T be given, then $L \propto F$; namely, the lengths of pendulums vibrating in the *ſame time* vary as the *forces* which *accelerate* them.

The accelerative force, added to the natural force of gravity in a clock, is uſually called the *maintaining power*, and may be derived either from an appended weight immediately through the eſcapement, or from a ſpring or ſprings wound up by a weight at ſhort intervals; and in either caſe the quantity and mode of its action will be limited by the nature of the eſcapement adopted; and frequently the ſpring of ſuſpenſion is ſo tapered upwards or downwards, as to become a modifier of the maintaining power, in conjunction with the eſcapement, all which circumſtances muſt be taken into the account when the nature and quantity of the extraneous force is eſtimated.

Before we conclude our account of the ſimple pendulum, we will juſt mention that ſuch pendulum vibrating in a given time, in a given ſituation on the Earth's ſurface, has been propoſed as a perpetual ſtandard of linear meaſure. Hatton, we underſtand, firſt propoſed this ſubject for the conſideration of the Society of Arts at the Adelphi buildings, (vol. i.) and Mr. Whitehurſt took up the ſubject after him, and contrived a method of aſcertaining the difference between the lengths of a ſeconds and half-ſeconds pendulum, by means of an apparatus, which afterwards came into the poſſeſſion of Dr. George Fordyce, who improved it, by adding a compenſation to the pendulum, and gave an account of it, with drawings, in the Philoſophical Tranſactions of London, of the year 1794.

We may alſo mention here, that Huygens, in his "Horologium Oſcillatorium," propoſed a *circular* or *circulating pendulum*, to move in the concave ſide of a parabolic conoid, the circuits of which would be performed, he ſays, in equal times, when the mechaniſm is properly adjuſted; but it would lead us beyond our mark in this place, to purſue this part of the ſubject further. See Horol. Oſcil.

Of compound or compenſating Pendulums.—If any ſubſtance, ſufficiently long for a pendulum, could be found in nature, that has not its dimenſions enlarged or diminiſhed by heat or cold, ſuch ſubſtance would be the moſt ſuitable for a ſimple attached pendulum; but all attempts to diſcover ſuch a ſubſtance have hitherto been ineffectual. (See EXPANSION.) Hence different contrivances have been deviſed by ingenious men,

men, to counteract the effects of variable temperature in the pendulum, and ſome of them have ſucceeded in effecting this deſirable purpoſe. So long ago as the year 1648, the different expanſibilities of various metals was known by Wendelin to exiſt, and in 1715 that ingenious artiſt George Graham made ſeveral experiments with the pyrometer, invented by Muſſchenbroek, to aſcertain the relative quantities of expanſions of different metals, with a view of availing himſelf of the difference of the expanſions of ſome two or more metals, when oppoſed to each other, in the conſtruction of a compenſating pendulum, but the difference which he detected, between iron and braſs, for inſtance, was ſo ſmall, that he relinquiſhed all hope of being able to accompliſh his object in this way, and therefore gave up the project. (Phil. Tranſ. London, 1726.) However, it occurred to the ſame artiſt, that mercury was affected by changes of heat and cold ſufficiently to anſwer the purpoſe of keeping the centre of oſcillation of a pendulum always equi-diſtant from the point of ſuſpenſion, provided the mercury could be made to aſcend while the verge of the pendulum deſcended, or elongated, by any degree of heat; and *vice verſâ*. In 1721 the trial was made in a pendulum, which we ſhall preſently deſcribe, and in which the compenſation, after ſome adjuſtment, was found ſo complete, that its error in the extremes of temperature was reduced to $\frac{1}{8}$th of the quantity obſervable in an ordinary ſimple pendulum. In the mean time, the idea of arranging metallic bars of different expanſibilities in ſuch a manner, as to conſtitute a good compenſating pendulum, being once ſuggeſted by Graham's paper on the ſubject, rouſed the ingenuity of different mechanics, and Harriſon, a carpenter of Barton in Lincolnſhire, in the year 1726, ſucceeded in completing the requiſite arrangement. His rods of metal were placed in ſuch parallel ſituations as to reſemble thoſe of a grid-iron, from which circumſtance the pendulum, ſo conſtructed, has been denominated the *grid-iron pendulum*. Since Harriſon's time, there have been various modifications of the compenſating pendulums, both in England and France, the principal of which we ſhall deſcribe in their order of time, but none of them have been found to excel the prototypes of Graham and Harriſon, which are ſtill uſed, one or other of them, in the principal obſervatories. If any improvement has been made on the original pendulums, it is in the external appearance, and in the eaſier modes of adjuſtment.

Graham's Mercurial Pendulum.—Graham's mercurial pendulum, which may be conſidered as the firſt compenſating pendulum, is repreſented in *fig.* 4, of *Plate* XXXIX. of *Horology*; A B is the verge conſiſting of a rod of ſteel, carrying at its lower extremity a ſtirrup, B C, of the ſame metal, holding the glaſs cylindrical veſſel, E F, on a circular plate, ſurrounded by a rim, to keep the veſſel from falling out; this veſſel was filled with pure mercury up to the letter F, near the top; and above there was ſpace enough to receive more, in caſe the compenſation ſhould not prove on trial to be ſufficient. The ſliding piece of metal, D, was adjuſtable to any part of the verge, for fixing the rate of going, and the compenſation could be made more or leſs by adding to or deducting from the mercury. It does not appear certain, whether the glaſs veſſel was choſen by conjecture, as to its ſize and ſhape, or was calculated previouſly, from a conſideration of the relative expanſibility and abſolute ſpecific gravity of mercury; but we are diſpoſed to think, that it muſt have been from calculation, becauſe it very much reſembles the veſſel calculated by Mr. Lowe of Iſlington, for the mercurial pendulum recently made for the Royal Obſervatory of Greenwich, by Hardy. In this pendulum, as now made by Hardy, and others, the lower end of the verge is tapped, and deſcends through the top of the ſtirrup, where a milled and divided nut ſcrews on, by means of which the adjuſtment for rate is made, without the ſliding piece D, which is now rejected. One peculiarity of this pendulum is, that the mercury which is uſed as the means of compenſation, conſtitutes alſo the weight of the bob. We are not made acquainted with the quantity of mercury uſed by Graham; nor yet with the particulars of Mr. Lowe's calculation; but the veſſel which the latter determined, as the moſt ſuitable, was cylindrical, ſeven inches and a half long within, and two inches in diameter, excluſively of the thickneſs of the glaſs, which was one-eighth of an inch; the capacity propoſed was for twelve pounds, but by eſtimating the ſolid contents of ſuch a veſſel, and trying its bulk of mercury, from a table of ſpecific gravities, we find that it will not hold more than ten pounds of this fluid, if ſo much. The depth to which Hardy's veſſel was propoſed to be filled was $6\frac{7}{15}$ inches, but on trial at the Royal Obſervatory, it was found neceſſary to add $\frac{1}{15}$th more, and $6\frac{8}{15}$ inches are now found to compenſate the ſteel verge ſo admirably, that the greateſt deviation in the rate, during winter and ſummer, has never yet exceeded $\frac{4}{15}$ths of a ſecond *per* day. The eſcapement that acts with Hardy's pendulum is of the remontoire ſort, acting almoſt conſtantly on the pedulum, and therefore requiring it to be ſomewhat longer than an exact ſtandard.

Harriſon's Grid-iron Pendulum.—Though we have given a general deſcription of Harriſon's grid-iron pendulum, as made by Reid of Edinburgh, under the article Clock, (ſee *Aſtronomical Clock by Mr. Thos. Reid of Edinburgh*) to which we refer our readers, as well as to *Plate* XXII. of *Horology*, where it is repreſented; yet we feel it due to that ingenious artiſt, to reſume our obſervations on his contrivance in this place. When Harriſon had determined the relative expanſions of ſteel and braſs to be very nearly as 3 to 5, it is probable that he reaſoned on the ſubject in ſome ſuch way as this; let a rod of ſteel be ſuſpended by its upper end on a fixed point, and let it hang freely in a vertical poſition, then to the lower end rivet the lower end of a braſs rod, ſo as to aſcend parallel and contiguous to the iron rod, while it is ſupported by the rivet; if the iron rod be divided into five equal portions, and the braſs rod be made juſt equal to three of thoſe portions, what will be the conſequence? Why the tops of the two rods will always preſerve the ſame diſtance from each other, becauſe the downward expanſion of the longer iron rod will always be equal to the upward expanſion of the ſhorter braſs rod, conſequently, if the vertex of the iron rod be made ſtationary, by being fixed to a wall or otherwiſe, the vertex of the braſs rod will alſo be ſtationary as it reſpects the other. Let now the vertex of the iron rod be aſſumed as the point of a pendulum's ſuſpenſion, and let the vertex of the braſs rod be the place for the centre of oſcillation, then the centre of oſcillation of a ſingle rod, or of any number of ſimilar rods, of uniform metal, is at $\frac{2}{3}$ds of the length from the point of ſuſpenſion; but the common centre of oſcillation will be ſtill lower in a compound pendulum that has another rod, not of the ſame length attached to its lower end; therefore two rods ſo united cannot produce a proper compenſation, and at the ſame time have the immoveable point at the common centre of oſcillation. But the rods may be cut into any number of lengths that the due arrangement may require; we have ſeen that if the long iron rod be cut in two between the ſecond and third portion, we ſhall have three lengths, one of iron $\frac{2}{5}$ths of the long rod, one of braſs $\frac{3}{5}$ths, and another of iron $\frac{3}{5}$ths, or equal to the one of braſs; in every arrangement of the various lengths to form a compound pendulum, the

the rods of leaſt expanſion muſt neceſſarily expand downwards, and thoſe of greater expanſion muſt expand upwards, becauſe theſe rods will be fewer in number, and the difference of the total expanſions muſt compenſate the laſt rod of the combination, by which the ball is ſuſpended; accordingly the length of the verge muſt be ²ths of the longer rod, and the other two each $\frac{3}{5}$ths; but $\frac{3}{5}$ths is $\frac{1}{3}$d longer than the required pendulum; this arrangement therefore in a ſeconds· pendulum is inconvenient, making the pendulum too long: let the iron rod then have three lengths and the braſs rod two, ſo that there may be two pairs of equal length and one to ſpare for the verge, to be compenſated by the ſurplus of expanſion of the braſs over the iron equal pairs of rods; this diviſion in due proportions will make a pendulum of the requiſite length, and will alſo preſerve the centre of oſcillation unchanged as it regards the axis of ſuſpenſion; but there is yet an inconvenience to be ſurmounted; the two pairs of alternate iron and braſs rods are to be united at both their extremities, each to the next contiguous rod, firſt below and then above, till they are all attached to each other, and the ball of the pendulum muſt in this caſe be ſuſpended by the *extreme* rod, which will therefore make the vertical line fall diagonally acroſs the parallel rods, and pull them to a diſtance from each other; but to prevent this evil the ball muſt hang from the *middle rod*, and alſo the rods muſt be braced to prevent their flying out; theſe requiſites may be effected by employing two pairs more of ſimilar iron and braſs rods, at the oppoſite ſide of the fifth or ſpare iron rod, that carries the ball, for theſe additional pairs of bars will not otherwiſe affect the compenſation, than do the ſame thing over again which the others did without them; and they will thus conſtitute a counterpoiſe by their weight, afford the means of a longer horizontal bracing piece, and preſerve the coincidence of the vertical line with the middle rod at the quieſcent poſition. Thus *nine* rods became neceſſary to conſtitute the grid-iron pendulum; and when they are well proportioned and nicely fitted into the holes of the connecting pieces, ſo as not to touch one another, nor to have friction in the holes, the pendulum completely anſwers its purpoſe in all variations of temperature, and may be brought to time, like the ſimple pendulum, by a ſcrew of adjuſtment either at the cock of ſuſpenſion under the ball, or at the point of oſcillation. But the proportions of the alternate rods may not produce the exact compenſation; that is, the pendulum may be either over-compenſated or under-compenſated; the adjuſtment for compenſation, when that is neceſſary, is uſually made by altering the poſition of one of the horizontal connecting pieces of braſs, by taking out the pins and inſerting them again into other holes, drilled in ſome of the bars to anſwer this purpoſe. This adjuſtment, however, is troubleſome, as it cannot be made without detaching the pendulum, and putting it probably out of its rate of going; and this inconvenience conſtitutes one of the principal objections to this conſtruction of the pendulum, for which there appears to be no remedy yet deviſed. For the conſtruction of a grid-iron pendulum rods, drawn into wire, are preferable to ſquare rods, both on account of their having leſs friction, and of their being more expoſed to the air, when not quite in contact. In Harriſon's pamphlet, publiſhed in 1763, it is ſtated that one of his clocks did not deviate from its rate, as meaſured by his grid-iron pendulum, more than one ſecond in a month, for ten years together; and Berthoud has expreſſed his regret that all the particulars of Harriſon's conſtruction of a clock, have not been divulged, that the public may benefit from his contrivances. It ſhould ſeem that notwithſtanding the objections that have been made to the cycloidal cheeks, Harriſon, while unacquainted with Huygens's invention, introduced the uſe of them into at leaſt one of his clocks. But their excellence did not conſiſt of this addition. About the year 1798 the compoſer of this article ſaw, in a back lane at Lincoln, and purchaſed one of Harriſon's clocks, which he recollects had wheels and pinions of hard wood; and that the teeth of the pinions were cylindrical rollers turning looſely on as many pins of ſteel: but he could not examine the eſcapement. The old woman, who was the poſſeſſor, was a lineal deſcendent from Harriſon, and ſome of her wiſe neighbours perſuaded her, that if ſhe parted with Harriſon's clock, his *ghoſt would haunt her*, which ſuggeſtion made ſuch an impreſſion on the good woman's ſpirits, that ſhe begged ſhe might be allowed to retain her clock; which requeſt was accordingly granted.

Regnauld's Pendulum.—In the year 1733, Regnauld, a clock-maker of Chalons in France, contrived a metallic pendulum of braſs and ſteel, the expanſions of which were oppoſed to each other in ſuch a way, that they conſtituted a *partial* compenſation. This pendulum is repreſented by *fig.* 5. of the ſame plate that contains Graham's mercurial pendulum, and probably derived its origin from the experiments mentioned in his paper on the Philoſophical Tranſactions of London of the year 1726. The pendulum in queſtion was made before Harriſon's account was publiſhed; it was compoſed of three parallel rods, two of ſteel and one of braſs; the middle rod was of ſteel, and carried the horizontal croſs-piece E A attached to its lower extremity; one end of this croſs-piece ſupported the braſs rod A B, to the top of which, at B, was rivetted the horizontal pin which paſſes through an oblong hole in the middle rod; the third rod, D E, which was of ſteel, was ſuſpended by the pin B, and carried the bob L, attached to it by a tapped nut of adjuſtment for rate beneath: the whole pendulum muſt have been ſuſpended by the top of the middle rod. The action is thus; while the middle rod elongates downward by increaſed heat, the braſs rod of greater expanſibility, A B, elongates upwards, and carries with it the pin B, by which the third rod D E is ſuſpended; and this rod paſſing through the croſs-piece E A, again elongates as much as its fellow, or rod A B of the ſame metal; therefore the compenſation is effected on a ſuppoſition that one rod of braſs will compenſate two of ſteel, or that braſs expands double the quantity of ſteel with the ſame degree of heat, and alſo contracts in the ſame ratio, neither of which aſſumptions is true. The relative expanſions being in fact as 17 to 10, according to Berthoud, or, according to others, as 113 to 68, or very nearly as 5 to 3.

Deparcieux's Improvement of Regnauld's Pendulum—In the year 1739, Deparcieux, a teacher of mathematics, and a diſtinguiſhed member of the Academy of Sciences at Paris, examined the conſtruction of Regnauld's pendulum, and propoſed an improvement in its conſtruction, which rendered it capable of adjuſtment for compenſation as well as for time. This pendulum is given in *fig.* 6. of the ſame plate, and may be thus deſcribed: A R P and E F are two bars of ſteel paſſing through the perforated piece of metal S T, which forms a clamping-piece to hold them in their places; C D is a bar of braſs reſting with its inferior end on the toe, F, of the bar E F, ſo as to be ſupported by it. Q R is another clamping-piece, faſt either at Q or R to one of the ſteel bars, but ſo adjuſted, that the other two may not be reſtrained from paſſing without friction whenever they elongate. Through the clamping-piece, S T, paſſes a ſlit lever reſting at the end S, as a centre of motion, in a notch, or otherwiſe made moveable round a fixed

 pin,

pin, in the clamping-piece ; this lever has a triangular point refting on the top of the brafs bar C D in fuch a way, that it is at liberty to afcend with the bar whenever it elongates ; at the other end of the faid lever the rod A R is fufpended, which therefore has more or lefs motion upwards produced by the expanfion of C D, accordingly as the triangular point on the lever is nearer to S, the centre of motion, or T, the point of the third bar's fupenfion : for when the rod E F elongates, the rod of brafs C D elongates more in the proportion of 17 to 10, and the furplus 7 acts on bar A R, and, by the aid of the lever, compenfates it.

This pendulum was afterwards ftill further improved by Berthoud, who made the triangular point of bearing in the lever adjuftable for diftance from the centre of motion, by a fcrew that acted on it, and altered its fituation a little. (See Effai fur l'Horlogerie, tome ii. p. 123.) Berthoud afcertained the relative expanfions of feveral different metals by a pyrometer that indicated the $\frac{1}{360}$th part of a French line, and according to his experiments, made the 17th of March 1760, when he was contriving the alteration in Deparcieux's improved pendulum, the following Table was conftructed ; in which the verge was taken at 3 feet 2 inches and 5 lines French, or 461 lines, in each experiment.

Berthoud's Table of Metallic Expanfions.

A verge of foft fteel - - -	$\frac{69}{360}$ of a line.
——— of fteel hammered cold -	$\frac{74}{360}$ of ditto.
——— of fpring-tempered fteel -	$\frac{77}{360}$ of ditto.
——— of foft iron - -	$\frac{75}{360}$ of ditto.
——— of hardened iron - -	$\frac{78}{360}$ of ditto.
——— of copper - - -	$\frac{107}{360}$ of ditto.
——— of brafs - - -	$\frac{121}{360}$ of ditto.
——— of tin - - -	$\frac{160}{360}$ of ditto.
——— of lead - - -	$\frac{193}{360}$ of ditto.
——— of glafs - - -	$\frac{62}{60}$ of ditto.
——— of filver - - -	$\frac{119}{360}$ of ditto.
——— of gold foftened - -	$\frac{82}{360}$ of ditto.
——— of gold in hard wire - -	$\frac{94}{360}$ of ditto.
——— of mercury - -	$\frac{1235}{360}$ of ditto.

In this table the whole fraction of each metal gives its abfolute expanfion, and the numerators the relative expanfions, as the denominator is common to all. Thefe experiments were firft made with fimple rods, and again with 35 pounds fufpended to each, without any variation in the expanfions, though it appeared that the fubfequent contractions were not exactly correfponding when the weight was fufpended and when there was no weight ; and it is to be regretted, that the author did not repeat his obfervations on the comparative contractions under different circumftances, as he has done the comparative expanfions, in order that the refults might have been afcertained as they relate to the pendulum's verge ftretched by a weight ; for in all the experiments we have reported under our article EXPANSION, the fubfequent contractions are confidered as being exactly proportional to the expanfions in the fame degrees of oppofite temperature. In compofing a pendulum of rods of any two of thefe metals, it is perhaps unneceffary to remark, that the rods of each metal muft have their total lengths in an inverfe ratio of their expanfibilities, unlefs indeed fome mechanical advantage by a lever, or otherwife, be given in aid of the difference of the abfolute expanfions, as is done in the inftance before us.

Deparcieux's Compenfation of a Simple Pendulum.—In the fame year in which Deparcieux contrived an improvement of Regnauld's pendulum, he invented a compound compenfation frame to be fubftituted for the common cock of fufpenfion, without altering the fimple pendulum, which mechanifm comes next to be defcribed.

This contrivance is reprefented by *fig.* 7, of our laft named plate, where A E D F is a fteel bar bent into the fhape of a two-pronged fork with the ends pointing upwards, and fixed to a folid wall at the points A and F. G E I H is another fork, formed fmaller, of brafs, and is inverted, or made to ftand with its ends refting on the horizontal part, B D, of the outer, or fteel fork. The horizontal, or upper part, E I, of the brafs fork, conftitutes the cock of fufpenfion, and has a fquared hole in the middle, which juft admits the tapped wire K, from which the fufpenfion-fpring depends ; in order that the nut of this tapped wire may give the adjuftment for time. Underneath this crofs-piece of the fork of brafs is a detached cock, not fhewn, near C, which has a flit cut through it in the direction of the vertical line of the fpring of fufpenfion, through which the fpring paffes without play, and from the under fide of which the length of the pendulum is reckoned. This cock of limitation is fcrewed to the frame of the wheel-work, and therefore remains unmoved. It is eafy to fee, in this difpofition of the bars of fteel and copper, that if the expanfions of thofe two metals be as 10 to 17, the length of the fteel verge of the pendulum added to the length, A B, of the outer fork, muft be to E G, the length of the inner fork above, as 17 to 10, to make the compenfation complete ; and accordingly the length of the brafs fork was made $54\frac{5}{7}$ French inches, the fum of the other two being very nearly 93.

About the fame time that this pendulum was made, Caffini propofed fome different conftructions, fome with three, and fome with five rods, fteel and brafs alternately, in the pendulum itfelf, which he gave in a memoir in the Hiftory of the Academy of Sciences at Paris in the year 1741.

Compenfating Tube and Rod of Julien le Roy, applied to the Simple Pendulum.—According to Thiout, Julien le Roy contrived a mode of compenfating the fimple pendulum in the fame year as Deparcieux, above ftated, and on the fame principle, though the compenfating mechanifm of Le Roy was altogether above the pendulum. This contrivance, exhibited in *fig.* 8, of the fame plate, confifts of a vertical brafs tube, 42 French inches long, denoted by the letters *a b c d*, refting at its lower end on A B, the cock of fufpenfion, which is alfo the cock of limitation of the pendulum's length, by having a flit made in it juft large enough to admit the fpring of fufpenfion without play, which fpring hangs from the lower end of a fteel rod, attached to the top part or cap, *a b*, of the brafs tube; therefore, in this conftruction alfo, the fum of the fteel rods, *i. e.* of the verge of the pendulum, including its fpring, and of that included in the tube, fhould be to the tube alone as 17 to 10. Thiout has given the length of the tube only at 42 inches, but Berthoud has remarked, that it either is, or ought to be, $54\frac{5}{7}$, like the brafs fork of Deparcieux. A pendulum of Le Roy's conftruction was fpoken well of in Caffini's Memoir of 1741, and was adopted in the obfervatory of Cluny about 1748, after which time the clock, to which it was applied, was fpoken of with great eulogium.

The principal objections that, we conceive, apply to either of thefe two laft pendulums, are, that the fpring of fufpenfion is conftantly altering its length and ftrength; that the flit in the cock of limitation muft have the effect of of lengthening the pendulum in proportion to the play that the fpring has in it, to the right and left of the vertical line, as the pendulum is vibrating; and that the pendulum rod in motion will cool fooner than the compenfating parts remaining at reft.

Ellicott's Compenfating Pendulum.—We have ftated under our article EXPANSION, that Ellicott, of London, made very accurate experiments on the relative expanfions of feven different metals, which, on a comparifon, we find to differ, more or lefs, from Berthoud's table, as well as from the determinations of Smeaton, Ramfden, Julien le Roy, Borda, and others; fo that when a compenfation pendulum is conftructed of any defcription, it feems proper, that the relative expanfions of the metal ufed fhould not be taken from any table of refults, but from an examination of the identical materials of which the pendulum is to be conftructed, and that after tempering, hammering, filing, polifhing, &c. in order to gain the due proportions in their ftate as prepared for ufe. Ellicott, we doubt not, ufed this precaution, and his experiments were made, moft probably, with a view to the conftruction of the pendulum we are now going to defcribe. It had been objected to Harrifon's grid-iron pendulum, that the adjuftment of the rods for temperature was inconvenient, and accompanied by a confiderable change in the rate, and that the expanfion of the ball fupported at its lower edge was all in an upward direction, to avoid which objections, as well as the fuppofed jerks during the elongation, was probably Ellicott's object, at the time of his contriving a new pendulum: whether or not he had feen the application of the lever in Deparcieux's pendulum is not certain, but his contrivance is nearly allied to that, the principal difference being that in Ellicott's pendulum two levers are adopted inftead of one, and are applied at the bob inftead of the fuperior end of the verge. In defcribing Ellicott's pendulum we fhall avail ourfelves of the inventor's own account, and then fubjoin our own remarks. *Fig.* 1, of *Plate* XL. of *Horology*, reprefents the pendulum; in which *a b* is a bar of brafs made quite faft at the upper end by pins, and held contiguous at feveral equal diftances, by the fcrews 1, 2, &c. to the rod of the pendulum, which is a bar of iron; and, fo far as the brafs bar reaches, is filed of the fame fize and fhape, though it does not appear fo in the figure, but, a little below the end of the brafs, the iron is left broader, as at *d d*, for the conveniency of fixing the work to it, and is made of a fufficient length to pafs quite through the ball of the pendulum to *c*. The holes 1, 2, &c. in the brafs, through which the fcrews pafs into the iron rod of the pendulum, are filed, as in the drawing, of a fufficient length to fuffer the brafs to contract and dilate freely by heat and cold under the heads of the fcrews: *e e e e* reprefent the ball of the pendulum; *f, f,* two ftrong pieces of fteel, or levers, whofe inner centres, or pivots, turn in two holes drilled in the broad part of the pendulum rod, and their outer ones in a ftrong bridge, or cock, fcrewed upon the fame part of the rod, but omitted in the figure, to fhew the mechanifm: *g, g,* are two fcrews entering at the edges, and reaching into the cavity near the centre of the ball. The ends of the fcrews next the centre are turned into the form feen in the figure, which, preffing with the weight of the ball againft the longer arms of the levers, caufe the fhort ends to prefs againft the brafs bar at *b*. Things being in this fituation, let us fuppofe that the rod of the pendulum, and the brafs annexed to it grow longer by heat, and that the brafs lengthens more than the iron of the fame length; then the brafs, by the excefs of its dilation, will prefs the fhort ends of the levers downwards at *b*, and at the fame time neceffarily lift up the ball, which refts on the long ends of the fame lever, at *f f*, to any proportion neceffary for due compenfations. The calculated proportion of the fhort arms of the levers muft be to the long ones, as the excefs in the expanfion of brafs is to that of the whole length of iron; and if this calculation be found on trial not perfectly accurate, the fide fcrews *g, g*, will produce the exact compenfation at a few adjuftments. At *i k*, the inferior end of the iron verge, a ftrong double fpring is fixed to bear the major part of the ball's weight, by its preffure upwards againft two points of the ball, equidiftant from the vertical line. This is the defcription of the firft pendulum made in this way, and the only alteration that the inventor made afterwards was placing the fide fcrews in the body of the ball.

Cumming, after defcribing this pendulum in his Elements of Clock and Watch-work, propofes that the brafs bar fhould elongate between two iron bars, in order to keep it ftraight, and to prevent the jerks to which he conceives Ellicott's conftruction is ftill liable, from friction impeding the free elongation of the brafs bar, and from a confequent tendency to curvature before each jerk: he has alfo propofed to alter the ftructure of the fhort arms of the levers, by making them turn each on an axis of motion, and to change their places of preffure mutually, in order that their united action may be applied in the fame point, at the centre of the brafs bar, in direct oppofition to the line of downward expanfion; and we entertain no doubt but that the propofed alterations, once adopted, would be found to be real improvements in Ellicott's pendulum. See Cumming's Elements, page 106, and plate xi.

Rhomboidal Pendulum.—Mr. Hooke propofed to make a compenfating pendulum of rods of fteel and brafs arranged into the form of a rhombus, or rhomboid, the four fides of which were of fteel and the long diagonal of brafs: this conftruction, requiring the pendulum to be broader than long, was found to be very inconvenient, and was therefore never applied to a clock. But the principle was afterwards taken up by a clock-maker at Ulverfton in Lancafhire, named Williamfon, or Williams, who, about the time that Ellicott conftructed his pendulum, made feveral pendulums, not however of one rhombus each, but of five united by joints at the angular points, oppofed to the diagonals of brafs; fo that as the brafs elongated more than the fteel, the angles opened and kept the diagonals always at the fame diftance from the faid angular points, and from each other. Mr. Edward Troughton alfo, without knowing what had been done by Williamfon in this way, contrived and made a pendulum fimilar to his in all refpects, except that feven rhombi were adopted inftead of five, which narrowed the pendulum ftill more than Williamfon's conftruction, and which anfwered its intended purpofe very well. We believe this conftruction is very little known at the prefent day, and, therefore, we beg leave to give a fketch of its arrangement in *fig* 2. of *Plate* XL., which will be fufficiently underftood from infpection. We have not calculated the proportions neceffary for the fides and common bafes of the triangles which compofe thefe rhombi, fo that the compenfation may be complete; but any one competent to conftruct this pendulum will eafily afcertain by calculation the due proportions.

Compenfating Pendulum of F. Berthoud by Frames.—When Berthoud had finifhed his experiments on the relative expanfions of metals in 1760, which was three years before Harrifon had publifhed his pamphlet relating to his pendulum, he contrived a compound pendulum, much refembling Harrifon's grid-iron at the firft appearance, and which is conftructed to anfwer the fame purpofe. He had previoufly noticed

noticed that Deparcieux's improvement of Regnauld's pendulum, and alſo his own, where the lever was introduced, were not calculated to ſupport a heavy pendulum with ſteadineſs; he therefore employed his mind in contriving one, by which any weight might be ſuſpended, that might be found deſirable to move in ſmall arcs. This pendulum, or rather a ſketch of it, is contained in *fig.* 3. of *Plate* XL., in which the oblong frame, ſurrounding the other parts, and denoted by the letters *a, b, c, d,* is of ſteel; the upper end of which carries the knife edge ſuſpenſion at *s*; *e f g h* is a ſecond frame, open at the top, and made of braſs; the horizontal portion, *e f*, reſts on the interior part of the lower end of the ſteel frame above the ball; the upper ends do not reach the whole length of the ſurrounding frame, but allow another frame of ſteel, *i n m n*, to reſt on them, by means of their chin-pieces, at their upper extremities; again, *o* and *p* are two rods of braſs reſting on the inferior part of the horizontal connecting piece, *i n*, of the ſecond ſteel frame, and at their upper ends ſupport a ſingle ſteel rod, with a chin at each ſide, which rod deſcends in the middle of the pendulum, and holds on its lower end the ball, after it has paſſed through holes made in the three frames above deſcribed. In this conſtruction, as in Harriſon's, it is eaſy to ſee, that the difference of the aſcending expanſions of the braſs frame and rods, and of the deſcending expanſions of the ſteel frames and middle rod, conſtitutes the compenſation for the effects of variable temperature; and if the ſum of the lengths of ſteel are to the ſum of the lengths of braſs inverſely as their expanſions, the compenſation will be perfect.

For the particulars relating to the ſuſpenſion, and metallic thermometer of a pendulum, made agreeably to this arrangement, we beg to refer our readers to Berthoud's own account, and to the correſponding engravings. See fig. 1. of plate 28. of Eſſai ſur l'Horlogerie, and fig. 1. of plate 10. Hiſtoire de la Meſure du Temps.

Troughton's Mercurial Pendulum.—When Graham's mercurial pendulum had been many years in uſe, and had performed in the moſt ſatisfactory manner, ſome prejudices were without reaſon raiſed againſt it, originating more in fancy than from experience, and Mr. Edward Troughton, about the year 1790, with a view of doing away the moſt material prejudice, contrived a modification of the mercurial pendulum, which may be conſidered as a diſtinct pendulum, inaſmuch as that the materials are differently arranged both as to quantity, quality, and ſituation. An idea prevailed that the metallic rod of Graham, and the veſſel of mercury, would not be affected by changes of temperature in a contemporaneous manner, but that the mercury being below the rod, and having on that account a quicker motion, would *cool* ſooner than the metallic rod, and be moreover liable to more frequent changes from its greater ſuſceptibility, notwithſtanding the teſtimony of a long trial to the contrary; Mr. E. Troughton therefore ſubſtituted a ſtrong glaſs tube with a bulb at the lower extremity, to be filled to a certain height with mercury, which ſhould riſe and fall preciſely as in a thermometer, which inſtrument, in fact, his tube is; and to make it uſeful as ſuch, it carries a graduated ſcale. The tube in queſtion is about the ſize of a barometer tube, and the bulb large enough to contain 45 ounces of pure mercury, by the aid of one-half the tube. *Fig.* 4. of *Plate* XL. will explain Mr. Troughton's arrangement more clearly than a mere verbal deſcription. A B is the mercurial rod filled with mercury to the *middle point* of the rod; at A the ſuſpenſion is ſimilar to that deſcribed in Meſſrs. Brockbank's aſtronomical clock, (ſee CLOCK,) where Troughton's tubular pendulum is deſcribed, with a reference to *Plates* XXIII. and XXVII.; C D is the bulb guarded by the ſurrounding lenticular bob of braſs and lead made in the uſual way, except that it is made hollow at the centre; the metallic parts of the pendulum weigh about nine pounds excluſively of the mercury; the rim that ſurrounds the bulb, and by which the weight of the lenticular bob is ſupported, projects into two notches cut in the interior edges of the metallic lens, at each ſide of the cavity, and the nice fitting of the bulb to its rim, renders the glaſs capable of bearing the weight of metal ſupported by it. The compenſation of this pendulum is effected entirely by the aſcent and deſcent, alternately, of the mercury up and down the glaſs tube as the heat varies, and as the expanſion of glaſs is much leſs than that of any of the metals, the ſmall column of mercury contained in the tube is found ſufficient to anſwer its purpoſe completely, after a trial of more than twenty years. Indeed this pendulum has proved itſelf equal, if not ſuperior, to any of its predeceſſors, and the difficulty of the conſtruction is perhaps the only objection why it is not more generally adopted, particularly in ſituations where there is no riſk of carriage. We cannot diſmiſs our account of this pendulum, of which ſort only a ſmall number has been conſtructed, and of which no account has been before publicly given, without mentioning a curious property in it, known only perhaps to the contriver and to the writer of this account, which is, that if the column of mercury be made to reach either above or below the *middle* of the pendulum, at a mean temperature, the pendulum in either caſe will be *under-compenſated;* but when the column reaches to the *middle exactly,* at a mean temperature, the compenſation is *exact;* that is, the compenſation has the greateſt effect when the column reaches to the middle of the pendulum, which the inventor, we underſtand, foreſaw would be the caſe, and calculated his dimenſions accordingly.

Dr. G. Fordyce's Compenſation of the Simple Pendulum, and others of a ſimilar Conſtruction.—The compenſation deſcribed in the Philoſophical Tranſactions of 1794, as the invention of Dr. G. Fordyce, is not attached to the pendulum, but to the cock of ſuſpenſion, in this reſpect like that of Deparcieux, contrived in 1739, from which it probably was borrowed. The original pendulum of Dr. Fordyce, as altered from Whitehurſt's, is now in the poſſeſſion of Gen. Bentham, from which an univerſal ſtandard of linear meaſure was attempted to be taken, and from which, as a model, ſome few others have been made. In *fig.* 5, of our laſt plate, let A B repreſent the back of the clock caſe, to which a tube of braſs may be attached at its lower end, of ſuch length as to compenſate the verge of the pendulum, which is a long ſlender ſpring. Near the top of this tube is a bracing piece, C, in form of a clamping tube, through which it paſſes without play or friction, and by which it is kept faſt to the back of the caſe. The upper end, A, carries the cock of ſuſpenſion and a counterpoiſe to the pendulum, and under it is fixed faſt to the clock frame a ſecond cock of limitation, for the ſpring of ſuſpenſion to be received in, and by which the effective length of the pendulum is limited. (For a minute account of this apparatus, ſee Nicholſon's Journal, 4to. ed. vol. iii. p. 20.) The theory of this pendulum is, that as the fixed tube behind the clock caſe lengthens by expanſion in heat, the ſteel pendulum, though more expanded, will be ſhortened in the ſame proportion by the upward motion of its ſpring and cock of ſuſpenſion above the cock of limitation; and if the wood of the clock caſe were unalterable by heat or moiſture, this theory would be a good one, but as it takes no notice of the material to which the fixed rod is attached at its lower end, the actual compenſation amounts to no more than the difference between the expanſion of metal and wood gives; or or in other words, it does no more fo circumftanced than a wooden pendulum would do applied in the common way. But inaccurate as the compenfation we have defcribed muft neceffarily be, where the compenfation tube is refting on another body not taken into the account; yet Mr. John Crofthwaite of Dublin has defcribed a pendulum of his invention, in the Tranfactions of the Royal Irifh Academy, as being a fuperior pendulum, which differs in no refpect from Dr. Fordyce's, except that the fixed rod of compenfation is folid as well as the verge, and is fupported by a bracket carried by a large piece of marble, and like Whitehurft's clock, has cycloidal cheeks near its fpring of fufpenfion. Likewife the Rev. J. Pine of Eaftdown, near Barnftaple in Devonfhire, has defcribed in the 3d vol. of the Repertory of Arts, &c. a pendulum fimilar in all refpects to Mr. Crofthwaite's, except that the cycloidal cheeks are omitted, and *fpiral fprings* applied to raife and lower the cock of fufpenfion as the rod of compenfation directs, which muft be lefs fteady than the original method of Dr. Fordyce.

Berthoud has defcribed a method of compenfating a fimple pendulum in his "Hiftoire de la Mefure du Temps," tome ii. p. 84, fig. 2. plate 11, which is not liable to the fame objection as the preceding ones; for he has not allowed any fubftance to be a part of his mechanifm, if it is his own that he defcribes, but what is taken into the account; a frame of three bars, the middle one of fteel, and the two extreme ones of brafs, all faftened at the bottom to a crofs piece, and the brafs ones alfo at their tops to another crofs piece, conftitutes the compenfation in queftion; the upper crofs piece, or end of the frame, is fixed to the back plate of the frame that holds the wheels, and confequently the brafs rods elongate downwards, while the fingle fteel rod elongates upwards in a fmaller quantity; it is not however faft to the crofs piece, but paffes through it, and carries a pin acting againft one end of a double lever, that has its centre of motion on the cock of fufpenfion, while the other end of the double lever bears the pendulum. The confequence is, that while the brafs rods lengthen downwards, they take the fteel rod with them as far as is equal to the difference of their refpective expanfions, and the contiguous end of the lever, which has a counterpoife for the pendulum, falls with it, and lifts the pendulum verge at the remote end of the double lever the requifite quantity for compenfation of the fimple rod: and this quantity is determined by an adjufting fcrew, altering the centre of motion between the two oppofite ends of the double lever. This conftruction feems to poffefs all the requifites of a good compenfation, except that the frame is not in motion.

Troughton's Tubular Pendulum of different Metals.—We have already defcribed two pendulums of Troughton's contrivance, both of which anfwered their purpofe of being good regulators, and hence it may be matter of furprife, why the fame man fhould contrive a third, very different in its conftruction from either of the former. But pendulums which may be equally good regulators of a clock, when properly adjufted, are not equally capable of ready adjuftment, particularly for variations of temperature, after they are brought to a mean time. All the pendulums which we have hitherto defcribed required their adjuftments for temperature to be made from a trial of their going in fummer and winter, and eight or nine months waiting between each adjuftment was an interval inconveniently long for an aftronomer, who was anxious to purfue his obfervations; Mr. Troughton therefore, whofe numerous contrivances are generally fo many improvements in the mechanical arts, converted the grid-iron pendulum into a tubular one, which has all the accuracy of the old pendulum, and at the fame time all the fimplicity of the external fhape of the fimple pendulum. This pendulum has its compenfation examined, and its compenfating mechanifm adjufted by means of the inventor's new fpirit-level pyrometer, without any trial for rate in the oppofite extremes of temperature, and that as foon as it has been roughly adjufted to mean time. This faving of time, and of a repetition of adjuftments, is a great improvement in the art of clock-making, and when generally known and adopted will enable the maker to fend home his machine perfectly adjufted in the fhort fpace of a few days, whereas the old method of adjufting for temperature, from obfervations repeated at diftant intervals, required as many months, and in fome cafes even years. Our *Plate* XXVII. of *Horology*, exhibits a variety of figures illuftrative of the arrangement of the different parts of Troughton's tubular pendulum, and in our defcription of *An Aftronomical Clock with Troughton's Pendulum*, under the head CLOCK, a minute account is given of the conftruction, to which our readers are defired to revert. It is afferted in the Britifh Encyclopedia, that Finny of Liverpool, and Chandler of Seven Dials, London, both conftructed tubular pendulums, very fimilar to Troughton's, many years before 1804, which is the date of Mr. Troughton's publication of his, in Nicholfon's Journal; but as the arrangement of the two metals is not the fame as in Troughton's conftruction, nor was known to him previoufly, this circumftance, if true, does not detract from his merit, as it regards the invention in queftion. Indeed we happen to know, that no tubular pendulum is at prefent, nor ever was, at the Richmond obfervatory, formerly fuperintended by M. de Mainbray, as ftated in the article above referred to.

Ward's Compenfation Pendulum.—In March of the year 1806, Mr. Henry Ward, of Blandford in Dorfetfhire, communicated to the Society of Arts at the Adelphi, an account of a new compenfation pendulum, accompanied by a model, which gained him the filver medal from the fociety. This pendulum confifted of three bars, two of iron and one of zinc, arranged in a manner very fimilar to Deparcieux's improvement of Regnauld's old French pendulum, but without the lever of adjuftment. The reprefentation is given in *fig.* 6. of our laft plate, where *h h* and *i i* are two flat rods of iron or fteel, about half an inch wide and an eighth of an inch thick each; and *k k* is a rod of zinc, of nearly a quarter of an inch thick, interpofed; the two extreme or iron rods are cranked, one at the top to receive the fufpenfion fpring, and the other at the bottom to hold the ball in the fame vertical line with the zinc rod. Thefe three parallel rods are united by four fcrews, *l, l, l, l,* which paffing through oblong holes in the bars *h h* and *k k,* fcrew into the rod *i i,* while the rod, *h h,* is faftened to the zinc one, *k k,* by the fingle fcrew *m,* near its lower extremity, which fcrew is the fcrew of adjuftment for temperature, and requires feveral holes to be made in the bar *h h,* and alfo in *k k,* that the length of the zinc bar may be limited to its due proportion for exact compenfation. The iron bar, *i i,* refts by a chin-piece on the upper end of the zinc bar, and therefore the ball fupported by its lower tapped end ftretches all the bars equally. If the zinc bar had refted on the crank at the bottom of the bar *h h,* inftead of being held by the fcrew *m,* the arrangement would have been the fame as in Deparcieux's, but with zinc fubftituted for brafs, which confequently does not require the aid of a lever to increafe its effect. The proportional expanfions of zinc and iron were taken for this pendulum from Smeaton's table in the 48th vol. of the Phil. Tranf. of London, and were made for hammered zinc as 373 : 151; but fubfequent obfervations on the rate of the clock, to which the pendulum was applied, proved

proved that the expanſion of the hammered zinc was greater than Smeaton's table gives it, though the quantity was within the reach of the adjuſtment.

In Nicholſon's Journal, 8vo edit. (vol. vii. page 300.) there is a notice of a ſimilar pendulum by Benzenberg, where one rod of lead is introduced to compenſate two rods of iron, but no drawing is given. For the original ſee Voigt's Magazine, vol. iv. page 787.

Compenſation Pendulum by Adam Reid.—Another compenſation pendulum, made by Mr. Adam Reid of Woolwich, was preſented to the Society of Arts in April 1809, and the inventor was rewarded with fifteen guineas for his contrivance and model. This pendulum is given in *fig.* 7. of our laſt plate, and has its compenſation of a tube of zinc acting on the ball from the lower end of the ſteel verge thus; A B is the ſteel verge, made a little thicker where it enters the ball C, and alſo of a lozenge ſhape, to prevent the ball's turning, but above and below it is cylindrical; near the centre of the ball is a ſhoulder in the verge, againſt which the upper end of the zinc tube D. preſſes the croſs piece of the ball when the nut E is turned up cloſe to its lower extremity, but when the nut is turned back, in adjuſting for rate, the tube deſcends a little, and the ball with it, while it reſts on the upper extremity of the tube. When the compenſation is too much, the tube may be ſhortened till it is found of the exact length, by a trial of the rate in extremes of temperature. In the figure before us the length of the zinc tube does not appear ſufficiently long to compenſate an entire verge of ſteel. The inventor obſerves that platina might be ſubſtituted for ſteel, and ſteel for zinc, for the formation of this kind of pendulum, but that the expenſe would be enhanced by the dearneſs of platina.

Doughty's Compenſation Pendulum.—Mr. Thomas Doughty, of Charles-ſtreet, Somer's-town, favoured us, in January 1811, with a ſketch and ſhort deſcription of a compenſation pendulum invented by him on the principle of flexure, by a compound bar that is applied to draw up the pendulum's ball in hot weather, after the manner of Mr. W. Nicholſon's, which we ſhall notice preſently. *Fig.* 8, of the ſame plate to which we have lately referred, repreſents this pendulum in its reduced ſtate; A is a compound bar of ſteel and braſs ſoldered together, with the ſteel uppermoſt; the length of which is ten inches, the breadth half an inch, and thickneſs four-tenths; the ſteel portion being $\frac{1}{10}$th of the whole; it is curved with the concave face downwards, the point E above the ball being the centre of curvature. Above this bar, and connected with it at the middle, riſes the braſs bar B, ſix inches long, to which the ſpring of ſuſpenſion is faſt, and which receives the impulſe from the crutch at the point 8. C, C, are two ſimilar ſliding heads of braſs, tapped to receive each a ſcrew and ſaddle-piece to fix them in any given part of the compound bar: by means of two clock ſprings theſe heads are attached to the ſide rods D, D, of ſteel, each $\frac{4}{10}$ths of an inch wide, and $\frac{1}{10}$th thick, which rods are again attached to the iron flat piece E by ſimilar ſprings, while the flat piece paſſes through the ball and receives the nut of adjuſtment for rate. The compenſation is effected thus; while the braſs piece B, and the ſteel ſide rods D, D, elongate by heat, the curve of the compound bar, A, becomes more and more approaching towards a ſtraight line, by the ſuperior expanſion of the braſs portion, and this alteration of figure lifts up the ball of the pendulum a correſponding quantity; which quantity is regulated by the diſtance of the heads C, C, from the middle of the compound bar; the greater diſtance being always accompanied by a greater effect of the compenſation, and *vice verſâ;* but in making this adjuſtment care muſt be taken that the clock be not put out of beat, by moving one head more than the other; but if the clock is already out of beat, this adjuſtment may be uſed alſo as an adjuſtment for beat, at the time of adjuſting for temperature. The ſprings are uſed for the purpoſe of preventing friction during the variations of temperature, but we ſuſpect that there will be another conſequence, not ſo favourable to the regular motion of the pendulum as might be wiſhed; which is, that when the impulſe is given from the maintaining power at the point 8 of the braſs bar B, the ſprings, both near C C and E, will *yield,* or bend a little, before the heavy ball will be urged, and thereby require a greater power to keep the pendulum in motion than would be neceſſary without ſuch ſprings; and alſo the ball and the rods will not begin and end their vibrations exactly in a contemporaneous manner, which circumſtance, without very nice equipoiſe of the ſide rods, will produce a vacillation in the ball. Ritchie's pendulum, however, which follows, and which was intended to remedy the faults of this, is liable to the ſame objection, though his ſliding adjuſtment for temperature will not interfere with the adjuſtment for rate, as muſt be the caſe with the pendulum before us; for bringing the upper ends C, C, of Doughty's ſide rods nearer together lengthens the effective length of the pendulum, thereby requiring a new adjuſtment for rate. The inventor heated a ſmall room with ſteam till the adjuſtment for temperature was found complete, which turned out to require eight out of the ten inches of the compound bar to make this adjuſtment perfect.

Ritchie's Compenſation Pendulum, and Nicholſon's.—In March of the year 1812, Mr. David Ritchie, of Clerkenwell, laid before the Adelphi Society, alſo, a model of a compenſation pendulum, for which he received, as a reward, twenty guineas. The compenſation in this pendulum was effected by means of the flexure of two compound horizontal bars interpoſed between the verge and the ball, which, like Doughty's, act on the ſame principle as the compound bars in the ordinary compenſation balances of a chronometer. *Fig.* 9. repreſents the pendulum at preſent before us, in which A is the ſpring of ſuſpenſion, B the ball, and C the ſteel verge as uſual, with the nut of adjuſtment for rate at E, below the ball. The compenſating bars are F and G, the upper part of F and the lower part of G being ſteel, and the other parts braſs; ſo that any additional heat may bring the bars nearer together by the convexity of the braſs faces of the compound bars being contiguous. Theſe compound bars have each a ſliding piece above and below *e* and *f* reſpectively, by the ſliding of which the bars may have their effective lengths altered, in the adjuſtment for temperature; and are connected by the ſlender ſprings *e* and *f,* which bear a portion of the ball's weight; the remainder being borne by the ſprings *l* and *m* attached by ſcrews at *h,* one to the verge piece *d* and the other to the ball piece *b,* while they themſelves are united by ſcrews at *l* and *m.* The compenſation frame, ſo conſtructed, is adjuſted by ſliding the compound bars, before they are fixed, till the centre of the ball falls in the vertical line of the verge, where the ſcrews at *h* fix it for trial. The impreſſion on our mind, from a contemplation of this conſtruction, is, that when the crutch of the clock urges the verge, the ball will not move with it till the ſprings *e* and *f* are bent a little, for when the ball is heavy, as is now cuſtomary in ſeconds pendulums, its weight will place a great ſtreſs on the ſlender ſprings that connect it with the verge, and produce, probably, a vacillation, as in Doughty's, that muſt be very unfavourable to iſochroniſm,

ſochroniſm, even allowing the compenſation to be perfect. It was to avoid this evil, that Mr. Nicholſon placed a ſimilar compenſation bar on the cock of ſuſpenſion above the verge of his pendulum, and alſo above the cock of limitation, as ſeen in *fig.* 10, but without any ſpring, and in this ſituation the varying curvature of the horizontal compound bar, E E, altering the length of the ſuſpenſion ſpring, produced the alternate riſing and falling of the ball, as the verge carried it in an oppoſite direction, ſo that the centre of oſcillation was comparatively ſtationary. (See Phil. Journal, vol. i. 4to. ed. plate 5. fig. 3.) The alteration in the length and ſtrength of the ſpring of ſuſpenſion is, however, an objection to Nicholſon's conſtruction. The ſame author alſo contrived a compenſation pendulum compoſed of four rods of ſteel and one of a compound, or alloy, of zinc and ſilver, which is deſcribed in vol. ii. of his 4to. Phil. Journal. (See page 205, and plate 9.) But as there are eight figures neceſſary to explain the fittings of the different parts, and as the work is in every one's hands, we will ſatisfy ourſelves by referring our readers to the author's own deſcription.

Hardy's Compenſation Pendulum.—Mr. Hardy of Clerkenwell, whoſe ingenuity and good workmanſhip have given him a high rank among chronometer makers, has lately contrived a compenſation pendulum that partakes partly of the grid-iron claſs, and partly of Ellicott's, and that promiſes to be exempted from the jerks that have been imputed to the other conſtructions during the elongations and flexure of the bars. It was our intention to have given a drawing of this pendulum, but our plates of pendulums are full; and as the pendulum before us has not yet been applied to a clock, we will conſider it as a project rather than as an approved pendulum. It is, however, completely finiſhed. It conſiſts of five parallel bars, the three middle ones of ſteel, and the two extreme ones of braſs; there are two croſs pieces at top, and as many at bottom, to connect the longitudinal bars, and the loweſt carries a pair of ſmall levers, on the remote ends of which the braſs bars act as they elongate downwards, but the centres of their motions are ſprings, like thoſe in the inventor's compenſation balance, which balance performs very well; the ball is attached to the interior ends of the ſaid levers by other ſprings, and a ſpiral ſpring acts near the top of the middle ſteel lever to carry a part of the ball's weight, in order to let the ſprings near the ball act without having too much ſtreſs on them. The excellence or defect of this pendulum muſt be proved by future trials.

Recapitulation of the principal Properties of a Pendulum.—As the various properties of, and circumſtances relating to, the pendulum, have been treated of under different heads in this work, we will conclude our preſent account by recapitulating the principal of them in a ſuccinct manner, thereby condenſing the ſubſtance that is neceſſarily diffuſed over our work into regular order and a narrow compaſs.

1. The theory of a pendulum's motion is derived from the uniformly accelerative force of the gravity of its bob, conſidered as having all its weight concentrated into one point.

2. If ſuch bob could be made to deſcend down the chord of its ſemi-arc of vibration, and to aſcend up the chord of the other half, its long and ſhort ſemi-vibrations would be performed in the ſame time, namely, in the time that would be taken up in falling through half the length of the pendulum, and the velocity at the loweſt point would be as the length of the chord.

3. A pendulum cannot be made to move in the chords of a circle, but will neceſſarily move in the arcs when ſuſpended by a cord or rod.

4. The longer arcs of vibration require more time than the ſhorter ones, when thoſe arcs are circular.

5. In that peculiar curve called a *cycloid*, or by the French a *roulette*, a ſemi-vibration is performed in the ſame time, whatever be the length of the arc, namely, in the time that would be taken up in deſcending down the chord of a correſponding arc of the generating circle.

6. The axis of a cycloid is equal to the diameter of the generating circle; the ſtraight line paſſed over in the generation is equal to the circumference; and the whole cycloid equal to twice the circumference of the ſaid circle.

7. The time of a ſemi-vibration in a cycloidal arc is to the time of the bob's deſcent down the axis of the cycloid, as a *quadrant* of a circle is to its *radius*; therefore the time of a whole cycloidal vibration is to the time of the fall down the axis as a *ſemi-circle* is to *radius*; or as the *circumference* of a circle is to its *diameter*.

8. The involute of a cycloid is itſelf a cycloid in every reſpect ſimilar to the cycloid evolved.

9. Any cycloidal arc is equal to *twice* the correſponding chord of the generating circle; and therefore the axis of the generating circle is equal to half the arc of the whole ſemi-cycloid; hence a cord repreſenting the length of a pendulum, evolved from contact with a ſemi-cycloid, till it reſts at the vertex, muſt be *double* the length of the axis of the generating circle, and will move in a ſemi-cycloid.

10. Therefore the time of a pendulum's vibrating in a cycloidal curve is to the time of its bob's deſcent down *one-half* of its length, as the circumference of a circle is to its diameter.

11. Hence Huygens placed two portions of ſeparate ſemi-cycloids, with their curves oppoſed to each other, and with their vertices nearly in contact at the point of ſuſpenſion, that the thread of ſuſpenſion might be evolved, firſt from one, and then from the other alternately, in each ſemi-vibration, in order to make all the vibrations, of whatever length, iſochronal.

12. The theory of this contrivance was admirable; but it ſuppoſed the action of the pendulum's gravity to be ſufficient to maintain the continuance of the vibrations, as well as all the ponderous matter collected into one point; likewiſe the centre of oſcillation to be in the ſame point with all lengths of the cycloidal cord; and alſo the cord itſelf to be invariable in length, as well as perfectly free from rigidity; all which ſuppoſitions, not founded in nature, proved ſo many impediments to the practical adoption of the cycloidal cheeks for an *attached* or going pendulum, where external force is neceſſarily ſuperadded to that of gravity, and where the centre of oſcillation does not coincide with the centre of gravity.

13. But *ſmall circular arcs* are ſo nearly coincident with cycloidal arcs near the vertical line, or line of quieſcence, that they may be ſubſtituted for the latter, where they do not exceed 4°, without ſenſible error.

14. Hence the time of a vibration in *ſmall circular arcs*, is to the time of the bob's falling down half the pendulum's length, *very nearly* as the circumference of a circle is to its diameter, which, in cycloidal arcs, has been ſhewn to be exactly ſo.

15. An eſcapement for *ſhort arcs*, therefore, became neceſſary, when the cycloidal cheeks were laid aſide, which acted with the *crown*-wheel eſcapement (an eſcapement adapted for long arcs), and Clement or Hooke contrived the *anchor*-eſcapement, which gave the pendulum,

vibrating in ſhort arcs, the name of the *royal pendulum*, from the noble property it poſſeſſes as a regulator. [Query, was not the ſign of the *crown* and *anchor* of horological origin ?]

16. As the momentum of a pendulum is compoſed partly of ponderous matter, and partly of velocity, the reduction of the arc of vibration produced a neceſſary addition to the weight of the bob.

17. The times of vibration of pendulums of different lengths will be always directly to each other as the *ſquare roots* of their reſpective lengths, where the accelerative force is conſtant and unaltered.

18. When pendulums of the *ſame length* vibrate with *different forces*, the times will vary inverſely as the *ſquare roots* of the *accelerative forces*.

19. The lengths of pendulums, vibrating in the *ſame times*, vary directly as the *forces* which accelerate them.

20. The times of vibration of pendulums of *different lengths*, acted upon by *different accelerative forces*, will vary as the ſquare roots of the *lengths* directly, and the ſquare roots of the *forces* inverſely.

21. Hence the ſame pendulum will not vibrate in preciſely the ſame time in all latitudes, where the force of gravity varies.

22. Nor will an attached pendulum, continuing to vibrate by the aid of ſome external force added to gravity, vibrate, even in the ſame latitude, in the ſame time as a detached pendulum vibrating by the force of gravity alone for a limited period.

23. Pendulums acting with different eſcapements, or having otherwiſe different maintaining powers, will therefore vary a little in their *lengths*, though they meaſure the ſame time, and even move in the ſame arcs.

24. External force, from the maintaining power, may be added to gravity, either inſtantly or gradually, for a longer or a ſhorter time, uniformly or variably, and commencing or ceaſing to act at any given point of the vibration ; or it may be applied to act in conjunction with gravity during the whole vibration ; all which modes will require ſo many correſponding lengths of pendulum to vibrate in the ſame time.

25. Hence originated all the variety of eſcapements that act with the pendulum, and modify the maintaining power, by giving their impulſes differently, as to place, quantity, quality, and duration.

26. That mode of giving the impulſe is beſt, which either deranges the action of gravity leaſt, or acts moſt in conformity with it.

27. When the ſame pendulum is by any cauſe made to vibrate in *different arcs*, the daily error, or difference, in ſeconds of time, will be $\frac{1}{15304}$ths of the difference of the ſquares of thoſe arcs, reckoned in degrees and parts of a degree.

28. A vibrating pendulum may have its arcs altered, not only by a variation in the additional force, or maintaining power, but may be ſo attracted by a contiguous heavy body, particularly if homogeneous, that its arc of vibration may be either enlarged or diminiſhed by the poſition of the ſaid body : if placed near the line of quieſcence, the arc may be ſhortened ; but if placed beyond the extremity of the arc, it may be enlarged.

29. If the point of ſuſpenſion be not firmly and immoveably fixed, a ſlight motion in it will have the ſame effect on the time, as lengthening the pendulum, though the maintaining power be the ſame ; and if the points of ſuſpenſion of two pendulums be contiguous, and be attached to the ſame ſhelf or yielding ſubſtance, they will continue to vibrate *together*, though they are not exactly of the ſame length, nor adjuſted preciſely to the ſame time. This curious property is obſervable in either the detached or attached pendulums.

30. When the pendulum is taken as an univerſal ſtandard of *linear meaſure*, as well as of time, the experiments to aſcertain the exact length of a ſeconds, or half-ſeconds pendulum, muſt be made always in the ſame latitude, on the ſame level, and in a detached ſtate if poſſible, or otherwiſe the reſult will be affected by all or ſeveral of the variety of circumſtances, which we have had occaſion to conſider as influencing the length of the vibrating pendulum.

31. That compenſation-pendulum may be accounted the beſt, that changes its rate leaſt throughout the year, provided that its adjuſtments can be conveniently made.

BOB, or BALL, in *Horology*, is the metallic weight which is attached to the lower extremity of a pendulum rod, by means of a tapped adjuſting nut, at ſuch a diſtance from the point of ſuſpenſion as the time of a given vibration requires. (See the articles CENTRE of OSCILLATION and PENDULUM.) In fixing upon a proper bob for any pendulum, two things are particularly to be attended to: the ſhape which is beſt calculated for meeting with the leaſt reſiſtance from the air; and the weight which is beſt adapted for preſerving the iſochroniſm of the vibrations with a given maintaining power. Each of theſe conſiderations has been the ſubject of much inveſtigation. A ſphere is a ſolid, the ſurface of which bears the leaſt proportion to its ſolidity of any other, and a cube is one with great extent of ſurface compared to its ſolid contents; conſequently, the former ſhape has obtained in cannon balls, intended to paſs through the air with as little obſtruction as poſſible, and has ſometimes, for the ſame reaſon, been applied as a bob for a pendulum. Such a ſhape is better adapted for a large bob than a ſmall one; for the increaſe of the ſurface is in proportion to the ſquare of the diameter, whereas the increaſe of weight or quantity of matter is as the cube. Mr. B. Martin (Mathematical Inſtitutions, vol. ii. p. 417.) propoſed two equal fruſta of ſimilar cones, to be joined together at the baſes, as a figure approximating nearly to a ſolid of leaſt reſiſtance, and recommended it as that out of which the middle one of three ſegments, cut longitudinally, will conſtitute the beſt ſhape for an appended weight, to anſwer the purpoſe of avoiding reſiſtance. Such a ſection, he conceived, would diſplace but a ſmall quantity of air in one vibration, and the impulſes of that quantity, being made obliquely on the ſloping ſurfaces of the ends, would produce but a ſmall effect. The lenticular ſhape, however, compoſed of two ſegments of a large globe joined to the line of ſection, has been generally adopted by clock-makers, probably becauſe two plates of braſs of that ſhape can be eaſily ſoldered together, and left hollow, ſo as to be filled with lead, or other heavy metal, in order to obtain a due degree of weight without adding to the thickneſs, which is an advantage that braſs or copper cannot have of itſelf in any ſhape. M. Ferdinand Berthoud (Eſſai ſur l'Horlorgerie, tome ii. chap. xiii.) made ſome experiments with bobs of both a ſpherical and lenticular ſhape, of equal weight, attached ſucceſſively to the ſame free pendulum, from which it appeared, that, in the ſame temperature, the latter continued to vibrate ſeconds much longer than the former before they arrived at the quieſcent ſtate, which experiment was conſidered as a proof that the lenticular ſhape has the advantage in eſcaping the effect of reſiſtance; it was diſcovered, however, on a repetition of the experiment, with ſeconds and half-ſeconds pendulums, that the friction at the point of ſuſpenſion occaſioned by heavy weights, particularly when vibrating in long arcs, made conſiderable alterations in the reſults, and proved itſelf a ſecond ſource of reſiſtance to the free motion of the pendulum.

The ſecond conſideration to be attended to in making the bob, as has been ſaid, is the weight which a given pendulum requires, with a given maintaining power. No theory is adequate to determine this deſideratum of itſelf, becauſe the diminution of the maintaining power by the friction of the pivots and the ſtate of the oil, the duration of the impulſe on the pallets, the nature of the eſcapements in other reſpects, the reſiſtance of the air, and at the point of ſuſpenſion, and particularly the nature and quantity of momentum of the pendulum, muſt all enter into the calculation; and theſe are data, many of which are conſtantly varying. The momentum, or whole quantity of motion of any pendulum, is the weight multiplied into the ſquare of its velocity, ſo that a large arc with a ſmall weight, and a ſmall arc with a correſponding large weight, cæteris paribus, ought to have an equal effect upon the iſochroniſm of the ſame pendulum; but the theory is not perfect; for firſt, large circular arcs deviate conſiderably from cycloidal ones, which, it has been demonſtrated by Huygens and others ſince, are thoſe which have the iſochronal property in an uniformly denſe medium; and ſecondly, they require a greater maintaining power than calculation gives: for inſtance, where the arc is 10° from the point of reſt, the addition to be made to the maintaining power beyond calculation is 57 parts in 100, according to Berthoud's experiments, which conſideration induces him to conclude, that irregularities in the maintaining power will affect the momentum principally compoſed of velocity, more than the momentum principally compoſed of weight; each kind of momentum, however, has its peculiar diſadvantage; for great velocity is ſubject to great reſiſtance from the air, and great weight to much friction at the point of ſuſpenſion. Mr. Alexander Cummings was an advocate for a large arc with a ſmall bob; but modern practice is in favour of a large ball or bob with a ſhort arc of vibration. It ſeems to be generally allowed, that the momentum of a pendulum ought to be as great as poſſible for a given maintaining power, provided the latter be ſufficient to overcome all obſtacles to conſtant motion; but as the momentum is obtained in two different ways, it may be proper to ſubjoin here an illuſtration of them.

If a pendulum, moving in an arc of one degree from the quieſcent point, with a bob or ball of ſix pounds, have its momentum denominated by unity, then the ſquare of any given momentum will give its correſponding weight for another bob, moving in the ſame arc; or the ſquare root of the given momentum will give the arc of ſemivibration with the ſame weight; for inſtance, a momentum 4 will require $6 \times 4 = 24$ for the weight where the arc remains unaltered; but the root of $4 = 2$ will be the arc from the point of reſt with the ſame weight; again, if the momentum be required to be 9 for a given maintaining power, the weight muſt be 54, or the arc 3° from the point of reſt, and in the ſame proportion for any other momentum; but, in fact, both the weight and arc may be varied according to circumſtances, which latitude affords great variety in the adjuſtment of the bob, and matter for multiplied experiments to determine what weight and arc ſhall be moſt deſirable, taken conjointly under different circumſtances.

The moſt practicable method of adjuſtment of the momentum of a pendulum, where a weight is uſed as a maintaining power, ſeems to be, to vary this power inſtead of the weight of the bob or ball, which, when once finiſhed of the requiſite ſhape, is not ſo eaſily altered. In pendulum clocks, actuated by a ſpring as a maintaining power, ſuch adjuſtment cannot indeed be made in the power properly without changing the ſpring, or altering the fuſee after they are adjuſted to each other; but as theſe inſtruments are not intended to meaſure time with great nicety, the adjuſtment of the bob to the maintaining power is exactly done by gueſs. In Huygens's beſt clock, the maintaining power was equal to three pounds, falling about 90 inches in 24 hours, and the ball of its ſeconds pendulum, together with the pendulum itſelf, was alſo three pounds; but the nature of his eſcapement required the vibrations to be performed in long arcs. Cummings tried different weights for his pendulums from 6 to 16 pounds, and various arcs from 3° to 6° from the point of reſt; but it does not appear that any other ſtandard has been adopted by clock-makers than what accords with the individual opinion of each. See MAINTAINING POWER.

Power, *Maintaining*, or *Maintaining power*, in *Horology*, is that extraneous force which is applied in a clock or watch, to maintain or perpetuate the natural vibrations of a pendulum, or oscillations of a balance, which would otherwise come to rest, by reason of friction in the mechanism, and the resistance opposed by the air to the parts in motion. This power, in the larger clocks, is usually a suspended weight; but in the portable clocks and watches, it is a spring, coiled in a metallic box, that actuates the wheelwork, by gradually unbending itself: in the former of these cases, the weight is suspended by a cord, or chain, that is coiled round a cylinder when wound up, which cylinder, being of uniform diameter throughout its length, is acted on by the cord, when fast at the interior end, by a similar force in every situation; and, therefore, imparts through the train, that is connected with its great wheel, invariable impulses to the escapement-wheel, at every vibration of the pendulum; which pendulum receives therefrom such a slight push, as is just sufficient to restore the momentum that it loses from friction and resistance of the air; and thus the uniform motion of the pendulum is perpetuated, or would be perpetuated, if no impediment from dirt, from the thickening of oil, or from wear, &c. were to interfere with the regularity of its action. But when a spring is substituted for a weight, its agency is not uniform, without some mechanical correction to equalize it: this correction is effected by the Fusee, which we have already explained in its proper place, but which is seldom so completely proportioned to the varying forces of the opening spring, as to be at all times equal to a weight suspended from an uniform cylinder. Hence have originated various escapements in horological machines, either to correct or to obviate the inequalities of the maintaining power, after it has been modified by transmission through the train of wheelwork, that is necessary both for limiting its quantity, and lengthening the time of its action. The whole quantity of maintaining power that is expended in a given time, when a weight is used, is the product of the weight into its descent, in inches and parts, from its situation at the commencement of action, and therefore the slower the fall is, the greater must be the weight, and *vice versâ*, when a given quantity is necessary for maintaining the vibrations in an undiminished state for the requisite duration. Hence a clock may be made to go at one winding, any length of time, from a day to a year, or more; but the most usual time, and perhaps as unobjectionable as any, is something more than a week, for which a weight of from ten to twelve pounds is sufficient, with an ordinary cylinder and double cord. In general, an excess of maintaining power is injurious to the natural motion of a pendulum or balance; but as different escapements require different degrees of power to produce the requisite effect, it would not be an easy matter to calculate the exact quantity that would be best in all cases, even if the same power were requisite in all states of foulness of the machine. The best rule in practice is to adopt that weight which experience has pointed out as adequate to its purpose, and if, on trial, a little addition or diminution of the power is found to bring the vibration to the most desirable length; the weight so changed will be proper to make choice of. In those particular escapements which are called *isochronal*, an increase or diminution of the power, though they affect the total length of the vibration, will not alter the time; and, therefore, with respect to them, the exact quantity of power is not of so much importance as in other escapements, provided it be sufficient, under all circumstances, to make the total vibration exceed the arc of escapement, so as to prevent quiescence. In those escapements which are denominated *remontoir*, the maintaining power does not itself act on the escapement-wheel, but serves only to raise a pair of small weights, or to wind up a couple of small springs, that become substitutes for the maintaining power, by giving their small alternate impulses at each vibration or oscillation, without variation in the intensity of each impulse, which is a great improvement in the art of clock-making and watch-making, particularly where the escapement is of the detached description; for whatever irregularities might arise from the transmission of the modified power, through the train, in different states of the weather, or of foulness, these irregularities, not influencing the effective forces derived from the secondary weights or springs alone, do not affect the pendulum or balance, that is actuated solely by one or other of these invariable powers, though these powers are under the agency of the maintaining power themselves, at certain intervals, more or less distant.

It has been asserted, that if the weight of a pendulum-bob be increased inversely as the square of the arc of vibration is diminished, and *vice versâ*, the maintaining power must remain the same, but the experiments of F. Berthoud, as reported in his 14th chap. of his "Essai sur l'Horlogerie," prove that this is far from being the case in practice, and that no theory, which has yet been proposed, is competent to the calculation of the exact maintaining power, that shall make a given pendulum move in a given arc, with all the variety of escapements; it appears, however, to be agreed on, that a heavy bob, moving in a small arc, with a comparatively small maintaining power, is most conducive to invariable regulation in a clock; but that in a watch, a large arc of oscillation is necessary, in order to obtain sufficient momentum. See Escapement.

TROUGHTON'S PENDULUM.

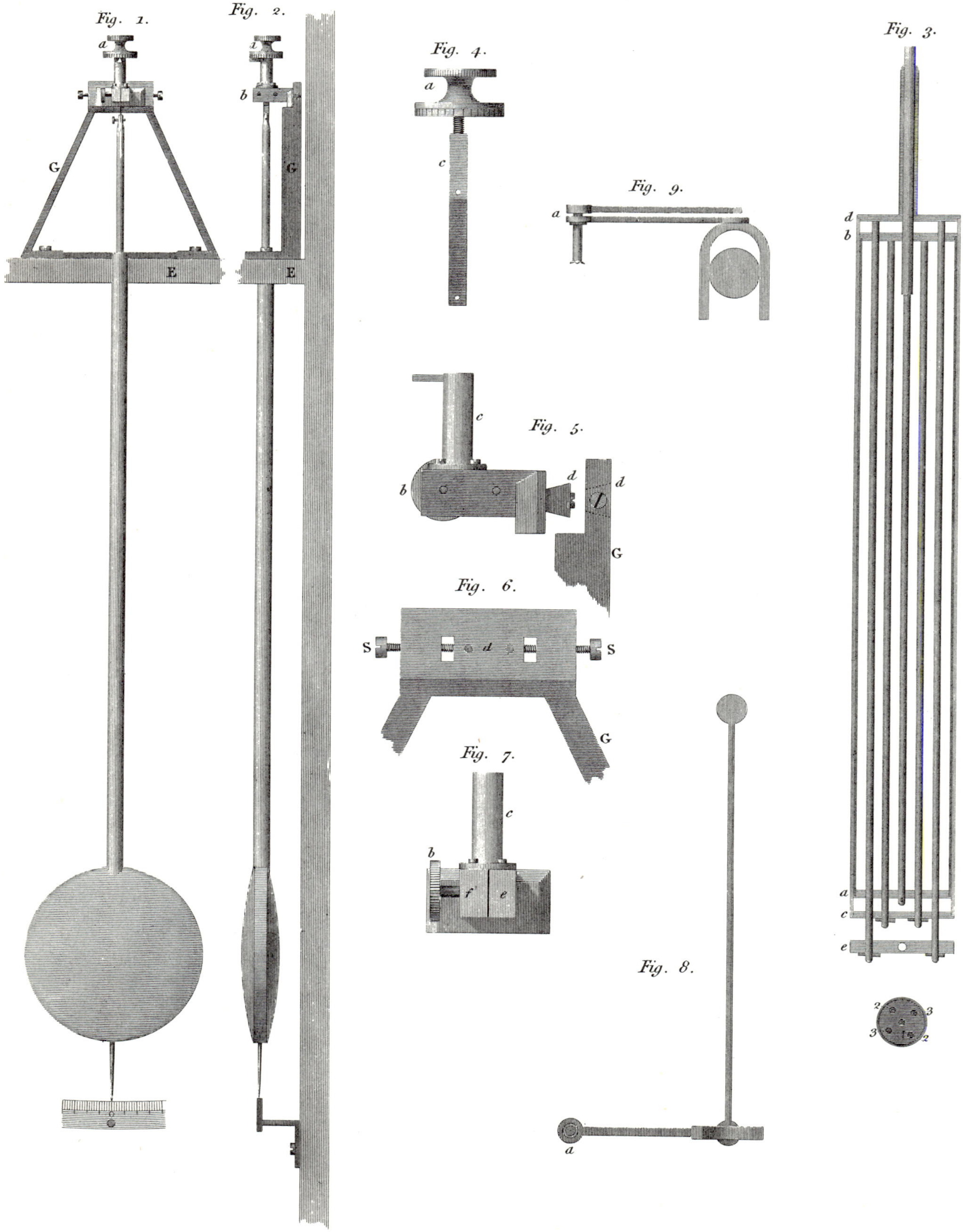

Drawn by J. Farey. Published as the Act directs, 1808, by Longman, Hurst, Rees & Orme, Paternoster Row. Engraved by Wilson Lowry.

COMPENSATING PENDULUMS.

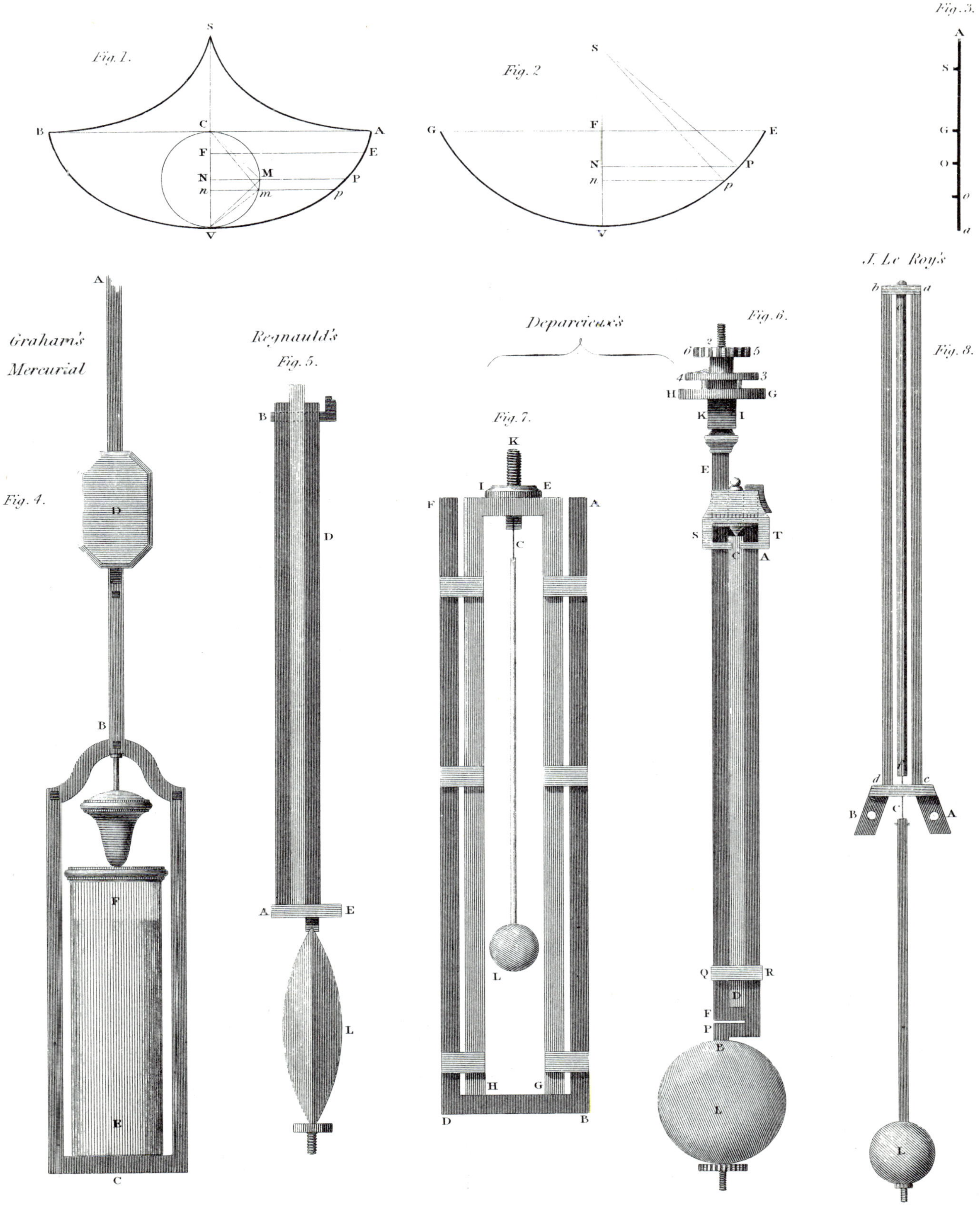

...n Jun. del. Published as the Act directs, 1815, by Longman, Hurst, Rees, Orme & Brown, Paternoster Row, London. Lowry sculp.

COMPENSATING PENDULUMS.

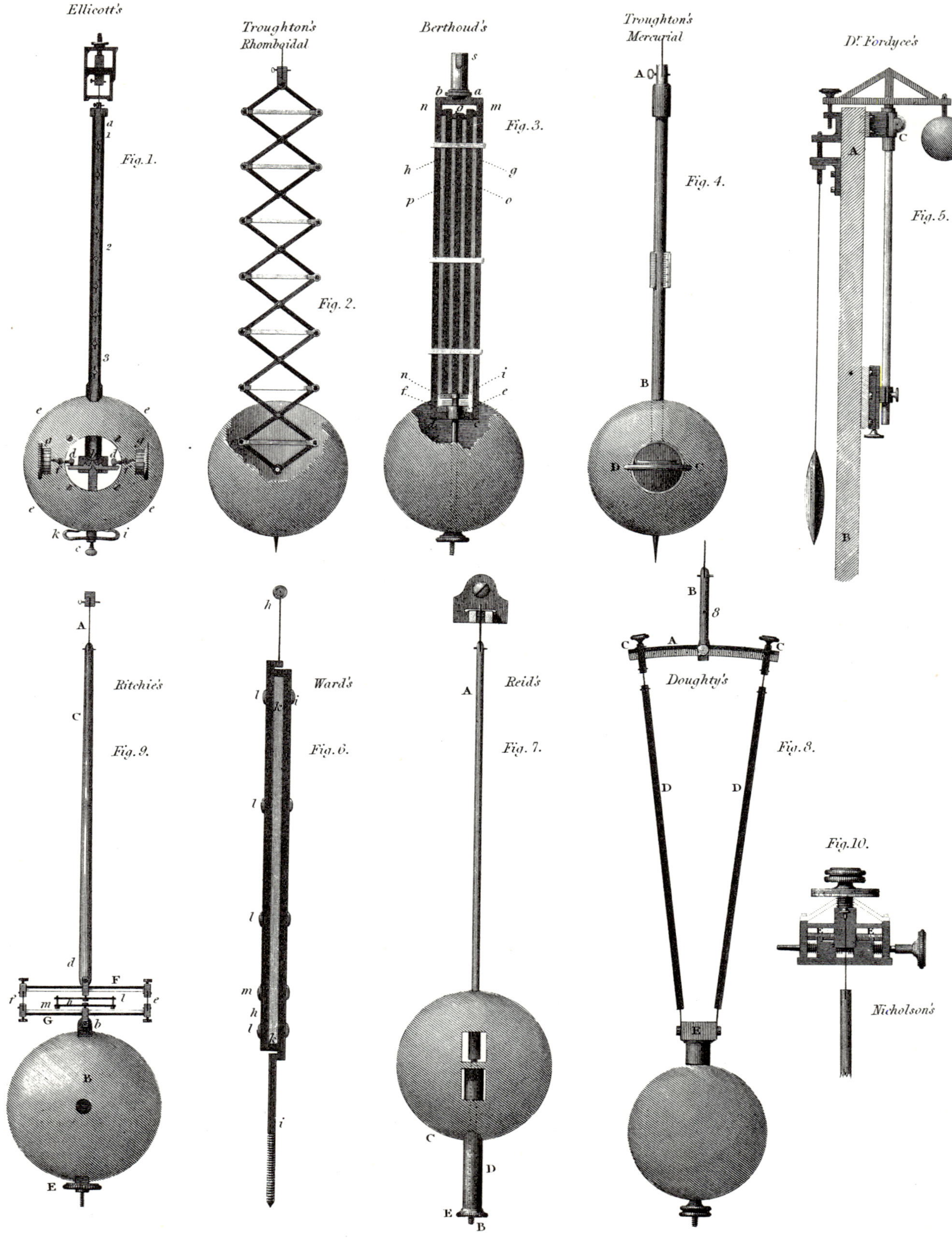

J. Farey Junr. delin.

Published as the Act directs, 1816, by Longman, Hurst, Rees, Orme, & Brown, Paternoster Row.

Lowry Sc.

Remontoir

REMONTOIR, or Remontoire, in *Horology*, is a ſpecies of eſcapement, in which a ſecondary ſpring frequently wound up, or a ſmall ſecondary weight frequently raiſed, by means of the maintaining power of a watch or clock, is ſubſtituted for the maintaining power itſelf, for the purpoſe of urging the balance or pendulum, at ſhort intervals, by more equable impulſes than can be conſtantly effected by the maintaining power alone, as varied by different degrees of friction in the train. When treating of Escapements in general, under their appropriate head, we deſcribed three only out of the four claſſes, and reſerved the fourth claſs, denominated *Remontoir*, till we arrived at our preſent article.

The firſt remontoir was invented and made by a German artiſt in 1600, according to Berthoud; but Huygens, who applied one to his marine clock, deſcribed it firſt in his "Horologium Oſcillatorium," page 17, and, in conjunction with his cycloidal cheeks, it promiſed to be a great improvement in his machine; but, as no compenſation had at that time been applied to the pendulum, and, as a pendulum is not calculated for a portable machine, particularly on the ſea, the utility of the contrivance remained to be proved by ſubſequent artiſts. The contrivance under our conſideration was a ſmall weight, ſuſpended by an endleſs well-made metallic chain, that was coiled round the crown-wheel of the eſcapement, and wound up a ſmall ſpace, at every vibration of a half-ſeconds pendulum, by the next wheel, which wheel took its motion from the maintaining power, through the medium of the train, as is uſual in common clocks; a ratchet and click, however, were a requiſite appendage, to act in the way that the endleſs cord was applied, in winding up the maintaining power without ſtopping the motion of the works; ſuch as we have already deſcribed in the fifth ſection of our article Clock-work. One half of the weight of the remontoir actuated the eſcapement wheel during its ſmall fall, and the other half was ſupported by the wheel that as often raiſed it again to its original height, while both parts of the folded chain were ſtretched alike by the ſaid weight. Leibnitz and Dr. Hooke alſo claimed the originality of a ſimilar invention, but do not appear to have put it in practice, as Sully afterwards did.

Mr. Harriſon and Mr. Mudge ſucceſſively applied remontoir ſprings, inſtead of ſuſpended weights, to the eſcapements of their time-pieces; but, as we have deſcribed theſe under our article Chronometer with ſufficient minuteneſs, it is not neceſſary to repeat here what we have there detailed of their conſtructions.

Mr. Cumming and Mr. Nicholſon, on the contrary, had recourſe again to weights inſtead of ſprings in their aſtronomical clocks; but as theſe weights did not act during the whole period of the vibration, we have already deſcribed them in another claſs, in the 21ſt and 38th ſections of our Escapements.

After Huygens and Sully, who left no plans of their mechaniſm behind them for the advantage of future workmen, who had not acceſs to the original machines, Gaudron contrived a remontoir, which performed its office very well, but which was applied in a wrong place, ſo as not to produce the deſired effect of equalizing the impulſes given to the regulator; for, inſtead of being applied to either the balance or balance-wheel, it was made to actuate the wheel preceding the minute-wheel; and, therefore, permitted the irregular friction of a conſiderable portion of the train to affect the motions of the balance, which fault was avoided in the conſtruction of Harriſon's and Mudge's time-keepers: the former of which had its ſecondary ſpring wound up eight times in every minute, and the latter had its two remontoir-ſprings alternately coiled at every correſponding oſcillation.

Haley's.—In the year 1796, Mr. Charles Haley, of Wigmore-ſtreet, Cavendiſh-ſquare, London, watch-maker, took out a patent for his invention of a new remontoir ſpring for a marine time-piece, or chronometer; the ſpecification of which is contained in the 6th volume of the Repertory of Arts and Manufactures. *Figs.* 1 and 2, of *Plate* XLI. of *Horology*, exhibit, the firſt a perſpective ſide view, and the ſecond a plan of Mr. Haley's remontoir eſcapement, as originally drawn; and we propoſe to retain the ſame letters of reference as are inſerted in the original deſcription. The utility of the invention is ſtated to conſiſt of its property of communicating an invariable force to the balance, which it does 150 times in the minute, in a train of 9000 beats in the hour. The ſame letters of reference apply to both figures, and indicate the correſponding parts, which will mutually illuſtrate each other. A B is the potance plate, and T the balance, the pivots of which, P, X, turn in the cock C and potance D; above the balance T is fixed a pendulum ſpring S, in the uſual way; on the axis of the verge, below the balance, are placed two ſmall ſteel collets I and K, by friction, having each a ruby pallet projecting a little way beyond their ſurfaces. I is called the diſcharging pallet, and K the impelled one; which pallets, together with the pendulum ſpring, all vibrate with the balance, whenever it is put in motion. E is the balance-wheel of the uſual form, moving juſt clear of the potance plate, and having its pivots ſupported by the cocks F and G. W V is the axis of the remontoir, which the inventor calls the *renovating* ſpring; and the three axes, or arbors, juſt deſcribed, ſtand in the ſtraight line in the direction A B. Below the remontoir ſpring the round ſteel pallet M is fixed, ſo as juſt

juſt to eſcape touching the ends of the balance-wheel's teeth in its reſting poſition, and the notch cut in this pallet is to receive the impulſe of any tooth of the ſaid wheel that may, at any time, act againſt it, in paſſing the line A B. Juſt above this large pallet is fixed, by friction, another ſmaller pallet N, of ſteel, in the form of a ſnail-piece, and having inſerted into it, near the centre of motion, and at right angles, a ſmall ruby pallet, which points directly to the radial end of the notch cut in the large pallet M. A ſmall collet, twiſted faſt to the axis of the remontoir, juſt above this ſnail-pallet, receives the lower end of the ſpring R, while the upper end is made faſt to a piece in the cock H, near W, in a manner ſimilar to that by which the pendulum-ſpring is fixed. On the arbor of the remontoir, and under the plate A B, is twiſted on a pallet I, which may be called the remontoir's impelling pallet, becauſe it gives motion to the balance by ſtriking the pallet K, which the author alſo called the *impelling* pallet; but is, as we have named it, more properly the *impelled* pallet, becauſe it *receives* the impulſe which pallet I imparts. From this detail of the pallets it is obvious, that whenever the balance-wheel impels the large pallet M, the remontoir-ſpring R, the ſnail-pallet N, together with its ſmall ruby pallet and impelling pallet I, muſt have a contemporary motion, and will deſcribe each its reſpective circle round the common arbor W V.

In *fig.* 2, *a* is a detent-ſpring, fixed by a ſcrew and ſteady pin to the upper face of the potance plate, and pointing directly to the axis of the verge, which it nearly approaches. Its ſhape and mode of being fixed will be better underſtood from an inſpection of *fig.* 3, which gives a ſide view of it, and from which it will be ſeen that it is placed high enough above the plate, to come in the way of the diſcharging pallet I, in each revolution of the latter. Upon the ſide of the ſaid detent-ſpring *a*, next to the balance-wheel, a ſecond, but very ſlender ſpring is pinned, ſo that its projecting end exceeds that of the detent-ſpring, and comes nearer to the balance verge than that of the detent, as repreſented at *m*, in *fig.* 3. To the detent-ſpring is made faſt a ſmall ruby pallet *r*, ſeen alſo in *fig.* 3. The cock *b* is ſcrewed to the potance, and the hole at *i* is tapped to receive the ſcrew *c*, the head of which, being turned towards the centre of the ſnail, forms a banking for the detent-ſpring *a*, when ſtruck by the ſnail-pallet. This ſcrew, *c*, is removed from its due place in the drawing, to avoid confuſion; and for a reaſon, which will preſently appear, this double ſpring *a* may be called the remontoir-detent. On another ſide of the balance-wheel a ſecond detent-ſpring *d* is fixed, pointing towards the centre of the remontoir axis, and forming nearly a right angle with the former detent: this ſecond detent is adjuſted by the ſcrew *y*. The ſhape and ſituation of this ſecond detent, which may be called the locking detent of the balance-wheel, is better ſeen in *fig.* 4, where its ſapphire pallet *s* may be diſtinguiſhed, againſt which the tooth 3 of the wheel is ſuppoſed to be reſting in *fig.* 2. The ſituation of this pallet *s*, as adjuſted by the ſcrew *y*, determines the reſting-places of the teeth 1 and 2, while they are equally free from the edge of the pallet M. The locking detent *d* is likewiſe made faſt by a ſcrew to the upper face of the potance plate; and the ſcrew *f*, placed in the cock *e*, but out of its place in the figure, forms a banking to the ſapphire pallet *s*, while the ſcrew *g*, in a ſmaller cock, limits the excurſion of the detent-ſpring *d* itſelf.

Having deſcribed the various pieces of mechaniſm that compoſe this remontoir eſcapement, we will now proceed to explain its mode of acting. Firſt, let us ſuppoſe all the parts at reſt with the wheel locked by the pallet *s* laſt deſcribed, but that the wheel is ready to move in the direction of the arrow Z, by the action of the main-ſpring, through the train, whenever the ſingle detent *d* is by any force impelled back to its banking. Let it alſo be conceived that, when the ſnail-pallet N is carried by any means in a direction contrary to that of the wheel till it reaches the pallet *r* of the remontoir or double detent-ſpring, the remointoir-ſpring is wound up by ſuch motion, and *vice verſâ*, and alſo that when the ſtroke of the ſnail-pallet has driven back this detent to its banking, by ſtriking the ſloped face of the ruby pallet *r*, the detent will inſtantly return by the force of its elaſticity, and the back part of the ſaid ruby pallet *r* will hold the ſnail-pallet locked at its return, till ſome other impulſe ſets it free again; but that when the unlocking takes place, the ſpring of the remontoir, now wound up, returns by its own force to its original ſituation; and that its axis brings back with it its affixed pallets. *Fig.* 1. will be of no uſe in deſcribing the action of this eſcapement, but in *fig.* 2. the parts are repreſented in a ſtate ready to commence motion; the wheel is locked by tooth 3 againſt the ſapphire pallet *s* of the ſingle detent *d*, and the remontoir-ſpring is wound up, and kept locked by the ſnail-pallet, reſting behind the ruby pallet *r* of the double detent: in this ſituation let the balance be wound round in ſuch direction that the diſcharging pallet I may ſtrike the end of the double or remontoir detent outwards, taking both its ſprings along with it, and thereby unlocking the ſnail-pallet; at this inſtant the remontoir-ſpring R begins to return, and brings all its four pallets with it, and during the return, pallet L, which we have called the impelling pallet, gives its ſtroke to pallet K on the verge, which we have called the impelled pallet, and through its medium to the balance itſelf, which now goes on in its oſcillation, till the pendulum or balance-ſpring is wound up; in the mean time the little unlocking ruby pallet, inſerted into the ſnail-pallet near its centre of motion, meets with the extreme end of detent *d*, and drives it back to its banking, thereby ſetting the tooth 3 of the wheel free from the ſapphire pallet *s*; the wheel, being urged by the train, now proceeds till tooth 2 falls into the notch of pallet M, and is checked, experiencing a little recoil; the balance-ſpring, being now wound up, returns, and alſo the wheel winds up the remontoir-ſpring, till it is again locked, by the ſnail-pallet, after its impulſe has driven back the detent: during this action the detent *d* returns by its own ſpring, and locks tooth 4 of the wheel, at the inſtant that tooth 2 eſcapes from the notch of pallet M, and the original ſituation of all the acting parts is now reſtored, except that the motion of the balance is not arreſted, the ſlender ſpring fixed to detent *a* allowing it to paſs in the return, without unlocking the ſaid detent; but when the backward oſcillation is finiſhed, and the balance returns, the ſame operation is repeated; *i. e.* the remontoir detent is unlocked, the pallets on its axis are brought back, the impulſe is next given to pallet K of the balance, to reſtore its loſs of momentum, and laſtly, the wheel is unlocked to wind up the remontoir-ſpring as before.

The obſervation we have to make on this ingenious remontoir eſcapement is this, that on enquiry from good authority we learn, that the theory could never be completely put into practice by the inventor, though he laboured to obtain his object full fourteen years. However well the parts of action were mechanically made, the locking of the remontoir-ſpring was never certain, for the *flirt* given to the ruby pallet *r* by the ſnail-pallet, drove back the detent ſo far, as frequently to enable the ſnail to be brought back by the remontoir-ſpring, before the detent returned to catch it, the conſequence of which was, that the pallets of the remontoir axis were all brought back, and the impulſe vainly expended.

pended by pallet L, before the pallet K of the balance returned to its place to receive it. Mr. Hardy, however, has lately contrived a fpecies of locking that is quite certain, as well as fafe, which we fhall prefently have occafion to defcribe.

Breguet's.—We have not been able to afcertain the exact date of the French watch-maker Breguet's invention of a remontoir efcapement, but as Berthoud has placed it after Haley's in his "Hiftoire de la Mefure du Temps," we muft conclude that it is of more recent origin. The mechanifm is fo complex, that the defcription given in the French work is divided into three parts, which plan we propofe to follow. *Figs.* 5. and 6. of *Plate* XLI. of *Horology*, will fuffice to explain the different parts of Breguet's remontoir; *fig.* 5. being a plan of the whole, and *fig.* 6. exhibiting a perfpective view of fuch parts as could not be well underftood without fuch reprefentation. The firft portion of the mechanifm confifts of two wheels of unequal numbers of teeth, with their planes in contact, and revolving on one common arbor concentrically, namely, the wheel B B′, and the wheel D, the firft having many teeth and a larger diameter, and the fecond having few, for a reafon which will appear hereafter: thefe wheels are actuated by a pinion *g*, connected with the train, and having juft fo many leaves as there are teeth in the large wheel B B′, contained in the fpace between two contiguous teeth of the fmall wheel D; on the arbor of this pinion, near the pivot, is inferted a fly with unequal arms, reprefented by the letters *i g h* in both figures, the longer arm of which, *g h*, is armed with a piece of fteel of peculiar fhape, feen in *fig.* 6. A fpring-detent *r* F, made faft to the cock at *r*, lies at right angles to the fly, when in a ftate of reft, and has a ruby pallet V at about one-third of its length from its interior end, which ftops the fly by oppofing its end *h* near *o*, and confequently prevents the train from urging the double wheel, fo long as no force unlocks the pallet from the end of the fly; but if any impulfe happen to bend the detent in a direction towards the pinion *g*, then the ruby pallet flips through a notch made in the fteel part of the fly, and the fly makes an entire revolution, while the pinion revolves and urges the large wheel the fpace of as many teeth as the pinion contains; that is, a fpace correfponding to the diftance between two contiguous large teeth in the fmall wheel, which wheel alfo moves the fame quantity, before the fly is again oppofed and ftopt.

The fecond portion confifts of the following parts; a fpring G, curved at the interior, or moveable end, ferves at the fame time for a remontoir-fpring, and alfo for a pallet to give the impulfe to the balance; it has confequently a heel-piece *n*, by which it is urged into a ftate of tenfion by the fmall wheel D, at every revolution of the fly, and alfo a ruby pallet *m*, with a locking notch cut in it; *a* H, made faft at *a*, is a detent-fpring, placed nearly at right angles to the fpring G, and having a flender fpring N made faft to its fide; this detent-fpring carries a fmall ruby *p*, which locks into the notch of the piece *m* in the fpring G, when this is in a ftate of tenfion. Another ruby, inferted into the extreme end of the detent *a* H at *s*, is fo placed, that the flender fpring N can move from right to left without impediment, but when urged in an oppofite direction againft the ruby *s*, it takes the large or detent-fpring along with it, and unlocks the remontoir-fpring G, by removing the fmall ruby *p* from the notch of *m*. The detent-fpring H has at its extreme end a claw, againft which the ruby *m* falls when this detent efcapes back from a tooth of wheel D; and this claw has a pin in it, againft which the flender fpring N refts; again, at the extreme end of the flender fpring N are fixed two parallel chamfered plates, fo contrived, that any preffure made on the upper flope, or inclined edge, will deprefs the fpring, but a fimilar preffure made on the under flope will elevate it again, the downward preffure being employed to difengage the ruby *m* from the claw, and the contrary.

The third portion confifts of the pieces or pallets K and *b*, borne by the fuperior end of the verge of the balance, and fixed at a quarter of a circle from each other. When the ofcillation of the balance is from right to left, or in the direction from *b* to K, the unlocking pallet piece K, in its motion, bends the flender fpring, and paffes over it, but as the pallet *b* is placed above the plane of the wheel B B′, and under the fpring H, this ofcillation from right to left is performed in freedom, except that the flender fpring N requires to be a little bent out of the way; but when the ofcillation is made from left to right, the pin at H preffes this flender fpring againft the ruby *s*, the fpring H then gives way, and bending, allows the ruby *p* to efcape from the notch at *m*, and the fpring G, being unlocked, is at liberty to perform its office, which we now proceed to defcribe.

The action of thefe three portions of the mechanifm may be thus explained: at the inftant that the ruby *p*, in the detent *a* H, is difengaged from the notch at *m* in the remontoir-fpring G, the pallet *b* of the balance is found pointing towards the common centre of the wheels, and ready to receive an impulfe from the extreme end *q* of the curved part of the remontoir-fpring G, which end now gives its ftroke to the balance through the medium of pallet *b*; immediately after which ftroke, the fame end *q* proceeds till it falls on the end F of the fly's detent, to which it now gives a pufh, and remains quiet; this pufh unlocks the fly, as has been before explained, and an entire revolution is immediately performed by it, while its pinion *g*, now at liberty to advance, urges on the double wheel, till another tooth in the fmall wheel, catching the heel-piece *n* of the remontoir-fpring, puts this fpring again into a ftate of tenfion; and the ruby *p*, falling into the notch of *m*, locks it as before, the balance in the interim completing its ofcillations: and in this manner the operation is repeated.

De Lafon's.—In the year 1801, the Adelphi Society for the Encouragement of Arts, &c. rewarded Mr. John de Lafons with thirty guineas for his invention of a new remontoir watch efcapement, which comes next under our notice. *Figs.* 1 and 2 of *Plate* XLII. of *Horology*, are a plan and fection of this efcapement, as given by the inventor in the Society's Tranfactions of the year above-mentioned. In both thefe figures, A is the efcapement wheel; B, the lever-pallet on an arbor with fine pivots, which has, at its lower end, the remontoir fpiral fpring C, fixed with a collar and ftud in the ufual way; D is the pallet of the verge, having a roller turning in fmall pivots, for the lever-pallet to act againft without friction; at E are the pallets for difcharging the locking, with a roller between, containing a fmall notch; F is the arm of the locking pallets, continued at the other end beyond the centre of motion, to preferve the equipoife, and having ftuds and fcrews for adjuftment of the banking; *a* and *b* are the locking pallets, being portions of circles faftened on an arbor, turning on fine pivots at the midway between the pallets; G is the triple fork, at the end of the arm of the locking pallets. In *fig.* 1, tooth 1 of the wheel having caught the interior end of the lever-pallet B, has urged it forwards and wound up the remontoir-fpring, and the inftant that the verge pallet D comes nearly in contact with the remote end of the lever-pallet, the difcharging pallet E, taking one prong of the fork, removes the arm F, thereby relieving the tooth 3 from the convex part of the claw *a*, that locks the wheels.

 The

The wheel now advances a little way, juſt enough to allow the interior end of the lever-pallet to paſs back again, as urged by the ſpiral ſpring, while the remote end of the ſaid lever gives an impulſe to the balance, through the pallet D; the tooth 4 is then locked on the concave ſide of the locking claw *b*, and the interior end of the lever-pallet partly ſupports the following tooth; in this ſituation the oſcillation is finiſhed, and on the return of the balance pallet E, ſtriking the prong of the fork in a contrary direction, again unlocks the wheel from claw *b*, while the force of this wheel is partly ſuſpended by the end of the lever-pallet; being now at liberty, the wheel proceeds to wind up the remontoir-ſpring again, till another tooth falls on the claw *a*, now brought inwards, when the operation is completed, and the preſſure of the inner end of the lever pallet againſt its contiguous tooth again relieves the claw *a* from a part of the preſſure of the wheel, and thereby renders the unlocking as eaſy as before, when claw *b* was unlocked. This eſcapement appears to us an improvement on Haley's, both as to its ſimplicity and ſafety of locking, and the inventor propoſes a ſtill further ſimplification, by ſubſtituting a ſtraight remontoir-ſpring for the ſpiral one, particularly in time-pieces intended to remain in a ſtationary ſituation, where the weight of the ſpring would form no objection; but it does not appear that ſuch conſtruction was ever adopted. The ſimplicity of the locking of the remontoir-ſpring by the wheel itſelf, where this ſpring and the maintaining power mutually re-act, requires, as might be foreſeen, an unuſually ſtrong maintaining power, which is, perhaps, one of the greateſt objections to this mode of applying the remontoir; unleſs, indeed, what was remarked by the ſociety's committee be true, that the balance muſt vibrate in a large arc before the piece will continue to perform.

Maſſey's.—Mr. Edward Maſſey, of Hanley, in Staffordſhire, received a reward of fifty guineas from the Adelphi Society, in the year 1803, for his invention of two different clock eſcapements, one of which we have already deſcribed, under our article Escapement, as being without a verge; and the other we ſhall make the ſubject of our preſent ſection. *Fig.* 3. of *Plate* XLII. repreſents ſo much of Maſſey's remontoir eſcapement as is ſufficient to explain all the eſſential parts; in which A is the ſwing-wheel; B, C, are two detached pallets, moving on ſeparate arbors, at oppoſite ſides of the ſwing-wheel; B is ſeen urged by the remontoir-ſpring E, by the aid of a tail-piece fixed on the arbor of pallet B, which receives the action; and the other pallet muſt be conceived to have a ſimilar ſpring and tail-piece, which cannot eaſily be repreſented in the drawing; F is the verge bearing two arms without pallets, which preſs under the pins of the detached pallets, and raiſe them alternately, at each vibration of the pendulum, from the teeth of the ſwing-wheel, which had been previouſly impelled by the maintaining power, and K is the pendulum ſuſpended in the uſual way, and having a crutch to communicate its force to the verge at the moment of withdrawing the pallets. All, therefore, that the pendulum has to do, is, to diſengage the pallets from the teeth of the ſwing-wheel that locks againſt the inclined planes of the pallets; for the remontoir-ſpring then oppoſes the aſcent of the pendulum, and aids its deſcent, by means of the connection between the pins of the pallets and the arms of the verge, which muſt, from the nature of the conſtruction, continue in contact during a large portion of each excurſion of the pendulum. It is not ſtated in the author's account (Tranſ. Adel. Soc. vol. xxi.) at what part of the arc of vibration the pendulum unlocks either of the pallets, but it ſeems to us capable of being made to effect this office at its point of *greateſt velocity*, if it does not do that as now conſtructed. An attention to this particular is of the utmoſt importance in any eſcapement, ſince the natural law of gravity is the leaſt deranged, when any force is given, or taken away, when the pendulum is at the *loweſt point* of its arc, for the velocity it has at *that point* determines the height to which the ball ſhall riſe; but any addition or diminution of the pendulum's momentum, beginning at any other point of the arc, either adds to, or diminiſhes the natural length of the vibration, and injures the iſochroniſm. Beſides, a ſpring commencing both its accelerating and retarding influence at the point of the pendulum's greateſt velocity, acts, not only by a law commenſurate with the law of gravity, but their various forces, thus exerted, are contemporary, and, therefore, act together as one; a conſideration which never ſhould be loſt ſight of in the conſtruction of an eſcapement of any denomination. The principal advantages that the author ſeems to inſiſt on in this eſcapement, are, that the friction is diminiſhed at the acting parts of the pallets, the impulſe being given by a direct *puſh*, without, or with very little, ſliding motion; and that a certain regular momentum is kept up in the pendulum, independently of any variation which may occur in the wheel-work, or in the acting part of the pallets during the ſhort time of unlocking; but, what is the principal object of ſuch a contrivance, he has not ſtated, perhaps not conſidered, how this *certain regular momentum* is, or ought to be, *modified.*—Indeed, in ſpeaking of his other eſcapement without arbors, he ſays, that one of its advantages over this is, that during a *part* of the vibration the pendulum is *diſengaged;* from which remark, it ſhould ſeem, that he thinks the conſtant action of this remontoir-ſpring, however modified, an objection to be avoided.

Antis'.—In vol. xxiii. of the Tranſactions of the Society laſt mentioned, it is ſtated, that Mr. John Antis of Fulneck, near Leeds, ſent this ſociety a model of a new clock eſcapement in 1805, with a correſponding deſcription, for which a reward was voted him of twenty guineas. The model is depoſited in the room at the Adelphi appropriated for the reception of models, but as the deſcription is not publiſhed in the Tranſactions of the Society, nor an engraving given of the model, we muſt infer, that the contrivance either is not new, or is not of ſuch importance as to merit the particular notice that has been given by them to the inventors of like contrivances, who both preceded and followed him. Indeed, in the letters of Mr. Antis to the Society, publiſhed with their account of the rewards, it ſeems that two eſcapements were ſent them, one a *detached* one, and the other, one that "*would equaliſe* the *power* of the *impulſe,*" which muſt, therefore, have been of the *remontoir* deſcription; but he confeſſes that theſe contrivances by him "may have been practiſed before." With reſpect to the latter eſcapement, he remarks, what is worthy of being recollected by clock-makers in general, that, by its means "a ſpring-clock will be as perfect as one which goes by weight, and more ſo if the latter has no remontoir." Should any of our readers have a wiſh to ſee and examine Mr. Antis' productions, which, he acknowledges, are done in a rough manner, and under diſadvantageous circumſtances, but which may not be the leſs ingenious on that account, there will be no difficulty in gaining admiſſion to the model room, where the original work is preſerved with a view to public inſpection, provided the applicant be properly introduced.

Mendham's.—It frequently happens that conſiderable improve-

improvements are made in mechanical contrivances, by men whoſe principal employment has no connection with the art to the improvement of which they contribute; this was probably the caſe with Mr. Antis, and is confeſſedly ſo with Mr. Mendham, who was rewarded with the ſilver medal of the Adelphi Society, in the year 1807, (ſee vol. xxv.) for a remontoir eſcapement for a chronometer, which next offers itſelf for our deſcription. *Figs.* 4, 5, and 6, of *Plate* XLII. are copied from the plate given by Mr. Mendham, *fig.* 4. being a ſection of the whole eſcapement, *fig.* 5. a plan of the wheel and locking pallet, and *fig.* 6. a detached view of the pallets and their common arbor. The letters *a a*, in *fig.* 4, are the two plates of the frame that contain the works; *b* is the balance, nearly of the uſual ſhape and ſize; *c* is the pallet that acts with the wheel, and *d* its arbor; *e* is the locking ſingle ſpring; and *f* is the unlocking pallet in the form of a pin, attached to *e*, and ſeen better in *fig.* 6. At the lower end of arbor *d* is placed *g*, the remontoir-ſpring, that gives force both to the ſmall pallet *f*, and alſo to the perpendicular pin *h*, that forms a part of the pallet *c*. The arbor *d* does not aſcend above the upper plate of the frame, but the pin *h* does, and the plate being formed, as we ſuppoſe, like a rim, *h* is permitted to move in the vacant ſpace within the rim, and ſufficiently high to fall in the way of the finer pin *l*, inſerted into and under the balance. The balance *i* has its verge immediately over the upper pivot of arbor *d*, point to point nearly, ſo that the motions of the balance-pin *i*, and of the pin *h*, that impels it, may be concentric. The cock *k* ſupports the upper pivot of the verge, and alſo holds the outer end of the balance or pendulum-ſpring. The action of theſe parts may be thus explained; before motion is given to the balance, we muſt conceive its pin *i* to be in contact, or nearly in contact with the pin *h* of the pallet *c*, and it muſt be remembered, that the wheel reſts on the pallet during the vibration of the balance: ſuppoſe now that the remontoir-ſpring *g* is in the act of being wound up by the balance-wheel, as actuated by the train; and that the pin *i* of the moving balance has kept in contact with the impelling pin *h*, borne by the arbor of the remontoir, it is evident, that the force of the remontoir-ſpring is here oppoſed to the momentum of the balance and at laſt ſtops it, thereby forming a banking of an unlimited extent; but the tooth of the wheel in contact with pallet *c*, at the point of ſhorteſt radius, aſſiſts to oppoſe or wind up the ſaid ſpring during a ſpace of one-fifth of a circle, till the tooth has paſſed and the wheel is locked; on the return of the balance the remontoir-ſpring exerts all its force on the balance, but without a jerk, the pins *i* and *h* being already in contact, till it comes to the ſtop; the balance then maintains its motion, and the ſmall pallet *f* having unlocked the wheel by forcing out the detent, the next tooth of the wheel falls on the pallet *c*, and waits the return of the balance, when the ſame operation is repeated.

Mr. Ramſay of Iſlington was conſulted on the merits of this eſcapement, and, on conſidering its properties, he gave a written report to this effect; that the impulſe is here given to the balance without friction, exactly as in Mudge's chronometer; and that the remontoir is wound up by the maintaining power in a ſimilar way, except that the ſhape of Mudge's tooth limited the quantity of tenſion of the ſpring, which here depends on the amount of the maintaining power; but that this eſcapement, having but one remontoir-ſpring, is more ſimple than Mudge's, and allows an unlimited arc of vibration, which advantages compenſate the foregoing diſadvantage in the ſhape of the tooth; and laſtly, that it is ſuperior to the detached eſcapement, inaſmuch as that it gives the impulſe without friction.

G. Prior's.—The ſilver medal and twenty-five guineas were voted to Mr. George Prior, junior, of Otley, in Yorkſhire, in the year 1809, by the Adelphi ſociety (vol. xxvii.) for his invention of a new clock eſcapement, in which an impelling (not however called a remontoir) ſpring actuates, or profeſſes to actuate, the pendulum at each alternate vibration. The account, particularly of the action, as printed in the ſociety's Tranſactions, ſeems to us ſo extraordinary, that we beg leave to tranſcribe it verbatim, in order to make ſome remarks on it, that may enable the reader to judge of the eſcapement's peculiar qualities, or at leaſt of the manner in which they are deſcribed. *Fig.* 7. of *Plate* XLII. already referred to, contains enough of this eſcapement to anſwer the purpoſe of the deſcription in queſtion, and therefore we will omit the references to *fig.* 1. in the original plate, and attend to its *fig.* 2. only. "*Fig.* (7) is a back view, which is ſuppoſed to be taken from behind the clock; *a* repreſents the axis of the ſwing-wheel, or laſt wheel of the train of the clock; *b d* is the ſwing-wheel fixed upon it, having thirty ſerrated teeth; it is turned round, in the direction from *b* to *d*, by the maintaining power of the clock; *g* is a ſpring-detent, which locks againſt one of the teeth of the ſwing-wheel, and this prevents its running down by the action of the maintaining power; *h* is another ſpring-detent, which is called the impelling ſpring; when left at liberty, it unlocks the former, by puſhing againſt the end of the ſmall arc *e*, faſtened to the detent *g*, and thus removing the end of the detent which obſtructed the wheel's motion; *l* is the rod of the pendulum, ſuſpended by a cock ſcrewed to the back plate of the clock; a ſmall piece of braſs, *k*, projects at right angles from the impelling ſpring *h*, ſo as to intercept the pendulum-rod in its vibration, and at this place a ſmall ſcrew is put through the pendulum-rod *l*, the point of which moves the impelling ſpring back: a ſmall pin is fixed to the frame, in a line between the point of ſuſpenſion of the pendulum, and the centre of the ſwing-wheel, againſt which the impelling ſpring ſtops when at liberty.

"Suppoſing the pendulum to be vibrating backwards and forwards, and the wheel locked as in the figure, the pendulum ſwinging from *m* to *n*, the impelling ſpring *h* follows by its elaſticity, until the pendulum *l* arrives at its perpendicular (or loweſt point of its arc); at this period the impelling ſpring comes *to reſt* againſt the end of the arc *e*, which it *puſhes back*, ſo as to releaſe the tooth of the wheel from the detent-ſpring *g*; the wheel now moves round a very ſmall ſpace before it meets the end of the impelling ſpring *h*, and *is ſtopped thereby*; in the mean time the pendulum continues its motion the extent of its vibration towards *n*, when it returns, and arriving at the *perpendicular*, it meets the *impelling ſpring h*, and *carries it along with it*, until the tooth of the wheel which reſts againſt it eſcapes from the end of it, and another tooth of the wheel comes to reſt againſt the ſpring-detent *g*. The ſucceeding vibration of the pendulum repeats the ſame operation." This is the whole account, from which the reader is induced to believe that the pendulum, and impelling ſpring *h*, act and re-act on each other *ad infinitum*, without any aid from the maintaining power through the medium of the wheel, which is never ſaid to raiſe, or otherwiſe to move the impelling (or remontoir) pallet, in order that it may, in its turn, impel the pendulum by its acquired force. Again, it is ſaid, the impelling ſpring unlocks the detent *g* when it arrives at the perpendicular, or nearly, if not quite, at the point of its quieſcence, where it has *no force*, and from that point, not the

the wheel, but the pendulum unbends it into a ſtate of tenſion, to reſume its operation; that is, the pendulum raiſes the ſpring, that the ſpring may drive the pendulum back again; and thus the vibrations are maintained without the aid, and conſequently without the neceſſity, of either a maintaining power or train; and what is equally extraordinary, the impelling ſpring locks the wheel by its *reſiſtance*, near the perpendicular line (within the ſpace of one tooth), where it has almoſt the ſmalleſt force, or nearly the ſame ſmall force that it has when it *unlocks* the *detent*. According to this account, even ſuppoſing the action of the wheel on the remontoir pallet to be *omitted*, the pendulum is under the influence of the impelling pallet during nearly the whole of one excurſion, and free from it in the other. The model, however, is ſaid to be preſerved in the ſociety's room, and will explain itſelf.

G. Prior's improved.—Mr. George Prior, jun. again preſents an improved remontoir eſcapement for a clock to the ſociety we have repeatedly had occaſion to name in this article, in the year 1811 (vol. xxix.), and again receives a reward at their hands, of twenty guineas, for his improvement. The deſcription given of this eſcapement, and of its action, is ſufficiently clear, and the alteration made in the conſtruction is in many reſpects an improvement, though we ſhall take occaſion to ſhew, in our remarks on it, that it is ſtill liable to objections. In the plate laſt referred to, *fig*. 8. contains the original *fig*. 1, which will explain the account with ſufficient preciſion. "The ſwing-wheel A," ſays the author, "has thirty teeth cut in its periphery, and is conſtantly urged forwards by the maintaining power; C, D, are two ſpring-detents, catching the teeth of the wheel alternately; theſe are, at the proper intervals, unlocked by the parts marked 2 and 3 upon the pendulum-rod H, intercepting ſmall pins *a*, *b*, projecting from the detents, as it vibrates towards the one or the other; E is the renovating or remontoir-ſpring, fixed to the ſame ſtud, F, as the detents; it is wound up by the higheſt tooth of the wheel, its poſition, when unwound, being ſhewn by the perpendicular dotted line. This being the caſe, ſuppoſe a tooth of the wheel caught by the detent D, which prevents the wheel from moving any further, and keeps the renovating ſpring from eſcaping off the point of the tooth; in this poſition the pendulum is quite detached from the wheel: now if the pendulum be cauſed to vibrate towards G, the part of it marked 2 comes againſt the pin *b*, projecting from the renovating ſpring E, and puſhes this ſpring from the point of the wheel's tooth; on vibrating a little further, it removes the detent D, which detained the wheel, by the part 3 ſtriking the pin *a*, which projects from the detent; the maintaining power of the clock makes the wheel, thus unlocked, to advance, until it is detained by a tooth reſting upon the end of the detent C, on the oppoſite ſide; by this means the renovating ſpring will be clear of the tooth of the wheel as it returns with the pendulum, and gives it an impulſe by its pin *b* preſſing againſt the part 2 of the pendulum, until the ſpring comes to the poſition ſhewn by the dotted line, in which poſition it is unwound, and reſts againſt a pin fixed in the croſs-bar of the plate; the pendulum continues vibrating towards I, nearly to the extent of its vibration, when the part 1 meets the pin in the detent C, and removes it from the wheel and unlocks it; the maintaining power now carries it forward, puſhing the renovating ſpring E before it, until another tooth is caught by the detent D, which detains the wheel in the poſition firſt deſcribed." Agreeably to this deſcription, the pendulum is oppoſed in its aſcent, and accelerated in its deſcent, in one of its excurſions from the centre, or perpendicular line, by both the impelling ſpring E, and detent D, after it is unlocked, though not the whole diſtance; but is alternately accelerated and retarded in the other excurſion by the detent C alone, and only after the unlocking; theſe unequal checks, given at different parts of the arc of vibration, we apprehend, are by no means favourable to the natural iſochroniſm of the pendulum.

Hardy's.—Mr. Hardy, chronometer-maker, of Coppice-row, Clerkenwell, London, whoſe ingenious improvements in clock and watch-making have placed him high in public eſtimation, on contemplating Mr. C. Haley's failure in the locking of his remontoir detent, contrived a remedy which fully anſwers its purpoſe, and which, on that account, merits our particular notice.

Fig. 7. of *Plate* XLI. of *Horology*, repreſents the plan of Hardy's new remontoir eſcapement, which, though it appears to differ conſiderably in conſtruction from Haley's, yet reſembles it ſo much in its action, as well as principle, that we will put the ſame letters of reference to the ſame correſponding parts, in order that the reader who has peruſed our deſcription of Haley's remontoir with attention, may the more clearly comprehend our deſcription of Hardy's, and perceive in what the improvement conſiſts. As in Haley's conſtruction, T T repreſents the balance, or rather the place of the balance not ſeen; E the eſcapement, or balance-wheel; P the pivot of the verge; S the balance-ſpring, ſuſpended as uſual; K the impelled pallet; I the unlocking pallet of the remontoir; N the remontoir detent, or locking pallet, on the ſame axis with *a*, the arm of the ſlender or unlocking ſpring, which lies parallel with it; *n* the cock, taking the pivot of the remontoir detent's arbor, and alſo the upper end of its ſpiral ſpring; *c* the head of the banking ſcrew, and *i* its ſtud; H the cock that takes the pivot of the remontoir, or cylindrical ſpring at W, and alſo one end of the ſaid ſpring, repreſented by the ſmall circle at R; L is the impelling lever or pallet, for giving motion to the pallet K of the balance; M is a lever, inſtead of the notch in Haley's large circular pallet, by means of which the wheel winds up the ſpring R of the remontoir; O is a lever, in place of the ſmall ruby unlocking pallet of Haley, fixed to the ſnail; *d* is the ſpring-detent of the wheel, and *s* its ruby, or ſapphire pallet, on which the wheel is locked, *f* and *g* being the banking ſcrew and its ſtud, and *y* the place where it is fixed by a ſcrew on the potance plate; and laſtly, the sectoral piece Q, in which the improvement chiefly conſiſts, is the addition introduced for rendering the locking of the remointoir-ſpring, when wound up, both certain and ſecure. The acting faces of the pallets *s*, M, K, N, and of the circular portion of Q, are of ruby, or other precious ſtone nicely poliſhed, and are properly ſhaped for their reſpective offices. From this deſcription of the mechaniſm before us, it will be perceived that the double ſpring *a*, and the detent with a claw pallet at N, are both faſt to the arbor of the ſpiral ſpring, which gently preſſes the detent at all times down upon the ſectoral piece Q, thereby preventing its being thrown back from its poſition for locking, as was the caſe with Haley's remontoir detent. It will alſo be ſeen that the arbor W, of the remontoir-ſpring R, has the ſaid ſectoral piece, the impelling lever L, the lever M impelled by the wheel, and the diſcharging lever O, all faſt to it at different heights, the two firſt being above, and the two laſt below, ſo that when one of theſe four pieces is moved out of its firſt ſituation by any external force, they all move together, as do alſo the remontoir detent N and its ſpring *a* on their common arbor, and as do likewiſe the pallets K and I on the verge of the balance. Theſe particulars being underſtood, and that levers in Hardy's conſtruction are uſed ufed for pallets in feveral inftances, we may proceed to explain the action of the different parts, and to fhew how they produce the defired effects of producing equable impulfes on the balance at all times, and of effecting certain and fecure locking of the remontoir-fpring. In the figure before us, tooth 1 of the wheel has juft proceeded far enough, by the action of the main-fpring and train, to wind up the remontoir-fpring, and its detent N, urged by its own fpiral fpring againft the curved face of the fectoral piece, after fliding fmoothly along it, has locked it; that is, by oppofing it at right angles, prevents its return, which would take place when the wheel leaves the arm M, or, which is the fame thing, when the remontoir-fpring, on the fame arbor with this arm, is wound up. The balance is now fuppofed to have commenced its motion, carrying the pallets K and I outwards from the wheel, but the fmall pallet I has not yet arrived at the flender fpring of the remontoir detent, though it is approaching it; the flender fpring, however, foon yields to the outward impulfe of this fmall pallet, without difturbing its arm or the detent N, therefore the remontoir-fpring remains locked: at the return of the balance, which takes place when its fpring is wound up, the faid little pallet I difplaces not only the flender fpring but its arm alfo, which lies next towards the wheel, the one not being liable to move in this inward direction without the other; the confequence is, that the detent N is now lifted by its connection with the arm of the flender fpring, and at that inftant the remontoir-fpring, being wound up, and at liberty to act, throws back the fectoral piece Q nearly its whole breadth, but not quite, for the claw of the detent in queftion refts on the curve of the fector during the motion we have defcribed, till the unlocking lever O ftrikes the end of the fingle detent-fpring *d*, and forces the ruby pallet *s* out of the wheel, which wheel inftantaneoufly refumes its motion from the train, and meeting with lever M winds up the remontoir-fpring a fecond time; in the mean time the impelling lever L has given its impulfe to the pallet or lever K, and confequently to the balance, which has thus had its momentum increafed; and on its returning vibration exactly the fame procefs is repeated. Thus the force of the maintaining power is expended in winding up at each alternate ofcillation the remontoir-fpring, which fpring, fo reinforced, impels the balance in its turn by quantities of force that are always the fame, whatever may be the irregularities of the force tranfmitted through the train that winds up the remontoir-fpring; therefore, properly fpeaking, the remontoir-fpring is the maintaining power that reftores, at each alternate ofcillation, the lofs of momentum that the balance has experienced from the refiftance of the air, and friction of the parts in action. This efcapement is, notwithftanding, of the detached kind, feeing the balance performs the greateft part of its ofcillations without any connection of even the auxiliary force derived at intervals from the remontoir-fpring; and a chronometer of this conftruction, made by Hardy for his royal highnefs the duke of Suffex, anfwers the maker's moft fanguine expectation, and will probably become a model for other makers when its merit is proved, and made public. The fame artift has made feveral experiments on the fame principle, with this difference, that the remontoir-fpring is retained by the lever refting on the tooth of the wheel, in place of being locked by the fectoral piece, as in the one before us.

Befides the preceding efcapement, Hardy invented one for the aftronomical clock he made for the Royal Obfervatory at Greenwich, which at firft fight appears to refemble Maffey's and Prior's improved one, with fpring pallets, but on clofe examination we difcovered that it has the following defirable properties peculiar to itfelf: 1ft, it gives the impulfe not only from remontoir-fprings, but ftrikes and unlocks the two feparate detents alternately, at the inftant when the pendulum has the *greateft velocity*; and, 2dly, the accelerating and retarding power imparted to the pendulum co-operates with the force of gravity, thereby deranging the natural law of gravity as little as poffible. We feel not at liberty to enter more minutely into the defcription of this efcapement, becaufe we underftand its inventor propofes giving an account of his machine himfelf, to be read before the Royal Society, as conftituting the companion to Troughton's tranfit circle, already defcribed by the aftronomer royal. We have permiffion, however, to fubjoin the rate of the clock in queftion for three quarters of a year, as taken at the Royal Obfervatory, which will afford the beft proof of its pretenfions to public notice. We have further to add, that the inventor has applied the fame principle of movement to a chronometer of a large fize, in which the balance has two metallic rods to adjuft it for temperature, in place of the metal being fluxed on the rim of the balance, as is common. See COMPENSATION.

Trial of Mr. Hardy's clock at the Royal Obfervatory at Greenwich in the year 1811.

	Daily Rate.	Mean of Thermometer.
	″	°
April 12 to May 13 - -	— 0.86	61
May 13 to June 9 - -	— 0.74	63
June 17 to July 12 - -	— 0.74	64
July 18 to Auguft 17 - -	— 0.83	68
Auguft 17 to September 14 -	— 0.77	65
September 14 to October 4 -	— 0.61	60
October 17 to November 17 -	— 0.60	51
November 17 to December 10	— 0.50	46
December 10 to December 30	— 0.47	39

From the above ftated authentic document it appears, that for the nine months of trial, and in oppofite extremes of temperature, the daily rate never varied quite *four-tenths* of a fecond; and when the quantity of mercury proper for the compenfation of the pendulum is exactly afcertained, it may be expected that the performance will be ftill more accurate; feeing that the deviation in the rate depends evidently on the change of temperature, and not on the efcapement.

REMONTOIR ESCAPEMENTS.

Fig. 1. Haley's.

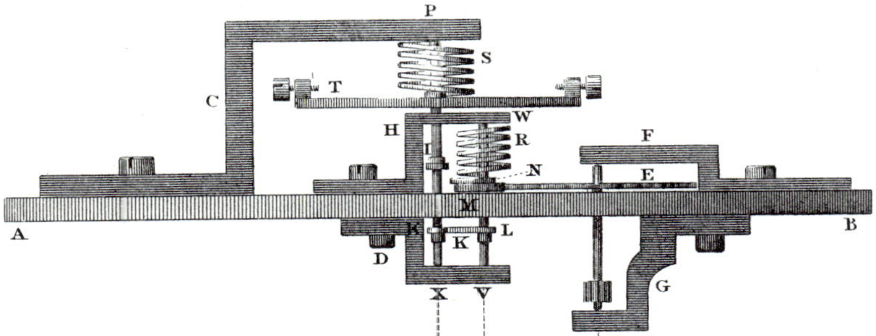

Fig. 7. Hardy's.

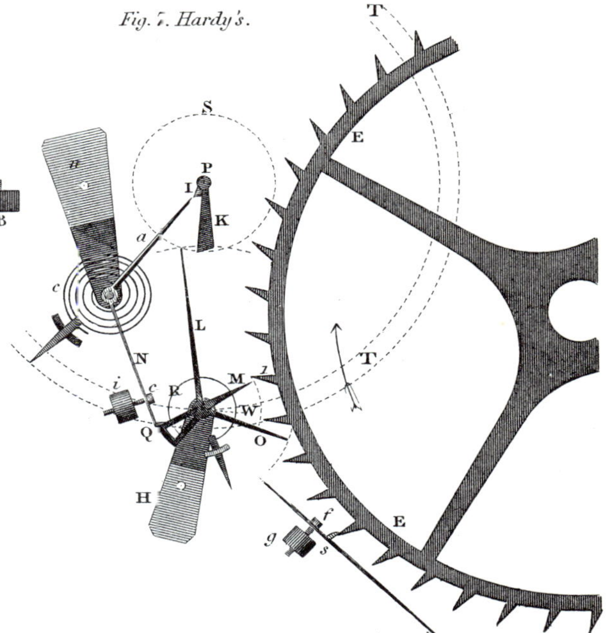

Fig. 3.

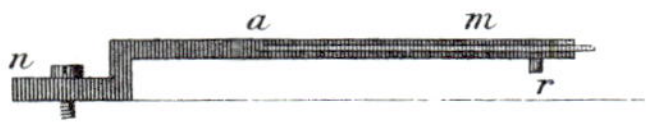

Fig. 2.

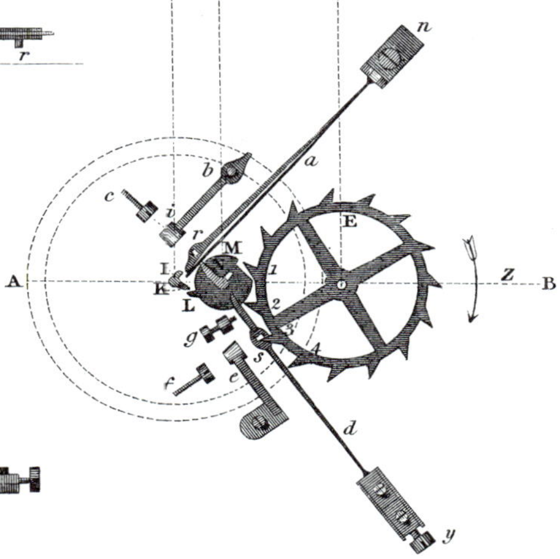

Fig. 4.

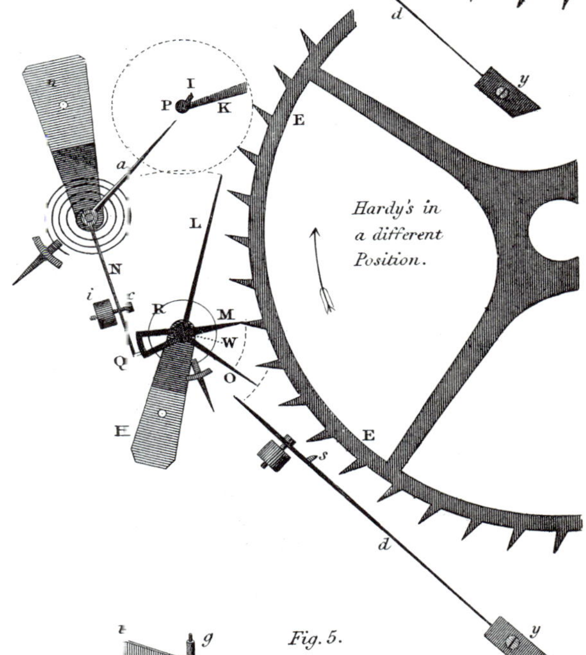

Fig. 6.

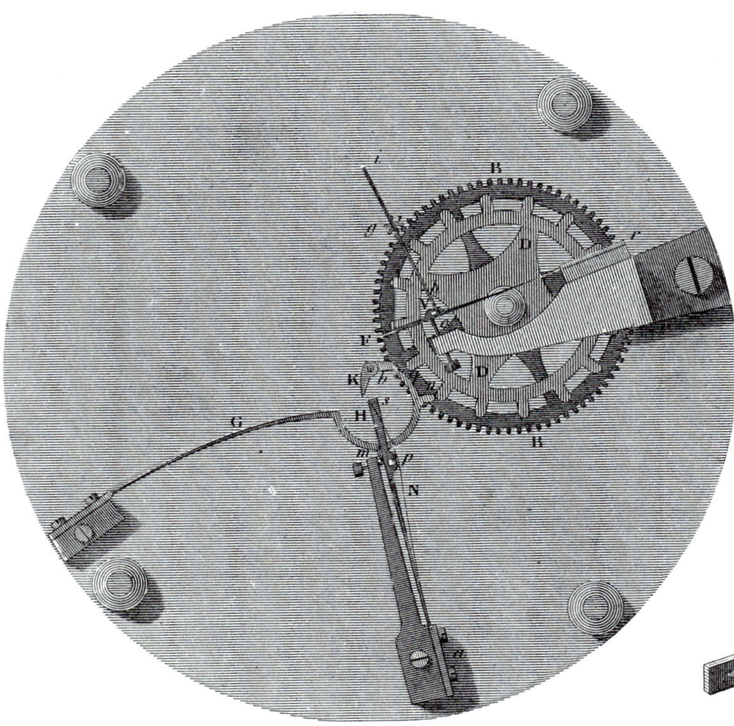

Fig. 5.

Breguet's

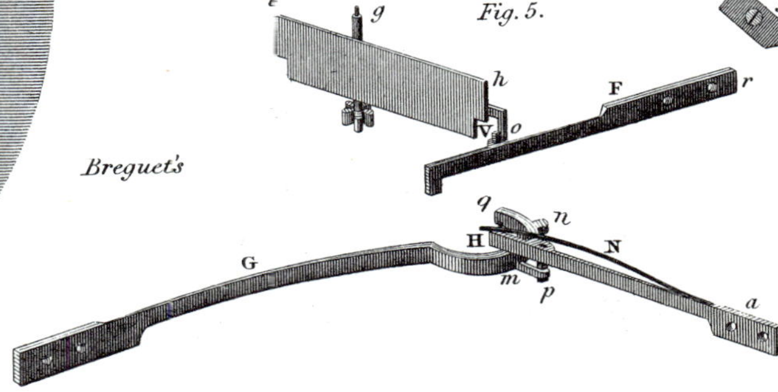

rey del. Published as the Act directs, 1816, by Longman, Hurst, Rees, Orme, & Brown, Paternoster Row, London. Lowry sculp.

Remontoire Escapements

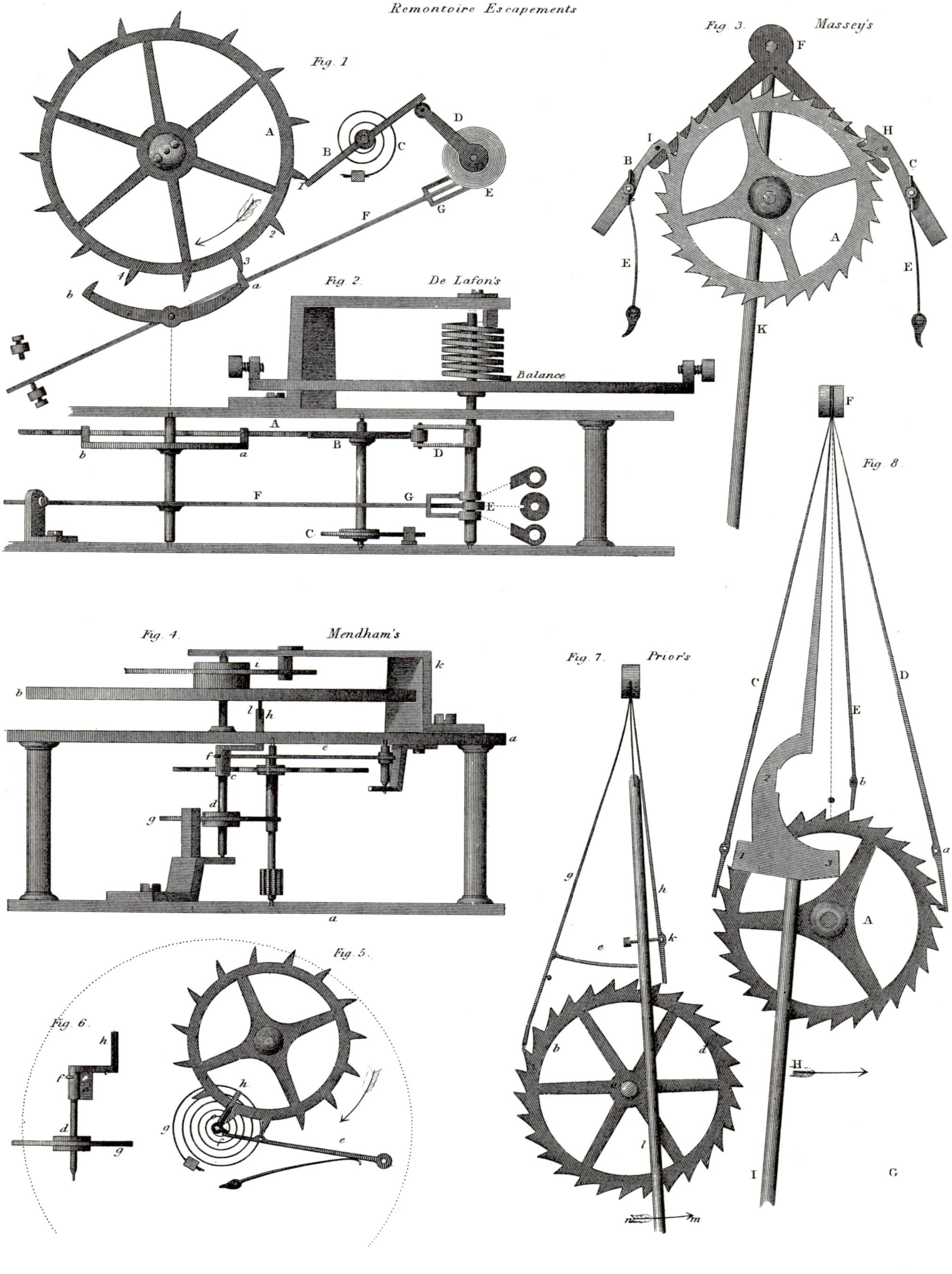

I. Farey delin. Published as the Act directs, 1816, by Longman, Hurst, Rees, Orme, & Brown, Paternoster Row. Lowry Sc.

Watch

Watch, in *Horology*, is a portable machine that meaſures and indicates the ſucceſſive portions of tranſient time. This uſeful piece of mechaniſm, when planned on the beſt ſcientific principles, and executed in the moſt perfect manner, contains within itſelf a collection of inventions, that have exerciſed the ſkill of the moſt ingenious mechaniſts through a ſucceſſion of three if not four centuries; and when we contemplate the curiouſly-contrived and nicely-adjuſted means by which the never-varying period of our globe's rotation on its axis is divided and ſubdivided into hours, minutes, and ſeconds, we need not be ſurpriſed that a *Paley* has ſelected this curious machine as a ſtriking ſpecimen of human ingenuity.

It is a matter of difficult reſearch to aſcertain what artiſt firſt reduced the portable ſpring-clock into the ſize of a watch, which is ſuppoſed to have been firſt effected in Germany; but it is evident that watches had become common in France before the year 1544, in which the corporation of maſter clock-makers in Paris had a ſtatute enacted, to enſure to themſelves the excluſive privilege of making, and of cauſing to be made, clocks, alarums, and watches, large or ſmall, within the precinct of the ſaid city.

The ſmall clocks and watches, however, which were made antecedently to the time of Huygens and Dr. Hooke, were very imperfect performers, and profeſſed not to ſubdivide the hour into minutes and ſeconds; the double lever, and the balance ariſing out of it, were very imperfect regulators of the motion, produced in the train of wheel-work by the maintaining power, inaſmuch as they were under the influence of various oppoſing agents, ſuch as friction ariſing from coarſe workmanſhip, the inertia of matter, reſiſtance of the air, &c.; the conſequence of which was, that the weight of the moving balance was to be determined by experiments, ſuch as would be a proper counterpoiſe to the agency of the main-ſpring on the moving train, and at the commencement of each returning oſcillation, a conſiderable pauſe took place, which made a part of the meaſure of time to be indicated. Theſe inconveniences at length were obviated by the introduction of a balance-ſpring, which became to the balance what gravity is to the pendulum; and the acceleration given to the moving balance during the firſt half of the oſcillation, is thus ſufficient to overcome the reſiſtance oppoſed to its motion during the ſecond half; and when the ſhape, length, and ſtrength of the regulating ſpring are duly proportioned, its iſochronal performance approaches very nearly to the regularity of the pendulum. The conteſt for the honour of this uſeful invention was warmly diſputed between Huygens and Dr. Hooke, for ſeveral years ſubſequently to 1658; but if priority of publication can be conſidered as a proof of priority of invention, the palm is due to our ingenious countryman.

In our articles Chronometer, Clock, Compensation-

Balance, Dial-*Work*, Escapement, Power, *Maintaining*, and Remontoir, we have anticipated the confideration of the moft material parts of a watch, and have given fuch a detailed account of moft of the varieties that occur in the practical conftruction of this machine, that little remains to be done in this place, but to defcribe an ordinary watch, without reference to its hiftory, fcientific principles, compenfations, or fuperior workmanfhip; all which have been amply difcuffed, and the conftituent parts explained by accurate engravings of the correfponding mechanifm. We may, however, comprehend in this article the appendages which have been applied to or actuated by the common watch, for the purpofe of either amufement or utility in civil fociety, fuch as chimes, alarums, ftriking-work, and repeating mechanifm, moft of which operate as a drag upon the works, and are therefore never introduced in chronometers, and feldom in watches of the moft perfect conftruction.

Fig. 1. *Plate* XLIV. of *Horology*, reprefents the interior works of an ordinary watch with the crown-wheel efcapement, as they remain on the pillar-plate when the upper plate of the frame, fhewn by *fig.* 5, is unpinned and removed; and *fig.* 2, which is a fection of the whole frame and of its contents, fhews the connection of all the parts, as though the calliper were in one right line. Thefe two figures, by having the fame letters of reference, mutually explain each other. The main-fpring, which actuates all the wheels and pinions that are called in one general term the movement, is contained in the circular box *a*, feen in different views in the feparate figures 1, 2, and 8, in the laft of which its parts are given in their detached ftate, *viz.* the box; the relaxed fpring immediately above, lying in a fpiral form; the arbor with its pin, on which the interior end of the fpring is hooked; and the lid through which the pivot of the arbor penetrates: this fpring is forced into the box by a tool on purpofe, when it is ftrong, and then the exterior end is hooked to a pin in the circular edge of the box, fo that if the box is made to turn round while the arbor is held faft, the fpring begins to coil at the centre, and is thereby wound clofe round the arbor, and is by this action faid to be *wound up*. The fame effect would be produced if the box were held faft, and the arbor only were turned; but in the latter cafe the chain, which requires to be uncoiled from the fpring-box as this fpring is wound up, would remain unmoved; it is neceffary therefore that the box be turned while the arbor is at reft, which is thus effected: one end of the chain is made faft to the fide of the fpring-box, and the other to the fufee *b*, after being coiled feveral times round the circumference of the box; then as the fquare end of the fpring-box arbor is held by the fmall ratchet and click *c*, feen on the reverfed face of the pillar-plate in *fig.* 7, fo that it cannot revolve, it is obvious that inferting a key on the fquare of the fufee-arbor, and turning it in a proper direction, will wind the chain upon the fpiral-groove of the fufee, while it is unwound from the box; and during this operation the fpring will be coiled up to the centre of the box, or be put into its ftate of greateft tenfion for pulling the fufee back again. The rapid motion which the fufee would have in a retrograde direction, when pulled by the whole force of the coiled fpring, is prevented by the train of wheel-work and balance thus; the great wheel *d* is not faft to the thick end of the fufee, as appears in the drawings, but carries a click and click-fpring 2, as feen in *fig.* 3, while the ratchet-wheel, feen in *fig.* 4, is made faft to the fufee; the confequence of which contrivance is, that while a key applied to the fufee-arbor winds up the watch and fills the fufee-groove with the chain, until the guard driven by it catches the beak at the fmall end of the fufee; the click in *fig.* 3. flides over the floping teeth of the ratchet in *fig.* 4, without acting on them, and thus leaves the great wheel *d* at reft, in connection with the pinion *e*, on the centre or minute-wheel arbor; but when the fpring acts on the fufee in a contrary direction, the click attached to the great wheel is laid hold of by the teeth of the ratchet, which thus makes it faft to the end of the fufee, fo long as the chain is unwinding from the fufee; or, in other words, till the fpring wants winding up again, which happens ufually once in 28 or 30 hours; but it is commonly wound up once in every 24 hours, more or lefs. The action of the great wheel *d* on the pinion *e*, is that of a long lever driving a fhort one; or this wheel may be faid to act under a mechanical difadvantage, where an increafe of velocity, but a lofs of power, is experienced by the pinion; again, on the fame central arbor of this pinion *e* is riveted the centre-wheel *f*, which revolves in an exact hour, as we fhall fee prefently, and this wheel drives the pinion *g*, on the arbor of the third wheel *h*, alfo with a mechanical difadvantage, for the force it imparts to the pinion *i*, on the arbor of the contrate-wheel, is again diminifhed in the ratio of the diameter of the wheel to that of its pinion; thus, the force of the main-fpring is continually diminifhing, as it is tranfmitted through the train, and when the contrate-wheel comes to be actuated, it has juft force enough to drive the horizontal pinion on the balance-wheel *l*, fo that the alternate impulfes given by its teeth to the pallets of the balance-verge are juft fufficient to perpetuate the ofcillations to the right and left, under all the obftacles of friction, dirt, wear, and the air's refiftance. It is a curious fact that this crown-wheel efcapement, though the *oldeft* that we know of, is ftill the moft in ufe in common watches, probably from the facility with which it is conftructed; for certainly it is more under the influence of the irregularities of the main-fpring's force than any other efcapement. The properties and action of this efcapement have been minutely explained under No. 1. of the article Escapement, with reference to *fig.* 6. *Plate* XXXI. of *Horology*, to which explanation and figure therefore we requeft our reader's attention.

In order that the force applied to the pallets of the verge at each ofcillation may not fenfibly vary, it was found neceffary to equalize, as much as poffible, the variable forces of the main-fpring in its different ftates of tenfion; and the moft practicable way of doing this has been found to convert the cylinder on the arbor of the great wheel, which would have been proper for a gravitating body, ufed as a maintaining power, into a figure of a parabolic form, that is, into a folid, generated by the revolution of a parabola, in order that, as the force of the fpring becomes greater by increafed tenfion, its action on the great wheel might be leffened in a fimilar proportion, by a gradual decreafe of the radius of the fufee, round which the chain is wound, to impart the force thus modified. Every feparate fpring, therefore, has not only its average force proportioned to the balance it is deftined to actuate, when diminifhed by tranfmiffion through a given train, but requires its *fcale* of varying forces to be nicely counteracted in every degree of tenfion by the *fhape* of the fufee; and this is done by means of a tool, called a fufee adjufting-tool, which is nothing more than a lever with a fliding weight attached to the fquared end of the fufee-arbor, as reprefented in *fig.* 10. *Plate* XXI. of *Horology*; for when the weight on the lever is an exact counterpoife to the force of the main-fpring in every part of the fucceffive revolutions of the fufee, as the fpring is wound up by the lever inftead of a key, then the fhape of the fufee is proper, but not otherwife. Hence, whenever a new main-fpring is put into a watch, the fufee ought to be adjufted in the

the fusee-engine accordingly as the adjusting-tool determines. The comparative forces of the spring at the two extreme ends of the fusee may be adjusted by the small ratchet *c*, on the back of the pillar-plate in *fig.* 7; but when the spring is put to a suitable degree of tension to act well at both extremities of the fusee, it must not be altered by the ratchet-click, but the intermediate forces must be equalized by a due shape given to the fusee. We have insisted the more on this part of the mechanism being attended to, because, as the *primum mobile*, it is the basis of all the other motions. The number of rounds that the spiral of the parabolic fusee may be cut into, depends on the length of the pillars of the frame, or, which is the same thing, on the shallowness of the watch. The French frequently leave out the fusee, and attempt to equalize the forces of the main-spring by tapering it, and with detached escapements this mode may sometimes answer tolerably, but with the crown-wheel escapement a fusee is indispensable. Again, the number of teeth in the great wheel, and in the centre pinion, depends on the number of rounds in the spiral of the fusee; in a 30 hours watch, with six turns of the fusee, the great wheel must have $\frac{30}{6}$, or 5 times as many teeth as the centre pinion; so that if this has 6 leaves, the wheel must have $5 \times 6 = 30$ teeth; but if 8, then $5 \times 8 = 40$; if the spiral has 7 turns, the great wheel 48, and the pinion 12, then the time of going will be $\frac{48}{12} \times 7 = 28$ hours; also if there be $5\frac{1}{2}$ turns on the fusee, 50 teeth in the wheel, and 10 leaves in the pinion, the period of going will be $27\frac{1}{2}$ hours, or $\frac{50}{10} \times 5\frac{1}{2} = 5 \times 5\frac{1}{2} = 27\frac{1}{2}$; but if 24 hours only were required as the period, with 6 turns and a pinion of 12, the great wheel would be required to have 48. Thus, when an alteration is made in either the pinion, the wheel, or the turns in the fusee, a corresponding variation may be made in the others, to produce the same period of going, but still the centre-wheel revolves once in an hour. In the commonest watches the pinions have only six leaves each, which do not act so well as pinions of higher numbers; but in the best watches, and in all chronometers, the leaves and teeth are more numerous. The pivot-holes, particularly of the verge and escapement-wheel arbor, have jewels for the purpose of diminishing the friction in the best watches; but detached and remontoir escapements are the best correctives of the unequal impulses given through the medium of the train in the different states of its foulness. The potance *m*, and small or counter potance *n*, that hold the pivots of the balance-wheel, are small cocks seen in *fig.* 2. both in their attached and detached states, and are screwed to the top or upper plate within the frame, but the springs, buttons, and joints of the case are not exhibited, as forming no part of the movement. *Fig.* 5. represents the outer face of the upper plate, with the balance *p*, cock *o*, and balance-spring *s*, called the pendulum-spring, from its having the properties of the pendulum; by means of this spring not only is the regulation made steady, but the adjustment for *time* is effected. In every balance-spring there is a certain length, to be taken as the effective length, by which the going of the watch, to which it is applied, is limited to exact performance, and when this length is determined by experiment, a pin is put in the stud that holds the exterior end, as at 4, in *fig.* 5, to prevent its being altered; but as the variation of temperature will alter the momentum of the moving balance, the effect thereby produced is a loss of time, in the rate, in hot weather, and a gain in cold, by an alternate increase and decrease in the dimensions of the balance itself, as well as by some alteration in the spring: to remedy this defect, in an ordinary watch, the contrivance shewn in *fig.* 6. is introduced; the wheel *t* is placed under the graduated circle *r*, seen in *fig.* 5, and a circular rack *u*, *fig.* 6, that holds the curb or slit-piece 5, seen in both figures, is moved by a sliding motion given to it, when a key is applied to the squared arbor of the figured circle, and thus the effective length of the spiral spring is limited by the position of the curb 5; and accordingly as the key is turned forwards or back, towards the words *fast* or *slow*, engraved on the cock, the shortened or lengthened spring alters the rate of going, till the proper length is found, that suits the season in question. In Harrison's time-piece the curb was moved by an expansion-lever of two metals, that acted by means of the change of temperature; but in the best chronometers of more recent date, the compensating levers constitute the three portions into which the rim of the balance is divided, and the adjustment for time, as well as compensation for temperature, are by means of heavy screws, which form a part of the moving balance. In these more perfect machines, the length of the spring, which is now made helical, or cylindrical, is first determined such, that the long and short vibrations are performed in the same time, and this is called the isochronal length, which is not afterwards altered by subsequent adjustments. But of these niceties we have treated more fully under our long article CHRONOMETER. The last portion of the watch, which demands our explanation, is the dial-work, for producing the hours and minutes; this will be easily understood by a reference to figures 2 and 7: when the pinion, called the cannon-pinion, seen near the minute-hand in *fig.* 2, is inserted on the arbor of the hour or centre-wheel, to which it fits rather tight by friction, it revolves therewith in an hour, and receives the minute or long hand on its protruding squared end; then this pinion drives the wheel *x* round a stud on the pillar-plate, and with it a pinion *w* made fast to its centre; which pinion again drives a second wheel *v* round the tube of the cannon-pinion in twelve hours, and to this the hour-hand is attached. This diminution of twelve revolutions from the cannon-pinion to the hour-wheel might be effected by one pinion driving a single wheel of twelve times its number of teeth; but as the motion must be brought back to the centre of the dial again, two more wheels, or a wheel and pinion, are necessary to be introduced, and these are therefore made a part of the train, and no large wheel or small pinion is wanted, for the ratio 12 : 1 may be more conveniently obtained by two factors, *viz.* 4 : 1 and 3 : 1; thus, suppose the cannon-pinion to have 15 leaves, its wheel may have $4 \times 15 = 60$ teeth for wheel *x*, and if wheel *v* be made the same its pinion will be $\frac{60}{3} = 20$, and the train $\frac{60}{15} \times \frac{60}{20} = \frac{360}{30} = \frac{72}{6}$ or $\frac{60}{5} = \frac{12}{1}$ or 12; so that when the pinions are fixed upon for the dial-work, the wheels are readily determined, and *vice versâ*. Under our term CLOCK-*Movement*, we have given three tables, containing the several varieties of the three different portions of a clock-movement, which are equally applicable to a watch-movement, and we presume will be found useful to the practical workman, who is disposed to vary his construction to shew seconds, or for other particular purposes. The following table, somewhat differently arranged, was published by W. Shirt, balance-wheel and fusee-cutter, No. 25, Coleman-street, Bunhill-row, London, with which we will conclude this division of our article.

A TABLE

A **Table** of Trains for Watches, ſhewing the Number of Turns on the Fuſee and Teeth in the Balance-wheel, with the Beats in an Hour, and the Number of Seconds in which the Contrate or Fourth Wheel revolves, for the eaſy Timing of Watches by the Vibrations of the Pendulum.

9 Teeth in the Balance-Wheel.

Second Wheel 58 6 Third Wheel Pinion	60 8	60 6	60 6	60 6	60 6	64 6	64 8
Third Wheel 56 6 Contrate Pin.	56 7	58 6	58 6	60 6	60 6	60 6	60 8
Contrate Wheel 54 6 Balance Pin.	80 6	52 6	56 6	54 6	60 6	54 6	80 6
Beats 14616 in an Hour	14400	15080	16240	16200	18000	17280	14400
Seconds $39\frac{9}{10}$ in which the 4th Wheel revolves	60	$37\frac{1}{4}$	$37\frac{1}{4}$	36	36	$33\frac{3}{4}$	60

11 Teeth in the Balance-Wheel.

Second Wheel 48 6 Third Wheel Pin.	54 6	54 6	56 7	56 6	56 6
Third Wheel 45 6 Contrate Pin.	45 6	50 6	45 6	54 6	56 6
Contrate Wheel 70 6 Balance Pin.	65 6	60 6	78 6	54 6	55 6
Beats 15400 in an Hour	16087	16500	17160	16632	17567
Seconds 60 in which the 4th Wheel revolves	$53\frac{1}{3}$	48	60	$42\frac{2}{3}$	$41\frac{1}{4}$

58 6	58 6	58 6	58 6	58 7	60 6	60 6	60 6	60 6	60 6	60 6
52 6	54 6	54 6	56 6	56 6	50 6	52 6	54 6	54 6	54 6	55 6
52 6	52 6	54 6	54 6	56 6	52 6	52 6	50 6	52 6	54 6	52 6
15973	16588	17226	17817	15879	15888	16520	16500	17160	17820	17477
$42\frac{3}{4}$	$41\frac{1}{3}$	$41\frac{1}{3}$	$39\frac{3}{4}$	$54\frac{1}{3}$	43	$41\frac{1}{2}$	40	40	40	39

60 6	60 7	60 8	60 8	60 6	60 8	60 7	62 6	62 7	63 6	63 6
56 6	56 6	56 7	56 7	60 6	60 6	60 7	54 6	58 6	54 6	56 7
50 6	56 6	74 6	78 6	48 6	56 6	60 6	52 6	52 6	50 6	56 6
17111	16426	16280	17160	17553	15400	16163	17935	16324	17325	17248
$38\frac{1}{2}$	40	60	60	36	48	49	$38\frac{1}{4}$	45	38	$42\frac{3}{4}$

64 6	64 6	65 7	70 8	70 7	72 8	72 7	80 8	75 10	72 9	72 9
50 6	52 6	62 7	54 7	63 7	63 7	64 7	72 8	72 9	66 8	60 8
50 6	52 6	59 7	68 6	58 7	54 6	58 7	68 8	66 8	60 6	54 6
16296	17625	15250	16830	16408	16035	17142	16830	13200	13200	11880
$40\frac{1}{2}$	39	$43\frac{3}{4}$	$53\frac{1}{3}$	40	$44\frac{1}{2}$	$38\frac{1}{4}$	40	60	66	60

13 Teeth in the Balance-Wheel.

Second Wheel 48 6 Third Wheel Pin.	48 6	52 6	54 6	54 6	54 6
Third Wheel 45 6 Contrate Pin.	45 6	52 6	50 6	52 6	52 6
Contrate Wheel 66 6 Balance Pin.	68 6	52 6	50 6	48 6	50 6
Beats 17160 in an Hour	17680	16925	16274	16224	16900
Seconds 60 in which the 4th Wheel revolves	60	$46\frac{1}{2}$	48	46	46

54 6	54 6	55 6	56 7	56 6	56 6	56 6	56 6	56 6	58 6	58 6
52 6	52 6	51 6	45 6	50 6	50 6	52 6	52 6	54 6	48 6	50 6
51 6	52 6	51 6	66 6	50 6	51 6	48 6	50 6	49 6	52 6	50 6
17238	17576	17219	17160	16851	17188	16824	17525	17836	17425	17453
46	46	$46\frac{1}{4}$	60	$46\frac{1}{2}$	$46\frac{1}{2}$	$44\frac{1}{2}$	$44\frac{1}{2}$	$42\frac{3}{4}$	$46\frac{3}{4}$	$44\frac{2}{3}$

60 6	60 8	60 6	60 6	60 7	60 6	60 7	60 8	60 7	60 8	60 6
48 6	48 6	50 6	50 6	54 6	54 8	56 7	56 7	58 7	58 6	60 7
48 6	66 6	46 6	48 6	52 6	60 6	56 6	66 6	56 6	56 6	48 6
16640	17160	16611	17333	17382	17550	16640	17160	17234	17593	17828
45	60	43	43	$46\frac{2}{3}$	54	$52\frac{1}{2}$	60	$50\frac{2}{3}$	$49\frac{1}{2}$	42

60 8	60 6	62 7	63 7	63 7	64 7	64 7	64 8	64 8	65 7	70 8
60 6	60 7	56 7	52 6	60 7	52 6	60 7	60 8	64 8	62 7	60 7
54 6	56 7	56 6	51 6	60 7	50 6	60 7	66 6	72 7	58 7	52 6
17550	17828	17194	17238	17191	17168	17464	17160	17115	17717	16900
48	42	$50\frac{1}{4}$	$46\frac{1}{4}$	$46\frac{2}{3}$	46	$45\frac{3}{4}$	60	$56\frac{1}{4}$	$43\frac{1}{2}$	48

70 8	72 8	72 8	74 8	74 8	75 10	75 10	80 10	96 12	96 12	90 10
66 8	52 6	70 8	64 8	68 8	72 9	72 9	60 8	75 10	75 10	90 10
64 7	52 6	68 8	63 7	68 8	70 7	72 9	60 8	80 8	88 8	90 10
17160	16673	17403	17316	17400.	15600	12480	15600	15600	17160	18954
50	$44\frac{3}{4}$	$52\frac{1}{2}$	$48\frac{2}{3}$	60	60	60	60	60	60	$44\frac{1}{2}$

TABLE *continued.*

15 Teeth in the Balance-Wheel.

Second Wheel 48 6 Third Wheel Pinion Third Wheel 45 6 Contrate Pin. Contrate Wheel 54 6 Balance Pin. Beats 16200 in an Hour Seconds 60 in which the 4th Wheel revolves						48 6 45 6 58 6 17400 60	48 6 45 6 60 6 18000 60	54 6 48 6 46 6 16560 50	54 6 48 6 48 6 17280 50	54 6 48 6 64 8 17280 50
54 6 50 6 48 6 18000 48	56 7 45 6 56 6 16800 60	56 7 45 6 58 6 17400 60	56 7 45 6 60 6 18000 60	56 6 48 6 46 6 17173 48	56 7 60 8 60 6 18000 60	58 6 48 6 46 6 17786 46¼	58 6 50 8 58 6 17520 59½	60 8 48 6 58 6 17400 60	60 8 48 6 60 6 18000 60	60 8 56 7 48 6 14400 60
60 8 56 7 56 7 14400 60	60 7 56 7 58 7 17044 52½	60 8 56 7 58 6 17400 60	60 8 56 7 60 6 18000 60	60 8 56 7 60 7 15386 60	60 6 60 8 48 6 18000 48	60 6 60 10 48 6 14400 60	60 6 60 8 56 7 18000 48	60 6 60 10 58 6 17400 60	60 10 60 6 60 6 18000 60	60 6 60 10 64 8 14400 60
60 8 64 8 66 7 16971 60	60 8 64 8 70 7 18000 60	62 8 60 8 60 6 17437 61⅔	63 7 54 7 50 6 17356 51¾	63 7 56 7 56 7 17280 50	64 8 45 6 56 6 16800 60	64 8 60 8 58 6 17400 60	64 8 60 8 60 6 18000 60	64 6 60 10 70 8 16800 56¼	65 7 56 7 56 7 17828 48⅓	70 6 60 10 48 6 16800 51⅓
70 7 60 10 70 7 18000 60	70 8 64 8 50 6 17500 51⅓	70 8 64 8 58 7 17400 51⅓	70 10 65 8 60 6 17062 56¾	72 6 60 10 48 6 17280 50	72 8 64 8 50 6 18000 50	72 8 64 8 54 7 16662 50	72 8 64 8 64 8 17280 50	72 8 65 8 64 8 17550 49	75 8 64 8 64 8 18000 48	81 9 72 9 72 9 17280 50

17 Teeth in the Balance-Wheel.

Second Wheel 48 6 Third Wheel Pin. Third Wheel 45 6 Contrate Pin. Contrate Wheel 50 6 Balance Pin. Beats 17000 in an Hour Seconds 60 in which the 4th Wheel revolves	56 7 45 6 53 6 18020 60	60 8 56 7 52 6 17828 60	64 8 60 8 60 7 17485 60

GW.	SWP.	TNS.	GW.	SWP.	TNS.	GW.	SWP.	TNS.
48	10	6¼	60	10	5	55	12	$6\frac{6}{11}$
50	10	6	62	10	4[illegible]	56	12	$6\frac{3}{7}$
52	10	5¼	64	10	4⅔	58	12	6⅕
54	10	$5\frac{5}{9}$	48	12	7½	60	12	6.
55	10	$5\frac{6}{11}$.	50	12	7[illegible]	62	12	5⅚
56	10	5⅓	52	12	$6\frac{1}{1}\frac{2}{3}$.	64	12	5[illegible]
58	10	5⅙	54	12	6⅔			

If we divide *double* the product of all the four wheels by the product of all the three pinions, the quotient will be the number of *beats* as given in any of the trains contained in this table; also, if we take the second and third wheels and their pinions respectively, as a compound fraction of an hour, they will give the *seconds* in which the contrate-wheel, attached to the latter pinion, will revolve; thus, $\frac{8}{60}$ of $\frac{7}{15}$ of 60^m = 1^m, or 60^s, which numbers are consequently proper for a watch that indicates seconds; and if the beats be 18000, or 14400, there will be five or four beats respectively in a second, which are the best trains for measuring fractional parts of a second.

French Repeater.—The mechanism which constitutes the repetition portion of a French, and also of a Swiss watch, was originally employed by Tompion, Quare, and other English artists, and is represented by the various figures contained in *Plate* XLV. of *Horology;* it is easier of construction than the repetition-motion of Stockten, which follows, but is not considered so perfect. We have put the same letters of reference to the detached parts, that stand near them in the larger figures, where they occupy their respective situations; and that the reader may be able to accompany us through our description of the action of the relative parts, we will explain previously the appellations by which the workmen designate these acting pieces. In *figs.* 1. and 2. A denotes the pendant-bow, carried at the end of a cylindrical piece, called the pendant, and the hollow piece, into which it is occasionally pushed, is the pendant-socket; B C D the triple lever is called the crémaillère; E is a fixed pulley; and F the hour-snail, by which the number of hours to be struck by the hour-hammer is limited; H is the star-wheel, to which the hour-snail F is fixed fast; I K is the *tout-ou-rien* and G its spring lying on its plane; L and N are the two sets of teeth, that take hold of the hammer-tails, which strike quarters by double blows; O is one of the quarter hammer-mer-tails, and Q, or Q 5, the other, which is attached to the hammer that ſtrikes alſo the hours; S is the quarter-ſnail that determines the number of quarters to be ſtruck at any time, when the hour-hammer has ſtruck the hours, and has three ſteps or arcs of different radii, preſented ſucceſſively to the part that acts on, or rather reſts on it; 7, 8, is the looſe piece attached to, and ſecured under the quarter-ſnail S; *a* is the end of the middle lever of the cremaillère; *b* the jumper, that makes the ſtar-wheel jump a whole ſpace when a tooth, in raiſing it, has arrived at its angular point, ſo as to give its ſpring *d* its full tenſion; *e* is the chain made faſt at one end to the crémaillère at D, and after paſſing round the pulley E, attached at the other to a ſecond pulley Z, which is inſerted on the arbor of the repeating main-ſpring, *f* is the quarter-piece ſpring, preſſing on a pin in the quarter-piece M; *h* is the quarter hammer-ſpring; *i* its counter-ſpring; *g* the quarter hammer-tail ſpring; *p* the hour-hammer ſpring, and *o* its counter-ſpring; *q* is the hour-hammer quarter tail-ſpring, and *r* the gathering piece or arm fixed on the arbor of the great wheel of the repeating train of wheels, over the pulley Z, that cauſes the quarter-piece to act on the hammer-tails, and is the ſame arbor which we have before called the arbor of the repeating main-ſpring. Theſe are the pieces of mechaniſm that lie under the face of the watch, and appear above the frame when the face is removed, together with the dial-work repreſented by the dotted circles in *fig.* 1, but by unſhaded wheels and pinions in *fig.* 2, that the other parts might not be concealed below them. *Fig.* 3. contains the works under the upper plate of the frame, of which the repetition-train (*petit rouage*) only is ſhaded, the ordinary movement being given in outline. The connection between the pieces exhibited in *figs.* 1. and 2, and the repetition train in the frame, ſhewn in *fig.* 3, is by means of the arbor of the great wheel and its circular rack G, ſeen in this figure, for this arbor protruding above the upper plate of the frame receives on its ſquare the main-ſpring of the repeating mechaniſm, and alſo the gathering-piece *r*, ſo that whenever this main-ſpring, exhibited in *fig.* 4, is wound up, the ratchet on the great wheel, ſeen in *fig.* 5, allows the great wheel to move with it without the reſt of the repeating train; but when the ſpring unbends itſelf, and pulls the chain and attached crémaillère back, the click of the ratchet catches and actuates the whole train, which terminates with a fly on the laſt pinion-arbor, as in the ſtriking part of a clock, and thus regulates the velocity with which the hammers reſpectively ſtrike.

We will now proceed to explain the action of the repeating mechaniſm, which we have deſcribed above, and ſee how the effect is produced by means that are thus apparently complex. When the pendant has been puſhed in ſlowly a ſhort ſpace, the end *a* of the middle prong of the crémaillère, being kept down by the ſmall cock Y, approaches one of the ſteps of the hour-ſnail F, and at the ſame time pulls the chain, by means of the prong D, round the pulley or friction-roller E, and winds up the main-ſpring coiled in the box of pulley Z, at the ſame time making the gathering-piece *r* retrograde from its pin, inſerted into the quarter-piece: in *fig.* 1. this motion is juſt commencing from a ſtate of quieſcence; ſuppoſe now, the retrograde motion of the gathering-piece to take place, while the pendant is puſhed very ſlowly in; and conceive the circular rack on the face of the large wheel within the frame, *viz.* G in *fig.* 3, to be retrograding alſo, as being on the ſame common axis; preſently the end *a* of the third prong of the crémaillère meets with one of the ſteps of the hour-ſnail, and puſhes againſt it; this ſnail, and its attached ſtar-wheel, having their common pivot borne by the *tout-ou-rien* at H, communicate the puſh received by them to this piece, which turning on its centre of motion at I, has its remote or looſe end K carried from its quieſcent poſition, notwithſtanding the oppoſing action of its ſpring; and when conſiderable force is applied to puſh the pendant home, this end K, which forms a detent to the quarter-piece at the points of their contact, quits its hold, and leaves the quarter-piece at liberty to be urged by its ſpring *f*, till its heel-piece *c* drops upon one of the ſteps of the quarter-ſnail, as in *fig.* 2, where it is ſeen reſting on the third ſtep, or ſhorteſt arc. At this inſtant the repeating main-ſpring begins to relax itſelf, and brings forward the concealed rack G, (*fig.* 3.) which had retrograded as many teeth only as the hour-ſnail permitted, before the *tout-ou-rien* was diſplaced; its neareſt tooth to the tail-piece 2, 3, of the hour-hammer R R, catches now this tail-piece, and makes the hammer ſtrike on the circular rim of ſteel, which is ſubſtituted for a bell, and as many blows are given in ſucceſſion, as there are teeth in the rack to fall againſt the hammer-tail, while the repeating train is running down; and during the time in which theſe ſtrokes are going on, the little pin between the hammer-tail ſpring *p*, and its counter-ſpring *o*, may be ſeen moving backwards and forwards, as though it gave the ſtrokes on the counter-ſpring. No ſooner are the hours limited by the hour-ſnail ſtruck, than the gathering piece *r* returns with the relaxing ſpring, till it catches the pin of the quarter-piece, which piece is moveable round a pivot at M, and is now gradually brought back by its pin till one of its teeth N catches O, the tail-piece of the quarter-hammer P, *fig.* 3, and then one of the teeth at L, at the oppoſite end of the quarter-piece, catches Q the upper tail-piece of the hour-hammer, which inſtantly repeats the blow with the hour-hammer, and thus as many double blows are given by the two hammers in immediate ſucceſſion for the quarters, as there are teeth to act on the ſaid tail-pieces, when the quarter-piece begins to return; and this number entirely depends on the ſtep of the quarter-ſnail S, on which the heel-piece falls, when the *tout-ou-rien* is diſplaced; hence if *any* blow is given, *all* the blows that the two ſnails limit will be given, from which neceſſity, the piece *tout-ou-rien*, (all-or-nothing,) takes its name. But leſt the quarter-piece ſhould return by a jerk before the tout-ou-rien has produced its full effect, the angular point *m* of the quarter-piece, in its return, ſlides down the interior face of the *tout-ou-rien*, in oppoſition to the action of its ſpring G, while the ſtrokes of the quarters are making, and arrives at the point K, at the termination of the ſtrokes, thus performing the office of a train and fly, after which the *tout-ou-rien* reſumes its quieſcent poſition, and its end becomes a detent to the quarter-piece. It is not neceſſary to deſcribe the action and re-action of the two hammer-tail pieces, which perform their operations, as in the ſtriking work of a clock heretofore deſcribed; but it may be proper to ſhew how the lower tail-piece 3, 4, of the hour-hammer is detached from the rack G, *fig.* 3, while the quarters are ſtruck, by means of the upper tail-piece Q acting with the teeth of the quarter-piece at L only; this will be beſt underſtood by a reference to the detached figures in the group denoted by *fig.* 6, as will alſo the action of ſeveral other parts, which we have deſcribed and explained as being in their places in *figs.* 1 and 2; the three pieces marked Q 5 are the ſame quarter tail-piece ſeen in different views, 2 C 1 R, is a part of the hour-hammer, and its arbor 6 paſſes the ſocket of the piece 3, 4, which we have called the lower tail-piece of the hour-hammer, before it receives the upper tail-piece Q on its ſquare;

 now

now the part 3, 4, is that which takes hold of the circular rack G in *fig.* 3, and is thus occasionally detached from the said rack; *viz.* its pin 3 ascends through the upper plate of the frame, so that its superior end is visible at 3 both in *fig.* 1. and *fig.* 2, and falls in the way of the extreme end of the quarter-piece, which, on its return from the quarter-snail, catches it and turns the piece 3, 4, round the central arbor 6 of the hour-hammer, and thereby takes the end 4 out of the circular rack, while the quarters are being struck, but whenever the hours are to be struck, the proper spring restores the due position of the tail-piece. As the repeating train of five wheels, and as many pinions, are introduced to give motion to the regulating fly, it is of no consequence what the numbers of their teeth be, provided they be duly proportioned to act smoothly, and to produce the requisite velocity for the proper intervals between the successive strokes. The dial-work for hours and minutes is the same as in any ordinary watch, except that the quarter-snail is attached to the cannon-pinion, and lies under it, so as to partake of its hourly motion, together with that of the minute-hand, shewn in dots, as being above the face, the piece, however, in *fig.* 6, denoted by the figures 7, 8, and called the loose-piece, (or surprise,) is also fast by friction to the same hour-arbor, and revolves contemporaneously with the snail and minute-hand, and when its pin 8 meets with one of the points of the star-wheel H, it moves it forwards until the angular point on the face of the jumper *b* has passed an opposite point of the star, when it will jump or move at once the remainder of the space; in this star are twelve points, and as the hour-snail, which has twelve steps, is made fast to it, the snail also jumps to the succeeding step once every hour, while the three steps of the quarter-snail follow one another by a constant slow motion, keeping pace with the minute-hand. Hence the *times* at which the respective hours and quarters are to be struck, corresponding with the positions of their proper snails, are guided by the common dial-work, and when once they are duly adjusted, a motion given to the minute-hand, by a suitable key, will always keep both the snails in their requisite positions for regulating the number of hour and quarter strokes, that the face of the clock has indicated by the hour and minute-hands; and in the same way the quarters even of the minute might be repeated, if such addition were deemed desirable. In our *figs.* 1 and 2, we have put the characters of the hours on the rim or edge of the case, as the face is removed, merely to shew how the hands, pointing to the divided spaces, are connected with the snails of the repeating mechanism. From this explanation, it will appear that the movement of the watch is not at all affected by pushing in the pendant, nor yet by the motions of the repeating parts, otherwise than as the heel of the quarter-piece falls against the quarter-snail carried by the cannon-pinion, and at the moments when the loose-piece, under this snail, moves the star at its hourly period; but trifling as these obstacles may appear in a common watch, they are seldom, if ever, introduced into a chronometer.

In the construction which we have here described, a large semi-circular rack and pinion are sometimes substituted for the chain and pulleys, in which case the rack is attached to the crémaillère, and the pinion to the arbor of the repeating main-spring; and it is obvious that such a substitution will remove the objection to the liability of the chain's breaking, and the consequent derangement of the works.

English Repeater.—The construction of the repeating motion, called, after the name of its inventor, a *Stockten* motion, differs in many respects from and is very superior to the original repeating motion, which, with a very little alteration in the mode of arranging its parts, is now generally known by the term *French* motion: this appellation is however improper, since the repeating watch is indisputably an English invention, the merit of which was disputed by Tompion, Quare, and other English watch-makers of celebrity of the last century.

Stockten was an Englishman, but his history, like that of many other ingenious men, is so little known to posterity, that even his sponsorial appellation is lost, and his birth-place unknown; all that can with any certainty be recorded respecting him is, that he lived in London, and worked for the celebrated Mr. George Graham, watch-maker and F.R.S. the successor to Mr. Tompion, the motions of whose repeaters were always made upon this construction. The houses of Mudge and Dutton, Fleet-street; of Ellicott, Royal Exchange; and of Vulliamy, Pall-mall; who, for many years after the death of Mr. Graham, were the principal makers of repeaters in London, constantly employed this construction of motion; and it is now generally made use of by all the London watch-makers in the manufacture of their best repeaters.

To enumerate very briefly the advantages of this repeating work, its main-spring is wound up by a rack and pinion; from its construction the action of the parts is uniform and gradual, and not depending on several springs, whose actions interfere with each other; and moreover it is susceptible of being made to strike the half-quarters, without inconvenience or additional work.

To understand the general action of this repeating motion, it is necessary to be well acquainted with its construction, the detail of its parts, the manner in which the different pieces come together, and their separate actions. The pieces composing the pendant work, and the action of the pushing-piece upon the crémaillère, or, as it is generally termed in this construction of a repeating motion, the *rack;* the brass edge, and its use and mode of being fixed to the pillar-plate; the dial, and the manner of fixing it to the brass edge, and the hands; the repeating main-spring and its barrel; the barrel arbor, and the mode of hooking the spring into the barrel; the repeating train of wheels, or, as they are frequently called, the running train, and the situations of the hammers, relative to the wheels between the plates, and their respective uses and actions are all so nearly the same as in the common motion, previously described, that any further description becomes unnecessary: the ratchet, click, and click-spring of the great wheel of the repeating train, are also the same; but the ratchet-wheel G, of twelve teeth (*fig.* 3. *Plate* XLV.) which in the common motion acts upon the hour-hammer, is in this construction of the repeating motion omitted.

In *Plate* XLVI. *fig.* 1. represents the repeating work complete in its quiescent state, with the wheel Q of 48, the hour-snail F, retrograding ratchet P, and wheel of communication R; and also the lantern-pinion *w*, quarter-snail G, and hour-wheel S; by which parts the motion is communicated to the hour-hand, and the hands carried round, perfectly detached from the other parts of the motion.

Fig. 2. represents the repeating work complete, as it appears at the instant of unlocking; the arm *f* of the piece D bearing on the hour-snail F, the arm *y* of the quarter-rack on the quarter-snail G, the little all-or-nothing piece I disengaged from the piece K, and the hammer-tails L and M in a proper situation to be acted upon by the hour and quarter ratchets N and O.

Fig. 3. represents merely the rack A B and pinion C; the piece D; and the two all-or-nothing pieces H and I, in their relative situations to one another on the pillar-plate, the rest of the work being supposed to be taken off.

Fig. 4.

Fig. 4. reprefents the pinion C, and the hour and quarter ratchets N and O, feparate in plan, and in profile; and alfo a fection of the three together.

Fig. 5. reprefents the cannon-pinion *w* feen from above, as well as below; and alfo in perfpective, with the quarter-fnail G attached to it.

Fig. 6. reprefents the wheel Q of 48, the hour-fnail F, the retrograding ratchet P, and the wheel of communication R, feparate in plan, and in profile; and alfo a fection of them together.

Fig. 7. reprefents the hour-fnail F on the wheel Q of 48, and the little fpring *z* in the notch of the focket of the hour-fnail.

The rack A B (fee *Plate* XLVI. *figs.* 1, 2, and 3.) is the piece firft put in motion when the watch is made to repeat, and is that by the action of which with the pinion C of twenty teeth, concealed in *figs.* 1. and 2. by the hour and quarter ratchets N and O, the main-fpring of the repeating part is wound up; this rack may be confidered as a portion of a very large wheel, whofe axis or centre of motion is placed, as near as it conveniently can be, to the edge of the watch: this axis is a hollow tube, paffing through a well fitted hole in the pillar-plate, and pivoted into the upper plate; (the reafon of its being a tube will be prefently fhewn;) it is fet upright in the frame, and confequently at right angles to the face of the pillar-plate; hence it follows, that the rack, which is alfo fixed at right angles to its axis, moves parallel to the pillar-plate, and is placed as near the plate as it can be, to move freely without rubbing it; the rack is cut into twenty-two teeth. Immediately connected with the rack are the two pieces D and E, called, the piece D the unlocking-arm, and the piece E the quarter-rack: the ufe of the piece D is two-fold; firft, to determine the number of blows to be ftruck by the great or hour-hammer, by means of its arm *f*, which, when the watch is made to repeat, comes to bear upon one of the fteps of the hour-fnail F; fecond, to unlock, or, as it is commonly termed, difcharge the ftriking. This effect is produced, as will be explained hereafter, by a motion of the piece D peculiar to itfelf. This piece D is attached to the rack, at its greateft diftance from its centre of motion, by the fcrew 1, which fcrew is tapped into the rack up to a fhoulder, leaving a plain part equal to the thicknefs of the piece D, and as much more as is neceffary for the piece not to be bound, between the under fide of the fcrew head and the upper furface of the rack; and the hole in the piece D, through which this fcrew 1 paffes, is made to fit very correctly on the plain part of the fcrew, upon which it moves as its centre of motion. The quantity of motion of the piece D is determined by a circular hole at its other extremity, through which the axis of the rack paffes, fomewhat larger than that axis, which, in order to pafs through the end of this piece D, and for it to act againft, is purpofely prolonged above the furface of the rack, a very little more than the thicknefs of this fame piece D: in *fig.* 3. the quarter-rack and the cock *a* are omitted to fhew the fhape of this piece.

The quarter-rack E, fituated above the piece D, has its centre of motion within the centre of motion of the rack A B, or confidering the centres of motion of both the pieces as lines, they may, under that fuppofition, be confidered as poffeffing one common centre of motion: this rack is alfo fixed at right angles to its axis, which paffes through the tube that forms the axis of the rack A B, the whole length of that axis to the upper plate; the under fide of this quarter-rack bears upon the top of the tube, or centre of motion of the great rack, which terminates a little above the centre of motion of the piece D, as has been before mentioned; and the upper extremity *e* of the axis of this piece is pivoted into the cock *a*, *figs.* 1 and 2, which cock is fcrewed faft to the pillar-plate Y Z: in this manner is the quarter-rack confined in its place between the upper end of the hollow arbor of the rack A B, prolonged through the piece D, as before defcribed, and the under fide of the cock *a*. The ufe of this piece is to determine the number of quarters, if any, or the half-quarter, as the cafe may be, to be repeated after the hour: this effect is produced by the action of one, and one only, of the eight teeth at the end of the rack, on the little *all-or-nothing* piece I; and according to the tooth which fo acts, the watch, after having repeated the hour, repeats the half-quarter, the quarter, or the quarter and half-quarter, &c. as fhewn by the hands; or if the minute-hand has not paffed the 7^{m} and 30^{s} after the hour, the firft tooth of the eight caufes the repeating to ceafe immediately after the repeating of the hour is completed. Which of the teeth fhall act on the little all-or-nothing piece is determined by the advance of the quarter-rack, which is regulated by the ftep on the quarter-fnail G, upon which the arm *y* comes in contact, when the watch is made to repeat. The arm *y* is made a feparate piece from and fixed to the quarter-rack E, by means of the fcrew 2, on which fcrew it moves as its centre of motion, in the fame manner as the piece D moves on the fcrew 1, and is kept in its place by the fpring *g*: the reafon of this piece being thus made, is to prevent the poffibility, in the cafe of the watch being made to repeat exactly at the quarter, of the repeating work ftopping the watch, by the arm *y* holding back the quarter-fnail, during the ftriking of the hours: the arm *o*, which is a portion of the quarter-rack, by its action with the retrograding-ratchet P, brings the hour-fnail F into its proper place to receive the arm *f*, of the piece D. The quarter-rack E is kept in its place by the pin 3, tapped into the piece D, which bears againft its edge; and is carried forward when the watch is made to repeat, by the action of its fpring *h*, which is fcrewed to its extremity the fartheft from the centre of its motion, and fet up by its other extremity being confined in a notch in the cock *a*.

The total furface of the piece D refts on the rack A B, and, confequently when, in the act of unlocking, it moves on its centre, or fcrew 1, its under face rubs on the furface of the rack A B; but the quarter-rack E, on the contrary, is entirely detached from, and does not touch the furface of the piece D, its under fide bearing upon the prolonged arbor of the rack A B, through the hole in the piece D; and is kept down by the cock *a*, which bears againft the fhoulder of its upper pivot, as has been before mentioned. There is what is termed, in the peculiar dialect of watch-makers, a *light* between the under fide of the rack and the pillar-plate, and between the under fide of the quarter-rack and the piece D; and from their conftruction, it is evident that they move in planes parallel to one another, and to the pillar-plate.

Having defcribed the rack, and the parts connected with it, we will next in order proceed to the defcription of the pinion C, and the hour and quarter ratchets N and O attached to it, and alfo their mode of connection, by which the hammers are raifed to caufe them to hit the blows, or to *ftrike*. The larger of the two ratchet-wheels, N, the one next the pinion, has originally been cut into twenty teeth, of which twelve confecutive teeth are left; then three more teeth, at an interval apart from each other, and from the laft of the twelve teeth, equal to the fpace of a tooth; the remainder of the teeth, that occupied the fpace from the laft of the three to the firft of the twelve teeth, are taken away. The upper ratchet, O, which acts on the fmall hammer, has been originally cut into ten teeth, of which only four confe-

cutive

cutive are left. The pinion C fits by a long ſquare hole on the arbor of the great wheel of the running train, prolonged through the pillar-plate, which arbor is pivotted into the upper plate, and the cock *b*, which is ſcrewed to the pillar-plate. The hour ratchet, of fifteen teeth, is fitted to this pinion C, by means of a ſocket, attached to and forming part thereof; and the upper ratchet, of four teeth, is attached by a hole through its centre, in a ſimilar manner to the under ratchet; and the two are firmly held together, and alſo to the pinion, by means of two ſcrews which paſs through them both, and are tapped into the pinion, as repreſented in ſection in *fig.* 4; conſequently, whatever portion of a circle the pinion is made to revolve by the action of the rack, the two ratchet-wheels neceſſarily do the ſame. It may not be amiſs to notice in this place, that the arbor of the great wheel paſſes through the barrel which contains the repeating main-ſpring; and that the barrel is ſcrewed to the under ſide of the pillar-plate, and is connected with it exactly in the ſame manner as in the caſe of the common repeating motion, and its operation is the ſame; but the barrel arbor is not pivotted between the upper and pillar-plates, as in the caſe of the common motion, but into the upper plate and the cock *b*, ſcrewed to the pillar-plate; and the connection between both the hammers and the motion is entirely above the pillar-plate, through the medium of the hammer-tails, as will be explained.

The pieces next to be deſcribed are the two hammers, and their connection with the motion. The two pivots or centres of motion of the hammers are generally placed at equal diſtances from the arbor that carries the pinion C, and perpendicular to the plates, and, conſequently, parallel to that arbor, and to one another; they are pivotted into the frame, and their upper pivots are prolonged through the pillar-plate, to receive the hammer-tails, L and M: there is a pin planted in each hammer, at a ſhort diſtance from its centre of motion, which alſo paſſes through the pillar-plate, in which there are notches or ſlits forming portions of circles concentric with the centres of motion of the hammers, for theſe two pins to paſs through, as ſeen in *fig.* 3; it is by means of theſe pins 4 and 5, *figs.* 1 and 2, which paſs into notches, made on purpoſe to receive them, in the hammer-tails L, M, that theſe tails, when acted upon by the two ratchets, in their turn act upon the hammers; were it not for this contrivance, the hammer-tails, when impelled by the ratchets, would turn on the pivots of the hammers, as their centres of motion, without raiſing the hammers. The hammers are kept to their places by the two hammer-ſprings *v* and *p*, which preſs againſt the pins 4 and 5, that paſs through the hammers, below the hammer-tails, and as cloſe to the ſurface of the pillar-plate as they can be, to be free from the ſurface of the plate: it is by theſe ſprings that the hammers, raiſed by the action of the ratchet-wheels, are impelled forward to ſtrike; and the ſtronger they are, provided they are not too ſtrong for the power of the main-ſpring, the louder will be the blow ſtruck. Immediately connected with the hammers are the hammer-tails L, M, through the agency of which the hammers are raiſed to ſtrike, by means of the two ratchets N and O; theſe tails require to be extremely well fitted on the upper pivots of the hammers, prolonged through the pillar-plate, which are their centres of motion, though not ſo tight but that they will move with eaſe up and down on thoſe pivots; for on this action depends the whole performance of the motion: the part of the hammer-tail, which, by way of diſtinction, is called the acting lever or pallet, is that againſt which the teeth of the ratchets act to raiſe the hammers, the upper ratchet being the ſmaller of the two; the lever of the quarter hammer-tail is left as much longer than the lever of the other hammer-tail, as is neceſſary for it to reach the ratchet. To prevent the hammer-tails, when acted upon by the ratchets, from turning on their centres of motion, there is, as has been noticed in deſcribing the hammers, a notch in each hammer-tail, to receive a pin planted in each hammer, parallel to its centre of motion, and which pin for this purpoſe is made to paſs through a circular notch in the pillar-plate.

The next piece to be deſcribed is the croſs-piece, or piece K. The ſurface of this piece, when at reſt, is in a plane parallel to the pillar-plate, and the piece moves upon two pivots, which act in holes in the two ſmall cocks *c* and *d*, ſcrewed, the cock *c* to the foot of the cock *b*, and the cock *d* to the pillar-plate Y Z. This piece K is maintained in its place, when the motion is at reſt, by the end of the little all-or-nothing piece I, which preſſes againſt its arm 8, on the one ſide of its centre of motion, and by the ſpring *q*, which preſſing againſt the under ſide of its arm 9, on the other ſide of its centre of motion, cauſes it to bear againſt the little all-or-nothing piece I; the hammer-tail L is maintained in its place, depreſſed, or raiſed on its centre of motion ſolely by the action of this piece K; the hammer-tail M is alſo depreſſed in the ſame manner, but is raiſed on its centre of motion by the ſpring *u*, the end of which is made to bear on the under ſide of the hammer-tail for that purpoſe; and by the power of this ſpring it is impelled upwards, and made to follow the piece K, when it is raiſed by the little all-or-nothing piece I, as before deſcribed.

The uſe of the piece K, when the unlocking takes place, is to depreſs the hammer-tails on their centres of motion, (the prolonged upper pivots of the hammers) to bring them into the ſame planes as the hour and quarter ratchets N and O, they being, when the motion is at reſt, ſituated conſiderably above thoſe planes; and as ſoon as the ſtriking is finiſhed, to raiſe up again the hammer-tail L, and by relieving the other tail M from its preſſure, ſuffer it to be raiſed by its ſpring *u*. For this purpoſe the arm 7 enters into a groove, made on purpoſe to receive it in the hammer-tail L, and the arm 6 is made to bear upon the top of the other hammer-tail M. The action of the piece I will be more fully explained preſently.

The pieces next to be deſcribed are the large *all-or-nothing* piece H, and the little *all-or-nothing* piece I. The upper ſurface of the piece H is in the ſame plane, and it is of the ſame thickneſs with the piece D, by which it is acted upon: the upper ſurface of the piece I is on the ſame plane with the upper ſurface of the quarter-rack, and as thick as it can be for its underſide to be perfectly free from the ſprings *v* and *p* on the pillar-plate: it is neceſſarily required to be of a certain thickneſs, being acted upon by the two pieces E and H. Both the *all-or-nothing* pieces move on ſtuds, ſcrewed into the pillar-plate, which thus become their centres of motion, the piece H on the ſtud *r*, the piece I on the ſtud *i*: it is indiſpenſable that theſe ſtuds ſhould be perpendicular to the face of the pillar-plate, and the pieces perfectly well fitted upon them: it is moreover abſolutely neceſſary that both the *all-or-nothing* pieces ſhould, in all ſituations, be parallel to the plane of the pillar-plate; on which account, the piece H, being a thin piece, is fixed to a ſocket, the hole through which ſocket is perfectly well fitted to the ſtud, and the ſtud is left as high as it can be, to avoid touching the underſide of the dial. The unlocking is effected by the action of the piece D on the piece H, in the following manner: when the pendant is puſhed in to make the watch repeat, the arm *f* of the piece D comes into contact with the hour-ſnail, and cauſes caufes this piece D to move on its centre of motion, (*viz.* the fcrew 1); its oppofite end is then preffed againft the extremity of the *all-or-nothing* piece H, and caufes that end of the piece to move in a direction outwards; confequently the other end of the piece H, which preffes againft the little *all-or-nothing* piece I, is moved inwards: the neceffary effect of this motion in the *all-or-nothing* piece H, is, from the fhape of the two pieces, to caufe the little *all-or-nothing* piece to move outwards from under the arm 8, of the piece K; and this action is called the *unlocking* of the motion. The piece I being thus, by the act of unlocking, difengaged from under the arm 8, of the piece K, this piece K, by the power of the fpring *q*, depreffes the hammer-tails, and brings them into the fame planes as the two ratchets, and confequently into a fituation to be acted on by thefe ratchets: after the ftriking is completed, the piece I is returned by the quarter-rack into its former fituation, and with it the piece K, and the two hammer-tails neceffarily refume their fituations. It is to be obferved, that the *all-or-nothing* piece H acts upon the little *all-or-nothing* piece I, very nearly at its centre of motion *i*, and on the end oppofite to that which paffes under the arm 8, of the piece K.

We come now to defcribe the wheels that carry the hands, commonly called the dial-work, and their mode of communication, together with the two fnails attached to them, which determine the hour and the quarter, or half-quarter, if any, to be ftruck.

The arbor of the centre-wheel, which makes one revolution in an hour, is prolonged through the pillar-plate, and on this arbor the pinion *w* is fitted, fufficiently tight to be carried round by the wheel, but not fo tight as to prevent its turning on the arbor, the upper end *w* only of the focket of this pinion is feen in *figs.* 1 and 2, but the pinion is fully reprefented, in *fig.* 5, detached from the pieces with which it is connected. The centre-wheel arbor is turned with a fhoulder to receive the bottom of the cannon-pinion's focket, and for it to bear againft, in order to prevent the extremities of its four teeth, that ftand parallel to the arbor of the wheel from rubbing on the furface of the pillar-plate: the upper part of its focket, above the dial, is fquared to receive the minute-hand, and the hand fits down to the fhoulder, formed, by the reduction of the original cylinder, into a fquare. At the bottom of this pinion's focket is a collar turned out of the fame piece, of which the pinion is made, purpofely to receive the quarter-fnail; and to this collar in the fame plane with the quarter-rack, is the quarter-fnail G fixed by two fcrews. This fnail is cut into eight fteps, by which the number of blows to be ftruck is regulated: if the arm *y* reaches the ftep next the centre, the ftriking ceafes with the repeating of the hour; if it only defcends to the next, or fecond ftep, the watch, after having repeated the hour, repeats the half-quarter, which is invariably defignated by a fingle faint blow; if it defcends to the third ftep, the quarter only is repeated; if to the fourth ftep, the quarter and half-quarter; if to the fifth ftep, the half hour; if to the fixth ftep, the half hour and half-quarter; if to the feventh ftep, the three-quarters; and if to the eighth ftep, the three-quarters and half-quarter. Under the fnail is the cannon-pinion, which communicates the motion to the wheel Q of 48, and through it to the hour-hand, and which, from its fingular fhape, obtains the name of lantern-pinion. This pinion is of a very peculiar conftruction, and confifts of four upright, equidiftant, cylindrical teeth, attached at one end to the under fide of the collar, and made out of the fame piece of fteel that forms the focket and collar above-mentioned, as feen in *fig.* 5.

Above the cannon-pinion, and refting on it, is the hour-wheel S, which makes one revolution in twelve hours; and this wheel carries the hour-hand by means of its focket, which fits on the wheel's focket prolonged through the hole in the centre of the dial. To explain the mode by which this wheel is made to perform one revolution in twelve hours, it will be neceffary to defcribe the wheel Q of 48, fo called from being cut into 48 teeth; the hour-fnail F, with its ratchet P; and the wheel R of communication to the hour-wheel. Thefe four pieces have one common centre of motion, which is a ftud *x*, fcrewed perpendicularly into the pillar-plate, fimilar to the ftuds *r* and *i*; the wheel of 48, to which are attached the other three, is placed as clofe as it can be to the plate to turn freely. To the centre of the wheel is fixed a long focket, well fitted to the ftud, by which means the perfect parallelifm of the wheel's plane to the furface of the plate is preferved: from the face of the wheel upwards, equal to the thicknefs of the hour-fnail, this focket is left of a confiderable diameter, equal to the hole in the centre of the hour-fnail; for on this part of the focket it is that the hour-fnail fits, but not fo tight as to prevent its turning eafily on the focket, while its underfide bears on the upper furface of the wheel: above the furface of the fnail the focket is reduced in diameter to the fize of a common focket. The ratchet P is faftened to the hour-fnail by two fcrews, and thus becomes one piece with it; the centre of the ratchet being cut out a quantity equal to about half its diameter, (fee *fig.* 6,) and it is fo placed on the hour-fnail, that the fmall circle thus cut out is concentric with its centre of motion. The wheel of communication R (fee *fig.* 6.) forms the laft of the four; it fits tight, by a hole through its centre, on the focket of the wheel of 48, and on its under fide a collar is left, which fits, but not tight, into the fpace formed by the cutting out of the centre of the ratchet, and which collar bears on the fhoulder of the large focket above-mentioned, that is formed by the reducing it from the fize, at which it paffes through the hour-fnail, to the fmaller fize: in this manner the hour-fnail and ratchet are confined between the wheel of Q 48 and the wheel of communication R. To enfure the fnail being carried round once in twelve hours by the wheel of 48, and its being preferved in its relative fituation with refpect to the hour-hand, fo that the hour ftruck fhall accord with the hour fhewn; and alfo to enfure the fafe bearing of the arm *f*, of the piece D, on the ftep of the fnail, there is a notch in the thick part of the ftud on which the fnail is fitted, and a correfponding long flit, commencing at the centre of the fnail, and extending nearly to its circumference; in which flit is a ftraight fpring *z*, that takes into the notch in the focket, (fee *fig.* 7.); the effect of this contrivance is twofold: firft, by the action of the fpring in the notch it preferves the fnail, when not acted upon by the arm *o* of the quarter-rack E, conftantly in the fame fituation relatively to the wheel of 48; and fecondly, it allows of its being moved on its centre when required, and at the fame time regulates the quantity of that motion by the width of the notch in the focket. The cafe in which the hour-fnail is required to turn on its centre, independently of the wheel of 48, is, when the minute-hand having juft paft the 60^{m}, and confequently the hour only being required to be ftruck, there would, without this peculiar additional motion of the fnail, be danger of the arm *f*, which cannot be a line, but muft poffefs width as well as thicknefs, not coming down fafe on the ftep of the fnail it ought to reach, but, by being ftopped by the preceding ftep, of its caufing the watch to repeat one hour lefs than it fhews; to prevent this, the arm *o*, of the quarter-rack, when the watch is made to repeat exactly at the hour, and for

for fome few minutes after, coming into contact with the ratchet, which is fituated in the fame plane as the arm *o*, caufes it, and alfo the fnail, to advance a fufficient quantity to enfure the fafe bearing of the arm *f* on the fnail, and confequently the correct ftriking of the watch: this effect, from the conftruction of the parts as juft defcribed, is produced, without difturbing either the wheel Q of 48, or the wheel of communication R, and confequently without moving the hands. But it yet remains to explain the mode in which motion is communicated to the hour-hand: the cannon-pinion of four teeth, or pins, revolving with the centre-wheel once in an hour, neceffarily caufes the wheel of 48, in which it acts, and with it the wheel of communication, to revolve once in $\frac{48}{4}$, or twelve hours; and the wheel of communication acting on the hour-wheel, fo called becaufe it carries the hour-hand, is of the fame fize and number of teeth as itfelf, and therefore caufes it and the hour-hand to make one revolution in twelve hours.

The feveral parts of the motion and their refpective ufes being now defcribed in detail, it remains laftly that we defcribe the general action of the whole.

Let us fuppofe the watch made to repeat in the ufual manner, by pufhing in the pendant at the three-quarters and a half-quarter paft feven o'clock, the rack A B, *fig.* 2. by the preffure of the pufhing-piece, is carried forward until its further advance is ftopped by the arm *f*, of the piece D, reaching the feventh ftep from the exterior point of the hour-fnail F; at the fame time the rack, by its action on the pinion C, winds up to a certain degree the main-fpring; and the hour and quarter ratchets N and O, being attached to this pinion, are carried forward with it, or rather, relative to the fhape of the teeth of thofe ratchets, are carried backwards with it: the piece D, neceffarily advancing with the rack, being immediately attached to it, difengages the quarter-rack E, previoufly kept in its place by the pin 3, which quarter-rack being difengaged, is by the power of its fpring *h* immediately brought forward after the rack A B, until it is ftopped by its arm *y*, reaching the quarter-fnail G; but in this cafe, the watch being to ftrike the three-quarters and the half-quarter, the advance of the quarter-rack is almoft immediately ftopped by its arm *y*, coming againft the ftep, the fartheft removed from the centre of the quarter-fnail G: the inftant the arm *f*, of the piece D, has, by its preffure againft the hour-fnail F, acted fufficiently upon the large *all-or-nothing* piece H, the unlocking is effected. By unlocking is meant that portion of the total action of the motion, every time the watch is made to repeat, which confifts in the hammer-tails being fet at liberty to defcend into the fame planes the two ratchet-wheels are in. By the action of the piece D, with the *all-or-nothing* piece H, which in its turn acts on the little *all-or-nothing* piece I, this piece I is thrown from under the piece K; and this piece K, being no longer fupported by the little *all-or-nothing* piece I, has its two arms 6 and 7, by which the fituation of the hammer-tails, on their centres of motion is determined, depreffed by the action of the fpring *q*, on its arm 9, until the two hammer-tails are brought into the fame planes as the two ratchets N and O, and are thus placed in a fituation to be acted upon by thofe ratchets, and through them the hammers. The power of the blows ftruck by the hammers depends on the ftrength of the hammer-fprings *o* and *p*; which power is communicated to the hammers, by the fprings bearing againft the pins 4 and 5, planted in them, which pafs through the pillar-plate into the hammer-tails: when by the action of the ratchet-wheels on the hammer-tails the hammers are raifed, thefe pins prefs againft the hammer-fprings and continue fo to do, thus gradually increafing the power of the fprings until the hammer-tail efcaping from the tooth of the ratchet, the hammer is by the ftrength of the fpring fuddenly thrown forward, and made to ftrike. The friction of the running-train and of the parts of the motion, added to the refiftance of the two hammer-fprings, is the whole power the repeating main-fpring has to overcome; and the ftronger thefe fprings are, provided they are not too ftrong for the power of the main-fpring, the louder will be the blows ftruck. The number of the teeth of the ratchets that are brought to act on the hammer-tails, is the fame as of the hour to be ftruck, which are determined by the two fnails. The unlocking being thus effected, the act of ftriking commences; the power by which the repeating main-fpring was wound up, being removed, the fpring immediately uncoils, and returns into its former ftate, carrying with it the pinion C, the two ratchets N and O, and the rack A B; the hammer-tails L M, having been depreffed, and from the action of the arms 6 and 7 of the piece K upon them, continuing in the fame planes with the ratchets, are neceffarily acted upon by them, and thus is the ftriking of the hours and quarters effected. The inftant the ftriking is completed, the rack A B continuing to return overtakes the quarter-rack E, and by means of its pin 3, carries it with it: at the moment the quarter-rack begins to move, one of its teeth, which tooth is regulated by the quarter or half-quarter that has been ftruck, acting on the pallet of the little *all-or-nothing* piece I, this piece is brought back to its former fituation, and through it the piece K, and the hammer-tails; and all the pieces of the motion return by the power of the repeating main-fpring, into the fame fituations they were in before the watch was made to repeat.

The *mode* by which, in this conftruction of repeating work, the proper number of quarters, and the half-quarter, if any, as fhewn by the hands, is ftruck, is peculiar to itfelf, and requires to be particularly explained. The hour-fnail and the great ratchet are fo proportioned to one another, that, as has been before obferved, according to the ftep on the fnail, the arm *f* of the piece D defcends upon; fo is the number of teeth of the great ratchet fent back, previoufly to the unlocking taking place, to engage as foon as the unlocking fhall be effected, in the hammer-tail of the hour-hammer: thus, fuppofe the watch to repeat *twelve*, the arm *f*. defcends to the twelfth or loweft ftep of the hour-fnail, and at the fame time the ratchets revolve a fufficient quantity for the twelve fucceffive teeth, that act on the great hammer-tail, to come into action: if, on the contrary, the watch were required to ftrike *one*, the arm of the rack would defcend to only the firft or exterior ftep of the fnail, and the ratchet would revolve a fufficient quantity for the firft only of the twelve teeth to come into action with the hour hammer-tail. It is evident, from the above defcription of the action of the twelve teeth of the great ratchet that act on the hour hammer-tail, that the three teeth in the fame wheel which, conjointly with the fmall ratchet of four teeth, are employed to ftrike the quarters, muft be fo fituated relatively to the twelve teeth that ftrike the hours, that when the watch is made to repeat they will precede thofe teeth, in their advance towards the hammer-tails; otherwife the quarters would be repeated before the hours; for if the action of thefe pieces is attended to, it will be feen, that, in the act of winding up the repeating main-fpring, as many teeth of the two ratchets will pafs under the hammer-tails, as are to act on them before the unlocking is effected: confequently, the tooth that firft paffes under the hour hammer-tail would be the laft to act upon it in returning; and every time the watch is made to repeat, the three teeth of the great ratchet,

ratchet, and the four teeth of the ſmall ratchet, muſt, by being in advance of the other twelve teeth of the great ratchet, be carried forward, and ſo ſituated, if not prevented by ſome external cauſe, as invariably to occaſion the watch to repeat the three-quarters and half-quarter, after the hour, every time the watch is made to repeat. Such effect is prevented taking place by the action of the quarter-rack: the extremity of this rack is cut into eight teeth; (the acting face only of the firſt tooth being cut, and the remainder of that tooth left ſolid, to inſure the ſafe action of the little *all-or-nothing* piece I, when the watch has finiſhed ſtriking,) and theſe eight teeth correſpond with and are proportionate to the eight ſteps in the quarter-ſnail; if the hour only is to be ſtruck, the quarter-ſnail is ſo ſituated with reſpect to the quarter-rack, that the arm *y*, of the latter, deſcends to the deepeſt ſtep in the ſnail, allowing the firſt tooth of the eight juſt to paſs the ſhort lever, or, as it is frequently called, *pallet* of the little *all-or-nothing* piece I, and, from the conſtruction of the parts, the quarter-rack is in ſuch a ſituation, relative to the great rack, that the latter, through the medium of the pin 3, coming into contact with it, at the inſtant the laſt blow of the hour is ſtruck, carries the quarter-rack with it, and cauſes it to return into its original ſituation: now the effect produced by the great rack, carrying the quarter-rack back, is, to occaſion the latter, by the action of its firſt tooth on the little *all-or-nothing* piece I, to bring that piece ſuddenly into the ſame poſition it was in before the ſtriking was diſcharged, and thus to raiſe the piece K, and conſequently the hammer-tails, and to diſengage them from the three teeth of the great ratchet that ſtrike the quarters, and alſo from all the teeth of the ſmall ratchet, which along with the rack-pinion, all now return into their original ſituation.

We are indebted to Mr. Benjamin Lewis Vulliamy, of Pall Mall, clock and watch-maker to the King and Prince Regent, for this minute deſcription and accompanying plate of the Stocken repeater, which has never before been deſcribed, and for which, therefore, we beg to acknowledge our obligation to him.

Elliot's repeating Watch.—A new, cheap, and ſimple apparatus for repeating the hours and quarters was contrived by J. M. Elliot of Ayleſbury-ſtreet, Clerkenwell, and publiſhed in Nicholſon's Journal (vol. vii. 8vo. Series, p. 157.), as being applicable to either clocks or watches. The ſketches that are given in the original drawings are not calculated to give a clear conception of the connection of the parts employed; and, therefore, we have given ſuch a new arrangement of the figures, as we truſt will enable our readers to comprehend the conſtruction and action of Elliot's mechaniſm, as applied to a watch. *Fig.* 1. of *Plate* XLVII. of *Horology*, exhibits the repeating work of Elliot's watch, publiſhed in 1804; and *figs.* 2, 3, 4, and 5, ſhew the parts detached, with the ſame letters of reference as in *fig.* 1. In this conſtruction, the uſual apparatus of wheels, pinions, chains, pulleys, and racks, are diſpenſed with; and the acting pieces, inſtead of being ſpread over the ſurface of the upper plate of the frame, are arranged concentrically on the axis of the pendant, which is not puſhed in, but made to turn round to the right or left, accordingly as the hours or quarters are required to be ſtruck; either of which may precede, or one only may be ſtruck, at the option of the wearer. A A B, in *fig.* 2, is called the repeating potance, ſcrewed to the upper plate at *x*, in *fig.* 1, and bears the works ſurrounding the axis of the pendant C D. The portion for ſtriking the quarters, ſeen in *fig.* 3, is attached to the axis *b*; but the part ſeen in *fig.* 2, for ſtriking the hours, is on a tube through which the axis paſſes, ſo that each part will revolve ſeparately; a circumſtance not adverted to in the original deſcription. The quarter-ſnail, S, in *fig.* 1, with its looſe piece, the ſtar-wheel and its hour-ſnail H, the jumper G, and its ſpring I, with the dial-work Y, are ſuch as we have already deſcribed. But the levers or detents M and N, with their ſprings *r* and *t*, act here with the circular racks on the common axis of the pendant, thus; the pendant ſocket D, in *fig.* 2, has a connection with the axis, ſeen in *fig.* 5, by means of the pin *f*, on the axis, taking hold of its projecting piece *a*, which may be called the hour-pallet; *e* is the hour-locking ſnail, in *figs.* 2 and 4, with its projecting pin, placed without or beyond the repeating potance, and fixed on a ſocket that ſurrounds the axis; on which ſocket alſo are fixed the repeating wheel *g*, and its ratchet R, with a ſpiral ſpring, exhibited in *fig.* 4; then while the tail-piece *x*, of detent N, of the hour-ſnail, falls on the proper ſtep of ſnail *e*, to regulate the number of ſtrokes to be made for the hours, the tail-piece *k*, in *fig.* 1, of the hammer V, is caught by the ſloping teeth of the ſtriking-wheel *g*, and raiſes the hammer under the plate of the frame, that ſtrikes a circular rim of ſteel ſurrounding the works, inſtead of a bell; but before theſe ſtrokes will be made, the pendant muſt be turned round by hand gently and regularly, and continued till all the blows are given: this manual turning, therefore, ſuperſedes the neceſſity of a repeating train of wheels and pinions. The ſtriking mechaniſm for the quarters is ſimilar in conſtruction to that for the hours, and is ſeen detached in *fig.* 3, where *l* is the pallet; *m*, the locking-ſnail, with its projecting pin to be caught by the pallet in the retrograde motion of the pendant; *o*, the ratchet-wheel, and *n*, its ſpiral ſpring; and laſtly, *p*, the contrate wheel for ſtriking the quarters, by means of the ſecond elongated hammer-tail S, while the hammer has its centre of motion at Z. Thus, when the hours and quarters are both ſtruck by the ſame hammer, as regulated by their reſpective ſnails, the ſprings *n* and *b*, connected with the concentric ſnails and their ratchets, bring back the acting parts to their original ſituations, for repeating the ſame ſtrokes as many times as may be wiſhed, for the purpoſes of either utility or curioſity. It may be neceſſary to notice further, that the rim ſubſtituted for a bell has a notch cut into it, to admit the arbor of the pendant to paſs without obſtruction; and that we give this conſtruction without having ſeen the watch itſelf, and, therefore, without making any remark on its merits or demerits, further than that it appears to have the recommendations of ſimplicity and cheapneſs.

In the ſame year and month in which this watch was firſt deſcribed, the inventor preſented the model of another repeating watch to the Society of Arts at the Adelphi, an account of which is publiſhed in the 22d volume of their Tranſactions, for which he received their bounty of thirty guineas. We have not given a drawing of this ſecond repeater, as being acceſſible to all ſcientific men properly introduced to the Society's rooms, and particularly as the mode of uſing it does not differ from what we have juſt deſcribed, ſo far as the rotatory motions to be given to the pendant, direct and retrograde, are concerned. In this watch the ſnails for the hours and quarters, the ſtar-wheel, dial-work, jumper with its ſpring, and locking detents, are nearly the ſame as in the other; but inſtead of the ſtriking-wheels, ratchets, and ſprings, being on the arbor of the pendant, they are placed on a flat circular rim of ſteel, that revolves, by means of friction-rollers, round the dial-work on the ſame plane, to about one quarter of a revolution. This rim is indented about a quarter of its circumference, and is actuated by a beveled pinion, placed on the inner extremity of the

the pendant's axis, which, by being turned to the right or left, will cause either the hours or quarters to strike first, according to the direction of motion; and pins inserted into the plane of the rim, at the side opposite to the indentation, lift the hammer-tail to strike as many blows as the hour and quarter notches made on the inner edge of the rim, and acting with the second arms of the detents, respectively determine. In this mechanism, a chain wound round a barrel containing a spring, and fixed on the pillar-plate, brings the steel rim back again to its original situation, which entirely depends on the position of the snails.

A patent was taken out for a repeating watch by the same ingenious man, dated 30th October, 1806; but as we have not seen the description thereof, we are unable to say how it differs from either of those which we have just described.

Berrollas' Repeater.—Joseph Anthony Berrollas of Denmark-street, in the parish of St. Giles in the Fields, London, took out a patent for what he calls an *infallible* repeating watch, bearing date 31st October, 1808, of which watch we shall give a brief description in this place. In the plate to which we last referred, *fig.* 6. shews the repeating portion of Berrollas' watch, in a state of action; *fig.* 7. shews the calliper of the common movement, hammer, and ring used for the bell; and the detached pieces of the repeating portion are represented singly in the group of *fig.* 8. Like Elliot, Berrollas founds his pretensions to public approbation on the simplicity of his contrivance, and on the consequent cheapness, where the repeating motion by wheels and pinions is dispensed with, and where one hammer only is necessary for striking both the hours and quarters. Though we have not seen the watch we now undertake to describe, yet as it has some peculiarities in its construction, not quite so intelligible as could be wished in the description given in the Repertory of Arts, vol. xiv. p. 364, we will venture to deviate a little from the author's own account. The outside of the watch resembles common watches, except in the pendant, which is mounted with a button, consisting of two parts, C and X: the lower one, X, does not move, and the upper one, C, having an endless screw annexed to it, turns round and comes out to the extent of four turns, and is cut into four turns and a half. The upper part of the button C, being turned to the right, screws off from the lower part X, and operating upon the hour-rack A, can be continued to be unscrewed until it has struck the hour which the hand indicates, when it cannot be further unscrewed. The same part C, being afterwards screwed to the left, to bring it back again to join the lower fixed part X, operates upon the quarter-rack B, and quarters are struck in the same manner as the hours, until the part C is completely joined to the part X. The piece W draws piece B back to its former station. The motion is composed of three principal parts, A, B, and C: A contains the hour-rack, B the quarter-rack, and C the pendant and endless screw. The piece C, turning on itself, ascends perpendicularly, and is kept in a proper direction by the piece E, which performs two objects. The interior of it forms the catch-work of the screw, whilst the exterior is fixed by two screws on the pillar-plate; and when the piece C is turned, it acts upon the piece A, and gives it a circular motion, first by means of the piece D D, whose interior is caught in the notch at the extremity of the piece C, while the exterior part of it is caught in the piece A; secondly, by the piece F, which holds the piece D D in a groove; thirdly, by the piece G, which is fixed to the pillar-plate with three screws, and under which the piece A is fixed by means of a pivot, on which it moves. The piece A, being thus moved, catches by means of twelve teeth, cut in its interior part, the piece H, which puts in action the hammer Q, that strikes on the bell-spring R, fixed to the pillar-plate S, as seen in *fig.* 6.

The piece A passes under the piece K, which is a brass bar with two screws to keep piece A from rising. In order to give a free and a steady motion to the piece A, it is operated upon by a pivot fastened to a spring U, placed on the inside of the pillar-plate; which pivot, passing through a hole in the pillar-plate, causes a steady friction under the teeth of the piece A. The piece A is regulated by the star N and hour-snail M, in the common way, with a jumper and spring, when the hours are struck; also the quarter-snail P, the quarter-piece B, and its spring O, regulate the same, when the quarters are struck, through the agency of the forked piece J, which is on the same arbor as the tail-piece H and hammer. The tooth V, on the piece A, falls on one of the steps of the hour-snail, and determines the number of strokes for the hours, when A is turned one way round, and the heel-piece of the quarter-piece near P falls on the quarter-snail, when the motion is given to A in the contrary direction, and thus determines the strokes given for the quarters, while two springs keep the tail-pieces H and J in their respective places; one of which springs, L, is seen upon the plate, and the other is fast to the piece K, not seen. Thus the striking in this watch is produced by manual pressure, as is done in Elliot's watches, and the mechanism differs from that at the rooms of the Adelphi Society only in these respects:—that the motion is produced by a screw, instead of a pinion; and the piece A moves round a pivot at the end of the cock G, instead of being formed into an exact ring to move within friction-rollers.

Alarum, 'larum, or *warning Watch.*—The watches which we have above described under the denomination of *repeating* watches, can be useful only to persons who are awake, and, therefore, do not answer the purpose of giving previous notice of the approach of any particular hour and minute, at which it may be required to be roused from sleep. This purpose is usually effected by a 'larum clock; but we will now describe how the same thing has been done by some additions to a pocket-watch. We will first describe the former method of adding the warning mechanism to a watch, and then explain the construction lately adopted by Berrollas, and secured to himself by a patent. The old 'larum watch has been so well described by Berthoud, in the first volume of his "Histoire de la Mesure du Tems," p. 66, &c. that we cannot do better than give a similar drawing and description, after omitting his account of the ordinary movement.

In the frame of the watch a spring-box is made fast to the under side of the upper-plate, and has a great wheel, ratchet, and click, to wind up by; but the winding is performed by the key inserted on the square end of the spring-arbor, while the box remains at rest. When the great wheel is made to revolve by the ratchet, it drives two other wheels and pinions, which, with it, constitute the warning train, and is in every respect similar to a repeating train, except that there is an escapement-wheel instead of a fly. This train, therefore, will be the more easily understood, from our preceding description of the striking or repeating train. *Fig.* 8. of *Plate* XLVI. exhibits so much of the warning mechanism as appears on the exterior face of the upper-plate, and will suffice to explain the construction and action of all the essential parts, if we bear in mind that a part of the warning train and the spring-box are within the frame, of which this plate forms the cover. The arbor of the concealed spring-box is seen at A, bearing a finger that acts with three teeth cut in the semicircle of the circular piece F, pivotted

pivotted into the cock G. In the preſent poſition, the finger carried by the ſpring-arbor is reſting on the plain or unindented part of piece F, and is keeping the ſpring to its required degree of tenſion. When the key for winding is applied to the ſquare of this arbor, the teeth of piece F regulate the number of turns that the ſpring requires to be wound, before it comes to its due tenſion for driving the warning-train. The wheel *n*, under the cock H, is on the arbor of the pinion driven, within the frame, by the great wheel, and may be called the ſecond wheel of the warning-train; and the pinion driven by it is on the ſame arbor as the eſcapement-wheel R, (pivotted alſo into cock H,) which is the third or laſt wheel of the train. The teeth of the eſcapement-wheel, when put in motion by the main-ſpring urging it through the train, acts with the two pallets *a* and *b* alternately, which are connected by the portions of two ſmall wheels: that repreſented by *a* is faſt to the ſquare end of the hammer-arbor, concealed in the frame; and the other, *b*, turns on a fixed ſtud on the plate. The forked piece or angular lever 1 2, of the piece *b*, embraces the angular end *d* of the warning-detent *d f* 4, which is moveable at *f*, while the remote end 4 preſſes on a ſpiral plate *p*, made faſt to the wheel of the hours C, by the action of the ſpring *q*. Now as plate *p* revolves once in twelve hours, and has only one ſtep at *o*, the end of the ſpiral, it is obvious that, when this ſtep comes to the angular point 4 of the detent, this point will drop ſuddenly towards the centre of the hours-wheel, and at the ſame inſtant the angular end *d* will quit the fork 1 2 of the piece *b*, which will now be at liberty to obey the force of the eſcapement-wheel R, exerted on the pallets *a* and *b* alternately; and the concealed ſemi-circular hammer, on the arbor of pallet *a*, will ſtrike at both ends alternately againſt the ſonorous ring that produces the warning noiſe, until the finger of A has gathered up all the teeth of piece F, and is again arreſted on the oppoſite point of the diameter beyond the laſt tooth. At *x* the piece *b* has a tail-piece, which vibrates between the elaſtic prongs of the fixed fork P, and thus brings back the hammer after each ſtroke to the right and left; and the frequency and loudneſs of theſe reiterated ſtrokes are competent to rouſe any perſon, not too lethargic, from a ſtate of ſleep. A ſmall dial-plate lying over the centre of the watch-face, and divided into twelve hours and parts, is ſo connected with the ſpiral-plate *p*, that turning this dial to a pointer, made on the ſhort end of the hour-hand, will put the ſtep *o* into a proper ſituation for making the point 4 of the detent fall at the time required, and will conſequently produce the alarm at the time for which the ſmall dial is thus adjuſted.

Warning Watch by Berrollas.—In the year 1810 we find J. A. Berrollas reſiding in Coppice-row, Clerkenwell, and taking out a patent, on the 26th of May of that year, for a warning watch of a new conſtruction, which we will next proceed to deſcribe; but the deſcription given of this invention in the Repertory of Arts, &c. is ſo imperfect, at leaſt to us ſo unintelligible, that we have been obliged to alter both the drawings and explanation, before we could make the mechanical contrivances underſtood. The reaſons which ſeem to have induced Berrollas to attempt a new conſtruction of a warning watch, were the inconveniences attending winding up, ſetting to time, and turning the ſmall dial-plate, all which he profeſſes to have obviated. We have ſhewn the different parts of this mechaniſm in ſeveral figures in *Plate* XLIV., which we ſhall now proceed to explain in our own way. In *fig.* 10, *a* ſhews the place of the main-ſpring, and *b* the fuſee of the ordinary works, which are conſtructed in the common way, but which are not ſeen in *fig.* 11, that repreſents an elongated ſection of the warning mechaniſm and dial-work only. At *c* is the arbor of the warning-ſpring box, of the great wheel of 60 teeth, and of a ratchet-wheel, which is made of ſteel with 33 teeth, that catch the tail of the hammer *d*, and make it ſtrike againſt the circular rim of ſteel, while the ſpring *e* brings back the hammer after each ſtroke. This part of the mechaniſm is not given in the original drawings, nor yet *fig.* 11, which explains the action. As the ſtrokes are made by a wheel on the arbor of the ſpring-box, it was neceſſary that it ſhould wind five times round, that the blows might be ſufficiently numerous and loud for giving the alarm: hence 165 (33 × 5) ſtrokes are given at once winding, and the firſt is the loudeſt, as being urged by the warning-ſpring, without a fuſee, at its greateſt degree of tenſion; an advantage which the inventor ſeems to have overlooked in his own account. Theſe parts, and alſo the pinion *f*, are planted within the frame, as ſeen in *fig.* 11; but the parts ſhewn in *figs.* 9, 13, and 14, are on the exterior face of the upper-plate, agreeably to the calliper given in *fig.* 9, but acting together, as more clearly repreſented in *fig.* 11, where the pivot-holes are ſuppoſed to be in a right line, for the ſole purpoſe of explanation. The arbor of pinion *f* aſcends through the upper-plate of the frame, and has the wheel *g* attached to it, which drives a ſecond pinion on the arbor of an eſcapement-wheel *h*; which two wheels are pivotted above into a long cock, ſcrewed to the upper plate; all which poſitions are clearly ſeen in *fig.* 11, as well as the mode by which the motion and force are tranſmitted from the main-ſpring *c* to the pallets *i*, *i*. The wheel *g* has 45 teeth, the eſcapement-wheel 20, and the two pinions each ſix leaves. This aſſemblage of wheels and pinions conſtitutes the warning-train; but the warning *detent*, on which much ſtreſs is laid by the inventor, remains yet to be deſcribed. This detent *k* is ſeen in two detached poſitions, in *fig.* 14, where the parts 1, 2, and 3, are taken off, to render their uſes more obvious. The dial-work conſiſts of a cannon-pinion of 12 leaves, the wheel *q* of 36 teeth on the ſtud, its pinion *r* of 10, and the hours-wheel *n* of 40, which are common numbers, that may be thus expreſſed: *viz.* $\frac{36}{12} \times \frac{40}{10} = \frac{1440}{120} = 12.$ Between the hours-wheel and the cannon-pinion the ring of the detent *k* ſurrounds the arbor of the centre-wheel, or rather the tube of the cannon-pinion, but not ſo as to be tight upon it: on the hours-wheel *n* is a pin projecting above and below its plane, as ſeen in *fig.* 13, againſt which pin the rim of the ſaid ring is preſſed by the ſpring *t*, as ſeen in *fig.* 9. Now as the ſcrew 1 takes into the cock 3, made faſt to the plate, after it has paſſed through a hole in the lever of the detent, this ſcrew becomes the centre of motion of the detent, and the preſſure of the ſpring *t* at one end elevates the ring at the other, and with it the perpendicular bar *l*, which reaches to the teeth of the eſcapement-wheel *h*; ſo that when the elevation of this bar *l* exceeds the plane of the eſcapement-wheel, the train will be at liberty to run on, and the hammer will be made to ſtrike a repetition of blows; but while the elevation of the ſaid bar *l* lies in the ſame plane with the teeth of the eſcapement-wheel, it will operate as a detent to the train, and ſilence will be preſerved. Above the hours-wheel *u*, and on its tube, revolves a large additional wheel *m*, bearing a circular piece of ſteel, with an oblong notch cut through it, as ſeen in *fig.* 13; and the preſſure of the upper end of the pin in the hours-wheel is againſt this circular ſteel piece at all times, except when it arrives at the ſaid notch, and then it aſcends into it by the downward action of the ſpring *t*, beyond the centre of motion 1; and at this inſtant the bar *l*, aſcending alſo, quits the teeth of the eſcapement-wheel, and ſuffers the alarm to commence, and to continue until the warning-ſpring is relaxed. It does not, however, appear by what means the pin of the hours-wheel is diſengaged from the oblong hole in the ſteel circle of wheel *m*, when the blows of the hammer are finiſhed, unleſs the ſecond edge of the ſlit be made ſloping; and its continuance there muſt obviouſly ſtop the watch, by preventing the free progreſs of the hours-wheel. Either the mechaniſm or the original deſcription muſt be defective in this part. Within the ring of the pendant is a button, in form of an acorn, which conſtitutes the head of a ſteel arbor, terminating with a pinion *s*, that actuates a contrate-wheel *o*, which has teeth alſo on the edge, that drive the double pinion *p*; the upper portion of which pinion again impels the large wheel *m*, that carries the ſmall hand, and alſo the circular piece of notched ſteel. This train is introduced for the purpoſe of ſetting the warning-hand, which is the ſhorteſt of the three ſeen in *fig.* 12, to the requiſite hour on the ſmall fixed dial, without opening the caſe, and deranging the other hands for ſhewing hours and minutes in the uſual way. At the bottom of the acorn there is a ſmall ratchet-wheel, with a click and ſpring, ſeen detached between *figs.* 11 and 14, the uſe of which is to allow the acorn to turn only in one direction, while the warning-hand is ſet to its place. It is a condition to be obſerved in ſetting the hour-hand, that it lie exactly over the *pin* in the hours-wheel, and alſo that the warning-hand lie parallel to the *ſlit* of the ſteel circle, attached to the wheel *m*. Laſtly, the warning mechaniſm may be put into a ſtate of action, or of inaction, by moving the button *v*, in the caſe, to the right or left, ſo as to fall in the way of the end of the detent *k*, or to eſcape it.

Rolling Watch.—After having deſcribed various conſtructions of watches adapted for the pocket, and contributing to the punctuality with which ſocial intercourſe is carried on in civilized life, we proceed laſtly to deſcribe the mechaniſm of a watch, which will meaſure time only by its deſcent down an inclined plane, and which therefore is more curious than uſeful. During the time that various experiments on the laws of moving bodies were made, and applied to the regulation of horological machines by Dr. Wallis, ſir Chriſtopher Wren, Dr. Hooke, Huygens, Leibnitz, Dr. Halley, ſir Iſaac Newton, and others, M. de Gennes and the marquis of Worceſter contrived watches, the former of which would *aſcend*, and the latter *deſcend* along an inclined plane, by means of a ſpring coiled up at the centre of the frame, which relaxed as the rolling motion proceeded; but as no explanation had been given of theſe contrivances in the year 1684, Mr. Maurice Wheeler publiſhed an account of a rolling watch, invented by him, in the firſt volume of Lowthorp's Abridgment of the Philoſophical Tranſactions of London, p. 468, et ſeq., which account has been copied into other works of ſcience; but which we will abridge, agreeably to our own plan of deſcription. In *fig.* 15. *Plate* XLIV., let *a b* be the diameter of the circle *a f b h*, ſtanding on the dotted horizontal baſe *b m*, on the point *b*. In this ſituation, ſuppoſing the circle to be an uniform plate of metal, it will remain at reſt when placed truly vertical, while the line *b m* remains horizontal; but make this line to coincide with the inclined line *n b*, and the circle will roll down this inclined plane, becauſe the vertical line, or line of direction, *d e*, raiſed from the point *d*, which will be the new point of contact, falls behind *c*, the centre of gravity of the uniform circular plate; ſo that the portion *e f d* becoming ſmaller than the portion *e h d*, the centre of gravity will be before the bearing point *d*, and the plate will roll down the plane; and the motion thus produced will have the greater velocity, the more the plane *n b* is inclined to the horizontal line. But if ſuch a piece of metal *g* be attached to the portion *e f d* as will form a counterpoiſe therewith to the larger portion *e h d*, then the plate will have no tendency to roll, but become ſtationary, ſo long as the inclination of the plane, and the poſition of the additional piece *g*, remain unaltered; but leſſen the angle of inclination, or remove the weight *g* towards *f*, and, in either caſe, the plate will actually aſcend, till the weight *g*, in its new poſition, balances the new angle of inclination. Alſo, if the angle be increaſed, or the weight *g* be brought nearer to *d*, in either caſe the plate will deſcend; but as the deſcent of the plate down the increaſed inclination, by a rolling motion, throws the weight *g* farther from the point of contact *d*, the rolling motion will ſtop as ſoon as the retrograde motion of the weight ſhall have produced a counterpoiſe to the portion *e h d* of the plate in the new inclined plane. Theſe premiſes being granted, we are now prepared to ſhew how the train of a watch in motion may be made to change the poſition of an appended weight in ſuch way as to render that weight a maintaining power during the whole time that a cylindrical box, which contains the weight and movement, ſhall gradually and ſlowly deſcend down a correſponding inclined plane, while a pendulous hand or index ſhall point out the ſucceſſive hours and parts during the ſaid deſcent. Let the four wheels and pinions, ſhewn in *fig.* 17, be placed like the common train of a watch, with the arbor *x* of the great wheel in the centre of the box, and conceive a balance and eſcapement to allow one tooth of the laſt wheel to eſcape at every alternate oſcillation of the moving balance; then let the weight *w*, at the end of the lever *x w*, be made faſt to the ſaid arbor at the hole *x*, *fig.* 16, as is ſeen in *fig.* 17; and let its poſition be between *d* and *f*; and the tendency of the weight *w* to come to *d*, will draw round the great wheel, pivotted to the ends of the box, and give ſuch motion to the train as will keep up the oſcillations of the balance; but this motion of the great wheel will be ſo ſlow, that it will be ſcarcely perceptible when the angle of inclination is ſmall; but increaſe this angle, or alter the poſition or the magnitude of the weight, and the force will be increaſed by which the train is actuated; ſo that by one alteration or other, or all, ſuch a final adjuſtment may be made, that the box will revolve exactly in twenty-four hours. But before this can be duly effected, the train of wheels muſt be counterpoiſed by a load, attached to the box at the oppoſite ſide of the centre, ſo that there may be one common centre of gravity of all the materials, excluſive of the weight *w*, falling in every poſition of the box, at the centre of motion of the weight *w*; *i. e.* at the arbor of the great wheel round which the box revolves. When this is the caſe, and the angle of elevation of the plane is nicely adjuſted by a ſcrew A, as in *fig.* 18, the regulated train of the watch will allow the weight *w* to approach *d*, juſt as faſt as the rolling motion of the box will throw it back, and the equipoiſe of theſe two contrary actions of gravity, and of the rotation of the box produced thereby, will keep up an uniform ſlow motion down the inclined plane. In ſome conſtructions detailed in the "Recueil d'Ouvrages curieux de Mathematique et de Mechanique; ou Deſcription du Cabinet de Monſieur de Serviere," à Lyon, 1733, the hours are drawn on the face of the inclined plane, and indicated by the point of contact of the box; but in the watch before us, one end of the box contains the figures from I. to XII. twice over, and a pendulous hand, made heavy below, and revolving looſely on the pivot of the great wheel, always

always points upwards to the paſſing hour, as repreſented in *fig.* 18.

In the work of which we have juſt given the double title, there are various devices for making balls deſcend on ſpiral and zigzag planes in a given time, which are again elevated by a ſpring, and which indicate the time by the number of their deſcents; but theſe matters of curioſity are no longer uſeful as horological machines in the preſent ſtate of the ſcience of horology.

Muſical Watch.—The works of a watch may be applied moreover to give motion to various devices and pieces of machinery, at the ſame time that the watch performs its own operations; ſuch as actuating the handle of a planetarium, or orrery, exhibiting the motions of any of the celeſtial bodies by means of its dial-work, or urging the barrel of ſmall muſical chimes: but after the detailed accounts we have given of PLANETARY *Machines*, it will only be neceſſary here to explain how a tune may be played by a muſical watch, from which our reader will perceive that the application of ſimilar means to other amuſing purpoſes is equally practicable. In *Plate* XLII. of *Horology*, *fig.* 4. exhibits, on an enlarged ſcale, the internal diſpoſition of a watch-movement, as ſeen on the pillar-plate from above, when the other plate of the frame is removed, and the watch laid down on its face; the pillars being ſuppoſed to be at the circles N, N, N, and N; the box A contains the main-ſpring; B is the fuſee, with the chain or cord winding round it, as it comes from the circumference of the ſaid box; D is the great wheel, and within it are the ratchet, click, and ſpring, as uſual; E is the pinion of the centre wheel, or hour-wheel, F, and is driven by the great wheel as ſoon as the main-ſpring is wound up; G is the pinion on the arbor of the ſecond wheel, and is driven by F; and H, that revolves in the ſame time with the pinion G, is the third wheel, which wheel again actuates a pinion, I, on the arbor of the fourth wheel, K; which here is not a contrate-wheel, becauſe the balance-wheel I is made for the cylinder eſcapement, which we deſcribed under the article ESCAPEMENT, N° 8. The arbors of theſe wheels and pinions pointing upwards to meet the eye, appear only in plan; and the dial-work, lying under the plate, is concealed from view. The parts done only in outline exhibit the calliper of the watch, independently of the muſical train, barrel, ſpring, and other appendages, which are *ſhaded*, for the purpoſe of diſtinguiſhing this portion from the ordinary movement of the watch. In this figure, the calliper of the muſical train and ſpring-barrel may be obſerved to lie on one ſide of the frame; but the mode of their action will be better explained by *fig.* 5, which is a ſection of the frame, in which the calliper is ſo altered into a ſtraight line, for the purpoſe only of explanation, that the effect to be produced may be clearly comprehended. In both the figures, 4 and 5, the great wheel attached to the ſpring-barrel is denoted by the unit 1, and the wheels that follow, with their reſpective pinions in the muſical train, are denoted by 2, 3, 4, and 5, till we come to the regulating fly, 6, placed on the arbor of the laſt pinion. This fly performs the ſame office as in the ſtriking part of a clock, or repeating train of a watch; that is, it regulates the velocity with which the main-ſpring ſhall unbend itſelf, and give motion to the barrel in which it is contained; ſo that if a quick motion be required, a few wheels and pinions only are neceſſary, and a light fly; but when the motion is required to be ſlow, there muſt be a longer train, or a heavier fly, proportioned to the ſtrength of the main-ſpring. The interior end of the ſpring is, as in the common main-ſpring boxes, hooked to a pin on the arbor *d*, and the exterior end to the ſide of the box, ſo that turning the arbor *d* round by a key, coils the ſpring into its ſtate of greateſt tenſion, which is adjuſted by the notched piece, or ratchet, *e*, which is held to its place by the click and ſpring in the uſual way, as repreſented in *fig.* 6. Upon the rim or cylindrical ſide of the box *a*, containing the main-ſpring, are inſerted various pins in parallel lines, but at unequal diſtances, according to the frequency of the occurrence of the reſpective notes to be ſounded by the correſponding prongs of the forked piece of ſteel *b c*, in any given tune which is to be played; and the number of prongs muſt be equal to the number of muſical tones and ſemi-tones to be produced. In the drawing before us the fork has eight prongs, and the notes are produced by the catching of the pins, inſerted into the revolving barrel, on the ends of the prongs, which are elaſtic and tapered, as well as tempered, to produce the requiſite ſucceſſion of tones that are required in the tune to be performed.

The upper pivots of all the arbors of the muſical train are callipered in the cock *h h*, while the lower pivots have their holes on the pillar-plate; and upon this cock *h h* the bent detent, or double lever *f e*, with a hook at *e*, is placed ſo as to be moveable round a ſcrew at the angular point in the middle: the hook of this detent is kept cloſe to the revolving barrel by the preſſing ſpring *i*, and when the tune is finiſhed, a hole is caught on the ſide of the barrel, free from the pins, by the hook *e*, which ſtays the motion till the button *g*, in the caſe of the watch, is puſhed in againſt the tail-piece of the detent, and frees the hook again from its hole, when the tune is repeated nearly in the ſame manner that chimes are uſually performed. (See CHIMES.) The parts drawn in perſpective in *fig.* 6. repreſent a conſtruction in which the elaſtic prongs of the muſical fork are bent into a curve, ſo as to occupy leſs ſpace than in *figs.* 4 and 5, or to produce more powerful tones where the ſpace will admit of an enlargement of their dimenſions. But inſtead of a barrel containing the main-ſpring, ſometimes a cylinder, A, revolving in the cocks *o* and *p*, contains the pins, as is ſeen in *fig.* 7, where a pinion on the projecting arbor of the cylinder is actuated by the great wheel attached to the ſpring-box; and this conſtruction is beſt calculated for a fork with more tones, and conſequently for a tune of greater compaſs. *Fig.* 8. ſhews how the prong of the fork is caught by the pins in ſucceſſion, and *fig.* 9. explains how a number of double-pronged forks, *b c*, are ſeparately ſcrewed into a frame, *k*, where the difference in the tones is produced by a correſponding difference in the dimenſions of the prongs. The hole in which the hook *e* falls is here in the end of the cylinder, but the motion is not ſtayed thereby; for the ſlender ſpring, preſſing againſt it, lays hold of the fly-pinion, when it follows the detent, and thus ſtops the part that has the greateſt velocity, and leaſt power; whereas conſiderable ſtrain takes place on the barrel when its motion is ſtopped by the hook of the detent, as in *figs.* 4 and 6. The arrangement in *fig.* 4. is beſt ſuited for a ſmall watch, but does not produce the moſt audible tones; and is that which is uſually concealed in the Swiſs *muſical ſeals*, that have been lately introduced into England: but the cylinder in *fig.* 7. is that which the *muſical boxes* contain, and which, from its length, is capable of containing two tunes, as well as notes on both the treble and baſs clefs. When the elaſtic prongs are tempered, they are brought to an exact muſical ſcale by grinding with oil-ſtone duſt, and the prongs that require to have their tones much flattened, are made more ſlender at the end moſt remote from the cylinder or barrel, where their reſiſtance to motion is diminiſhed, ſo that the vibrations are rendered leſs frequent, and the tones leſs acute. When a ſecond

second tune is played, the cylinder is pushed into and held in a new situation, as in a barrel-organ.

WATCH-*Maker* is a term that might originally denote the *maker* of a *watch* out of the rough materials; but this was the case only in the infancy of the art; for when the demand for pocket-watches became so great as to render the manufacture an object of commerce, the consequent division of labour was soon found to facilitate the execution; and each branch of the trade had its appropriate tools and artisans; so that thousands of families have at length been employed and maintained by their respective manipulations, without interfering with the business of their neighbours, who are equally occupied in their respective departments. In this art, the materials are of very little value in their original state; but the workmanship is so various, and requires such delicacy in the management, that, as has been stated to a committee in the house of commons, materials that did not cost sixpence in the state of ore, have been and frequently are manufactured into a watch that is worth 100*l.* and upwards. This view of the subject induced ministers to withdraw their projected plan of taxing watches, as being, in fact, a tax on ingenuity and industry.

We have anticipated much of the subject of watch-making under the articles we have already referred to in horology, for a watch may be considered as a spring-clock in miniature; and what we have said about calculating, callipering, making, and adjusting the various portions of a small clock, is therefore equally applicable to a watch, except that the balance must be substituted for the pendulum. It may gratify the curiosity of our readers, however, to explain how the labour, we have adverted to, is divided and subdivided into various branches of manufacture; and how persons situated at different parts of the kingdom conspire, without any mutual knowledge of each other, to bring this small piece of mechanism into its finished state.

The best watch-movements are made at Prescot, in Lancashire, by persons called movement-makers, who furnish the movement complete to the London watch-makers.

The following is a list of the principal workmen employed in manufacturing a movement, previously to its coming into the hands of the London watch-maker.

1. The frame-maker, who makes the frame; that is to say, the two plates, the bar, and the potance.

2. The pillar-maker, who turns the pillars, and makes the stud for the stop-work.

3. The cock-maker, who makes the cock and the stop-work.

4. The barrel and fusee-maker, who makes the barrel, great wheel, fusee, and their component parts.

5. The going fusee-maker, who makes the going fusee, (the means by which the watch is kept going while winding up,) when made use of.

6. The centre wheel and pinion-maker, who makes the same.

7. The small pinion-maker, who makes it of wire, previously drawn by another workman, called pinion-wire; the third and fourth wheels, and escapement-wheel pinion; and in the case of repeaters, the pinions of the repeating train of wheels: these are all finished in the engine.

8. The small wheel-maker, who makes the third and fourth wheels, and the wheels of the repeating train for repeating movements, and rivets them to their pinions.

9. The wheel-cutter, who cuts the wheels.

10. The verge-maker, who makes the verge of vertical watches.

11. The movement-finisher, who turns the wheels of a proper size previously to their being cut, forwards them to and receives them from the wheel-cutter, examines all the parts as they are made, to see that they are as they should be; and finally completes the movement, and puts it together.

12. The balance-maker, who makes the balance of steel or brass.

Note.—The brass balance is preferred to the steel balance by some watch-makers, in consequence of the latter being subject to the influence of magnetism: but others prefer the steel to the brass balance, in consequence of the latter being more influenced by variation of temperature than the former.

13. The pinion wire-drawer, who prepares the pinion-wire; this, however, may be considered as only a branch of the trade of wire-drawing.

The plates and wheels are now all made out of rolled brass; but formerly, when it was to be had, they were made of Dutch brass, it being considered preferable to the English.

The movement, in the state in which it is sent to the London watch-maker, consists of the frame, composed of two plates, connected together by four or five pillars, as the case may be, which pillars are riveted to one of the plates, called the pillar-plate; the wheels, consisting of the great wheel attached to the fusee, the second or centre wheel, the third and fourth wheels, the fusee and barrel, potance and stop-work, which latter are attached to the upper plate, (so called in contra-distinction to the pillar-plate,) but the potance screwed to it as between the plates; and lastly, the cock screwed to the outside of the upper plate.

The following is a list of the principal workmen employed on a watch to complete it from the state in which the movement is received from the country.

1. The slide-maker, who makes the slide.

2. The jeweller, who jewels the cock and potance, and, in a more forward state of the watch, any other holes that are required to be jewelled.

3. The motion-maker, who makes the brass edge; and, after the case is made, joints and locks the watch into the case, and makes the motion-wheels and pinions.

4. The wheel-cutter, who cuts the motion-wheels for the motion-maker.

5. The cap-maker, who makes the cap.

6. The dial-plate maker, who makes the dial.

7. The painter, who paints the dial.

8. The case-maker, who makes the case.

9. The joint-finisher, who finishes the joint of the case.

10. The pendant-maker, who makes the pendant.

11. The engraver, who engraves the name of the watch-maker on the upper plate; and also engraves the cock and slide, or index, as the case may be.

12. The piercer, who pierces the cock and slide for the engraver, and afterwards engraves them.

13. The escapement-maker, who makes the horizontal, duplex, or detached escapements; but the escapement of a vertical watch is made by the finisher.

14. The spring-maker, who makes the main-spring.

15. The chain-maker, who makes the chain.

16. The finisher, who completes the watch, and makes the pendulum-spring, and adjusts it.

17. The gilder, who gilds the watch.

18. The fusee-cutter, who cuts the fusee to receive the chain, and also balance-wheel of the vertical escapement.

19. The hand-maker, who makes the hands.

20. The glass-maker, who makes the glass.

21. To these must be added the pendulum-spring wire-drawer,

drawer, who draws the wire for the pendulum-fprings, which is almoft a diftinct trade.

The above are the principal workmen employed in the manufacture of a plain watch; in the manufacture of a repeater, the fame workmen are employed, with the addition of the repeating motion-maker, in the place of the maker of the plain motion, who makes the brafs edge and motion, and joints and locks or fixes in, which ever way it is done, the watch into its cafe; and when the repeater is made to ftrike on bell-fprings, the motion-maker makes the fprings. Formerly, repeaters were made to ftrike what was termed dumb; that is to fay, a dead blow againft the cafe or cap, or elfe on a bell, which bell was made by a feparate workman, entitled the bell-maker. But the bell-fprings have now entirely fuperfeded the old bells. The fame fpring-maker who makes the main-fpring of the watch, alfo makes the main-fpring of the repeating train of wheels.

The fprings of a hunting-cafe are made by a feparate workman, called a fecret fpring-maker. Single cafes (not hunting-cafes) are frequently made to open with fprings: pairs of cafes (the old-fafhioned box and cafe) are fprung, lined, and polifhed by a workman called a fpringer and liner; the better defcription of fingle cafes and hunting-cafes are polifhed by a perfon fimply called the polifher: this is fometimes done by women, particularly by the wives of fome of the cafe-makers; and this is the only branch of the trade, probably, in which women are employed in this country.

The compenfation-balances for chronometers are fometimes made by the efcapement-maker, and fometimes by a feparate workman, who confines himfelf entirely to making compenfation-balances.

Plain watch-movements are made of all prices, from 2*s.* 6*d.* to 2*l.* 12*s.* 6*d.*; but repeating movements coft from 2*l.* to 4*l.* 4*s.* each, according to their quality. There are many inferior movements, made as low as 21*s.* a dozen. The loweft prices at which the movements called Lancafhire movements are fold, is 7*s.* for plain, and 2*l.* 10*s.* for repeating movements.

The principal London watch-makers order the movements, as above defcribed, of the movement-makers of Prefcot, who make them according to the callipers they receive from each maker with their orders. But the ordinary defcription of movements may be purchafed at moft of the watch-tool fhops in London; one of the chief of which is Fenn's, N° 105, Newgate-ftreet, where every defcription of clock and watch-makers' tools and engines may alfo be procured at moderate prices.

At and near Geneva, but chiefly at Locle and Chaux de Fond, in the principality of Neufchatel, the Swifs watches are manufactured in great numbers. In thefe manufactories women are very generally employed, and the fubdivifion of labour is carried ftill further than in our's; and this concurs with the poverty of the workmen, and other caufes, to render thefe watches cheaper than the Englifh manufacturers can make them. The Swifs watches have accordingly fupplanted the Englifh in many countries of Europe. In general, the workmanfhip of the Swifs watches is exceedingly flight.

WATCH-*Making* is the art of making watches. See WATCH-*Maker*.

WATCH-*Tools* are the tools with which watches are ufually made, fuch as vices, pliers, files, hammers, drills, gravers, turning-tools, broaches, turn-benches, balance-tools, fpring-tongs, gauges, fpring-tools, fufee-adjufting tools, pitching-tools, callipers, fcrew-plates, burnifhers, wire-nippers, fcrew-drivers, and various others, which would require feveral plates to reprefent them, but which greatly refemble the *clock-tools* that we have defcribed and explained by a reference to *Plates* XIX., XX., and XXI., of *Horology*.

WATCH-*Glafs*, the concavo-convex portion of a glafs-fphere ufually employed to cover the dial of a watch. The fpheres out of which the watch-glaffes are cut are blown of various dimenfions, according to the degree of convexity required: the edges are then ground to fit the groove of the cover of the watch-cafe. There is a fuperior defcription of watch-glaffes, technically called *bottoms*, which are not portions of fpheres, but are flat on the top, the edges only being concavous. Each of thefe is cut from a feparate piece of hollow glafs, blown in the fhape of a cone, of which the watch-glafs forms the bottom; whence it takes its name. The wafte of glafs is, therefore, confiderable, and the flat glaffes are confequently much more expenfive than the fpherical. The method of converting a circular piece of plain glafs into a concave, by a heated convex piece of iron, as recommended by Boyle, (fee Works Abr. vol. i. p. 135.) is no longer practifed in the conftruction of watch-glaffes; but large convex glaffes for clock-faces are frequently made in this way.

Recordon's Renovator.

Fig. 1.

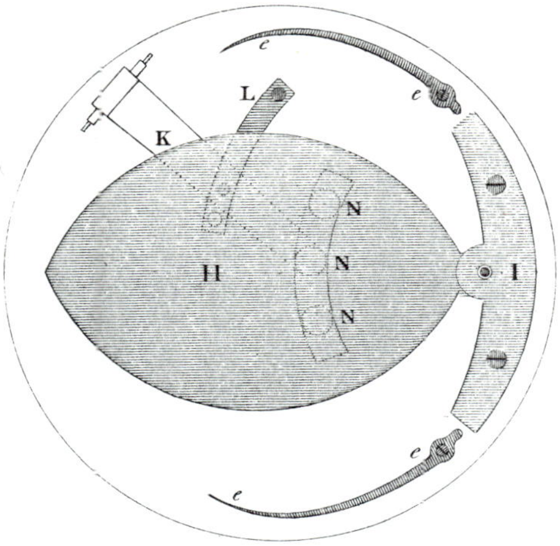

Fig. 3.

Fig. 2.

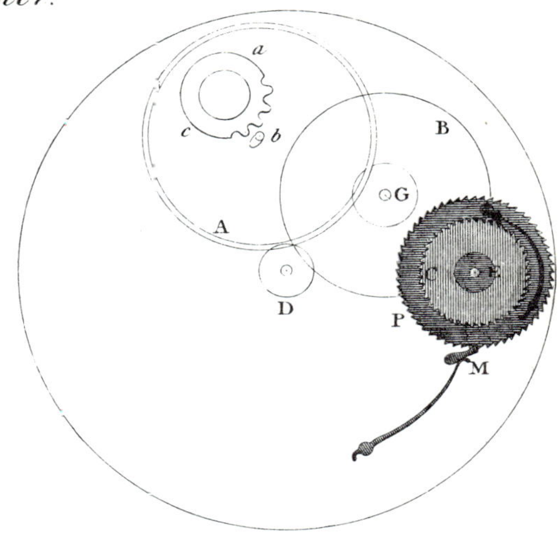

WATCH with MUSIC.

Fig. 4.

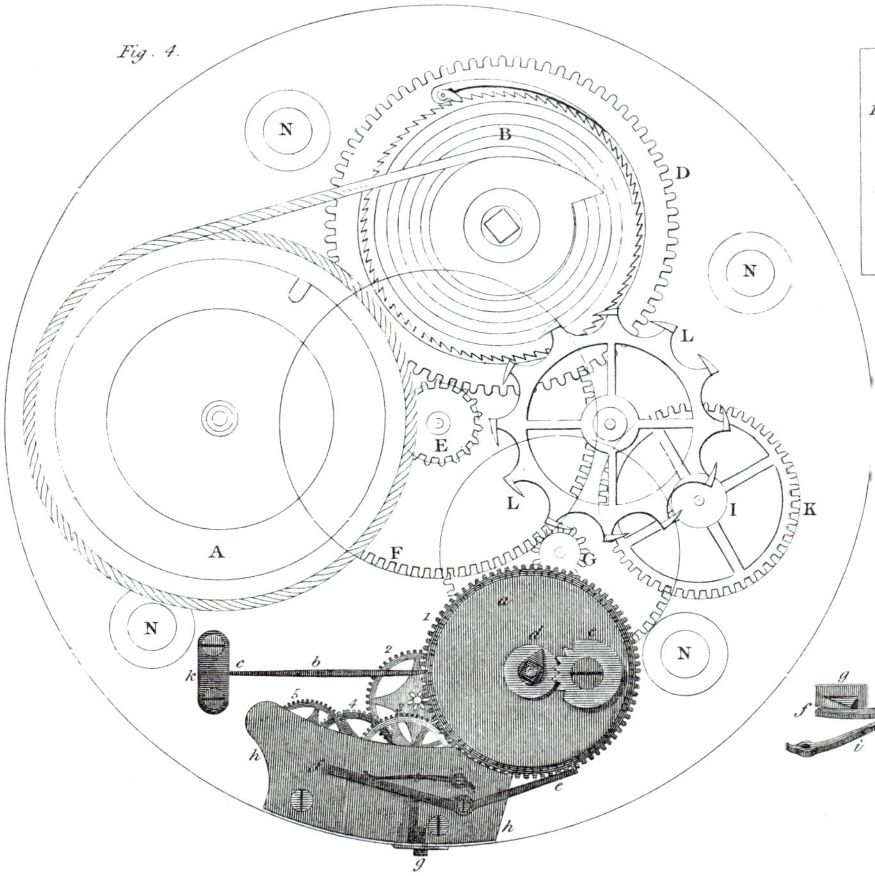

Fig. 7.

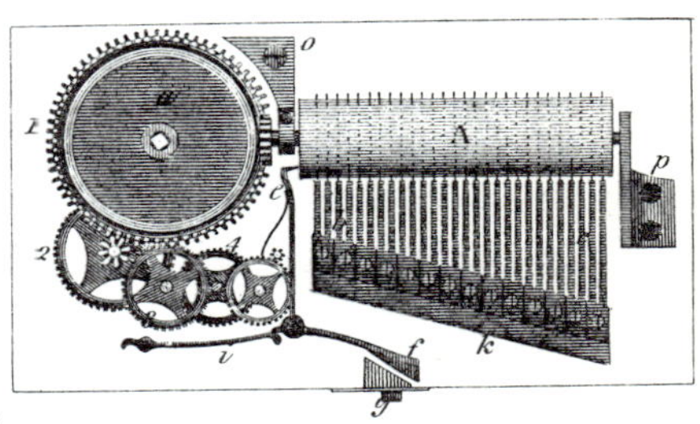

Fig. 8.

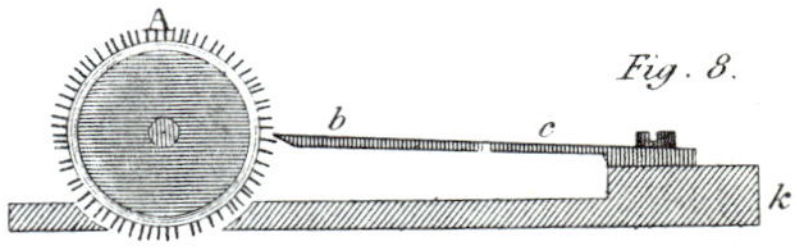

Fig. 9.

Fig. 6.

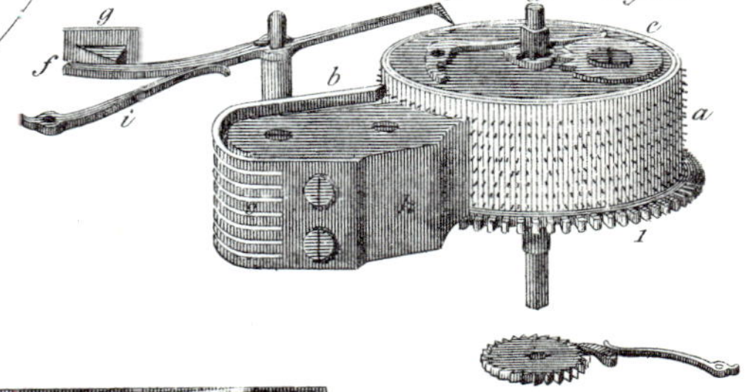

Fig. 5.

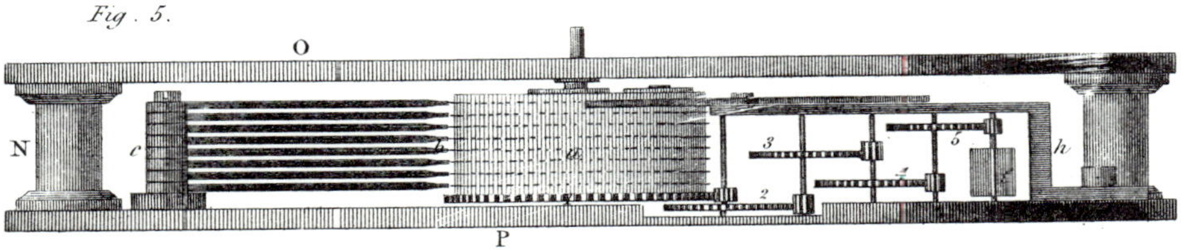

Farey delin. Published as the Act directs, 1817, by Longman, Hurst, Rees, Orme & Brown, Paternoster Row. Lowry Sc.

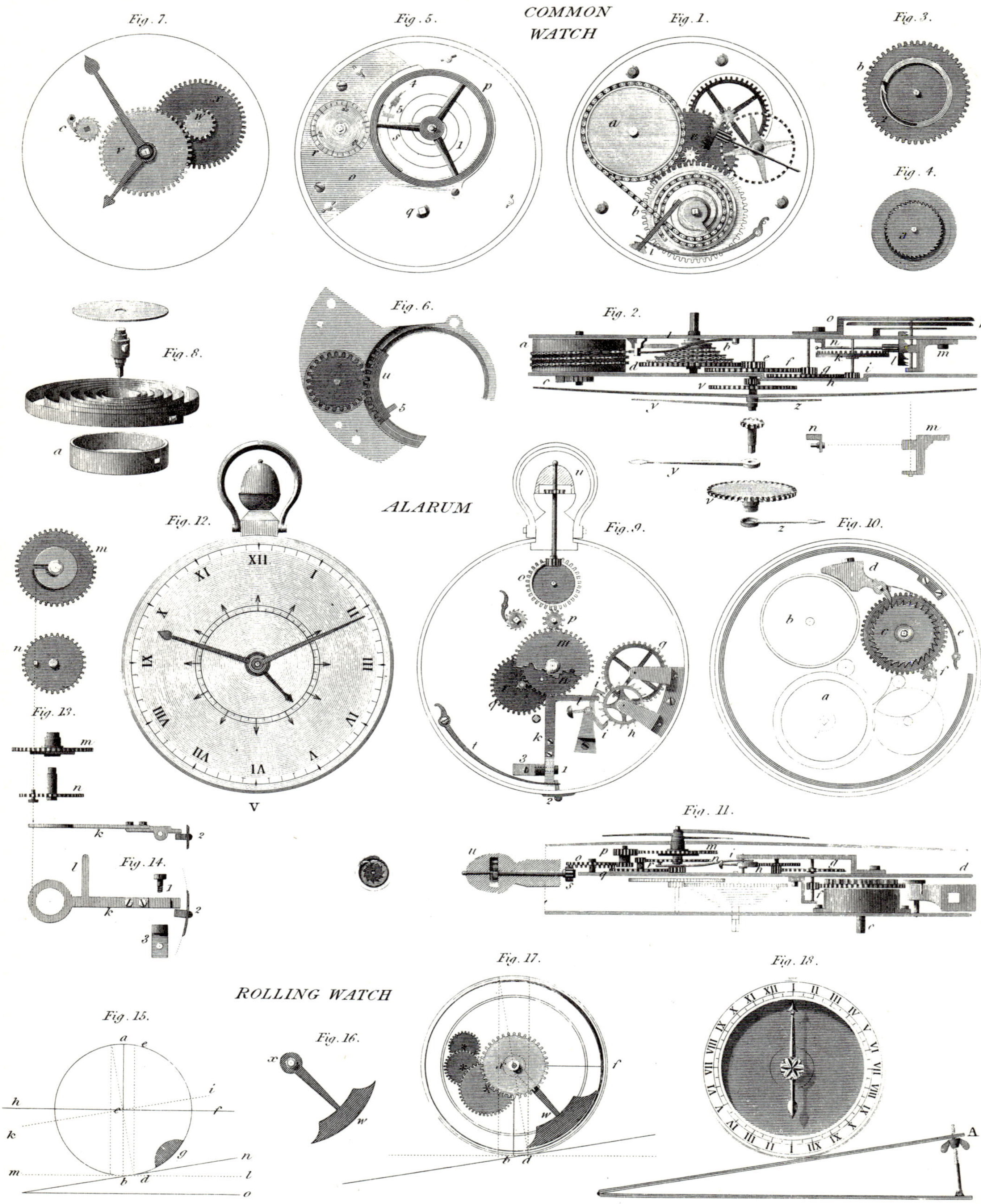

Published as the Act directs, 1818, by Longman, Hurst, Rees, Orme & Brown, Paternoster Row. *Lowry Sc.*

REPEATING WATCH.

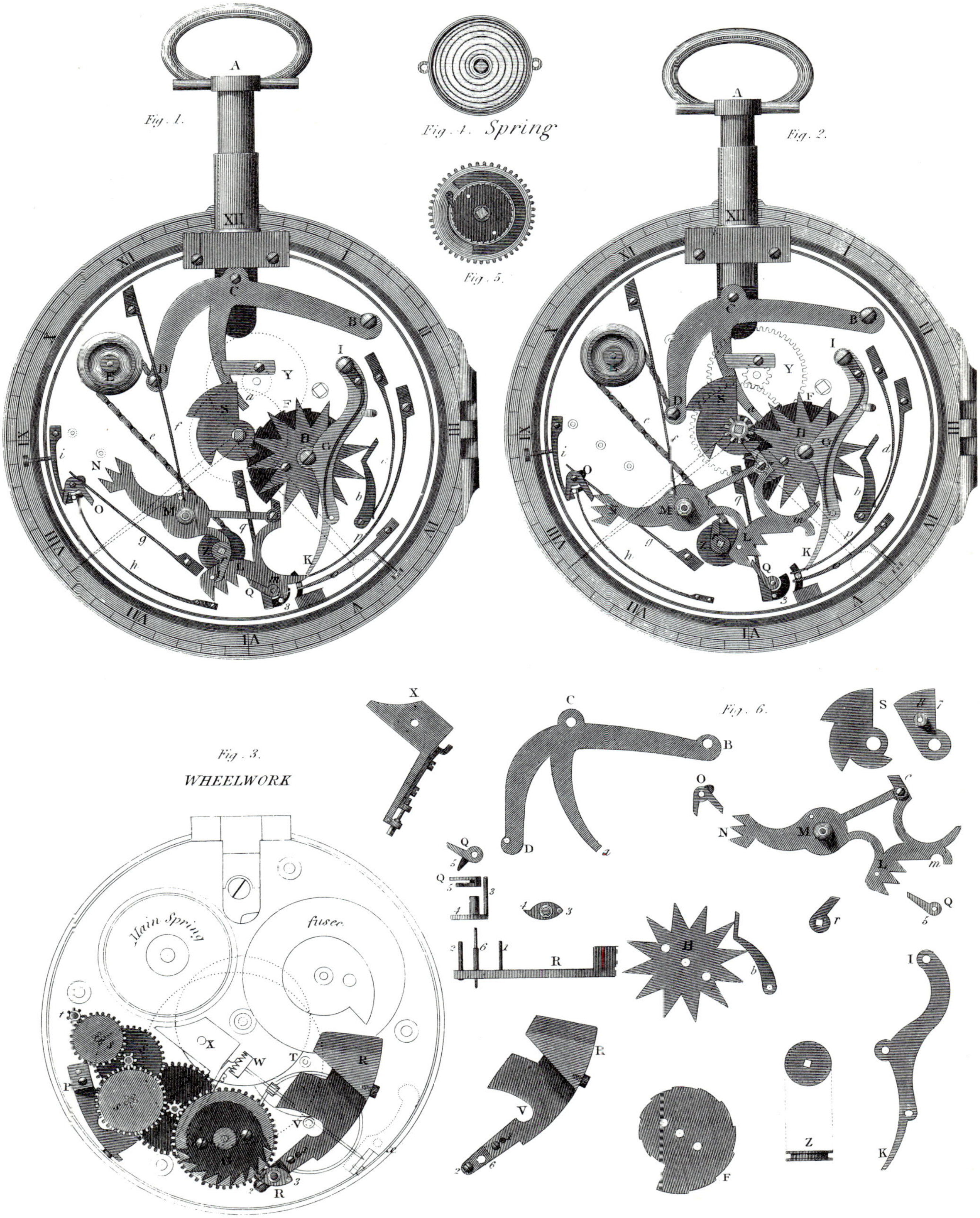

J. Farey del. Fn. Published as the Act directs, 1817, by Longman, Hurst, Rees, Orme & Brown, Paternoster Row. Lowry Sc.

HOROLOGY.

REPEATING and ALARUM WATCHES.

Fig. 1.

Fig. 5.

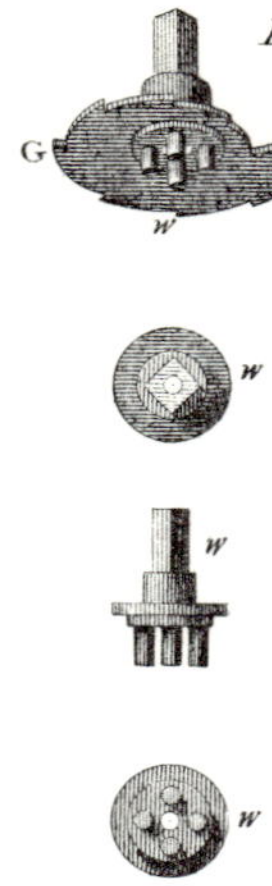

Fig. 2.

Fig. 7.

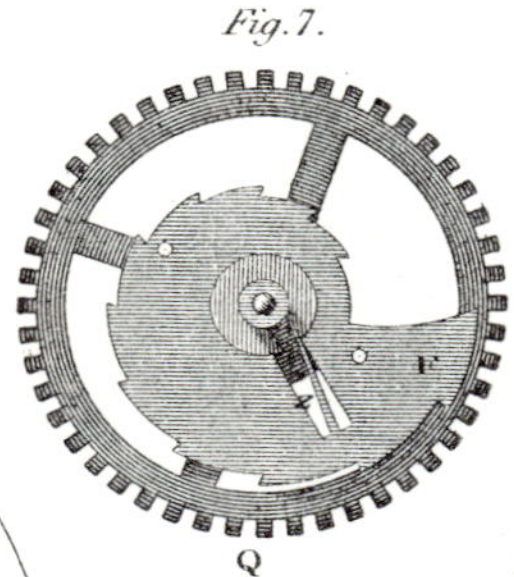

Fig. 8.

Fig. 3.

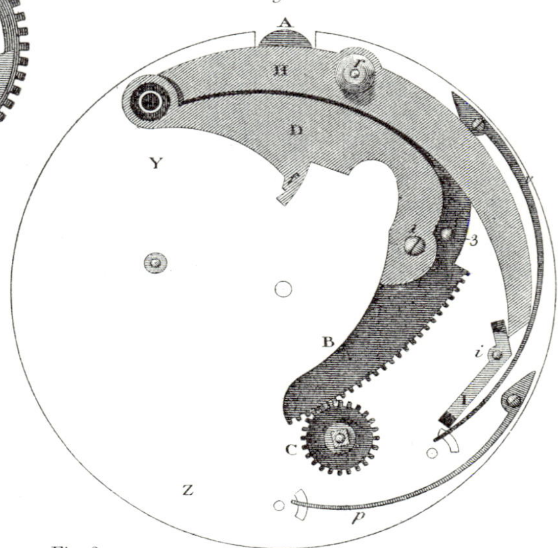

Fig. 4.

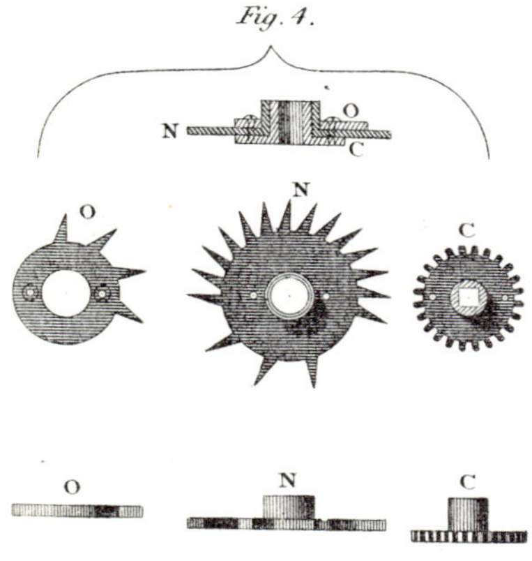

Fig. 6.

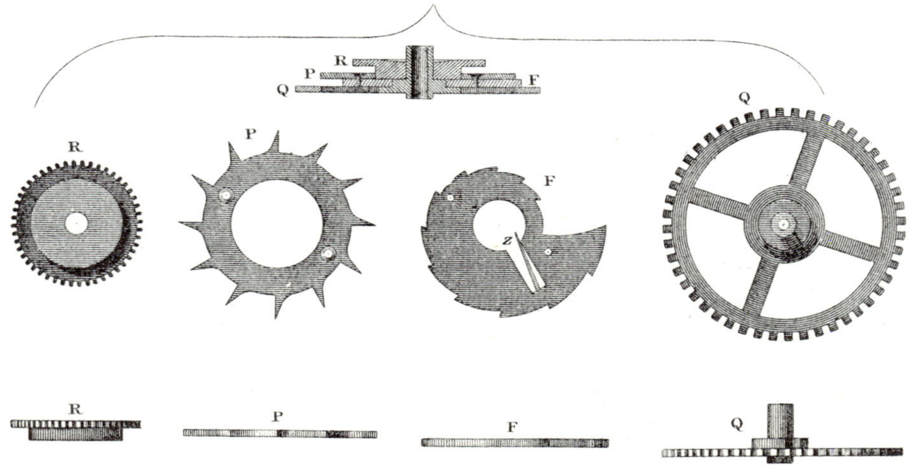

B. L. Vulliamy, Clock & Watch Maker, to the King, & Prince Regent, delin^t.

Published as the Act directs, 1819, by Longman, Hurst, Rees, Orme, & Brown, Paternoster Row, London.

Lowry sculp.

HOROLOGY.
REPEATING WATCHES.

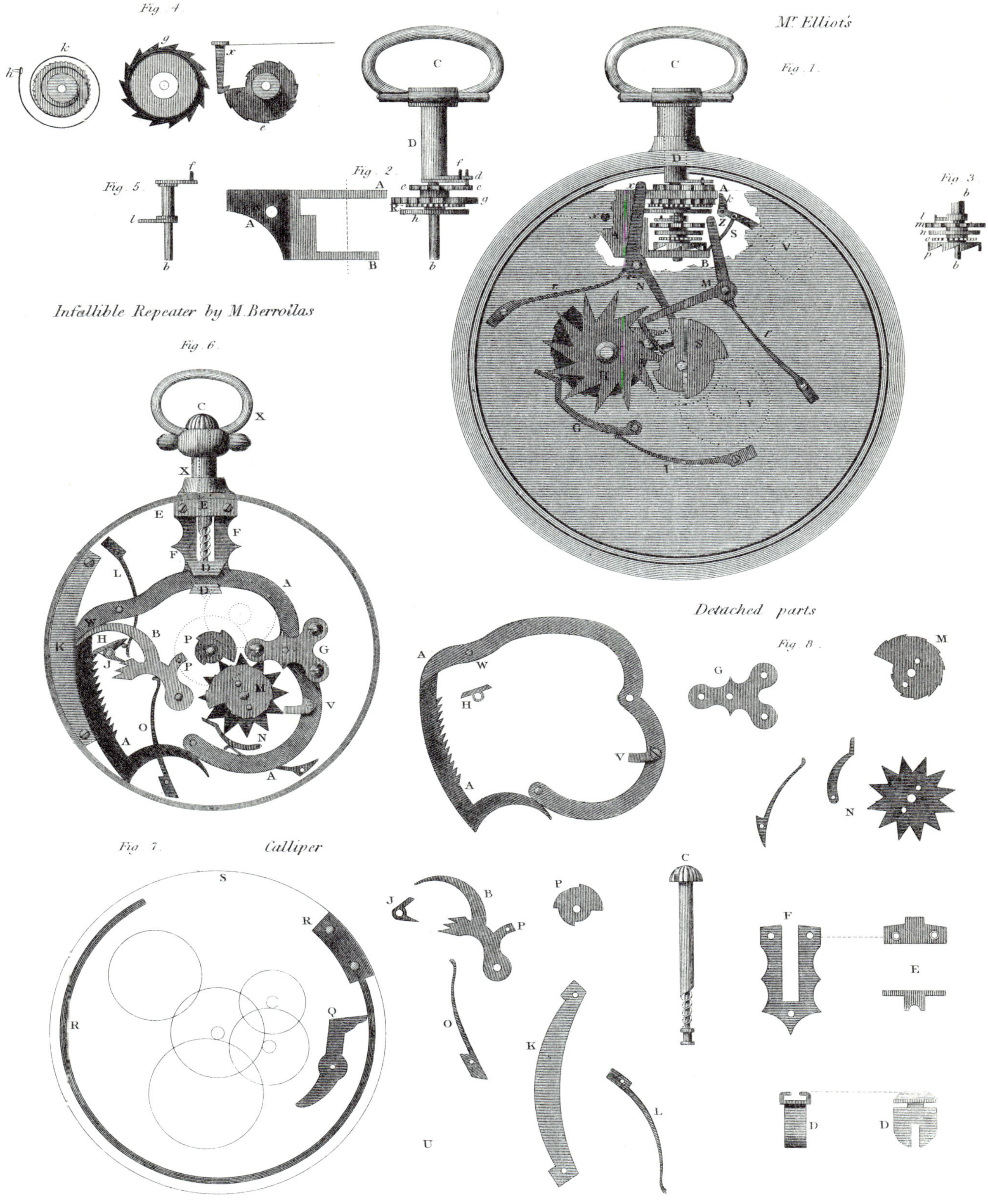

J. Farey Junr. delin. Published as the Act directs, 1816, by Longman, Hurst, Rees, Orme, & Brown, Paternoster Row. Lowry Sc.

Index

Bold figures denote plate pages